ORIGINS AND EVOLUTION OF LIFE

Devoted to exploring questions about the origin and evolution of life in our Universe, this highly interdisciplinary book brings together a broad array of scientists. Thirty chapters assembled in eight major sections convey the knowledge accumulated and the richness of the debates generated by this challenging theme. The text explores the latest research on the conditions and processes that led to the emergence of life on Earth and, by extension, perhaps on other planetary bodies. Diverse sources of knowledge are integrated, from astronomical and geophysical data, to the role of water, the origin of minimal life properties and the oldest traces of biological activity on our planet. This text will appeal not only to graduate students but also to the large body of scientists interested in the challenges presented by the origin of life, its evolution, and its possible existence beyond Earth.

MURIEL GARGAUD is a Research Scientist at the Laboratoire d'Astrophysique de Bordeaux, CNRS - Université Bordeaux 1, and is vice-president of the Société Francaise d'Exobiologie, which is associated with the NASA Astrobiology Institute.

PURIFICACIÓN LÓPEZ-GARCÍA is a Research Director at the CNRS, Université Paris-Sud, and leads a research team exploring microbial diversity and evolution in different ecosystems, including extreme environments.

HERVÉ MARTIN is a Professor at the Laboratoire Magmas et Volcans at the Université Blaise Pascal in Clermont-Ferrand. He has been in charge of several international research programmes on the geochemistry and geodynamic processes on early Earth.

Cambridge Astrobiology

Series Editors

Books in the series

1. Planet Formation: Theory, Observations, and Experiments
   Edited by Hubert Klahr and Wolfgang Brandner
   ISBN 978-0-521-86015-4

2. Fitness of the Cosmos for Life: Biochemistry and Fine-Tuning
   Edited by John D. Barrow, Simon Conway Morris, Stephen J. Freeland and Charles L. Harper, Jr.
   ISBN 978-0-521-87102-0

3. Planetary Systems and the Origin of Life
   Edited by Ralph Pudritz, Paul Higgs and Jonathan Stone
   ISBN 978-0-521-87548-6

4. Exploring the Origin, Extent, and Future of Life
   Edited by Constance M. Bertka
   ISBN 978-0-521-86363-6

5. Life in Antarctic Deserts and other Cold Dry Environments
   Edited by Peter T. Doran, W. Berry Lyons and Diane M. McKnight
   ISBN 978-0-521-88919-3

6. Origins and Evolution of Life: An Astrobiological Perspective
   Edited by Muriel Gargaud, Purificación López-García and Hervé Martin
   ISBN 978-0-521-76131-4

ORIGINS AND EVOLUTION OF LIFE

An Astrobiological Perspective

MURIEL GARGAUD
Université Bordeaux 1

PURIFICACIÓN LÓPEZ-GARCÍA
Université Paris-Sud 11

HERVÉ MARTIN
Université Blaise Pascal, Clermont-Ferrand

CAMBRIDGE UNIVERSITY PRESS
Cambridge, New York, Melbourne, Madrid, Cape Town, Singapore,
São Paulo, Delhi, Dubai, Tokyo, Mexico City

Cambridge University Press
The Edinburgh Building, Cambridge CB2 8RU, UK

Published in the United States of America by Cambridge University Press, New York

www.cambridge.org
Information on this title: www.cambridge.org/9780521761314

First published 2011

Printed in the United Kingdom at the University Press, Cambridge

A catalogue record for this publication is available from the British Library

Library of Congress Cataloguing in Publication data
Origins and evolution of life : an astrobiological perspective / [edited by]
Muriel Gargaud, Purificación López-García, Hervé Martin.
p. cm. – (Cambridge astrobiology ; 6)
Includes bibliographical references and index.
ISBN 978-0-521-76131-4 (hardback)
1. Exobiology. 2. Life–Origin. 3. Evolution (Biology) I. Gargaud, Muriel.
II. López-García, Purificación. III. Martin, H. (Hervé)
QH326.O75 2010
576.8′3–dc22
2010037097

ISBN 978-0-521-76131-4 Hardback

Contents

Contributors

Kristin BARTIK
Matières et Matériaux CP 165/64
Faculté des Sciences Appliquées
Université Libre de Bruxelles
50 Av. F. D. Roosevelt
1050 Brussels
Belgium

Karim BENZERARA
IMPMC
Batiment 7
140 Rue de Lourmel
75015 Paris
France

Hugues BERSINI
IRIDIA
Université Libre de Bruxelles
CP 194/6
50 Av. Franklin Roosevelt
1050 Bruxelles
Belgium

Jean-Pierre BIBRING
Institut d'Astrophysique Spatiale
Centre Universitaire d'Orsay
Bat 120-121
91405 Orsay Cedex
France

Bernard BILLOUD
UPMC
Univ Paris 06
UMR7139
Station Biologique
F29682 Roscoff Cedex
France

Laurent BOITEAU
Institut des Biomolécules Max Mousseron
UMR CNRS 5247
Universités Montpellier 1 & 2
Groupe Dynamique des Systèmes Biomoléculaires Complexes (DSBC)
CC 1706
Université Montpellier 2
Place Eugène Bataillon
F-34095 Montpellier Cedex 5
France

Céline BROCHIER-ARMANET
Université de Provence
Aix-Marseille I
Laboratoire de Chimie Bactérienne (UPR CNRS 9340) IFR88
31 Chemin Joseph Aiguier
13402 Marseille Cedex 20
France

Gilles BRUYLANTS
Matières et Matériaux CP 165/64
Faculté des Sciences Appliquées
Université Libre de Bruxelles
50 Av. F. D. Roosevelt1050 Brussels
Belgium

Jean CADET
Laboratoire Lésions des Acides Nucléiques
INAC/SCIB
CEA/Grenoble
38054 Grenoble Cedex 9
France

Cecilia CECCARELLI
Laboratoire d'Astrophysique de Grenoble
Université Joseph Fourier/CNRS
BP53
F-38041 Grenoble Cedex 9
France

José CERNICHARO
Laboratory of Molecular Astrophysics
Department of Astrophysics
Centro de Astrobiología
INTA
Ctra de Torrejón a Ajalvir, km 4
28850 Torrejón de Ardoz
Madrid
Spain

John CHAMBERS
Carnegie Institution of Washington
5241 Broad Branch Road, NW
Washington DC 20015
USA

Thierry DOUKI
Laboratoire Lésions des Acides Nucléiques
INAC/SCIB
CEA/Grenoble
38054 Grenoble Cedex 9
France

Emmanuel DOUZERY
Lab. de Paleontologie, Phylogenie & Paleobiologie
CC064 (RDC bat. 22)
Institut des Sciences de l'Evolution (UMR 5554 CNRS)
Universite Montpellier II
Place E. Bataillon
34 095 Montpellier Cedex 5
France

Ana DURBÁN
Institut Cavanilles de Biodiversitat i Biologia Evolutiva
Universitat de València
Poligon de la Coma s/n
46980 Paterna
Spain

Thérèse ENCRENAZ
LESIA
Observatoire de Paris
F-92195 Meudon
France

Muriel GARGAUD
Laboratoire d'Astrophysique de Bordeaux
Université Bordeaux1/CNRS UMR 5804
BP 89
33270 Floirac
France

Matthieu GOUNELLE
LMCM
UMR7202

Muséum National d'Histoire Naturelle
Case Postale 52
57 Rue Cuvier
75 231 Paris Cedex 05
France

Manuel GÜDEL
University of Vienna
Department of Astronomy
Tuerkenschanzstr. 17
A-1180 Vienna
Austria

Emmanuelle JAVAUX
Geology Department
University of Liège
17 allée du 6 Août B18
Sart-Tilman Liège 4000
Belgium

James KASTING
443 Deike
Penn State University
University Park
PA 16802
USA

Dominique LAMBERT
FUNDP
Faculté des Sciences
Université de Namur
61 rue de Bruxelles
B-5000 Namur
Belgium

Amparo LATORRE
Institut Cavanilles de Biodiversitat i Biologia Evolutiva
Universitat de València
Poligon de la Coma s/n
46980 Paterna
Spain

Antonio LAZCANO
Facultad de Ciencias
Universidad Nacional Autonoma de Mexico
Apdo Postal 70-407, Cd Universitaria
04510 Mexico D. F.
Mexico

Richard LEVEILLE
Canadian Space Agency
6767 route de l'Aéroport
St-Hubert,
Québec
J3Y 8Y9
Canada

Emanuela LOCCI
Dipartimento di Scienze Chimiche
Cittadella Universitaria di Monserrato
S.S. 554 Bivio per Sestu
09042 Monserrato (CA)
Italy

Purificatión LÓPEZ-GARCÍA
Unité d'Ecologie, Systématique et Evolution
UMR CNRS 8079
Université Paris-Sud
Bâtiment 360
91405 Orsay Cedex
France

Jonathan LUNINE
1629 E University Blvd.
LPL/University Arizona
Tucson AZ 85721
USA
and
Dipartimento di Fisica
Università degli Studi di Roma
Tor Vergata Rome
Italy

Hervé MARTIN
Laboratoire Magmas et Volcans
CNRS-UMR 6524
Université Blaise Pascal
5 Rue Kessler
63038 Clermont-Ferrand
France

Jennyfer MIOT
11 Lotissement les Aulnes
191 Route du Mas Rillier
Les Echets
01700 Miribel
France

Thierry MONTMERLE
Institut d'Astrophysique de Paris
98 bis Bd Arago
75014 Paris
France

Michel MORANGE
Centre Cavaillès
Ecole Normale Supérieure
29 Rue d'Ulm
75230 Paris Cedex 05
France

David MOREIRA
Unité d'Ecologie
Systématique et Evolution
UMR CNRS 8079
Université Paris-Sud Bâtiment 360
91405 Orsay Cedex
France

Andrés MOYA
Institut Cavanilles de Biodiversitat i Biologia Evolutiva
Universitat de València
Poligon de la Coma s/n
46980 Paterna
Spain

Robert PASCAL
Institut des Biomolécules Max Mousseron
UMR CNRS 5247
Universités Montpellier 1 & 2
Groupe Dynamique des Systèmes
Biomoléculaires Complexes (DSBC)
CC 1706
Université Montpellier 2
Place Eugène Bataillon
F-34095 Montpellier Cedex 5
France

Juli PERETÓ
Institut Cavanilles de Biodiversitat i Biologia Evolutiva
Universitat de València
Poligon de la Coma s/n
46980 Paterna
Spain

Sandra PIZZARELLO
Dept. of Chemistry & Biochemistry
Arizona State University
Tempe
AZ 85287-1604
USA

Nicolas PRANTZOS
Institut d'Astrophysique de Paris
98bis Bd Arago
75014 Paris
France

Gilles RAMSTEIN
Laboratoire des Sciences du Climat et de l'Environnement
IPSL CEA/CNRS/UVSQ
D. S. M.
Orme des Merisiers
Bat. 701
C. E. Saclay
91191 Gif-sur-Yvette
France

François RAULIN
LISA
UMR CNRS 7583
IPSL
Universités Paris Est
Créteil & Denis Diderot
61 Avenue du Général de Gaulle
F 94010 Créteil Cedex
France

Jacques REISSE
Matières et Matériaux
CP 165/64
Faculté des Sciences Appliquées
Université Libre de Bruxelles
50 Avenue F. D. Roosevelt
1050 Brussels
Belgium

Suzanne SOMMER
Institut de Génétique et Microbiologie
Bâtiment 409
Université Paris-Sud
91409 Orsay Cedex
France

Ken TAKAI
Subsurface Geobiology Advanced Research (SAGAR) Project &
Precambrian Ecosystem Laboratory
Japan Agency for Marine-Earth Science & Technology (JAMSTEC)
2-15 Natsushima-cho
Yokosuka 237-0061
Japan

Magali TOUEILLE
Institut de Génétique et Microbiologie
Bâtiment 409
Université Paris-Sud
91409 Orsay Cedex
France

Frances WESTALL
Centre de Biophysique Moléculaire
CNRS Rue Charles Sadron
45071 Orléans Cedex 2
France

Giuseppe ZACCAI
Institut Laue Langevin
6 rue Jules Horowitz
BP 156 38042 Grenoble Cedex 9
France

Foreword

William M. Irvine
University of Massachusetts and Goddard Center for Astrobiology

Astrobiology, also known as bioastronomy or exobiology, is the study of the origin, evolution and distribution of life in the Universe. These are subjects which have been of interest to mankind throughout recorded history. Although questions of origins have most frequently invoked divine beings, non-supernatural speculation on these fundamental issues dates back at least to the Ionian school of pre-Socratic Greek philosophers. Anaximander, the successor to Thales, is reported as saying that all living creatures arose from the moist element (water) through the action of the Sun (Freeman, 1966), a prescient insight given current ideas that life as we know it requires water, that radiation acting on inorganic matter can produce the molecular components of life (amino acids, nucleic acids, etc.) and that the Sun is the ultimate energy source for almost all life on Earth. In fact, Anaximander seems to have gone further and suggested that human beings arose from fish-like creatures (presumably a natural result of life having originated in water).

Speculation about life beyond the Earth has also had a long tradition. Although Pythagoras himself is not known to have recorded his teachings, his school (in particular, Philolaus, *ca.* 400 BCE) is said to have written that the Moon appears Earth-like because it is inhabited with animals and plants (Dreyer, 1953). At roughly the same time the atomist school of Leucippus and Democritus taught that the Universe is infinite and contains innumerable worlds. Since Democritus is quoted as saying that 'There are some worlds devoid of living creatures or plants', presumably he believed some are in fact inhabited, and this view was explicitly stated by his later follower Epicurus (*ca.* 300 BCE). The atomist ideas are best known from the Roman poet and philosopher Lucretius (*ca.* 99–55 BCE), who firmly embedded the idea of an infinity of worlds in the atomist tradition. Also during Roman times Plutarch, better known for his biographies, raised in an essay the distinction between habitability and the actual presence of life; a distinction of fundamental importance in modern astrobiology (Dick, 1982).

Aristotle's rejection of the atomist theories ended most Western discussion of life beyond the Earth for the next millennium, although some medieval scholars such as William of Ockham (of the famous razor; *ca.* 1280–1347) argued that the omnipotence of God certainly allowed for the possible existence of other worlds like ours. Then, as the Renaissance began, Nicholas of Cusa (1401–1464) argued that 'Rather than think that so many stars and parts of the heavens are uninhabited and that this earth of ours is peopled ... we will suppose that in every region there are inhabitants'. Subsequently Johannes Kepler, arguing on the basis of its newly discovered moons, 'deduce[d] with the highest degree of confidence that Jupiter is inhabited' (Dick, 1982).

Islamic science had a considerable history of speculation about the evolution of species. Al-Jahiz (real name Abu Uthman Amr ibn Bahr al-Fuqaimi al-Basri) (*ca.* 780–*ca.* 869), an Afro-Arab descendant of an African slave, wrote that the effect of the environment can cause animals to develop new characteristics and can thus lead to new species (Sarton, 1975; Bayrakdar, 1983). Later, Nasir al-Din al-Tusi (born in 1201 in what is now Iran) apparently held an atomist-like view of the origin of life and also propounded ideas on the evolution of species (Alakbarov, 2001). Fakr al-Din al-Razi (1149–1209, in Iran) was an atomist as well and proposed that there are possibilities for other beings and other universes (A. Ragab, Harvard University).

In modern times ideas concerning extraterrestrial life have been expressed by many, including Huyghens and Fontenelle, while Percival Lowell built the Lowell Observatory in the USA primarily to investigate Mars, where he was convinced that the 'canals' were the work of an intelligent species. Modern scientific study of the origin of life perhaps began with the theoretical work of Oparin and Haldane and the laboratory experiments by Miller and Urey. Governmental funding for what was initially called exobiology was initiated in the USA shortly after the formation of NASA in 1958, with the aim of exploring the origin, evolution and distribution of life, and life-related molecules, in the Universe. The Exobiology Program included the Viking missions, intended specifically to search for evidence of life on Mars. At present the International Astronomical Union has a Commission (51) on Bioastronomy, there is an active International Astrobiology Society (ISSOL) and astrobiology societies or institutes exist in Spain, the USA, Japan, the United Kingdom, Australia, France, Italy and more generally in Europe.

Modern astrobiology encompasses the search for extant life, evidence of past life or evidence of prebiotic chemistry on Solar-System bodies; the search for and characterization of planets around other stars; the study of biologically relevant molecules in the interstellar medium and in primitive Solar-System objects such as comets, undifferentiated asteroids and some meteorites; the study of the origin, evolution and environmental constraints for life on Earth; and the search for intelligent signals of extraterrestrial origin. This book addresses all of these questions except the last one and also probes the complex issue of the definition of life. The authors are experts in the field, so that their work here will be a valuable resource for both students and established scientists in the many disciplines which contribute to astrobiology.

References

Alakbarov, F. (2001). A 13th-century Darwin? Tusi's views on evolution. *Azerbaijan International*, **9**, 48.

Bayrakdar, M. (1983). Al-Jahiz and the rise of biological evolutionism. *The Islamic Quarterly*, 3rd quarter, 149.

Dick, S. J. (1982). *Plurality of Worlds: The Origins of the Extraterrestrial Life Debate from Democritus to Kant*. Cambridge: Cambridge University Press.

Dreyer, J. L. E. (1953). *A History of Astronomy from Thales to Kepler*. New York: Dover Publications, p. 46.

Freeman, K. (1966). *The Pre-Socratic Philosophers: A Companion to Diels, Fragmente der Vorsokratiker*. Cambridge, MA: Harvard University Press, p. 62.

Sarton, G. (1975). *Introduction to the History of Science*. Huntington, NY: R. E. Krieger Publ. Co.

Preface

This book aims at exploring several crucial issues related to the origin(s) and evolution of life in the Universe, starting from the only example of life known so far: terrestrial life. It is clear, though, that many of the circumstances that surrounded the emergence of life on Earth may have occurred, are occurring or will occur in other regions of our Galaxy or in other galaxies of our Universe. Therefore, the critical exploration of those conditions and the elaboration of models explaining the transition from the organic chemistry of the Universe to the biochemistry of terrestrial living forms are relevant at a much more global scale.

Just as with this volume, the field of astrobiology is by nature multidisciplinary. Astrophysicists, geologists, chemists, biologists, computer scientists and philosophers, as well as scientists working at the different interfaces between those disciplines, can all contribute to a better understanding of the processes and conditions that led to the emergence of life. The points of view and approaches of those different disciplines should not only superimpose, but also converge towards a unified explanation of the phenomenon of life in our Universe.

This book is an attempt to contribute to such an ambitious objective. It summarizes a series of lectures presented by selected speakers during two successive summer courses sponsored by the French Research Council (CNRS, Centre National de la Recherche Scientifique): *Exobio'05 and Exobio'07, Ecole d'exobiologie du CNRS*, which were respectively held in September 2005 and September 2007 in Propriano, Corsica (http://www.u-bordeaux1.fr/exobio07/).

The different chapters condense the animated discussions held in Propriano by a community of astronomers, geologists, chemists, biologists, computer scientists, philosophers and historians of science, all sharing the common goal of critically assessing potential scenarios for the origin of life on Earth and in the Universe. This book will attempt to convey the enthusiasm and richness of the debates that took place among those different specialists that gathered their strength to address a specific and challenging issue with an open mind. Under such an atmosphere, long-standing assumptions may be put into question, and lead to a stronger interdisciplinary basis, where the astronomer learns to reason as a biologist, or the chemist as a geologist. The ambition of this book is to reflect such broad scientific trespassing and to make it accessible for the broader public as well as for teachers and master and PhD students.

The book is divided into eight major sections. Part I is introductory to the heart of the problem and gathers a series of chapters addressing the fundamental question of what life is and what kind of proxies are used to approach, in practice, a minimum life system by various disciplines. Michel Morange (Chapter 1) critically revisits the problem of the definition of life in its historical context and takes into account the epistemological challenges that it implies. Along a similar epistemological perspective, Dominique Lambert carries out a logical analysis of the cosmological anthropic principles, making a clear dissection of the underlying premises and assumptions. The following three chapters by, respectively, Céline Brochier-Armanet, Hugues Bersini and Bernard Billoud discuss the standards used to define the minimal cell from both biology and from computer sciences as well as the approach taken by computer scientists to mimic life.

Part II deals with the astronomical and geophysical constraints that led to the emergence of life on Earth and, perhaps, elsewhere in the Universe. It contains a series of chapters reviewing organic chemistry in the interstellar medium (Cecilia Ceccarelli and José Cernicharo) and meteorites (Sandra Pizzarello), as well as the astronomical setting allowing the right conditions for life to evolve (Matthieu Gounelle and Thierry Montmerle), including the formation of habitable planets (John Chambers) and galactic habitable zones (Nicolas Prantzos). Two additional chapters are devoted more specifically to the Solar System, with a chapter by Manuel Güdel and James Kasting on the influence of the young Sun on planets' atmospheres and a chapter by Gilles Ramstein about the evolution of climates on planet Earth.

One essential requisite for life as we know it is liquid water, which has incomparable properties, making it a solvent of choice for biochemical reactions. The three chapters of Part III explain why those properties are essential to life (Kristin Bartik *et al.*), how water intervenes in planet formation and evolution (Thérèse Encrenaz) and what the history of water on Mars has been like (Jean-Pierre Bibring).

The three chapters of Part IV are devoted to the very transition from inert matter to true living organisms, a fading frontier that is explored from both the bottom-up approach of prebiotic chemistry (Robert Pascal and Laurent Boiteau) and from the top-down approach of comparative genomics of extant organisms to deduce the features of more ancestral living forms (Antonio Lazcano). Juli Peretó details the fundamental, though complex, issue of the origin of metabolism and of how the first cells made a living, i.e. obtained free energy and organic matter to sustain themselves and reproduce.

Part V includes three key chapters to understand the molecular phylogenetic tools that allow for the retrieval of the genetic history of the extraordinary diversity of lineages of living organisms on our planet (Emmanuel Douzery) as well as crucial mechanisms with an increasingly recognized primordial role in evolution, namely, horizontal gene transfer (David Moreira) and symbiosis (Amparo Latorre *et al.*).

The three chapters of Part VI describe particular adaptations of life to some of the most challenging conditions: life under ionizing radiation and desiccation (Magali Toueille and Suzanne Sommer), which cause important damage to biological macromolecules, notably nucleic acids (Jean Cadet and Thierry Douki); and high salt (Giuseppe Zaccai).

Part VII deals with the kinds of traces that biological organisms can leave directly (their remains) or indirectly (through their activity on the immediate surrounding environment) in the fossil record and how these can be used to trace back and attempt dating of the oldest occurrences of life on our planet (chapters, respectively, by Frances Westall, Emmanuelle Javaux, Karim Benzerara and Jennyfer Miot, and Ken Takai).

Finally, Part VIII closes the book with two chapters by Jonathan Lunine and François Raulin and Richard Leveille on the interest in studying other systems different from Earth to tackle the possibility of finding life elsewhere in the Solar System or in the rest of the Universe.

We hope that the reader will enjoy this book as much as we have by putting together these insightful chapters on the origin of life from a multitude of scientific perspectives.

Part I

What is life?

of life. If the properties necessary and sufficient to define 'living objects'

1

Problems raised by a definition of life[1]

Michel Morange

Introduction

Looking for a definition of life raises various issues, the first being its legitimacy. Does seeking such a definition make sense, in particular to scientists? I will successively refute the different arguments of those who consider that looking for such a definition makes no sense, and then propose good reasons to do just that, but also add some caveats regarding what sort of definition is sought. After considering definitions proposed in the past, I will examine various present-day definitions, what they share and how they differ. I will show that the recent suggestion that viruses are alive makes no sense and obscures discussions about life. Finally, I will emphasize two important recent transformations in the way life is defined.

Philosophical and scientific legitimacy of a definition of life

Two questions immediately emerge. Are we seeking a definition of life or a definition of organisms? And what kind of definition should be sought? Two types of definition are, in fact, traditionally distinguished. A definition may aim to give the essential characteristics that causally explain the existence of the category of objects considered. Or a definition may be of more limited scope: to establish a list of properties that are necessary and sufficient to define this category of objects and to distinguish them from objects belonging to other categories. If one adopts the first kind of definition it will be possible to define life. If one opts for the second, one will look for a definition of organisms. I will support the idea that these two kinds of definition are somehow equivalent, at least in the case of a definition of life. If the properties necessary and sufficient to define ‘living objects’ are well chosen they will constitute a ‘causal’ definition of life.

More serious are the philosophical objections raised against the possibility of finding such a definition. The main argument is that organisms do not constitute a natural kind of category, but rather one constructed by humans. This was expressed by Norman Pirie in 1957: ‘My attitude, which might be labelled empirical nihilism, is that the statement that

[1] This contribution was inspired by a series of essays on defining life published in OLEB vol. 40 (2010).

a system is or is not alive is a statement about the speaker's attitude of mind rather than about the system, and that no question is scientifically relevant unless the questioner has an experiment in mind by which the answer could be approached.' (Pirie, 1957). It was also recently formulated again by the philosopher Evelyn Fox Keller (Keller, 2002) and found support from some biologists working in the new discipline of synthetic biology. It is frequently proposed that in a more or less near future, biologists will be able to synthesize 'objects' intermediary between non-life and life, and that the decision to call them living will be a purely human one. The idea that there is a 'minimal' form of life is an illusion. There are many different minimal forms of life, but also subliminal forms intermediary between life and non-life.

Doubts about the possibility of defining life are supported by historical studies. Michel Foucault argued that life did not exist before the 'invention of biology' at the beginning of the nineteenth century (Foucault, 1966). Less affirmative, the philosopher Georges Canguilhem has simply remarked that the question of life was not seriously discussed before the last decades of the eighteenth century. He added that the discussions about the question of life also disappeared at the end of the twentieth century, after the development of the new molecular vision of life (Canguilhem, 1982).

I consider that these arguments are of limited value and should not prevent us looking for a definition of life. I will try to deconstruct them, starting with the historical ones.

Foucault's statement is a clear deformation of historical facts. Aristotle discussed the question of life. What is true, and rightly underlined by Canguilhem, is that the intensity of the discussion has varied greatly from one period to another. But that answering this question was no longer a priority does not mean that it had fully disappeared. The question of life is a historical question, linked in a complex way with changes in the study of organisms.

The philosophical argument is ontological, and for this reason it is impossible to demonstrate that it is wrong. But there is also no way to demonstrate that it is right. Is it really necessary to address such ontological issues and to make of organisms a 'natural kind'? What is necessary for our purpose – in searching for a definition of life – is to consider that organisms have a set of properties that unambiguously distinguish them from other objects of the natural world. And that this set of properties can be sufficiently well defined for the search for objects on other planets having these properties to be a valuable goal and for such newly discovered objects unambiguously to be called 'living'. There is a recurrent confusion stemming from misinterpretation of the word 'convention', which is viewed as equivalent to 'arbitrary'. A definition of life will be conventional, in the sense that it will be an intersubjective agreement. This does not mean that this agreement is not based on rational arguments.

The existence of intermediary stages between life and non-life, either natural or created by synthetic biologists, is not an objection to the definition of life. If one admits – and it seems difficult not to! – that life appeared from non-life, there were surely objects that were intermediary between life and non-life at one or another time in the process. The decision to call these objects 'alive' or not will clearly be a human decision. But not because the set of properties that they have to possess to be called 'alive' is ill-defined, but rather because these objects probably did not have all these properties in a stable way. The existence of categories is not incompatible with the existence of intermediary states. The problem has been nicely described by Radu Popa under the name of 'the dilemma of endless gradualism' (Popa, 2004): it is easy to designate objects that are alive and objects that are not, but it is probably possible to go from one to the other by a gradual transformation.

This is a general philosophical difficulty, encountered also, for instance, when attempting to define consciousness or a species. There are different ways to try to solve this difficulty. To conclude that the adjectives 'living' and 'non-living' have no significance is not the right conclusion. A second possibility is to suppose that, in a gradual continuous process, there are nevertheless 'thresholds' or 'phase transitions'. The problem with this solution is that the latter expressions are metaphoric and do not provide any clue to the nature of the change. Another possibility would be to abandon traditional logic for fuzzy logic, in which between being A (1) and not being A (0) there is a continuum of values between 1 and 0. The problem with fuzzy logic is that everything is both true and false and the most absurd propositions such as 'viruses are alive' cannot be shown to be wrong. My feeling is that a more qualitative solution has to be sought, in which an object can be A, or non-A or intermediate between A and non-A. The limits to our logic are not the fact that A and non-A cannot be clearly distinguished, but rather that in our Universe a historical process can generate A from non-A. To be alive means to be alive for a relatively long period of time. It seems reasonable to admit that the first living systems were alive for very short periods of time. To be alive requires a certain stability, but there is no way to define a minimal duration below which a system might not be called alive and beyond which it might be called alive. This tight association between 'to be alive' and 'duration' explains why the capacity to reproduce is an obligatory component of a definition of life (see later). Reproduction is the only way to provide a living system with a sufficient stability.

Scientific objections to a search for a definition of life are also diverse, but of a different nature. The first one is historical. To search for a definition of life is likened by many scientists to a search for a principle of life, and therefore to the vitalist movement. Historians have long shown that the vitalist movement was diverse and that many of its supporters in the eighteenth and nineteenth centuries were simply opposed to the mechanistic–reductionist programme of Descartes and Galileo; not in principle for many of them, but in practice, because this reductionist project was premature and ill-adapted (Canguilhem, 1994). Most contemporary biologists nevertheless consider vitalism as a crime against science. It is obvious that looking for a definition of life, looking for a list of properties associated with life, does not mean that organisms are not natural objects, nor that behind their properties a 'principle of life' is hidden somewhere.

The second objection seems stronger, and has been advocated in particular by François Jacob (Jacob, 1982). Science has always progressed by focusing on questions of a limited amplitude – in contrast with, for instance, philosophy. By so doing, scientists have been able to construct a solid form of knowledge. The question of life is too large to be a scientific one.

This argument is true in the sense that the search for a definition of life is not comparable to the work that scientists do every day at the bench. And finding a good definition cannot be a PhD project! But it does not mean that finding a definition of life is not a scientific objective. It only means that it is an objective of a different nature, distinct from the day-to-day questions raised by scientists; a question which nevertheless must remain in the minds of biologists.

Another argument, closely linked with the previous one, is that looking for a definition of life is a waste of time; it is far more useful to ask 'small' questions, and to answer them by precise experiments. Once again, there is a confusion between experimental questions and questions that belong to a science, without being questions that are directly addressed by experiments.

Another objection is the apparent heterogeneity of the answers that are presently proposed (Popa, 2004). These answers are specific to one discipline or subdiscipline of biology, the answer of a biochemist being different from the answer of a geneticist. Such differences have been shown by historians to have persisted throughout the twentieth century (Kamminga, 1988). Another slightly different version of the same objection is to consider that the question of life will be asked for different purposes, for instance by astrobiologists constructing a device to be put aboard a space probe, or chemists studying prebiotic reactions, and that the answers will always be, for this reason, contextual and different. The fact that the answers are biased by the scientific training or field within which they are posed is obvious, but the error is to consider that these biases will remain forever barriers in a search for a common definition of life. The limits of these provisional definitions do not mean that the production of a consensual definition, including all the aspects that have been underlined by these partial definitions, is not possible. What is clear is that a valid definition of life cannot belong to a single discipline. It has to be shared by different disciplines, biochemistry as well as genetics, chemistry and physics.

The last objection appears radical. Discussing the nature of life makes no sense. What is needed is the elaboration of an experimental device thanks to which the distinction between living and non-living objects would be unambiguous. This was one of the arguments of Norman Pirie (Pirie, 1957). The only way for scientists to answer questions is by doing experiments. The logical flaw in this reasoning is that, to construct such experimental devices, the researcher must have a theoretical idea about what it is 'living'. What will be inserted in this experimental device are these preconceived ideas, and the results of the experiments will only confirm them.

Good scientific and societal reasons to search for a definition of life

In contrast to the fallacy of the previous objections, there are excellent reasons to try to answer the question 'What is life?'. The first and major reason, which I will develop later when specifically examining the case of viruses, is that the absence of a definition leaves us defenceless when absurd statements are made on the living or non-living character of such and such objects. A definition, however imperfect it may be, is a framework to which it is possible to refer.

A second good reason is that scientists constantly answer this question: not explicitly, but implicitly. A simple example will clearly demonstrate that it is impossible not to adopt an implicit definition of life. A meeting was organized by the Royal Society in London for 3–4 December 2003. The title was: 'The molecular basis of life: is life possible without water?'. The answer to this question is crucial in selecting the planets and satellites of the Solar System to be explored as a priority in the search for traces of life. How is it possible to answer the question on the place of water in living systems without an idea of the main characteristics of a living system? Since a definition of life is clearly necessary, in particular in astrobiology, it is better for the discussion to make it explicit instead of keeping it implicit.

It is also crucial to have an answer, however imperfect, to the question 'What is Life?' for the relations between scientists (and, in particular, biologists) and the lay public. The latter would not understand that this question is not high on the list of biologists' priorities. The right attitude for scientists is not to reject this question as non-scientific: this would create complete misunderstanding. It is to explain why the answer is difficult and what characteristics are considered today as essential for a system to be called 'alive'.

By doing this, biologists will occupy a place too often left to 'mavericks', journalists, amateurs, people without a satisfactory scientific training who propose categorical and frequently wrong answers to the question 'What is life?' and also a door for creationists to give their own non-scientific answer.

The tendency, common among scientists, to consider that this is a 'philosophical' question, and therefore one that does not belong to science, has the same consequence. Philosophers propose answers not informed by recently acquired scientific knowledge and for this reason these answers are at odds with scientific practices. Scientists must not dictate to philosophers what they have to say about life, but they have to do their best to ensure that this philosophical discourse is not in conflict with current scientific knowledge.

An additional argument in favour of a 'natural' place for the question of life within science is the need to attract to science the best minds of the next generations. They will not be seduced by the technological achievements of science, but by the ambition to address the most fundamental issues at the boundary between science and philosophy. When they receive their scientific training they will rapidly learn to delay seeking answers to these fundamental questions and that the progress of science results from focusing on small limited questions, accessible to experimental practice. These fundamental questions will not be abandoned, but left to one side, ready to be taken up again as soon as important scientific breakthroughs have been accomplished. In such a way, the dreams that support any scientific career will not be dashed, but simply tempered by the strict rules that any scientific work has to obey.

While the search for a definition of life has its place in science, this does not mean that any definition is acceptable. Such a definition has to be 'open', in the sense that it must not be limited by our present knowledge, which there is no reason to consider complete. When definitions of life were proposed in the nineteenth century, proteins seemed to be its essential components. For molecular biologists of the 1960s, DNA was placed at the pinnacle. Today, one considers that living organisms, using other macromolecules, may have preceded life on Earth and exist on other planets. To define precisely the nature of the macromolecules present in organisms in a definition of life would therefore be a mistake.

Looking for a definition of life must not be considered to reflect the belief that life is common in the Universe. It does not mean that the process of formation of organisms was a deterministic one: it may be that the conditions favourable to the emergence of life are rarely encountered in the Universe.

Successive answers to the question of life

My objective will not be to provide a full description of the answers that have been provided so far, but to point to some of their characteristics. Two strategies can be used in a search for a definition of life. The first is to observe the characteristics of organisms present on Earth and to try to isolate the most fundamental ones. The other is more abstract, less tied to observations made on extant organisms. It aims to define abstract characteristics that organisms ought to possess, independently of their precise material constitution. This second approach has been favoured by researchers working on artificial life (Varela, 1979; Ganti, 2003). Most of the historical definitions of life were the product of the former approach.

The most traditional one is that of Aristotle: 'By life we mean self-nutrition and growth (with its correlative decay)' (Aristotle, 1941, 555). One can retrospectively see in this definition an emphasis on two characteristics of life that we will show are still essential in our present vision: the fact that organisms permanently exchange with their environment and that life – in organisms as well as at the global level – is a historical phenomenon.

As we previously saw, the second half of the eighteenth century and the first years of the nineteenth century were a crucial period for the question of life.

One of the most popular definitions was proposed by Bichat in 1802: 'Life is the totality of functions that resist death' (Bichat, 1994). Its weakness is its rhetorical nature, the answer being sought in its apparent opposite, death, the definition of which is unfortunately no more obvious than the definition of life!

More important, and significant for future developments, is the impetus given at the same time by Immanuel Kant, Georges Cuvier and many others in considering organisms as systems, in which the explanation of the functions of the different parts cannot be reached independently of the role they fulfil in the whole organism.

The systemic vision of life has persisted up to the present time, but is complemented by the emphasis placed on reproduction. Buffon was the first to place reproduction at the core of living phenomena, but it was through Darwin's contribution that reproduction became the central mechanism in the evolution of life: the occurrence of chance variations and the differential reproduction of organisms harbouring these variations is the mechanism that drives the evolution of organisms. The progressive description of the chemical components present in organisms – proteins, DNA – in the middle of the twentieth century pressed biologists to give these informational macromolecules a pre-eminent place in the definition of life.

The result of this complex history is that, as we have already outlined, there are many current definitions of life. However, two characteristics of life clearly predominate (Joyce, 1994). The first is that organisms are 'autopoietic systems', chemical systems able to maintain themselves by permanently synthesizing all or most of their components (Varela, 1979). The notion of autonomy is essential to characterize these self-maintaining systems (Ruiz-Mirazo and Moreno, 2009). The second is the capacity of organisms to reproduce. These two characteristics are the legacy of the two historical traditions that we described previously, enriched by improved knowledge of the mechanisms involved. Throughout the twentieth century, these two traditions were clearly supported by two groups of disciplines, chemistry and biochemistry on one side, genetics and evolutionary biology on the other.

Most biologists would agree that these two characteristics are important, but would disagree on their relative importance. Biochemists, for instance, would consider that the existence of a self-sustaining chemical system is the basis of life and that the capacity to reproduce is a consequence. Conversely, geneticists would emphasize that the capacity to reproduce came first, maybe in the simple form of self-replicating macromolecules, and that the metabolic side of life emerged later to support the reproductive system.

Disagreements also emerge as to whether or not it is necessary to combine other properties to generate a living system. Are the complex macromolecular structures present in organisms part of a definition of life? Is it possible to consider that the formation of these complex macromolecular structures was the consequence of the improvement of primitive, already living, chemical reproducing systems or, conversely, that the existence of self-maintaining and reproducing systems would not have been possible without the invention of these complex macromolecular structures? In particular, the existence of a self-maintaining chemical system is due to the presence within organisms of highly efficient catalysts, which all are macromolecules.

Similarly, it is obvious that a system has, in one way or another, to be insulated from its environment. The formation of cellular membranes can be considered as the necessary condition for the existence of such systems, or as a last step in the evolution of complex systems that already existed thanks to the confinement of their components in space, for instance on mineral surfaces. The formation of membranes will be considered as a fundamental step, if they are believed to have been the primeval place where energy was produced.

The invention of genetic information can also be seen as a way to reproduce pre-existing systems more faithfully; or the *sine qua non* for the emergence of a system able to reproduce. The existence of genetic information and the complex interactions between organisms and their environment make it reasonable to attribute a certain form of cognition to any organism. Is cognition a part of the definition of life (Bitbol and Luisi, 2004)? There is a long philosophical tradition that the capacities exhibited by 'higher organisms' have to be included in a minimal definition of life.

Is it also necessary to emphasize that the reproduction of organisms has to be imperfect to generate variants susceptible to be screened by natural selection? It seems difficult to imagine what 'perfect' reproduction would be. More significant, but difficult to be precise about, is the necessity for the rate of variation to be intermediate between two extreme values: too low a value will prevent the generation of enough variations and too high a value will rapidly lead organisms to utter disorganization of their most fundamental functions.

The discussions are not closed, but some trends are clearly visible. The 'informational' vision of many molecular biologists is not as dominant as it was some years ago; dominant to the point of considering that the riddle of life had been definitively solved with the discovery of genetic information and of the genetic code.

The fading of the informational vision is due to the immediately disappointing results of genome sequencing; and also to the hypothesis that an RNA world preceded the present DNA and protein world – in which RNA is not informational in the same sense as DNA is. The contributions of specialists of artificial life, who emphasize the functioning of organisms as systems, did not abolish the role of information in organisms, but put an end to the identification of life with the existence of genetic information.

Similarly, the vision of organisms as self-maintained chemical systems, which was put aside during the development of the informational vision of molecular biology and the dominant place of genes, made a comeback with the increasing number of projects in astrobiology. For the detection of extraterrestrial forms of life will be totally dependent on the capacity of organisms to exchange matter and energy with their environment, not on their capacity to reproduce. Personally, I consider that the two properties – a self-maintaining chemical system and reproduction – are necessary and sufficient to define life; necessary, because a self-maintaining chemical system can be transiently alive, but will not be stable enough to support life; sufficient, at least provisionally, because the addition of other properties depends on scenarios that have not yet been validated. A useful distinction was introduced by Tibor Ganti (Ganti, 2003) between a criterion considered as absolute – the capacity of organisms to self-maintain – and one considered as actual – the capacity to reproduce. The first allows definition of a system as alive, whereas the second is required for the existence of life.

Are viruses alive?

After they were distinguished from bacteria, viruses were considered in the first decades of the twentieth century as alive. They were the simplest living organisms, and for this reason models to characterize the most fundamental properties of organisms (Morange, 1998). Otherwise, it would be impossible to understand the role that bacteriophages and the tobacco mosaic virus played in the rise of molecular biology (Creager, 2002).

But difficulties accumulated in the 1930s: it was impossible to cultivate viruses except in the presence of living cells. Viruses were progressively considered as parasites. At the end of the 1950s, André Lwoff stated the differences between viruses and organisms: the absence of a metabolism – the lack of a molecular machinery able to translate the information contained in their DNA or RNA genomes (Lwoff, 1957).

There have been recent attempts to give life back to viruses. The first reason for this is the increasing evidence that horizontal gene transfer, due in most cases to viruses, played an important role in the evolution of life. Such genetic transfer, by viruses or plasmids, still plays an important part in the rapid adaptation of microorganisms to new environments, such as the addition of antibiotics. It is admitted that at the early stages of life, at the time when the last universal common ancestor (LUCA) of all extant living forms existed, these genetic exchanges might have been so extensive that it would be more appropriate to talk about a community of genes shared by organisms than of a collection of independent organisms. The recent systematic search for viruses and bacteriophages has revealed not only their abundance, but also the diversity of the genes they harbour, with no related sequences found in extant organisms. A possible interpretation of these data is that viruses are the remnants of a living world that pre-dated LUCA.

The way of considering viruses has also changed. During its cycle a virus may replicate within cells in particular structures called viral factories, which are sometimes surrounded by a membrane. It has been suggested that these viral factories represent the true nature of the virus, whereas the virions are only inactive states during their cycle. Hence the hypothesis that cells invaded by viral factories, in which the virus has reoriented all the synthetic capacities of the cells towards their own propagation, might be called 'virocells' and have a living status comparable to that of 'traditional' cells, now called 'ribocells' (Raoult and Forterre, 2008).

Large viruses of sizes comparable to bacteria have been discovered (Raoult *et al.*, 2004). These mimiviruses have large genomes containing hundreds of genes. In addition, they can themselves be parasitized by other viruses (La Scola *et al.*, 2008). These observations led their authors to propose that mimiviruses are alive and that the world of life contains a fourth kingdom, in addition to Archaea, eucarya and eubacteria.

I consider with others that, despite their importance, viruses remain strict parasites (Moreira and López-García, 2009). There is no symmetry between a virocell and a ribocell. The first needs the second: the reverse is not true. To consider viruses as living creates much useless confusion. Nothing in the definition of life or in the non-living nature of viruses has to be changed to acknowledge the importance viruses had in the evolution of organisms.

A crucial time for the definition of life

Since antiquity, multiple definitions have been proposed. Despite the accumulation of observations, the present ones are not utterly different from those that preceded them. Is there real progress in this desperate search for a definition of life?

My conviction is that the search for a definition of life cannot be considered independently from two other issues – the origin of life and the capacity to 'master' living phenomena. Both have experienced a dramatic evolution in the twentieth century. The fears generated by the development of genetically modified organisms bear witness to this new capacity to modify organisms. Researchers working in synthetic biology also have the ambition to synthesize artificial organisms in a more or less distant future; and this ambition does not seem unreasonable.

The study of the origin of life also dramatically changed during the second part of the twentieth century. A marginal activity for retired scientists has evolved into a serious discipline, the scientific status of which is no longer denied.

These changes put the question of life into a different context. The question of life is no longer a riddle and a general definition of life can be provided. Nothing fundamental is missing in our understanding of the phenomena of organisms. Such a statement is the acknowledgement that the molecular descriptions obtained from the mid twentieth century onwards constitute a dramatic breakthrough in our understanding of organisms. It is a vain hope that the discovery of new laws, such as the laws of complexity, will illuminate the question of life. Organisms are complex systems and a better description of complex systems will obviously cast light on the functions of organisms. But it will not be the solution to the question of life for the simple reason that complexity is not a property unique to organisms.

As Oparin (1953) remarked, every characteristic of organisms is shared with objects of the non-living world. Life is simply the gathering of these different properties in a single object. What has to be explained is not the realization of each of these characteristics, but the historical process by which they were linked. Simple self-sustaining chemical systems exist in the inorganic world; the capacity to reproduce also. What remains mysterious is the complex process by which these different characteristics were progressively combined in a stable form in organisms and the roles of self-organization and Darwinian evolution in this process (Weber and Depew, 1996). There was a diachronic emergence of life: what is exactly meant by this expression constitutes the work that has still to be accomplished. The question of life is no longer the search for a definition, but a historical question lacking a precise scenario.

Acknowledgements

We are indebted to David Marsh for critical reading of the manuscript and to Purificación López-García for fruitful comments.

References

Aristotle. (1941). De Anima 2.1.14–15. In *The Basic Works of Aristotle*, ed. R. McKeon. New York: Random House.

Bichat, X. (1994). *Recherches Physiologiques sur la Vie et la Mort*, ed. A. Pichot. Paris: Flammarion.

Bitbol, M. and Luisi, P. L. (2004). Autopoiesis with or without cognition: defining life at its edge. *Journal of the Royal Society Interface*, **1**, 99–107.

Canguilhem, G. (1982). *Vie*. Paris: Encyclopaedia Universalis.

Canguilhem, G. (1994). The vitalist imperative. In *A Vital Rationalist: Selected Writings from Georges Canguilhem*, ed. F. Delaporte. New York: Zone.

Creager, A. N. H. (2002). *The Life of a Virus: Tobacco Mosaic Virus as an Experimental Model*. Chicago: The University of Chicago Press.

Foucault, M. (1966). *Les Mots et les Choses*. Paris: Gallimard.
Ganti, T. (2003). *The Principles of Life*. Oxford: Oxford University Press.
Jacob, F. (1982). *The Possible and the Actual*. New York: Pantheon.
Joyce, G. F. (1994). Foreword. In *Origins of Life: The Central Concepts*, eds. D. W. Deamer and G. R. Fleischaker. Boston: Jones and Bartlett.
Kamminga, H. (1988). Historical perspective: the problem of the origin of life in the context of developments in biology. *Origin of Life and Evolution of the Biosphere*, **18**, 1–11.
Keller, E. F. (2002). *Making Sense of Life*. Cambridge: Harvard University Press.
La Scola, B., Desnues, C., Pagnier, I., Robert, C., Barrassi, L., Fournous, G. Merchat, M., Suzan-Monti, M., Forterre, P., Koonin, E. and Raoult, D. (2008). The virophage as a unique parasite of the giant mimivirus. *Nature*, **455**, 100–4.
Lwoff, A. (1957). The concept of virus. *Journal of General Microbiology*, **17**, 239–53.
Morange, M. (1998). *A History of Molecular Biology*. Cambridge: Harvard University Press.
Moreira, D. and Lopez-Garcia, P. (2009). Ten reasons to exclude viruses from the tree of life. *Nature Reviews in Microbiology*, **7**, 306–11.
Oparin, A. I. (1938). *The Origin of Life* (S. Morgulis, translation). 2nd ed., 1953. New York: Macmillan.
Pirie, N. W. (1957). The origin of life: the Moscow symposium. *Nature*, **180**, 886–8.
Popa, R. (2004). *Between Chance and Necessity: Searching for the Definition and Origin of Life*. New York: Springer-Verlag.
Raoult, D. and Forterre, P. (2008). Redefining viruses: lessons from Mimivirus. *Nature Reviews in Microbiology*, **6**, 315–19.
Raoult, D., Audic, S., Robert, C., Abergel, C., Renesto, P., Ogata, H., La Scola, B., Susan, M. and Claverie, J.M. (2004). The 1.2-megabase genome sequence of Mimivirus. *Science*, **306**, 1341–50.
Ruiz-Mirazo, K. and Moreno, A. (2009). New century biology could do with a universal definition of life. In *Information and Living Systems: Essays in Philosophy of Biology*, eds. G. Terzis and R. Arp. Cambridge: MIT Press.
Varela, F. J. (1979). *Principles of Biological Autonomy*. New York: Elsevier.
Weber, B. H. and Depew, D. J. (1996). Natural selection and self-organization: dynamical models as clues to a new evolutionary synthesis. *Biology and Philosophy*, **11**, 33–65.

2

Some remarks about uses of cosmological anthropic 'principles'

Dominique Lambert

The aim of this contribution is to underline some problems related to what is called the cosmological anthropic principle. There are several statements (weaker or stronger) of this 'principle', which was initially introduced by Brandon Carter (cf. Demaret and Barbier, 1981; Barrow and Tipler, 1986; Demaret, 1991; Demaret and Lambert, 1996). Today, there are a huge number of references defining and discussing these statements and we do not want to enter such a discussion here. In fact, for our purpose, we can simply say that the 'weak' version expresses simply the causality principle: if human life exists in the Universe, then there exist precise constraints that render the emergence of such life possible. This can also be presented as an observational constraint. If, as human beings, we are observing the Universe now, the latter cannot be arbitrary. It has to be such that human life is possible. These 'weak principles' are in fact a translation of the fact that each empirical event or each phenomenon can be characterized by a set of necessary conditions. And the weak versions of what one called the anthropic principle are then nothing more than a logical implication: human life (H) implies necessary conditions for human life to exist (NCH). Or, if we are considering the observational constraint approach: the existence of human observers implies necessary constraints on the Universe that render this existence possible. The 'strong' versions of the anthropic principle reverse these implications by saying that the Universe has to be such that it renders the emergence of human life possible or even necessary or that the Universe has the right properties allowing for the emergence of human observers.

The statements of the anthropic principle have generated many comments, critics and even fears. We will assess in the following in which sense this 'principle' is relevant or irrelevant for scientists and for philosophers.

Logical and empirical problems arising from anthropic statements

The first logical problem and risk related to the anthropic statements is the reversing of the implication arrows. It is perfectly relevant, according to a scientific point of view, to express the fact that if human life exists then there are some necessary conditions for that kind of existence. It can even be interesting because this can lead to the study of necessary conditions that are surprising or very far from the interest of biologists. For example, it is now well known that necessary conditions for life to exist are linked to some very special properties of the strong interaction (which explain the existence, studied first by Fred Hoyle, of precise energy levels of the carbon nucleus that facilitate the production of carbon nuclei from beryllium ones and that forbid the transmutation of all carbon nuclei to oxygen ones[1]) and to a precise spectrum of values for the cosmological constant (which controls the acceleration of the Universe's expansion and galaxy formation). It is really non-trivial to realize that there are some connections between human existence on Earth and constraints occurring at the (sub)microscopic level (the atomic nucleus) and also on a very large scale (galaxies, the Universe as a whole). Studying how the existence of humans or any other organism is affected by variations of some micro-level or large-scale-level parameters is also interesting. But what is illegitimate is to reverse the arrow of a logical inference to transform necessary conditions into sufficient conditions! In order to get human life we need carbon, hence stars in which carbon is generated, and thus galaxies where star birth occurs. And finally, we need a Universe with some precise cosmological constant values rendering large-scale structure formation possible. All the previously quoted constraints and implications are scientifically relevant, but from these it is not legitimate to infer that the cosmological constant is such that human life had to emerge in the Universe. From a logical point of view we cannot transform into a necessary condition that of an existence which is only 'possible'[2]:

$$(\lozenge H) \wedge (H \Rightarrow NCH) \tag{2.1}$$

into a necessary condition of an existence which is assumed to be 'necessary'[3]:

$$(\square H) \wedge (H \Rightarrow NCH) \tag{2.2}$$

or worse, into a 'sufficient' condition for this existence:

$$NCH \Rightarrow H \tag{2.3}$$

Empirically, today, the only thing we can state is that human life was 'possible' (because it exists now) and that this possibility implies some very specific necessary conditions that are now (at least partially) known at nearly each phenomenological level: from the elementary particle level to the cosmological level. In fact the discovery of a set of existing sufficient conditions (SCH) for human life to exist would be equivalent to a proof of the necessity for human life to emerge in the Universe, because usually, in modal logic, due to the rule of 'necessitation'[4], if we have proved SCH $\Rightarrow$ H we have also proved $\square$H. In principle, we cannot exclude the possibility of an a-priori proof that physical laws have to lead 'necessarily' to human existence. The simulations of self-reproducing and self-evolving

[1] The carbon nucleus is produced by the following reactions: $^4He + ^4He \rightarrow ^8Be$ and $^4He + ^8Be \rightarrow ^{12}C$. The latter reaction is favoured by a 'resonance' phenomenon but the following reaction is not favoured: $^4He + ^{12}C \rightarrow ^{16}O$. Therefore all the carbon produced is not transformed into oxygen.

[2] $\lozenge$H means: the existence of human being (H) is possible; $\wedge$ is the conjunction operator 'AND'.

[3] $\square$H means: the existence of human beings is necessary. The expression $\square$H is equivalent to the following $\neg\lozenge\neg$H and we also have: $\square H \Rightarrow \lozenge H$ (what is necessary is also possible).

[4] The 'necessitation rule' accepted in the so-called normal modal logics says: if t is a theorem then $\square$t is also a theorem. In other words, the theorems are necessities of the formal language defined by this logic.

automata (like those imagined in 'The Sims') and theories of complexity are positive steps in this direction, but up to now we can only shed some light on the necessary conditions for human existence and we cannot transform them into sufficient conditions.

In fact another logical problem involving the use of the weak anthropic principle is the risk of confusing the implication of $H \Rightarrow NCH$ with $H \Rightarrow (NCH_1 \vee NCH_2)$, where NCH_1 and NCH_2 are two possible sets of necessary conditions. In the former case, for life to exist it is necessary that NCH conditions must exist. In the latter case, for this existence, it is necessary to have either NCH_1 or NCH_2 or both. Then if either NCH_1 or NCH_2 is not satisfied, it does not imply that H is not possible due to the contraposition[5]:

$$\neg (NCH_1 \vee NCH_2) \Rightarrow \neg H \tag{2.4}$$

and due to the de Morgan law:

$$\neg (NCH_1 \vee NCH_2) \Leftrightarrow (\neg NCH_1) \wedge (\neg NCH_2) \tag{2.5}$$

which leads to:

$$(\neg NCH_1) \wedge (\neg NCH_2) \Rightarrow \neg H \tag{2.6}$$

and not to $(\neg NCH_1) \Rightarrow \neg H$ or to $(\neg NCH_2) \Rightarrow \neg H$. The problem is now that it may be very difficult to prove that a set of constraints is a (exhaustive or not) conjunction of necessary conditions or a (complete or not) disjunction of necessary conditions of H. But the discrimination of both cases could be very important. In the first case we have:

$$H \Rightarrow (NCH_1 \wedge NCH_2 \wedge \ldots \wedge NCH_n)$$

and thus

$$(\neg NCH_1 \vee \neg NCH_2 \vee \ldots \vee \neg NCH_n) \Rightarrow \neg H \tag{2.7}$$

which means effectively that the absence of one of the NCH_j conditions forbids human existence. Here, we could write for each $j = 1,\ldots n$, $H \Rightarrow \Box NCH_j$. But then the difficulty is to prove that $\Box NCH_j$ is empirically true. The second case can be translated as follows:

$$H \Rightarrow (NCH_1 \vee NCH_2 \vee \ldots \vee NCH_n)$$

and then

$$(\neg NCH_1 \wedge \neg NCH_2 \wedge \ldots \wedge \neg NCH_n) \Rightarrow \neg H \tag{2.8}$$

This shows that in this situation human life could exist even if $(n - 1)$ were not satisfied, because the logical expression means that at least one (but not all) NCH_j condition is necessary for human life to exist. The difference with respect to the first case is now obvious, because now we can write $H \Rightarrow \Diamond NCH_j$ for $j = 1,\ldots n$, which means that it is possible that one of the NCH_j conditions was a necessary condition of the process leading to human

[5] $\neg$ is the operator of negation 'NOT' and $\vee$ is the disjunction 'OR' operator.

existence. The conclusion of this analysis is that it is scientifically perfectly legitimate to focus on the conditions allowing for life existence (human or other: bacteria etc.) but it is very difficult to prove that some conditions are really necessary ($H \Rightarrow \Box NCH_j$) because, as in the second case we just mentioned, the absence of one condition can in fact be compensated for by the presence of another, as we only have: $H \Rightarrow \Diamond NCH_j$. The very important and interesting problem of the sensibility of H to a variation in several physical conditions is logically nothing but the study of the 'extension' of the disjunction in the inference $H \Rightarrow (NCH_1 \vee NCH_2 \vee \ldots \vee NCH_n)$.

If we adopt the interpretation of the weak anthropic principle as an observational bias we encounter some empirical problems. Let us consider, for example, the problem of explaining the present value of the cosmological constant Λ, which occurs in Einstein field equations: $R\mu\nu - 1/2\ R\ g\mu\nu + \Lambda\ g\mu\nu = -\ \kappa\ T\mu\nu$. In the Lemaître Universe (1931) this constant can be interpreted as describing a kind of repulsive gravity related to what we now call the Dark Energy. A variation of the numerical value of Λ has crucial consequences on the formation process of galaxies. For example, if the repulsive gravity is too strong, the Universe is accelerating too much, hindering or preventing the collapse of the matter clouds necessary for the galaxies' formation. The anthropic statement is used to fix some bounds on the numerical values of Λ implying that if we are now observing the Universe, it is necessary that carbon nuclei exist, which in turn implies that stars and galaxies exist. The existence of the latter implies some constraints on Λ. In particular, the ratio between the dark-matter energy density (related to Λ and responsible for the acceleration of the Universe) and the energy–matter density (responsible for the usual gravitational attraction) has to be not too high in order to allow the matter clouds to collapse. To give a precise meaning to the anthropic constraint on Λ, we can try to evaluate the (conditional) probability that Λ has a precise value if it is observed by a human observer in a galaxy (Peter and Uzan, 2005; Starkman and Trotta, 2006; Bernardeau, 2007). But this probability cannot be computed without knowledge of the a-priori probability for Λ to have a given numerical value. This is very difficult because of the lack of a precise, unique fundamental theory giving the a-priori probability distribution of the value of Λ. As a matter of fact, it is very difficult to give a rigorous expression of the observational bias that the weak anthropic statements are referring to. And now the situation is very strange, because if we search this fundamental theory in the framework of superstring theory, we get a huge number of possible 'fundamental theories', and in order to choose the right one, we are invited by some physicists to also use an anthropic statement (Susskind, 2006; Becker *et al.*, 2007). We are here facing an annoying difficulty. In order to explain the values of several physical parameters (Λ, the Planck constant, the speed of light, the fine-structure constant of electromagnetic interaction and space–time dimensions) we generally do not have at our disposal a fundamental theory allowing us to deduce these parameters a-priori. That is the reason why we try to constrain the value of these numbers by the fact that we are observing them. However, for that reasoning to be meaningful, we have to compute the probability for a human to observe a particular value of the parameter, which in turn requires knowledge about the probability distribution of the parameter value, the definition of which is rooted in an unknown fundamental theory.

We have to be careful when using the weak anthropic arguments and anthropic observational bias because, as we have already mentioned above, we have some possible (as yet unknown) compensations that allow us to extend the admissible spectrum of values of some fundamental parameters when modifying values of others. To illustrate this, let us consider an example where we are trying to explain the precise value of the speed of light, c, by saying that this value is related to an anthropic bias. We will remark that our existence as biochemical organisms implies a 'least bound' on the intensity of electric forces. A decrease in c would imply an immediate decrease in the Coulomb force due to the relations:

$$F = 1/4\pi\varepsilon_0 \, (Q.Q'/r^2) \text{ and } c = (1/\mu_0\varepsilon_0)^{1/2} \qquad (2.9)$$

where μ_0 and ε_0 are constants characterizing the intensity of the magnetic and electric forces, Q and Q′ are the electric charges and r the distance between them. It is true that a decrease in the Coulomb force would be fatal for life to exist due to a destabilization of the atoms and molecules. However, according to the above formula, c can decrease its value without changing the intensity of the Coulomb force if μ_0 increases. Of course, increasing μ_0 might have some deleterious consequences for the existence of biological organisms. Nevertheless, this model shows simply that we cannot jump too rapidly to conclusions concerning the lethal variation of fundamental parameters due to the possible existence of some unknown compensation phenomena.

We face a similar problem in the context of superstring theory, to which we alluded before. Every possible fundamental theory can be represented as a minimum, a 'valley' in a landscape. However, an incredibly huge number of such valleys exist. Of course not all of these are admissible, because many of them do not lead to universes in which life can possibly emerge and evolve. Nevertheless, among this huge number of fundamental theories, many are life-admissible. Even if the parameters and the formal structure of the physical laws change, life remains possible. In this context it is impossible to choose a unique theory and to fix univocally the value of the fundamental constants and parameters using simply the statement that life had to emerge in our Universe. This is precisely what Susskind (2006) said:

> … les considérations anthropiques […] ne suffisent pas à elles seules pour déterminer ou tout prédire. Ce qui est inévitable si le paysage contient plus d'une vallée favorable à notre vie. Avec 10^{500} vallées, on peut être sûr que c'est bien le cas. Disons qu'il s'agit de vides (Becker *et al.*, 2007) anthropiquement acceptables. La physique et la chimie ordinaires des électrons, des noyaux, de la gravité, des galaxies et des planètes ressemblent sans doute beaucoup à celles de notre univers. Les différences tiennent peut-être seulement à ce qui peut intéresser les physiciens des hautes énergies. Par exemple, il existe une multitude de particules dans la nature – le quark top, le lepton tau, le quark bottom et autres – dont les différentes propriétés n'ont pratiquement aucun effet sur notre monde ordinaire […] Y a-t-il moyen d'expliquer pourquoi nous vivons dans l'un de ces vides anthropiquement acceptables? Manifestement, le principe anthropique ne peut pas nous permettre de prédire celui où nous vivons – n'importe quel vide pourrait convenir.

Epistemological problems concerning the use of weak anthropic statements

A first immediate epistemological remark that could be made concerning the use of the 'anthropic principles' is that they are not at all 'principles' in the actual sense of other well-established principles: the covariance principle in general relativity, the second principle of thermodynamics, the principle of least action in classical mechanics etc. A principle is a kind of fundamental axiom that can be used in order to frame the construction of a theory and to derive, for example, fundamental equations, theorems etc. As an example, let us consider the principle of least action in field theory. It is built using some general principles of symmetry. We can impose, for instance, that the 'action' is invariant to a particular group undergoing transformation. This leads to a special mathematical form for the action. Then we apply the principle of least action by saying that the fundamental equation of the theory expresses the stationarity of this action (the fact that the action is an extremum-maximum or -minimum on the space of trajectories), namely, the fact that the first variation of this action is zero. A very simple example is the one where the action is simply the integral with respect to the time of the classical kinetic energy[6]. This action is invariant under the transformations of the Galileo group (for example translations in time). The principle of least action leads to the equations of Euler–Lagrange[7]. In this case, the latter is simply the Newton equation without any force. We see here clearly the framing role of a physical principle. An anthropic principle is not at all a principle in this sense. It can be considered as a heuristic tool that sheds some light on the value of fundamental parameters or on the formal structure of certain laws referring to some observational bias or to the a-posteriori coherence of the theories with the existence of some systems existing in the Universe. Epistemologically, a principle is the root, the foundation of a scientific explanation. In science, starting with general principles we try to explain particular empirical facts. However, 'weak anthropic principles' do not actually explain a fact (the precise value of a constant, for example). They express simply the 'coherence' of a theoretical framework with respect to certain facts (the existence of a human observer, for example). It is thus a manifestation of the 'coherence' of an explanatory scheme, but not a real explanation. Using Aristotle's way of speaking we could say that the weak anthropic principle 'cannot be an explanation' of some empirical reality because it does not help to progress in the direction of unravelling what is the 'cause' of such a reality.

The dream of those who use anthropic statements too naïvely would be that the formalism describing the Universe could be univocally determined by the study of one phenomenon in this Universe: the human being, for example. Philosophically, this dream refers to Plato's belief that the human being is, in fact, a 'microcosm', a reality reflecting the structure of the whole cosmos. But is this idea acceptable? Is empirical knowledge of a local part of the cosmos, as complex as we want it to be, sufficient to determine the formal structure of the theory describing the cosmos as a whole? The answer is no. Not only

[6] In one dimension the action is: $\int 1/2\ m\ (dx/dt)^2\ dt = \int L(x,t)\ dt$.

[7] $d/dt(\partial L/\partial v) - \partial L/\partial x = 0$ with $v = dx/dt$. This equation gives: $m\ d^2x/dt^2 = 0$.

because of the fact that there are some empirical limitations to getting this local knowledge but also, speaking the language of logicians, because in general, the knowledge of one particular model of a sufficiently rich formal language cannot determine univocally the latter. Let us explain this fact. According to logic, a model of a formal language[8] is a set of 'objects' upon which the axioms of this language are satisfied, i.e. are true. The usual plane with its points and straight lines is, for example, a model for the formal language defining Euclidean plane geometry. A discrete infinite set of points is a usual model for the formal language of Peano's arithmetic, the usual rules governing elementary computations on natural numbers. But this is not the only model for such a language. If you add to the infinite sequence of 'natural' numbers (0, 1, 2, ...) another infinite sequence of 'transfinite numbers' (ω, $\omega + 1$, $\omega + 2$, $\omega + 3$, ...) the set (0, 1, 2, 3, ..., ω, $\omega + 1$, $\omega + 2$, $\omega + 3$, ...) is still a model of the Peano arithmetic axioms expressing, essentially, the fact that there is a first natural number (0) and that each natural number 'n' has a successor 'n + 1'. And you can go on in order to get plenty of non-standard models of natural numbers. In fact, model theory proves that every formal language that has a model with an infinite number of elements has models of all cardinalities[9]. This simply means that 'the' formal theory of the Universe (a formal language) implies, in fact, an infinity of non-equivalent universes described by this theory! The idea arising from model theory is now that, in general, the complete knowledge of one model whose theoretical description is 'supposed' to be given by a formal language is not sufficient to describe adequately and univocally this formal language. Therefore, we cannot expect to adequately fix all the fundamental parameters of physics and to determine univocally the form of its fundamental laws by focusing only on complex human phenomena. The microcosm, in principle, cannot even reflect the deep formal structure of the cosmos. Incorporating Plato's ideas into weak anthropic statements is blocked by logic. Some locally gained theoretical bounds to the empirical knowledge about the Universe as a whole can add to the logical limitations. An example of such bounds is the so-called Beckenstein bound, which can be derived using general relativity, quantum mechanics and black-hole thermodynamics. If you are observing the Universe through the surface of a sphere centred at your position, you cannot measure more than 1/4 of the area of this surface in Planck units (Smolin, 2000). In conclusion, we can say that it is very interesting to check the coherence of a fundamental theory against the fact that human beings exist. However, this consistent checking does not mean that we would be able to determine in detail (by a kind of 'anthropic bootstrap') the formal description of our Universe only by studying the features of the complex human system!

[8] We recall that a formal language is a language based on a set of symbols that are combined together, according to some precise rules of syntax, in order to get well-formed expressions. The language is then characterized giving a set of axioms which are particular well-formed expressions considered as the starting point of deductions, the latter being controlled by precise rules of inferences. A set of symbols endowed with such rules of syntax, axioms and rules of inferences determine what the logicians call a formal language or system. Fundamental theoretical physics, yet not completely axiomatized, can be considered as a quasi-formal system.

[9] This is the essence of results related to the so-called Löwenheim–Skolem theorem (Cameron, 2008).

Metaphysical problems concerning the use of the strong anthropic principle

The fears concerning the use of anthropic arguments are mainly related to the strong versions of these arguments. In fact, the strong anthropic principle introduces a kind of finality, which is legitimate at the level of a philosophical approach, but which is excluded from the scientific methodology. A strong version could say: 'The Universe is such that human life had to emerge.'

Let us note first that such a statement is not necessarily connected to a teleology in a metaphysical sense. It is true that we can complete this statement by saying: 'The Universe is such that human life had to emerge according to a meta-natural purpose,' however, you would have to complete the statement as follows: 'The Universe is such that human life had to emerge according to the necessity of the physicochemical laws.' Then the strong anthropic principle is simply a version of a kind of strong determinism (this determinism is defended, for example, by Christian de Duve in his wonderful book *Vital Dust: Life as a Cosmic Imperative*, 1995): the origin of life and of human beings was not only possible, it was a necessity due to the structure of fundamental laws. The strong version is not necessarily a teleological one. However, if it does not have teleological commitment it is an expression of a strong determinism that has to be justified by studies proving that an evolution towards human complex systems was unavoidable on some protected cosmological site, from the existence, in the past, of elementary biochemical building blocks.

For a philosopher, such radically different interpretations of the same strong anthropic principle – a teleological one and a strongly deterministic one (and thus non-teleological) – is not a surprise. Several great philosophers have shown how some naïve conceptions of teleology are, in fact, very close to a purely deterministic vision. We can quote Henri Bergson here, who brought near-determinism and finalism into his famous book *L'Evolution Créatrice* in order to shed light on their inner limits (Bergson, 1907):

> L'erreur du finalisme radical, comme d'ailleurs celle du mécanisme radical, est d'étendre trop loin l'application de certains concepts naturels à notre intelligence [...] Qu'on se figure la nature comme une immense machine régie par des lois mathématiques ou qu'on y voie la réalisation d'un plan, on ne fait, dans les deux cas, que suivre jusqu'au bout deux tendances de l'esprit qui sont complémentaires l'une de l'autre et qui ont leur origine dans les mêmes nécessités vitales.
>
> C'est pourquoi le finalisme radical est tout près du mécanisme radical sur la plupart des points. L'une et l'autre doctrines répugnent à voir dans le cours des choses ou même simplement dans le développement de la vie, une imprévisible création de forme [...] Bref, l'application rigoureuse du principe de finalité, comme celle du principe de causalité mécanique, conduit à la conclusion que 'tout est donné'. Les deux principes disent la même chose dans leurs langues parce qu'ils répondent au même besoin.

The philosophical conception of Bergson is neither purely deterministic nor radically finalistic. For the French philosopher, physical and chemical processes are necessary conditions for biological evolution, but the emergence of new biological forms is not completely predictable; there is a fundamental indeterminacy, stochasticity. This way of thinking inspired

Ilya Prigogine (Prigogine and Stengers, 1979). It can also be found in cosmologist Georges Lemaître's work when he wrote (1967), concerning his famous primeval atom hypothesis:

> On peut retourner la phrase célèbre de Laplace et dire que celui qui connaîtrait l'atome primitif ou les premiers stades de sa division ne pourrait en aucune manière en déduire les particularités de l'univers qui commence. Dans le déterminisme laplacien tout est écrit, l'évolution est semblable au déroulement d'une bande magnétique enregistrée ou des spirales gravées sur un disque de phonographe. Tout ce que qu'on entendra aurait pu se lire sur la bande ou sur le disque. Il en est tout à fait autrement de la conception de la physique moderne et, dans la théorie actuelle, cette conception s'applique à l'univers, du moins au début de son évolution. Ce début est parfaitement simple, insécable, indifférenciable, atomique au sens grec du mot. Le monde s'est différencié au fur et à mesure qu'il évoluait. Il ne s'agit pas du déroulement, du décodage d'un enregistrement; il s'agit d'une chanson dont chaque note est nouvelle et imprévisible. Le monde se fait et il se fait au hasard. Tel est du moins tout ce que peut dire la physique ou l'astronomie. Il n'en est pas moins vrai que physique et astronomie n'épuisent pas toute réalité.

The critical approach of Bergson is interesting for us because it explains why a strong anthropic principle is not exclusively connected to a final cause or to a naïve conception of 'meta-natural' design, but that it can also express a purely 'natural' strong determinism. The reason is that naïve teleology is nothing but a reverse determinism. However, along the same lines, radical determinism is also a kind of reverse finalism! Indeed, logically, when you are assuming radical determinism, you have to impose that the whole history of the Universe is already written. Then we can say that the Universe is such that it will evolve necessarily towards such and such final or future states. For naïve finalism or for radical determinism the Universe is fine-tuned and it behaves according to a meta-natural (in the first case) or to a natural (in the second case) design. Many criticisms of the strong versions of the anthropic principle can be rechannelled into arguments against radical determinism.

We realize that defenders of the strong anthropic principle are allied to those supporting radical determinism due to the fact that their notion of teleology is nothing but the concept of a 'mechanical' cause directing all historical trajectories towards goals fixed a priori in the future. The problem that arises here is that, from a philosophical point of view, the concept of final cause is independent of the notions of a fixed design, a strong determinism or the execution of a programme. Coming back to Aristotle, the formal cause and final cause can be identified and finality can be conceived from a strictly immanent point of view. Imposing the existence of global regularity (physical laws whose legitimacy extends into the future as well as into the past) and formal structures has, in this context, already a flavour of finality because *telos* and *logos* meet. The history of philosophy offers many teleological systems that maintain a crucial place for contingency. In this case, the action of a *telos* is precisely to generate many degrees of freedom, allowing plenty of possible and non-necessary evolutions. Let us quote here an example coming from medieval philosophy. Thomas Aquinas in his famous *Summa Contra Gentiles* defends the thesis that God, whom he considers as the final cause (*causa causarum*), neither imposes a complete necessity on the Universe nor excludes pure contingency in immanent realities. At the end of his 'article' '*Quod divina providentia non exlcudit contingentiam a rebus*' of his *Summa*

(Lib. III, LXXII) he states: '*Non igitur divina providentia necessitatem rebus imponit, contingentiam a rebus universaliter excludens.*' (Divine providence does not thus impose a law of necessity on all the beings by pushing aside totally from them the contingency.) Then, logically, one cannot exclude, at a philosophical level, the existence of a meta-physical finality, which raises up and supports, at the physical level, free, plastic and nondeterministic evolutions. This precise position was the one adopted by Georges Lemaître who, at the same time, believed in a metaphysical finality but refused, as explicitly stated in the extract quoted above, the strong determinism of a fixed a-priori design.

The defenders of the strong anthropic principle believe that it is the actual expression of a cosmological finality. The opponents of this principle (which is not a principle!) are against this principle for the same reason. We have shown that both are wrong. Both a metaphysical finality without the execution of an a-priori fixed and rigid design and the absence of a metaphysical (meta-natural) teleology coupled to a global principle leading to the emergence of complex systems like human beings can be rightly considered.

Conclusion

Anthropic 'principles' are in fact incorrectly termed as they are not really principles in the true sense of the word, and thus it may be more appropriate to call them anthropic 'statements'. Weak anthropic statements are neither able to determine univocally the structure of a fundamental theory nor are they able to fix, without ambiguity, the values of physical parameters and constants. Weak anthropic statements do not involve, as such, any explanatory power. Nevertheless, these statements are not at all useless. In fact, they can be used to check the coherence of a fundamental theory, which must be compatible with the existence of life as it exists today. They can also be interesting as they can show the hidden links that relate biological phenomena with the properties of the cosmos at every level, from elementary particles to large-scale structures, and even to the cosmos as a whole. The interest in such statements is in their heuristic power, as a way to discover and to focus on some part of the phenomenological reality, and there is certainly no reason to exclude this heuristic method from scientific activity.

Strong anthropic statements can be considered as a commitment to a metaphysical finality. It is not purely science that, methodologically, can exclude any final cause. Yet the statement can also be considered as a position of radical determinism: the physical laws of our Universe are such that human life had to emerge. In this case, it is either a philosophical commitment or something that has to be proved, showing that effectively human life was unavoidable if we accept the formal structure of physical laws.

Philosophically, defence of the anthropic 'principles' or opposition to them is very often based on an incorrect conception of finality. Fine-tuning of parameters and constants or the exceptionality of the Universe are not necessarily linked with a philosophical teleology. A Laplacian Universe without any metaphysical goal and intelligent design can be perfectly fine-tuned for a fundamental immanent reason! On the contrary, a final cause can raise up multiverses, non-deterministic evolution, regressive evolutionary phyla and generate many degrees of freedom. A final cause could have highlighted evolution as a mere tinkering, leading to a history where determinism became supple.

The real danger in the use of anthropic statements is thus to mix science and metaphysics. However, it is useless to spend time fighting against arguments thought to be metaphysically loaded when they are only heuristic ones. From the end of the 1930s and until the 1960s, some physicists spent time fighting against the Big Bang models because they assimilated that kind of physical beginning to the Universe, the initial singularity, to a creation. At the same time, some physicists and philosophers believed that it was important to defend the primeval atom hypothesis due to its theological value. Fortunately, Georges Lemaître counteracted this by insisting that a natural beginning (from an initial singularity) had not to be confused with a creation in a metaphysical sense (a theistically created Universe can have no initial singularity and a materialistic cosmos can allow for a physical beginning). During the nineteenth century some philosophers fought against the second principle of thermodynamics because they were convinced that it was incompatible either with the eternity of matter or with Christian eschatology. Fortunately, some philosophers remarked that the second principle was not a metaphysical principle and that several philosophical readings of this principle were possible. The moral of this story is to be applied here: it is legitimate to search for some philosophical interpretation of scientific content, but it is epistemologically illegitimate to immediately transform scientific concepts into philosophical ones or to directly apply some extrinsic philosophical principle onto scientific knowledge in order to transform science into an ideological instrument. We have to articulate science and philosophy but without confusing both.

References

Barrow, J. D. and Tipler, F. J. (1986). *The Anthropic Cosmological Principle*. Oxford: Oxford University Press.

Becker, K., Becker, M. and Schwartz, J. H. (2007). *String Theory and M-Theory: A Modern Introduction*. Cambridge: Cambridge University Press, pp. 522–6.

Bergson, H. (1907). *L'Evolution Créatrice*. (Paris: Quadrige/P.U.F., 1981). Paris: Alcan.

Bernardeau, F. (2007). *Cosmologie: Des Fondements Théoriques aux Observations*. Paris: EDP Sciences, pp. 668–9.

Cameron, P. J. (2008). *Sets, Logic and Categories*. Berlin: Springer, pp. 95–112.

de Duve, C. (1995). *Vital Dust: Life as a Cosmic Imperative*. New York: Basic Books.

Demaret, J. (1991). *Univers: Les Théories de la Cosmologie Contemporaine*. Aix en Provence: Le Mail.

Demaret, J. and Barbier, C. (1981). Le principe anthropique en cosmologie. *Revue des Questions Scientifiques*, **152**, pp. 181–222, and 461–509.

Demaret, J. and Lambert, D. (1996). *Le Principe Anthropique. L'homme Est-il le Centre de L'Univers?* Paris: Armand Colin.

Lemaître, G. (1931). Expansion of the Universe: the expanding Universe. *Monthly Notices of the Royal Astronomical Society*, **91**, 490–501.

Lemaître, G. (1967). L'expansion de l'Univers: réponses à des questions posées par Radio-Canada le 15 avril 1966. *Revue des Questions Scientifiques*, t. CXXXVIII (5e série, t. XXVIII), Avril 1967, n°2, pp. 153–62.

Peter, P. and Uzan, J. P. (2005). *Cosmologie Primordiale*. Paris: Belin.
Prigogine, I. and Stengers, I. (1979). *La Nouvelle Alliance: Métamorphose de la Science*. Paris: Gallimard.
Smolin, L. (2000). *Three Roads to Quantum Gravity. A New Understanding of Space, Time and the Universe*. London: Phoenix, pp. 169–178.
Starkman, G. D. and Trotta, R. (2006). Why anthropic reasoning cannot predict lambda. *Physical Review Letters*, **97** 201301 (4 pages).
Susskind, L. (2006). *Le Paysage Cosmique: Notre Univers en Cacherait-il des Millions D'autres?* Paris: Robert Laffont.

3

Minimal cell: the biologist's point of view

Céline Brochier-Armanet

Introduction

The word 'cell' was first used by the scientist Robert Hooke in the seventeenth century in his book *Micrographia* (1665), where he described observations made with his own handmade microscope (Hooke, 1665). In particular, he noticed small cavities in a piece of cork delimitated by cell walls of cellulose and suber that he called cells. However, neither Hooke nor his contemporaries realized the importance of this discovery and it was only during the nineteenth century that the cell stood out as a central dogma in biology: the cellular theory. This theory is based on the work of Matthias Schleiden and his friend Theodor Schwann, who showed that all living organisms (plants, animals, moulds) are made of microscopic building units: cells (Schwann and Schleyden, 1847). The cellular theory is based on two central ideas. First, the cell is the unit of structure, physiology and organization in all living beings. Thus, the cell retains a dual existence – as a distinct entity and as a building block in the construction of organisms: unicellular organisms correspond to a single and autonomous cell, whereas multicellular organisms are made of two up to several billions of often highly differentiated cells. The second concept of the cellular theory is that 'every cell stems from another cell', and was formulated by Rudolf Virchow in 1855 (Virchow, 1855). In its modern form, cellular theory can be summed up as follows: (1) all living beings are made of cell(s) and are the descendants of pre-existing cells, and (2) each cell possesses all three of the following three properties: self-reproduction (i.e. they are able to produce systems having those three properties), self-maintenance (i.e. they are able to import and transform nutrients from the extra-cellular milieu to sustain their activities) and self-regulation (i.e. they are able to control and to regulate their metabolic activities) (see Luisi, 1997 for additional discussion concerning life and cell properties).

The concept of the minimal cell emerged in the late 1950s (Maniloff, 1996) (see also Morowitz, 1984 and references therein) and has been discussed by scientists belonging to various fields (e.g. artificial life, evolution, exobiology, cellular biology, prebiotic chemistry etc.). This concept can be formalized as the smallest number of properties and elements that are required to allow cellular life. Given that the number of genes required by each of them is not known, most researchers of the minimal cell try in fact to identify the minimal set of genes that is necessary and sufficient to sustain an autonomous cellular life.

Properties of cells

Significant progress concerning the definition of the minimal cell has emerged from cell biology works realized during the twentieth century that allowed the identification of a number of properties shared by all cells, without any exception. Consequently, these properties are expected to be present in the minimal cell.

The first feature common to all cells is the use of double-stranded DNA (deoxyribonucleic acid) as support for their genetic information. Cells transmit their genetic information (or a part of it) to their offspring. This implies that during cell reproduction, i.e. the mechanism by which a mother cell gives birth to two daughter cells, the DNA material is duplicated. This process – called replication – involves numerous proteins (e.g. DNA polymerases, topoisomerases etc.) and uses each of the two strands composing the DNA molecule as a template to synthesize the complementary one. At the end of replication the genetic material is present in two identical copies, each composed of one of the two original strands and of a newly synthesized one. Importantly, even if mechanisms assuring the fidelity of the replication exist, mutations can occur, leading to small variations in the genetic information that are transmitted to the daughter cells.

The second characteristic shared by all cells concerns the mechanism involved in the expression of the genetic information that is encoded in their genomes. In fact, the genetic information stored in DNA cannot be used directly by cells. It must be first converted to a single-stranded RNA (ribonucleic acid) by means of a process called transcription. The main differences between RNA and DNA are the type of sugar constituting the backbone of the molecules (i.e. ribose versus deoxyribose, respectively), and more importantly, the type of bases used to code the genetic information (i.e. A, U, G and C in RNA and A, T, G and C in DNA). Transcription is realized by a complex apparatus that involves many proteins, such as RNA polymerases, transcription factors etc. If a gene codes for an RNA, e.g. ribosomal RNA (rRNA), transfer RNA (tRNA) or small RNA (sRNA), its transcript can be used directly by the cell. On the other hand, if a gene codes for a protein, an additional step is required where the genetic information carried by the transcript (called messenger RNA, mRNA) is converted into protein. This process is called translation and is realized by complex cellular machineries called ribosomes made of rRNA and from 60 up to 130 different proteins. During translation, ribosomes ensure the conversion of the genetic information of mRNAs, based on a 4-character alphabetical code, to proteins that are composed of 20 types of amino acid. Ribosomes read mRNAs by triplets of bases (codons) and associate to each triplet a single amino acid. The correspondence between codons and amino acids is determined by the genetic code. Because all organisms use the same genetic code, it is called universal. However, a few organisms (e.g. the bacteria of the genus *Mycoplasma*) exhibit slight deviations of the universal genetic code, but these have been proven to result from recent and punctual evolution. Finally, the expression of the genetic information is coordinated through a great diversity of regulatory proteins able to bind DNA to activate or to repress gene transcription. Other regulatory mechanisms acting at the translation level exist and involve proteins or sRNA.

A third and important common feature is that all cells require an input of energy to sustain their integrity and activities (division, metabolism etc.). Energy can be obtained from various sources, but not all cells are able to use all of them: phototrophs use light, chemo-organotrophs the degradation of organic matter, and chemo-lithotrophs chemical reactions involving inorganic molecules. In addition to energy, all cells require raw materials to synthesize their own components. In fact, even if all cells use similar basic molecules (e.g. sugars, lipids, nucleic acids, amino acids, cofactors etc.), few organisms are able to synthesize all of them *de novo*. For example, autotrophs such as plants are able to use inorganic CO_2 as a sole source of carbon, whereas heterotrophs use the carbon present in organic matter for the biosynthesis of their components. Finally, some organisms such as bacterial symbionts (i.e. mutualistic, commensalist or pathogens according to the fitness of each partner) have lost many (and sometimes most) of their biosynthetic capacities and are thus dependent on their hosts for their supply of small molecules.

Finally, all cells are delimited by a semi-permeable membrane, called a cell membrane, primarily (but not exclusively) made of lipids and proteins. It acts as a selective barrier that allows the importation of raw materials and the excretion of waste molecules resulting from metabolic activities, and prevents the dispersion of cell components. More than that, cell membranes are essential to the production of energy via the creation of a proton gradient that is exploited by the universally conserved ATPases to generate ATP, the universal energy currency within cells.

Minimal genomes in present-day cells

An intuitive approach to tackle the concept of a minimal cell is to search for present-day 'minimal cells' (i.e. the simplest cells, with small genomes, a limited set of metabolic activities etc.). Indeed, this could provide a good approximation of the functions and genes a minimal cell should harbour. Present-day prokaryotic genomes range from a few hundred kilo base pairs (kbp) up to 13 mega base pairs (Mbp), corresponding for example to 9384 genes in the delta-proteobacterium *Sorangium cellulosum* (Galperin, 2008) (Figure 3.1). By contrast, eukaryotic genomes are most often significantly larger, having sizes ranging from a few Mbp to 670 000 Mpb in the amoebozoan *Amoeba dubia* (by comparison the human genome only has 2900 Mpb) (Gregory and Hebert, 1999). The smallest prokaryotic and eukaryotic genomes are harboured by symbionts. In eukaryotes, the smallest genome known (2.9 Mpb encoding 1997 genes) comes from the microsporidian *Encephalitozoon cuniculi* (a parasitic fungus) (Katinka *et al.*, 2001). In Archaea, the smallest sequenced genome (491 Kpb encoding 536 genes) is that of *Nanoarchaeum equitans*, a tiny mutualistic–parasitic symbiont that lives at the surface of another archaeon, *Ignicoccus hospitalis* (Waters *et al.*, 2003). Concerning bacteria, for a long time the smallest known genomes have been those of parasites from the genus *Mycoplasma* (Firmicutes) with genomes ranging from 0.58 Mpb to 1.36 Mpb (Fraser *et al.*, 1995), as well as those of animal mutualistic gammaproteobacterial symbionts such as *Buchnera* (0.42 to 0.66 Mpb), *Baumannia* (0.69 Mpb) and *Wigglesworthia* (0.71 Mpb) (see Gil *et al.*, 2004; Klasson and Andersson 2004; McCutcheon, 2010, and

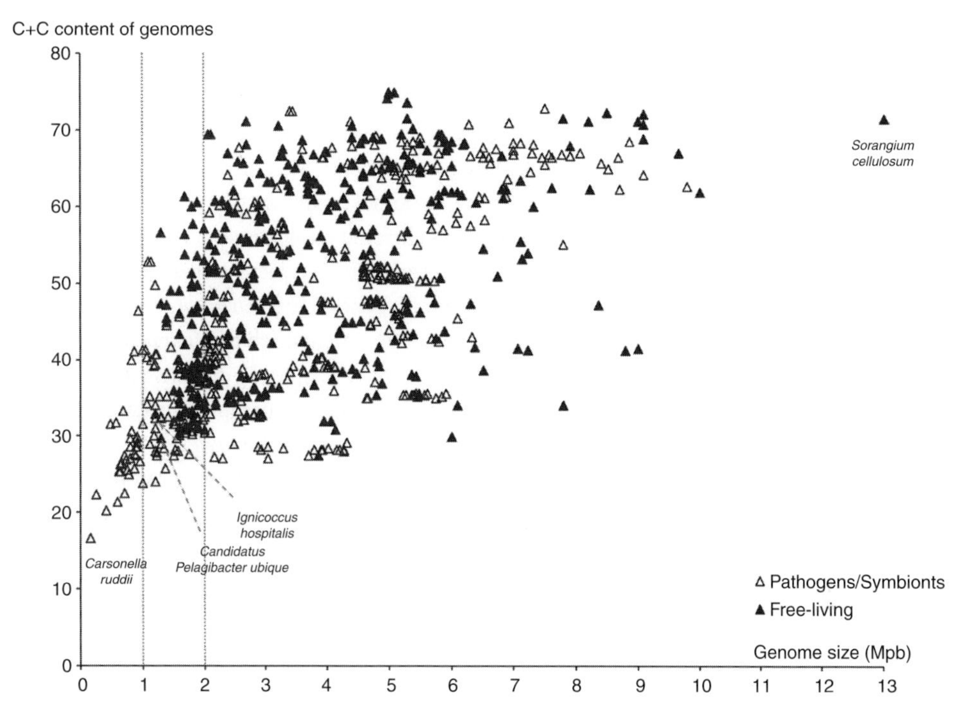

Figure 3.1. Graphic showing the genome size distribution of the 881 completely sequenced prokaryotic genomes (http://www.ncbi.nlm.nih.gov/genomes/lproks.cgi). Filled and empty triangles correspond to genomes from free-living and from parasitic/symbiotic organisms, respectively.

references therein). The observation of highly reduced genomes in various lineages among the three domains of life indicates that extreme genome reduction events are not rare and have occurred several times and independently during evolution.

For a long time the threshold of 400 Mpb was considered as the minimal size compatible with cellular life. However, this limit was recently overcome with the publication of the genomes of two bacterial symbionts of animals: the Bacteroidetes *Sulcia muelleri* (0.25 Mpb containing 261 genes) (McCutcheon and Moran, 2007) and the gamma-proteobacerium *Carsonella ruddii*, (0.16 Mb corresponding to 213 genes) (Nakabachi *et al.*, 2006). These are by far the most streamlined genomes known to date, and pose even the question of the cellular nature of these organisms. Indeed, a survey of their gene repertoires shows that most biological processes essential for an autonomous cellular life are not encoded in these genomes (Tamames *et al.*, 2007). The extreme genome reduction observed in *S. muelleri* has been explained by a tripartite mutualistic symbiosis involving, in addition to this organism, the animal host and a co-symbiont (the bacterium *Baumania cicadellinicola*) (McCutcheon and Moran, 2007). Each of these organisms specifically encodes at least one essential function that is missing in the two others, ensuring that neither of them can exist without the others. For example, *B. cicadellinicola* can produce nucleotides, vitamins and cofactors, but not amino acids, whereas *S. muelleri* is able to synthesize all essential amino acids, but not nucleotides, vitamins or cofactors, whereas the host cells cannot produce either of them (McCutcheon and Moran, 2007). In the case of *C. ruddii*, the situation is different. First, in contrast to *C. muelleri*, the gene complement of *C. ruddii* is not sufficient to replicate, transcribe and synthesize proteins, thus questioning its status of an independent living entity (Tamames *et al.*, 2007). Second, because no other symbionts are present in the host, the hypothesis that missing functions can be compensated for cannot be retained to explain this extreme reduced genome (Nakabachi *et al.*, 2006). Thus, it has been proposed that – as in the case of typical organelles (i.e. mitochondria and chloroplasts) – the genes missing in the genome of *C. ruddii*, but coding for essential functions, have been transferred to the host genome, which has taken control over their expression (Nakabachi *et al.*, 2006). Alternatively, it has been proposed that *C. ruddii* may compensate for its missing functions by importing and using the components of the mitochondrial machineries encoded into the host nucleus (Tamames *et al.*, 2007). This suggests that this bacterium might be a new sub-cellular entity, intermediate between living cells and organelles (Tamames *et al.*, 2007) or a degenerating symbiont bound to extinction, rather than a bona fide living organism.

Interestingly, a careful examination of the size distribution of complete prokaryotic genomes indicates that genomes smaller than 1 Mbp are found exclusively in obligate parasites and symbionts, whereas the 1- to 2-Mbp range corresponds to both symbionts and parasites and to free-living organisms (Figure 3.1). Accordingly, it was predicted that free-living minimal cells (i.e. those that do not require an established and continuous association with another species and can replicate independently) should have genomes within this range, which correspond approximately to 1000–2000 genes (Podar *et al.*, 2008). Among them, the smallest genomes of free-living organisms (i.e. those that do not require an established and continuous association with another species and can replicate independently) are found in the archaeon *I. hospitalis* (1.29 Mpb coding for 1434 proteins) (Podar *et al.*, 2008) and in the bacterium *Pelagibacter ubique* (1.31 Mpb, 1354 proteins). The smallest sequenced genomes of free-living eukaryotes correspond to yeasts (Fungi) with 12 Mbp, the green algae *Ostreococcus tauri* (12.5 Mpb) and the thermophilic red algae *Cyanidioschyzon merolae* (16.5 Mpb).

The survey of present-day genomes, although indicative, is far from defining minimal cells. Indeed, even in the highly reduced genomes mentioned above, some genes may be dispensable (see below). Thus, experimental approaches are required to better define the minimal-cell concept. Two approaches were developed in the twentieth century. The first is based on *in-vivo* genetic experiments that consist of the systematic inactivation of each gene present in the genome of an organism. The second approach is instead based on *in-silico* comparisons of genomes from various organisms. Both types of analysis assume that, as in present-day living organisms, proteins and RNA are the functional entities of the minimal cell, and that the instructions for their synthesis are encoded in genes (Koonin, 2003). However, even if the minimal cell should include an integrated and coherent network that controls gene expression and regulates cellular processes, most of the published studies are restricted to the identification of the minimal set of essential proteins. In consequence, what they reconstruct is a minimal and static genome rather than a minimal cell. Finally these two approaches use either genomic data from modern organisms or present-day organisms as experimental models. This implies de facto that the minimal genomes inferred (and thus the minimal cell) will be, in fact, a minimal modern-like genome and a minimal modern-like cell. It is important to notice that these entities are probably very different to the first genomes and first cells. Based on results from *in-silico* and *in-vivo* approaches, theoretical approaches have also tried to deduce the genome and cellular properties of the first cells (Luisi *et al.*, 2002).

In-vivo genetic experiments

The goal of these approaches is the identification of genes that are indispensable for cell growth in particular environmental conditions. The assumption made is that the set of genes that is essential for cell growth forms the minimal core of genes that define a cell. The dispensability or indispensability of each gene is measured by the growing defect in the corresponding mutant. If the mutant is unable to grow, this means that the mutation is lethal and, accordingly, the corresponding gene is classified as essential. On the other hand, if no significant growth defect is observed, the mutation is classified as non-lethal, and the corresponding gene is considered as dispensable. Various genetic methods have been used for this type of analysis (Table 3.1), but all make the strong assumption that the experimentally defined minimal cells live in an ideal and stable environment where all nutrients are available in abundance. It is important to notice that these experimental conditions are very different from the natural habitats of the organisms. Thus, some genes detected as non-essential under laboratory conditions may be essential in natural environments.

Mutagenesis by transposon insertion is one of the most popular experimental approaches to tackle the minimal-cell study (Table 3.1). Transposons are short DNA sequences that can randomly insert into genomes and disrupt genes, leading to their inactivation. The mutagenesis by transposon experiments is designed so that on average only one single transposition event occurs in each cell. The analysis of many thousands of cells allows for the expectation that each gene has been disrupted at least once. After mutagenesis, only mutants where transpositions occurred in non-essential genes are expected to grow. By contrast, genes where insertions are never observed are supposed to be essential because their mutants are not viable.

Such a study has been realized in the bacterium *Mycoplasma genitalium*, a human pathogen (Firmicutes) whose genome sequence was published in 1995 (Fraser *et al.*, 1995; Hutchison *et al.*, 1999). *M. genitalium* possesses one of the smallest cellular genomes known (540 Kpb coding for 537 genes) and accordingly was considered as a good model to tackle the question of the minimal cell (Fraser *et al.*, 1995). Indeed, it was believed that this genome was mainly composed of essential genes (for example, around one-third of its genes code for proteins involved in protein biosynthesis). Surprisingly, transposon insertions have estimated that 180–215 genes (~1/3 to 1/2 of its genes) are non-essential in *M. genitalium*, whereas 265 to 350 are essential (Hutchison *et al.*, 1999). The set of non-essential genes was later scaled down to 100 genes by the authors (Glass *et al.*, 2006). However, both studies showed that the different functional categories are not equally represented in

Table 3.1. *In vivo* experimental reports focused on the minimal genome.

Year	Organism used	Experimental procedure	Genome size	Minimal set of genes	Reference
1995	*Bacillus subtilis*	Site-directed mutagenesis	4188	318	(Itaya, 1995)
1999	*Mycoplasma genitalium*	Mutagenesis by transposons	537	265–350	(Hutchison *et al.*, 1999)
1999	*Mycoplasma pneumoniae*	Mutagenesis by transposons	733	265–350	(Hutchison *et al.*, 1999)
1999	*Haemophilus influenzae*	Mutagenesis by transposons	1709	990–1000	(Reich *et al.*, 1999)
1999	*Saccharomyces cerevisiae*	Single-gene deletion	6925 (2026 tested)	344	(Winzeler *et al.*, 1999)
2001	*Mycobacterium bovis*	Mutagenesis by transposons	3953	658	(Sassetti *et al.*, 2003)
2001/2002	*Staphylococcus aureus*	Antisense RNA inihibition	2509	150/658	(Ji *et al.*, 2001; Forsyth *et al.*, 2002)
2002	*Haemophilus influenzae*	Mutagenesis by transposons	1709	478–670	(Akerley *et al.*, 2002)
2002	*Saccharomyces cerevisae*	Single-gene deletion	6925	>1000	(Giaever *et al.*, 2002)
2002	*Escherichia coli* K-12	Large genomic deletions	4254	$$$	(Kolisnychenko *et al.*, 2002)
2002	*Escherichia coli* K-12	Genomic deletion + mutagenesis by transposons	4254	###	(Yu *et al.*, 2002)
2003	*Bacillus subtilis*	Non-replicating plasmid	4106	271	(Kobayashi *et al.*, 2003)
2003	*Pseudomonas aeruginosa*	Mutagenesis by transposons	5566	667–300	(Jacobs *et al.*, 2003)
2003	*Escherichia coli* K-12	Genetic footprinting	4254	620	(Gerdes *et al.*, 2003)
2004	*Escherichia coli* K-12	Mutagenesis by transposons	4254	2312	(Kang *et al.*, 2004)
2004	*Helicobacter pylori*	Mutagenesis by transposons	1566	344	(Salama *et al.*, 2004)
2006	*Mycoplasma genitalium*	Mutagenesis by transposons	537	487	(Glass *et al.*, 2006)

$$$: In this study 8.1% of the genome of *E. coli* K-12 was successfully removed without diminution of growth rate on minimal medium.

###: In this study single deletions of 59–117 Kpb were generated. The combination of some of these deletions in a 0.31 Mpb 'cumulative-deletion-strain' (corresponding to 287 ORF) exhibited normal growth under standard laboratory conditions.

terms of number of essential genes. In fact, and as expected, these genes are over-represented in categories corresponding to fundamental cellular processes such as translation, transcription, replication, lipid metabolism, energy conversion, cell reproduction, transport etc. (Figure 3.2A). By contrast, essential genes are under-represented in other categories such as interaction with host cells, cell envelope, gene regulation, nucleotide, small molecule and cofactor biosynthesis etc. (Hutchison *et al.*, 1999).

However, as mentioned above, some of the genes that are dispensable under laboratory conditions may be important in natural environments. This was well illustrated in *Mycobacterium*, where a total of 194 genes essential for growth *in vivo* have been identified, whereas most of these genes are non-essential under laboratory conditions (Sassetti and Rubin, 2003). This may explain why most genes involved in cell-envelope biosynthesis (and are thus important for host–pathogen interactions during infection) are non-essential in laboratory growth conditions. Similarly, most experimental studies use quite unrealistic ideal conditions where small molecules (amino acids, nucleotides, cofactors etc.) are available in the growth medium. Accordingly, genes involved in their biosynthesis may turn out to be dispensable if the cells are able to import those molecules. A most unexpected discovery was that 67 out of 111 genes of unknown function appeared to be essential (Hutchison *et al.*, 1999). This was quite surprising because it assumed that all genes involved in essential functions were already known and, accordingly, it was expected that all the genes of unknown function should represent species-specific non-essential genes. The high number of essential genes with unknown function may imply that they are involved in fundamental but as yet unidentified functions, or that they are somehow involved in essential known functions.

Similar studies, but using different approaches and/or different organisms as models, showed very similar results to those obtained with *M. genitalium* from a qualitative point of view (see references in Table 3.1). For example, 658 essential genes have been identified in the bacterium *Staphylococcus aureus* (Forsyth *et al.*, 2002). Among these, 168 genes have homologues in the highly reduced genome of *M. genitalium* and 146 (87%) have been identified as essential in this bacterium (Hutchison *et al.*, 1999; Forsyth *et al.*, 2002). This indicates that most genes predicted as essential in *M. genitalium* are also essential in *S. aureus*. However, it seems to be quite surprising that most of the essential genes (490) in *S. aureus* have no counterpart in *M. genitalium*. This points to the fact that gene essentiality is a subjective notion because it is strongly dependent on the considered organism: some genes essential in a given species can be non-essential (or even missing) in other organisms. For example, in contrast to *S. aureus*, *M. genitalium* has no cell wall. Accordingly, genes required for the synthesis and maintenance of this structure are fully dispensable in *M. genitalium* and may have been lost during evolution.

Despite the efficiency of *in-vivo* inactivation experiments, they suffer from a number of technical limitations that may over- or under-estimate the proportion of essential genes in genomes, which may explain the great variability in the size of the essential gene sets predicted by these experiments, even when the same organism is used (Table 3.1) (see Gil *et al.*, 2004; Feher *et al.*, 2007, and references therein). The main problem is linked

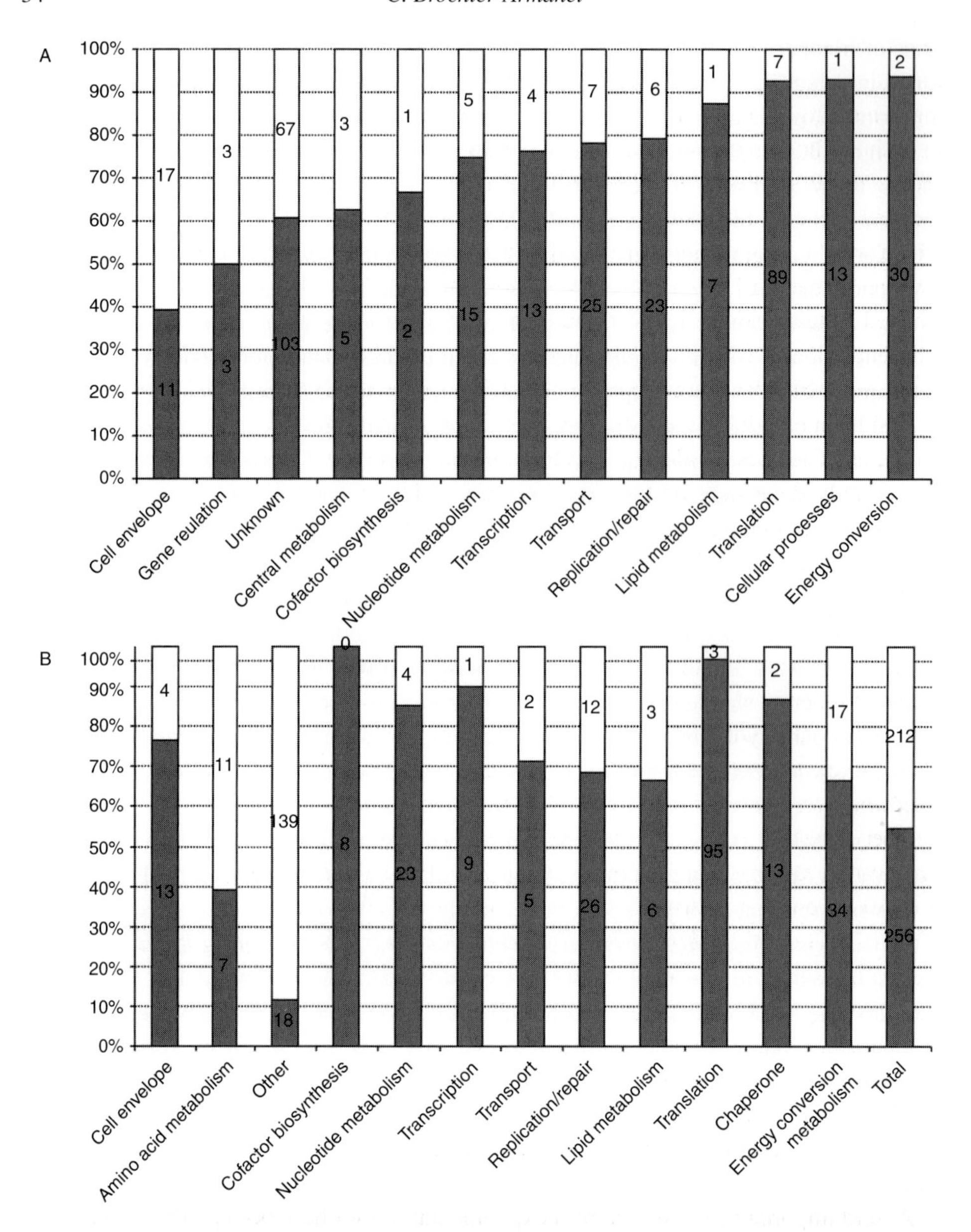

Figure 3.2. A: Histogram showing the proportion of essential (in grey) and non-essential (in white) genes among main functional classes identified by transposon mutagenesis experiment (Hutchison *et al.*, 1999). The corresponding numbers of genes are indicated within each bar. B: Histogram showing the proportion and the function of *M. genitalium* genes having orthologues in *H. influenzae* (in grey) and without orthologue in *H. influenzae* (in white) (Mushegian and Koonin, 1996). The corresponding numbers of genes are indicated within each bar.

to difficulties in the interpretation of mutant phenotypes. For example, some mutations can slow down the growth of cells without arresting it and, accordingly, the corresponding genes might be classified as essential, whereas they are not. Other biases are specific to each methodology. For example, in transposon mutagenesis some essential genes may escape detection because there is no guarantee that each gene of the genome has been mutated. Second, in some cases transposon insertions may fail to fully inactivate genes, which will be thus wrongly classified as non-essential. Finally, many studies are affected by a phenomenon known as 'cross-feeding'. In fact, because mutants are not grown in isolated cultures, some of them can obtain nutrients that they are unable to synthesize from adjacent colonies where the corresponding gene has not been inactivated. Similarly, horizontal gene-transfer events between cells in the same culture may rescue some mutants.

Besides methodological bias, *in-vivo* approaches suffer from a number of conceptual problems. The most evident is that the minimal-gene complements required for growth are strongly dependent on the experimental conditions used. For example, some genes may be non-essential in some environmental conditions (e.g. aerobiosis) but may be indispensable in others conditions (e.g. anaerobiosis). Similarly, if the medium contains amino acids, the genes required for their biosynthesis may be dispensable, whereas if amino acids are not present in the medium, the same genes will be indispensable. Second, most studies used bacterial models, so that the essential genes (and by extension the minimal genomes) identified are de facto 'bacterial'. Thus, it is likely that similar experiments in Archaea or eukaryotes will lead to the identification of different sets of genes. This was shown in the case of eukaryotes, where single-gene deletions were performed in the yeast *Saccharomyces cerevisiae* (Table 3.1). Even if yeasts have a highly reduced genome, the number of genes to test is significantly larger than in prokaryotes, making systematic gene-inactivation experiments very challenging. The two analyses performed suggest that more than 1000 genes are indispensable in this organism (Winzeler *et al.*, 1999; Giaever *et al.*, 2002). This relatively high number of essential genes predicted in yeast compared to bacteria is not surprising because eukaryotic cells harbour a more complex organization than their prokaryotic counterparts. In particular, they possess a number of essential cellular structures and processes such as the nuclear envelope (Mans *et al.*, 2004), the nuclear pore complex (Bapteste *et al.*, 2005), the spliceosome (Roy and Irimia, 2009), mitochondria (Embley, 2006) etc., that have no equivalent in prokaryotic cells, and that probably require numerous eukaryotic-specific genes (Makarova *et al.*, 2005), among them some that are essential.

A third important issue is that most experimental studies have been performed on organisms with relatively small genomes, which are often considered to be close to minimal genomes. The main consequence of the use of these very specialized organisms is that the essential genes identified will not be compatible with a minimal but autonomous cellular life. This problem may be bypassed by the use of free-living organisms for experiments. However, such approaches will be confronted by other problems, such as genomic redundancy. Indeed, genomes of free-living organisms are often large and contain numerous paralogues (Glass *et al.*, 2006). These genes result from duplication events and paralogues often display similar functions. A direct consequence is that the proportion of genes detected as essential in these genomes will be underestimated because their function might be rescued by paralogues. Thus, paralogues can be individually non-essential but collectively essential. Such bias is expected to be marginal in the case of small genomes because the proportion of paralogues is negligible (e.g. ~ 6% in *M. genitalium*) (Glass *et al.*, 2006). By contrast, this problem is not trivial in the case of large genomes because paralogues may represent up to 50% of the genes (Kobayashi *et al.*, 2003; Glass *et al.*, 2006). This outlines the concept of 'mutually inclusive' and 'mutually exclusive' genes. The latter designates genes that can be individually but not collectively deleted. This was highlighted in the case of *E. coli*, where several deletions were successfully combined in a single strain (including one strain that lacks 287 genes), whereas other deletions were incompatible (Kolisnychenko *et al.*, 2002; Yu *et al.*, 2002). For example, the deletion of either *nrdA* (essential for aerobic growth) or *nrdD* (essential for anaerobic growth), two genes involved in deoxyribonucleotide synthesis, is viable, whereas their simultaneous deletion would eliminate the ability of *E. coli* to grow both aerobically and anaerobically and is thus lethal (Garriga *et al.*, 1996). By contrast, mutually inclusive deletions encompass cases where genes cannot be individually deleted, suggesting that they are essential, whereas they can be deleted in combination with other genes. For example, cell-killing systems in *E. coli* composed of toxin and anti-toxin coding genes have been described (Gotfredsen and Gerdes, 1998). Accordingly, deletion of the anti-toxin genes leads to cell death whereas deletion of both genes is viable. These few examples illustrate the fact that the set of essential genes is not equivalent to the minimal genome. In fact, single-gene-inactivation experiments are only the first step through the characterization of essential genes in genomes. The next step would be the systematic combination of data from single-gene-inactivation experiments with gene product interactions, in order to have a clearer picture of the functional relationships between genes within a genome and gain insight into the definition of the minimal genome.

Despite serious limitations and biases, *in-vivo* approaches to the minimal-genome problem have provided a number of very important results. Comparison of these studies allows the drawing of an emerging portrait of a minimal bacterial cell. It consists of a cell compartment delimited by a membrane that is able to perform most of the universal cellular functions, such as replication, cell division, transcription, translation and energy production through a very simple metabolism (glycolysis). By contrast, this minimal bacterium includes few genes involved in regulatory networks, cell envelope and cell shape, cell division, respiration and synthesis of small components (e.g. nucleotides, amino acids etc.). This is not surprising because these functions are often very species-specific or non-essential under laboratory conditions. The minimal genome associated with this minimal cell ranges from two hundred to one thousand genes. This reflects differences in the genetic background used for *in-vivo* experiments: minimal estimates are obtained in the case where model organisms already have highly reduced genomes, whereas the number of essential genes is more important in the case of free-living organisms because their genomes encode more functions. Finally, an unexpected discovery was that most of the essential genes are present in a single copy in genomes (e.g. ribosomal proteins, RNA polymerases etc.). By contrast, a higher redundancy is observed in non-essential genes. This was quite surprising because the presence of multiple copies of essential genes would be advantageous for cells, because it would prevent the loss of essential functions. Different explanations have been proposed. First, the presence of several copies of essential genes will allow for the accumulation of slightly deleterious mutations, thus leading to a progressive decrease of the fitness of these genes in cell populations. By contrast, the presence of these genes in one copy ensures the maintenance of the integrity of their sequence, because cells carrying mutated genes would be eliminated by natural selection from cell populations. On the other hand, a redundancy of genes considered as non-essential under laboratory conditions has been proposed to increase the fitness of organisms because it allows them to cope better with changes in environmental conditions, each copy allowing subtle advantages in different situations.

In-silico approaches

These approaches are based on the simple idea that all present-day living organisms are cells, a feature that has been inherited from the last universal common ancestor (LUCA), which was probably a free-living unicellular organism (Forterre *et al.*, 2005; Ouzounis *et al.*, 2006). It seems thus reasonable to suppose that LUCA already had all the genes necessary to sustain a typical cellular life and that these genes have been transmitted to its offspring, generation after generation, up to present-day cells. This implies that although present-day genomes harbour very different sizes and gene repertoires, they are expected to be composed of a conserved core of genes (CCG) inherited from LUCA (encompassing the genetic information required to sustain cellular life) and of a variable shell of accessory genes involved in more species-specific activities. According to the importance of gene loss in prokaryotes, only genes involved in fundamental cell functions are expected to belong to the CCG. Thus, in theory, the identification of the CCG inherited from LUCA is a feasible task through the comparison of present-day genomes. In fact, this idea is not new and was put forward even before the sequencing of the first cellular genomes, and applied when the sequencing of the first two cellular genomes came out in 1995 (Fraser *et al.*, 1995).

The first two genomes compared were those of *H. influenzae*, a gamma-proteobacterium, and a very distant relative *M. genitalium* (Fraser *et al.*, 1995; Mushegian and Koonin, 1996). This analysis allowed the identification of 256 genes shared by both genomes (Figure 3.2B). The corresponding genes encompass most of the fundamental functional categories suggesting a minimal cell with a complete translation, transcription and replication apparatus, a rudimentary machinery involved in DNA repair and recombination, several chaperones, a limited anaerobic central metabolism restricted to glycolysis and a secretory system. Interestingly, most of the genes belonging to the CCG were predicted to have homologues in the two other domains, Archaea and Eucarya (an intriguing exception concerned genes involved in DNA replication; see below). This suggested that a universal CCG inherited from LUCA still exists in present-day cells and would progressively emerge with the inclusion of data from future genome-sequencing projects.

Table 3.2. Proteins conserved in 100 bacterial and archaeal genomes (Koonin, 2003).

Functional category	Number of genes	Gene missing in one or two genomes
Translation	41	3
Transcription	4	1
Nucleotide metabolism	2	2
Secretion	4	0
Repair/recombination	1	1
Central metabolism	0	3
Unknown	0	1
Total	52	11

The main conclusions of those works were that the CCG identified was considered as sufficient and necessary to sustain a modern cellular life. Moreover, the CCG identified by this *in-silico* approach was very similar to the minimal genome identified by *in-vivo* genetic experiments, both in terms of conserved functional categories and number of genes involved (Figures 3.2A and 3.2B). This suggested that both were close to the theoretical minimal genome. However, additional experiments showed that for ~15% of the 256 genes composing the CCG, viable knockouts were obtained in *M. genitalium*, meaning that some of the genes of the CCG were, in fact, non-essential (see Koonin, 2000, and references therein). This outlines the main difference between the *in-vivo* and *in-silico* approaches: the first identifies genes that are essential for growth in individual species whereas the second identifies a set of conserved genes among diverse taxa. Accordingly, it is not surprising that some genes of the CCG may be successfully inactivated by *in-vitro* experiments because they are not (or are no longer) essential in some organisms. Reciprocally, some genes inferred to be essential in some organisms by *in-vitro* experiments may be of recent origin and thus do not belong to the CCG defined by *in-silico* approaches. Some approaches have tried to combine the results of both types of analysis. Among them the analysis of Gil *et al.* (2004) allowed the definition of a minimal genome containing 206 genes. The minimal cell that emerged from this study is able to synthesize genes involved in DNA and RNA metabolism, protein synthesis and processing, energetic and intermediary metabolisms and cellular processes (Gil *et al.*, 2004).

Similar studies published afterwards have shown that, depending on the type of organisms compared, the CCG is highly variable. In fact, the inclusion of an increasing number of genomes leads to a progressive diminution in the number of genes shared by all genomes. For example, a similar analysis but including 21 genomes (1 eukaryote, 4 Archaea and 16 bacteria) led to a CCG that contained only 80 genes (instead of 256 in the previous study) (Koonin, 2003) and this number falls to 52 when 100 genomes are considered (Table 3.2). This new CCG is mainly composed of genes that encode for proteins involved in transcription and translation, but the detected genes are not even sufficient to code for a functional ribosome or a functional transcription apparatus. Thus, in contrast to previous expectations, the increase of genomic data leads to a CCG that is incompatible with any form of cellular life (and that is probably not representative of the genome of LUCA) because most essential cellular functions are not represented. This is a consequence of several factors. First, some universal genes escape detection because they do not share sufficient similarity to be detected as homologues. Second, even essential genes can be lost or replaced in some lineages during evolution, following, for example, the emergence of more recent genes with the same function. Similarly, symbionts often have very simplified biochemistries (compared to their free-living ancestors) resulting from recent and massive gene-loss events that may also include genes that are essential in free-living organisms. Moreover, key cellular functions that are conserved in reduced genomes may vary from one organism to another, reflecting highly specialized but different lifestyles (Klasson and Andersson, 2004). Thus, because their inclusion in such *in-silico* analyses will strongly and artificially constrain the size of the CCG, these genomes should be discarded. Accordingly, the use of less stringent criteria, such as the absence of conserved genes in one or two genomes, allows the inclusion of eleven protein-coding genes to the CCG (Table 3.2), and cellular functions such as glycolysis or nucleotide metabolism, that were not represented or were poorly represented in previous estimates (Koonin, 2003). Nevertheless, this set of CCG is still incompatible with a cellular life.

To limit the above-mentioned biases, alternative *in-silico* approaches have been proposed. These are based on the identification of core conserved functions (CCF) instead of CCG (Ouzounis and Kyrpides, 1996). These studies use the presence of protein families as characters and link them to biochemical pathways, allowing the inference of cellular processes in LUCA. Such analyses have inferred the presence of a set of CCF containing 77 protein families in LUCA among the 944 analysed. Without surprise, and similarly to approaches focused on CCG, the authors infer that a modern-like translation apparatus was present in LUCA based on CCF. However, the analysis of other components of the CCF provided a very different picture of LUCA. It was, in fact, predicted to harbour a complex network of reactions including amino acid, nucleotide, fatty acid, sugar and coenzyme biosynthesis pathways. By contrast, other processes such as DNA replication and repair, or transcription, were predicted to have undergone important structural changes. This approach was limited by the scarcity of data available for Archaea at that time. An update of this analysis has increased the size of the CCF to 246, representing at least 324 proteins (Kyrpides *et al.*, 1999). This new CCF encompasses nearly all of the most important cellular functions (e.g. metabolism of amino acids and nucleotides, energy conversion and electron transfer, transporters, carbohydrate metabolism, transcription and translation, protein modification and degradation), whereas a few functions are still under-represented (e.g. cell division, intracellular signalling, transcription factors). This estimate of the CCF is in agreement with a study published by Tatusov *et al.*, who identified more than 300 universal functions probably present in LUCA (Tatusov *et al.*, 1997).

The functional categories absent from minimal genomes inferred in those studies are very intriguing. This is the case for instances of enzymes involved in DNA replication. This was unexpected because all present-day cells use DNA to store their genetic information, but reflects the fact that the replication machineries are not homologous in the three domains of life. More precisely, most archaeal and eukaryotic proteins involved in replication are non-homologous to their bacterial counterparts. Similarly, only a few proteins involved in DNA repair are homologous among the three domains of life. Three hypotheses have been proposed to explain these surprising results (Mushegian and Koonin, 1996; Forterre *et al.*, 2005). The first hypothesis assumes that a non-homologous replacement of the ancestral replication machinery present in LUCA occurred during evolution (i.e. in the bacterial lineage or in a specific ancestor of Archaea and eukaryotes). The second hypothesis is that DNA and DNA replication arose after LUCA, meaning that this organism had an RNA genome. This hypothesis also implies that at least two independent RNA to DNA replacements occurred during evolution (i.e. one in the bacterial lineage and one in an ancestor of Archaea and eukaryotes). Finally, a third scenario assumes a LUCA with a DNA genome but with a very primitive replication apparatus that was subsequently modernized by recruitment of non-homologous proteins in the three domains.

More sophisticated *in-silico* analyses have recently been developed. Briefly, they compare genome contents in the light of their evolutionary relationships. The number of genes or protein families in ancestors of the studied organisms is inferred using sophisticated models that take into account gene losses and/or gene transfers that may bias these analyses if not correctly considered (Snel *et al.*, 2002). Using those new approaches, a recent analysis of 37 402 protein families across 184 genomes inferred that a CCF composed of ~ 1400 gene families was present in LUCA (Ouzounis *et al.*, 2006). This estimate challenged the widely held view of a minimal genome containing ~ 300 genes supported by previous *in-silico* or *in-vivo* analyses. A careful examination of the predicted CCF shows a number of protein families involved in previously poorly represented important functions such as replication, recombination, repair and modification. This suggests that most aspects of DNA metabolism were already present in LUCA and therefore that LUCA may have had a DNA genome. Concerning other processes, LUCA was predicted to harbour functions involved in various aspects of cell division, thermoprotection, signalling and proteolysis. Very importantly, LUCA appears to have been a complete cell with a well-established membrane system and with membrane-associated proteins involved in transport and key electron-transport systems. Finally, a large set of metabolic enzymes was also detected, including enzymes involved in amino acid, nucleotide, sugar and lipid biosynthesis and a few degradation pathways. The most important point reached by this study is that it shows that LUCA had a relatively large genome, i.e. greater in size than a number of modern cellular genomes. This indicates that in contrast to early hypotheses, LUCA was far from being a minimal cell because its genome was far from a minimal genome. This reflects the fact that LUCA was a relatively complex modern-like cell and was probably the result of a relatively long evolutionary process that began with the first cell (Forterre *et al.*, 2005).

Conclusion

Throughout life history, the evolution of genomes has been punctuated by gains and losses of genetic material. Thus, genomes should be viewed as highly dynamic objects that provide temporary residence for genes. A direct consequence is that the gene repertoires harboured by present-day cells are very variable in size and content and this, in turn, leads to an abundant diversity of cell types. Accordingly, the basic formulation of the problem of the minimal cell (i.e. the smallest number of genes required and sufficient to sustain a cellular life) is too vague. In fact, it does not pinpoint with accuracy the type of cell that has to be considered. It is evident that the minimal genome for a phototrophic cell able to use inorganic CO_2 will be very different from the one defined for a basic heterotrophic cell that grows in a medium containing a unique organic carbon source, or for a more complex heterotrophic cell requiring an enriched medium, or for a eukaryote, or for an endosymbiont etc. The situation can then be further complicated by the definition of the environment where this minimal cell is supposed to have lived. Virtually, it exists as many minimal genomes depending on cellular types and environmental conditions.

Nonetheless, all cells share a number of functions that define their cellular nature, meaning that their genomes have retained this information during evolution. *In-silico* analyses based on the comparison of gene contents have shown that the proportion of genes shared by all genomes is nearly inexistent and clearly insufficient to define a viable cell. Paradoxically, *in-silico* analyses studying conserved functions have shown that a number of important functions are universal. This highlights the importance of replacements of genes sharing the same functions during evolution. As a consequence, for a clearly defined minimal cell type (e.g. a minimal phototroph), many solutions can be proposed. For example, if the minimal cell is supposed to use DNA as support for genetic information, what kind of DNA replication machinery will be considered: a bacterial or an archaeal/eukaryotic machinery? The same choice is likely to occur for each function. This point illustrates the main difference between real and inferred minimal cells. Indeed, the former are the result of a long evolutionary process. As a consequence, their genomes have been shaped by interactions between their genes and by interactions of cells with the environment. These complex interactions during millions of years have strongly constrained the evolution of genomes. For example, the loss of one or several genes can provoke irreversible changes that will condition the number and the type of genes that will be conserved, acquired or lost, and thus the evolutionary paths that the cell will be able to follow. By contrast, inferred minimal cells and minimal genomes can be compared to *de novo* constructions realized by an engineer that would artificially gather together genes from different origins (i.e. that appeared and evolved in different lineages).This will allow for the experimentatal combination of genes that may never occur naturally.

The two points mentioned above underline the fact that the minimal-cell concept is, behind its apparent simplicity, a very complex question for which no really satisfactory answer can be proposed and that will probably never be fully resolved. However, this does not imply that research on this question is useless. Indeed, the problem of the minimal cell has stimulated very interesting research in biology since its inception. The diversity of the experimental approaches developed has allowed for important technical progress and discoveries in genetics, comparative genomics etc. Finally, the recent report on the artificial synthesis of an *M. genitalium* genome (Gibson *et al.*, 2008) opens up a new avenue because an important technical step has been overcome, allowing for the testing of practical solutions for minimal-cell research.

Acknowledgements

I would like to thank Simonetta Gribaldo and David Moreira for stimulating discussions and their critical reading of the manuscript.

References

Akerley, B. J., Rubin, E. J. V., Novick, L., Amaya, K., Judson, N. and Mekalanos, J. J. (2002). A genome-scale analysis for identification of genes required for growth or survival of *Haemophilus influenzae*. *Proceedings of the National Academy of Sciences of the USA*, **99**, 966–71.

Bapteste, E., Charlebois, R. L., MacLeod, D. and Brochier, C. (2005). The two tempos of nuclear pore complex evolution: highly adapting proteins in an ancient frozen structure. *Genome Biology*, **6**, R85.

Embley, T. M. (2006). Multiple secondary origins of the anaerobic lifestyle in eukaryotes. *Philosophical Transactions of the Royal Society of London B Biological Sciences*, **361**, 1055–67.

Feher, T., Papp, B., Pal, C. and Posfai, G. (2007). Systematic genome reductions: theoretical and experimental approaches. *Chemical Reviews*, **107**, 3498–513.

Forsyth, R. A., Haselbeck, R. J., Ohlsen, K. L., Yamamoto, R. T., Xu, H., Trawick, J. D., Wall, D., Wang, L., Brown-Driver, V., Froelich, J. M., King, P., Kedar, G. C, McCarthy, M., Malone, C., Misiner, B., Robbins, D., Tan, Z., Zhu, Z. Y., Carr, G., Mosca, D. A., Zamudio, C., Foulkes, J. G. and Zyskind, J. W. (2002). A genome-wide strategy for the identification of essential genes in *Staphylococcus aureus*. *Molecular Microbiology*, **43**, 1387–400.

Forterre, P., Gribaldo, S. and Brochier, C. (2005). Luca: the last universal common ancestor. *Médecine Sciences (Paris)*, **21**, 860–5.

Fraser, C. M., Gocayne, J. D., White, O., Adams, M. D., Clayton, R. A., Fleischmann R. D., Bult, C. J., Kerlavage, A. R., Sutton, G., Kelley, J. M., Fritchman, R. D., Weidman, J. F., Small, K. V., Sandusky, M., Fuhrmann, J., Nguyen, D., Utterback, T. R., Saudek, D. M., Phillips, C. A., Merrick, J. M., Tomb, J. F., Dougherty, B. A., Bott, K. F., Hu, P. C., Lucier, T. S., Peterson, S. N., Smith, H. O., Hutchison, C. A. 3rd and Venter, J. C. (1995). The minimal gene complement of *Mycoplasma genitalium*. *Science*, **270**, 397–403.

Galperin, M. Y. (2008). Social bacteria and asocial eukaryotes. *Environmental Microbiology*, **10**, 281–8.

Garriga, X., Eliasson, R., Torrents, E., Jordan, A., Barbe, J., Gibert, I., and Reichard, P. (1996). nrdD and nrdG genes are essential for strict anaerobic growth of *Escherichia coli*. *Biochemical and Biophysical Research Communications*, **229**, 189–92.

Gerdes, S. Y., Scholle, M. D., Campbell, J. W., Balazsi, G., Ravasz, E., Daugherty, M. D., Somera, A. L., Kyrpides, N. C., Anderson, I., Gelfand, M. S., Bhattacharya, A., Kapatral, V., D'Souza, M., Baev, M. V., Grechkin, Y., Mseeh, F., Fonstein, M. Y., Overbeek, R. L., Barabasi, A., Oltvai, Z. N. and Osterman, A. L. (2003). Experimental determination and system level analysis of essential genes in *Escherichia coli* MG1655. *Journal of Bacteriology*, **185**, 5673–84.

Giaever, G., Chu, A. M., Ni, L., Connelly, C., Riles, L., Veronneau, S., Dow, S., Lucau-Danila, A., Anderson, K., Andre, B., Arkin, A. P., Astromoff, A., El-Bakkoury, M., Bangham, R., Benito, R., Brachat, S., Campanaro, S., Curtiss, M., Davis, K., Deutschbauer, A., Entian, K. D., Flaherty, P., Foury, F., Garfinkel, D. J., Gerstein, M., Gotte, D., Guldener, U., Hegemann, J. H., Hempel, S., Herman, Z., Jaramillo, D. F., Kelly, D. E., Kelly, S. L., Kotter, P., LaBonte, D., Lamb, D. C., Lan, N., Liang, H., Liao, H., Liu, L., Luo, C., Lussier, M., Mao, R., Menard, P., Ooi, S. L., Revuelta, J. L., Roberts, C. J., Rose, M., Ross-Macdonald, M. P., Scherens, B., Schimmack, G., Shafer, B., Shoemaker, D. D., Sookhai-Mahadeo, S., Storms, R. K., Strathern, J. N., Valle, G., Voet, M., Volckaert, G., Wang, C. Y., Ward, T. R., Wilhelmy, J., Winzeler, E. A., Yang, Y., Yen, G., Youngman, E., Yu, K., Bussey, H., Boeke, J. D., Snyder, M., Philippsen, P., Davis, R. W. and Johnston, M. (2002). Functional profiling of the *Saccharomyces cerevisiae* genome. *Nature*, **418**, 387–91.

Gibson, D. G., Benders, G. A., Andrews-Pfannkoch, C., Denisova, E. A., Baden-Tillson, H., Zaveri, J., Stockwell, T. B., Brownley, A., Thomas D. W., Algire, M. A., Merryman, C., Young, L., Noskov, V. N., Glass, J. I., Venter, J. C., Hutchison, C. A. 3rd and Smith, H. O. (2008). Complete chemical synthesis, assembly, and cloning of a *Mycoplasma genitalium* genome. *Science*, **319**, 1215–20.

Gil, R., Latorre, A. and Moya, A. (2004). Bacterial endosymbionts of insects: insights from comparative genomics. *Environmental Microbiology*, **6**, 1109–22.

Gil, R., Silva, F. J., Pereto, J. and Moya, A. (2004). Determination of the core of a minimal bacterial gene set. *Microbiology and Molecular Biology Reviews*, **68**, 518–37.

Glass, J. I., Assad-Garcia, N., Alperovich, N., Yooseph, S., Lewis, M. R., Maruf, M., Hutchison, C. A. 3rd, Smith, H. O. and Venter, J. C. (2006). Essential genes of a minimal bacterium. *Proceedings of the National Academy of Sciences of the USA*, **103**, 425–30.

Gotfredsen, M. and Gerdes, K. (1998). The *Escherichia coli* relBE genes belong to a new toxin–antitoxin gene family. *Molecular Microbiology*, **29**, 1065–76.

Gregory, T. R. and Hebert, P. D. (1999). The modulation of DNA content: proximate causes and ultimate consequences. *Genome Research*, **9**, 317–24.

Hooke, R. (1665). *Micrographia: Or Some Physiological Descriptions of Minute Bodies Made by Magnifying Glasses with Observations and Inquiries Thereupon.* London: Martyn Allefiry.

Hutchison, C. A., Peterson, S. N., Gill, S. R., Cline, R. T., White, O., Fraser, C. M., Smith, H. O. and Venter, J. C. (1999). Global transposon mutagenesis and a minimal *Mycoplasma* genome. *Science*, **286**, 2165–9.

Itaya, M. (1995). An estimation of minimal genome size required for life. *FEBS Letters*, **362**, 257–60.

Jacobs, M. A., Alwood, A., Thaipisuttikul, I., Spencer, D., Haugen, E., Ernst, S., Will, O., Kaul, R., Raymond, C., Levy, R., Chun-Rong, L., Guenthner, D., Bovee, D., Olson, M. V. and Manoil, C. (2003). Comprehensive transposon mutant library of

Pseudomonas aeruginosa. *Proceedings of the National Academy of Sciences of the USA*, **100**, 14339–44.

Ji, Y., Zhang, B., Van, S., Horn, F., Warren, P., Woodnutt, G., Burnham, M. K. and Rosenberg, M. (2001). Identification of critical staphylococcal genes using conditional phenotypes generated by antisense RNA. *Science*, **293**, 2266–9.

Kang, Y., Durfee, T., Glasner, J. D., Qiu, Y., Frisch, D., Winterberg, K. M. and Blattner, F. R. (2004). Systematic mutagenesis of the *Escherichia coli* genome. *Journal of Bacteriology*, **186**, 4921–30.

Katinka, M. D., Duprat, S., Cornillot, E., Metenier, G., Thomarat, F., Prensier, G., Barbe, V., Peyretaillade, E., Brottier, P., Wincker, P., Delbac, F., El Alaoui, H., Peyret, P., Saurin, W., Gouy, M., Weissenbach, J. and Vivares, C. P. (2001). Genome sequence and gene compaction of the eukaryote parasite *Encephalitozoon cuniculi*. *Nature*, **414**, 450–3.

Klasson, L. and Andersson, S. G. (2004). Evolution of minimal-gene-sets in host-dependent bacteria. *Trends in Microbiology*, **12**, 37–43.

Kobayashi, K., Ehrlich, S. D., Albertini, A., Amati, G., Andersen, K. K., Arnaud, M., Asai, K., Ashikaga, S., Aymerich, S., Bessieres, P., Boland, F., Brignell, S. C., Bron, S., Bunai, K., Chapuis, J., Christiansen, L. C., Danchin, A., Debarbouille, M., Dervyn, E., Deuerling, E., Devine, K., Devine, S. K., Dreesen, O., Errington, J., Fillinger, S., Foster, S. J., Fujita, Y., Galizzi, A., Gardan, R., Eschevins, C., Fukushima, T., Haga, K., Harwood, C. R., Hecker, M., Hosoya, D., Hullo, M. F., Kakeshita, H., Karamata, D., Kasahara, Y., Kawamura, F., Koga, K., Koski, P., Kuwana, R., Imamura, D., Ishimaru, M., Ishikawa, S., Ishio, I., Le Coq, D., Masson, A., Mauel, C., Meima, R., Mellado, R. P., Moir, A., Moriya, S., Nagakawa, E., Nanamiya, H., Nakai, S., Nygaard, P., Ogura, M., Ohanan, T., O'Reilly, M., O'Rourke, M., Pragai, Z., Pooley, H. M., Rapoport, G., Rawlins, J. P., Rivas, L. A., Rivolta, C., Sadaie, A., Sadaie, Y., Sarvas, M., Sato, T., Saxild, H. H., Scanlan, E., Schumann, W., Seegers, J. F., Sekiguchi, J., Sekowska, A., Seror, S. J., Simon, M., Stragier, P., Studer, R., Takamatsu, H., Tanaka, T., Takeuchi, M., Thomaides, H. B., Vagner, V., van Dijl, J. M., Watabe, K., Wipat, A., Yamamoto, H., Yamamoto, M., Yamamoto, Y., Yamane, K., Yata, K., Yoshida, K., Yoshikawa, H., Zuber, U. and Ogasawara, N. (2003). Essential *Bacillus subtilis* genes. *Proceedings of the National Academy of Sciences of the USA*, **100**, 4678–83.

Kolisnychenko, V., Plunkett, G. 3rd, Herring, C. D., Feher, T., Posfai, J., Blattner, F. R. and Posfai, G. (2002). Engineering a reduced *Escherichia coli* genome. *Genome Research*, **12**, 640–7.

Koonin, E. V. (2000). How many genes can make a cell: the minimal-gene-set concept. *Annual Reviews in Genomics and Human Genetics*, **1**, 99–116.

Koonin, E. V. (2003). Comparative genomics, minimal gene-sets and the last universal common ancestor. *Nature Reviews in Microbiology*, **1**, 127–36.

Kyrpides, N., Overbeek, R. and Ouzounis, C. (1999). Universal protein families and the functional content of the last universal common ancestor. *Journal of Molecular Evolution*, **49**, 413–23.

Luisi, P. L. (1997). About various definitions of life. *Origin of Life and Evolution of the Biosphere*, **28**, 613–22.

Luisi, P. L., Oberholzer, T. and Lazcano, A. (2002). The notion of a DNA minimal cell: a general discourse and some guidelines for an experimental approach. *Helvetica Chimica Acta*, **85**, 1759–77.

Makarova, K. S., Wolf, Y. I., Mekhedov, S. L., Mirkin, B. G. and Koonin, E. V. (2005). Ancestral paralogs and pseudoparalogs and their role in the emergence of the eukaryotic cell. *Nucleic Acids Research*, **33**, 4626–38.

Maniloff, J. (1996). The minimal cell genome: 'on being the right size'. *Proceedings of the National Academy of Sciences of the USA*, **93**, 10004–6.

Mans, B. J., Anantharaman, V., Aravind, L. and Koonin, E. V. (2004). Comparative genomics, evolution and origins of the nuclear envelope and nuclear pore complex. *Cell Cycle*, **3**, 1612–37.

McCutcheon, J. P. (2010). The bacterial essence of tiny symbiont genomes. *Current Opinion in Microbiology*, **13**(1), 73–8.

McCutcheon, J. P. and Moran, N. A. (2007). Parallel genomic evolution and metabolic interdependence in an ancient symbiosis. *Proceedings of the National Academy of Sciences of the USA*, **104**, 19392–7.

Morowitz, H. J. (1984). The completeness of molecular biology. *Israel Journal of Medical Sciences*, **20**, 750–3.

Mushegian, A. R. and Koonin, E. V. (1996). A minimal gene set for cellular life derived by comparison of complete bacterial genomes. *Proceedings of the National Academy of Sciences of the USA*, **93**, 10268–73.

Nakabachi, A., Yamashita, A., Toh, H., Ishikawa, H., Dunbar, H. E., Moran, N. A. and Hattori, M. (2006). The 160-kilobase genome of the bacterial endosymbiont *Carsonella*. *Science*, **314**, 267.

Ouzounis, C. and Kyrpides, N. (1996). The emergence of major cellular processes in evolution. *FEBS Letters*, **390**, 119–23.

Ouzounis, C. A., Kunin, V., Darzentas, N. and Goldovsky, L. (2006). A minimal estimate for the gene content of the last universal common ancestor-exobiology from a terrestrial perspective. *Research in Microbiology*, **157**, 57–68.

Podar, M., Anderson, I., Makarova, K. S., Elkins, J. G., Ivanova, N., Wall, M. A., Lykidis, A., Mavromatis, K., Sun, H., Hudson, M. E., Chen, W., Deciu, C., Hutchison, D., Eads, J. R., Anderson, A., Fernandes, F., Szeto, E., Lapidus, A., Kyrpides, N. C., Saier, M. H. Jr., Richardson, P. M., Rachel, R., Huber, H., Eisen, J. A., Koonin, E. V., Keller, M. and Stetter, K. O. (2008). A genomic analysis of the archaeal system *Ignicoccus hospitalis–Nanoarchaeum equitans*. *Genome Biology*, **9**, R158.

Reich, K. A., Chovan L. and Hessler, P. (1999). Genome scanning in *Haemophilus influenzae* for identification of essential genes. *Journal of Bacteriology*, **181**, 4961–8.

Roy, S. W. and Irimia, M. (2009). Splicing in the eukaryotic ancestor: form, function and dysfunction. *Trends in Ecology and Evolution*, **24**, 447–55.

Salama, N. R., Shepherd, B. and Falkow, S. (2004). Global transposon mutagenesis and essential gene analysis of *Helicobacter pylori*. *Journal of Bacteriology*, **186**, 7926–35.

Sassetti, C. M. and Rubin, E. J. (2003). Genetic requirements for mycobacterial survival during infection. *Proceedings of the National Academy of Sciences of the USA*, **100**, 12989–94.

Sassetti, C. M., Boyd, D. H. and Rubin, E. J. (2003). Genes required for mycobacterial growth defined by high density mutagenesis. *Molecular Microbiology*, **48**, 77–84.

Schwann, T. and Schleyden, M. S. (1847). *Microscopical researches into the accordance in the structure and growth of animals and plants*. London: Printed for the Sydenham Society.

Snel, B., Bork, P. and Huynen, M. A. (2002). Genomes in flux: the evolution of archaeal and proteobacterial gene content. *Genome Research*, **12**, 17–25.

Tamames, J., Gil, R., Latorre, A., Pereto, J., Silva, F. J. and Moya, A. (2007). The frontier between cell and organelle: genome analysis of *Candidatus* Carsonella ruddii. *BMC Evolutionary Biology*, **7**, 181.

Tatusov, R. L., Koonin, E. V. and Lipman, D. J. (1997). A genomic perspective on protein families. *Science*, **278**, 631–7.

Virchow, R. (1855). *Collected Essays on Public Health and Epidemiology*. Canton, MA: Science History Publications.

Waters, E., Hohn, M. J., Ahel, I., Graham, D. E., Adams, M. D., Barnstead, M., Beeson, K. Y., Bibbs, L., Bolanos, R., Keller, M., Kretz, K., Lin, X., Mathur, E., Ni, J., Podar, M., Richardson, T., Sutton, G. G., Simon, M., Soll, D., Stetter, K. O., Short, J. M. and Noordewier, M. (2003). The genome of *Nanoarchaeum equitans*: insights into early archaeal evolution and derived parasitism. *Proceedings of the National Academy of Sciences of the USA*, **100**, 12984–8.

Winzeler, E. A., Shoemaker, D. D., Astromoff, A., Liang, H., Anderson, K., Andre, B., Bangham, R., Benito, R., Boeke, J. D., Bussey, H., Chu, A. M., Connelly, C., Davis, K., Dietrich, F., Dow, S. W., El Bakkoury, M., Foury, F., Friend, S. H., Gentalen, E., Giaever, G., Hegemann, J. H., Jones, T., Laub, M., Liao, H., Liebundguth, N., Lockhart, D. J., Lucau-Danila, A., Lussier, M., M'Rabet, N., Menard, P., Mittmann, M., Pai, C., Rebischung, C., Revuelta, J. L., Riles, L., Roberts, C. J., Ross-MacDonald, P., Scherens, B., Snyder, M., Sookhai-Mahadeo, S., Storms, R. K., Veronneau, S., Voet, M., Volckaert, G., Ward, T. R., Wysocki, R., Yen, G. S., Yu, K., Zimmermann, K., Philippsen, P., Johnston, M. and Davis, R. W. (1999). Functional characterization of the *Saccharomyces cerevisiae* genome by gene deletion and parallel analysis. *Science*, **285**, 901–6.

Yu, B. J., Sung, B. H., Koob, M. D., Lee, C. H., Lee, J. H., Lee, W. S., Kim, M. S. and Kim, S. C. (2002). Minimization of the *Escherichia coli* genome using a Tn5-targeted Cre/loxP excision system. *Nature Biotechnology*, **20**, 1018–23.

4

Minimal cell: the computer scientist's point of view

Hugues Bersini

Introduction to artificial life

Would a theoretical biologist be surprised to be told that computer use and software developments should help him make substantial progress in his discipline? It is doubtful. There is a long tradition of software simulations in theoretical biology to complement pure analytical mathematics which are often limited to reproducing and understanding the self-organization phenomena resulting from non-linear and spatially grounded interactions of the huge number of diverse biological objects. Nevertheless, proponents of artificial life would bet that they could help them further by enabling them to transcend their daily modelling/measuring practice by using software simulations in the first instance and, to a lesser degree, robotics, in order to abstract and elucidate the fundamental mechanisms common to living organisms. They hope to do so by resolutely neglecting much materialistic and quantitative information deemed as not indispensable. They want to focus on the rule-based mechanisms making life possible, supposedly neutral with respect to their underlying material embodiment, and to replicate them in a non-biochemical substrate. In artificial life, the importance of the substrate is purposefully understated for the benefit of the function (software should 'supervene' to an infinite variety of possible hardware). Minimal life begins at the intersection of a series of processes that need to be isolated, differentiated and duplicated as such in computers. Only software development and usage make it possible to understand the way these processes are intimately interconnected in order for life to appear at those crossroads.

Artificial life obviously relates to exobiology, this other recent scientific discipline equally centred on life and the study of its origins, not only on the obvious environment of Earth, but also throughout the Universe. Exobiology cannot restrict itself to a mere materialistic view of life in order to attempt to detect it elsewhere, as the material substrate could be something totally different. This substrate could be as much singular on a distant planet as it could be in the RAM (random access memory) of a computer, somewhere in a computer lab. The presence of life might be suspected through its functions, much before being able to dissect it. Artificial life does not attempt to provide an extra thousandth attempt at a definition of life, any more than do most biologists. 'Defining' is a sociological endeavour which consists in grounding something semantically rather weak on a stronger semantic support. How to assess the strength of this semantic support is far from easy. It solidifies with the possibility of connecting the expression to define with a perceived reality although, for the philosopher Wittgenstein and for most of the common expressions, the usage will finally decide the final meaning. As a matter of fact, the concept of 'life', as opposed to 'gravity' or 'electromagnetism' or the 'quantum reduction of a wave packet' had already been in widespread existence prior to any scientific reading or reification. Its semantic limits are now definitely out of control, allowing James Lovelock (Lovelock, 2000) the latitude to assimilate the planet Earth into a living organism and many protagonists of artificial life to see life everywhere as soon as an amusing little animation appears on their screen or a mechanical dog wags its tail. The rejection of an authoritative definition of 'life' is often compensated for by a list of functional properties which never finds unanimity among its authors. Some demand more properties, others require fewer of those properties that are often expressed in terms of a vague expression such as 'self-maintenance', 'self-organization', 'metabolism', 'autonomy', 'self-replication' and 'open-ended evolution'. A first determining role of artificial life consists in the writing and implementing of software versions of these properties and of the way they do connect, so as to disambiguate them, making them algorithmically precise enough so that, in the end, the only reason for disagreement on the definition of life would lie in the length or the composition of this list and on none of its items.

The biologist obviously remains the most important partner; but what may he expect from this 'artificial life'? What can he expect from these new 'Merlin hackers', whose ambitions seem, above all, disproportionally naïve? These computer platforms could be useful in several ways, presented in the following in terms of their increasing importance or by force of impact. First of all, they can open the door to a new style of teaching and advocating of the major biological ideas like, for example, Richard Dawkins (1986), who, bearing the Darwinian good news, did so with the help of a computer simulation where creatures known as 'biomorphs' evolve on a computer screen by means of genetic algorithms. These same platforms and simulations can, insofar as they are sufficiently flexible, quantifiable and universal, be used more precisely by the biologist, who will find in them a simplified means of simulating and validating a given biological system under study. Cellular automata, Boolean networks, genetic algorithms and algorithmic chemistry are excellent examples of software to parameterize and use in order to reproduce the natural phenomenon required. Their predictive power varies from very qualitative (their results just reproduce very general trends of the real world) to very quantitative (the numbers produced by the computer may be precisely compared to those in the real world). Although being at first very qualitative, a precise and clear coding is already the guarantee of an advanced understanding accepted by all. Algorithmic writing is an essential stage in formalizing the elements of the model and making them objective. The more the model allows for the integration of what we know about the reality reproduced, the detailed structures of objects and relationships between them, the more the predictions will move from qualitative to precise and the easier the model will be validated according to the Karl Popper ideal falsifying process.

Finally, through systematic software experiments, these platforms can lead to the discovery of new natural laws, whose impact will be as great as the simulated abstractions will be that are present in many biological realms. In the 1950s Alan Turing (Turing, 1952) discovered that a simple diffusion phenomenon, propagating itself at a different speed, depending on whether it was subject to a negative or positive influence, produces zebra or alternating motifs; this had a considerable effect on a whole section of biology studying the genesis of forms (animal skins, sea shells, etc.) (Meinhardt, 1998). When some scientists discovered that the number of attractors in a Boolean network or a neural network has a linear dependency on the number of units in these networks (Kauffman, 1993, 1995), these results can equally well apply to the number of cells expressed as dynamic attractors in a genetic network or to the quantity of information capable of being memorized in a neural network. Entire chapters of biology dedicated to networks (neural, genetic, protein, immune, hormonal) had to be re-written in the light of these discoveries. When some scientists recently observed a non-uniform connectivity in many networks, whether social, technological or biological, showing a small number of key nodes with a large number of connections and a greater number of nodes with far fewer connections, and when, in addition, they described the way in which these networks were built in time (Barabasi, 2002) by preferential attachment, again biology has been clearly affected. Artificial life is, of course, at its apogee when it reveals new biological facts, destabilizing biologists' presuppositions or generating new knowledge, rather than simply illustrating or refining the old. At the moment, the fact that this discipline is rather young and has a certain immaturity in comparison with biology has led several observers to remain in want of information in the face of the current discrepancy between promises and reality. In my opinion, they have tended to underestimate the importance of the results already obtained, as they are too riveted to their microscope. They should show less reticence, less coldness and arrogance towards these new computer explorers and more curiosity and conviviality for these people who, like them, have set out on the conquest of life while remaining in front of their computer screens.

In this chapter I shall attempt to set out the history of life as the disciples of artificial life understand it, by placing these different lessons on a temporal and causal axis, showing which one is indispensable to the appearance of the next and how each one does connect to the next. This history will certainly be very incomplete and full of numerous unknowns, but most people involved in artificial life will be in agreement. They will mainly disagree on the number of these functions and on the sequence of their appearance, acknowledging, however, that the appearance of any will have been conditioned by the presence and the functioning of the previous ones. The task of artificial life is to set up experimental software platforms where these different lessons, whether taken in isolation or together, are tested, simulated, and, more systematically, analysed. I will sketch some of these existing software platforms whose running delivers interesting take-home messages to open-minded biologists.

The history of life as seen by artificial life

Appearance of chemical reaction cycles and autocatalytic networks

In order to emerge and maintain itself inside a soup of molecules which are potentially reactive and that contain several different components (which could correspond to the initial conditions required for life to appear, i.e. the primordial soup), a reactive system must form an internally cycled network or a closed organization, in which every molecule is consumed and produced back by the network. Above all, in order for life to begin, all of the components must have been able to stabilize themselves in time. These closed networks of chemical reactions are thus perfect examples of systems which, although heterogeneous, are capable of maintaining themselves indefinitely, despite the shocks and impacts that attempt to destabilize them. This comes about through a subtle self-regeneration mechanism, where the molecules end up producing those molecules that have produced them. It may be obtained on a basic level in a perfectly reversible chemical reaction but can be obtained more subtly in the presence of many intermediary molecules and catalysts. By this reactions-based roundabout in which they all participate, all molecules contribute to maintaining themselves at a constant concentration, compensating and re-establishing any disruption in concentration undergone by any one of them. The bigger the network, the more stable it should be and the more molecules it will contribute to maintain itself in a concentration zone which would vary very little, despite external disruptions.

A network of this kind will be materially closed but energetically open if none of the molecules appear in or disappear from the network as a result of external factors, whereas energy, originating in external sources, is necessary for the reactions to start and take place. The presence of such an energy flux, maintaining the network far from the thermodynamic equilibrium, is needed, since without it no reactive flow would be possible circulating through the entire network. A molecular end of the cycle must be re-energized back in order to start the whole circular reaction process again. This cycle thus acts as a chemical machine, energetically driven from the outside. As soon as one of the molecules is being produced in the network without, in its turn, producing another of the molecules making up the network, it absorbs and thus destroys the network. In the presence of molecules of this kind, produced but non-productive, the only way of maintaining the network consists of feeding it materially and making it open to material influx. The network acts on the flow of material and energy as an intermediate ongoing stabilization zone, made up of molecules which may be useful to other vital functions (such as the composition of enclosing membranes or catalysing self-replication), to be described in the following chapters. It transforms, as much as it 'keeps on' all the chemical agents which it recruits. Biologists generally agree that a reactive network must exist prior to the appearance of life, at least to catalyse and make possible the other life processes such as genetic reading and coding; it is open to external influences in terms of matter and energy, but necessarily contains a series of active cycles. They are most often designated 'metabolism' or 'proto-metabolism'. The most popular and active advocates of this 'metabolism-first' hypothetical scenario of the origin of life are S. Kauffmann (1993), J. Maynard-Smith and E. Szathmary (1999), F. Dyson (1999), C. De Duve (2002), T. Ganti (2003) and R. Shapiro (2007).

In my laboratory (Lenaerts and Bersini, 2009), we favour the study of chemical-reaction networks, viewing them as key protagonists in the appearance of life. The elements of these networks are the molecules participating in the reactions and the connections are the reactions linking the reacting molecules to the molecules produced. The networks are generally characterized by fixed-point dynamics, the chemical balances during which the producers and the products mutually support each other. The attractors in which these networks fix themselves are as dynamic – the concentrations slowly stabilize – as they are structural – the molecules participating in the network are chosen and 'trapped' by the network as a whole. These networks are perfect examples of systems which combine dynamics (the chemical kinetics in this case) and metadynamics (the network topological change), as new molecules may appear as the result of reactions while some of the molecules in the network may disappear if their concentration vanishes in time. Both the structure of the network and the concentration of its constituents tend to stabilize over time. Kauffman (1993, 1995) and Fontana (1992) were the forerunners in the study of the appearance and properties of these networks. Figure 4.1 illustrates the work of these two artificial-life pioneers, dedicated to the study of prebiotic chemistry, limiting the reactions studied to polymerization, such as: aa + bb → aabb or inversely, depolymerization or hydrolysis: abaa → ab + aa.

Kauffman showed that provided the probability that a reaction takes place is affected by the presence of a catalyst which is itself produced by the network (in such a case the whole network is called by him an autocatalytic set), a process of percolation or phase transition, characteristic of this type of simulation, is produced. For probabilities much too low, the network does not pop up because the reactions are too improbable, but as soon as a threshold value is reached for this same probability, the network 'percolates', giving rise to multiple molecules produced by multiple reactions. Kauffman grants a privileged status to this threshold value and to the giant 'explosive' network resulting from it in his scenario of the origin of life, without really arguing the reason as to why such a status should exist, but passing the immense interest and enthusiasm which the phenomena of phase transition arouse among physicists onto the world of biology. Fontana, for his part, is concerned with the inevitable appearance of reaction cycles (such as that illustrated in Figure 4.1). All the molecules produced by these cycles in the network in turn produce molecules of the network. He is among those many biologists who see these closed networks or organizations as forming a key stage in the appearance of life, due both to their stability and to the fact that they form structural and dynamic attractors for the system. They induce a stabilization and internal regulation zone together with an energetic motor in a chemical soup that is continually crossed by a flow of matter and energy. Fontana went on to show how these networks are also capable of self-regeneration and self-replication.

With Tom Lenaerts (Lenaerts and Bersini, 2009), we programmed the genesis of these chemical reaction networks by adopting the object-oriented (OO) programming paradigm. The OO simulator aims to reproduce a chemical reactor and the reaction network which emerges from it (as shown in Figure 4.2). This coevolutionary (dynamics + metadynamics)

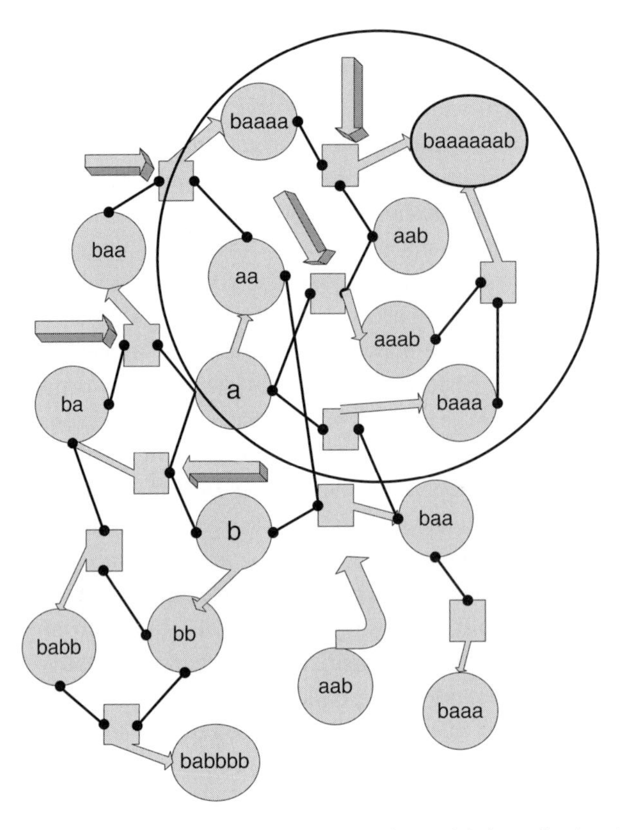

Figure 4.1. Representation of a network of chemical reactions of polymerization (a + b → ab) and depolymerization (ab → a + b) taking place in a simulated chemical reactor. Molecules are represented by circles and reactions by squares. Each reaction can be catalysed, as the arrows pointing to the squares show, by a molecule of the network (giving rise to an autocatalytic network). Some molecules can appear (such as the molecule 'aab') or simply disappear from the network. Reaction cycles can appear, such as the one circled in Figure 4.1 (aa → baaaa → baaaaaab → baaa → aa).

model incorporates the logical structure of constitutional chemistry and its kinetics on the one hand and the topological evolution of the chemical reaction network on the other hand. The network topology influences the kinetics and the other way round, since only molecules with a sufficient concentration are allowed to participate in new reactions. Our model is expressed in a syntax that remains as close as possible to real chemistry. Starting with some initial molecular objects and some initial reaction objects, the simulator allows us to follow the appearance of new molecules and the reactions in which they participate as well as the evolution of their concentration over a period of time. The molecules are coded as canonical graphs. They are made up of atoms and bonds which open, close or break during the reactions. The result of the simulation consists of various reaction networks unfolding in time, and whose properties can be further studied (for instance, the presence of cycles or of a particular topology such as scale-free or random).

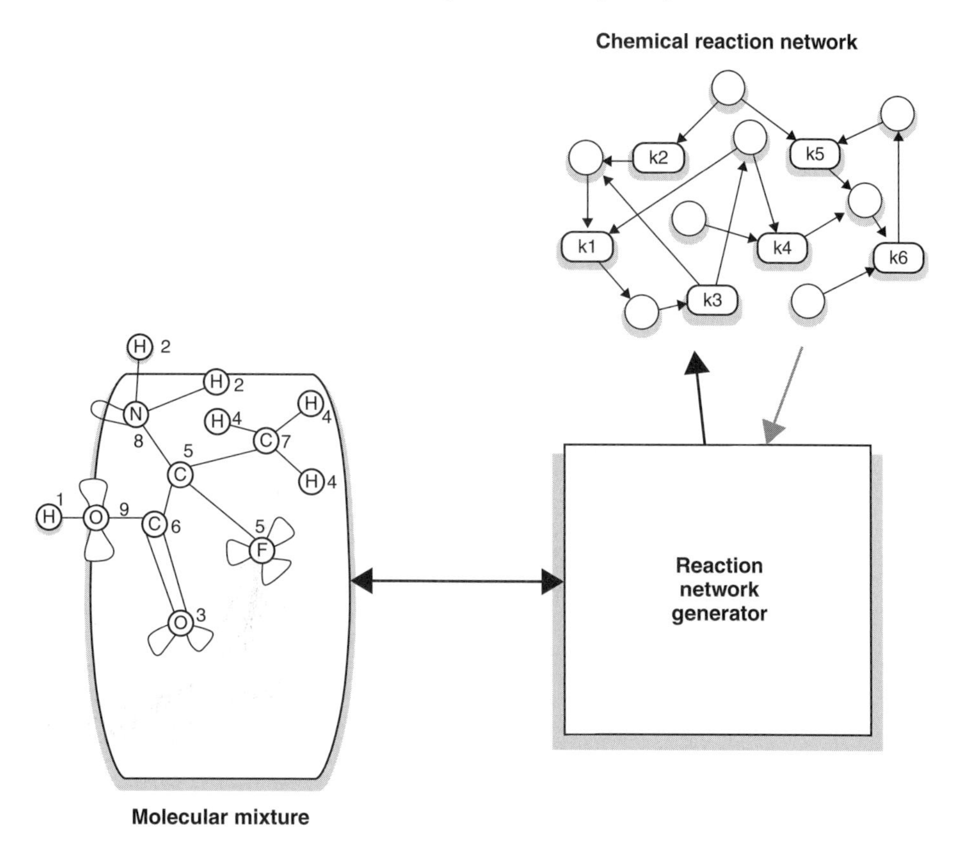

Figure 4.2. The OO chemical simulator developed by Lenaerts and Bersini (2009). On the left, the molecules are represented as canonical graphs. On the right, the outcome of the simulator is an evolving reaction network that can be studied in its own right (the presence of cycles, the type of topology etc.).

One of these reaction schemes which is more than just cycling, can further be autocatalytic when a product of the reaction cycle is double in concentration of one of the reactants: $a + b \rightarrow a + a$. It can happen either in a direct or very indirect way. This is, for instance, the case of the so-called Formose reaction (that Ganti and Szatmary have discussed at length in Ganti, 2003), during which a two-carbon molecule, reacting twice with a carbon monomer, leads to a four-carbon molecule, which subsequently splits, thus duplicating the original molecule. This is the chemical variant of genetic self-replication, since in both cases the original is duplicated. As will be discussed later, Ganti (2003) was the first to connect and synchronize these two replication processes: chemical and genetic, in order for the cell to simultaneously duplicate its boundary, its metabolism and its informational support. In the presence of autocatalysis, the kinetics of some molecules exponentially increases and, more interestingly, when various autocatalytic cycles enter into antagonistic interaction, they turn out to be responsible for symmetry breaking (one of the cycles, randomly favoured initially,

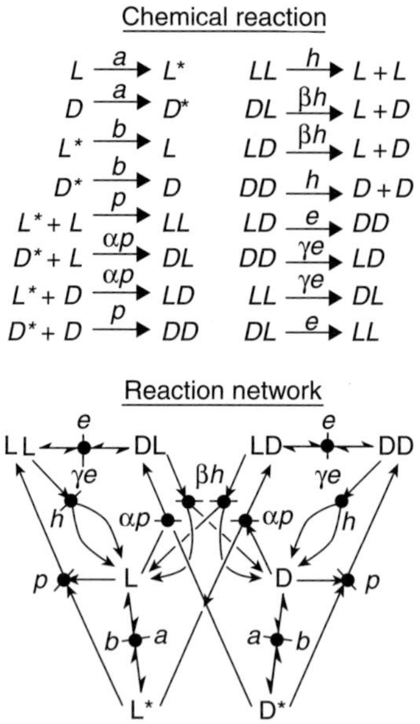

Chemical reaction

$$
\begin{array}{ll}
L \xrightarrow{a} L^* & LL \xrightarrow{h} L+L \\
D \xrightarrow{a} D^* & DL \xrightarrow{\beta h} L+D \\
L^* \xrightarrow{b} L & LD \xrightarrow{\beta h} L+D \\
D^* \xrightarrow{b} D & DD \xrightarrow{h} D+D \\
L^*+L \xrightarrow{p} LL & LD \xrightarrow{e} DD \\
D^*+L \xrightarrow{\alpha p} DL & DD \xrightarrow{\gamma e} LD \\
L^*+D \xrightarrow{\alpha p} LD & LL \xrightarrow{\gamma e} DL \\
D^*+D \xrightarrow{p} DD & DL \xrightarrow{e} LL
\end{array}
$$

Reaction network

Figure 4.3. The prebiotic chemical reactor system responsible for a homochiral steady state studied by Plasson *et al.* (2007). The complete set of reactions is indicated containing activation (the necessary energy source), polymerization and hydrolysis (which together shape the cycles) and epimerization (which induces the competition between the enantiomers).

wins and takes it all). The early origin of life should not be studied without taking into account the self-organization of chemical networks, the emergence and antagonism of autocatalytic cycles and how energy flows drive the whole process. Such chemical networks are, for instance, interesting in order to understand the onset of biological homochirality as the destabilization of the racemic state resulting from the competition between enantiomers and from amplification processes relating to both autocatalytic competitors (one left-oriented and the other right; see Plasson *et al.*, 2007). The chemical reaction network under study (shown in Figure 4.3) is made up of the same types of polymerization and depolymerization reactions as the ones studied by Fontana (1992). In the additional presence of epimerization reactions allowing the transformation of a right-hand monomer into a left-hand one and vice versa, the concentration of one family of monomers (for instance the left one) vanishes in favour of the other. The flux of energy is transferred and efficiently distributed through the system, leading to cycle competitions and to the stabilization of asymmetric states.

Production by this network of a membrane promoting individuation and catalysing constitutive reactions

The emergence of a reaction network of this kind undeniably creates the stability necessary for exploiting its constituents in many reactive systems such as the ones dedicated to the construction of membranes or the replication of molecules carrying the genetic code. This network also acts as a primary filter as it can accept new molecules within it, but can equally well reject other molecules seeking to be incorporated within it. They would be rejected as they do not participate in any of the reactions making up the network. Can we see a primary form of individuation in this network? No, because by definition it can only be unique as no spatial frontier allows it to be distinguished from another network. Although it is roughly possible to conceive of an interpenetration of several chemical networks, establishing a clear separation between them would remain a problem.

It would seem fundamental that a living organism of any kind can be differentiated from another. We know that the production of a second organism from a first one is a fundamental mechanism of life that can only operate if the 'clone' elaborates something to spatially distinguish itself from its 'original'. The best way of successfully completing this individuation and distinguishing between these networks is to revert to a spatial divide, which can only be produced by some form of container capable of circumscribing these networks in a given space. Biochemists are well acquainted with an ideal type of molecule, a raw material for these membranes in the form of lipidic/amphiphilic molecules or fatty acids, the two extremities of which behave in an antagonistic way – the first hydrophilic, attracted to water, and the second hydrophobic, repulsed by it. Quite naturally these molecules tend to assemble in a double layer (placing the two opposing extremities opposite to each other), formed by the molecules lining up and resulting in the shape of a sphere, which protects the hydrophobic extremities from water. Like soap bubbles, these lipid spheres are semi-permeable and imprison the many chemical components trapped during their formation. They do, however, actively channel in and out the most appropriate chemicals for maintaining themselves.

In assimilating living organisms to autopoietic systems, Varela *et al.* (1974) were the first who insisted that this membrane should be endogenously produced by the elements and the reactions making up the network (for example, lipids would come from the reactions of the networks themselves) and would in return promote the emergence and self-maintenance of the network. The membrane can help with the appearance of the reactive and growing network by the frontiers that it sets up, the concentration of certain molecules trapped in it or by acting as a catalyst to some of the reactions due to its geometry or its make-up. Basically, autopoiesis requires a cogeneration of the membrane and of the reactive network which it 'walls up'. The network presents a double closure – one chemical, linked to the cycling chain of its reactions, and the other physical, due to the frontiers produced by the membrane. In the cellular automata model of Varela (Varela *et al.*, 1974; McMullin and Varela, 1997), illustrated in Figure 4.4, there are three types of particle capable of moving around a two-dimensional surface: 'substrates', 'catalysts' and 'links'. The working and updating rules of this cellular automaton are as follows:

- If both of two substrates are near a catalyst, they disappear to create one single link where one of the two was located.
- If two links are near each other they link up and attach themselves to each other. Once attached these links become immobile.

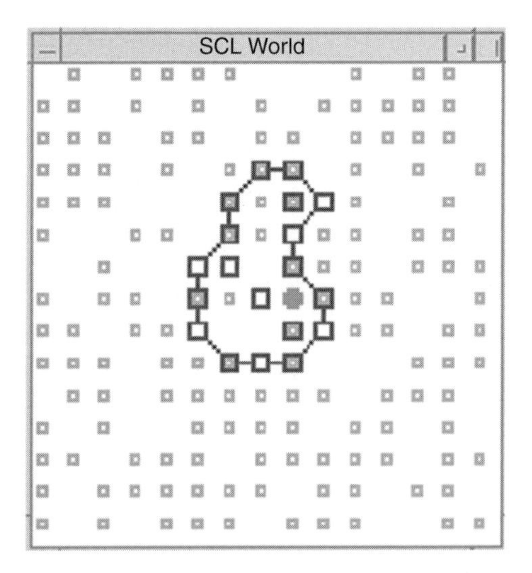

Figure 4.4. Simulation by means of a cellular automaton of the autopoietic model originally proposed by Varela *et al.* (1974). The minimal cell can easily be identified and recognized together with the catalysts and the substrates that it encapsulates.

- Each link is only allowed to attach itself to two other links at the most; which allows the links to form chains and to be able to make up a closed membrane.
- The substrates can diffuse through the links and their attachments, while the catalysts and the other links cannot. We can therefore understand how the product of the cogeneration discussed previously comes about. The membranes shut in the catalysts and the links, which in turn support the membrane by being essential to its formation and regeneration when it is destroyed.
- The reactions creating the links are reversible, as the links can re-create the two original substrates (and thus cause the membrane to deteriorate), but at a lower speed. When this happens, the attachment between the links also disappears.

Continuous updating and execution of these rules produces minimal versions of reactive systems, physically closed and confined by means of a membrane, which is itself produced by the reactive system. For Varela and the others following him, this turns out to be an essential stage in the road to life. Running the software, many difficulties are encountered such as the simple attainment of a closed cell on account of the many more possibilities for the membranes to unfold in a straight way. As a matter of fact, the original simulations departed from an 'already in place' minimal cell, the rules then consisting essentially of stabilizing the cell despite the intermittent disappearance of the membrane parts. And even this is far from obvious, as reported in McMullin and Varela (1997). This difficulty has been exploited by urging the different researchers interested in programming minimal life to respect good programming practices such as OO patterns, to render their code available, understandable and as well-commented on as possible. Only software simulations can interconnect the physical compartment played by the membrane with the generating metabolism, and further show how far from obvious it is for these two systems to mutually sustain each other.

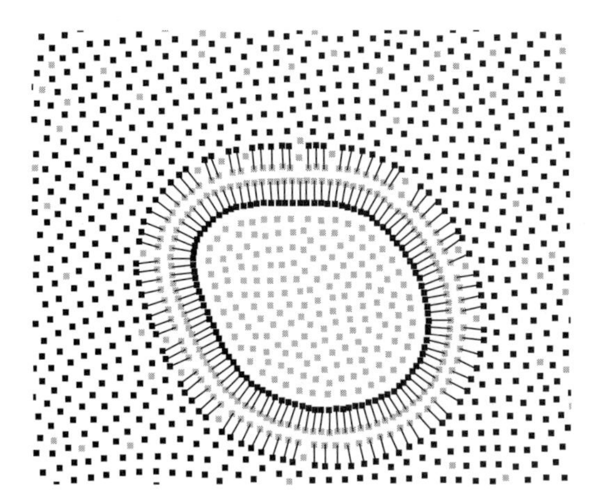

Figure 4.5. Simulation of a minimal cell based on A–B (A is hydrophobic and B hydrophilic) and water molecules. All molecules move in reaction to repulsive forces of different intensity and thermal agitation. The light dots are the As, the grey dots are the Bs that connect to give A–B by a simple chemical reaction.

The whole interactive 'metabolism and membrane' prefigures a minimal elementary cell, which already seems capable both of maintaining itself and detaching itself from its environment and to give rise to cells similar to it. This is at this stage of our successive conceptual additions on the way to establishing a better and more exact characterization of life, that the definition given by Luisi (2002) takes on its full meaning (restating the idea of autopoeisis in more biological terms): 'Life is a system which can be self-maintaining by using external energy and nutritional sources to the production of its internal constituents. This system is spatially circumscribed by a semi-permeable membrane of its composition.' In the footsteps of Varela, considering life impossible without a way for individuation and compartmentalization, the constitution of the membrane by simple self-organization or self-assembly processes of bipolar molecules (hydrophilic and hydrophobic) has become a very popular field for the study of artificial life in its own right. It is indeed rather simple to reproduce this phenomenon in software (as illustrated in Figure 4.5). We need water molecules that just randomly move. We need two kinds of sub-molecules (called A and B) which when they meet, form, through the only authorized additive chemical reaction, an A–B molecule (A is hydrophobic and B hydrophilic), whose two poles are connected by a small string. We also need to adjust the degree of repulsion between A and water, of attraction between B and water, the strength of the string of the A–B molecule and the random component (akin to the thermal noise) to add at each of the intermolecular forces. Nevertheless, the final outcome turns out to be rather robust. The bi-layer of B–A/A–B molecules will very naturally and spontaneously form just as for real cells.

Again, as for Varela's minimal cell, the closure turns out to be quite difficult to achieve. One very simple way to obtain it consists of locating the source of A sub-molecules at a singular point so that the closed membrane will simply surround that source, the circular shape being the local minimal of the mechanical energy connecting all A–Bs together. As in Varela's model, and somewhat paradoxically, the source needs to be circumscribed by the membrane for that same membrane to close on itself. However, in contrast to this autopoietic model, once in place the membrane cannot deteriorate and thus no further internal chemistry is required to endogenously produce what would be needed to fix it. Ultimately this membrane must exhibit some selective channelling in and channelling out (akin, for some authors such as Luisi [2002], to a very primitive form of cognition) providing its internal metabolism with the right nutrients and the right evacuating way out so as to facilitate the cell's self-maintaining. These two software models raise exciting questions for biologists such as: how are the molecular parts of the membrane generated (endogenously or exogenously) and is this cogeneration of the membrane and the internal metabolism the signature of minimal life?

Self-replication of this elementary cell

Self-replication is the ability of a system to produce a copy of itself on its own; it is one of the essential characteristics which has most intrigued and impassioned disciples of artificial life. Biology, and in particular the faculty of self-replication, fascinated von Neumann to the extent that he devoted the majority of his final years in science to it (see von Neumann, 1966). The comparison of a cell to a computer and of a genome to a code requires an understanding of how the computer itself is able to be created out of this code. Let us follow, step by step, the reasoning of von Neumann, as it is the perfect illustration of an 'artificial life' type of approach with no material instantiation but just pure functions or rules. To a sequence of purely functional questions and showing an almost complete ignorance of biological materiality, his reasoning led to a logical explanation, the content of which retraces astonishingly closely those lessons we have since learned about the way biology functions. Von Neumann works on the principle that a universal constructor C must exist, which is based on the plan that some kind of machine P_M (P the plan, M the machine) must be capable of constructing the machine M^P. This idea may be simply translated by $C(P_M) = M^P$. The question of self-replication which is then raised is 'Is this universal constructor capable of constructing itself?' In order to do so, it must, following the example of other construction products, have a plan of what it wants to construct; in this specific case, it is the constructor's plan P_C. The problem is then expressed as follows: can $C(P_C)$ give $C(P_C)$ such that it would be a perfect replication of the original? Von Neumann therefore realized that the question at issue is that of the fate of the construction plan, because if the constructor constructs itself, it has to add the plan itself to the product of the construction. Von Neumann proposed then allotting two tasks to the universal constructor; first, that of constructing the machine according to the given plan and thus adding the original plan to this construction. The constructor's new formula then becomes: $C(P_M) = M^P(P_M)$. If the constructor applies itself to its own plan, then the replication will be perfect: $C^P(P_C) = C^P(P_C)$.

The fascinating aspect of von Neumann's solution is that it anticipated the two essential functions that, as we have since discovered, are the main attributions of the protein tools

Figure 4.6. Langton's self-replicating cellular automata (see explanation in the text).

constituting the cell; constructing and maintaining this cell, but also duplicating the code in order for this construction to be able to prolong itself for further generations. Starting with DNA, the whole protein machinery first of all builds the cell then, by an additional procedure, duplicates this same DNA. Von Neumann did not stop at duplication, because, at the same time he imagined how this same machinery could evolve and become gradually more complex as a result of random mutations taking place while the plan recopies itself. Von Neumann also gave a cellular automaton solution of the problem in which each cell of the automaton possessed 5 neighbours and 29 states, and about 200 000 cells were necessary for the phenomenon of self-replication to take place. Many years later, Langton (1984, 1989) proposed an extremely simplified version of this automaton (8 states, but 219 rules remain necessary), although it still follows the pattern mapped out by von Neumann. This automaton, shown in Figure 4.6, incessantly reproduces a small motif shaped as a loop. However, the departure of von Neumann's version was a system capable of universal computation and universal construction, whereas Langton made a key short cut, simply producing a spatial pattern capable of replication.

For many biologists, as opposed to Varela, Luisi, Ganti and Maynard Smith, life is not simply inseparable from, but also essentially reducible to, this capacity for self-replication. Nevertheless, they still need to explain how life can actually reproduce without entire pre-existent metabolic chemical machinery. Departing from the elementary cell introduced in the preceding chapter, let us imagine a simpler scenario leading to self-replication. The closed circuit of chemical reactions could be destabilized by some kind of disturbance, causing a concentration increase of some of its constituents, including those involved in the formation of membranes. This would also be the case provided all the reactions of the metabolism turn out to be autocatalytic, entailing the exponential growth in concentration of all its molecular elements (including again the membrane constituents). The membrane and the elements that it captures begin to grow (as illustrated in Figure 4.7) until they reach the fatal point where the balance is upset. This is followed by the production of a new cell generated by and from the old one. When the new one comes, it quickly grows fast enough to catch up with the 'generator' and 'nursing' cell, as a chemical network is capable of some degree of self-regeneration due to its intrinsic stability; each molecule 'looks around' for another that it can couple up to.

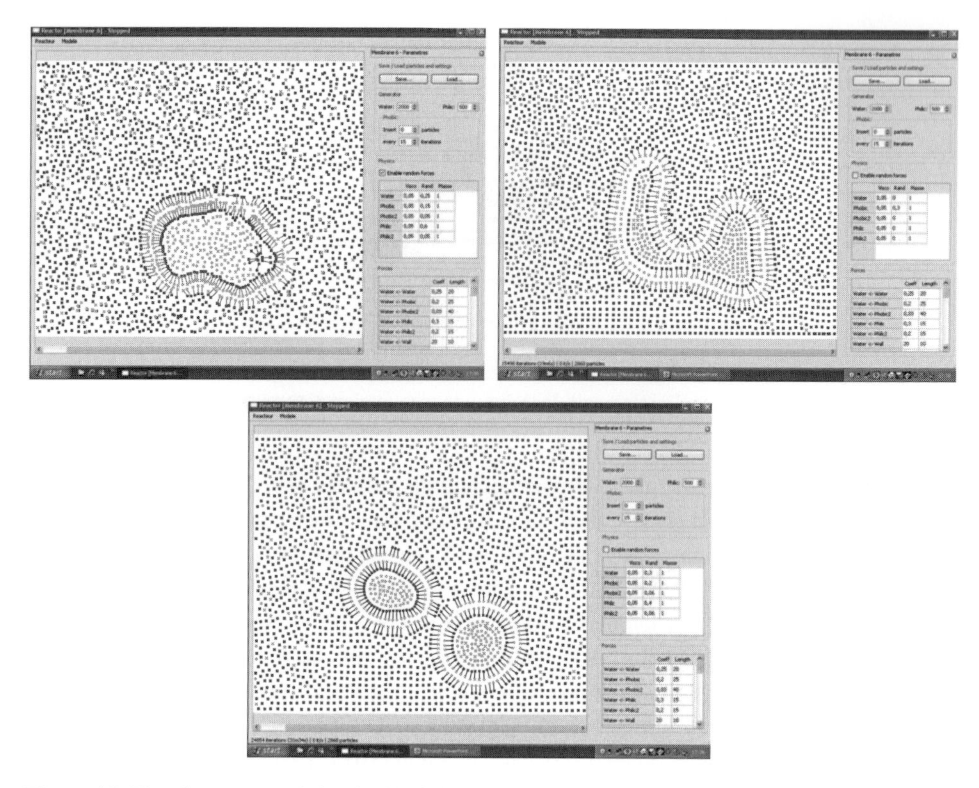

Figure 4.7. The elementary minimal cell of Figure 4.5 in a process of self-replication induced by the growing and division of the chemical metabolic network together with the membrane individuating it. There is much random thermal noise here, which is indispensable in destabilizing the initial cell (symbols are the same as in Figure 4.5).

This reconstitutes the natural chain reaction of the whole. The new membrane and the new chemical network reconstitute on their own by helping each other. Again, obtaining such duplication is far from obvious since with any cell being intrinsically stable only a thermal but quite unnatural agitation would do the job. Similarly, hackers ought to find some more convincing mechanisms, including the genetic template, in the future.

Rather than this elementary form of chemical self-replication coupled to the physical self-replication induced by the growth and division of the membrane, life has opted for a more sophisticated physico–chemical version of it, more promising for the forthcoming evolution: self-replication by interposing an 'information template'. Each element of the template can exclusively couple itself with one complementary element. The new coming elements will, as a whole, naturally reconstitute the template they were attracted to, causing then the replication of the entire template. In biology, it is the extraordinarily emblematic double helix of DNA that acts as a template, shouldering the major role in the history of life – that of the first known replicator. Our elementary cell must now be internally equipped with this information template. In (2003) Ganti proposed a first-minimum mathematical system, named

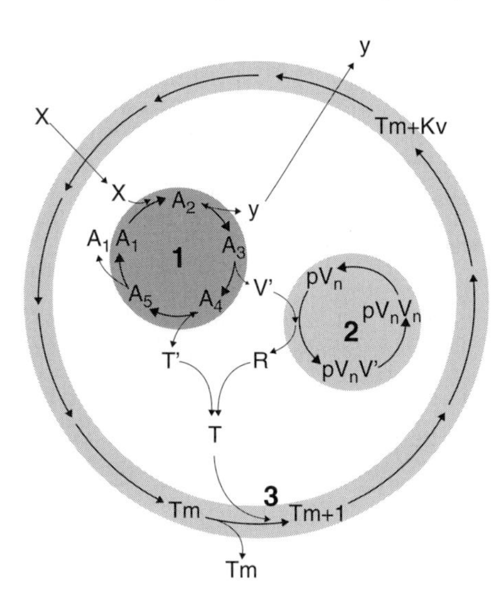

Figure 4.8. The schematic representation of Ganti's (2003) chemoton. One can easily see the three autocatalytic subsystems: the metabolism, the membrane and the information template, chemically coupled.

'chemoton', represented in Figure 4.8. This is the oldest known abstract computational protocell, construed by Ganti as the original ancestor of living organisms. It possesses three autocatalytic chemically linked subsystems: a metabolic network, a membrane and an information template responsible for scheduling and regulating self-replication. They all three grow exponentially until they reproduce and essentially depend on each other for their existence and their stability. The metabolism feeds the membrane and the template, the membrane concentrates the metabolites and the template mechanism dictates the reproduction of the whole. Once the concentration of the metabolites has doubled, template reproduction is initiated, consuming this concentration excess. The triadic ensemble is indeed capable of a whole self-replication and tries to computationally answer questions about the three subsystems and their interdependency, such as 'how does self-replication of the template automatically accompany the self-replication of the whole?'. This complex software object, the 'chemoton', has recently become the topic of several software development and experimentation projects, such that it is emblematic of artificial life at its best.

To a large extent, Ganti's work initiated a full campaign of research dedicated to the conception of the minimal cell (or protocell), a very active field of scientific investigation these days, both with real biochemical materials and with software (Rasmussen *et al.*, 2008).

Genetic coding and evolution by mutation, recombination and selection

The information template introduced in the previous section can in no way be described as 'information' by chance, given that, apart from its self-replication, each letter constituting

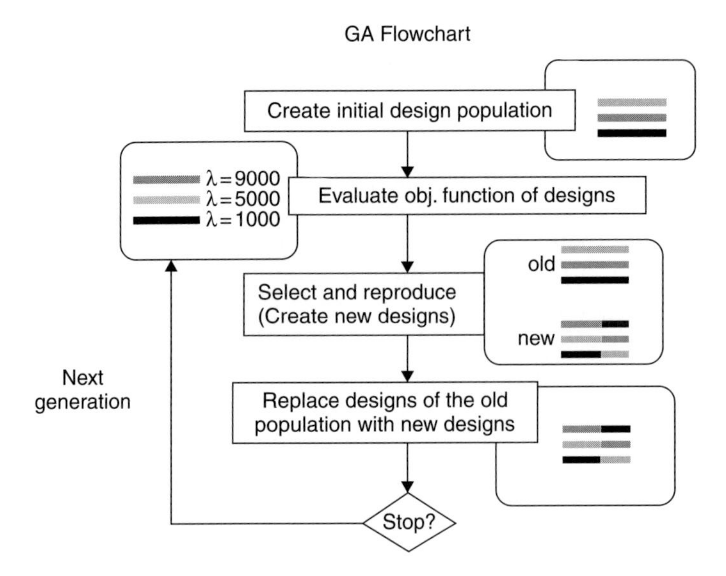

Figure 4.9. Illustration of the genetic algorithm: the recurrent iterated sequence of selection, mutation and recombination easily lead to an interesting solution of a complex optimization problem.

it contributes to the code of a functional component essential to the cell and designed on the basis of that code – a protein. As soon as he hears anyone talking about code, the software specialist has, quite legitimately, to put his head in through the window, because it is to him and him alone that we in fact owe the metaphor of the genetic code. Since Darwin and thereafter, throughout all evolutionary science, we have a good idea of what the last chapter of the history of life is. Doubtless what has stimulated the most development in 'artificial life' (primarily from the point of view of engineering) is the fact that the genetic code can evolve through mutation and sexual crossing between the old machines, evolving so as to produce new machines that are more and more efficient. Over the past twenty years, many of those developments in artificial life have been eager to show how beneficial this idea is for the research and the automated discovery of sophisticated solutions to complex problems. As illustrated in Figure 4.9, this research can take place though a succession of mutations and recombinations operating at the level of the code, with the best solutions proposed being preserved in the next generation in order to be used for a new cycle of these same operations. The brute force of the computer is used to its full effect.

These are the same genetic algorithms which Dawkins (1986) used in his Darwinian crusade, when he developed his biomorphs. It should also be stressed that another element in Dawkins' program is that when it is finally evaluated, the phenotype is not directly obtained from the genotype, as would be the case for classical optimization in a real or combinatorial space. In his work, the biomorphs are the result of a recursive sophisticated program which is carried out starting from a given genotype to give a phenotype. A great 'semantic distance' is maintained between these genotypes and phenotypes, which reflects the long process of cell construction from the genetic code and the need for a sophisticated metabolism building the machine out of the code. Similarly, a very prized derivative of these algorithms is genetic programming (Koza, 1992), where the individuals are software codes.

Conclusions

Parallelism, functional emergence and adaptability are the conditions necessary to allow these new biologically inspired artefacts to emerge to 'face the world'. We are jumping straight into the robotics' branch of artificial life (Brooks, 1990). The interfacing with the real world required by these robots needs a parallel information-reception mechanism, because the environment subjects it to a constant bombardment of stimuli. They have to learn to organize and master this avalanche falling on their perceptions. They have to learn to build their own concepts, fed and stimulated by this environment, and which, in return, allows them to master it. The conceptual high-level cognitive processes are born out of motor–sensory interactions and serve to support them. Cognitive systems extend to new levels what the minimal cell in the primitive soup does, with a flow of matter and energy crossing straight through, maintaining itself by selectively integrating this influx to form a closed reactor network and the membrane enclosing it. As with the simplest membrane, they do not passively receive a predetermined world, but integrate it in a way which is adapted to their structure and their maintenance in the world. Biological systems initially function in a closed way before opening themselves to the outside, and what they do with this 'outside' is first and foremost a result of the concern for stabilizing themselves.

My conclusions are addressed to the three partners: the biologist, the engineer and the philosopher. To the first, the outcomes of artificial life consist of bringing out what the computer and biology share intimately: an elementary way of working at the ultimate lowest level, but which by the brute force of parallelism and incessantly repeated iterations, can make unknown and sophisticated phenomena emerge at higher levels. These processes give rise to many different 'ways of being', which will be sorted by a further phase of selection aimed at a higher 'well-being' or viability of organism. The qualitative aspect of these simulations can give them new roles in the vast scientific register: use it for education, illustrate biological principles which are already understood, open up possible experiences of thought, play and replay multiple biological scenarios very quickly, stimulate the imagination by on-screen representations and call into question some of the ambiguously interpreted but commonly accepted facts.

The second partner, the engineer, is vigorously encouraged to use the computer for what it is best at doing – this infinite possibility of trial and error. He must exclude from the loop what he cannot take in; the time for the computer to propose this immense range of solutions to the problem facing him. Then, similar to a child playing at 'hot and cold', he will be able to guide the computer as he wishes, because he remains the only one who will know, *in fine*, the nature of the problem and who will be capable of appreciating the quality of the solutions. It is a perfect synergy, where both participants ideally complement each other. While the engineer must bow to the computer in terms of calculating power, this is compensated for by his judgement. Genetic algorithms, ant colonies, neural networks and reinforced learning have enriched the engineer's toolbox. It is rare for new problems to be completely created piece by piece and we are still left with the usual data processing, with situations requiring either optimization or regulation – but these new algorithms coming from artificial life have the singular advantage of adding simplicity to performance and leaving the computer, by its calculating power, to soften the torment undergone by the engineer in his inferential progression to better solutions.

Finally, for the philosopher, at each attempt to define life, artificial life makes a real attempt to achieve a computerized version consistent with this definition. For the sceptic, unhappy with this computerized 'lining', the question now becomes how to refine his definition, to complete it or to renounce the possibility that there is no definition which cannot be computerized. The other possibility, doubtless more logical but more difficult for many philosophers to accept, would be that life poses no problem for a computer snapshot since it is computational at its roots. The beneficial effect of such an attitude is to help de-sanctify the idea of life in its most primitive form, when it is the privilege of the most elementary organisms, and not, as in more evolved organisms, when it takes root and becomes indistinguishable from manifestations of consciousness; in substance this means to keep life and consciousness of life clearly separate. By referring to a philosophical article that has become famous in the artificial intelligence community (Nagel, 1974), if a computer cannot know what it is like to be a bat, living like one could be much more within its reach.

References

Barabasi, L-A. (2002). *Linked: The New Science of Networks*. New York: Perseus Books Group.
Brooks, R. (1990). Elephants don't play chess. In *Designing Autonomous Agents*, ed. P. Maes. Cambridge, MA: MIT Press.
Dawkins, R. (1986). *The Blind Watchmaker*. New York: W. W. Norton & Company, Inc.
De Duve, C. (2002). *Life Evolving: Molecules, Mind, and Meaning*. (Oxford: Oxford University Press.
Dyson, F. (1999). *Origins of Life*, 2nd edn. Cambridge: Cambridge University Press.
Fontana, W. (1992). Algorithmic chemistry. In *Artificial Life II: A Proceedings Volume in the SFI Studies in the Sciences of Complexity*, vol. 10, eds. C. G. Langton, J. D. Farmer, S. Rasmussen and C. Taylor. Reading, MA: Addison-Wesley.
Ganti, T. (2003). *The Principles of Life*. Oxford: Oxford University Press.
Kauffman, S. (1993). *The Origins of Order: Self-Organization and Selection in Evolution*. Oxford: Oxford University Press.
Kauffman, S. (1995). *At Home in the Universe: The Search for the Laws of Self-Organisation and Complexity*. New York: Oxford University Press.
Koza, J. (1992). *Genetic Programming*. Cambridge, MA: MIT Press.
Langton, C. G. (1984). Self-reproduction in cellular automata. *Physica D*, **10**, 135–44.
Langton, C. G. (1989). *Artificial Life I*. Redwood City, CA: Addison-Wesley.
Lenaerts T. and Bersini H. (2009). A synthon approach to artificial chemistry. *Artificial Life Journal*, **15**(1), 89–103.

Lovelock, J. (2000). *Gaia: A New Look at Life on Earth*. Oxford: Oxford University Press.
Luisi, P. L. (2002). Some open questions about the origin of life. In *Fundamentals of Life*. Paris: Elsevier, pp. 287–301.
Maynard Smith, J. and Szathmary E. (1999). *The Origins of Life: From the Birth of Life to the Origin of Language*. Oxford: Oxford University Press.
McMullin, B. and Varela F. R. (1997). Rediscovering computational autopoiesis. In *Proceedings of the 4th European Conference on Artificial Life*, eds. P. Husband and I. Harvey. Cambridge, MA: MIT Press, p. 38.
Meinhardt, H. (1998). *The Algorithmic Beauty of Sea Shells*, 2nd enlarged edn. Heidelberg, New York: Springer.
Nagel, T. (1974). What is it like to be a bat? *Philosophical Review*, **83**, 435–50. Reprinted in his *Mortal Questions*. New York: Cambridge University Press, pp. 165–80.
Plasson, R., Kondepudi, D. K., Bersini, H., Commeyras, A. and Asakura, K. (2007). Emergence of homochirality in far-from-equilibrium systems: mechanisms and role in prebiotic chemistry. *Chirality*, **19**, 589–600.
Rasmussen, S., Beday, M. A., Chen, L., Deamer, D., Krakauer, D. C., Packard, N. H. and Stadler, P. F., eds. (2008). *Protocells: Bridging Nonliving and Living Matter*. Cambridge: MIT Press.
Shapiro, R. (2007). A simpler origin for life. *Scientific American*, **296**, 46–53.
Turing, A. M. (1952). The chemical basis of morphogenesis. *Philosophical Transactions of the Royal Society of London B*, **237**, 37–72; In *The Collected Works of A. M. Turing: Morphogenesis*, ed. P. T. Saunders. Amsterdam: North-Holland (1992).
Varela, F. R., Maturana, H. R. and Uribe, R. (1974). Autopoiesis: the organisation of living systems, its characterization and a model. *Biosystems*, **5**, 187–96.
von Neumann, J. (1966). *Theory of Self-Reproducing Automata*, ed. A.W. Burks. Urbana, IL: University of Illinois Press.

5

Origins of life: computing and simulation approaches

Bernard Billoud

Introduction

The scope of life models and simulations is as broad as that of biology. It encompasses studies on cell metabolism, intercellular communication, immunology, physiology, development, cognitive processes, molecular evolution, population genetics, epidemiology etc. However, for the purpose of the present book, we will focus on the use of computing and simulation approaches as tools for studying the origins of life. Under the global denomination of 'automata', a large number of different frameworks have been used to implement and test models accounting for the emergence of life.

Despite their diversity, these approaches call upon the same fundamental grounding: bottom-up model building. Instead of identifying state variables of the whole system and formulating their relationships by equations, the idea is to start with elementary components and then specify how they interact with one another and with the environment. The whole system behaviour is therefore not an *a priori* descriptor of the model, but rather emerges from within the system. This kind of reasoning is obviously appropriate for research on the origins of life, where the main question is precisely to find out how a property (life) which is valid for the system (a living being) has emerged out of multiple elements (chemical components), which are not individually endowed with this property.

Birth of a new research field

The interest of computer scientists regarding the question of the origins of life dates back to the origins of computer science. One century before the time of electronic machines, Charles Babbage (1791–1871) simulated models of evolution using the premises of automatic computation (Bullock, 2000). Later, during and after World War II, the initiators of the first electronic computers were committed to metaphorically speak about their creations as some (primitive) living beings. These machines were indeed able to perform complex tasks, and they readily differed even from the most sophisticated mechanical devices in that their interaction with their environment was based on information. The nascent interdisciplinary field of research called 'cybernetics' focused on 'control and communication in the animal and the machine', to quote the famous Wiener's book title (Wiener, 1948). Thus, automatic information processing intended to be bio-inspired with an emphasis on the key concept of feedback, a common feature in living organisms, as used in various functions like metabolism or cognitive abilities. Meanwhile, if the pioneers of modern computer science regarded life studies as a provider of examples to imitate, they also wanted to contribute to the understanding of life processes by a constructivist approach: using computer programs to implement and test models of life.

Alan Turing (1912–1954) was the first to propose the theoretical design of a universal computer (Turing, 1936). With a very similar concern, John von Neumann (1903–1957) elaborated a theory of automata in the late 1940s. His project aimed at identifying the upper bounds of automatic computation. In particular, he established the conditions necessary for an automaton to be able to perform any possible computation. As a limit case, he showed that a self-reproducing automaton was possible, and defined five different models for it (von Neumann, 1966). Among them is found the cellular model (see Figure 5.1A), initially proposed by Stanislaw Ulam (1909–1984), who formalized it in collaboration with von Neumann (Beyer *et al.*, 1985). Their purpose was to design an automaton able to be self-reproductive, but they wanted to make a clear distinction between self-reproduction akin to that of living beings and a simple, repetitive structured accretion process such as the growth of crystals. The latter is achieved as soon as a set of transition rules, applied on an initial pattern, produces more and more instances of this pattern (Figure 5.1A). Conversely, the lifelike reproduction must be a property of a cell configuration, which, given the transition rules, can drive the creation of a copy of itself. To this purpose, von Neumann introduced the requirement for an internal representation of the being, in other words, an informational content which can be transmitted to the next generation. As a model, he proposed a homogeneous 2D cell automaton with 29 states and a neighbourhood of 4 cells. This cell automaton is a universal constructor, able to build any configuration of cell states. In particular, it can become a self-reproducing machine, provided it is given an appropriate structure: it must contain a 'blueprint', which has to be embedded in the automaton, and alternatively used in two modes. It is either decoded in order to build an organism (less the blueprint), or fully copied, the copy being then included in the new organism. The separation between these two usages is essential to prevent entering an infinite logical loop: build a new organism containing the building plan of a new organism containing the building plan … The 'self-replicator' of von Neumann was purely theoretical at this time and was fully implemented only in 1994 (Pesavento, 1995). It represents, however, a big success of the automata theory, because its replicative cycle is similar to the way actual cells divide and duplicate their DNA. Let us remember that in 1949 the structure and functioning of DNA had not been elucidated. Theoretical considerations had allowed predicting a yet unknown biological mechanism.

These works initiated the synthetic approach in life studies and were immediately understood as a powerful source of ideas and arguments for the questions related to the origins of life. By starting with a minimalist set of assumptions, including nothing more than low-level processes with a physical basis, and showing that lifelike properties arise in the whole system, these simulations assess the likelihood of the scenarios that they model.

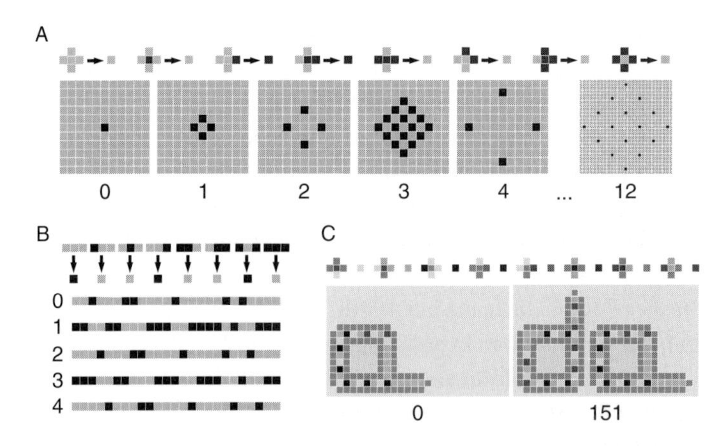

Figure 5.1. Examples of cell automata. A: Ulam's 'recursively defined geometrical objects'. The rules are shown above: each configuration of one cell and its four neighbours defines the state of the central cell at the next time step (the next state of the other cells is not represented because it also depends on cells of their respective neighbourhood). These rules can be comprehensively expressed as: a cell having exactly one black neighbour at time t is black at time $t + 1$; otherwise it remains or becomes clear. Starting with one black cell at time 0, successive applications of this rule result in an endlessly growing pattern of regularly spaced black cells (state at time 12 is represented on a smaller scale in order to show all the black cells). Note that the mechanism is not a simple replication at each time step, as the pattern undergoes a cyclic increase and decrease of black cell density. B: A 1D Cell Automaton, with two states and a neighbourhood of three cells. The eight transitions are represented above, and their application is shown on a randomly chosen initial state, during four time steps. In this example, we use a closed cell line, i.e. the leftmost and rightmost cells are connected. C: Langton's loop is a 2D cell automaton, with eight states (different shades of grey) and a neighbourhood of four cells. Only some of the transition rules are represented above. The initial configuration is a self-reproducing automaton: after 151 time steps, a new loop has been produced by extending the 'arm'. This loop is ready to reproduce on its right side. Meanwhile, the parental loop has produced a new arm, pointing upwards, where another loop will be formed etc. The self-reproduction is not a property of the transition rules, but of the loop configuration which, given these rules, works as an 'embedded automaton'. Similar to a real living being, this loop is not a universal constructor, but it is able to make copies of itself.

A definition of 'A-Life'

Although very diverse, these approaches share a common methodological idea, which is now known as the 'artificial life' approach. Christopher Langton promoted 'A-Life' as a new discipline in 1987, by launching a series of conferences called 'artificial life'. He defined the field as 'the study of man-made systems that exhibit behaviours characteristic of natural living systems' (Langton, 1989). The approach is therefore clearly synthetic and focuses on the processes rather than the objects. These objects (extant living organisms made of carbon-based molecules) are only one instance of life, called 'life as we know it', or 'B-Life' (B for biotic).

The expression 'exhibit behaviours characteristic of natural living systems' deserves a more precise definition. It is indeed essential to prevent the confusion between a true reconstruction of life properties and a realistic representation of living beings. For instance, ray-tracing 3D models which move and make noises like real animals, however impressive the result could be, is not relevant to an A-Life approach, because the underlying methodology is definitively different: A-Life requires the construction of several objects (usually parts of a computer program) which are endowed with basic properties, and prove able to interact in such a way that a new higher-level property will be observed. Indeed, a 3D cartoon could be (and is sometimes) the result of an A-Life experiment, provided the behaviour of the characters results from the interactions of their low-level components, instead of the execution of a predefined scenario. In other words, the behaviour in question must be an emergent property of the system.

We also ought to clarify which behaviours are 'characteristic of natural living systems'. Trying to make an exhaustive list of life-specific properties would be similar to asking 'What is life?', a notoriously difficult question (Lazcano, 2008). Biologists themselves cannot answer this question. Most of them consider that the definition of life is, in fact, the ultimate goal of their research (see Chapter 1). Nevertheless, many definitions have been proposed, at least as parts of other programs, like the search for extraterrestrial life forms. So, no one among the scientists involved in A-Life would claim that he or she has the definitive definition of life, which should serve as a goal for all research in the field. Instead, the different approaches have to be considered as different paradigms, each of them including its own list of characteristic behaviours to reconstruct. But all the recent definitions agree on one essential point: if life has to be defined as a concept, then the definition should not be related to any explicit chemicals. Instead, the definition should only refer to properties.

Cell automata

Many variants of cell automata were proposed during the 1960s to 1980s, with special attention placed on the search for a minimal automaton able to perform universal computation, i.e. to execute any algorithm, provided it is encoded in a suitable form. For instance, in two dimensions with a neighbourhood of 5 cells – one cell and its north, south, east and west neighbours – universal computation is possible with as few as 4 states (Banks, 1971). It is also possible in a 2-state cell automaton, provided the neighbourhood is increased to 9 – adding the four nearest cells diagonally (Smith, 1976). An example of this category is the famous Conway's game of life (Gardner, 1970). Another simple class of automata was designed in one dimension (Figure 5.1B) with two states. The neighbourhood is made of one cell, its immediate neighbours and a variable number of non-immediate neighbours, usually resulting in a neighbourhood of 3 to 7 cells (Wolfram, 1983). Various self-reproducing automata were also produced, among which is found Langton's loop (Figure 5.1C) (Langton, 1984).

Although the rules in cell automata that we have seen so far are simple and perfectly deterministic, it is difficult to predict the future state of a given initial configuration: the only way to determine a long-term evolution is to explicitly apply the transition rules and repeatedly build the next state of the automaton. In other words, there is no analytical solution to the problem of the automaton state at time t, and the only approach is to simulate the automaton throughout all the successive time steps. However, it is possible to study the question of the global behaviour of a certain set of rules. By systematically studying the transition rules in one-dimensional automata, Wolfram proposed a qualitative classification based on their usual outcome from 'disordered' initial states. Four classes were thus defined (Wolfram, 1984): (I) homogeneous configuration (all cells have the same state); (II) inhomogeneous fixed configurations or oscillators; (III) chaotic succession of configurations; and (IV) complex localized spatio–temporal structures that are 'sometimes long-lived' and form propagating structures. The property of non-trivial reproduction, related to the ability to perform universal computation, was expected to be found among the class IV automata. This is, however, not the case, and the relationship between universal computation and belonging to a given class appeared to be loose. Moreover, classifying a given rule into one of the four classes is not straightforward. By trying to quantitatively characterize Wolfram's classes, Langton defined a parameter called λ, a value between 0 and 1, which increases among the four classes, in the order I-II-IV-III. The complex behaviours are associated with a critical value, λ_c, which defines the frontier between ordered and chaotic average outcomes, thus placing the ability to perform universal computation, and hence life, at the 'edge of chaos' (Langton, 1992). This notion has been subject to a rich debate (Crutchfield *et al.*, 1993). The topic of the cell-automaton classification has been further developed (Dubacq *et al.*, 1999), especially with proposals of classification schemes allowing for the deduction of the classification directly from the transition rules.

From another viewpoint, the fact that cell automata with self-reproducing properties have necessitated remarkable intuitions and/or a long and patient development has raised the question of their relevance for the question of the origins of life. Is it likely that a simple and locally connected system would 'discover' how to construct copies of itself? Is it possible that this feature has emerged spontaneously? Indeed, systematic simulations starting from randomly distributed initial states showed that replicators often appear spontaneously and grow (Chou and Reggia, 1997).

Cycles, coupling and hypercycles

A natural cognate challenge of these studies was to propose a mechanism by which a self-reproducing system could appear in a medium devoid of this property. An approach to this problem consists of finding the conditions allowing for the appearance of autocatalytic cycles, which represent the most basic form of self-reproduction. By modelling the dynamics in a network made by potential reactions and catalytic activities of a set of chemically active molecules, it was shown that such cycles spontaneously appear when the number of links becomes sufficiently large (Kauffman, 1993): when reaching a threshold of about 6000 reactions, the system, which is initially a disorganized soup of reactants, undergoes a 'phase transition' and becomes suddenly structured in cycles. These cycles allow the system to stabilize in a dynamic steady state, able to last as long as an energy source remains available. The fact that self-maintenance is likely to appear in purely chemical systems is of major importance for the question of the origins of life, as this phenomenon represents a first step towards a complex self-reproducing metabolism.

A lifelike model has been designed on the basis of coupled cycles: the 'chemoton' (Gánti, 1971). Three autocatalytic cycles maintain the major characteristics of life: metabolism, information and boundary, linked by regulation and control processes (Figure 5.2). With this structure, the cell is depicted as a fluid (chemical) automaton, which differs from a mechanical (hardware) automaton or a cellular (software) automaton in that properties in the fluid state are not tamed by geometric constraints. The arithmetic analysis was conducted by an original approach, called 'cycle stoichiometry', in which the catalysts are incorporated into the reaction equations. Thus, fed-back networks and autocatalytic processes are amenable to arithmetic analysis. On this basis, numerical computations and simulations have shown that the dynamics of the system can be sustained (Munteanua and Sole, 2006) and can account for divisions (Fernando and Di Paolo, 2004). The initial chemoton model depicted a fully featured cell, and seemed, therefore, too complex to be a good candidate as a prebiotic system. It would rather represent a model for a minimal unit of life (i.e. the unit of life that could exist without requiring another unit of life for its maintenance). But a chemoton-like system was shown to be able to emerge out of a purely chemical set of reactions (Gánti, 1997). Thus, fluid automaton represents a seductive scenario for the origins of life, which is further supported by the possibility of actually performing the theoretical cycles with realistic chemical reactions. The chemoton theory, however, remained totally unknown by the international scientific community for over thirty years, because Tibor Gánti wrote it in Hungarian and published it from behind the Iron Curtain.

Nevertheless, it appears that in the 1970s the minds and techniques were ready to develop models based on coupled cycles. On the Western side of the world, the theory of self-reproduction has been greatly enriched by the discovery of an organization scheme called the 'hypercycle' (Eigen and Schuster, 1977). A hypercycle appears when autocatalytic cycles are coupled to one another, and their coupling closes in a higher-order cycle (Figure 5.2). In other words, it is a cooperative system made of self-reproducing units. Up to this point, the model remains abstract, and the nature of the replicative units does not have to be defined. In fact, the hypercycle is versatile enough to apply to various models of early life. We will discuss two of them, namely 'replicators' and 'embodied catalysts', corresponding, in the B-Life implementation, respectively, to scenarios known as 'RNA first' and 'protein/lipid first'.

The seminal work on hypercycles was conducted under the assumption that the basic self-reproducing units are informative molecules (Eigen and Schuster, 1977) with an explicit reference to the 'RNA world' hypothesis. Under this hypothesis, a paradox appears: with a realistic error rate in replication only a limited sequence length can be maintained. This length (in the order of several hundreds of nucleotides) is not sufficient to encode complex functions, and in particular an error-correcting system. To overcome this paradox, the hypercyclic organization was shown to maintain the stability of information through cooperation between replicators, while allowing for high replication rates. However, the hypercycle was soon shown to be vulnerable to parasite invasion (Smith, 1979), a topic which was further developed across numerous studies. Among the solutions to this problem is the arrangement

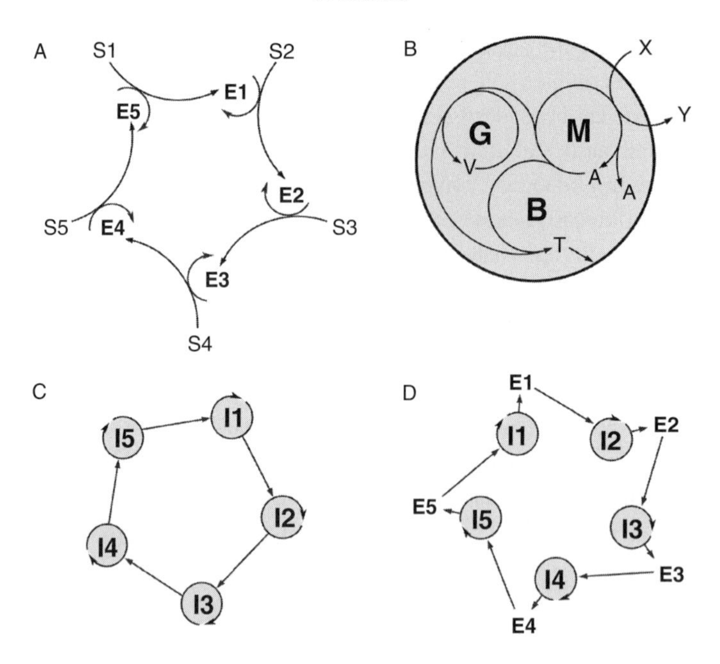

Figure 5.2. Coupled cycles. A: A catalytic cycle of size *n* (here *n* = 5) is formed when in successive reactions, the reaction *i*, using S*i*, produces E*i*, which is a catalyst for the reaction *I* + 1, and this process loops with reaction *n* producing a catalyst for reaction 1. B: The chemoton is made of three stoichiometrically coupled cycles, corresponding to metabolism (M), genetic information (G) and boundary formation (B). The M cycle uses substrates and energy from the outside (represented by X), and produces matter (A) and sub-products (Y). Its products are used by the two other cycles. The G cycle maintains and duplicates the information (V). It instructs the B cycle to form monomers (T) which are included in the membrane. C: When a catalyst I*i* has both an autocatalytic activity, promoting its own synthesis (circular arrows) and a catalytic activity on reaction *I* + 1 (straight arrows), then the loop is a catalytic hypercycle (for simplicity, the input of substrates and energy is not represented). I*i* cycles can represent the behaviour of informational molecules. D: A second-order hypercycle is achieved when the coupling between the autocatalytic cycles I*i* and I*i* + 1is mediated by intermediate catalysts E*i*. A, C and D are adapted from Eigen and Schuster (1977); B is adapted from Gánti (1971).

of the replicators in a given dynamic spatial organization. Assuming the reactions occur in a well-mixed solution, modelled by a stochastic cell automaton, the spatial arrangement takes the form of a rotating spiral (Boerlijst and Hogeweg, 1991). This requirement was not obvious a priori, but emerged out of numerical and simulation studies (Cronhjort, 1995). Simulations were also able to address the question of the onset of a hypercyclic system, and revealed, for instance, that symbiosis occurs spontaneously in multi-species systems, and helps stabilize them (Kim and Jeong, 2005), suggesting that prebiotic models should take this phenomenon into account. In contrast to the well-mixed-solution hypothesis, another spatial arrangement which allows a hypercycle system to sustain errors is achieved by compartmentalization. The first models involved pre-existing vesicles encompassing hypercycles, but a more pertinent approach in the context of prebiotic models implies taking the dynamics of vesicle formation and their self-organization into account (Hogeweg and Takeuchi, 2003). Once established within vesicles, the replication cycles can undergo evolution, and incorporate new actors, for instance, a type of encoded effector, for which the genetic code can appear spontaneously, as shown by simulation experiments (Kuhn, 2005).

In the 'protein/lipid first' version, the replicative units are supposed to be made of molecules with a given catalytic activity, such as oligopeptides, able to link or unlink peptides together. The efficiency and selectivity of the catalysis can be very low as long as there is no competition with a better-fitted system: the laws of dynamics apply, although with very long characteristic times. Provided the critical number of reactions mentioned above has been attained, replicative cycles are formed, and the whole system is able to reach a steady state. The second prerequisite is the isolation of replicative units into vesicles. Such vesicles are known to self-assemble spontaneously when lipids are mixed in water. If peptides are present in the aqueous medium, they are captured into the vesicles, and their concentration can increase if the membrane is permeable to monomers (amino acids) but not to oligomers. With peptides inside membranes, or 'protocells', the system is now ready to evolve, without any informational molecule, according to a mechanism called 'inherited efficiencies' (New and Pohorille, 2000). Simulations show that the mean efficiency of catalysis increases with time in the community of protocells. The use of multi-agent systems allows us to expand the model by setting up a more complex and realistic evaluation of the fitness of protocells (Gupta *et al.*, 2006). The non-genomic scenario therefore represents a likely starter for life, allowing for adaptive evolution of abiotic 'metabolisms', before the emergence of a type of coupling called 'information replication and expression'. Self-organization and autocatalysis are indeed sufficient to allow for information transmission from one generation to the next (Monaco and Rateb de Montozon, 2005).

Autopoiesis

The concept of autopoiesis (Maturana and Varela, 1980) is also relevant to the category of the non-genetic models. It refers to a constructivist viewpoint on the question of defining life. A living system is defined by its ability to build its own body in space and maintain it. This definition makes the distinction between living systems and non-living growing systems: a crystal, for instance, does 'make use' of its pre-existing structure to build its newly formed 'body', but it does not regenerate its pre-existing components. On the other hand, the components of a living system continuously change their interactions in a way that produces and maintains the organization of the system. The essence of life resides neither in the nature of the components of the being (molecules, organs etc.) nor in its functional properties (metabolism, replication, evolution etc.) but in the relational properties between the components. Thus, for a chemical system, becoming autopoietic means bridging the gap from inert matter to life. Importantly, this gap does not involve replication, which is viewed as a side-effect and not as an essential condition for life. Reproduction of autopoietic units can occur as a simple process of mechanical splitting. Unequal repartition of components

Table 5.1. Examples of artificial-life software.

Program	Author(s)	Institution	URL	Purpose
[no name]	C. Fernando	University of Sussex, Brighton, UK	www.cogs.susx.ac.uk/users/ctf20/dphil_2005/dphil.htm	Chemoton-like fluid automaton
Avida	C. Adami	California Institute of Technology, CA, USA	dllab.caltech.edu/avida	A digital world in which simple computer programs mutate and evolve
COSMOS	T. Taylor, J. Hallam	University of Edinburgh, UK	tim.taylor@ed.ac.uk	Self-replicating programs in cellular organisms
ECHO	J. Holland, T. Jones	Santa Fe Institute, USA	www.santafe.edu/~pth/echo	Complex adaptive systems
NetLogo	U. Wilensky	Northwestern University, IL, USA	ccl.northwestern.edu/netlogo	Multi-agent programmable modelling environment
PyCage	E. M. Francis	Alcyone Systems, CA, USA	www.alcyone.com/software/cage	Complete cellular automata simulation engine
SCL	B. McMullin	Dublin City University, UK	elm.eeng.dcu.ie/~alife/Research.html	Autopoiesis
Tierra	T. Ray	Department of Zoology, University of Oklahoma, USA	life.ou.edu/tierra/	Synthetic organisms using CPU time as energy and memory as material

can account for some sort of evolution, without any requirement for informative molecules. A late-coming genetic material may provide an important increase in fitness, which explains why it is widespread, but it is not a necessary feature of a living being.

In contrast to the chemoton and hypercycle, there are only a few computer simulations of autopoietic systems (Table 5.1). Two of them were co-authored by the initiator of the concept, Francisco Varela. The first one was issued soon after the birth of the concept itself (Varela *et al.*, 1974). It was then extended (McMullin and Varela, 1997) in order to better fit with the structure of the model.

Evolvability

Whatever the initial assumptions and the details of the approach, all the models of life emergence involve a mechanism of self-reproduction. Besides, once a basic form of

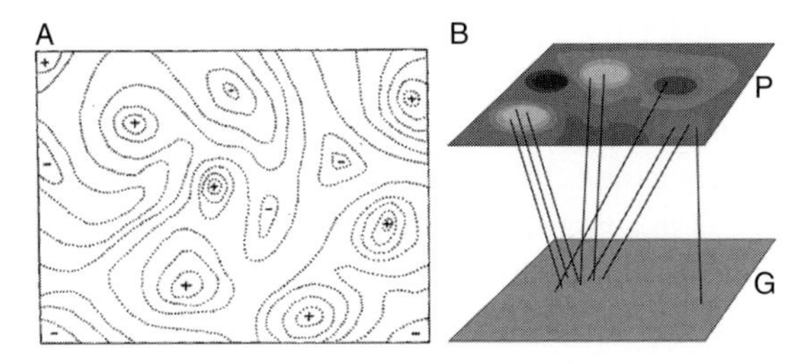

Figure 5.3. Fitness landscapes. A: A simplified fitness landscape, as shown in Wright (1932). The dimensions of the fitness space (parameters used to define the fitness) have been reduced to two in order to allow visualization of high (+) and low (−) fitness regions as peaks and valleys. Contour plots denote lines of equal fitness. B: The sequence–structure space of RNA, adapted from Schuster *et al.* (1994). It is made of two spaces (dimension reduced to two for visualization): genotype (G) where mutations occur and phenotype (P) where the selective pressure applies. The mapping between the two spaces is complex, with similar sequences (close in G space) mapping to very different structures (distant in P), and distantly related sequences mapping to close structures.

self-reproduction does exist, then the most fascinating phenomenon of life can occur, namely evolution. Evolution plays two roles in the theories of origins of life: it is one of several processes which are characteristic of living beings, and therefore, a property to be reconstructed in a model-based approach. But it is also a powerful means of accounting for the acquisition of new features by a system, this system being already alive or not (Szathmary and Smith, 1995). It is therefore not surprising that even models of prebiotic systems account for evolution, involving both stochastic and deterministic processes, in other words: '*hasard et nécessité*' (Monod, 1972). The transition from pre-life to life can be mathematically characterized (Nowak and Ohtsuki, 2008). Simulations involving random events in a chemical world show that the ability to perform replication can occur within reasonable timescales. Once initiated, replication unavoidably promotes more efficient replication (Ma *et al.*, 2007). The evolutionary process is strikingly bare in these conditions. Chemical cycles competing for template molecules or energy draw a pure sketch of 'descent with errors and selection through reproductive success', free of any complex mechanism which would blur the interplay between genotype and phenotype.

An important tool for the computational studies of evolution is termed the 'fitness landscape' (Wright, 1932). This expression refers to an abstract space with one dimension for each variable parameter. Thus, each point in this space represents a combination of parameters for which a fitness value can be computed (Figure 5.3A). If a genotype is depicted by such parameters, then a mutation is a move from one point to one of its neighbours, which may result in a change in fitness. Within this framework, a model of evolution is mainly specified by a fitness function, a definition of neighbourhood and a mechanism of selection. The associated fitness landscapes became by themselves a subject of study. In particular, there is more to say about the structure of the fitness landscape itself than what was thought initially. Actual fitness landscapes are seldom like geographical landscapes because their respective juxtapositions of peaks and valleys result from very different constraints. Among some important descriptors, they can be rugged or smooth, with important implications in the stepwise history of adaptation, random genetic drift or population diversity *vs.* homogeneity. This tool is of special interest for the computational approach in evolution studies because it allows for quantifying and parameterizing the relationships between genotype, phenotype and selective pressure. For instance, the prebiotic evolution of RNA has been studied as a random walk in a fitness landscape shaped by the secondary-structure energy of RNA sequence variants (Fontana *et al.*, 1993). One of the main outcomes is that in such a space, any sequence has many neighbours with the same folding energy, which means that evolving according to a random neutral drift allows for exploring a significant part of the sequence space. Moreover, the neighbourhood of any sequence contains sequences which fold in many different common secondary structures. This property is a consequence of the relation between the sequence space and the structure space (Schuster *et al.*, 1994). This mapping has a great impact on the evolution process in structured molecules: the mutations occur in the sequence space, which defines the neighbourhood of the sequences; the selection, if any, occurs in the structure space, which defines the fitness function (Figure 5.3B). So the RNA fitness landscape as a whole is a complex space, in which evolution driven by the fitness of secondary structures can occur efficiently with limited exploration of sequence variations.

Among other tools, fitness landscapes are also used to analyse results of *in-silico* experiments which could not be performed *in-vivo*, such as modifying the genetic alphabet. This allowed us to see that the four-letter alphabet was not optimal in extant living systems, but was the most robust in systems with a lower copy fidelity, and was thus suitable in an early RNA world (Gardner *et al.*, 2003).

Evolving programs

While computer scientists, as we already noticed, use the living being as a source of inspiration to describe and even conceive their machines and programs, the reverse metaphor is also extensively used. The cell has been compared to a computer, in which the genetic material is a program. Indeed, DNA is strikingly similar to the tape which encodes programs and data in the Turing machine (Adelman, 1998), and the encoding of the primary structures of proteins is digital: the genetic code is based on a set of symbols and their combinations. However, the similarity between computers and cells has often led to overestimating the role of the genetic message in cell properties and behaviour (Feitelson and Treinin, 2002).

Nevertheless, if explored with care, the parallel between the genetic message and a computer program provides an interesting way to study the evolutionary process (see Table 5.1). The idea is to use programs as models for living beings (Ray, 1991). A program can be subject to slight random modification upon copying, and these modifications may improve or, most of the time, decrease its efficiency. If the role of a program is to make copies of itself, then programs will need to duplicate with a low energy and/or CPU time consumption. On the other hand, having a longer code allows it to perform more complex tasks, like error checking … a reminder of the motivation for the creation of hypercycles (see above).

What is interesting within the framework of evolvable software is that the possibilities of a program are virtually infinite, making it possible to observe how phenomena that are not defined a priori will appear (Ray, 1994). Thus, more than imitating extant life forms, artificial-life systems exhibit an essential property of living beings: the ability to undergo open-ended evolution.

Conclusions

Biology tries to uncover the mechanisms of life by studying the life forms which can be observed, or 'life-as-we-know-it'. The purpose of artificial life is to include the biology into the more general question of the essence of life, or 'life-as-it-could-be' (Langton, 1989). Artificial life is therefore a natural component of research domains such as the search for extraterrestrial life forms or the elucidation of the origins of life. Even if a consensual definition of life is still lacking, the researchers in each community rely on a more or less explicit way of distinguishing between the living and the non-living (or, for the origins of life: the not-yet living). Chris Adami (1995) proposed that a means of understanding life is precisely to try to exhibit the simplest model that displays the attributes necessary to be recognized as living. It is possible that several different models would be able to fulfil this requirement, but in any case, extant life forms are far too sophisticated, and only artificial systems can be eligible.

As a limit consequence, some researchers involved in artificial life claim that computer programs are more than models of life, and that we can speak of literally creating living organisms without any materials, just in the form of information processing (Emmeche, 1994). The different opinions on this topic would deserve a chapter to themselves. Let us simply mention that according to many philosophers, such a proposition seems to suffer from weakness, at least from an ontological point of view (Olson, 1997).

By generalizing life to any of its conceivable implementations, artificial life allows us to address questions at a conceptual level. The biotic life that we know here and now is no more than an instance of these concepts. No more, but no less. If a concept is general enough, then it must apply in particular to this precise instance of life. For example, cell automata are theoretical constructions which have been the subject of a great variety of mathematical investigations and applications in many fields of knowledge (Bandini *et al.*, 2001). The transposition of these mathematical results to the question of the origins of life requires a mapping of the automaton's constituents and properties to some physical objects and their interactions. This is not a trivial task, and to our knowledge there is no complete model of such a mapping. For instance, one can consider that the cells represent positions in a medium and that a cell state stands for a molecule present in this place. But the transition rules accounting for the Brownian motion of molecules would be difficult to express in the cell automata formalism. Nevertheless, the main value of cell automata is to exemplify how local interactions (each cell is aware of the state of some of its immediate neighbours) result in global properties at the level of the whole automaton. Moreover, quantitative studies, especially those concerning the classification of cell automata, have allowed us to precisely define the conditions necessary for the emergence of lifelike properties.

Consequently, artificial-life models do not pretend to be accurate representations of actual living beings. Instead, they can be seen as 'thought experiments', to which the computer adds the power of simulation (Bedau, 1998). This is especially true when considering evolution. Due to the central role of chance in the evolutionary process, the simulation approach has proved to be of outstanding value. Of course, both analytical and numerical computations can incorporate probabilities in their formulas and account for variability within populations. Their predictive power is far beyond the computation of averages and standard errors. But evolution is a highly heuristic process involving bifurcations and dead ends. The occurrence of a single event often changes the global outcome for a whole population. The broad trends of all evolvable systems towards a better adaptation to their environment should not mask the fact that we never see the same movie when we play the tape many times. By repeatedly running a simulation, it becomes possible to highlight the main properties of an evolutionary process and to decipher the respective contributions of the driving forces at work in the model as a whole versus the singular history occuring in any specific run.

The use of repeated simulations has been encountered in many different contexts. In fact, they are present everywhere throughout the origins-of-life theories. Obviously, the role of the simulations is not to bring a final ending to the controversies (Peretó, 2005), either by summoning *ex nihilo* the synthesis between divergent viewpoints, or by definitely proving the validity of one theory against that of another. For instance, many models focus on the question of self-reproduction, somehow adopting the 'genetic first' party. But at the same time, the tenants of the 'metabolism first' position also have their models and simulations. Both types of models (and others) are supported by experimental data and theoretical arguments. In each of these models, computational approaches, including mathematical analyses and simulations, have assessed the feasibility of some crucial points, and allowed for quantitative elucidation of the conditions necessary for the occurrence of key events. In several cases, unexpected phenomena have emerged from simulations and have initiated new ideas. This is probably the way simulations will participate in the emergence of a synthetic theory of the origins of life. In addition, expressing a model so as to make it run in a simulation requires an effort to precisely define the role of every component of the system and to clarify every assumption. As just reward, simulations will show with an incomparable expressive power how a set of basic hypotheses gives rise to new and integrated properties in a system.

Acknowledgements

The author wishes to thank Bénédicte Charrier (CNRS – UMR7139) for her support and proofreading of the manuscript, and Guillaume Achaz for his candid comments and friendly discussion.

References

Adami, C. (1995). On Modeling Life. *Artificial Life*, **1**, 429–438.

Adelman, L. M. (1998). Computing with DNA. *Scientific American*, 54–61.

Bandini, S., Mauri, G. and Serra, R. (2001). Cellular automata: from a theoretical parallel computational model to its application to complex systems. *Parallel Computing*, **27**, 539–53.

Banks, E. R. (1971). Information processing and transmission in Cellular Automata. PhD thesis, Massachusset Institute of Technology.

Bedau, M. A. (1998). Philosophical content and method of artificial life. In *The Digital Phoenix: How Computers are Changing Philosophy*, ed. T. W. Bynam and J. H. Moor. Oxford: Basil Blackwell.

Beyer, W. A., Sellers, P. H. and Waterman, M. S. (1985). Stanislaw M. Ulam's contributions to theoretical theory. *Letters in Mathematical Physics*, **10**, 231–42.

Boerlijst, M. and Hogeweg, P. (1991). Spiral wave structure in pre-biotic evolution: hypercycles stable against parasites. *Physica D*, **48**, 17–28.

Bullock, S. (2000). What can we learn from the first evolutionary simulation model? In *Artificial Life vii*. Cambridge, MA: MIT Press, pp. 477–86.

Chou, H. and Reggia, J. A. (1997). Emergence of self-replicating structures in a cellular automata space. *Physica D*, **110**, 252–76.

Cronhjort, M. B. (1995). Hypercycles versus parasites in the origin of life: model dependence in spatial hypercycles systems. *Origins of Life and Evolution of the Biosphere*, **25**, 227–33.

Crutchfield, J. P., Haber, P. T. and Mitchell, M. (1993). Revisiting the edge of chaos: evolving cellular automata to perform computations. *Computing Systems*, **7**, 89–130.

Dubacq, J., Durand, B. and Formenti, E. (1999). Kolmogorov complexity and cellular automata classification. *Theoretical Computer Science*, **259**, 271–95.

Eigen, M. and Schuster, P. (1977). The hyper cycle. A principle of natural self organization. Part A. Emergence of the hyper cycle. *Naturwissenschaften*, **64**, 541–65.

Emmeche, C. (1994). *The Garden in the Machine: The Emerging Science of Artificial Life*. Princeton, NJ: Princeton University Press.

Feitelson, D. G. and Treinin, M. (2002). The blueprint for life? *IEEE Computer*, **35**, 213–22.

Fernando, C. and Di Paolo, E. (2004). The chemoton: a model for the origin of long RNA templates. In *Artificial Life ix*. Cambridge, MA: MIT Press, pp. 1–8.

Fontana, W., Stadler, P., Bornberg-Bauer, E., Griesmacher, T., Hofacker, I., Tacker, M., Tarazona, P., Weinberger, E. and Schuster, P. (1993). RNA folding and combinatory spaces. *Physical Review E*, **47**, 2083–9.

Gánti, T. (1971). *The Principles of Life*. Budapest: Gondolat.

Gánti, T. (1997). Biogenesis itself. *Journal of Theoretical Biology*, **187**, 583–93.

Gardner, M. (1970). The fantastic combinations of John Conway's new solitaire game 'life'. *Scientific American*, **223**, 120–3.

Gardner, P. P., Holland, B. R., Moulton, V., Hendy, M. and Penny, D. (2003). Optimal alphabets for an RNA world. *Proceedings of the Royal Society of London B*, **270**, 1177–82.

Gupta N., Agogino, A. and Tumer, K. (2006). Efficient agent-based models for non-genomic evolution. In proceedings of the Fifth International Joint Conference on Autonomous Agents and Multi-Agent Systems, 58–64.

Hogeweg, P. and Takeuchi, N. (2003). Multilevel selection in models of prebiotic evolution: compartments and spatial self-organization. *Origins of Life and Evolution in the Biosphere*, **33**, 375–403.

Kauffman, S. A. (1993). *The Origins of Order*. New York, USA: Oxford University Press.

Kim, P. and Jeong, H. (2005). Spatio-temporal dynamics in the origin of genetic information. *Physica D*, **203**, 88–99.

Kuhn, C. (2005). A computer-glimpse of the origin of life. *Journal of Biological Physics*, **31**, 571–85.

Langton, C. G. (1984). Self-reproduction in cellular automata. *Physica D*, **10**, 135–44.

Langton, C. G. (1989). Artificial life. In *Artificial Life*, ed. C. G. Langton. Massachusetts: Addison-Wesley.

Langton, C. G. (1992). Life at the edge of chaos. In *Artificial Life ii*. Cambridge, MA: MIT Press, pp. 41–91.

Lazcano, A. (2008). Towards a definition of life: the impossible quest? *Space Science Review*, **135**, 5–10.

Ma, W., Yu, C. and Zhang, W. (2007). Monte Carlo simulation of early molecular evolution in the RNA World. *BioSystems*, **90**, 28–39.

Maturana, H. R. and Varela, F. J. (1980). *Autopoiesis and Cognition: The Realisation of the Living*. London: Reidel.

McMullin, B. and Varela, F. J. (1997). Rediscovering computational autopoiesis. *ECAL IV*, 38–47.

Monaco, R. R. and Rateb de Montozon, F. (2005). Self-organization, autocatalysis and models of the origin of life. In *Complex Systems Summer School*, Santa Fe.

Monod, J. (1972). *Chance and Necessity: An Essay on the Natural Philosophy of Modern Biology*. London, UK: Collins.

Munteanua, A. and Sole, R. V. (2006). Phenotypic diversity and chaos in a minimal cell model. *Journal of Theoretical Biology*, **240**, 434–42.

New, M. H. and Pohorille, A. (2000). An inherited efficiencies model of non-genomic evolution. *Simulation Practice and Theory*, **8**, 99–108.

Nowak, M. A. and Ohtsuki, H. (2008). Pre-evolutionary dynamics and the origin of evolution. *Proceedings of the National Academy of Sciences of the USA*, **105**, 14924–7.

Olson, E. (1997). The ontological basis of strong artificial life. *Artificial Life*, **3**, 29–39.

Peretó, J. (2005). Controversies on the origin of life. *International Microbiology*, **8**, 23–31.

Pesavento, U. (1995). An implementation of von Neumann's self-reproducing machine. *Artificial Life*, **2**, 337–54.

Ray, T. S. (1991). An approach to the synthesis of life. In *Artificial Life ii*. Cambridge, MA: MIT Press, pp. 367–86.

Ray, T. S. (1994). Evolution, complexity, entropy and artificial reality. *Physica D*, **75**, 239–63.

Schuster, P., Fontana, W., Stadler, P. F. and Hofacker, I. L. (1994). From sequences to shapes and back: a case study in RNA secondary structures. *Proceedings of the Royal Society B*, **255**, 279–84.

Smith, A. (1976). Introduction to and survey of polyautomata theory. In *Automata, Languages, Development*. The Netherlands: North Holland Publishing.

Smith, J. M. (1979). Hypercycles and the origin of life. *Nature*, **280**, 445–6.

Szathmary, E. and Smith, J. M. (1995). The major evolutionary transitions. *Nature*, **374**, 227–32.

Turing, A. M. (1936). On computable numbers with an application to the entscheidungsproblem. *Proceedings of the London Mathematical Society*, **42**, 230–65.

Varela, F., Maturana, H. and Uribe, R. (1974). Autopoiesis: the organization of living systems, its characterization as a model. *BioSystems*, **5**, 187–96.

von Neumann, J. (1966). *Theory of Self-Reproducing Automata*. Champaign, IL: University of Illinois Press.

Wiener, N. (1948). *Cybernetics, or Control and Communication in the Animal and the Machine*. Cambridge, MA: MIT Press.

Wolfram, S. (1983). Statistical mechanics of cellular automata. *Review of Modern Physics*, **55**, 601–44.

Wolfram, S. (1984). Universality and complexity in cellular automata. *Physica D*, **10**, 1–35.

Wright, S. (1932). The roles of mutation, inbreeding, crossbreeding and selection in evolution. In *Proceedings of the Sixth International Congress of Genetics*, pp. 356–66.

Part II

Astronomical and geophysical context of the emergence of life

6

Organic molecules in the interstellar medium

Cecilia Ceccarelli and José Cernicharo

Introduction

At one end, the Earth, we have complex molecular structures giving rise to life; at the other end, the diffuse interstellar medium (hereinafter ISM), we have atoms floating in an almost empty space. How, when and where did the transition from unbound atoms to complex molecular structures occur? Had it occurred already in the dense molecular-cloud phase of the ISM and/or during the formation and evolution of the protosolar nebula? These are the questions whose answers will help in understanding whether, as the Nobel prizewinner C. De Duve wrote, 'The building blocks of life form naturally in our galaxy and, most likely, also elsewhere in the cosmos. The chemical seeds of life are universal' (De Duve, 2005). What we know for sure is that a long process, of a few billions years, brought matter from the diffuse state of the ISM to the condensed state of planets (Earth), comets and meteorites (see Chapter 7 of this volume). We also know that primitive meteorites, the oldest fossils we have from the Solar-System formation aeons, contain the 'seeds of life' that De Duve alluded to: amino acids.

In this chapter we will show that the formation process of solar-type stars, while bringing matter from a diffuse to a condensed state, also leads to increasing molecular complexity. Although solar-type star-forming regions are not the only places in the ISM where organic molecules are found, two reasons lead us to focus here on them: (1) they are among the places with the richest harvest of organic molecules; and (2) they are regions similar to our Solar-System progenitor, so that the organic chemistry observed there is directly linked to the possible inheritance of terrestrial life from the ISM.

The chapter is organized as follows. We will initially describe the observed increasing molecular complexity as matter evolves from the diffuse state to form a star and planetary system similar to the Solar System, and then briefly review the mechanisms thought to be at the origin of the molecular complexity and outline the difficulties of the present theories. This is followed by a discussion as to how the observed molecular deuteration helps us to understand the heritage passed throughout the various phases of star formation. We then focus on the issue of amino acids in the ISM and discuss the similarities and differences so far observed in the organic material found in the ISM and meteorites (as described in Chapter 7 of this volume) prior to concluding the chapter.

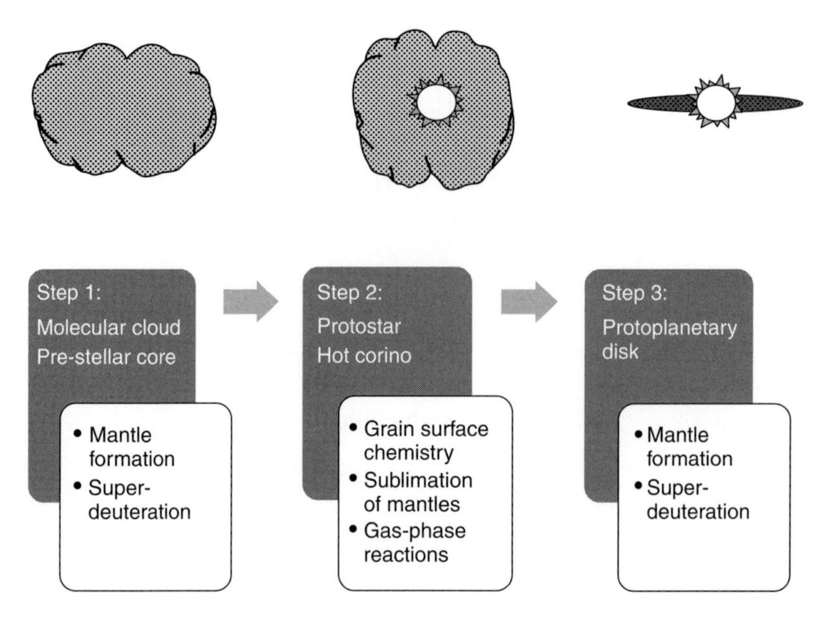

Figure 6.1. Schema of the three major steps in solar-type star formation and chemical complexity.

Star formation and molecular complexity

Star formation and molecular complexity go hand in hand. Figure 6.1 shows schematically the three major steps in the formation of a solar-type protostar, which also correspond to three major steps in increasing molecular complexity.

Step 1: Molecular clouds and pre-stellar cores

Stars like our Sun and planetary systems like our Solar System form out of cold (~ 10 K) and dense (> 10^6 particles per cm^{-3}) condensations, called 'pre-stellar cores' (see the recent reviews by Di Francesco *et al.*, 2007; Ceccarelli *et al.*, 2007; Bergin and Tafalla, 2007), inside the densest regions of the ISM, the 'molecular clouds'. As their name indicates, hydrogen, the most abundant element in the Universe, is predominantly in the molecular form in molecular clouds. Similarly, the heavier elements in the gas are prevalently bound in a variety of molecules: from simple diatomic molecules like CO to long chains like the cyanopolyynes (HC_nN with n = 3, 5, 7, 9 and 11) or unsaturated chains like propylene (CH_3CHCH_2), just to mention a few examples that are important in the context of this chapter (for a list of all detected molecules in the ISM see www.en.wikipedia.org/wiki/List_of_molecules_in_interstellar_space). We know that the chemical reactions leading to the above molecules occur both in the gas phase and on the surfaces of the dust grains. Indeed, although dust constitutes only 1/100 of the mass of the ISM, it plays a key role in the gas chemistry. First, molecular hydrogen is formed on the grain surfaces (Hollenbach and Salpeter, 1971). Second, given the low temperatures, heavy atoms and molecules tend to freeze out onto the dust grains forming icy mantles. As an important example, at the centre of the pre-stellar cores (in the inner few thousand astronomical units [AUs]), the CO molecules are expected to freeze out onto the dust grains (disappearing from the gas phase) in less than 10^4 yr. This phenomenon leads to a triple effect: (1) the disappearance of gaseous molecules limits the importance of gas-phase chemistry at this stage; (2) atoms and molecules are oxidized and hydrogenated on the grain surfaces and form molecules such as H_2O, CO_2, CH_4, NH_3, H_2CO and CH_3OH (Tielens and Hagen, 1982); (3) the low temperature, coupled with the disappearance of CO (and all heavy-element-bearing molecules), triggers the super-deuteration phenomenon and the formation of multiply deuterated molecules with abundances enhanced up to 13 orders of magnitude with respect to the elemental deuterium/hydrogen abundance ratio.

Observations show that the grain mantles in molecular clouds are composed prevalently of H_2O, CO and CO_2 (e.g. Knez *et al.*, 2005). Although similar observations are impossible towards the centre of the pre-stellar cores, where all CO and heavy-element-bearing molecules freeze out onto the grain mantles, much indirect evidence suggests that hydrogenation and oxidation of atoms and simple molecules takes place. This is the first step towards the molecular complexity process, even though it is not clear yet how big the step is. Certainly, a relatively complex molecule such as formic acid (HCOOH), with two atoms of oxygen and one of carbon, is present in the grain mantles with an abundance of about 10^{-5} with respect to hydrogen, namely containing an important fraction of the gaseous carbon not locked in CO or CO_2 (Knez *et al.*, 2005).

Step 2: Protostars and hot corinos

Once the collapse starts, the envelope surrounding the forming star is heated by radiation energy produced from the gravitational energy released by the infalling matter. This leads to three effects: (1) in the warm dust, atoms and molecules of the mantles become mobile and probably react to form more complex molecules (e.g. Garrod and Herbst, 2006); (2) in the region of the envelope where the dust temperature reaches 100 K, called the 'hot corino', and whose sizes are estimated to be similar to those of the Solar System (Maret *et al.*, 2004), the grain mantles sublimate and their components are released into the gas phase (Ceccarelli *et al.*, 2000a); (3) in the warm (> 100 K) gas, some endothermic reactions between the neutral molecules (including those sublimated from the grain mantles) become efficient, forming more complex molecules (Rodgers and Charnley, 2003).

From a chemical point of view, therefore, solar-type protostars consist of two regions (see the recent review by Ceccarelli *et al.*, 2007): (1) the cold envelope, where the dust temperature is lower than 100 K: the chemistry is rather similar to that in pre-stellar cores and dominated by the super-deuteration phenomenon (step 1 chemistry); (2) the hot corino region, where the dust temperature exceeds 100 K: the chemistry is totally dominated by the sublimation of the grain mantles. Table 6.1 lists all detected complex (with more than 4 atoms) organic molecules observed in hot corinos and their abundances. Given the difficulty in observing these objects (smaller than ~ 1″), only a few molecules have been detected in a few objects so far. However, from those observations it is clear that a substantial fraction (at least 1%) of carbon (whose majority is locked into gaseous CO or grains) is contained in organic molecules. It is worth noticing that the abundance of the detected complex organic molecules normalized to the abundance of formaldehyde or methanol (both molecules represent starting points for the synthesis of more complex molecules) is greater in hot corinos than in hot cores, analogous to hot corinos in massive protostars (Bottinelli *et al.*, 2007). In this sense, hot corinos can be considered the 'retail shops' of complex organic molecules in the ISM, despite the fact that they are more difficult to observe than their more massive cousins. Furthermore, solar-type protostars differ from massive protostars in the observed degree of deuteration. In the former, multiply deuterated molecules have been detected with abundances of a small percentage with respect to their relative hydrogenated species (D_2CO and CD_3OH: Ceccarelli *et al.*, 1998; Parise *et al.*, 2004), whereas the same molecules are often undetected in massive protostars (Ceccarelli *et al.*, 2007). For whatever reason, a different previous history or a longer lifetime, hot corinos are not simply a scaled version of hot cores, as much as solar-type protostars are not a scaled version of massive protostars. Therefore, it would be more appropriate to compare Solar-System objects to what is found in solar-type protostars and hot corinos, rather than ISM or massive protostars in general.

Step 3: Protoplanetary disks

'Protoplanetary disks' are the last step towards the formation of planets and the leftovers of the process, namely comets and asteroids. The dust slowly coagulates into 'planetesimals' and settles onto the equatorial plane while the disk cools down. Similarly to protostars (step 2) from the chemical point of view, protoplanetary disks are formed by two regions: (1) the inner disk, where the equatorial dust temperature exceeds 100 K, at around 10–20 AU from the central star: the chemistry is dominated by the sublimation of the grain mantles; and (2) the outer disk, where the lower temperature again triggers the phenomena seen in the pre-stellar cores, namely mantle formation and super-deuteration. A key point here is that the grain mantles may now contain all molecules formed during the previous phases, from the simple ones (H_2O, CO, CO_2, NH_3, CH_4 etc.) to the more complex ones formed in the hot-corino phase (see Table 6.1). During the dust-coagulation period those mantles may be (entirely or partially) trapped and conserved throughout the whole process up to the planet-, comet- and meteorite-formation stages. Therefore, the latter may have the chemical imprint of the whole star-formation process. As mentioned several times, molecular deuteration may be the hallmark of the imprint, even though its interpretation is not straightforward, as it may originate in several and in different phases.

In the inner disk, which is warm and dense, a very active chemistry takes place (e.g. Agúndez *et al.*, 2008). Recent observations have detected large quantities of acetylene (H_2C_2) and hydrogen cyanide (HCN) in the inner few tens of AUs of some protoplanetary disks (Lahuis *et al.*, 2006; Carr and Najita, 2008). Unfortunately, detection of more

Table 6.1. List of the complex (with 5 or more atoms) organic molecules observed in hot corinos.

Source	Species	Abundance/H_2	Reference
IRAS16293–2422	CH_3OH	1×10^{-7}	Maret *et al.*, 2005
	HCOOH	6×10^{-8}	Cazaux *et al.*, 2003
	CH_3CHO	5×10^{-8}	
	CH_3OCHO	4×10^{-7}	
	CH_3OCH_3	2×10^{-7}	
	$HCOOCH_3$	4×10^{-7}	
	CH_3CN	1×10^{-8}	
	C_2H_5CN	1×10^{-8}	
	CH_3CCH	3×10^{-7}	
	NH_2CH_2COOH	$< 10^{-10}$	Ceccarelli *et al.*, 2000b
NGC1333-IRAS4A	CH_3OH	2×10^{-8}	Maret *et al.*, 2005
	HCOOH	5×10^{-9}	Bottinelli *et al.*, 2004
	$HCOOCH_3$	7×10^{-8}	
	CH_3CN	2×10^{-9}	
NGC1333-IRAS4B	CH_3OH	3×10^{-6}	Maret *et al.*, 2005
	$HCOOCH_3$	2×10^{-6}	
	CH_3CN	1×10^{-7}	Bottinelli *et al.*, 2007
NGC1333-IRAS2A	CH_3OH	2×10^{-7}	Maret *et al.*, 2005
	CH_3CN	1×10^{-8}	Bottinelli *et al.*, 2007
	CH_3OCH_3	3×10^{-8}	Jørgensen *et al.*, 2005

complex molecules is hampered by the sensitivity of the available instruments. Similar limitations apply for the detection of complex molecules in the outer disk, where the largest detected molecule is H_2CO (Dutrey *et al.*, 2007). Finally, measures of the molecular deuteration in protoplanetary disks are scarce (DCO^+, H_2D^+ and HDO: van Dishoeck *et al.*, 2003; Ceccarelli *et al.*, 2004, 2005), but they confirm the occurrence of the super-deuteration phenomenon.

Theories of organic chemistry in the ISM

There are three known mechanisms for the formation of organic molecules in the ISM: (1) in the gas phase; (2) on the grain surfaces; and (3) by UV irradiation and/or energetic bombardment of grain mantles. What mechanism is important and when are matters of intense debate and it certainly depends on the molecules and the time of evolution. At one end there are the observations (reviewed in the previous section) and at the other end there are the predictions of the astrochemical models that aim to reproduce the observations. In between, there are quantum chemistry calculations and laboratory experiments to identify the reactions, the reaction products and the rates of the reactions.

Quantum chemistry calculations are complex and often require huge amounts of computing time, so that we are far from having an exhaustive coverage of the various possibilities. Laboratory experiments are also complex, as they need to simulate the ISM conditions, namely vacuum, low temperatures and various doses of UV irradiation and energetic bombardment, for instance, by protons from cosmic rays. While several experiments exist on the gas-phase reactions (e.g. Chastaing *et al.*, 2001) and UV irradiation and/or energetic bombardment (e.g. Muñoz Caro *et al.*, 2002; Bennet and Kaiser, 2007), experiments of surface reactions are scarce. To illustrate the situation, the hydrogenation of CO and formation of H_2CO and CH_3OH (step 1) has been a source of controversy for a long period of time and only recently have experiments agreed on the results (e.g. Hidaka *et al.*, 2009).

In the absence of theoretical computations and laboratory experiments, astrochemical models assume 'reasonable' reaction rates. One has also to notice that the methodology to deal with surface chemistry is itself a matter of controversy. Conversely to gas-phase chemistry, where the products of a reaction only depend on the number density of the reactants and the reaction rate, this deterministic approach is not suitable for describing reactions on the grain surfaces. In this case, the exact method involves taking into account the probability that the two reactants meet on the grain surface (which is not necessarily so). Solving the exact method is, in practice, only possible for a small chemical network (with less than a dozen species: Barzel and Biham, 2007). However, various approximate methods have been recently developed and have started to give some interesting results (see below and Herbst and van Dishoeck, 2009, for a recent detailed review).

In general, the 'old' gas-phase astrochemical models fail to reproduce the observations of complex organic molecules, so that the present models assume that the majority of these species, for example, methyl formate ($HCOOCH_3$) or dimethyl ether (CH_3OCH_3), are formed on the grain surfaces (as described in step 2 of the previous section) (e.g. Horn *et al.*, 2004). The latest generation of astrochemical models assume that those molecules are formed during the warming up of the dust because of reactions of radicals formed on the grains by the cosmic rays (Garrod and Herbst, 2006). Depending on the various assumed histories and durations of the dust warm-ups, those models are able to reproduce the available observations (Garrod *et al.*, 2008). However, as discussed above, many aspects of the models need confirmation, so predictions of the abundance of molecules that have not yet been observed cannot be considered as reliable, even when assuming the correctness of the assumptions made within those models.

Molecular deuteration

As mentioned earlier, the low temperature coupled with the disappearance of CO and heavy-element-bearing molecules from the gas phase give rise to the super-deuteration phenomenon. Briefly, in molecular gas the main reservoir of deuterium is HD and (a fraction of) deuterium atoms are (mainly) transferred from HD to other low-abundance molecules via various reactions involving the deuterated isotopologues of the ion H_3^+, namely H_2D^+, HD_2^+ and D_3^+. These isotopologues are, in fact, formed by reactions with

HD: e.g. $H_3^+ + HD^- \rightarrow H_2D^+ + H_2$. In practice, the molecular deuteration depends on the abundance of H_2D^+, HD_2^+ and D_3^+ with respect to the H_3^+ abundance. In normal conditions, H_2D^+ reacts with H_2 and gives back HD, so that the H_2D^+/ H_3^+ ratio is set by the elemental D/H ratio, namely $H_2D^+/ H_3^+ \sim 10^{-5}$ (similar considerations apply to $HD_2^+/ H_3^+ \sim 10^{-10}$ and $D_3^+/ H_3^+ \sim 10^{-15}$) and the deuteration of all other molecules also reflects the D/H elemental abundance ratio. However, at low temperatures (< 30–40 K) the situation changes because the reaction of H_2D^+ with H_2 is inhibited, so that the H_2D^+/H_3^+ ratio is enhanced with respect to the elemental D/H ratio. The enhancement is limited because of the competition of CO (rather than HD, which gives H_2D^+) in reacting with and destroying H_3^+. When CO (and the other heavy-element-bearing molecules) disappear from the gas phase the reaction with HD becomes the major route of destruction of H_3^+, increasing the abundance of H_2D^+, whose reaction with HD forms HD_2^+, which again reacts with HD to form D_3^+. To illustrate the point, Figure 6.2 shows the H_2D^+/ H_3^+, HD_2^+/ H_3^+ and D_3^+/ H_3^+ ratios as a function of the CO abundance in the gas. For extremely low CO abundances, the ratios can become much larger than unity and D_3^+ becomes orders of magnitude more abundant than H_3^+. The consequence is the formation of doubly and triply deuterated molecules with abundances of a small percentage, as mentioned previously. The theory has been basically confirmed by the detection of abundant H_2D^+ and HD_2^+ (Caselli *et al.*, 2003; Vastel *et al.*, 2004).

It is now clear why a large molecular deuteration is a hallmark of the first phases of the formation of the Solar System, even though it may be difficult to pin down exactly when and where the deuteration occurred. Table 6.2 summarizes the deuteration ratios observed during the various phases in different molecules. Note the case of water: the deuteration of this species is lower than for other molecules, notably formaldehyde and methanol. The reason for that is not entirely clear: it is probably due to the different mechanisms and times of formation of the three molecules. Therefore, when comparing the deuteration observed in Solar-System objects (comets and meteorites) it is important to compare the

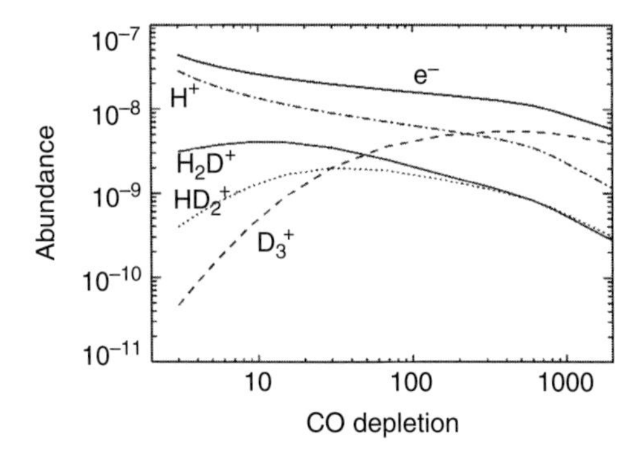

Figure 6.2. Abundance ratio of the isotopologues of the H_3^+ ion as a function of CO depletion in the gas, namely the factor by which CO is under-abundant with respect to the normal abundance in molecular clouds of 10^{-4} (adapted from the model by Ceccarelli and Dominik, 2005). e^- = the abundance of electrons.

Table 6.2. Molecular deuteration D/H, namely the abundance ratio between D-bearing and H-bearing molecules in the ISM and Solar System.

Object	Molecule	D/H	References
Oceans	H_2O	1.6×10^{-4}	De Witt *et al.*, 1980
Comets	H_2O	3×10^{-4}	Villanueva *et al.*, 2009
Protoplanetary disks	H_2O	0.01?	Ceccarelli *et al.*, 2005
Ices of protostars	H_2O	~ 0.03	Parise *et al.*, 2005
Carbonaceous chondrites	Amino acids	up to 0.001	Pizarello, this volume
Ices of protostars	H_2CO, CH_3OH	0.01–0.80	Ceccarelli *et al.*, 2007

same molecules. Furthermore, the situation may also be complicated by the exchange of deuterium and hydrogen atoms between the frozen molecule and the water–ice matrix, as observed in laboratory experiments (Ratajczak *et al.*, 2009; Weber *et al.*, 2009). Depending on the molecule and the functional group, the time of residence of the frozen molecule in the ice, the ice temperature and UV irradiation, this mechanism may significantly alter the original deuteration.

Ultimate molecular complexity and amino acids in the ISM

For a long time the amino acids observed in meteorites have been thought to be synthesized by aqueous alteration on the parent body. The observations of a large deuteration, up to 0.001, in the amino acids found in carbonaceous meteorites have raised the question as to whether (some) amino acids are formed in the ISM (see the discussion in Chapter 7 in this volume). Based on what has been voiced above, while testifying to an early origin, a large deuteration of the amino acids alone does not ensure formation in the ISM. First, D-atoms may be 'acquired' from the deuterated water–ice during the long period of co-existence of frozen amino acids and water. Both thermal alteration and UV irradiation during that period may lead to D-enrichment of the amino acids at the expense of the deuterated ice (Ratajczak *et al.*, 2009; Weber *et al.*, 2009). Second, the large deuteration may simply come from the (deuterated) amino-acid precursors, as also discussed by Pizarello in this volume (Chapter 7).

Evidently, the ultimate proof would be to detect amino acids in the ISM. Unfortunately, this is not easy and no detection exists so far of even the simplest of the amino acids, glycine, despite several attempts. The first observations targeted the most luminous star-forming regions of the Galaxy but with no success (e.g. Combes *et al.*, 1996). Later observations aimed at solar-type protostars, but the results were similarly negative (Ceccarelli *et al.*, 2000b). Recently, the detection of glycine was first claimed and then definitively disproved (Jones *et al.*, 2007). All these observations put an upper limit to the glycine abundance of about 10^{-10} with respect to H_2.

The reason for this 'failure' is that the larger the molecule the more numerous the modes of rotation and vibration of the molecule are, giving rise to a large number of spectral lines, often thousands. This has two effects: (1) the energy is spread over several lines, making them weak and difficult to detect; and (2) the large number of weak lines makes a 'grass' of lines, increasing the probability of mere coincidence (confusion limit). Although observations with high spatial resolution may alleviate the problem, the basic problem remains and detection of amino acids in the ISM may remain a chimera.

The alternative to the observational approach is the theoretical one. Even though amino acids are not detectable in the ISM, a reliable theory of the formation of organic molecules may apply to observations. Unfortunately, as discussed earlier, we do not have a fully reliable theory for the formation of complex organic molecules yet. But the huge efforts in that direction should soon give reliable answers. We should add the possibility, reproduced in laboratory experiments, that amino acids are formed on the grain mantles exposed to UV photons (Muñoz Caro *et al.*, 2002; Lee *et al.*, 2009). However, also in this case, the results have to be regarded with some caution, as the detection is made after a heavy hydrolysis process with HCl, a disputable 'natural' process. Finally, the recent detection of amino acids in comets may greatly contribute to understanding whether amino acids are formed in the ISM and how. If confirmed, the recent claim of the detection of extraterrestrial amino acids in the STARDUST samples from the comet WILD2 (Elsila *et al.*, 2009) is a first important step in that direction, even considering the huge differences between the comet's formation environment and the ISM.

In summary, while it is possible, it is still unproven that amino acids are formed in the ISM. What is proven is that all the precursors of amino acids are synthesized during the star-formation process, and that they have the imprint of the first phases, namely a large deuteration. The recent observations of molecules such as propylene (CH_3CHCH_2: Marcelino *et al.*, 2007), ethylene glycol ($HOCH_2CH_2OH$: Hollis *et al.*, 2002), aminoacetonitrile (NH_2CH_2CN: Belloche *et al.*, 2008) or cyanoformaldehyde (CNCHO: Remijan *et al.*, 2008) in the massive clouds towards the Galactic centre suggest that we are just seeing the tip of the iceberg and support the possibility that many more complex organic molecules, including amino acids, are formed in the ISM. Nonetheless, it is worth noticing that the upper limits on the abundance of glycine in the ISM ($< 10^{-10}$) play against the hypothesis that the glycine found in meteorites is prevalently formed in the ISM. In fact, in meteorites, the abundance of glycine is much higher than that of other organic molecules, whereas the opposite is true in the ISM (see Table 6.1 of this chapter and Table 7.1 of Chapter 7).

Conclusions: from ISM to meteorites

The formation of stars like the Sun is accompanied by an increase of the complexity in the organic chemistry. Several and abundant complex organic molecules have been detected, notably in the hot corinos. Despite the huge progress of the last decade in this field, we do not yet have a comprehensive theory for exactly how molecular complexity develops in the ISM in general, and during the star-formation period in particular. We are also far from being able to affirm that the molecular complexity achieved is such that amino acids are certainly formed in the diffuse matter during the star-formation process. Nonetheless, in spite of all difficulties and limitations in our knowledge, it is clear that a rich organic chemistry takes place in the ISM, and that it may have had a role in what we find today in meteorites. 'Ariadne's thread' throughout the various phases, from the very first to the last one, is provided by molecular deuteration.

In summary, given the present limitations in the observations and in the theory, we are far from knowing what is the ultimate attained complexity: it is possible but not certain that molecules as complex as amino acids are formed during the first phases of systems similar to the Solar System, even though probably not in the large quantities observed in carbonaceous meteorites. Whether those molecules survive in comets and asteroids and their harsh environment during and after the planetary system formation and then reach the surfaces of forming or formed telluric planets undestroyed, is still a largely open question.

References

Agúndez, M., Cernicharo, J. and Goicoechea, J. R. (2008). Formation of simple organic molecules in inner T-Tauri disks. *Astronomy and Astrophysics*, **483**, 831–7.

Barzel, B. and Biham, O. (2007). Efficient stochastic simulations of complex reaction networks on surfaces. *Journal of Chemical Physics*, **127**, 4703.

Belloche, A., Menten, K. M., Comito, C., Müller, H. S. P., Schilke, P., Ott, J., Thorwirth, S. and Hieret, C. (2008). Detection of amino acetonitrile in Sgr B2(N). *Astronomy and Astrophysics*, **482**, 179–96.

Bennett, C. J. and Kaiser, R. I. (2007). On the formation of glycolaldehyde and methyl formate in interstellar ice analogs. *The Astrophysical Journal*, **661**, 899–909.

Bergin, E. and Tafalla, M. (2007). Cold dark clouds: the initial conditions for star formation. *Annual Review of Astronomy and Astrophysics*, **45**, 339–96.

Bottinelli, S., Ceccarelli, C., Williams, J. P. and Lefloch, B. (2007). Hot corinos in NGC 1333-IRAS4B and IRAS2A. *Astronomy and Astrophysics*, **463**, 601–10.

Bottinelli, S., Ceccarelli, C., Lefloch, B., Williams, J. P., Castets, A., Caux, E., Cazaux, S., Maret, S., Parise, B. and Tielens, A. G. G. M. (2004). Complex molecules in the hot core of the low-mass protostar NGC 1333 IRAS 4A. *The Astrophysical Journal*, **615**, 354–8.

Carr, J. S. and Najita J. R. (2008). Organic molecules and water in the planet formation region of young circumstellar disks. *Science*, **319**, 1504–6.

Caselli, P., van der Tak, F. F. S., Ceccarelli, C. and Bacmann, A. (2003). Abundant H_2D^+ in the pre-stellar core L1544. *Astronomy and Astrophysics*, **403**, L37–41.

Cazaux, S., Tielens, A. G. G. M., Ceccarelli, C., Castets, A., Wakelam, V., Caux, E., Parise, B. and Teyssier, D. (2003). The hot core around the low-mass protostar IRAS 16293–2422: scoundrels rule! *Astrophysical Journal Letters*, **593**, L51–5.

Ceccarelli, C. and Dominik, C. (2005). Deuterated H_3^+ in protoplanetary disks. *Astronomy and Astrophysics*, **440**, 583–93.

Ceccarelli, C., Caselli, P., Herbst, E., Tielens, A. G. G. M. and Caux, E. (2007). Extreme deuteration and hot corinos: the earliest chemical signatures of low-mass star formation. In *Protostars and Planets*, eds. V. B. Reipurth, D. Jewitt and K. Keil. Tucson, AZ: University of Arizona Press, pp. 47–62.

Ceccarelli, C., Castets, A., Loinard, L., Caux, E. and Tielens, A. G. G. M. (1998). Detection of doubly deuterated formaldehyde towards the low-luminosity protostar IRAS 16293–2422. *Astronomy and Astrophysics*, **338**, L43–6.

Ceccarelli, C., Dominik, C., Caux, E., Lefloch, B. and Caselli, P. (2005). Discovery of deuterated water in a young protoplanetary disk. *The Astrophysical Journal*, **631**, L81–4.

Ceccarelli, C., Dominik, C., Lefloch, B., Caselli, P. and Caux, E. (2004). Detection of H_2D^+: measuring the midplane degree of ionization in the disks of DM Tauri and TW Hydrae. *The Astrophysical Journal*, **607**, L51–4.

Ceccarelli, C., Loinard, L., Castets, A., Faure, A. and Lefloch, B. (2000b). The search for glycine in the solar type protostar IRAS 16293–2422. *Astronomy and Astrophysics*, **362**, 1122–6.

Ceccarelli, C., Loinard, L., Castets, A., Tielens, A. G. G. M. and Caux, E. (2000a). The hot core of the solar-type protostar IRAS 16293–2422: H_2CO emission. *Astronomy and Astrophysics*, **357**, L9–12.

Chastaing, D., Le Picard, S. D., Sims, I. R. and Smith, I. W. M. (2001). Rate coefficients for the reactions of $C(^3P_J)$ atoms with C_2H_2, C_2H_4, $CH_3C{\equiv}CH$ and $H_2C{=}C{=}CH_2$ at temperatures down to 15 K. *Astronomy and Astrophysics*, **365**, 241–7.

Combes, F., Q-Rieu, N. and Wlodarczak, G. (1996). Search for interstellar glycine. *Astronomy and Astrophysics*, **308**, 618–22.

De Duve, C. (2005). *In Singularities: Landmarks on the Pathways of Life*. Cambridge: Cambridge University Press.

De Witt, J., van der Straaten, C. and Mook, W. (1980). Determination of the absolute D/H ratio of VSMOW and SLAP. *Geostandards Newsletter*, **4**, 33–6.

Di Francesco, J., Evans, N. J., 2nd, Caselli, P., Myers, P. C., Shirley, Y., Aikawa, Y. and Tafalla, M. (2007). An observational perspective of low-mass dense cores: internal physical and chemical properties. In *Protostars and Planets*, eds. V. B. Reipurth, D. Jewitt and K. Keil. Tucson, AZ: University of Arizona Press, pp. 17–32.

Dutrey, A., Guilloteau, S. and Ho, P. (2007). Interferometric spectroimaging of molecular gas in protoplanetary disks. In *Protostars and Planets*, eds. V. B. Reipurth, D. Jewitt and K. Keil. Tucson, AZ: University of Arizona Press, pp. 495–506.

Elsila, J. E., Glavin, D. P. and Dworkin, J. P. (2009). Cometary glycine detected in samples returned by stardust. *Meteoritics and Planetary Science*, **44**(9), 1323–30.

Garrod, R. T. and Herbst, E. (2006). Formation of methyl formate and other organic species in the warm-up phase of hot molecular cores. *Astronomy and Astrophysics*, **457**, 927–36.

Garrod, R. T., Weaver, S. L. W. and Herbst, E. (2008). Complex chemistry in star-forming regions: an expanded gas-grain warm-up chemical model. *The Astrophysical Journal*, **682**, 283–302.

Herbst, E. and van Dishoeck, E. F. (2009). Complex organic interstellar molecules. *Annual Review of Astronomy and Astrophysics*, **47**, 427–80.

Hidaka, H., Watanabe, M., Kouchi, A. and Watanabe, N. (2009). Reaction routes in the CO-H_2CO-d_n -CH_3OH-d_m system clarified from H(D) exposure of solid formaldehyde at low temperatures. *The Astrophysical Journal*, **702**, 291–300.

Hollenbach, D. and Salpeter, E. E. (1971). Surface recombination of hydrogen molecules. *The Astrophysical Journal*, **163**, 165–74.

Hollis, J. M., Lova, F. J., Jewell, P. R. and Coudert, L. H. (2002). Interstellar antifreeze: ethylene glycol. *The Astrophysical Journal*, **571**, L59–62.

Horn, A., Møllendal, H., Sekiguchi, O., Uggerud, E., Roberts, H., Herbst, E., Viggiano, A. A. and Fridgen, T. D. (2004). The gas-phase formation of methyl formate in hot molecular cores. *The Astrophysical Journal*, **611**, 605–14.

Jones, P. A., Cunningham, M. R., Godfrey, P. D. and Cragg, D. M. (2007). A search for biomolecules in Sagittarius B2 (LMH) with the Australia telescope compact array. *Monthly Notices of the Royal Astronomical Society*, **374**, 579–89.

Jørgensen, J. K., Bourke, T. L., Myers, P. C., Schöier, F. L., van Dishoeck, E. F. and Wilner, D. J. (2005). Probing the inner 200 AU of low-mass protostars with the submillimeter array: dust and organic molecules in NGC 1333 IRAS 2A. *The Astrophysical Journal*, **632**, 973–81.

Knez, C., Boogert, A. C. A., Pontoppidan, K. M., Kessler-Silacci, J., van Dishoeck, E. F., Evans, N. J., Augereau, J. C., Blake, G. A. and Lahuis, F. (2005). Spitzer mid-infrared spectroscopy of ices toward extincted background stars. *The Astrophysical Journal*, **365**, L145–8.

Lahuis, F., van Dishoeck, E. F., Boogert, A. C. A., Pontoppidan, K. M., Blake, G. A., Dullemond, C. P., Evans, N. J., Hogerheijde, M. R., Jørgensen, J. K., Kessler-Silacci, J. E. and Knez, C. (2006). Hot organic molecules toward a young low-mass star: a look at inner disk chemistry. *The Astrophysical Journal*, **636**, L145–8.

Lee, C., Kim, J., Moon, E., Minh, Y. C. and Kang, H. (2009). Formation of glycine on ultraviolet-irradiated interstellar ice-analog films and implications for interstellar amino acids. *The Astrophysical Journal*, **697**, 428–35.

Marcelino, N., Cernicharo, J., Agundez, M., Roueff, E., Gerin, M., Martin-Pintado, J., Mausberg, R. and Thum, C. (2007). Discovery of interstellar propylene: missing link in interstellar gas-phase chemistry. *The Astrophysical Journal*, **665**, L127–30.

Maret, S., Ceccarelli, C., Tielens, A. G. G. M., Caux, E., Lefloch, B., Faure, A., Castets, A. and Flower, D. R. (2005). CH_3OH abundance in low mass protostars. *Astronomy and Astrophysics*, **442**, 527–38.

Maret, S., Ceccarelli, C., Caux, E., Tielens, A. G. G. M., Jørgensen, J. K., van Dishoeck, E., Bacmann, A., Castets, A., Lefloch, B., Loinard, L., Parise, B. and Schöier, F. L. (2004). The H_2CO abundance in the inner warm regions of low mass protostellar envelopes. *Astronomy and Astrophysics*, **416**, 577–94.

Muñoz Caro, G. M., Meierhenrich, U. J., Schutte, W. A., Barbier, B., Arcones Segovia, A., Rosenbauer, H., Thiemann, W. H. P., Brack, A. and Greenberg, J. M. (2002). Amino acids from ultraviolet irradiation of interstellar ice analogues. *Nature*, **416**, 403–6.

Parise, B., Castets, A., Herbst, E., Caux, E., Ceccarelli, C., Mukhopadhyay, I. and Tielens, A. G. G. M. (2004). First detection of triply-deuterated methanol. *Astronomy and Astrophysics*, **416**, 159–63.

Parise, B., Caux, E., Castets, A., Ceccarelli, C., Loinard, L., Tielens, A. G. G. M., Bacmann, A., Cazaux, S., Comito, C., Helmich, F., *et al.* (2005). HDO abundance in the envelope of the solar-type protostar IRAS 16293–2422. *Astronomy & Astrophysics*, **431**, 547.

Ratajczak, A., Quirico, E., Faure, A., Schmitt, B. and Ceccarelli, C. (2009). Hydrogen/deuterium exchange in interstellar ice analogs. *Astronomy and Astrophysics*, **496**, L21–4.

Remijan, A. J., Hollis, J. M., Lovas, F. J., Stork, W. D., Jewell, P. R. and Meier, D. S. (2008). Detection of interstellar cyanoformaldehyde (CNCHO). *The Astrophysical Journal*, **675**, L85–8.

Rodgers, S. D. and Charnley, S. B. (2003). Chemical evolution in protostellar envelopes: cocoon chemistry. *The Astrophysical Journal*, **585**, 355–71.
Tielens, A. G. G. M. and Hagen, W. (1982). Model calculations of the molecular composition of interstellar grain mantles. *Astronomy and Astrophysics*, **114**, 245–60.
van Dishoeck, E. F., Thi, W. F. and van Zadelhoff, G. J. (2003). Detection of DCO^+ in a protoplanetary disk. *Astronomy and Astrophysics*, **400**, L1–4.
Vastel, C., Phillips, T. G. and Yoshida, H. (2004). Detection of D_2H^+ in the dense interstellar medium. *The Astrophysical Journal*, **606**, L127–30.
Villanueva, G. L., Mumma, M. J., Bonev, B. P., Di Santi, M. A., Gibb, E. L., Böhnhardt, H. and Lippi, M. (2009). A sensitive search for deuterated water in comet 8p/Tuttle. *The Astrophysical Journal*, **690**, L5–9.
Weber, A. S., Hodyss, R., Johnson, P. V., Willacy, K. and Kanik, I. (2009). Hydrogen–deuterium exchange in photolyzed methane–water ices. *The Astrophysical Journal*, **703**, 1030–3.

7

Cosmochemical evolution and the origin of life: insights from meteorites

Sandra Pizzarello

Introduction

Meteorites have so far provided our best analytical window into the cosmochemical evolution of the elements that make up Earth's complex chemical systems. They are for the most part fragments of asteroids, i.e. of the small-sized and odd-shaped planetesimals that orbit the Sun in great numbers between Mars and Jupiter. According to the regular spacing of inner planets from the Sun (the Titius–Bode law), this orbit should be occupied by a planet, but it appears that the smaller lumps of early Solar-System material on their way to form a planet in this region fell under the strong gravity of the gas-giant Jupiter, which slung many away and left the rest unable to fully coalesce. Asteroids, therefore, are the remnants of a planet that never was and, just like comets and other smaller bodies in the Solar System that avoided the geological reprocessing of planet formation, may offer a pristine record of early Solar-System material. Yet, they have the important distinction of being concentrated in a crowded orbit and, subjected still to the gravitational pull of planets nearby and many collisions, regularly send their fragments to the Earth as meteorites, whose direct laboratory analyses secure unequivocal data of their extraterrestrial material.

In the case of carbonaceous meteorites, the delivery has taken on an astrobiological significance because this subgroup of meteorites contains abundant and diverse organic material, including compounds having identical counterparts in the biosphere, such as amino acids (e.g. Pizzarello *et al.*, 2006). Organic compounds have their largest molecular representation in terrestrial living systems, where they attain structures, functions and selection possibilities that allow and sustain life. Not surprisingly, therefore, the discovery and study of these organic compounds in meteorites have become part of the discourse regarding the origin of life. How life began is simply not known, nor can we infer with any certainty the specific compounds, physico–chemical processes or contingencies that might have contributed to these origins. However, it is reasonable to conceive that the long cosmic history of the biogenic elements might have merged with the terrestrial molecular evolution that yielded life and, through yet unknown processes, contributed to it. To this hypothesis, carbonaceous meteorites provide unique analytical insights into abiotic organic materials, as they came to be in a planetary setting of the Solar System and ahead of the onset of life.

Carbonaceous chondrites (CCs) take their name from containing various percentage abundances of total carbon, about 1–4%, in addition to round glassy inclusions, the chondrules. These meteorites have a primitive elemental composition that is similar to that of the Sun and, since the Sun is a rather average star, to that of the cosmos overall. They have the appearance of aggregate rocks, i.e. materials packed together without signs of having been transformed by extreme heat or pressure, and their major constituent is a fine-grained silicate 'matrix', which holds larger mineral inclusions and, interspersed within it, a complex and heterogeneous organic material. Hydrous silicates and other mineral features also present in CC matrices give evidence of an aqueous phase having affected their asteroidal parent bodies (Brearley, 2006).

Carbon-containing meteorites have been known and analysed since the beginning of the nineteenth century but their observed falls, after the one first recorded in 1806 for the Alais meteorite, have been few, and less than fourty-five are known to date. CCs are, in fact, not very dissimilar from some Earth rocks in porosity or clay content and, if not retrieved upon fall, easily conceal themselves and crumble in terrestrial environments where they also incorporate terrestrial contaminants, including bacteria and biomolecules, and lose their indigenous organic content with time. The first unequivocal analyses of extraterrestrial organic material were those of the Murchison meteorite shortly after its fall in 1969. At the time, many laboratories in the USA were readying themselves for the possibility that lunar samples, soon to return, would contain organics and NASA scientists decided to analyse at once a pristine Murchison sample as a probable analogue of Moon rock (Kvenvolden *et al.*, 1970). This meteorite fell weighing 100 kg, has been studied now for forty years and is the source of our understanding of the chemistry as well as formative history of this type of meteorite. Recently, new and surprisingly different data were obtained from stones recovered in Antarctica (referred to as finds, e.g. Pizzarello *et al.*, 2008), a region where in-falling meteorites are quickly covered by snow, buried within the ice and resurface unspoiled with the ice sheets' flow. In the following brief review section, we will describe separately the data obtained from the Murchison meteorite and these Antarctic finds, not only because the latter are as yet partial, but also to convey our current understanding of the abiotic organic chemistry of meteorites, which appears to be wide-ranging in composition, distribution, and possibly, synthetic history.

The Murchison meteorite

Organic composition

Murchison organic material represents about 90% of the meteorite's total carbon; its composition is complex as well as diverse and can be found as soluble compounds and insoluble macromolecules (30% and 70% of total, respectively). Both appear important as markers of the extent and distribution of chemical evolution: the insoluble macromolecular material gives a record of the range and complexity of the organic molecular structures attained through abiotic chemistry, while the soluble compounds allow us to compare this chemistry

Figure 7.1. Transmission electron microscope image of Murchison macromolecular organic material. 1: Unstructured form; 2: Nanoglobules; 3: Microscope supporting grid. Courtesy of Laurence Garvie, Arizona State University.

with the production of compounds we know well from extant biochemistry. Individual soluble compounds are readily obtained by extraction of whole meteorite powders with water or solvents while the insoluble organic matter (IOM) can only be analysed after the dissolution of the inorganic components of the meteorite and, being insoluble, has to be studied *in toto* by various spectroscopic methods or by analysis of its molecular fragments after pyrolysis. By its bulk, the macromolecular material can also be seen by microscopy (Figure 7.1) and has given some indirect hints as to its location in the stone, which appears to be intermixed with the hydrous silicates of the matrix (e.g. Pearson *et al.*, 2007). However, details of the relationship between organic and inorganic components of CC matrices have been challenging to obtain and are largely unknown, notwithstanding their possible significance.

Murchison insoluble organic material

Meteoritic IOM has been referred to as kerogen-like because, similar to terrestrial kerogens, it is composed of other biogenic elements in addition to carbon and hydrogen, with the elemental formula: $C_{100}H_{70}N_3O_{12}S_2$. Microscopically (Figure 7.1), it appears to be made up of both an unstructured and heterogeneous 'fluffy' material without any clear resolution in its continuity and distinct nanometre-sized entities such as flakes, solid or hollow spheres and tubes. The main unstructured portion is itself heterogeneous and, although not well defined chemically, consists mainly of condensed aromatic, hydroaromatic and heteroaromatic macromolecules that contain alkyl branching, functional groups, such as OH and COOH, and are bridged by alkyl chains, with an overall aromatic to aliphatic ratio of approximately two. The more structured portions of IOM nanospheres and tubes also have a large aromatic component as well as varying composition from close to pure graphitic carbon (C) (> 99%) to containing several per cent of oxygen, nitrogen and sulphur. The relationship between the many heterogeneous components of the IOM has been difficult to assess and, in general terms, has led to the conclusion that this material is the complex end-product of formative conditions, chemical regimes and cosmic environments that varied extensively.

Despite the IOM insolubility and overall analytical intractability, some findings have pointed to its possible astrobiological value. For example, the IOM releases soluble compounds in hydrothermal conditions not too dissimilar from those of terrestrial volcanic vents (300°C, 100 MPa). These compounds are mostly simple alkyl and aromatic molecular species such as branched phenols and benzothiophenes or biphenyls and terphenyls, however, and so far inexplicably, this hydrothermolytic treatment also leads to the release of a series of dicarboxylic acids of up to C_{17} chain length (Yabuta *et al.*, 2007). These acids are amphiphilic at such chain lengths, i.e. carry both a hydrophobic hydrocarbon chain and a hydrophilic group at its opposite ends, and could be useful in the formation of membrane-like bilayer enclosures in water. Interestingly, it was also found that the IOM triggers an asymmetric response of the Soai autocatalytic reaction (Soai *et al.*, 1995) but fails to do so after hydrothermal treatment (Kawasaki *et al.*, 2006), suggesting the presence in the IOM of yet undefined labile chiral entities, whose asymmetry could be molecular-based as well as structural.

It is probable that our current analytical methodologies have missed large and complex molecules of unexpected composition present in the meteorite. There is an example in the complex and not fully identified amphiphilic material that can be extracted from whole Murchison powders by chloroform (Deamer, 1985). It forms hydrated gels that assemble into membranous vesicles in alkaline solutions and appears to contain aromatic hydrocarbon polymers, phenolic and carboxyl groups as well as a tell-tale fluorescent yellow pigment, also of unknown composition, that can be observed directly in the matrix of the cut meteorite.

Murchison soluble organic compounds

Murchison indigenous compounds make up an abundant suite of various molecular species that, although smaller and simpler than IOM macromolecules, are not less complex or varied. As Table 7.1 shows, they range from water-soluble amino acids or polyols to non-polar poly-aromatic and long-chain hydrocarbons. For all classes of compounds, except for purines and pyrimidines, their abundance and numbers are sometimes significantly larger than the table shows, because many molecular species can often be identified analytically (e.g. by mass spectrometry) but reference standards are not available. Overall, their molecules show large structural heterogeneity, preference for branched alkyl chains and a broad isomeric range, with all isomers of a given class often found present up to the limit of their solubility; concentrations also decline in homologous series with increasing carbon number.

Table 7.1. Soluble organic compounds in the Murchison meteorite.

Compound class	n[1]	ppm[2]	Example molecule and structure	
Carboxylic acids	31	300	acetic acid	$H_3C{-}COOH$
Amino acids	74	60	alanine	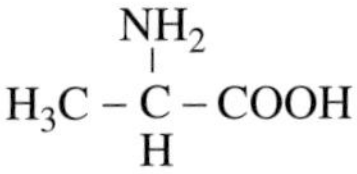 $H_3C-CH(NH_2)-COOH$
Hydroxy acids	15	7	lactic acid	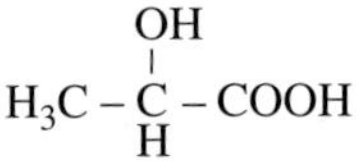 $H_3C-CH(OH)-COOH$
Ketoacids	5	nd[3]	pyruvic acid	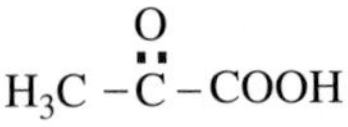 $H_3C-C(=O)-COOH$
Dicarboxylic acids	17	30	succinic acid	$HOOC-(CH_2)_2-COOH$
Sugar alcohols and acids	19	30	glyceric acid	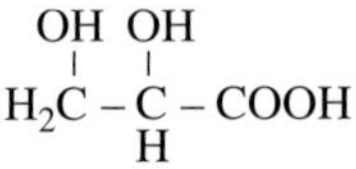 $H_2C(OH)-CH(OH)-COOH$
Aldehydes and ketones	18	20	acetaldehye	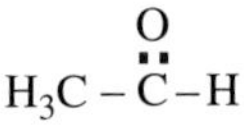 $H_3C-C(=O)-H$
Amines and amides	20	13	ethyl amine	$H_3C-CH_2{-}NH_2$
Pyridine carboxylic acids	7	7	nicotinic acid	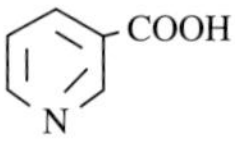 COOH, N
Purines and pyrimidines	5	1.5	adenine	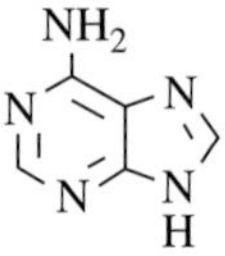 NH_2, N, N, N, N, H
Hydrocarbons:				
Aliphatic	140	35	propane	$H_3C-CH_2-CH_3$
Aromatic	25	30	naphthalene	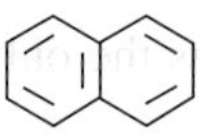
Polar	10	120	isoquinoline	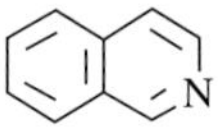 N

[1] Number of compounds identified (represents minimum numbers, see text).

[2] Parts per million. [3] Not determined.

Murchison amino acids offer a good example of these general molecular characteristics, as they number over one hundred and have C_2 through C_8 alkyl chain lengths that may be linear, branched or cyclic and comprise all the possible relative positions of the amino and carboxyl functions along those chains. These distributions contrast starkly with extant biochemistry; taking again amino acids as an example, it should be remembered that all terrestrial proteins, numbering in the tens of thousands, are made up of just twenty amino acids of the same general structure (with the amino group at the α-C).

The complex and diverse composition of Murchison organics points to abiotic synthetic processes that were randomly governed, lacked selective influences such as catalysis and may have involved chain growth by addition of single C units (Yuen *et al.*, 1984); these compositional traits and the contrast with the selectivity and functional specificity displayed by biomolecules also validate the indigeneity of Murchison compounds. At the same time, as Table 7.1 also shows, several classes of Murchison organics can be found in terrestrial biochemistry and some of their individual compounds are actually identical to biomolecules; e.g. eight Murchison amino acids are components of terrestrial proteins (glycine, alanine, valine, leucine, isoleucine, proline, aspartic acid, glutamic acid). Furthermore, a subgroup of Murchison chiral amino acids with limited or no known terrestrial distribution (the 2-methyl-2-amino acids) display small but significant L-enantiomeric excesses, i.e. carry the essential biochemical trait of chiral asymmetry (Cronin and Pizzarello, 1997). A great number of biomolecules contain one or more asymmetric carbons and are chiral[1], for example, only the smallest of the twenty protein amino acids, glycine ($CH_2(NH_2)COOH$), is not chiral. We know also that life builds its polymers with monomers of just one configuration, L-amino acids for proteins and D-sugars for RNA, DNA and polysaccharides. This chiral homogeneity (homochirality) governs syntheses, interactions and functions of biopolymers as well as the many stereochemical reactions of extant metabolism and appears indispensable to life as we know it (e.g. Cronin and Reisse, 2005). However, it is unknown outside the biosphere because, although the natural world offers numerous examples of chirality and separation of chiral forms (e.g. left- and right-handed circularly polarized light or D- and L-crystals such as quartz), as of to date, there are only two known cases of quantitative differences between one chiral form over the other: the subatomic energy difference between enantiomers, which is due to parity violation in weak-force interactions (PVED) and the enantiomeric excesses (*ee*) we find in the amino acids of meteorites.

The *ee* of meteoritic amino acids, therefore, take on a unique and profound significance as they stand as the only natural example of molecular asymmetry measured outside the biosphere. So far, *ee* appear to be confined to a subgroup of six C_5–C_8 α-amino acids having a methyl branching at the α-C and the dicarboxylic α-methylglutaric acid; they are also

[1] Chirality (from the Greek for hand) is a general property of objects that, like the left and right hands, come in two forms made up of equal components and are mirror images of each other but do not superimpose. Chirality is an attribute of organic molecules because the carbon atom has tetrahedral configuration and, when bonded to four different substituents, becomes asymmetric and imparts chirality to molecules containing it. The two configurations of molecules having one asymmetric carbon are called enantiomers.

broad ranged, 1–18%, with variability even for the same molecule within the meteorite. How and when these amino acids acquired their *ee* has been debated but is not known. Their indigeneity has been proven by carbon and hydrogen isotopic analyses, by which we also learned that asymmetry-carrying amino acids display, as a group, the largest δD values of any Murchison compound, i.e. that their formation was probably related to cold, deuterium-enriching cosmic environments (*vide infra*). This suggestion of a pre-planetary locale for *ee* formation would appear to agree with the hypothesis of their production by asymmetric photolysis of meteorite organics by UV circularly polarized-light (CPL) irradiation during their synthesis (Rubenstein *et al.*, 1983); asymmetry-causing UV radiation, in fact, would have probably been shielded in the meteorite parent bodies. Analyses, however, have revealed that Murchison amino acids' *ee* can be larger than would be theoretically allowed if UV CPL photolysis had been the direct source of their asymmetry (Pizzarello *et al.*, 2003). Another often-debated possibility is that of the effect of PVED, however, how the basic asymmetry of matter at the subatomic level could relate to the chemistry of chiral molecules is poorly understood. PVED has not yet been measured and is estimated to be exceedingly small, 10^{-11} J mole^{-1}, i.e. less than necessary to be detected analytically over statistical fluctuations during reactions (Soai *et al.*, 2003). *Ab initio* calculations have also led to contradicting results, being hindered by questions on the theoretical approach to be used as well as the fact that organic molecules such as amino acids and sugars can take numerous conformations (e.g. Quack, 2002). How PVED might have influenced the formation of molecules in cosmic environments is even less known and remains a question of vast importance. As of today, therefore, the physico–chemical processes and/or cosmic locales that may have produced *ee* in extraterrestrial molecules such as meteoritic amino acids and their possible distribution beyond the Earth remain unknown.

Syntheses and locales involved in the formation of Murchison compounds

The question of how Murchison organic compounds were formed was raised shortly after the meteorite's first analyses revealed the presence of biomolecule-like compounds. Clearly, this knowledge would allow insights into the syntheses and locales where biomolecules, or biomolecule precursors, can be synthesized abiotically as well as on how common or rare their cosmic distribution may be. The understanding came, at least in general terms, with the learning of Murchison's isotopic composition, where several soluble compounds as well as fractions of the IOM show enrichment in the heavy isotopes of carbon, hydrogen (Figure 7.2) and nitrogen compared to terrestrial standards. It should be noted why these findings have been important in the context of meteorite organics' formation. The stable isotopes of the biogenic elements may, like other isotopes, display mass-dependent isotopic fractionation during reactions that derives from differences in energy of bond formation between isotopes. Because such fractionation is not only dependent on isotopomers' mass difference but is also inversely proportional to the reaction temperature, i.e. is also sensitive to environmental conditions, it may become a good diagnostic tracer of a molecule's formative history and allow us to infer physico–chemical conditions during syntheses.

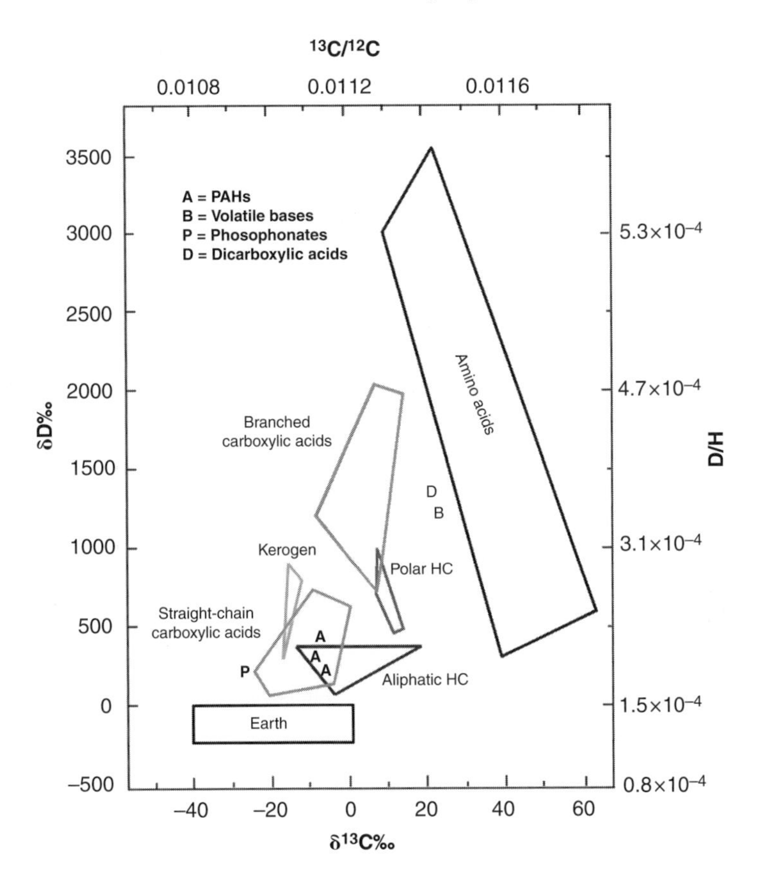

Figure 7.2. Isotopic distributions of Murchison insoluble organic material and soluble compounds. The δ values used in geochemical and biochemical studies give the difference between the heavy/light isotope ratio of a sample and that of a chosen standard as ‰ of the standard ratio. For carbon: $\delta^{13}C$ ‰ $= (^{13}C/^{12}C_{sample} - ^{13}C/^{12}C_{standard} / ^{13}C/^{12}C_{standard}) \times 10^3$. Astrochemical data are usually reported as the direct isotopomers' ratio instead.

This has been dramatically evident in the study of interstellar molecules and the understanding of their probable formation. The dense clouds of the interstellar medium (ISM) contain many gas-phase organic molecules that are characterized by large deuterium enrichment, with measured D/H ratios as high as 0.06 (for CDH_2OH/CH_3OH) compared to the average terrestrial value of 1.5×10^{-4}. This fractionation is made possible by the extremely low ISM temperatures (5–10 K), where reactions are confined to those that do not require activation energy and are exothermic by the zero-point energy difference between the D and H isotope bonds (D/H ratio is 2, the largest of the stable isotopes, compared to ≈ 1.08 for C and ≈ 1.07 for N). Carbon fractionation is also observed and measured in ISM molecules but is less pronounced due to the smaller differential between ^{13}C and ^{12}C atomic masses. An example is the $^{13}CO/^{12}CO$ reported value of 3.1×10^{-2} compared to the terrestrial standard of 1.12×10^{-2} (e.g. Langer *et al.*, 1984).

Murchison compounds' heavy-isotope enrichments vary significantly between classes and probably represent the variety of cosmic environments involved in their formation; for amino acids, the group of compounds studied in most detail, this heterogeneity is observed between and within their various molecular subgroups. For example, the 2-methylamino acids display δD values of up to + 3600‰, near the range of those determined for ISM molecules, suggesting that these amino acids, or their direct precursors, probably formed in cold cosmic environments (Pizzarello and Huang, 2005). The 2-H-amino acids, less enriched in deuterium (δD values 400–2500‰), may have been synthesized in asteroidal parent bodies during a transient aqueous phase, e.g. by the reaction with other condensed ISM volatiles such as HCN, ammonia, aldehydes and ketones. The carbon isotopic composition of Murchison compounds varies as well: only 2-amino acids (of both 2-H and 2-methyl series) show declining ^{13}C values with increasing alkyl chain length while dicarboxylic amino acids and 3-, 4- and 5-amino acids do not. The first trend indicates a possible growth of the compounds' homologous series by single-moiety addition while its absence for the latter groups suggests that C-additions were probably not their synthetic route. Since all these molecules also display deuterium enrichment, albeit of different levels, the isotopic diversity of Murchison compounds clearly indicates that cold chemical regimes must be able to efficiently perform syntheses by diverse pathways.

The deuterium and ^{13}C enrichment determined for many Murchison soluble organic compounds, when interpreted on the basis of astrochemical findings, has led to the hypothesis that their formation may have been multistaged and spanned several cosmic environments, where precursor molecules were synthesized by ion–molecule reactions in the cold molecular clouds of the ISM, carried through the subsequent stages of stellar as well as nebular processes and reacted further in the asteroidal parent body during an aqueous phase. This 'interstellar-parent body hypothesis' is broadly outlined, however, and does not account for specific synthetic pathways and locales, e.g. between dense clouds and subsequent prestellar cores, or explain the large isotopic diversity shown between the various compound groups within meteoritic organic classes.

Antarctica finds

The organic composition of two pristine CR2 meteorites

As mentioned in the Introduction, the Antarctica ice fields have been good curators for CCs and several of the meteorite finds from the region have shown little trace of terrestrial weathering or contamination. The large number of meteorites collected has also allowed for the classification of new families of CCs, as in the case of chondrites of the Renazzo family (CR, meteorites that have a petrology closely similar to that of the Renazzo meteorite, which fell in 1864 and had long remained unclassified). Two such meteorites, GRA95229 and LAP02342 (collected in Antarctica Graves Nunataks and LaPaz ice fields, respectively) were analysed recently and found to be exceptionally pristine (Pizzarello *et al.*, 2008; Pizzarello and Holmes, 2009). They have offered a yet unknown view of the

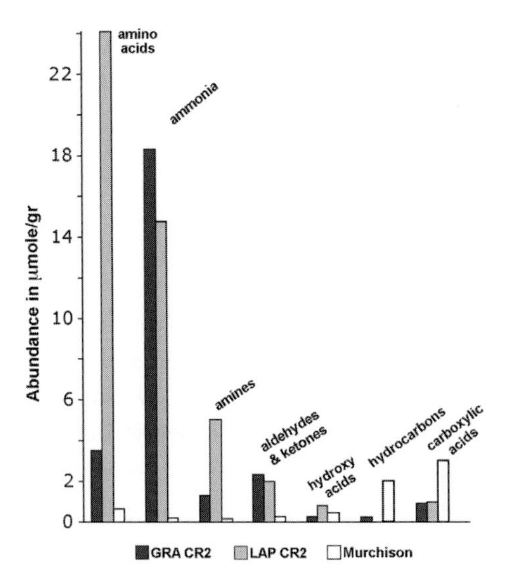

Figure 7.3. Plot of selected soluble-compound abundances in CR2 and the Murchison meteorites, in parts per million.

synthetic capabilities of extraterrestrial environments by revealing soluble organic suites that, unlike that of Murchison, are made up of mainly water-soluble compounds, with predominantly N-containing species. Figure 7.3 gives an overview of the novelty of the two CR2[2] organic compositions, where the amounts of amino acids, ammonia, amines and carbonyl-containing compounds are considerably larger than Murchison's while those of carboxylic acids and hydrocarbons are lower. Other differences were observed within the groups of these compounds, for example, compound abundances within groups do not decrease linearly with increasing carbon chain length, as seen in Murchison, but lower molecular-weight compounds are dominant over higher homologues, resulting in less variety but a far larger quantity of individual molecular species. In the case of amino acids, these are the terrestrial biomolecules glycine and alanine. Intriguing also was the finding in both meteorites of the hydroxy amino acids serine, threonine, *allo*threonine and tyrosine as well as of phenylalanine, which brought to twelve the total of protein amino acids detected in extraterrestrial samples. Is it possible that these hydroxy amino acids, which are reactive molecular species, were not observed in Murchison because of their further aqueous-phase reactions in that meteorite? This would also explain the large abundances of unreacted CR2 aldehydes and ketones (see plot, Figure 7.3) and suggests a short or colder aqueous phase for these meteorites.

The *ee* for 2-methyl amino acids were found to be much lower (3%, GRA95229) or absent (LAP02342) in the two CR2 meteorites than in the Murchison meteorite. Given the rapid abundance decline with increasing molecular size for CR2 compounds, only the

[2] The number represents a classification of petrographic type and estimates asteroidal secondary processes.

first isomer of the CR2 2-methyl-2-amino acid series, isovaline ($CH_3CH_2CH(CH_3)(NH_2)COOH$), was abundant enough to be measured. The divergence in molecular abundances between Murchison and CR2 2-amino acid subsets could possibly relate to the results. The 2-methyl-2-amino acids often represent the most abundant Murchison amino acid subgroup and their displaying of significantly higher δD values than 2-H-2-amino compounds has led to the hypothesis that their formation was distinct and must have involved cold cosmic regimes. That CR2 isovaline (and its lower C_4 homolog 2-aminoisobutyric acid) also have high δD values (*vide infra*) would point to pathways of formation similar to those that produced Murchison amino acids of the same type. If so, their differences in abundance with Murchison would seem unlikely to have affected their enantiomeric ratios. On the other hand, the fact that δD values for these two amino acids are so much higher in GRA95229 than in Murchison could indicate diverse ISM synthetic regimes for their formation, where those that lead to the formation of CR2 2-amino-2-methyl amino acids offered lesser exposure to symmetry-breaking effects.

An unusual and interesting finding came from the study of CR2 diastereomer amino acids, which suggested that their precursor aldehydes might have carried *ee* during the aqueous-phase reactions that took place in the meteorites' asteroidal parent bodies. The understanding of this assumption requires a few reminders. Several water-soluble chiral compounds having *ee* may, in this medium, 'flip' from one configuration to the other with time, leading to a racemic mixture (i.e. having equal amounts of D- and L-enantiomers). Protein amino acids all have a hydrogen attached to the alkyl chain second carbon that, being slightly acidic due to its proximity to the electron-withdrawing carboxyl, readily loses and reacquires this H in water, becoming racemic with time (if H is replaced by an alkyl group, as is the case for the α-methylamino acids, racemization would not be possible). These properties would lead to the conclusion that, if the conditions of the parent body were such that they allowed racemization, no amino acids with an α-H would show *ee* in meteorites after a water phase.

On the other hand, when a molecule has two chiral carbons instead of one, such as some meteoritic and protein amino acids do, they are present in two sets of enantiomers called diastereomers. In water, these do not racemize to their enantiomers but 'epimerize' to the diastereomers of opposite configuration (only C_2 readily racemizes). The best-known example is the conversion with time of L-isoleucine to D-*allo*isoleucine (Figure 7.4), which has been used to determine the age of fossil protein samples because isoleucine is a protein amino acid and *allo*isoleucine is formed in biological materials only by epimerization of isoleucine. In nature, *ee* between diastereomers could also result from asymmetric syntheses or effects, for example, during the formation of amino acids from chiral aldehydes, HCN and ammonia in water (cyanohydrin synthesis). As shown in Figure 7.5, for the formation of the isoleucine diasteromers, an *ee* of the aldehyde precursor, e.g. of (S) configuration, would appear in the isoleucine and *allo*isoleucine enantiomer products that carried the S-part of the molecule through the synthesis. These two amino acids had been found to be non-racemic in Murchison but their *ee* could not be recognized as indigenous, in spite of isotopic data suggesting so, because of the possibility of contamination by terrestrial isoleucine and its subsequent epimerization.

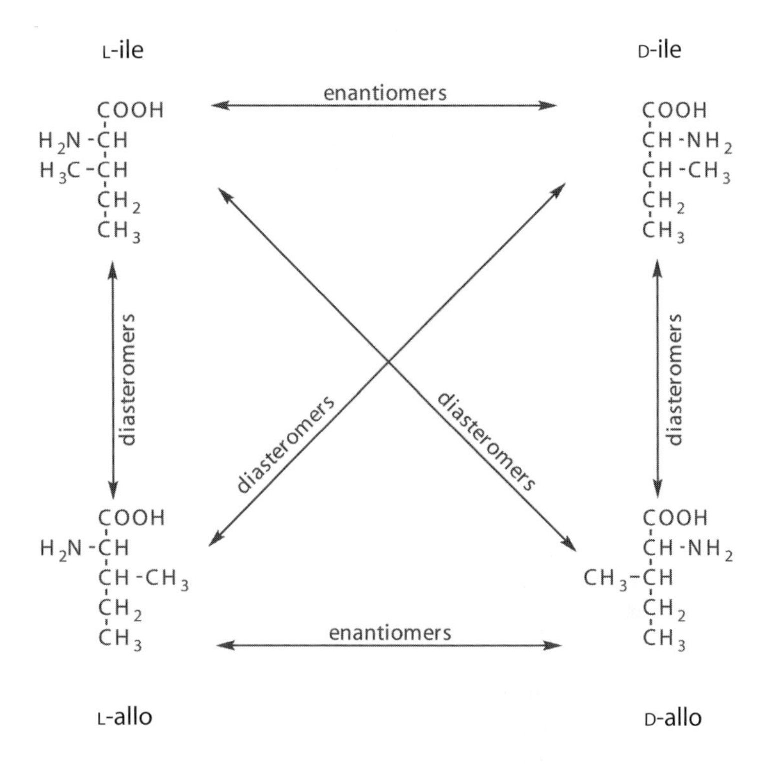

Figure 7.4. Example of racemization of *allo*isoleucine; see explanation in the text.

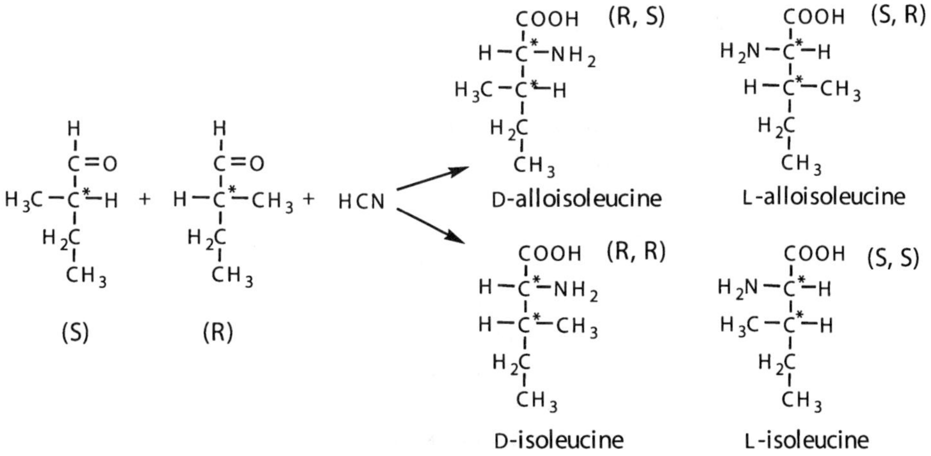

COOH (R, S)
H−C*−NH2
H3C−C*−H
H2C
CH3
D-alloisoleucine

COOH (S, R)
H2N−C*−H
H−C*−CH3
H2C
CH3
L-alloisoleucine

H
C=O
H3C−C*−H + H−C*−CH3 + HCN
H2C H2C
CH3 CH3
(S) (R)

COOH (R, R)
H−C*−NH2
H−C*−CH3
H2C
CH3
D-isoleucine

COOH (S, S)
H2N−C*−H
H3C−C*−H
H2C
CH3
L-isoleucine

Figure 7.5. Scheme illustrating the formation of the isoleucine diasteromers by addition of HCN to chiral 2-butyraldehyde (in the presence of NH_3 and H_2O, not shown. *C denotes chiral carbon).

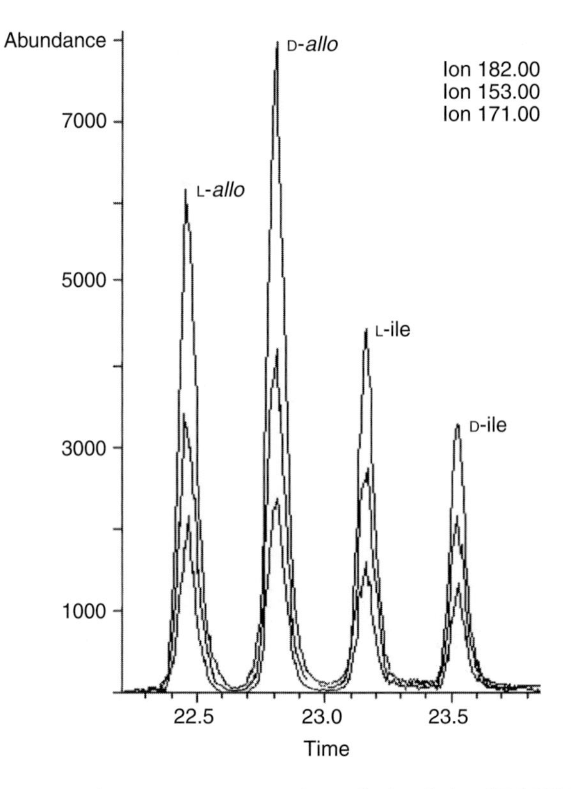

Figure 7.6. Gas chromatography–mass spectrometric analysis of the GRA95229 isoleucine and *allo*isoleucine diastereomers; single ion traces on a β-cyclodextrin stationary phase.

However, the same diastereomer distributions seen in Murchison were also measured in GRA95229 (Figure 7.6) and, in this case, the pristine nature of the meteorite vouched for their indigeneity. The finding was further validated by ^{13}C isotopic analyses and interpreted to signify an L-*ee* of the precursor aldehyde (Pizzarello *et al.*, 2008). These data would add, albeit indirectly, a new class of known interstellar molecules such as the aldehydes to the list of compounds carrying abiotic molecular asymmetry.

CR2 isotopic composition and the large variety of possible cosmochemical precursor environments

The isotopic composition of CR2 amino acids has revealed both similarities and differences with that of Murchison in that the isotopic distinction between subgroups of amino acids was confirmed but also magnified in the GRA95229 meteorite, with δD values for 2-H amino acids ranging from 890 to 2890‰ and those for the branched isomers reaching ≥ 7200‰ (Pizzarello *et al.*, 2008), which are the highest ever recorded for an extraterrestrial molecule by direct analysis, and fall well within the range of D/H values established spectroscopically for ISM molecules (5800–45 000‰ in the δD notation). Remarkable also were the data obtained from $\delta^{15}N$ measurements of several amino acids and ammonia, which were prompted by the large abundance of N-containing molecular species in CR2 meteorites. The $\delta^{15}N$ values determined for each amino acid set of the two meteorites showed a distribution between molecular subgroups that is opposite to the one of their δD values, with 2-H amino acids having higher $\delta^{15}N$ values, 122–130‰ for four amino acids measured, than for 2-methyl amino acids, 75–90‰ (Pizzarello and Holmes, 2009).

These results had to be interpreted against a near absence of molecular $\delta^{15}N$ values available for cosmic environments and far less detailed knowledge of nitrogen isotope cosmic distributions than for the D/H or $^{13}C/^{12}C$ isotopes. In fact, it was the finding in meteorites and interplanetary dust particles (IDPs) of elevated $\delta^{15}N$ values (dramatically so in the case of IDPs, e.g. Messenger, 2000), that spurred theoretical studies of cosmic nitrogen ratios (Terzieva and Herbst, 2000; Charnley and Rodgers, 2002; Rodgers and Charnley, 2004; Rodgers and Charnley, 2008). According to the first study, which predicts low nitrogen fractionation by ion–molecule reactions in the ISM, we would conclude that the values determined for the two CR2 meteorites are all higher than those that could form in hydrogen-fractionating molecular clouds. The studies by Charnley and Rodgers offer further insights; they describe a mechanism for higher nitrogen fractionations in regions of the ISM that preceded star formation. Here, the enhanced density and pressure of the core would cause the freeze out of most carbon- and oxygen-containing molecules and, with their disappearance, a disruption of N_2-formation pathways employed in less dense clouds and the result of gas-phase atomic nitrogen prevalence. In turn, this would lead to the efficient production of ammonia and $^{15}NH_3/^{14}NH_3$ ratios higher than the cosmic $^{15}N/^{14}N$ ratios (by up to as much as 80%).

These predictions are interesting in that they appear to match, if in broad terms, the findings in meteorites and the current interpretation of meteoritic amino-acid formation. As the distinctly higher δD values of 2-methyl amino acids and lower values of 2-H amino acids had suggested that their syntheses, or that of their precursors, would have taken place in the ISM and parent body, respectively, so the respective $\delta^{15}N$ distributions would fit, according to Charnley and Rodgers, with earlier (molecular clouds) and later (pre-stellar cores) ISM cosmochemical processes, removing a parent-body environment from the overall scheme.

Very little is known of the molecular sequence of events that would have taken place in a pre-stellar core, however, we can expect that several stages of temperature, pressure and ensuing chemical regimes followed the initial collapse phase described by these authors and others (e.g. Ceccarelli *et al.*, 2007). We could hypothesize (Pizzarello and Holmes, 2009), therefore, that some of the warmer stages of star formation might have allowed selected environments, where the desorption, mixing and reactions of radical precursor molecules, water and ammonia led to the synthesis of higher ^{15}N amino acids and favoured shorter molecular-species formation. It also appears that such locales and the kinetic processes they allow us to envision could, rather than parent-body reactions, explain some of the molecular distributions seen in these two CR2 meteorites such as: the far from unity diastereomer ratios seen for the thermodynamically similar amino acids *allo*isoleucine and isoleucine, their erratic levels of enantiomeric excesses as well as the abundance of unreacted carbonyl-containing molecules.

Cosmochemical evolution and the origin of life

When seeking insights from meteorites about the chemistry that preceded life's origin, we are looking for support of a chembiogenesis hypothesis (von Kiedrowsky, 2005), i.e. that this origin resulted from a process with evolutionary roots in abiotic chemistry. The hypothesis is not new: when in 1871 Darwin mused with a friend about the first production of a living organism in the now-famous writing about a warm little pond '...But if (and oh! What a big if!) we could conceive in some warm little pond, with all sort of ammonia and phosphoric salts, light, heat, electricity, etc ..., present', he foretold of both the idea of extending evolution to prebiotic chemistry as well as our struggle in proving it. The proposal appears valid because, inasmuch as there is exceedingly little knowledge about the actual origin(s) of life, it seems reasonable to extend to life's beginnings the same fundamental evolutionary nature that has been recorded throughout its cellular history, and the concept that this evolution could have reached back into prebiotic chemistry is worth exploring. As we conceive it, there would be three progressive steps involved in this prebiotic evolutionary process: the gathering of necessary chemical building blocks from an abiotic pool; the building of a molecular scaffolding with structural and functional properties; and the emergence of a self-sustaining and reproductive system. Unfortunately, to provide analytical evidence of prebiotic relevance, or progression, for any experiment that would mimic these hypothetical evolutionary stages is not easy or direct. There are inherent difficulties in the fact that emergent-life processes may have differed in their core molecular organization and function from even the simplest of extant life forms, from which they are so far removed. Furthermore, while reasonably acknowledging a prebiotic dominance to the chemistry of carbon in the appearance of life, it is not known to what extent early Earth's physical and chemical environments affected this chemistry with influential contingencies.

We encounter the same predicaments when trying to interpret the relevance of the extraterrestrial organic chemistry of meteorites. The unique advantage in this case is that CC represents the reality of a natural sample and provides an unambiguous, even if partial, analytical database of attainable abiotic organic compounds. From an overall astrobiological perspective, meteorites tell us that abiotic syntheses are capable of producing a large variety of organic species of remarkable complexity, including compounds having identical counterparts in terrestrial biochemistry. Regardless of the actual involvement of these abiotic molecules in the origin of terrestrial life, therefore, the formation of several molecules important to life in cosmic environments clearly preceded the chemistry of life on the Earth. The significant compositional differences observed within and between several types of carbonaceous meteorites also allow us to conclude that such environments were diverse and differed in elemental and isotopic composition as well as in their effective selection of molecular species and/or their traits, such as molecular asymmetry. The understanding through the study of meteorites of how this organic material came to be present, survived and evolved in the Solar System may help constrain the conditions that would allow for planetary life on Earth and elsewhere.

Compounds suggestive of prebiotic potential

As to the origin of terrestrial life, the details of meteoritic compound distributions should allow us to explore systematically whether chemical evolution displayed compositional traits that could have aided in subsequent prebiotic evolutionary stages. Of the several organic constituents of carbonaceous meteorites with compositional correspondence to extant-life components, two types appear to have the most opportunities for further evolution: the vesicle-forming amphiphilic compounds and macromolecules described in the Murchison meteorite and the amino acids found in several families of CC.

Amphiphilic compounds

Murchison amphiphilic compounds may have had a prebiotic importance that is readily explained by their ability to form vesicular enclosures. On the assumption that life evolved in a water-based environment, cell-like compartments to separate and protect useful evolutionary chemistry from the bulk of the environment appear to be indispensable intermediates for the onset of life (e.g. Luisi *et al.*, 1999). Membrane-like vesicles have been grown experimentally from lipids as well as from meteoritic materials and, in lipid bilayers, have been shown to replicate under opportune physico–chemical conditions. This is a cell-like ability that would carry the evolutionary advantage of assuring the survival of certain sets of molecules having better reproducing enclosures (e.g. Szostak *et al.*, 2001). Meteoritic amphiphiles are also released under hydrothermal conditions and give water solutions a sustained increase in surface tension (Mautner *et al.*, 1995); this property might have worked well in combination with early Earth endogenous processes towards enclosure formation and encapsulation of relevant organics.

Amino acids

Meteoritic amino acids might have carried an advantage as biomolecule precursors over other single molecules provided to the early Earth by meteorites and, considering their demonstrated properties to form proteins and act as individual catalysts, we can imagine simple analogous pathways by which these compounds might have aided terrestrial molecular evolution. As of now, it seems that CR2 meteorites could have provided their bulk delivery. Compounds of Murchison-type meteorites have been, in fact, often regarded as too complex in isomeric composition for any particular molecular species to gain significance in prebiotic chemistry (e.g. Cronin and Chang, 2003), while the new Antarctica CR2 finds point to the possibility of abundant and selective deliveries of low molecular-weight compounds, amino acids in particular. For example, the abundance of C_2 glycine relative to the estimated total of organic compounds released in water from Murchison is 1/46 while for the LAP02342 meteorite it is about 1/3; the value of these ratios for C_3 alanine would be even lower in Murchison and about the same as for glycine in the CR2 meteorite. Assuming that early evolutionary steps of accreted amino acids on the Earth may have included their condensation, possible under certain scenarios (Leman *et al.*, 2004), these abundances may have led to less heterogeneous peptide forms; α-aminobutyric acid (aiba), the third largest CR2 amino acid and known helix stabilizer (e.g. Toniolo and Benedetti, 1991), could have provided early helical structures that, in turn, may have benefited from the *ee* of other meteoritic deliveries.

As mentioned previously, one important prebiotic feature of several amino acids from Murchison-type meteorites is that they are chiral and carry *ee* that, if not as sizable, have the same configuration (L-) as all those comprising terrestrial proteins. Although we do not know how biological homochirality originated, and many theories have been postulated on whether it was the product of prebiotic processes or the result of selection brought about by life itself, we do know that it is a pervasive functional requirement of extant biomolecules and we may postulate that a chiral homogeneity of early molecules or polymers was an attribute necessary for the origin and/or development of life. The finding of *ee* in the amino acids of meteorites, therefore, has raised the reasonable suggestion that extraterrestrial abiotic processes might have provided the early Earth with a 'primed' inventory of exogenous organic molecules that held an advantage in prebiotic molecular evolution.

One possible way this could have occurred is during the condensation of *ee*-carrying amino acids into peptides. Just as the interactions between enantiomers (or hands) of the same kind differ from those between opposite enantiomers (or hands), so if the coupling of two chiral amino acids favoured a homochiral pair, this would influence their successive additions during condensation. If secondary structures such as helices are achieved in the process, then these structures are also chiral and would aid in the further development of chirality. The same as aiba, all amino acids found with *ee* in meteorites are helix formers and their helices are stable in water.

As individual molecules or small peptides, the non-racemic amino acids from meteorites could have had an inductive effect on the development of homochirality by the catalytic transfer of their asymmetry to other biomolecules. Amino acids are known catalysts (e.g. Weber, 2000) and the hypothesis that they could also act as asymmetric catalysts in water has been tested on a reaction that involves the condensation of aldehydes to give sugars. This formose reaction, which in its simplest form uses formaldehyde, is considered a model of possible early Earth chemistry, where smaller aldehydes could have been formed by various conditions (e.g. Miller and Schlesinger, 1984). When the self-condensation reaction of C_2 glycolaldehyde was carried out in the presence of either non-racemic amino acids (Pizzarello and Weber, 2004) or enantiomerically pure dipeptides (Weber and Pizzarello, 2006), it was found that the resulting four-carbon sugars carried considerable *ee*, up to 12% and 85%, respectively; with an L-valine dipeptide catalyst, the highest *ee* belonged to the D-erythrose product, i.e. a sugar having the same OH– configuration along the alkyl chain as D-ribose.

Conclusions

To conclude, the analyses of carbon-containing meteorites have provided a wealth of analytical detail about the abiotic organic chemistry that preceded life's origin in the Solar

System. These studies are still ongoing and the recent analyses of pristine CR2 meteorites have shown how incomplete our overall outlook of this chemistry is. However, they do allow several inferences on the general properties of this chemistry and whether it could possibly evolve into biochemistry. The first, and somewhat pessimistic conclusion is that abiotic chemistry has a path to complexity that is far different from that of life's molecular assemblages and appears to follow random aggregation of heterogeneous components, possibly accumulated through diverse environments and revealing no sign of selective influences. Yet, even if the insoluble material of meteorites does not seem amenable to evolutionary processes, this amorphous material may still have provided useful hydrophobic associations with water-soluble aggregates upon interaction with early Earth hydrothermal environments. We have to think, however, that soluble compounds such as those seen in meteorites represented the bulk of material having prebiotic possibilities that chemical evolution was capable of producing in the Solar System. Some of these molecules and their traits, such as amino acids and molecular asymmetry, are currently fundamental attributes of extant biomolecules. In spite of the absolute lack of certainty about the actual processes that allowed the emergence of life, therefore, we must consider plausible the idea that, once delivered to the early Earth and concomitant with Earth physico–chemical features, these exogenous molecules intervened, facilitated and eventually carried out the first biochemical processes.

References

Brearley, A. J. (2006). The action of water. In *Meteorites and the Early Solar System* II, eds. D. Lauretta and H. Y. McSween. Tucson, AZ: University of Arizona Press, pp. 587–624.

Ceccarelli, C., Caselli, P., Herbst, E., Tielens, A. G. G. M. and Caux, E. (2007). Extreme deuteration and hot corinos: the earliest chemical signatures of low-mass star formation. In *Protostars and Planets*, eds. V. B. Reipurth, D. Jewitt and K. Keil. Tucson, AZ: University of Arizona Press, pp. 47–62.

Charnley, S. D. and Rodgers, S. B. (2002). The end of interstellar chemistry and the origin of nitrogen in comets and meteorites. *The Astrophysical Journal*, **569**, L133–7.

Cronin, J. R. and Chang, S. (2003). Organic matter in meteorites: molecular and isotopic analyses of the Murchison meteorite. In *The Chemistry of Life's Origins*, eds. J. M. Greenberg, C. X. Mendoza-Gómez and V. Pirronello. The Netherlands: Kluwer Academic Publishers, pp. 209–58.

Cronin, J. R. and Pizzarello, S. (1997). Enantiomeric excesses in meteoritic amino acids. *Science*, **275**, 951–5.

Cronin, J. R. and Reisse, J. (2005). Chirality and the origin of homochirality. In *Lectures in Astrobiology Vol. 1*, eds. M. Gargaud, B. Barbier, H. Martin and J. Reisse. Berlin: Springer-Verlag, pp. 473–514.

Deamer, D. (1985). Boundary structures are formed by organic components of the Murchison carbonaceous chondrite. *Nature*, **317**, 792–4.

Kawasaki, T., Hatase, K., Fujii, Y., Jo, K., Soai, K. and Pizzarello, S. (2006). The distribution of chiral asymmetry in meteorites: an investigation using asymmetric autocatalytic chiral sensors. *Geochimica et Cosmochimica Acta*, **70**, 5395–402.

Kvenvolden, K., Lawless, J., Pering, K., Peterson, E., Flores, J., Ponnamperuma, C., Kaplan, J. R. and Moore, C. (1970). Evidence of extraterrestrial amino acids and hydrocarbons in the Murchison meteorite. *Nature*, **228**, 923–6.

Langer, W. D., Graedel, T. E., Frerking, M. and Armentrout, P. B. (1984). Carbon and oxygen isotope fractionation in dense interstellar clouds. *The Astrophysical Journal*, **277**, 581–604.

Leman, L., Orgel, L. and Gadhiri, M. R. (2004). Carbonyl sulfide-mediated prebiotic formation of peptides. *Science*, **306**, 283–6.

Luisi, P. L., Walde, P. and Oberholtzer, T. (1999). Lipid vesicles as possible intermediates in the origin of life. *Current Opinion in Colloid and Interface Science*, **4**, 33–9.

Mautner, M. N., Leonard, R. L. and Deamer, D. W. (1995). Meteorite organics in planetary environments: hydrothermal release, surface activity, and microbial utilization. *Planetary and Space Science*, **43**, 139–47.

Messenger, S. (2000). Identification of molecular-cloud material in interplanetary dust particles. *Nature*, **404**, 968–71.

Miller, S.L. and Schlesinger, G. (1984). Carbon and energy yields in prebiotic syntheses using atmospheres containing CH_4, CO and CO_2. *Origins of Life*, **43**, 83–90.

Pearson, V. K., Kearsley, A. T., Sephton, M. A. and Gilmour, I. (2007). The labeling of meteoritic organic material using osmium tetroxide vapour impregnation. *Planetary and Space Science*, **55**, 1310–18.

Pizzarello S. and Holmes W. (2009). Nitrogen-containing compounds in two CR2 meteorites: ^{15}N composition, molecular distribution and precursor molecules. *Geochimica et Cosmochimica Acta*, **73**, 2150–62.

Pizzarello, S. and Huang, Y. (2005). The deuterium enrichment of individual amino acids in carbonaceous meteorites: a case for the presolar distribution of biomolecule precursors. *Geochimica et Cosmochimica Acta*, **69**, 599–605.

Pizzarello, S. and Weber, A. L. (2004). Meteoritic amino acids as asymmetric catalysts. *Science*, **303**, 1151.

Pizzarello, S., Cooper, G. W. and Flynn, G. J. (2006). The nature and distribution of the organic material in carbonaceous chondrites and interplanetary dust particles. In *Meteorites and the Early Solar System* II, eds. D. Lauretta and H. Y. McSween. Tucson, AZ: University of Arizona Press, pp. 625–51.

Pizzarello, S., Huang, Y. and Alexandre, M. D. R. (2008). Molecular asymmetry in extraterrestrial chemistry: insights from a pristine meteorite. *Proceedings of the National Academy of Sciences of the USA*, **105**, 7300–4.

Pizzarello, S., Zolensky, M. and Turk, K. A. (2003). Non racemic isovaline in the Murchison meteorite: chiral distribution and mineral association. *Geochimica et Cosmochimica Acta*, **67**, 1589–95.

Quack, M. (2002). How important is parity violation for molecular and biomolecular chirality? *Angewandte Chemie International Edition*, **41**, 4618–30.

Rodgers, S. B. and Charnley, S. B. (2004). Interstellar diazenylium recombination and nitrogen isotopic fractionation. *Monthly Notices of the Royal Astronomical Society*, **352**, 600–4.

Rodgers, S. B. and Charnley, S. B. (2008). Nitrogen superfractionation in dense cloud cores. *Monthly Notices of the Royal Astronomical Society*, **569**, L48–52.

Rubenstein, E., Bonner, W. A., Noyes, H. P. and Brown, G. S. (1983). Supernovae and life. *Nature*, **306**, 118.

Soai, K., Shibata, T., Morioka, H. and Choji, K. (1995). Asymmetric autocatalysis and amplification of enantiomeric excess of a chiral molecule. *Nature*, **378**, 767–8.

Soai, K., Sato, I., Shibata, T., Komiya, S., Hayashi, M., Matsueda, Y., Imamura, H., Hayase, T., Morioka, H., Tabira, H., Yamamoto, J. and Kowata, Y. (2003). Asymmetric synthesis of pyrimidyl alkanol without adding chiral substances by the addition of diisopropylzinc to pyrimidine-5-carbaldehyde in conjunction with asymmetric autocatalysis. *Tetrahedron Asymmetry*, **14**, 185–8.

Szostak, J. W., Bartel, D. P. and Luisi, P. L. (2001). Synthesizing life. *Nature*, **409**, 387–90.

Terzieva, R. and Herbst, E. (2000). The possibility of nitrogen fractionation in interstellar clouds. *Monthly Notices of the Royal Astronomical Society*, **317**, 563–8.

Toniolo, C. and Benedetti, E. (1991). The polypeptide 3_{10}-helix. *Trends in Biochemical Science*, **16**, 350–3.

von Kiedrowsky, G. (2005). Coined this term for *Chembiogenesis 2005: a conference on prebiotic chemistry and early evolution*, Venice, Italy, 2005.

Weber, A. L. (2000). The sugar model: catalysis by amines and amino acid products. *Origins of Life and Evolution of the Biosphere*, **31**, 71–86.

Weber, A. L. and Pizzarello, S. (2006). The peptide-catalyzed stereospecific synthesis of tetroses: a possible model for prebiotic molecular evolution. *Proceedings of the National Academy of Sciences of the USA*, **103**, 12713–7.

Yabuta, H., Williams, L. B., Cody, G. D., Alexander, C. M. O. D. and Pizzarello, S. (2007). The insoluble carbonaceous material of CM chondrites: a possible source of discrete organic compounds under hydrothermal conditions. *Meteoritic and Planetary Science*, **42**, 37–48.

Yuen, G., Blair, N., Des Marais, D. J. and Chang, S. (1984). Carbon isotopic composition of individual, low molecular weight hydrocarbons and monocarboxylic acids from the Murchison meteorite. *Nature*, **307**, 252–4.

8

Astronomical constraints on the emergence of life

Matthieu Gounelle and Thierry Montmerle

Before the Solar System

The formation of the Sun and stars

The Sun is somewhat a late-comer in our Galaxy, the Milky Way. It was born 4.6 Gyr ago (4.5685 Gyr ± 0.5 Myr, to be precise, from the decay of specific radioactive heavy elements in the most primitive meteorites – see below). This is to be compared with the age of the Universe, constrained by the best theoretical fits to the observed spatial fluctuations of the 'cosmic background radiation' to be 13.7 Gyr after the Big Bang, within 2%. When galaxies form is less certain, but current estimates give a time lapse of less than 1 Gyr after the Big Bang – implying that our own Galaxy has an age of over 12.7 Gyr and that the Sun was born over 8.1 Gyr later. So at the time the Sun formed, our Galaxy was already sufficiently evolved by successive generations of stars that it presented no major differences with the one we observe today. Therefore, we can safely derive conclusions about the distant birth of the Sun from observations of contemporary young stars.

In a nutshell, from various observations we know that bright nebulae, including some famous ones like Orion, the Eagle or Carina nebulae, are 'stellar nurseries' (Figure 8.1), where stars like the Sun form in clusters of thousands of low- to intermediate-mass stars. A few massive stars, like the Orion Trapezium, for which the highest mass is of order 45 $M_\odot$ also form in these stellar nurseries. In addition, except for the brightest (and most massive) ones, all stars appear surrounded by dense dark circumstellar disks (Figure 8.1). There are strong observational and theoretical arguments to claim that these disks are 'protoplanetary', i.e. future sites where planetary systems will form. In 'OB' associations, which are clusters of hundreds or thousands of stars, named after their most massive brightest members of spectral types O and B, such disks exist in a quite hostile environment. Indeed, during their lifetime, these stars, which are very hot (having surface temperatures of order several 10 000 K), emit very strong UV radiation, and generate very intense 'winds' (velocities of several 1000 km s^{-1}, mass loss rate on the order of several $M_\odot$ per Myr). Stars more massive than 8 $M_\odot$ end their evolution in 'supernova explosions'; due to their short lifetimes, the most massive ones will explode within the association in which they were born.

But we also see stars forming in relative isolation, i.e. sufficiently far apart from one another that, contrary to the case of OB associations, they do not mutually influence each

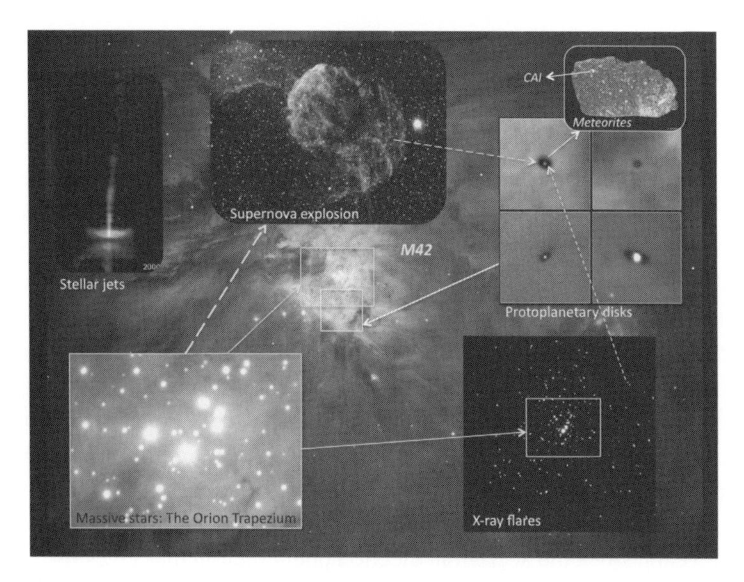

Figure 8.1. The 'pre-planetary world' in a star-forming region. We have chosen the Orion Nebula (M42), the closest massive star-forming region, as an example of the proposed 'cradle of the Sun'. Background: the nebula, seen with the Hubble Space Telescope (HST). It is illuminated and heated by a small cluster of very massive stars, the Orion 'Trapezium' (near-infrared VLT image). Around them, and almost everywhere in the nebula, protoplanetary disks, having typical radii of up to 500 AU (or 10 times the radius of the present-day Solar System), can be seen in silhouette against the bright background of the nebula (HST images). All the ~ 2000 stars present in the nebula, the majority of them having masses comparable with that of the Sun, emit strong X-ray flares (Chandra satellite image), which irradiate the circumstellar disks. Young low-mass stars are also associated with luminous jets (upper left corner: the HH30 star in Taurus; HST image). After a few million years only, the most massive stars explode as supernovae (upper middle panel: optical image of the IC443 supernova remnant, associated with the IC444 star-forming region), which back-react on the lower-mass stars and their disks (of the same generation or of later generations). During that time, dust grains in circumstellar disks aggregate in less than a million years, forming km-sized planetesimals. Meteorites which now fall on Earth are fragments of these primitive planetesimals. These meteorites contain small 'calcium–aluminium-rich inclusions' (CAIs) (white dots on the upper-right photograph of the Allende meteorite; Muséum National d'Histoire Naturelle). Their isotopic composition reveals a unique record of the formation of the Solar System, in particular traces of both early irradiation by flares from young stars and of contamination by supernovae.

other: their formation and early evolution can be considered as those of a single star. Most notably, there are essentially no stars more massive than a few solar masses in their vicinity. In this case, observations of young stars in nearby (less than 500 light years) dark clouds reveal very conspicuous 'jets' of bright (hence hot) gas, visible in opposite directions against a dark background. This is especially the case in the near-infrared band, where we can observe resolved circumstellar disks perpendicular to the jets (Figure 8.1).

As a result, theoretical developments have been ongoing in two main directions. The formation of single stars, with their protostellar envelopes, disks and jets has been intensively and successfully modelled (e.g. Shu *et al.*, 1987). More recently, and in parallel with refinements in single-star formation and early evolution theory, a more global approach to the formation of multiple stars of various masses has been developed using the fastest computers. Even in dark clouds, the stars, taken globally, are not really isolated, but seem to form in loose clusters. Therefore, it is thought that a large-scale mechanism must be at work to trigger star formation in a broadly generic way, sometimes in a dispersed fashion (as in dark clouds), sometimes in a more collective fashion (as in OB associations). The key word for this mechanism is 'turbulence'. Observations show that the gas in molecular clouds has a velocity distribution (on the order of a few km s^{-1}, typically) consistent with eddies transferring energy from the large scales to the small scales, where it is eventually dissipated and allows collapse and star formation. The details are, however, far from being clear. For instance, the source of the large-scale turbulence in molecular clouds is still controversial, but it is thought that jets and outflows from young stars play an important feedback role to stir the gas out of which they were born (see Elmegreen and Palous, 2007).

However, in low-mass star-forming regions this feedback mechanism slows down after a few million years, since the accretion–ejection phenomenon that gives rise to jets eventually stops because circumstellar disks disappear (see next section). On the other hand, in massive star-forming regions, the feedback accelerates, since, over the same timescale, the most massive members generate intense stellar winds, creating large pc-sized hot bubbles, and then explode as supernovae, leading to the disruption of the parent molecular cloud.

From young suns to the young Sun: the preplanetary phase

Observations and theory tell us that newly born stars are not hot enough in their centres to burn hydrogen: they are called 'pre-main sequence stars' and derive their luminous energy from gravitational contraction only, on timescales of order of millions of years. 'Pre-main sequence' suns, or more simply 'young suns', are known under the name of 'T-Tauri stars'. In dark clouds, for typical low-mass clusters, their mass range runs from a few solar masses down to 0.1 $M_{\odot}$, and their ages from 1 Myr or less, to several 10 Myr. In OB associations, the stellar masses can be as high as several 10 $M_{\odot}$, up to nearly 100 $M_{\odot}$.

To understand the early stages of the evolution of our Sun (thus during its T-Tauri phase) and the first steps towards planet formation, two key facts derived from observations of young low-mass stars must be taken into account: their enhanced 'magnetic activity' and the short lifetime of their circumstellar disks.

Their enhanced magnetic activity is generated by convection. Because they are convective, wholly or within their outer layers in the same way that the present-day Sun is, young stars generate magnetic fields (via the so-called dynamo mechanism i.e. by the creation of magnetic fields by moving electric charges). On the Sun, these magnetic fields are seen directly in the form of large loops of various shapes and sizes (arcades, protuberances etc.) in the vicinity of dark sunspots. The ubiquitous presence of magnetic fields is inferred from the X-ray variability of T-Tauri stars, in the form of hour-long powerful flares, where the stellar atmosphere is locally heated to million-degree temperatures (Figure 8.1) by the reconnection of magnetic-field lines, and temporarily confined in magnetic loops, as we see on the Sun (e.g. Güdel, 2004).

Basically, because their outer layers undergo the same phenomenon (convection), T-Tauri stars and the present-day Sun share qualitative similarities: their magnetic fields are on the same order (a few kGauss in the co-called active regions), they have spots and they show flares, especially in X-rays. Quantitatively, however, the 'magnetic activity' of T-Tauri stars is enormously enhanced with respect to the Sun: their starspots cover at least 50% of their surface (as opposed to 5% on average for the Sun); the X-ray luminosity (scaled to the total luminosity) is elevated by 3 to 4 orders of magnitude with respect to the Sun, and attributed to a hot plasma trapped in magnetic loops of sizes at least as large as the stellar radius, sometimes several times larger than that. T-Tauri X-ray flares are much more frequent: typically, they generate one strong flare per day equivalent to the same flares emitted by the Sun once a year. For the young Solar System, as discussed below, this has a very important consequence: since solar flares also emit energetic particles (mainly protons and helium nuclei – in two varieties, the usual ^{4}He, or α-particles, and ^{3}He, that we see in the interplanetary medium), we infer that T-Tauri stars must inject in their environment per unit time something like 100 000 times more energetic particles than the present-day Sun.

The lifetime of their circumstellar disks is comparatively short. Infrared studies, both imaging and spectroscopic of clusters of young stars, have shown that fewer and fewer stars possess detectable circumstellar disks as they age along the T-Tauri phase (Figure 8.1). Most stars of a very young cluster (less than 1-Myr old, say), provided they are not too massive, are surrounded by disks, whereas in a 10-Myr cluster most stars are found to be diskless. In between, one finds, on average, that the fraction of stars surrounded by disks declines rapidly with age with a characteristic 'lifetime' of about 3 to 5 Myr, which is found to differ appreciably from cluster to cluster, but is significantly less than the duration of the T-Tauri phase (Cieza *et al.*, 2007).

Why do disks 'disappear' so rapidly? At this early stage, circumstellar disks are made of gas and dust grains of size up to a millimetre or so. The first idea is disk evaporation by the combined effect of external UV radiation and winds in regions of massive star formation (see above). However, this will affect only the T-Tauri stars in close proximity to OB stars, not the more distant ones; and it cannot operate in low-mass star-forming regions where the stars are cooler and more 'isolated' from each other. Another possibility is evaporation caused by the UV radiation from the central star itself. Indeed, even though T-Tauri stars are 'cool', a consequence of their magnetic activity is that, like the Sun, they possess a 'chromosphere', i.e. a hot (10^5 K) upper-atmosphere region which is a strong UV emitter. Quantitatively, this mechanism is efficient if the UV flux is strong enough: in this case, the disk material may be dissipated in a few Myr only. Another competing explanation for the apparent 'disappearance' of disks is not that they physically disappear, but that they undergo a radical internal evolution: their dust grains somehow aggregate and form large bodies which will eventually become 'planets'.

This is, of course, the mechanism that led to the formation of the Solar System. According to recent theoretical work, the initial phases of planet formation are characterized by a high frequency of collisions between small bodies (km-sized planetesimals, planetary embryos; see Chapter 9). Amid the debris of these collisions, a large number of boulder-sized or smaller bodies may orbit the Sun for billions of years, and eventually produce fragments which fall on the Earth: the meteorites. As the following sections will show, some meteorites contain a unique tell-tale record of the early stages of the formation of the Sun and the Solar System.

Chondrites and the origin of the Solar System

Chondrites

There exist three main groups of meteorites: stones (94% of the observed falls), stony-irons (1%) and irons (5%). Though some of them might have an impact origin, iron and stony-iron meteorites are believed to represent the core and the core-mantle boundary, respectively, of asteroids large enough to have undergone silicate–metal differentiation as the Earth did. We will not discuss these further.

Stony meteorites can be divided into chondrites and achondrites. The former are characterized by a large abundance of chondrules (millimetre-sized beads of silicates and metal, see below) and have a chemical composition similar to that of the Sun for rock-forming elements. They are primitive rocks. Achondrites do not contain chondrules and they have a strongly fractionated chemical composition relative to that of the Sun. They are evolved rocks which cannot tell us much about early Solar-System processes.

Chondrites are made of calcium–aluminium-rich inclusions (CAIs), chondrules and matrix. These three components were made in the solar protoplanetary disk. The relative abundance of these three components varies between chondrite groups. In most chondrites, chondrules represent the dominant component. CAIs are very abundant in carbon-rich carbonaceous chondrites while they are virtually absent from other chondrites. Studies of CAIs during the last 30 years are heavily biased towards CAIs found in the CV3 chondrites epitomized by the Allende meteorite which fell in Mexico in 1969. The reason for that bias is that CAIs in CV3 chondrites are large (up to a cm) and abundant (~ 10% volume). It is, however, important to note that CAIs in CV3 chondrites are quite peculiar when compared to CAIs in other chondrite groups and may record specific events in the solar protoplanetary disk.

CAIs are an assemblage of calcium and aluminium oxides and silicates. Some of them show evidence of melting. CAIs are the oldest rock pieces we have at hand with high-precision absolute Pb–Pb ages ranging from 4567.2 to 4568.7 Myr (Amelin *et al.*, 2006; Bouvier and Wadhwa, 2009). Note that only CAIs from CV3 chondrites have been dated with that method so far.

Chondrules are an assemblage of iron–magnesium silicates plus metal and sulphides. Chondrules usually show an igneous texture indicating they were, at some point, fully melted in the solar protoplanetary disk. Chondrules from the CR2 and CV3 chondrites have respective ages of 4564.7 ± 0.6 Myr (Amelin *et al.*, 2002) and 4565.45 ± 0.45 Myr (Connelly *et al.*, 2007). It seems, therefore, that there is an age difference of ~ 2 Myr between CV3 CAIs and chondrules. It should, however, be kept in mind that experimental difficulties might lead to some biases in the Pb–Pb absolute ages. In addition, different oxidation states of CAIs and chondrules might lead to different initial $^{238}U/^{235}U$ ratios in these objects. As Pb–Pb ages assume an identical $^{238}U/^{235}U$ ratio of 137.88 for the 2 kinds of objects, a revision of that assumption might lead to a revision of the ages.

The matrix is made of fine-grained (< 1 μm) iron–magnesium silicates, metal and sulphides. The matrix has not been dated yet due to analytical challenges. Indeed, for most chondrite groups, the matrix has been modified on the parent asteroid due to secondary geological processes such as thermal metamorphism or hydrothermalism. Though processed, the matrix contains a significant amount of presolar grains, i.e. submicron-sized oxides and silicates which were made in stars anterior to our Solar System and survived high-temperature events in the early Solar System.

An important property of CAIs and chondrules is that they contained, when they formed, short-lived radionuclides (SRs). SRs are radioactive elements whose half-life is low compared to that of the Solar System (4.568 Gyr assuming CAIs define a time zero). SRs have now entirely decayed; they are said to be extinct. Their past presence in the Solar System is inferred from the excess of their daughter isotopes relative to the terrestrial isotopic composition. SRs are usually found in a larger abundance in CAIs than in chondrules. Because CAIs are supposed to have formed before chondrules the initial abundance of SRs in CAIs is taken to be that of the Solar System (Table 8.1).

It is important to note that there is a key experimental bias in determining the initial abundance of SRs in both CAIs and chondrules. To positively demonstrate that an SR was 'alive' in an object (CAI or chondrule), one needs to characterize the isotopic composition of phases having a high father/daughter ratio. Such phases do not always exist, or can be too small to be the target of a high-precision isotopic measurement. For example, no high Be–B phase has been identified so far in chondrules, preventing the discovery of the past presence of ^{10}Be (which decays to ^{10}B with a half-life of 1.5 Myr) in chondrules (Table 8.1).

When the abundance of an SR is unknown in CAIs, its initial abundance is measured in other phases/meteorites and the abundance in CAIs is calculated using another isotopic system (another SR or the Pb–Pb age) for which data are available both in CAIs and in the other phases/meteorites under investigation. This method assumes that the two isotopic systems under consideration have similar closure temperatures – this assumption being usually unproven. In some other cases, the initial Solar-System ratio is identified with the initial value measured for bulk carbonaceous chondrites (see Table 8.1 and text below), assuming that their precursors separated from the accretion disk at time zero (i.e. the CAI-formation time).

Secondary processes on parent asteroids can blur the determination of the initial abundance of SRs. Thermal metamorphism or hydrothermalism can mobilize elements and compromise the interpretation of isotopic data.

Table 8.1. Initial abundances of short-lived radionuclides. Most numbers come from Wadhwa *et al*. (2007) unless discussed in the text.

Radionuclide (R)	Half-life (Myr)	Daughter isotope	Reference stable isotope (S)	R/S
^{7}Be	52 days	^{7}Li	^{9}Be	6.1×10^{-3}
^{41}Ca	0.1	^{41}K	^{40}Ca	1.5×10^{-8}
^{36}Cl	0.3	^{36}S	^{35}Cl	$> 1.6 \times 10^{-4}$
^{26}Al	0.74	^{26}Mg	^{27}Al	4.5×10^{-5}
^{10}Be	1.5	^{10}B	^{9}Be	$5–10 \times 10^{-4}$
^{60}Fe	2.6	^{60}Ni	^{56}Fe	$< 4 \times 10^{-7}$
^{53}Mn	3.7	^{53}Cr	^{55}Mn	$3–10 \times 10^{-5}$
^{107}Pd	6.5	^{107}Ag	^{108}Pd	4.5×10^{-5}
^{182}Hf	9	^{182}W	^{180}Hf	1.1×10^{-4}
^{129}I	16	^{129}Xe	^{127}I	1×10^{-4}
^{92}Nb	36	^{92}Zr	^{93}Nb	1.1×10^{-3}
^{244}Pu	81	Fission products	^{238}U	0.007
^{146}Sm	103	^{142}Nd	^{144}Sm	5.0×10^{-3}

Since the discovery of ^{129}I, many SRs have been identified in CAIs and chondrules (Table 8.1). Given the difficulties due to analytical measurements and to secondary events, the numbers in the table above (Table 8.1) should be viewed with caution. Detailed analysis of initial abundance is discussed in Wadhwa *et al*. (2007) and Gounelle and Meibom (2008). On that note, it is also important to realize that it might be incorrect to assume there was one initial value in the early Solar System. This postulates implicitly that SRs were homogeneously distributed in the Solar-System protoplanetary disk, which is unproven so far (Gounelle and Russell, 2005).

The origin of short-lived radionuclides

SRs in the early Solar System are generally believed to originate either from the galactic background, from protoplanetary disk-dust irradiation by protosolar cosmic rays or by the last minute injection of a supernova (Wasserburg *et al*., 2006).

On-going nucleosynthesis by a diversity of stars (supernovae, novae, AGB stars etc.) in the Galaxy continuously replenishes the interstellar medium with freshly made SRs. At a given time in the history of the Galaxy, the background abundance of a given short-lived radionuclide R relative to its stable reference isotope S will depend on a diversity of parameters such as the number and nature of nucleosynthetic events responsible for the production of R over the last few half-lives of R, the number and nature of nucleosynthetic events responsible for the production of S during the history of the Galaxy, the respective yields of R and S in the nucleosynthetic events aforementioned, astration, the mixing timescales and processes of the different phases of the interstellar medium, the rate of decay of R etc. Final isolation of the average interstellar medium from nucleosynthetic events introduces an extra parameter, the isolation time Δ, during which R decays without further addition of freshly made matter. Galactic background can account for the SRs with the longer half-lives – from ^{53}Mn to ^{244}Pu (Meyer and Clayton, 2000; Gounelle, 2006).

Irradiation of nebular dust by flare-accelerated hydrogen or helium nuclei can result in the production of SRs. Beryllium-10 has an irradiation origin since it cannot be made in stars (McKeegan *et al.*, 2000). Beryllium-7, which has been identified in 'one' Allende CAI (Chaussidon *et al.*, 2006), is an unambiguous tracer that some irradiation took place in the early Solar System. As discussed in the first section above, ubiquitous, thousand-fold enhanced and flare-like X-ray activity of T-Tauri stars provides firm evidence for the existence of accelerated particles in the vicinity of the early Sun (e.g. Wolk *et al.*, 2005). In addition to ^{7}Be and ^{10}Be, some models have established that it is possible to produce ^{26}Al, ^{36}Cl, ^{41}Ca and ^{53}Mn at abundances in line with those of the early Solar System (Leya *et al.*, 2003; Gounelle *et al.*, 2006) provided that proton and helium nuclei are accelerated by impulsive events. Assuming that ^{26}Al was ubiquitous and homogeneously distributed in the entire protoplanetary disk, Duprat and Tatischeff (2007) proposed that irradiation could not account for the initial ^{26}Al/^{27}Al ratio. There is, however, no positive evidence for ^{26}Al ubiquity and homogeneous distribution in the protoplanetary disk. The claim for homogeneous distribution of ^{26}Al (Villeneuve *et al.*, 2009) relies on the observation that CAIs formed with a deficit of ^{26}Mg relative to chondrites and the Earth (Jacobsen *et al.*, 2008), indicating that ^{26}Al decay happened in the chondrites and the Earth's forming region. This reasoning, however, assumes that the small ^{26}Mg deficit (~ 0.003%) observed in CAIs is not the result of nucleosynthetic anomalies. Given that nucleosynthetic anomalies are widespread and of the order of the observed deficit, it is difficult to prove ^{26}Al was homogeneously distributed in the Solar System. In addition, Duprat and Tatischeff's (2007) calculations rely on the use of the minimum-mass solar nebula (MMSN), a concept which modern astrophysics has shown to be irrelevant due to planetary migration (Crida, 2009). It remains true, however, that it is difficult for irradiation models to account for the high initial abundance of ^{26}Al in the solar protoplanetary disk (Fitoussi *et al.*, 2008).

Injection of SRs by a nearby AGB star or supernova in the nascent Solar System has long been considered an attractive possibility (Cameron and Truran, 1977; Wasserburg *et al.*, 2006). It has, however, been shown that the probability of encounter between an AGB star and a star-forming region is of the order of 10^{-6} (Kastner and Myers, 1994), ruling out these stars as a reasonable source for SRs. The case for a nearby supernova (SN) will be discussed below.

The cradle of the Sun

Iron-60 cannot be produced by energetic particle irradiation (Lee *et al.*, 1998). It necessarily has a stellar origin. Its origin can shed light on the astrophysical context of our Sun's birth.

The initial abundance of ^{60}Fe in the Solar System

The initial abundance of ^{60}Fe in the early Solar System is poorly known. Measurements of Ni isotopes in CAIs are challenging because CAIs are rich in Ni nucleosynthetic anomalies, which can blur isotopic effects due to ^{60}Fe decay (Birck, 2004). An upper limit on the $^{60}Fe/^{56}Fe$ ratio of 1.6×10^{-6} was given by Birck and Lugmair (1988) for an Allende CAI. No isochron was reported, and the authors expressed caution about the ^{60}Fe decay origin of the small (~ 0.01%) Ni excess detected. A two-point internal isochron for an Allende CAI measured by Quitté *et al.* (2007) gave a lower limit for the initial $^{60}Fe/^{56}Fe$ ratio of 3×10^{-7}.

Measurements on the silicate portion of chondrules from primitive meteorites, including Semarkona, yielded $^{60}Fe/^{56}Fe$ ratios of between 1.7 and 3.2×10^{-7} (Tachibana *et al.*, 2006; Tachibana *et al.*, 2007). If a hypothetical time delay of 1.6 Myr is assumed between the formation of CAIs and the closure of the Fe–Ni system in ordinary chondrites (Connelly *et al.*, 2007), initial $^{60}Fe/^{56}Fe$ ratios of between 2.6 and 4.0×10^{-7} can be calculated for CAIs, using a half-life of 2.6 Myr for ^{60}Fe (Rugel *et al.*, 2009). A measurement from a Semarkona troilite (FeS) gives $^{60}Fe/^{56}Fe \sim 9 \times 10^{-7}$ (Mostefaoui *et al.*, 2005). These high $^{60}Fe/^{56}Fe$ initial ratios in troilite are probably due to Fe–Ni redistribution in the sulphides during later alteration processes (Chaussidon and Barrat, 2009).

Recently, several studies performed with multi-collector–inductively coupled mass spectrometers (MC–ICPMS) failed to detect any ^{60}Ni excess due to the decay of ^{60}Fe in a large suite of samples (Dauphas *et al.*, 2008; Regelous *et al.*, 2008; Chen *et al.*, 2009). These authors put respective upper limits on the $^{60}Fe/^{56}Fe$ ratio of 6×10^{-7} (Dauphas *et al.*, 2008) and 3×10^{-7} (Chen *et al.*, 2009; Regelous *et al.*, 2008). It is important to note that using carbonaceous chondrites, Regelous *et al.* (2008) found an upper limit of 1×10^{-7} for the $^{60}Fe/^{56}Fe$ ratio. Despite a few early reports claiming $^{60}Fe/^{56}Fe$ ratios as high as 1×10^{-6} (Cook *et al.*, 2006; Mostefaoui *et al.*, 2005; Tachibana *et al.*, 2006), it is doubtful that the initial Solar-System ratio of $^{60}Fe/^{56}Fe$ was higher than a few 10^{-7}.

A single nearby supernova origin for ^{60}Fe

Elaborating on the pioneering work of Cameron and Truran (1977), two different quantitative scenarios with a 'nearby single' SN have been proposed whereby ^{60}Fe is injected either into the solar protoplanetary 'disk' (e.g. Ouellette *et al.*, 2005) or into the molecular cloud (MC) 'core' progenitor of our Solar System (e.g. Cameron *et al.*, 1995). In the case of the core model it is assumed that the supernova shockwave triggers the gravitational collapse of the core.

In the case of a single SN the distance r at which it has to lie to inject enough SRs is dictated by solid-angle considerations and is given by the expression (e.g. Cameron *et al.*, 1995):

$$r = \frac{r_0}{2}\sqrt{\eta \frac{Y_{SN}}{M_{SS}} e^{-\frac{\Delta}{\tau}}} \tag{8.1}$$

where r_0 is the size of the receiving phase (disk or core), η the injection efficiency, Y_{SN} the yield of ^{60}Fe, M_{SS} the mass of ^{60}Fe present in the receiving phase and Δ the time delay between nucleosynthesis and injection. With $^{60}Fe/^{56}Fe = 3 \times 10^{-7}$, an $^{56}Fe/^{1}H$ ratio of

3.2×10^{-5}, and a metallicity of 0.7 (Lodders, 2003), it is calculated that the concentration of ^{60}Fe in the ESS was of 0.4 ppb.

SN injection into a disk

Using Equation 8.1 and assuming that the protoplanetary disk had a mass of 0.01 $M_{\odot}$, Looney *et al.* (2006) as well as Ouellette *et al.* (2005) calculated that the nearby SN had to lie at a minimum distance from the protoplanetary disk of a few tenths of a parsec to receive SRs at the early Solar-System level.

Williams and Gaidos (2007) as well as Gounelle and Meibom (2008), however, showed that it is very unlikely that a protoplanetary disk lies that close to a massive star ready to explode as a supernova. This is mainly because the vast majority of low-mass stars in clusters lie further away than 1 pc from the most massive star (Reach *et al.*, 2004), and because disks so close to a massive star evaporate due to the massive star's UV radiation (Johnstone *et al.*, 1998).

To exemplify the disk contamination by nearby SN ejecta, beautiful HST images showing disks within a few tenths of a parsec of the massive (40 $M_{\odot}$) star θ^1 C Ori in the Orion Nebula Cluster were often used (Hester *et al.*, 2004; see Figure 8.1). These images are, however, misleading as the ONC is only ~ 1-Myr old. By the time the most massive star in the Orion Nebula Cluster C explodes as an SN (~ 4 Myr from now), the disks will have photo-evaporated or formed giant planets (Gounelle and Meibom, 2008).

The requirement for the disk to lie at a few tenths of a parsec from the massive star is calculated assuming an injection efficiency of 1. This assumption can be seriously questioned. Indeed, Ouellette *et al.* (2007) showed that the supernova ejecta bounces back on the dense disk and proposed that SRs are injected as micron-sized grains. However, as noted by Boss *et al.* (2008), supernova dust grains are essentially smaller than 0.1 μm and are sputtered to even smaller sizes in the shock (Bianchi and Schneider, 2007). The injection efficiency is therefore probably $<< 1$, imposing the disk to lie even closer to the SN, and lowering, accordingly, the probability that this model works.

SN injection into a core

In the case of the core model, it is required that the SN shockwave triggers the core gravitational collapse at the same time as SRs are injected (Cameron and Truran, 1977; Boss *et al.*, 2008). In their simulations of such a process, Boss *et al.* (2008) consider a 1 $M_{\odot}$ core having a size $r_0 = 0.058$ pc. For such a core, and given the 0.4 ppb abundance of ^{60}Fe, it means that 4×10^{-10} $M_{\odot}$ of ^{60}Fe needs to be delivered by the SN.

Boss *et al.* (2008) calculated that the injection efficiency of their model is $\eta = 0.003$. Assuming conservatively $\Delta = 0$ (Boss *et al.*, 2008), and with Y_{SN} in the range 1×10^{-5} to 6×10^{-5} (Rauscher *et al.*, 2002), Equation 8.1 gives an SN distance varying between 0.8 and 2 pc. To trigger the collapse of the core, the SN shockwave needs to be of ~ 20 km s^{-1} (Boss *et al.*, 2008). For an SN shockwave to have slowed from 1000 km s^{-1} (its approximate starting value) to 20 km s^{-1}, it needs to have travelled at least 8 pc (assuming an SN of 25 $M_{\odot}$ and an interstellar density of 1 molecule cm^{-3}) (Cameron *et al.*, 1995). Injection of SRs and gravitational collapse of the core are, therefore, incompatible with each other.

To conclude, it seems very unlikely that injection of ^{60}Fe in the nascent Solar System by a nearby single SN is compatible with basic observations of star-forming regions.

Multiple distant supernovae as a source for ^{60}Fe

It has been recently proposed that ^{60}Fe could have originated from multiple distant supernovae which seeded the protomolecular cloud where our Sun formed. This proposition solves many of the problems met by the single SN models. Called SPACE (supernova propagation and cloud enrichment), this model is based on the most recent observations of star-forming regions.

A new paradigm concerning the formation mechanisms and lifetimes of MCs has emerged during the last few years (e.g. Hennebelle *et al.*, 2007). In this new paradigm, referred to as the 'turbulent convergent-flow model', MCs result from the collision of coherent flows and large-scale shocks in the ISM driven by winds from massive stars and SN explosions. Such collisions compress the interstellar atomic gas and after 10–20 Myr of evolution, the gas is dense enough to be shielded from the UV radiation and to become molecular. Star formation follows immediately after the formation of the dense molecular gas. The turbulent convergent-flow model elegantly accounts for the division of OB associations in subgroups of different ages (Lada and Lada, 2003). A famous example is the Scorpio–Centaurus region made of the Lower Centaurus Crux (LCC ~ 16 Myr), the Upper Centaurus Lupus (UCL ~ 17 Myr) and the Upper Scorpius (Upper Sco ~ 5 Myr) subregions (Preibisch and Zinnecker, 2007). If the turbulent convergent-flow model is correct, relatively high concentrations of ^{60}Fe and other radioisotopes with half-lives > 1 Myr are expected in MCs. This is because SN ejecta, whose compression effects build MCs, also carry large amounts of radioactive elements such as ^{60}Fe. Although it can take as long as 20 Myr to build an MC, depending on the starting density of the atomic gas, 'live' ^{60}Fe is continuously replenished in the second-generation MC by the supernovae originating from the first episode of star formation, which explode every few Myr (see Figure 8.2).

In the context of the SPACE model, Gounelle *et al.* (2009) made a quantitative evaluation of the amount of ^{60}Fe expected in a protomolecular cloud. They concentrated on a fiducial case where a set of 5000 stars belonging to a first-generation molecular cloud (MC1) enrich in ^{60}Fe a molecular cloud of second generation (MC2). To account for the stochasticity of star formation they simulated this process 100 times.

They find that on average, MC2 made by turbulent convergent flows created by MC1 contain $3 \times 10^{-6}\, M_{\odot}$ of ^{60}Fe. Given that the early Solar-System ^{60}Fe concentration is 0.4 ppb, it means that a molecular cloud of second generation of mass $0.8 \times 10^4\, M_{\odot}$ would have an abundance ratio of $^{60}Fe/^{56}Fe \sim 3 \times 10^{-7}$, similar to that of the Solar System. The question is to know whether it is reasonable to have such a cloud created by turbulent convergent flows from massive stars belonging to an association containing 5000 stars. The answer is yes, as it appears that supernovae from the association UCL–LCC whose total number of stars is ~ 5000 built up the younger Upper Sco star-forming region, whose progenitor molecular cloud had a mass in the range $0.8–4.7 \times 10^4\, M_{\odot}$ (Gounelle *et al.*, 2009). Both the abundance

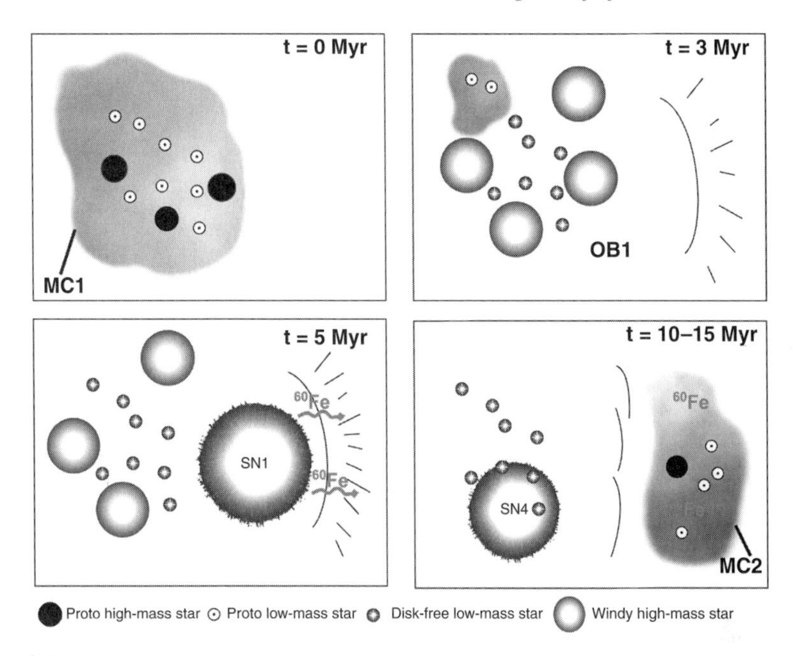

Figure 8.2. Sketch of the SPACE model for the origin of ^{60}Fe in meteorites: Figure 8.1 in time. At t = 0 Myr, star formation starts in MC1. At t = 3 Myr, high-mass stars emit powerful winds which dissipate the molecular gas and start to accumulate interstellar gas further away. At t = 5 Myr, the first SN in OB1 explodes and mixes ^{60}Fe in the swept-up gas. At t = 10–15 Myr, the swept-up gas becomes dense enough to become molecular and star formation starts in MC2. As a result of sequential MC formation, MC2 contains a relatively high abundance of ^{60}Fe produced by the previous SNe (four in that particular case), which contributed to its formation. (From Gounelle *et al.*, 2009.)

of ^{60}Fe and the size of the second-generation molecular cloud scale with the number of stars in the first-generation molecular cloud. In the case of the SPACE model, an astrophysical setting coherent with the observations of star-forming regions therefore exists.

Not only is the astrophysical setting robust and self-consistent, but the SPACE model is also supported by recent work. The ^{60}Fe half-life was recently revised towards larger values, 2.6 Myr instead of 1.5 Myr (Rugel *et al.*, 2009). This helps to build up larger quantities of ^{60}Fe in MC2. Second, d'Orazi *et al.* (2009) have discovered that, in the Orion star-forming regions, older sub-regions (see above) have lower metallicities than younger ones, demonstrating supernova enrichment in heavy metals such as iron. Finally, we note that the $^{60}Fe/^{56}Fe$ ratio adopted in the model of Gounelle *et al.* (2009) is compatible with the upper limit presently observed (Chen *et al.*, 2009). Future work might establish that the actual $^{60}Fe/^{56}Fe$ ratio in the early Solar System was significantly lower than 3×10^{-7}, reducing the need for delivering large quantities of that SR.

Gounelle *et al.* (2009) noted that the SPACE model is not necessarily tied to the turbulent convergent-flow model. It could occur in giant molecular clouds (GMCs). Such an astrophysical setting relies on the observation that during the lifetime of a GMC the regions where active star formation is taking place are changing with time. Iron-60 made in supernovae from a first generation of stars can pollute a region of the GMC which becomes active several million years later. In such a context, it is still the 'collective' explosion of correlated supernovae which helps to build up an important ^{60}Fe background.

The SPACE model differs from previous models in several important points which all contribute to the fact that it successfully accounts for the ^{60}Fe in the nascent Solar System. First, ^{60}Fe is not injected into a dense phase ($n_{H2} \sim 10^5$ molecules cm^{-3} and $\sim 10^{14}$ molecules cm^{-3} for the core and disk, respectively) isolated from the rest of the ISM, but is delivered into a relatively diffuse ISM phase interacting with other ISM components leading to high mixing efficiency. Second, mixing efficiency is also promoted by the longer mixing timescale compared to nearby SN models (10 Myr vs. a few tenths of Myr). Third, the delivery in the whole molecular cloud ensures that all protostars formed in that molecular cloud are endowed with ^{60}Fe, in contrast to the unlikelihood inherent in other models (see above). Finally, such a model might account for the special Solar-System oxygen-isotopic composition, when compared with nearby stars (Young *et al.*, 2008).

Conclusions

The Sun is a low-mass star born in a star-forming region. There are, however, different types of star-forming regions exemplified by the Taurus cloud, the Orion Nebula Cluster or the bright nebula Carina. Furthermore, in large star-forming regions such as the former, the stellar environment varies widely depending on the distance to the most massive stars.

Complementary to observations of star-forming regions at a variety of wavelengths (IR, mm, visible, UV, X-ray, γ-ray) are the laboratory studies of meteorites. These complex rocks are leftovers from our own Solar-System formation. Though identifying processes which took place during the first million years of the solar protoplanetary disk is complicated by geological processes on their parent bodies (asteroids and comets), analytical breakthroughs in a diversity of fields (chemistry, mineralogy, isotope cosmochemistry etc.) have made the message of meteorites clearer and clearer over the last twenty years.

Because the formation of low-mass stars in general and of our Sun in particular is better understood, it becomes possible to address a diversity of important questions. How typical is our Solar System? Did it form under special conditions? Are its properties which seem necessary for life to appear likely to be encountered elsewhere?

Short-lived radionuclides are important tools helping to answer these questions for a variety of reasons. First, because some of them were made within the 20 Myr preceding the formation of our Solar System, providing unique insights into its birth environment. Second, because some of them record very energetic events that occurred during the very first million years of our Solar System's existence. These energetic events might also help to process organic matter inherited from the interstellar medium (e.g. Remusat *et al.*, 2009). Third, some SRs (^{26}Al, ^{53}Mn, ^{182}Hf) can help build a chronology of early Solar-System events which can now be linked to timescales inferred from the observation of protoplanetary disks. Fourth, ^{26}Al and ^{60}Fe decay via γ-ray emission might have provided an important heat source for differentiating planetesimals, and therefore might have played an important role in the geological evolution of planetary bodies and the emergence of life (Gargaud *et al.*, 2006).

Beryllium-10, and possibly ^{36}Cl, ^{41}Ca and ^{53}Mn, were made by energetic-particle irradiation when the Sun was a young T-Tauri star. Proton fluxes inferred from their abundances point to a very active early Sun. This activity, also traced at X-ray wavelengths in all young low-mass stars, might provide clues to the evolution of organic matter found in the most primitive carbonaceous chondrites (see Chapter 7).

Iron-60 abundance seems incompatible with the presence of a supernova exploding very close to the nascent Solar System. It appears, however, compatible with supernovae from a previous generation of stars having mixed some of their nucleosynthetic products, such as ^{60}Fe and oxygen isotopes, with the molecular-cloud progenitor of our Solar System. This could indicate that the Sun was born in a molecular cloud formed by turbulent convergent flows created by massive stars belonging to a previous generation of stars. The exact proportion of stars which were born in such an environment is not exactly known but it might be as high as 20% (Hennebelle *et al.*, 2007), suggesting that endowment with ^{60}Fe might be a common property of planetary systems, and not of the Solar System in particular.

Fascinatingly enough, one of the most studied SRs, ^{26}Al, still has a mysterious origin. Given its short half-life (0.7 Myr), enrichment on a 20-Myr timescale by supernovae belonging to a previous generation of stars appears to be difficult. Even if one considers the unlikely scenario of enrichment by a nearby supernova, ^{26}Al is underproduced relative to ^{60}Fe. Winds from massive stars might be a promising possibility for explaining the high abundance of ^{26}Al in the nascent Solar System (Gaidos *et al.*, 2009; Montmerle *et al.*, 2009).

Constraints from SRs, put together with constraints on planetary formation (Adams and Laughlin, 2001) and oxygen isotopes (Young *et al.*, 2008), point to the Sun being born in a moderate-size cluster much like Orion (500 to a few thousand members). It seems also that some previous episodes of star formation were recorded in the nascent Solar System. Given the sequential nature of star formation, and that a significant number of stars were born in such moderate clusters (Adams and Laughlin, 2001), it makes our Solar System relatively typical. In other planetary systems, one would expect to find SRs such as ^{10}Be, ^{26}Al and ^{60}Fe. While the first is a tracer of high-energy events that might have processed organic matter, the others are efficient heat sources for early planetesimal differentiation and therefore for the development of geological processes, possibly essential for the emergence of life.

References

Adams, F. C. and Laughlin, G. (2001). Constraints on the birth aggregate of the Solar System. *Icarus*, **150**, 151–62.

Amelin, Y., Wadhwa, M. and Lugmair, G. W. (2006). Pb isotopic dating of meteorites using ^{202}Pb–^{205}Pb double spike: comparison with other high-resolution chronometers (abstract). *Lunar and Planetary Science Conference*, **37**, 1790.

Amelin, Y., Krot, A. N., Hutcheon, I. D. and Ulyanov, A. A. (2002). Lead isotopic ages of chondrules and calcium–aluminium-rich inclusions. *Science*, **297**, 1678–83.

Bianchi, S. and Schneider, R. (2007). Dust formation and survival in supernova ejecta. *Monthly Notices of the Royal Astronomical Society*, **378**, 973–82.

Birck, J.-L. (2004). An overview of isotopic anomalies in extraterrestrial materials and their nucleosynthetic heritage. In *Geochemistry of Non-traditional Stable Isotopes*, eds. C. M. Johnson, B. L. Beard and F. Albarède. Washington, DC 55: Mineralogical Society of America, pp. 25–64.

Birck, J. L. and Lugmair, G. W. (1988). Nickel and Chromium in Allende inclusions. *Earth and Planetary Science Letters*, **90**, 131–143.

Boss, A. P., Ipatov, S. I., Keiser, S. A., Myhill, E. A. and Vanhala, H. A. T. (2008). Simultaneous triggered collapse of the presolar dense cloud core and injection of short-lived radioisotopes by a supernova shock wave. *Astrophysical Journal*, **686**, L119–22.

Bouvier, A. and Wadhwa, M. (2009). Synchronizing the absolute and relative clocks: Pb–Pb and Al–Mg systematics in CAIs from the Allende and NWA 2364 CV3 chondrites. *Lunar and Planetary Science Conference*, **40**, 2184.

Cameron, A. G. W. and Truran, J. W. (1977). The supernova trigger for formation of the Solar System. *Icarus*, **30**, 447–61.

Cameron, A. G. W., Hoflich, P., Myers, P. C. and Clayton, D. D. (1995). Massive supernovae, Orion gamma rays, and the formation of the Solar System. *Astrophysical Journal*, **447**, L53–7.

Chaussidon, M. and Barrat, J. A. (2009). ^{60}Fe in Eucrite NWA 4523: evidences for secondary redistribution of Ni and for secondary apparent high ^{60}Fe/^{56}Fe ratios in troilite. *Lunar and Planetary Science Conference*, **40**, 1752.

Chaussidon, M., Robert, F. and McKeegan, K. D. (2006). Li and B isotopic variations in an Allende CAI: evidence for the in situ decay of short-lived ^{10}Be and for the possible presence of the short-lived ^{7}Be in the early Solar System. *Geochimica et Cosmochimica Acta*, **70**, 224–45.

Chen, J. H., Papanastassiou, D. A. and Wasserburg, G. J. (2009). A search for nickel isotopic anomalies in iron meteorites and chondrites. *Geochimica et Cosmochimica Acta*, **73**, 1461–71.

Cieza, L., Padgett, D. L., Stapelfeldt, K. R., Augereau, J-C., Harvey, P., Evans II, N. J., Merin, B, Koerner, D., Sargent, A., van Dishoeck, E. F., Allen, L., Blake, G., Brooke, T., Chapman, N., Huard, T., Shih-Ping, L., Mundy, L., Myers, P. C., Spiesman, W., and Wahhaj, Z. (2007). The Spitzer c2d survey of weak-line T Tauri stars II. New constraints on the timescale for planet building. *Astrophysical Journal*, **667**, 308.

Connelly, J. N., Amelin, Y., Krot, A. N. and Bizzarro, M. (2007). Chronology of the Solar System's oldest solids. *Astrophysical Journal*, **675**, L121–4.

Cook, D. L., Wadhwa, M., Janney, P. E., Dauphas, N., Clayton, R. N. and Davis, A. M. (2006). High precision measurements of non-mass-dependent effects in Nickel isotopes in meteoritic metal via multicollector ICPMS. *Analytical Chemistry*, **78**, 8477–84.

Crida, A. (2009). Minimum mass solar nebulae and planetary migration. *Astrophysical Journal*, **698**, 606–14.

d'Orazi, V. D., Randich, S., Flaccomio, E., Palla, F., Sacco, G. G. and Pallavicini, R. (2009). Metallicity of low-mass stars in Orion. *Astronomy and Astrophysics*, **501**(3), 973–83.

Dauphas, N., Cook, D. L., Sacarabany, C., Frohlich, C., Davis, A. M., Wadhwa, M., Pourmand, A., Rauscher, T. and Gallino, R. (2008). Iron-60 evidence for early injection and efficient mixing of stellar debris in the protosolar nebula. *Astrophysical Journal*, **686**, 560.

Duprat, J. and Tatischeff, V. (2007). Energetic constraints on in situ production of short-lived radionuclei in the early Solar System. *Astrophysical Journal*, **671**, L69–72.

Elmegreen, B.G. and Palous, J. (eds.) (2007). *Triggered star formation in a turbulent interstellar medium*. IAU Symposium 237, Cambridge University Press.

Fitoussi, C., Duprat, J., Tatischeff, V., Kiener, J., Naulin, F., Raisbeck, G., Assunção, M., Bourgeois, C., Chabot, M., Coc, A., Engrand, C., Gounelle, M., Hammache, F., Lefebvre, A., Porquet, M.-G., Scarpaci, J.-A., de Séréville, N., Thibaud, J.-P. and Yiou, F. (2008). Measurement of $^{24}Mg(^{3}He,p)^{26}Al$ cross section: implication for ^{26}Al production in the early Solar System. *Physical Review*, **C 78**, 044613, 15 pp.

Gaidos, E., Krot, A. N., Williams, J. P. and Raymond, S. N. (2009). ^{26}Al and the formaton of the Solar System from a molecular cloud contaminated by Wolf–Rayet winds. *Astrophysical Journal*, **696**, 1854–63.

Gargaud, M., Claeys, P., López-García, P., Martin, H., Montmerle, T., Pascal, R. and Reisse, J. (2006). *From Suns to Life: A chronological approach of the history of life on Earth*. Dordrecht: Springer, p. 370.

Gounelle, M. (2006). The origin of short-lived radionuclides. *New Astronomy Reviews*, **50**, 596–9.

Gounelle, M. and Meibom, A. (2008). The origin of short-lived radionuclides and the astrophysical environment of our Solar System formation. *Astrophysical Journal*, **680**, 781–92.

Gounelle, M., Meibom, A., Hennebelle, P. and Inutsuka, S. I. (2009). Supernova propagation and cloud enrichment: a new model for the origin of ^{60}Fe in the early solar system. *Astrophysical Journal Letters*, **694**, L1–L5.

Gounelle, M. and Russell, S. S. (2005). Spatial heterogeneity in the accretion disk and early solar chronology. In *Chondrites and the Protoplanetary Disk*, eds. A. N. Krot, E. R. D. Scott and B. Reipurth. San Francisco: ASP Conference Series, vol. 341, pp. 548–601.

Gounelle, M., Shu, F. H., Shang, H., Glassgold, A. E., Rehm, K. E. and Lee, T. (2006). The irradiation origin of beryllium radioisotopes and other short-lived radionuclides. *Astrophysical Journal*, **640**, 1163–70.

Güdel, M. (2004). X-ray astronomy of stellar coronae. *Astronomy and Astrophysic Reviews*, **12**, 71.

Hennebelle, P., Mac Low, M.-M. and Vazquez-Semadeni, E. (2007). Diffuse interstellar medium and the formation of molecular clouds. *arXiv*: 0711.2417v2.

Hester, J. J., Desch, S. J., Healy, K. R. and Leshin, L. A. (2004). The cradle of the Solar System. *Science*, **304**, 1116–17.

Jacobsen, B., Yin, Q. Z., Moynier, F., Amelin, Y., Krot, A. N., Nagashima, K., Hutcheon I. D. and Palme, H. (2008). ^{26}Al ^{26}Mg and ^{207}Pb ^{206}Pb systematics of Allende CAIs: canonical solar initial $^{26}Al/^{27}Al$ ratio reinstated. *Earth and Planetary Science Letters*, **272**, 353–64.

Johnstone, D., Hollenbach, D. and Bally, J. (1998). Photoevaporation of disks and clumps by nearby massive stars: application to disk destruction in the Orion nebula. *Astrophysical Journal*, **499**(2), 758.

Kastner, J. H. and Myers, P. C. (1994). An observational estimate of the probability of encounters between mass-losing evolved stars and molecular clouds. *Astrophysical Journal*, **421**, 605–14.

Lada, C. J. and Lada, E. A. (2003). Embedded clusters in molecular clouds. *Annual Review of Astronomy and Astrophysics*, **41**, 57–115.

Lee, T., Shu, F. H., Shang, H., Glassgold, A. E. and Rehm, K. E. (1998). Protostellar cosmic rays and extinct radioactivities in meteorites. *Astrophysical Journal*, **506**, 898–912.

Leya, I., Halliday, A. N. and Wieler, R. (2003). The predictable collateral consequences of nucleosynthesis by spallation reactions in the early Solar System. *Astrophysical Journal*, **594**, 605–16.

Lodders, K. (2003). Solar System abundances and condensation temperatures of the elements. *Astrophysical Journal*, **591**, 1220–47.

Looney, L. W., Tobin, J. J. and Fields, B. D. (2006). Radioactive probes of the supernova-contaminated solar nebula: evidence that the Sun was born in a cluster. *Astrophysical Journal*, **652**, 1755–62.

Meyer, B. S. and Clayton, D. D. (2000). Short-lived radioactivities and the birth of the Sun. *Space Science Reiews*, **92**, 133–52.

McKeegan, K. D., Chaussidon, M. and Robert, F. (2000). Incorporation of short-lived ^{10}Be in a calcium–aluminium-rich inclusion from the Allende meteorite. *Science*, **289**, 1334–7.

Montmerle, T., Augereau, J. C., Chaussido, M., Gounelle, M., Marty, B. and Morbidelli, A. (2006). From Suns to Life: A chronological approach to the history of Life on Earth 3: Solar System formation and early evolution: the first 100 Million years. *Earth Moon and Planets*, **98**, 1–4, 39–95.

Montmerle, T., Gounelle, M. and Meynet, G. (2009). Circumstellar disks in high-mass star environments: irradiation and nucleosynthetic products in the early Solar System. In *Highlights of Astronomy*, eds. I. Corbett, S. Alencar and J. Gregorio-Hetem. Cambridge: Cambridge University Press, vol. 14.

Mostefaoui, S., Lugmair, G. W. and Hoppe, P. (2005). ^{60}Fe: a heat source for planetary differentiation from a nearby supernova explosion. *Astrophysical Journal*, **625**, 271–7.

Ouellette, N., Desch, S. J. and Hester, J. J. (2007). Interaction of supernova ejecta with nearby protoplanetary disks. *Astrophysical Journal*, **662**, 1268–81.

Ouellette, N., Desch, S. J., Hester, J. J. and Leshin, L. A. (2005). A nearby supernova injected short-lived radionuclides in our protoplanetary disk. In *Chondrites and the Protoplanetary Disk*, eds. A. N. Krot, E. R. D. Scott and B. Reipurth. San Francisco: ASP Conference Series, vol. 341, pp. 527–38.

Preibisch, T. and Zinnecker, H. (2007). Sequentially triggered star formation in OB associations. In *Triggered Star Formation in a Turbulent ISM*, eds. B. Elemgreen and J. Palous. IAU Symposium 237, p. 270.

Quitté, G., Halliday, A. N., Markowski, A., Meyer, B. S., Latkoczy, C. and Günther, D. (2007). Correlated Iron 60, Nickel 62, and Zirconium 96 in refractory inclusions and the origin of the Solar System. *Astrophysical Journal*, **655**, 678–84.

Rauscher, T., Heger, A., Hoffman, R. D. and Woosley, S. E. (2002). Nucleosynthesis in massive stars with improved nuclear and stellar physics. *Astrophysical Journal*, **576**, 323–48.

Reach, W. T., Rho, J., Young, E., Muzerolle, J., Fajardo-Acosta, S., Hartmann, L., Sicilia-Aguilar, A., Allen, L., Carey, S., Cuillandre, J. C., Jarrett, T. H., Lowrance,

P., Marston, A., Noriega-Crespo, A. and Hurt, R. L. (2004). Protostars in the Elephant Trunk Nebula. *The Astrophysical Journal Supplement Series*, **154**, 385–90.

Regelous, M., Elliott, T. and Coath, C. D. (2008). Nickel isotope heterogeneity in the early Solar System. *Earth and Planetary Science Letters*, **272**, 330–8.

Remusat, L., Robert, F., Meibom, A., Mostefaoui, S., Delpoux, O., Binet, L., Gourier, D., and Derenne, S. (2009). Protoplanetary disk chemistry recorded by d-rich organic radicals in carbonaceous chondrites. *Astrophysical Journal*, **698**, 2087–92.

Rugel, G., Faestermann, T., Knie, K., Korschinek, G., Poutivtsev, M., Schumann, D., Kivel, N., Günther-Leopold, I., Weinreich, R. and Wohlmuther, M. (2009). New measurement of the ^{60}Fe half-life. *Physical Review Letters*, **103**, 072502.

Shu, F. H., Adams, F. C., and Lizano, S. (1987). Star formation in molecular clouds: observation and theory. *Annual Review of Astronomy and Astrophysics*, **25**, 23–81.

Tachibana, S., Huss, G. R. and Nagashima, K. (2007). ^{60}Fe-^{60}Ni systems in ferromagnesian chondrules in least equilibrated ordinary chondrites (abstract). *Lunar and Planetary Science Conference*, **38**, 1709.

Tachibana, S., Huss, G. R., Kita, N. T., Shimoda, H. and Morishita, Y. (2006). ^{60}Fe in chondrites: debris from a nearby supernova in the early Solar System. *Astrophysical Journal*, **639**, L87–90.

Villeneuve, J., Chaussidon, M. and Libourel, G. (2009). Homogeneous distribution of ^{26}Al in the Solar System from the Mg isotopic composition of chondrules. *Science*, **325**, 985–8.

Wadhwa, M., Amelin, Y., Davis, A. M., Lugmair, G. W., Meyer, B. S., Gounelle, M. and Desch, S. J. (2007). From dust to planetesimals: implications for the solar protoplanetary disk from short-lived radionuclides. In *Protostars and Planets* V, eds. B. Reipurth, D. Jewitt and K. Keil. Tucson: University of Arizona Press, pp. 835–48.

Wasserburg, G. J., Busso, M., Gallino, R. and Nollett, K. M. (2006). Short-lived nuclei in the early Solar System: possible AGB sources. *Nuclear Physics A*, **777**, 5–69.

Williams, J. P. and Gaidos, E. (2007). On the likelihood of supernova enrichment of protoplanetary disks. *Astrophysical Journal*, **663**, L33–6.

Wolk, S. J., Hardnen, F. R., Flaccomio, E., Micela, G., Favata, F., Shang, H. and Feigelson, E. D. (2005). Stellar activity on the young suns of Orion: COUP observations of K5–7 pre-main sequence stars. *Astrophysical Journal Supplement Series*, **160**, 423–49.

Young, E. D., Gounelle, M., Smith, R., Morris, M. R. and Pontoppidan, K. M. (2008). Solar System oxygen isotope ratios result from pollution by type II supernovae. *Lunar and Planetary Science Conference*, **39**, 1329.

9

Formation of habitable planets

John Chambers

Characteristics of a habitable planet

What is a habitable planet? There is no formal definition at present, but the term is generally understood to mean a planet that can sustain life in some form. This concept is of limited use in practice since the conditions required to support life are poorly constrained. A narrower definition of a habitable planet is one that shares some characteristics with Earth, and hence one that could support at least some of Earth's inhabitants. A commonly adopted minimum requirement is that a planet can sustain liquid water on its surface for geological periods of time. Earth is the only body in the Solar System that qualifies as habitable in this sense. One advantage of this definition is that it can be used to categorize hypothetical and observable planets in a relatively straightforward manner, and we will use it in the rest of this chapter. However, one should bear in mind that not all life-sustaining environments will be included under this definition. Tidally heated satellites of giant planets, like Europa, are likely to possess oceans of liquid water beneath a layer of ice (Cassen *et al.*, 1979), but these objects would not be 'habitable' according to the conventional usage.

Planets that can support liquid water at their surface must have an atmosphere, and surface temperatures and pressures within a certain range. These planets will occupy a particular range of orbital distances from their star that is commonly referred to as the star's habitable zone (HZ). For a planet undergoing crustal recycling the width of the HZ is enhanced by the carbon–silicon cycle (Walker *et al.*, 1981). The weathering of silicate rocks in the presence of liquid water can be encapsulated by the reaction:

$$SiO_3^{2-} + CO_2 \rightarrow SiO_2 + CO_3^{2-} \quad (9.1)$$

which proceeds more rapidly with increasing temperature. However, weathering removes carbon dioxide from the atmosphere, ultimately depositing it in the mantle, which lowers the temperature due to a weakened greenhouse effect. This negative-feedback loop keeps the surface temperature in the range for liquid water over a wide range of orbital distances. Kasting *et al.* (1993) estimate the Sun's habitable zone extends from 0.95 to 1.4 AU, while more recent calculations by Forget and Pierrehumbert (1997) suggest the outer edge may be > 2 AU from the Sun. Stars tend to grow more luminous as they age, so the HZ moves outwards over time. This has led to the concept of a continuously habitable zone, which is the range of orbital distances that lie in the HZ over a particular interval of time.

The precise extent of the HZ depends on a planet's albedo, which depends on the degree of cloud cover and whether the surface is covered in ice. For example, Earth currently has two stable climate modes: the current climate with a low albedo, and 'snowball Earth', in which surface water is frozen at all latitudes and the planetary albedo is high. Habitable-zone calculations implicitly assume the climate is in the former mode. The location of the HZ depends on the distribution of continents and oceans on a planet's surface, and thus can change over time as a result of plate tectonics (Spiegel *et al.*, 2008). Additional greenhouse gases, such as methane and sulphur dioxide, can also raise the surface temperature, extending the outer edge of the HZ (Pavlov *et al.*, 2000; Halevy *et al.*, 2007).

The HZ concept is a useful starting point, but a planet must have additional characteristics in order to support life. The calculations used to determine a star's HZ assume that a planet possesses an atmosphere and a reservoir of water to begin with. It is currently unclear how planets like Earth acquire their water and other volatiles, and how these inventories change over time. Venus is almost dry today but probably possessed more water in the past, while Earth may have acquired far more water during its formation than exists in its oceans today (Abe *et al.*, 2000). Earth is depleted in all atmosphere-forming elements compared to the Sun, and several of these same elements (C, N, H, O) are essential to terrestrial life. The sources and removal mechanisms for these elements during the accretion of terrestrial planets are poorly understood at present, but the abundances of these elements will have a profound impact on a planet's habitability.

A planet must be geologically active in order for the carbon–silicon cycle to operate. The nature and degree of crustal recycling probably depend on a planet's size and composition. Small planets like Mars lose their heat rapidly and are likely to be less active than larger bodies. Small planets are also less able to retain their atmospheres due to their weaker gravitational fields, and the lack of crustal recycling means that atmospheres are less likely to be replenished once lost. There may be a critical lower limit to the mass of a life-sustaining planet, somewhere between the mass of Mars and Earth, perhaps 0.2–0.3 Earth masses (Williams *et al.*, 1997; Raymond *et al.*, 2007a). A planet's mass and thermal evolution determine whether a magnetic dynamo operates in its core, generating a strong magnetic field that helps reduce atmospheric erosion caused by interactions with the surrounding stellar wind (Hutchins *et al.*, 1997; Stevenson, 2003).

The climate and habitability of a planet are affected by its obliquity ε (axial tilt relative to its orbit) and orbital eccentricity e (degree of non-circularity), since these quantities affect the spatial and temporal distribution of light from the star on the planet's surface. An increase in either of these quantities increases the degree of seasonal variation on the planet. Surface temperature variations of > 100 K are possible for planets in the HZ with ε near 90 degrees (Williams and Pollard, 2003). Both ε and e will vary on million-year timescales if giant planets are present in the same system. For Earth, these variations are small, because Earth's obliquity evolution is dominated by gravitational interactions with its large Moon (Laskar *et al.*, 1993), and because its orbit is not near an orbital resonance with Jupiter or Saturn. However, terrestrial planets in some systems may undergo large changes in e and ε.

The HZs of low-mass stars lie sufficiently close to the star that habitable planets on circular orbits will be tidally locked, with one face permanently lit and the other in darkness. However, the presence of a thick atmosphere should render such planets at least partially habitable (Joshi, 2003). Planets moving on eccentric orbits around low-mass stars are unlikely to be tidally locked. These planets can move significantly closer to their star on billion-year timescales due to stellar tides, changing their location with respect to the star's HZ (Barnes *et al.*, 2008).

Theories of planet formation

Information about how planets form comes from a variety of sources, including astronomical observations of young stars, cosmochemical analysis of meteorites, robotic space missions to planets, asteroids and comets in the Solar System, laboratory experiments and computer simulations. These data have given rise to a standard model of planet formation in which planets begin life as dust grains in orbit around young stars. These dust grains grow as a result of mutual collisions, aided by gravity, until a handful of planetary-mass bodies remain (Lissauer, 1993). This model is widely accepted for the formation of terrestrial planets – rocky bodies such as Earth and Mars. It is less clear how giant planets form. The popular 'core-accretion' model posits that solid cores form in the same way as terrestrial planets, and once these cores become large enough they accrete gaseous envelopes from their surroundings (Pollack *et al.*, 1996).

The standard model was developed in order to account for the origin of the Sun's planets. At the time of writing, several hundred planets have been discovered orbiting other stars. Most of these objects are believed to be gas giants on the basis of their mass and density. The orbital distribution of these extrasolar planets suggests the standard model may need to be modified somewhat in order to explain the variety of systems observed. In particular, it appears that some giant planets migrate substantial distances towards their star during their formation, something that apparently did not happen in the Solar System (Armitage, 2007).

This chapter will focus on the formation of terrestrial planets since these appear to be the most likely to support life. I will place particular emphasis on processes that might affect planetary habitability. To date, Earth-like planets have not been firmly identified orbiting other Sun-like stars. Three terrestrial-mass objects have been discovered in orbit around a pulsar (neutron star), but unfortunately little is known about these objects or how they formed (Wolszczan and Frail, 1992). As a result, the discussion in this chapter will be driven to a large extent by what we know about the terrestrial planets in the Solar System, with the obvious caveat that the standard model may be revised in light of future discoveries.

Formation of terrestrial planets

Protoplanetary disks

Most young stars are surrounded by disks of gas and dust. The presence of dusty material in orbit around young stars, and the flattened nature of disks, which mimics the planar arrangement of the orbits of the planets in the Solar System, suggests these disks provide the environment in which planets form. These are commonly called 'protoplanetary disks' as a result. The Sun's own disk is referred to as the solar nebula. The minimum mass of the solar nebula can be gauged from the amount of material of solar composition needed to form the planets, which is 1–2% of a solar mass, although this provides only a lower limit.

Infrared observations of protoplanetary disks show that they contain huge numbers of dust particles with sizes ranging from < 1 μm to > 1 mm (Natta *et al.*, 2007). Emission lines in the spectra of young stars indicate that they are accreting gas from their disks at rates of roughly 10^{-7} to 10^{-9} solar masses per year (Gullbring *et al.*, 1998). The mechanism driving this viscous accretion remains unclear, but it is likely to involve turbulence in the disk gas. A disk with a composition similar to the Sun will contain roughly 99% gas by mass, mainly hydrogen and helium, and 1% dust. The dust fraction depends on the stellar 'metallicity' (logarithm of the Fe/H ratio), so metal-rich stars have dust-rich disks, and may be more likely to form terrestrial planets as a result. A typical disk has a radius of order 200 AU (Andrews and Williams, 2007), which suggests the maximum radial extent of most planetary systems will be similar to this. Stars older than about 10 Ma (million years) appear to have lost their disks (Haisch *et al.*, 2001; Pascucci *et al.*, 2006; Cieza *et al.*, 2007). Disks probably dissipate due to a combination of several factors: accretion onto the star, acceleration of gas away from the star by interactions with ultraviolet photons (photo-evaporation) and accretion of material by planets. The final stage of disk dispersal appears to be especially rapid, taking a few times 10^5 years (Cieza *et al.*, 2007).

Temperature and pressure in a protoplanetary disk generally decrease with distance from the star. In the inner disk, only refractory materials such as silicates, oxides, metal and sulphides (collectively 'rock') remain in the solid phase. Further from the star, water ice is also present, first appearing at a distance that is referred to as the 'snow line'. Highly volatile ices such as methane and carbon monoxide can exist in the outermost regions of a disk. In the cooler regions of the solar nebula, the rock-to-water-ice ratio would have been very roughly 1:1 (Lodders, 2003). Temperatures decline over time as the disk loses mass and the central protostar becomes less luminous (Kennedy and Kenyon, 2008). As a result, volatile materials may be progressively depleted compared to refractory elements, since the former remain in the gas phase for longer (Cassen, 2001). The make-up of solid material in any given region of a disk will naturally influence the composition of large bodies that form there. To a first approximation, this explains why the inner planets of the Solar System are composed mainly of rock, while comets and the satellites of the outer planets contain large amounts of water ice.

The dynamics of dust grains is controlled by the gravity of the central star and aerodynamic drag forces from the surrounding disk gas. Gas tends to orbit the star more slowly than solid objects due to the outward pressure gradient. As a result, solid objects experience a headwind, losing angular momentum and drifting inwards. Inward radial drift is especially rapid for m-sized bodies, which can move inwards by 1 AU in 10^2–10^3 years (Weidenschilling, 1977). Particles also tend to settle towards the disk midplane due to the vertical component of the star's gravity. In a turbulent disk, vertical and radial motions are partially opposed by turbulent diffusion. A combination of turbulence and radial drift of solid particles can lead to large spatial and temporal variations in the solid-to-gas ratio, chemical composition and oxidation state of the disk, and these variations may be reflected in planets and asteroids that form subsequently (Ciesla and Cuzzi, 2006). Solid material can accumulate near the snow line or at local pressure maxima, possibly making these preferred sites for planet formation (Stevenson and Lunine, 1988; Haghighipour and Boss, 2003).

Planetesimals

The μm-sized grains observed in protoplanetary disks provide the starting point for the formation of terrestrial planets. Objects larger than about 1 km in size have appreciable gravitational fields and are able to attract and hold on to solid material gravitationally. Bodies this size and larger are referred to as 'planetesimals'. It is currently unclear how planetesimals form from dust grains. Several mechanisms have been proposed and we will examine each of these briefly. A key factor in determining how planetesimals actually form is the level of turbulence in the disk, which is poorly constrained.

Laboratory experiments show that when μm-sized grains collide in a microgravity environment, they stick together electrostatically forming fractal aggregates (Krause and Blum, 2004). Further collisions lead to compaction and to the formation of mm-to-cm-sized dust balls. In the solar nebula, many of these dust balls were melted in brief high-temperature events of unknown origin in the disk, forming round rocky beads called chondrules, which are the dominant component of most primitive meteorites (Ciesla and Hood, 2002; Desch and Connolly, 2002). Volatile elements preferentially escape during these heating events, which may explain why many primitive meteorites are progressively depleted in volatile elements (Alexander *et al.*, 2001; Cohen *et al.*, 2004). If chondrules are one of the main precursors of planets, these planets will inherit the chemical depletions that arose during chondrule formation.

In a non-turbulent disk, planetesimals can potentially form via continued sticking of smaller objects. Laboratory experiments show that mm-to-cm-sized dusty aggregates will embed themselves in larger ones, even at high collision speeds, leading to net growth (Wurm *et al.*, 2005). Chondrules can also stick together during collisions provided that they first sweep up rims of dust on their surface (Ormel *et al.*, 2008). Growth becomes more difficult for m-sized bodies, since these objects have short lifetimes with respect to radial drift. Growth probably stalls at this size if the disk is turbulent since collisions can become destructive in this case.

Small particles tend to sediment towards the disk's midplane. If the concentration of particles becomes high enough, their mutual gravitational attraction can render the particle layer unstable, causing regions to contract to form planetesimals (Goldreich and Ward, 1973). However, this gravitational instability is prevented if the disk is even weakly turbulent (Cuzzi and Weidenschilling, 2006).

In a turbulent disk, mm-sized particles are preferentially concentrated in stagnant regions between the smallest turbulent eddies. Large dense concentrations of particles can become gravitationally bound and slowly shrink to form solid planetesimals (Cuzzi *et al.*, 2008). Turbulent concentration provides a plausible mechanism for forming the parent bodies of primitive meteorites since these are primarily composed of mm-sized chondrules. Dense concentrations only occur rarely, however, so planetesimal formation is likely to be a prolonged process if this is the primary formation mechanism (Cuzzi *et al.*, 2008).

If a substantial fraction of the solid mass is contained in m-sized boulders, these will tend to collect at temporary pressure maxima in a turbulent disk as a result of gas drag. Further concentration takes place via the 'streaming instability' (Johansen *et al.*, 2007). Here, the headwind experienced by objects in a region containing many boulders is reduced, so these objects drift slowly inwards. Boulders further out in the disk drift rapidly as usual, and catch up with objects in the dense region, enhancing the density of boulders still further. This process may form gravitationally bound clumps that collapse to form planetesimals.

The efficiency with which dust is converted into planetesimals will determine the amount of solid material that is available to form terrestrial planets and thus the masses of these bodies. In addition, the timing of planetesimal formation will affect the degree of heating that these objects undergo due to the decay of short-lived radioactive isotopes, and this is likely to influence the degree to which volatile materials are retained and ultimately incorporated into rocky planets.

Formation of planetary embryos

The subsequent evolution of planetesimals is relatively well understood compared to the processes that led to their formation. Collisions and interactions between planetesimals are dominated by gravity rather than electromagnetic forces. The outcome of a collision between two planetesimals depends on their masses and the collision speed. Low-velocity collisions typically result in a merger to form a single body. High-velocity collisions can lead to substantial fragmentation, although some fragments will be travelling too slowly to escape, and these will reaccumulate onto the largest remnant of the collision. Gravitational reaccumulation becomes more important with increasing planetesimal mass, making large bodies better able to survive high-speed collisions (Benz and Asphaug, 1999).

Gravitational attraction means that a planetesimal can pull nearby objects towards it, increasing the chance of a collision. When this is taken into account, a planetesimal of mass M grows at a rate:

$$\frac{dM}{dt} \approx \frac{\pi R^2 v_{rel} \Sigma}{2H} \left(1 + \frac{v_{esc}^2}{v_{rel}^2} \right) \tag{9.2}$$

where R is the planetesimal's radius, $v_{esc} = (2GM/R)^{1/2}$ is its escape velocity, G is the gravitational constant, v_{rel} is the typical relative velocity of planetesimals, Σ is the surface (column) density of solid material in the disk and H ($\propto v_{rel}$) is the vertical scale height of the planetesimals. The growth rate depends mainly on the quantity in parentheses, called the 'gravitational focusing factor', which can become very large if $v_{rel} << v_{esc}$. Large planetesimals naturally have larger gravitational focusing factors, due to their greater escape velocities, so they grow the most rapidly; a process referred to as 'runway growth'.

Most close encounters between planetesimals do not result in a collision. However, the orbits of the planetesimals change due to their mutual gravitational interaction. Encounters typically increase v_{rel}, while gas drag tends to decrease v_{rel}. Once an object grows to about 10^{-6} Earth masses, it is likely to dominate the dynamics of the surrounding region, increasing the relative velocities of nearby planetesimals and moderating its own growth as a result (Ida and Makino, 1993; Thommes *et al.*, 2003). Simulations show that each portion of the disk becomes dominated by a single large body, called a 'planetary embryo', while most planetesimals remain small, a regime referred to as 'oligarchic growth' (Kokubo and Ida, 1998). Embryos grow mainly by absorbing planetesimals from a 'feeding zone' centred on their own orbit. As a result, each embryo is likely to have a unique composition, reflecting the local make-up of the disk. The radial zoning seen in asteroid spectral types, indicating differences in their composition and degree of thermal processing, probably arose at this stage (Gradie and Tedesco, 1982). Simulations suggest that runaway and oligarchic growths are rapid for single stars, taking < 1 Ma to form Mars-mass bodies in the terrestrial-planet region (Kokubo and Ida, 2002). However, the presence of a nearby stellar companion increases the relative velocities of planetesimals and can slow or prevent oligarchic growth (Thebault *et al.*, 2008).

There is abundant evidence that the solar nebula once contained short-lived radioactive isotopes with half-lives < 100 Ma. Several of these isotopes have been used as chronometers to place constraints on the timing of events during planet formation. Isotopes such as ^{26}Al also provide a powerful heat source as they decay. Many planetesimals that formed within the first 2 Ma of the Solar System were melted or thermally altered as a result of this heating (Woolum and Cassen, 1999). Iron meteorites come from the cores of planetesimals that melted and differentiated, with iron and other siderophile elements sinking to the centre. Isotopic dating shows that some differentiated planetesimals formed in < 1 Ma (Markowski *et al.*, 2006). The parent bodies of iron meteorites and other melted and thermally metamorphosed planetesimals probably lost most of their water, carbon and other volatile components as a result of heating soon after they formed.

Planetary embryos larger than Mars acquire substantial atmospheres of gas captured from the surrounding disk (Inaba and Ikoma, 2003). The composition of such an atmosphere reflects that of the disk itself, mostly hydrogen and helium. Dust entrained in the gas and released from collisions with planetesimals makes the atmosphere opaque, trapping heat and possibly raising surface temperatures enough to melt rock (Ikoma and Genda, 2006). The high temperatures mean that vigorous exchange of material takes place between the atmosphere and planetary surface. Some of the helium and neon outgassing from Earth's mantle today may be a remnant of such an early atmosphere (Porcelli *et al.*, 2001).

The presence of water-ice increases the amount of solid material available for building embryos in the outer disk. In addition, the width of an embryo's feeding zone increases with distance from the star due to the reduced stellar gravity. For both these reasons, embryos are likely to grow larger in the outer disk than the inner disk. If an embryo grows above a critical mass, roughly 10 times that of the Earth, gas will begin to flow onto the embryo from the surrounding disk (Ikoma *et al.*, 2000). Gas continues to flow onto this solid core at an accelerating rate until the supply is shut off by the dispersal of the gas disk, or by tidal perturbations from the growing body itself (Pollack *et al.*, 1996). This mechanism is the most widely accepted model for the formation of giant planets. Alternatively, giant planets may form when portions of the outer disk become gravitationally unstable and collapse; referred to as 'disk instability' (Boss, 1997). Each of these models suggests that giant planets typically form in the outer regions of a protoplanetary disk.

Late-stage growth

Oligarchic growth ends when planetary embryos contain roughly half of the solid mass in a particular region (Kenyon and Bromley, 2006). At this point, embryos in the inner solar nebula would have been roughly 0.01–0.1 Earth masses, while those in the outer nebula would have been 1–10 Earth masses. Gravitational interactions between neighbouring embryos increase their relative velocities and cause their orbits to overlap. Gravitational focusing becomes weak and growth rates slow by several orders of magnitude. This final stage of growth is characterized by the gradual sweeping up or removal of remaining planetesimals, and occasional giant impacts between planetary-mass embryos. Numerical simulations suggest it took roughly 100 Ma for Earth to reach its current mass, with the growth rate declining exponentially over time (Chambers, 2001; O'Brien *et al.*, 2006; Raymond *et al.*, 2006a). This timescale is comparable to estimates based on radioactive dating (Halliday, 2004; Touboul *et al.*, 2007).

Late-stage growth is highly stochastic since each planetary embryo typically undergoes many encounters with other large bodies and the orbital evolution depends sensitively on the details of each encounter. Numerical simulations find that a range of outcomes are possible for similar initial conditions (see Figure 9.1). As a result, the number of terrestrial planets, their masses and their orbits are likely to vary widely from one star to another, even for stars with similar protoplanetary disks. A substantial amount of radial mixing takes place during this stage, as embryos acquire eccentric orbits that cross those of many of their neighbours. Overall, embryos are more likely to collide with nearby objects than distant ones, so chemical gradients established during oligarchic growth are partially retained (Chambers, 2001). This is probably the reason why Mars is richer in many volatile rock-forming elements than Earth, and the two planets have different oxygen isotope ratios (Wanke and Dreibus, 1988; Clayton and Mayeda, 1996).

Simulations show that head-on low-velocity impacts between embryos typically result in a merger with little mass escaping in fragments (Agnor and Asphaug, 2004). Collisions between objects larger than the Moon lead to global melting and the differentiation of an iron-rich core and silicate mantle, if the bodies are not already differentiated (Tonks and Melosh, 1992). When two embryos of comparable size collide at an oblique angle, they often separate again and escape, rather than merge, exchanging some material and ejecting volatile

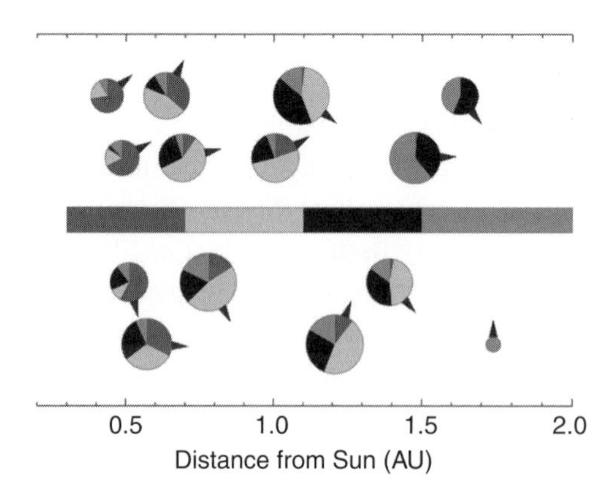

Figure 9.1. Outcome of four simulations of late-stage terrestrial planet formation with similar initial conditions. Each row of symbols shows the planets formed in one simulation, with symbol radius proportional to planetary radius (the largest objects have masses comparable to Earth). The pie-chart segments show the proportion of material originating in different regions of the solar nebula, indicated by the horizontal grey scale bar in the centre of the plot. Arrows indicate spin-axis orientation. Data from Chambers (2001).

materials in the process (Asphaug *et al.*, 2006). Impacts at very high speeds can strip away the outer silicate-rich layers of an embryo, leaving an intact iron core with a thin rocky mantle (Benz *et al.*, 1988). The high uncompressed density of Mercury compared to the other terrestrial planets suggests it experienced at least one disruptive impact late in its formation.

An oblique impact by a small embryo onto a large one can eject debris into orbit, forming a circumplanetary disk. It is likely that the Moon formed within such a disk, generated when a Mars-mass body hit Earth near the end of its formation (Canup and Asphaug, 2001). The Moon would have originally formed at a few Earth radii from the planet, but subsequent tidal interactions caused the size of its orbit to increase rapidly. Simulations of planet formation find that impacts capable of forming a large satellite like the Moon may be quite common in other systems (Agnor *et al.*, 1999).

Giant impacts tend to randomize the spin orientations of terrestrial planets and result in short spin periods (Agnor *et al.*, 1999; Chambers, 2001). Subsequent tidal interactions with the star, and friction between a planet's core and mantle reduce the spin rate and force the obliquity towards 0 or 180 degrees (Correia, 2006). These effects become progressively weaker with distance from the star, which probably explains why Mercury and Venus rotate very slowly and have spin axes almost perpendicular to their orbits, while Earth and Mars do not. Planets that lie close to their star can become tidally locked with one face permanently facing the star. Planets with eccentric orbits may be captured into more complicated spin-orbit resonances, such as that of Mercury, which rotates three times for every two orbits (Correia and Laskar, 2004).

The orbital eccentricities of embryos are excited by mutual encounters and damped by gravitational interactions with leftover planetesimals and tidal interactions with gas. The final eccentricities of terrestrial planets probably depend on how many planetesimals remain during the final stage of growth, and whether any gas is still present (O'Brien *et al.*, 2006; Kominami and Ida, 2004).

Gas-giant planets must form before the protoplanetary disk disperses since they are largely composed of hydrogen and helium, which do not condense and must be captured from the disk. Thus, the final stage of terrestrial-planet formation takes place in the presence of fully formed giant planets. Gravitational perturbations from giant planets affect the growth and stability of terrestrial planets. When giant planets are located in the inner disk, or the giants have eccentric orbits, the number of stable niches for terrestrial formation is reduced and fewer terrestrial planets are formed (Chambers and Cassen, 2002; Levison and Agnor, 2003; O'Brien *et al.*, 2006; Rivera and Haghighipour, 2007).

Perturbations from giant planets are particularly important at orbital resonances, which occur when the orbital period or precession period of an object is in a simple integer ratio with the corresponding quantity for one of the giant planets. In the Solar System, these resonances are associated with the Kirkwood gaps in the main asteroid belt. An asteroid in a resonance develops a highly eccentric orbit, and is typically ejected from the Solar System or collides with the Sun on a timescale of the order of 1 Ma (Gladman *et al.*, 1997). A planetary embryo in a resonance will suffer the same fate. Numerical simulations show that a combination of gravitational interactions between embryos and loss of bodies via resonances is likely to have removed all planetary embryos and roughly 99% of planetesimals from the asteroid belt on a timescale that is short compared with the age of the Solar System (Chambers and Wetherill, 2001; Petit *et al.*, 2001). In addition, some resonances swept radially across the asteroid belt as the solar nebula was dissipating (Nagasawa *et al.*, 2000). A combination of gas drag and sweeping resonances would have removed many planetesimals and deposited them in the region containing the terrestrial planets (Franklin and Lecar, 2000). It seems likely that other planetary systems will contain asteroid belts rather than terrestrial planets in regions that contain strong resonances with giant planets.

The origin of water and other volatile materials on terrestrial planets is unclear. Models suggest the solar nebula was too hot for water ice to condense at 1 AU when planetesimals were forming (Lecar *et al.*, 2006; Kennedy and Kenyon, 2008). Meteorites that are thought to come from the inner asteroid belt contain little water (Baker *et al.*, 2003). Since temperatures were higher closer to the Sun, planetesimals that formed at 1 AU probably contained little or no water. Some of Earth's water could have been generated when hydrogen captured from the solar nebula reacted with rock. However, the deuterium-to-hydrogen (D/H) ratio of water on Earth is 6 times higher than that of the Sun (Robert, 2001), and by extension the solar nebula. The nebula was unlikely to be a major source of Earth's water unless Earth's D/H ratio has been modified substantially, perhaps by the escape of an early hydrogen-rich atmosphere (Genda and Ikoma, 2008).

It seems more likely that Earth received much of its water via collisions with planetesimals and embryos from cold regions of the solar nebula. The outer asteroid belt is particularly favoured since meteorites from this region contain up to 10% water by mass in the form of hydrated silicates; this water has a similar D/H ratio to Earth's oceans, and objects in the asteroid belt have a relatively high chance of colliding with Earth (Morbidelli *et al.*, 2000;

Hartmann *et al.*, 2000; Robert, 2001). Cometary planetesimals also contributed, but their low collision probability with Earth and their high D/H ratio means they were probably a minor source, providing perhaps 10% of Earth's water (Morbidelli *et al.*, 2000; Hartmann *et al.*, 2000; Robert, 2001). Earth's water probably arrived before the planet finished forming and much of it would have been incorporated into the mantle (Drake and Righter, 2002). Water abundances are likely to vary from one terrestrial planet to another due to the stochastic nature of late-stage accretion. A collision with a single water-rich embryo can make a large difference to a planet's composition (Lunine *et al.*, 2003). Simulations suggest that the efficiency of water delivery depends sensitively on the orbits of the giant planets (Chambers and Cassen, 2002; Raymond *et al.*, 2004; O'Brien *et al.*, 2006; Raymond, 2006). When the giants are confined to circular orbits in the outer disk, terrestrial planets may acquire water fractions of 0.1–1% (Raymond *et al.*, 2007b). In systems containing giant planets moving on highly eccentric orbits, terrestrial planets may be essentially dry.

The high N/Ne ratio on Earth and the very low abundances of noble gases on the terrestrial planets in general, argues that Earth received its carbon and nitrogen in condensed form rather than captured from the nebula (Zahnle *et al.*, 2007). Some meteorites contain up to several per cent carbon by mass, and smaller amounts of nitrogen (Gibson *et al.*, 1971), so planetesimals from the asteroid belt may have supplied much of Earth's C and N as well as its water.

Hydrogen-rich atmospheres captured from a protoplanetary disk are probably lost at an early stage by 'hydrodynamic escape' due to heating by ultraviolet photons from the star. Subsequent outgassing from the mantle replenishes the atmosphere with heavier gases such as carbon dioxide and water. Giant impacts partially remove existing atmospheres, especially if oceans are present, since an ocean increases the efficiency with which the impact shock is transmitted to the atmosphere (Genda and Abe, 2005). Impacts such as the Moon-forming event will vaporize oceans and melt much of the mantle, but little water escapes as a result (Genda and Abe, 2005; Canup and Asphaug, 2001). Low-mass planets like Mars probably lose much of their atmosphere on geological timescales due to interactions with the solar wind (Jakosky and Phillips, 2001), and continued hydrodynamic escape (Tian *et al.*, 2009). Atmospheres are less strongly bound to a terrestrial planet than oceans and material in the crust and mantle, so rocky planets will naturally tend to become depleted over time in atmosphere-forming elements such as C, N, H and the noble gases.

Prospects for habitable planets outside the Solar System

Surveys show that at least 10% of main-sequence stars have giant planets (Cumming *et al.*, 2008). If giant planets form by core accretion, these systems are likely to contain terrestrial planets as well since the same physical processes are involved in each case. Giant planets are predicted to form in the outer regions of a protoplanetary disk in both the core-accretion and the disk-instability models. The small orbital radii of some extrasolar giant planets suggest that these objects migrated inwards during or after their formation. Migration can happen as a result of tidal interactions with the gas disk or perturbations between giant planets (Ward, 1997; Ford and Rasio, 2008). The efficiency of giant-planet formation may depend sensitively on timing (Thommes *et al.*, 2007). If a giant planet forms too early it can fall into the star as a result of migration. If a giant-planet core forms too late, there will not be enough mass left in the disk to form the planet's gaseous envelope. In disks with a low mass or a low metallicity there may not be enough solid material to form a giant-planet core at all. By contrast, even low-mass disks are likely to contain enough dust to form terrestrial planets. In addition, these planets can undergo most of their growth after the gas has dispersed, so low growth rates are not a problem. As a result, terrestrial planets are likely to be more common than giant planets.

Many young stars that have lost their gas-rich protoplanetary disk still possess low-mass disks of dust, referred to as 'debris disks' (Hillenbrand *et al.*, 2008). The lifetime of dust grains in these systems is typically short compared to the age of the star, which suggests that new dust grains are being generated continually from larger bodies such as asteroids or comets (Artymowicz and Clampin, 1997). This implies that at least the early stages of planet formation took place in these systems, and suggests that terrestrial planets may orbit many if not most stars.

Migrating giant planets pose a potential problem for the growth and stability of terrestrial planets in the star's HZ because the giant can dynamically remove much of the solid material from this region. However, simulations indicate that terrestrial planets can still form from material that survives the giant's migration, together with new material that moves inwards from outer disk (Raymond *et al.*, 2006b). These terrestrial planets are likely to be more volatile-rich than Earth, due to the large fraction of their mass that originates from the outer disk.

While terrestrial planet formation is likely to be commonplace, not all stars will possess planets capable of supporting life. Rocky planets will not be present in the HZ when this region is rendered unstable by perturbations from nearby giant planets. Similarly, if strong resonances are present in the HZ, this region will probably contain an asteroid belt rather than terrestrial planets. Even when there are no giant planets in the system, the stochastic nature of late-stage planet formation means that some systems will not have a large geologically active terrestrial planet in the HZ. The compositions of terrestrial planets are also likely to vary widely. The amount of water and other volatile material retained in planetesimals depends on how much they are heated, which in turn depends on the abundance of short-lived radioactive isotopes such as ^{26}Al in the disk and the timing of planetesimal formation. The amount of volatile material delivered to rocky planets from elsewhere in the disk also depends sensitively on the nature of any giant planets in the system and their orbital evolution (O'Brien *et al.*, 2006). Thus, while terrestrial planets are likely to be abundant, 'habitable' planets may be much less common.

Acknowledgements

I would like to thank Lindsey Chambers for helpful comments during the preparation of this article. This work was partially supported by the NASA Origins of Solar Systems programme.

References

Abe, Y., Ohtani, E., Okuchi, T., Righter, K. and Drake, M. (2000). Water in the early Earth. In *Origin of the Earth and Moon*, eds. R. M. Canup and K. Righter. Tucson, AZ: University of Arizona Press, pp. 413–33.

Agnor, C. B. and Asphaug, E. (2004). Accretion efficiency during planetary collisions. *Astrophysical Journal*, **613**, L157–60.

Agnor, C. B., Canup, R. M. and Levison, H. F. (1999). On the character and consequences of large impacts in the late stage of terrestrial planet formation. *Icarus*, **142**, 219–37.

Alexander, C. M. O'D., Boss, A. P. and Carlson, R. W. (2001). The early evolution of the inner Solar System: a meteoritic perspective. *Science*, **293**, 64–8.

Andrews, S. M. and Williams, J. P. (2007). High-resolution submillimeter constraints on circumstellar disk structure. *Astrophysical Journal*, **659**, 705–28.

Armitage, P. J. (2007). Massive planet migration: theoretical predictions and comparisons with observations. *Astrophysical Journal*, **665**, 1381–90.

Artymowicz, P. and Clampin, M. (1997). Dust around main-sequence stars: nature or nurture by the interstellar medium? *Astrophysical Journal*, **490**, 863–78.

Asphaug, E., Agnor, C. B. and Williams, Q. (2006). Hit-and-run planetary collisions. *Nature*, **439**, 155–60.

Baker, L., Franchi, I. A., Wright, I. P. and Pillinger, C. T. (2003). The oxygen isotopic composition of water extracted from unequilibrated ordinary chondrites. *Lunar and Planetary Science*, **34**, 1800.

Barnes, R., Raymond, S. N., Jackson, B. and Greenberg, R. (2008). Tides and the evolution of planetary habitability. *Astrobiology*, **8**, 557–68.

Benz, W. and Asphaug, E. (1999). Catastrophic disruptions revisited. *Icarus*, **142**, 5–20.

Benz, W., Slattery, W. L. and Cameron, A. G. W. (1988). Collisional stripping of Mercury's mantle. *Icarus*, **74**, 516–28.

Boss, A. P. (1997). Giant planet formation by gravitational instability. *Science*, **276**, 1836–9.

Canup, R. M. and Asphaug, E. (2001). Origin of the Moon in a giant impact near the end of the Earth's formation. *Nature*, **412**, 708–12.

Cassen, P. (2001). Nebular thermal evolution and the properties of primitive planetary materials. *Meteoritics and Planetary Science*, **36**, 671–700.

Cassen, P., Reynolds, R. T. and Peale, S. J. (1979). Is there liquid water on Europa? *Geophysical Research Letters*, **6**, 731–4.

Chambers, J. E. (2001). Making more terrestrial planets. *Icarus*, **152**, 205–24.

Chambers, J. E. and Cassen, P. (2002). The effects of nebula surface density profile and giant planet eccentricities on planetary accretion in the inner Solar System. *Meteoritics and Planetary Science*, **37**, 1523–40.

Chambers, J. E. and Wetherill, G. W. (2001). Planets in the asteroid belt. *Meteoritics and Planetary Science*, **36**, 381–99.

Ciesla, F. J. and Cuzzi, J. N. (2006). The evolution of the water distribution in a viscous protoplanetary disk. *Icarus*, **181**, 178–204.

Ciesla, F. J. and Hood, L. L. (2002). The nebular shock wave model for chondrule formation: shock processing in a particle gas suspension. *Icarus*, **158**, 281–93.

Cieza, L., Padgett, D., Stapelfeldt, K., Augereau, J. C., Harvey, P., Evans, N., Merin, B., Koerner, D., Sargent, A., Van Dishoeck, E., Allen, L., Blake, G., Brooke, T., Chapman, N., Huard, T., Lai, S. P., Mundy, L., Myers, P. C., Spiesman, W. and

Wahhaj, Z. (2007). The Spitzer c2d survey of weak-line T Tauri stars ii. New constraints on the timescale for planet building. *Astrophysical Journal*, **667**, 308–28.
Clayton, R. N. and Mayeda, T. K. (1996). Oxygen isotope studies of achondrites. *Geochimica et Cosmochimica Acta*, **60**, 1999–2017.
Cohen, B. A., Hewins, R. H. and Alexander, C. M. O'D. (2004). The formation of chondrules by open-system melting of nebular condensates. *Geochimica et Cosmochimica Acta*, **68**, 1661–75.
Correia, A. C. M. (2006). The core-mantle friction effect on the secular spin evolution of terrestrial planets. *Earth and Planetary Science Letters*, **252**, 398–412.
Correia, A. C. M. and Laskar, J. (2004). Mercury's capture into the 3/2 spin-orbit resonance as a result of its chaotic dynamics. *Nature*, **429**, 848–50.
Cumming, A., Butler, R. P., Marcy, G. W., Vogt, S. S., Wright, J. T. and Fischer, D. A. (2008). The Keck planet search: detectability and the minimum mass and orbital period distribution of extrasolar planets. *Publications of the Astronomical Society of the Pacific*, **120**, 531–54.
Cuzzi, J. N. and Weidenschilling, S. J. (2006). Particle-gas dynamics and primary accretion. In *Meteorites and the Early Solar System II*, eds. D. S. Lauretta and H. Y. McSween. Tucson, AZ: University of Arizona Press, pp. 353–81.
Cuzzi, J. N., Hogan, R. C. and Shariff, K. (2008). Toward planetesimals: dense chondrule clumps in the protoplanetary nebula. *Astrophysical Journal*, **687**, 1432–47.
Desch, S. J. and Connolly, H. C. (2002). A model of the thermal processing of particles in solar nebula shocks: application to the cooling rates of chondrules. *Meteoritics and Planetary Science*, **37**, 183–207.
Drake, M. J. and Righter, K. (2002). Determining the composition of the Earth. *Nature*, **416**, 39–44.
Ford, E. B. and Rasio, F. A. (2008). Origins of eccentric extrasolar planets: testing the planet–planet scattering model. *Astrophysical Journal*, **686**, 621–36.
Forget, F. and Pierrehumbert, R. T. (1997). Warming early Mars with carbon dioxide clouds that scatter infrared radiation. *Science*, **278**, 1273.
Franklin, F. and Lecar, M. (2000). On the transport of bodies within and from the asteroid belt. *Meteoritics and Planetary Science*, **35**, 331.
Genda, H. and Abe, Y. (2005). Enhanced atmospheric loss on protoplanets at the giant impact phase in the presence of oceans. *Nature*, **433**, 842–4.
Genda, H. and Ikoma, M. (2008). Origin of the ocean on the Earth: early evolution of water D/H in a hydrogen-rich atmosphere. *Icarus*, **194**, 42–52.
Gibson, E. K., Moore, C. B. and Lewis, C. F. (1971). Total nitrogen and carbon abundances in carbonaceous chondrites. *Geochimica et Cosmochimica Acta*, **35**, 599–604.
Gladman, B. J., Migliorini, F., Morbidelli, A., Zappala, V., Michel, P., Cellino, A., Froeschle, C., Levison, H. F., Bailey, M. and Duncan, M. (1997). Dynamical lifetimes of objects injected into asteroid belt resonances. *Science*, **277**, 197–201.
Goldreich, P. and Ward, W. R. (1973). The formation of planetesimals. *Astrophysical Journal*, **183**, 1051–61.
Gradie, J. and Tedesco, E. (1982). Compositional structure of the asteroid belt. *Science*, **216**, 1405–7.
Gullbring, E., Hartmann, L., Briceno, C. and Calvet, N. (1998). Disk accretion rates for T Tauri stars. *Astrophysical Journal*, **492**, 323–41.
Haghighipour, N. and Boss, A. P. (2003). On pressure gradients and rapid migration of solids in a nonuniform solar nebula. *Astrophysical Journal*, **583**, 996–1003.

Haisch, K. E., Lada, K. A. and Lada, C. J. (2001). Disk frequencies and lifetimes in young clusters. *Astrophysical Journal*, **553**, L153–6.

Halevy, I., Zuber, M. T. and Schrag, D. P. (2007). A sulphur dioxide climate feedback on early Mars. *Science*, **318**, 1903–7.

Halliday, A. N. (2004). Mixing, volatile loss and compositional change during impact-driven accretion of the Earth. *Nature*, **427**, 505–9.

Hartmann, W. K., Ryder, G., Dones, L. and Grinspoon, D. (2000). The time-dependent intense bombardment of the primordial Earth/Moon system. In *Origin of the Earth and Moon*, eds. R. M. Canup and K. Righter. Tucson, AZ: University of Arizona Press, pp. 493–512.

Hillenbrand, L. A., Carpenter, J. M., Kim, J. S., Meyer, M. R., Backman, D. E., Moro-Martin, A., Hollenbach, D. J., Hines, D. C., Pascucci, I. and Bouwman, J. (2008). The complete census of 70 micrometer-bright debris disks within 'the Formation and Evolution of Planetary Systems' Spitzer legacy survey of Sun-like stars. *Astrophysical Journal*, **677**, 630–56.

Hutchins, K. S., Jakosky, B. M. and Luhmann, J. G. (1997). Impact of a paleomagnetic field on sputtering loss of Martian atmospheric argon and neon. *Journal of Geophysical Research*, **102**, 9183–9.

Ida, S. and Makino, J. (1993). Scattering of planetesimals by a protoplanet: slowing down of runaway growth. *Icarus*, **106**, 210–27.

Ikoma, M. and Genda, H. (2006). Constraints on the mass of a habitable planet with water of a nebula origin. *Astrophysical Journal*, **648**, 696–706.

Ikoma, M., Nakazawa, K. and Emori, H. (2000). Formation of giant planets: dependencies on core accretion rate and grain opacity. *Astrophysical Journal*, **537**, 1013–25.

Inaba, S. and Ikoma, M. (2003). Enhanced collisional growth of a protoplanet that has an atmosphere. *Astronomy and Astrophysics*, **410**, 711.

Jakosky, B. M. and Phillips, R. (2001). Mars' volatile and climate history. *Nature*, **412**, 237–44.

Johansen, A., Oishi, J. S., MacLow, M.-M., Klahr, H., Henning, T. and Youdin, A. (2007). Rapid planetesimal formation in turbulent circumstellar disks. *Nature*, **448**, 1022–5.

Joshi, M. (2003). Climate model studies of synchronously rotating planets. *Astrobiology*, **3**, 415–27.

Kasting, J. F., Whitmire, D. P. and Reynolds, R. T. (1993). Habitable zones around main sequence stars. *Icarus*, **101**, 108–28.

Kennedy, G. M. and Kenyon, S. J. (2008). Planet formation around stars of various masses: the snow line and the frequency of giant planets. *Astrophysical Journal*, **673**, 502–12.

Kenyon, S. J. and Bromley, B. C. (2006). Terrestrial planet formation i. The transition from oligarchic growth to chaotic growth. *Astronomical Journal*, **131**, 1837–50.

Kokubo, E. and Ida, S. (1998). Oligarchic growth of protoplanets. *Icarus*, **131**, 171–8.

Kokubo, E. and Ida, S. (2002). Formation of protoplanet systems and diversity of planetary systems. *Astrophysical Journal*, **581**, 666–80.

Kominami, J. and Ida, S. (2004). Formation of terrestrial planets in a dissipating gas disk with Jupiter and Saturn. *Icarus*, **167**, 231–43.

Krause, M. and Blum, J. (2004). Growth and form of planetary seedlings: results from a sounding rocket microgravity aggregation experiment. *Physical Review Letters*, **93**, 021103.

Laskar, J., Joutel, F. and Robutel, P. (1993). Stabilization of the Earth's obliquity by the Moon. *Nature*, **361**, 615–17.

Lecar, M., Podolak, M., Sasselov, D. and Chiang, E. (2006). On the location of the snow line in a protoplanetary disk. *Astrophysical Journal*, **640**, 1115–18.
Levison, H. F. and Agnor, C. (2003). The role of giant planets in terrestrial planet formation. *Astronomical Journal*, **125**, 2692–713.
Lissauer, J. J. (1993). Planet Formation. *Annual Review of Astronomy and Astrophysics*, **31**, 129–74.
Lodders, K. (2003). Solar System abundances and condensation temperatures of the elements. *Astrophysical Journal*, **591**, 1220–47.
Lunine, J. I., Chambers, J., Morbidelli, A. and Leshin, L. A. (2003). The origin of water on Mars. *Icarus*, **165**, 1–8.
Markowski, A., Leya, I., Quitte, G., Ammon, K., Halliday, A. N. and Wieler, R. (2006). Correlated helium-3 and tungsten isotopes in iron meteorites: quantitative cosmogenic corrections and planetesimal formation times. *Earth and Planetary Science Letters*, **250**, 104–15.
Morbidelli, A., Chambers, J., Lunine, J. I., Petit, J. M., Robert, F., Valsecchi, G. B. and Cyr, K. E. (2000). Source regions and timescales for the delivery of water to the Earth. *Meteoritics and Planetary Science*, **35**, 1309–20.
Nagasawa, M., Tanaka, H. and Ida, S. (2000). Orbital evolution of asteroids during depletion of the solar nebula. *Astronomical Journal*, **119**, 1480–97.
Natta, A., Testi, L., Calvet, N., Henning, T., Water, R. and Wilner, D. (2007). Dust in protoplanetary disks: properties and evolution. In *Protostars and planets v*, eds. V. B. Reipurth, D. Jewitt and K. Keil. Tucson, AZ: University of Arizona Press, pp. 767–81.
O'Brien, D. P., Morbidelli, A. and Levison, H. F. (2006). Terrestrial planet formation with strong dynamical friction. *Icarus*, **184**, 39–58.
Ormel, C. W., Cuzzi, J. N. and Tielens, A. G. G. (2008). Co-accretion of chondrules and dust in the solar nebula. *Astrophysical Journal*, **679**, 1588–610.
Pascucci, I., Gorti, U., Hollenbach, D., Najita, J., Meyer, M. R., Carpenter, J. M., Hillenbrand, L. A., Herczeg, G. J., Padgett, D. L., Silverstone, M. D., Schlingman, W. M., Kim, J. S., Stobie, E. B., Bouwman, J., Wolf, S., Rodman, J., Hines, D. C., Lunine, J. and Malhotra, R. (2006). Formation and evolution of planetary systems: upper limits to the gas mass in disks around Sun-like stars. *Astrophysical Journal*, **651**, 1177–93.
Pavlov, A. A., Kasting, J. F. and Brown, L. L. (2000). Greenhouse warming by methane in the atmosphere of early Earth. *Journal of Geophysical Research*, **105**, 11981–90.
Petit, J. M., Morbidelli, A. and Chambers, J. (2001). The primordial excitation and clearing of the asteroid belt. *Icarus*, **153**, 338–47.
Pollack, J. B., Hubickyj, O., Bodenheimer, P., Lissauer, J. J., Podolak, M. and Greenzweig, Y. (1996). Formation of the giant planets by concurrent accretion of solids and gas. *Icarus*, **124**, 62–85.
Porcelli, D., Woolum, D. and Cassen, P. (2001). Deep Earth rare gases: initial inventories, capture from the solar nebula, and losses during Moon formation. *Earth and Planetary Science Letters*, **193**, 237–51.
Raymond, S. N. (2006). The search for other Earths: limits on the giant planet orbits that allow habitable terrestrial planets to form. *Astrophysical Journal*, **643**, L131–4.
Raymond, S. N., Mandell, A. M. and Sigurdsson, S. (2006b). Exotic Earths: forming habitable worlds with giant planet migration. *Science*, **313**, 1413–16.
Raymond, S. N., Quinne, T. and Lurine J. I. (2004). Making other Earths: dynamical simulations of terrestrial planet formation and water delivery. *Icarus*, **168**, 1–17.

Raymond, S. N., Quinn, T. and Lunine, J. I. (2006a). High-resolution simulations of the final assembly of Earth-like planets I. Terrestrial accretion and dynamics. *Icarus*, **183**, 265–82.
Raymond, S. N., Scalo, J. and Meadows, V. S. (2007a). A decreased probability of habitable planet formation around low-mass stars. *Astrophysical Journal*, **669**, 606–14.
Raymond, S. N., Quinn, T. and Lunine, J. I. (2007b). High-resolution simulations of the final assembly of Earth-like planets II. Water delivery and planetary habitability. *Astrobiology*, **7**, 66–84.
Rivera, E. and Haghighipour, N. (2007). On the stability of test particles in extrasolar multiple planet systems. *Monthly Notices of the Royal Astronomical Society*, **374**, 599–613.
Robert, F. (2001). The origin of water on Earth. *Science*, **293**, 1056–8.
Spiegel, D. S., Menou, K. and Scharf, C. A. (2008). Habitable climates. *Astrophysical Journal*, **681**, 1609–23.
Stevenson, D. J. (2003). Planetary magnetic fields. *Earth and Planetary Science Letters*, **208**, 1–11.
Stevenson, D. J. and Lunine, J. I. (1988). Rapid formation of Jupiter by diffusive redistribution of water vapour in the solar nebula. *Icarus*, **75**, 146–55.
Thebault, P., Marzari, F. and Scholl, H. (2008). Planet formation in Alpha Centauri A revisited: not so accretion friendly after all. *Monthly Notices of the Royal Astronomical Society*, **388**, 1528–36.
Thommes, E. W., Duncan, M. J. and Levison, H. F. (2003). Oligarchic growth of giant planets. *Icarus*, **161**, 431–55.
Thommes, E. W., Nilsson, L. and Murray, N. (2007). Overcoming migration during giant planet formation. *Astrophysical Journal*, **656**, L25–8.
Tian, F., Kasting, J. F. and Solomon, S. C. (2009). Thermal escape of carbon from the early Martian atmosphere. *Geophysical Research Letters*, **36**, L02205.
Tonks, W. B. and Melosh, H. J. (1992). Core formation by giant impacts. *Icarus*, **100**, 326–46.
Touboul, M., Kleine, T., Bourdon, B., Palme, H. and Wieler, R. (2007). Late formation and prolonged differentiation of the Moon inferred from W isotopes in lunar metals. *Nature*, **450**, 1206–9.
Walker, J. C. G., Hays, P. B. and Kasting, J. F. (1981). A negative feedback mechanism for the long-term stabilization of the Earth's surface temperature. *Journal of Geophysical Research*, **86**, 9776–82.
Wanke, H. and Dreibus, G. (1988). Chemical composition and accretion history of terrestrial planets. *Philosophical Transactions of the Royal Society of London A*, **325**, 545–57.
Ward, W. R. (1997). Protoplanet migration by nebular tides. *Icarus*, **126**, 261–81.
Weidenschilling, S. J. (1977). Aerodynamics of solid bodies in the solar nebula. *Monthly Notices of the Royal Astronomical Society*, **180**, 57–70.
Williams, D. M. and Pollard, D. (2003). Extraordinary climates of Earth-like planets: three-dimensional climate simulations at extreme obliquity. *International Journal of Astrobiology*, **2**, 1–19.
Williams, D. M., Kasting, J. F. and Wade, R. A. (1997). Habitable moons around extrasolar giant planets. *Nature*, **385**, 234–6.
Woolum, D. S. and Cassen, P. (1999). Astronomical constraints on nebular temperatures: implications for planetesimal formation. *Meteoritics and Planetary Science*, **34**, 897–907.

Wolszczan, A. and Frail, D. A. (1992). A planetary system around the millisecond pulsar PSR 1257+12. *Nature*, **355**, 145–7.
Wurm, G., Paraskov, G. and Krauss, O. (2005). Growth of planetesimals by impacts at ~25 m/s. *Icarus*, **178**, 253–63.
Zahnle, K., Arndt, N., Cockell, C., Halliday, A., Nisbet, E., Selsis, F. and Sleep, N. (2007). Emergence of a habitable planet. *Space Science Review*, **129**, 35–78.

10

The concept of the galactic habitable zone

Nicolas Prantzos

Introduction

The modern study of the 'habitability' of circumstellar environments started almost half a century ago (Huang, 1959). The concept of a circumstellar habitable zone (CHZ) is relatively well defined, being tightly related to the requirement of the presence of liquid water as a necessary condition for life-as-we-know-it; the corresponding temperature range is a function of the luminosity of the star and of the distance of the planet from it. An important amount of recent work, drawing on various disciplines (planetary dynamics, atmospheric physics, geology, biology etc.) has refined considerably our understanding of various factors that may affect the CHZ; despite that progress, we should still consider the subject to be in its infancy (e.g. Chyba and Hand, 2005; Gaidos and Selsis, 2007; and references therein).

Habitability on a larger scale was considered a few years ago by Gonzalez *et al.* (2001), who introduced the concept of the galactic habitable zone (GHZ). The underlying idea is that various physical processes, which may favour the development or the destruction of complex life, may depend strongly on the temporal and spatial position in the Milky Way (MW). For instance, the risk of a supernova (SN) explosion sufficiently close to represent a threat to life is, in general, larger in the inner Galaxy than in the outer one, and has been larger in the past than at present. Another example is offered by the metallicity (the amount of elements heavier than hydrogen and helium) of the interstellar medium, which varies across the Milky Way disk, and which may be important for the existence of Earth-like planets. Indeed, the host stars of the ~ 300 detected extrasolar giant planets are, on average, more metal-rich than stars with no planets in our cosmic neighbourhood; however, it must be noticed that Neptune-sized planets do not appear to share this property (e.g. Souza *et al.*, 2008). Several other factors, potentially important for the GHZ, are discussed in Gonzalez (2005).

The concept of GHZ is much less well defined than the one of CHZ, since none of the presumably relevant factors can be quantified in a satisfactory way. Indeed, the role of metallicity in the formation and survival of Earth-like planets is not really understood at present, while the 'lethality' of supernovae and other cosmic explosions is hard to assess. The first study attempting to quantitatively account for such effects was made by Lineweaver *et al.* (2004), with a detailed model for the chemical evolution of the Milky Way disk. They found that the probability of having an environment favourable to complex life is larger in a 'ring' (a few kpc wide) surrounding the Milky Way's centre and spreading outwards in the course of the Galaxy's evolution.

We repeat that exercise here with a model that reproduces satisfactorily the major observables of the Milky Way disk in the section entitled 'The evolution of the Milky Way disk'. In 'Probability of having stars with Earth-like planets' we discuss the role of metallicity and we quantify it in a different (and, presumably, more realistic) way than in Lineweaver *et al.* (2004), in the light of recent progress in our understanding of planetary formation. In 'Probability of life surviving supernova explosions' we discuss the risk of SN explosions and conclude that it can hardly be quantified at present, in view of our ignorance of how robust life really is. For comparison purposes, though, we adopt the same risk factor as the one defined in Lineweaver *et al.* (2004). In 'A GHZ as large as the whole Galaxy?' we present our results, showing that the GHZ may, in fact, extend to the whole galactic disk today. We conclude that, at the present stage of our knowledge, the GHZ may extend to the entire MW disk.

The evolution of the Milky Way disk

The evolution of the solar neighbourhood is rather well constrained, due to the large body of observational data available. In particular, the metallicity distribution of long-lived stars suggests a slow formation of the local disk through a prolonged period of gaseous infall (e.g. Goswami and Prantzos, 2000; see however, Haywood, 2006, for a different view). For the rest of the disk, available data concern mainly its current status and not its past history; in those conditions, it is impossible to derive a unique evolutionary path. Still, the radial properties of the Milky Way disk (profiles of stars, gas, star-formation rate, metallicity, colours etc.), constrain significantly its evolution and point to a scenario of 'inside-out' formation (e.g. Prantzos and Aubert, 1995). Such scenarios fit 'naturally' in the currently favoured paradigm of galaxy formation in a cold dark-matter Universe (e.g. Naab and Ostriker, 2006).

Some results of a model developed along those lines (Boissier and Prantzos, 1999; Hou *et al.*, 2000) are presented in Figure 10.1. Note that the evolution of the galactic bulge (innermost ~2 kpc) is not studied here, since it is much less well constrained than the rest of the disk; the conclusions, however, depend very little on that point. The model assumes a star-formation rate proportional to R^{-1} (where R is the galactocentric distance), inspired from the theory of star formation induced by spiral waves in disks. It satisfies all the available constraints for the solar neighbourhood and the rest of the disk; some of those constraints appear in Figure 10.1. Among the successes of the model, one should mention the prediction that the abundance gradient should flatten with time due to the 'inside-out' formation scheme (bottom right panel in Figure 10.1), a prediction that is quantitatively supported by recent measurements of the metallicity profile in objects of various ages (Maciel *et al.*, 2006).

It should be stressed, however, that the output of the model is only as good as the adopted observational constraints. For instance, there is still some uncertainty concerning the level

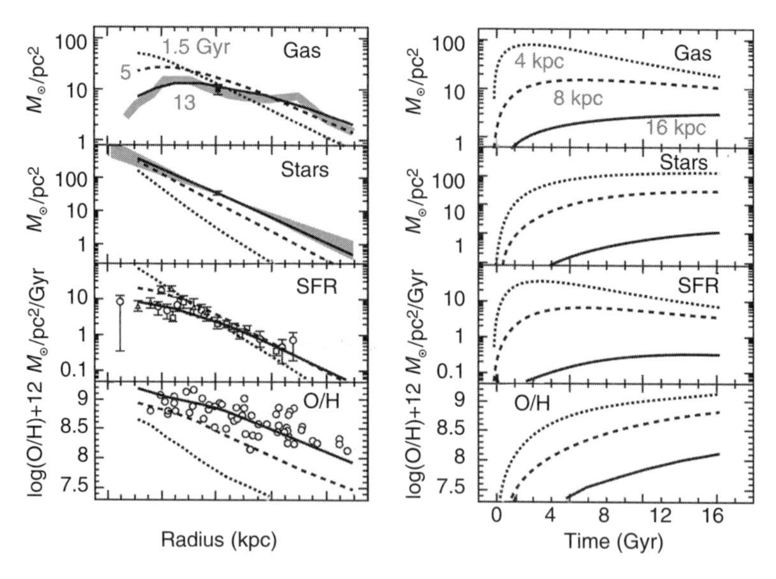

Figure 10.1. The chemical evolution of the Milky Way disk, obtained in the framework of a one-dimensional model with radial symmetry (see the 'Introduction' for a description of the model). Left, from top to bottom: radial profiles of the surface density of gas, stars, star-formation rate (SFR) and of the oxygen abundance. Profiles are displayed at three epochs, namely at 1.5, 5 and 13 Gyr; the latter (thick curve in all figures) is compared to observations concerning the present-day MW disk (grey shaded areas in the first two figures, data points in the last two). Right, from top to bottom: the same quantities are plotted as a function of time for three different disk regions, located at galactocentric distances of 4, 8 and 16 kpc; the second one (thick curves in all panels of the right column) corresponds to the solar neighbourhood.

of the oxygen-abundance gradient: the 'standard' value of dlog(O/H)/dR = –0.07 dex/kpc is challenged by the recent evaluations of Daflon and Cunha (2004), who find only half that value. If the latter turns out to be true, some of the model parameters (e.g. the adopted radial dependence of the infall rate and/or the star-formation rate) should have to be revised.

In any case, it appears rather well established now that the star-formation rate and the metallicity are more important in the inner disk than in the outer one, and that this trend has been even stronger in the past (see left panel in Figure 10.1). Some authors have suggested that those two parameters play an important role in the 'habitability' of a given region of the galactic disk (Gonzalez *et al.*, 2001; Lineweaver *et al.* 2004); we discuss briefly that role in the next two sections.

Probability of having stars with Earth-like planets

Since the first detection of an extrasolar planet around Peg 51 (Mayor and Queloz, 1995), more than 330 stars in the solar neighbourhood have been found to host planets. The masses of those planets range from 0.005 to 18 Jupiter masses and their distances to the host star lie in the range of 0.03–6 AU. These ranges of mass and distance, however, are at present due to selection effects, resulting from current limitations of detection techniques. Continuous improvement of those techniques may well reveal the presence of smaller Earth-like planets at small distances from the host stars (as well as massive ones further away). In fact, the mass distribution of the detected planets is $dN/dM \propto M^{-1.1}$ (Butler *et al.*, 2006) and suggests that Earth-like planets should be quite common (unless an as yet-unknown physical effect truncates the distribution from the low-mass end).

The unexpected existence of 'hot Jupiters' is usually interpreted in terms of planetary migration (e.g. Papaloizou and Terquem, 2006 for a review): those gaseous giants can, in principle, be formed at a distance of several AUs from their star (where gas is available for accretion onto an already rapidly formed rocky core); their subsequent interaction with the protoplanetary disk leads, in general, to loss of angular momentum and migration of the planets inwards. On their way, those planets destroy the disk and any smaller planets that may have been formed there. Thus, the presence of hot Jupiters around some stars implies that the probability of life (at least as we know it) in the corresponding stellar system is rather small.

A key feature of the stars hosting planets is their high metallicity (compared to stars with no planets). That feature was first noticed by Gonzalez (1997), who interpreted it in terms of accretion of (metal-enriched) planetesimals onto the star. However, subsequent studies found no correlation of stellar metallicity with the depth of the convective zone of the star, invalidating that idea (e.g. Fischer and Valenti, 2005, and references therein). Thus, it appears that the high metallicity is intrinsic to the star and it presumably plays an important role in the giant-planet formation. Fischer and Valenti (2005) quantified the effect, finding that the fraction of FGK stars with hot Jupiters increases sharply with metallicity Z, as $P_{HJ} = 0.03\ (Z/Z_\odot)^2$, where Z designates Fe abundance (here, and in the rest of this chapter). A similar relationship is found with the larger sample of Souza *et al.* (2008). Note that this function has lower values and is much less steep than the one suggested by Lineweaver (2001), which reaches a value of $P_{HJ} = 0.5$ at a metallicity of $2\ Z_\odot$; the latter function is used in the calculations of Lineweaver *et al.* (2004).

The impact of metallicity on the formation of Earth-like planets is unknown at present. Metals (in the form of dust) are obviously necessary for the formation of planetesimals, but the required amount depends on the assumed scenario (and its initial conditions, e.g. size of protoplanetary disk). Inspired by the early results on extrasolar-planet host stars, several authors argued for an important role of metallicity. Thus, in Lineweaver (2001) it is suggested that the probability of forming Earth-like planets should simply have a linear dependence on metallicity $P_{FE} \propto Z$. On the other hand, Zinnecker (2003) argues for a threshold on metallicity, of the order of $0.5\ Z_\odot$, for the formation of Earth-sized planets.

However, the situation may be more complicated than the one emerging from simple analytical arguments. Recent (and yet unpublished) numerical simulations by the Bern group find that at low metallicities, the decreasing probability of forming giant planets leaves quite a lot of metals to form a significant number of Earth-sized planets (Figure 10.2). Those simulations cover a factor of 3 in metallicity (from $0.6\ Z_\odot$ to $2\ Z_\odot$) and they roughly reproduce the observations concerning P_{HJ} (Z) (see the earlier discussion); their predictions

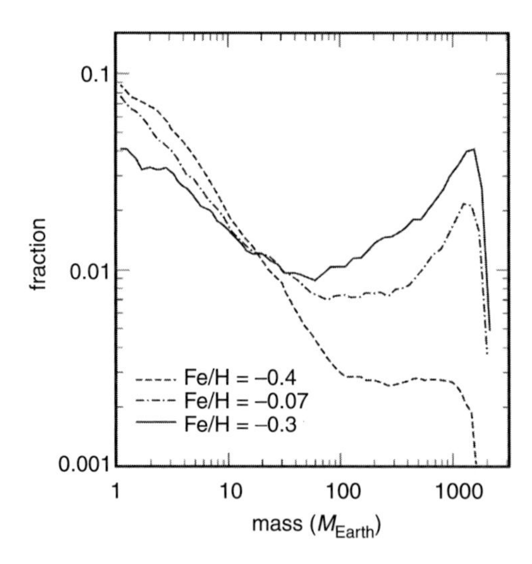

Figure 10.2. Fraction of planets around a 1-$M_\odot$ star as a function of their mass (from simulations of Mordasini *et al.*, 2008). Results are displayed for three different values of the metallicity of the star (and of the corresponding protoplanetary disk), as shown on the figure (where Fe/H is for [Fe/H]). It is seen that high metallicities favour larger fractions of massive planets (in rough agreement with observations), while at low metallicities the presence of Earth-like planets is enhanced rather than suppressed. (Courtesy Y. Alibert.)

for P_{FE} (Z) have to be confirmed by further simulations and, ultimately, by observations. They suggest, however, that the formation of Earth-like planets may be quite common, even in low-metallicity environments like the outer Galaxy and the early inner Galaxy.

Those theoretical arguments appear to be supported by the recent detection of Neptune-sized planets. Indeed, in the handful of cases of such detections, statistics (still limited to small numbers) show no correlation between frequency of detection and metallicity of host star (Souza *et al.*, 2008; Figure 10.3).

Assuming that P_{HJ} (Z) and P_{FE} (Z) are known, the probability of having stars with Earth-like planets (but not hot Jupiters, which destroy them) is simply: $P_E (Z) = P_{FE} (Z) [1 - P_{HJ} (Z)]$. However, it is clear from the previous paragraphs that neither of the terms of the right member of this equation can be accurately evaluated at present. In fact, it is not even certain that the migration of hot Jupiters inwards prohibits the existence of terrestrial planets in the habitable zone: based on recent simulations of planetary system formation, Raymond *et al.* (2006) found that 'about 34% of giant planetary systems in our sample permit an Earth-like planet of at least 0.3 M_{Earth} to form in the habitable zone'. Just for illustration purposes, we adopt in this work a set of metallicity-dependent probabilities different from the one adopted in Lineweaver *et al.* (2004), who assumed P_{FE} (Z) to be linearly dependent on Z, and P_{HJ} strongly increasing with Z: we assume that P_{FE} = const = 0.4 (so that the metallicity integrated probability is the same as in Lineweaver *et al.*, 2004) for $Z > 0.1\ Z_\odot$ and P_{HJ} from Fischer and Valenti (2005). The two sets appear in Figure 10.4, which also displays

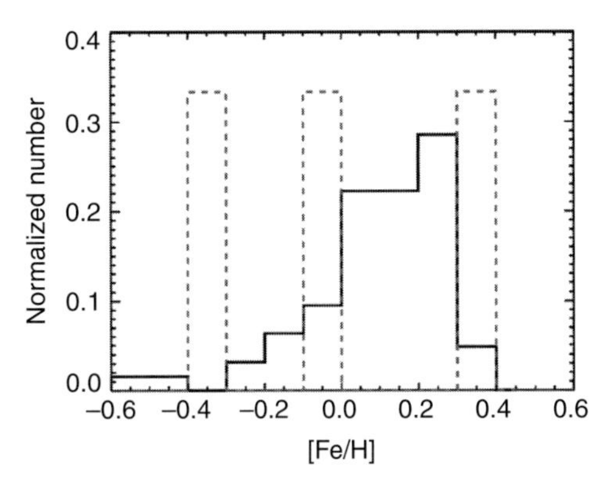

Figure 10.3. Metallicity dependence of Neptune-like hosts (dashed) and of Jupiter-like hosts (solid). Despite poor statistics in the former case, no metallicity dependence appears to exist, contrary to the latter case (from Souza *et al.*, 2008).

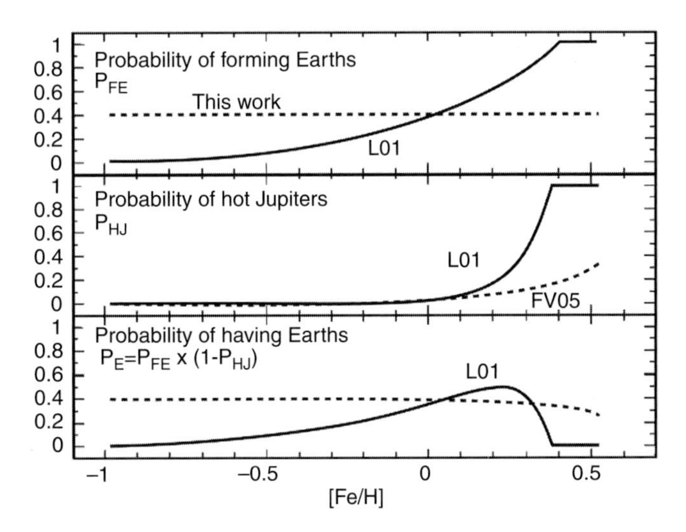

Figure 10.4. Role of metallicity of the protostellar nebula in the formation and presence of Earth-like planets around solar-mass stars, according to various estimates. Top: The probability of forming Earths has been taken as proportional to metallicity in Lineweaver (2001), while it is assumed to be quasi-independent of metallicity (at least for [Fe/H] > −1) here, after the results displayed in 'The evolution of the Milky Way disk'. Middle: The probability of forming hot Jupiters (= destroying Earth-like planets in circumstellar habitable zones) is larger in absolute value and steeper as a function of metallicity in Lineweaver (2001) than in Fischer and Valenti (2005); the latter function is adopted here. Bottom: The probability of having Earths, obtained from the two previous ones, in Lineweaver (2001) and in this work.

P_E (bottom panel): with our adopted set P_E extending to non-zero values at metallicities much lower and higher than in the work of Lineweaver *et al.* (2004). From that factor alone it is anticipated that any GHZ (found through a chemical-evolution model) will be much larger with our set of data than with the one of Lineweaver *et al.* (2004).

Probability of life surviving supernova explosions

The potential threat to complex life on Earth represented by nearby SN explosions was first studied by Ruderman (1974). He pointed out that energetic radiation from such events, in the form of hard X-rays, gamma-rays or cosmic rays, may (partially or totally) destroy the Earth's atmospheric ozone, leaving land life exposed to lethal doses of UV fluxes from the Sun. The paper went virtually completely unnoticed with just one citation for about 20 years. In the last few years, however, a large number of studies have been devoted to that topic (for reasons that are not quite clear to the author of this paper). Two factors are, perhaps, at the origin of that interest: the availability of complex models of Earth's atmospheric structure and chemistry; and the discovery of extrasolar giant planets. The latter suggests that Earth-like planets may also be abundant in the Galaxy; however, complex life on them may be a rare phenomenon because of various cosmic threats like SN explosions. This is the basic idea underlying the concepts of GHZ (Gonzalez *et al.*, 2001) and of the 'rare Earth' (Brownlee and Ward, 2000): complex life in the Universe may be rare, not for intrinsic reasons (i.e. improbability of development of life on a planet), but for extrinsic ones, related to the hostile cosmic environment.

Despite the simplicity of the idea, however, the studies devoted to the topic revealed that it is very difficult to quantify the SN threat to life (or any other cosmic threat). Studies of that kind evaluate the energetic particle flux/irradiation impinging on Earth, which could induce a significant number of gene mutations (based on our understanding of such mutations in various organisms on Earth). Knowing the intrinsic emissivity of energetic particles from a typical SN (which depends essentially on the configuration of the progenitor star and the energetics of the explosion), one may then calculate a lower limit for the distance to the SN for such fluxes/irradiations to occur. The distribution of the rates of various SN types in the Milky Way (presumably known to within a factor of a few) is then used to evaluate the frequency/probability of such events in our vicinity and elsewhere in the Galaxy.

Numerous studies in the last few years have explored several aspects of the scenario, with the use of models of various degrees of complexity for the atmosphere of the Earth (or Mars) and the transfer of high-energy radiation through it (see e.g. Ejzak *et al.*, 2006 and references therein). Thus, Smith *et al.* (2004) found that a substantial fraction (about 1%) of the high-energy radiation (X- and gamma-rays) impinging on a thick atmosphere (column density ~ 100 g/cm^2, similar to the Earth's) may reach the ground and induce biologically important mutations. Ejzak *et al.* (2006) found that the ozone depletion depends mainly on the total irradiation (for durations in the 0.1–10^8 s range) and only slightly on the received flux; this result allows one to deal with both short (gamma-ray bursts[1]) and long (SN explosions) events. They also found that the overall result depends significantly on the shape of the spectrum, with harder spectra being more harmful.

[1] Gamma-ray bursts are more powerful, but also much rarer events than supernovae. Their beamed energy makes them lethal from much larger distances than SN (several kpc in the former case, compared to a few pc in the latter) and several studies have recently been devoted to that topic (Ejzak *et al.*, 2006 and references therein). However, they are associated with extragalactic regions of low metallicity (in the few cases with available observations), implying that their frequency in the Milky Way has probably been close to zero in the past several Gyr; the formation of gamma-ray bursts (GRB) progenitors in low-metallicity environments is also favoured on theoretical grounds (e.g. Hirschi *et al.*, 2005).

Even if one assumes that such calculations are realistic (which is far from being demonstrated), it is hard to draw any quantitative conclusions about the probability of definitive sterilization of a habitable planet (which is the important quantity for the calculation of a GHZ). Even if 100% lethality is assumed for all land animals after a nearby SN explosion, marine life will certainly survive to a large extent (since UV is absorbed by a couple of metres of water). In the case of the Earth, it took just a few hundred million years for marine life to spread on to the land and evolve to dinosaurs and, ultimately, to humans; this is less than 4% of the lifetime of a G-type star.

Even if land life on a planet is destroyed from a nearby SN explosion, it may well reappear again after a few hundred Myr or so. Life displays unexpected robustness and a cosmic catastrophe might even accelerate evolution towards life forms that are presently unknown[2]. Only an extremely high frequency of such catastrophic events (say, more than one every few tens of Myr) could, perhaps, ensure permanent disappearance of complex life from the surface of a planet.

Gehrels *et al.* (2003) found that significant biological effects due to ozone depletion – that is, doubling of the UV flux on Earth's surface – may arise for SN explosions closer than $D \sim 8$ pc. Taking into account the estimated SN frequency in the Milky Way (a couple per century), the frequency of the Sun crossing spiral arms (~ 10 per Gyr), and the vertical to the galactic plane-density profile of the supernova progenitor stars (scale height $h \sim 30$ pc, comparable to the current vertical displacement of the Sun $h_{\odot} \sim 30$ pc), Gehrels *et al.* (2003) found that a SN should explode closer than 8 pc to the Sun with a frequency $f \sim 1.5$ per Gyr. Repeating their calculation with a realistic radial-density profile for SN resulting from massive stars (see profile of SFR in Figure 10.1), we find a number closer to $f \sim 1$ per Gyr. Note, however, the sensitivity of that number to the assumed critical distance ($f \propto D^3$): reducing that distance from 8 pc to 6 pc, would correspondingly reduce the frequency to 0.4 per Gyr. In any case, it appears that no such catastrophic event has occurred close to the Earth in the past Gyr or so.

In an attempt to circumvent the various unknowns, Lineweaver *et al.* (2004) quantified the risk for life represented by SN explosions by using the time-integrated rate of SN (TIR_{SN}) within 4 Gyr (see Figure 10.5). Adopting such a variable avoids considering the spiral-arm passage and allows for a rapid implementation of the SN risk factor in a galactic chemical-evolution model; however, the assignment of probabilities is rather arbitrary. For instance, it is assumed that $P < 1$ for $TIR_{SN} = 1$ local unit, a rather strange assumption in view of the fact that life on Earth has survived the local SN rate. Also, it is assumed that $P = 0$ for $TIR_{SN} = 4$ local units, implying that if the local SN time-integrated rate were just 4 times larger than what it has actually been, then our planet would have been permanently unable to host complex life. But such an increase in the SN rate would increase the frequency of 'lethal SN' (closer than 8 pc, according to Gehrels *et al.*, 2003), from 1 per Gyr to 4 per Gyr, still leaving 250 Myr between lethal events, which is probably more than

[2] An illustration is offered by the 'Cambrian explosion', with a myriad of complex life forms appearing less than about 40 Myr after the last 'snowball Earth' (which presumably occurred in the Neoproterozoic Era, 750 Myr ago).

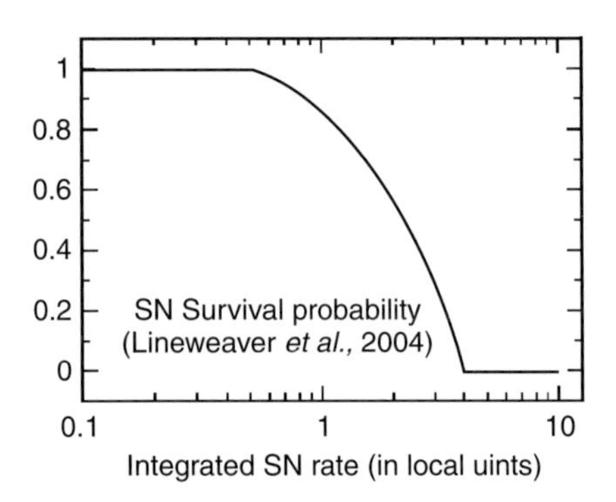

Figure 10.5. The probability that complex life on an Earth-like planet survives a nearby supernova explosion, according to Lineweaver *et al.* (2004). It is plotted as a function of the time-integrated supernova rate $TIR_{SN}(t) = \int_t^{t+4\,Gyr} SNR(t')\,dt'$; the latter is expressed in units of the corresponding time-integrated SN rate in the solar neighbourhood and in the last 4 Gyr. The probability is quite arbitrarily assumed to be $P_{SN} = 1$ for integrated SN rates less than 0.5 and $P = 0$ for integrated SN rates larger than 4 (a GHZ as large as the whole Galaxy?).

enough for a 'renaissance' of land life. Assuming $P = 0$ for substantially larger TIR values (say, for $TIR_{SN} = 10$ local units) seems a safer bet, but it still constitutes a very imperfect evaluation of the SN threat. For comparison purposes with Lineweaver *et al.* (2004), the 'SN risk factor' of Figure 10.5 is also adopted here.

A GHZ as large as the whole Galaxy?

Using the chemical evolution model of 'The evolution of the Milky Way disk', the probabilities for forming Earth-like planets that survive the presence of hot Jupiters from 'Probability of having stars with Earth-like planets' and the risk factor from SN explosions from 'Probability of life surviving supernova explosions', we have calculated the distribution of stars potentially hosting Earth-like planets with complex life in the Milky Way.

The distribution of various probabilities as a function of galactocentric radius appears in Figure 10.6 for five different epochs of the Galaxy's evolution, namely 1, 2, 4, 8 and 13 Gyr. Due to the inside-out formation scheme of the Milky Way, all metallicity-dependent probabilities peak early on in the inner disk and progressively increase outwards. The time-integrated SN rate is always higher in the inner than in the outer Galaxy. However, at late times its absolute value in the inner disk is smaller than at early times; as a result, the corresponding probability for surviving SN explosions, which is null in the inner disk at early times, becomes quite substantial at late times.

The bottom line is depicted in the bottom right panel of Figure 10.6. It displays the fraction of all stars having Earths (but no hot Jupiters) that survived supernova explosions. It is calculated as: $F(R,t) = \int_0^t SFR(R,t')\,P_E(R,t')\,P_{SN}(R,t')\,dt' / \int_0^t SFR(R,t')\,dt'$, where SFR (R,t′) is the star-formation rate at galactocentric distance R and time t′ and $P_E(R,t')$ and $P_{SN}(R,t')$ are the corresponding probabilities (respectively, Earth's without hot Jupiters and

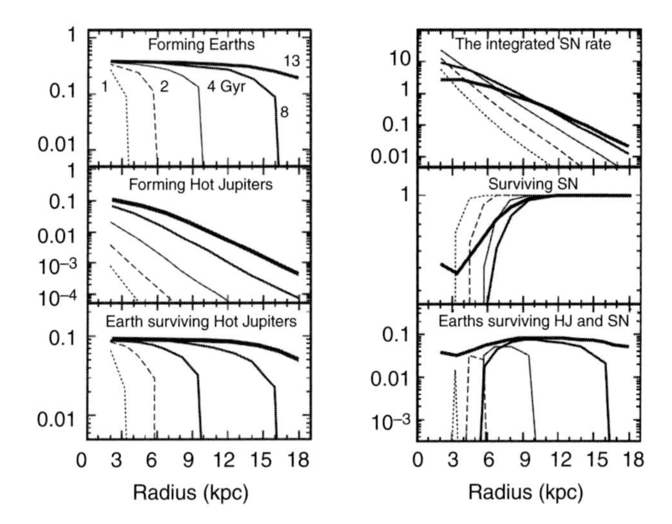

Figure 10.6. Probabilities of various events as a function of galactocentric distance at five different epochs of the Milky Way's evolution: 1, 2, 4, 8 and 13 Gyr, respectively (the latter is displayed with a thick curve in all figures). The probabilities of forming Earths P_{FE} and hot Jupiters P_{HJ} and of actually having Earths $P_E (Z) = P_{FE} (Z) [1 - P_{FE} (Z)]$ (left part of the figure) depend on the metallicity evolution (Figure 10.4), while the probability of life-bearing planets surviving SN explosions is obtained with the criteria of Figure 10.5. Finally, the overall probability for Earth-like planets with life (bottom right) defines a ring in the MW disk, progressively migrating outwards; that 'probability ring' is quite narrow at early times, but fairly extended today and peaking at about 10 kpc.

surviving supernova), as depicted in the bottom left and middle right panels of Figure 10.6. That fraction F (R,t) has indeed a ring-like shape, quite narrow early on (as it should, since there is basically no star formation in the outer disk at that time) and progressively 'migrating' outwards and becoming more and more extended. Today, that 'ring' peaks at ~ 10 kpc, but it is quite large, since even in the inner disk (the molecular ring) the SN risk factor is not much larger (just a factor of a few) than in the solar neighbourhood. Obviously, the GHZ extends to the quasi-totality of the galactic disk today. Figure 10.7 (left panel) shows the same result as Figure 10.6 (bottom right) in a different way (space–time diagram), comparable with Figure 10.4 of Lineweaver *et al.* (2004).

In the right panel of Figure 10.7, the Earth-hosting star fraction of the left panel is multiplied by the corresponding number of stars created up to time t. Since there are more stars in the inner disk, this latter quantity peaks in the inner Galaxy (in other words: the left panel displays the relative probability of having complex life around one star at a given position, while the right panel displays the relative probability of having complex life per unit volume, or surface density, in a given position). Thus, despite the high risk from SN early on in the inner disk, that place becomes relatively 'hospitable' later. Because of the large density of stars in the inner disk, it is more pertinent to seek complex life there than in the outer disk; the solar neighbourhood (at 8 kpc from the centre) is not particularly privileged in that respect.

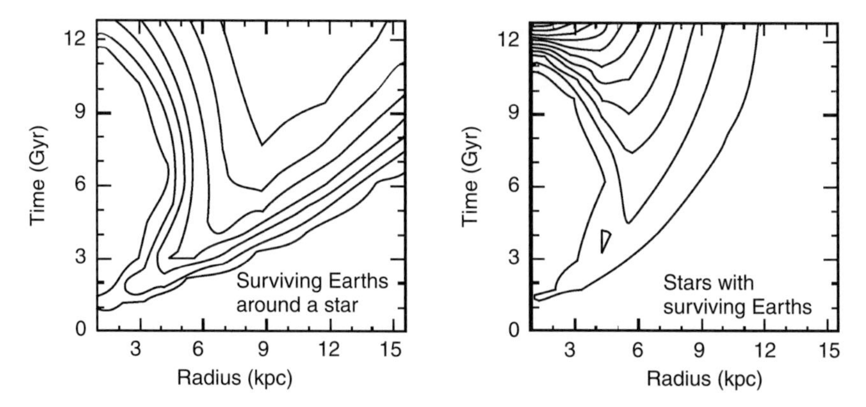

Figure 10.7. Left: Probability isocontours in the time vs. galactocentric distance plane of having around one star an Earth-like planet that survived SN explosions and migration of hot Jupiters; as in the previous figure, the probability peaks today at about 10 kpc from the galactic centre. Right: Multiplying that probability by the corresponding surface density of stars (much larger in the inner disk than in the outer one, because of the exponentially decreasing stellar profile), one finds that it is more likely to find a star with an Earth-like planet that survived SN explosions in the inner Galaxy, rather than in the outer one. In both figures, isocontours are on a linear scale, separated by intervals of 5%, i.e. the innermost ones are at 95% and the outmost ones at 5% levels of the corresponding maximum values.

The results obtained in this section depend heavily on the assumptions made in 'The evolution of the Milky Way disk' and 'Probability of having stars with Earth-like planets', which are far from being well founded at present. It is clear that the contours of a GHZ in the Milky Way cannot be, even approximately, defined, either in space or in time; it may well be that most of our Galaxy is (and has been) suitable for life. Thus, the concept of a GHZ may have little or no significance at all. It should be considered, at best, as a broad framework, allowing us to formulate our thoughts/educated guesses/knowledge about a very complex phenomenon such as life (origin, development and survival) in the Milky Way.

References

Boissier, S. and Prantzos, N. (1999). Chemo-spectrophotometric evolution of spiral galaxies: i. The model and the Milky Way. *Monthly Notices of the Royal Astronomical Society*, **307**, 857–76.

Brownlee, D. and Ward, P. (2000). *Rare Earth: Why Complex Life Is Uncommon in the Universe*. New York: Springer-Verlag.

Butler, R. P., Wright, J. T., Marcy, G. W., Fischer, D. A., Vogt, S. S., Tinney, C. G., Jones, H. R. A., Carter, B. D., Johnson, J. A., McCarthy, C. and Penny, A. J. (2006). Catalog of nearby exoplanets. *The Astrophysical Journal*, astro-ph/0607493.

Chyba, C. F. and Hand, K. P. (2005). Astrobiology: the study of the living Universe. *Annual Review of Astronomy and Astrophysics*, **43**, 31–74.

Daflon, S. and Cunha, K. (2004). Galactic metallicity gradients derived from a sample of OB stars. *The Astrophysical Journal*, **617**, 1115–26.

Ejzak, L., Melott, A., Medvedev, M. and Thomas, B. (2006). Terrestrial consequences of spectral and temporal variability in ionizing photon events. *The Astrophysical Journal*, **654**, 373, astro-ph/0604556.
Fischer, D. and Valenti, J. (2005). The planet-metallicity correlation. *The Astrophysical Journal*, **622**, 1102–17.
Gaidos, E. and Selsis, F. (2007). From protoplanets to protolife: the emergence and maintenance of life. In *Protostars and Protoplanets v*, eds. V. B. Reipurth, D. Jewitt and K. Keil. Tuscon, Arizona: University of Arizona Press.
Gehrels, N., Laird, C. M., Jackman, C. H., Cannizzo, J. K., Mattson, B. J. and Chen, W. (2003). Ozone depletion from nearby supernovae. *The Astrophysical Journal*, **585**, 1169–76.
Gonzalez, G. (1997). The stellar metallicity-giant planet connection. *Monthly Notices of the Royal Astronomical Society*, **285**, 403–12.
Gonzalez, G. (2005). Habitable zones in the Universe. *Origins of Life and Evolution of Biospheres*, **35**, 555–606.
Gonzalez, G., Brownlee, D. and Ward, P. (2001). The galactic habitable zone: galactic chemical evolution. *Icarus*, **152**, 185–200.
Goswami, A. and Prantzos, N. (2000). Abundance evolution of intermediate mass elements (C to Zn) in the Milky Way halo and disk. *Astronomy and Astrophysics*, **359**, 191–212.
Haywood, M. (2006). Revisiting two local constraints of the galactic chemical evolution. *Monthly Notices of the Royal Astronomical Society*, **371**(4), 1760–76.
Hirschi, R., Meynet, G. and Maeder, A. (2005). Stellar evolution with rotation. xiii. Predicted GRB rates at various Z. *Astronomy and Astrophysics*, **443**, 581–91.
Hou, J., Prantzos, N. and Boissier, S. (2000). Abundance gradients and their evolution in the Milky Way disk. *Astronomy and Astrophysics*, **362**, 921–36.
Huang, S. (1959). Occurrence of life in the Universe. *American Scientist*, **4**, 393–402.
Lineweaver, C. (2001). An estimate of the age distribution of terrestrial planets in the Universe: quantifying metallicity as a selection effect. *Icarus*, **151**, 307–13.
Lineweaver, C., Fenner, Y. and Gibson, B. K. (2004). The galactic habitable zone and the age distribution of complex life in the Milky Way. *Science*, **303**(5654), 59–62.
Maciel, W., Lago, L. and Costa, R. (2006). An estimate of the time variation of the abundance gradient from planetary nebulae. *Astronomy and Astrophysics*, **453**, 587–93.
Mayor, M. and Queloz, D. (1995). A Jupiter-mass companion to a solar-type star. *Nature*, **378**, 355–9.
Mordasini, C., Alibert, Y., Benz, W. and Naef, D. (2008). Giant planet formation by core accretion. Extreme Solar Systems. In *ASP Conference Series*, eds. D. Fischer, F. A. Rasio, S. E. Thorsett and A. Wolszczan, vol. 398, p. 235.
Naab, T. and Ostriker J. (2006). A simple model for the evolution of disc galaxies: the Milky Way. *Monthly Notices of the Royal Astronomical Society*, **366**, 899.
Papaloizou, J. and Terquem, C. (2006). Planet formation and migration. *Reports on Progress in Physics*, **9**, 119–80.
Prantzos, N. and Aubert, O. (1995). On the chemical evolution of the galactic disk. *Astronomy and Astrophysics*, **302**, 69–85.
Raymond, S., Mandell, A. and Sigursson, S. (2006). Exotic Earths: forming habitable planets with giant planet migration. *Science*, **313**, 1413–16.
Ruderman, M. (1974). Possible consequences of nearby supernova explosion for atmospheric ozone and terrestrial life. *Science*, **184**, 1079–81.

Smith, D. S., Scalo, J. and Wheeler, J. C. (2004). Transport of ionizing radiation in terrestrial-like exoplanet atmospheres. *Icarus*, **171**, 229–53.

Souza, S. G., Santos, N., Mayor, M., Udry, S., Casagrande, L., Israelian, G., Pepe, F., Queloz, D. and Monteiro, M. J. P. F. G. (2008). Spectroscopic parameters for 451 stars in the HARPS GTO planet search program: stellar [Fe/H] and the frequency of exo-Neptunes. *Astronomy and Astrophysics*, **487**, 373–81.

Zinnecker, H. (2003). Chances for Earth-like planets and life around metal-poor stars. In *Bioastronomy 2002: Life Among the Stars*, eds. R. Norris and F. Stootman. Proceedings of IAU Symposium 213. San Francisco: Astronomical Society of the Pacific, p. 45.

11

The young Sun and its influence on planetary atmospheres

Manuel Güdel and James Kasting

The young Sun: activity and radiation

Magnetic activity in the young Sun

The Sun's magnetic activity has steadily declined throughout its main-sequence lifetime. This is an immediate consequence of the declining dynamo as a star spins down by losing angular momentum through its magnetized wind. Along with the decline in magnetic activity, solar radiation ultimately induced by the magnetic fields declined as well, and hence the short-wavelength radiative input into planetary atmospheres diminished with time. (By contrast, solar radiation at visible wavelengths increased with time, as discussed below.) Similarly, the magnetically guided solar wind and high-energy particle fluxes were very likely to be different in the young solar environment compared to present-day conditions. A closer understanding of the magnetic behaviour of the young Sun is therefore pivotal for further modelling of young planetary atmospheres, their chemistry, heating and erosion.

Magnetic activity expresses itself in a variety of features, including dark photospheric, magnetic spots, photospheric faculae and chromospheric plage producing optical and ultraviolet excess radiation, and – most dramatically – magnetically confined coronae containing million-degree plasma that emits extreme-ultraviolet and X-ray emission. Occasional magnetic instabilities (flares) and shocks both in the corona and in interplanetary space accelerate particles to energies much beyond 1 MeV; related electromagnetic radiation (e.g. from collisions) is emitted in the hard X-ray and gamma-ray range (Lin *et al.*, 2002).

While important meteoritic evidence indicates that the Sun was magnetically much more active in its youth than at present (e.g. Caffee *et al.*, 1987), the study of the Sun's past magnetic behaviour is best pursued by looking at nearby 'solar analogues' with known ages, such as the well-studied 'Sun in Time' sample covering ages of 100–7000 Myr in the optical, ultraviolet (UV), far-ultraviolet (FUV), extreme ultraviolet (EUV), X-rays and also radio waves (e.g. Dorren and Guinan, 1994; Güdel *et al.*, 1997; Guinan and Ribas, 2002; Ribas *et al.*, 2005).

The optical and ultraviolet Sun

Stellar evolution calculations indicate that the young, zero-age main-sequence Sun was bolometrically about 30% fainter than now (Gough, 1980; Gilliland, 1989; Sackmann and Boothroyd, 2003). The reason has to do with the fact that hydrogen is slowly being converted into helium in its core, raising the density and causing corresponding increases in temperature and the rate of nuclear fusion. At the same time, the optical light was more strongly modulated by rotation due to surface magnetic spots, but also due to magnetic-activity cycles that induced light variations up to 0.3 mag in the course of 2–13 year spot cycles (Messina and Guinan, 2002). The presence of excess faculae outweighs the darkening by the spot area in old solar analogues, making these stars slightly brighter in optical light during activity maxima, while in young stars, spots dominate and make them fainter (Radick *et al.*, 1990). Cycles are not always present, while irregular variations abound among very active solar-like stars (Hempelmann *et al.*, 1996). The surface coverage with spot regions may reach several per cent in such cases, although the spot complexes appear predominantly at high latitudes or in the polar region itself (Strassmeier and Rice, 1998), perhaps due to the action of internal Coriolis forces (Schüssler and Solanki, 1992).

Ultraviolet emission from more active transition regions strongly enhances the short-wavelength UV part of the spectrum, in particular in emission lines forming at temperatures around 10^5 K. This is illustrated in Figure 11.1.

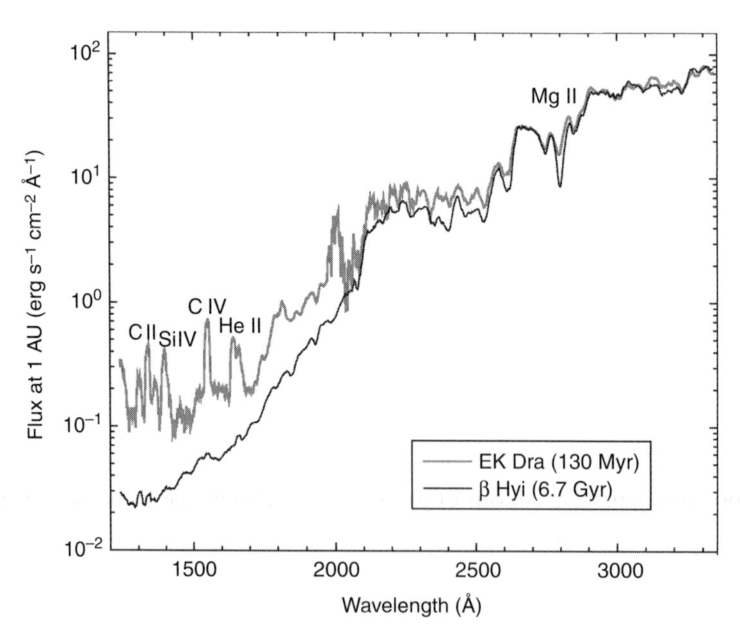

Figure 11.1. Comparison between UV irradiances at a distance of 1 astronomical unit (AU) from a young (EK Dra, grey) and an old (β Hyi, black) solar analogue (from Guinan and Ribas, 2002; reprinted with permission of ASP).

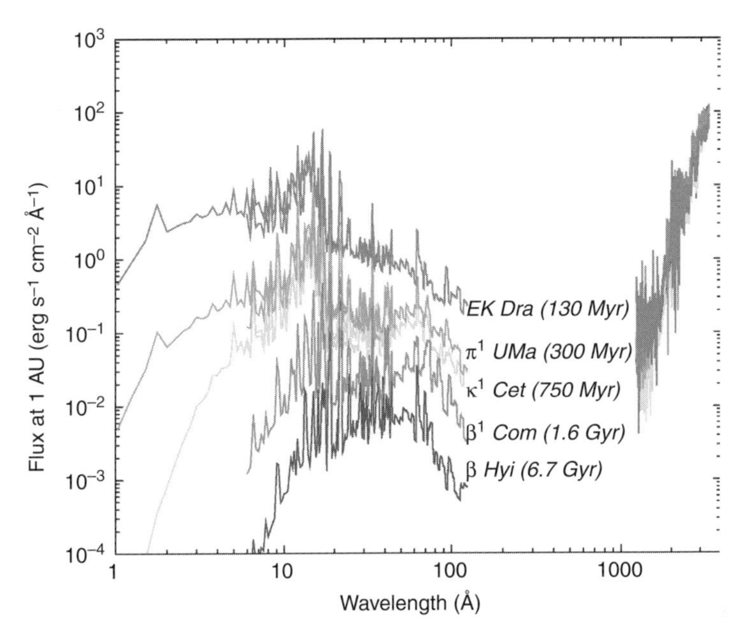

Figure 11.2. Spectral irradiance at 1 AU in the UV, EUV and X-ray ranges for various solar analogues at ages of 0.1–7 Gyr. Spectra have not been obtained in the ≈ 150–1000 Å range because of severe interstellar photoabsorption. Note the logarithmic flux scale (from Guinan and Ribas, 2002; reprinted with permission of ASP).

The extreme ultraviolet and X-ray Sun

Much more dramatic changes have occurred in the evolution of the short-wavelength radiation from the Sun. Studies of solar analogues in young stellar clusters and in the field show X-ray and EUV luminosities exceeding the present-day solar output by several hundred to approximately 1000 times at ages of order 1–100 Myr, including the T-Tauri phase of stellar evolution (Stauffer *et al.*, 1994; Preibisch *et al.*, 2005). The evolutionary changes in spectral irradiance from 100 Myr to 7 Gyr are depicted in Figure 11.2. High-pressure high-temperature plasma (T of order 10 MK) confined in magnetic fields predominates in young active stars (Giampapa *et al.*, 1996; Güdel *et al.*, 1997), producing much harder X-ray emission than the present-day Sun. The physics of coronal heating to such temperatures is unclear; an interesting possibility is heating by a high rate of flares occurring nearly continuously (refer to 'solar and stellar flares' below). 'Continuously flaring coronae' are crucially important as they also produce appreciable levels of gamma rays and high-energy particles that interact with planetary atmospheres.

The long-term evolution of solar irradiance from the optical to X-rays

The 'Sun in Time' stellar sample has been used to study the long-term changes of the solar output from optical light to soft X-rays, and thus to pinpoint its spectral irradiance. It is

Table 11.1. Enhancement factors of solar irradiance in solar history[a].

Solar age (Gyr)	Time before present (Gyr)	Enhancement in X-rays (1–20 Å)	Soft X (20–100 Å) EUV (100–360 Å) XUV (1–1180 Å)	FUV (920–1180 Å)	UV lines, transition region	UV lines, chromosphere
0.1	4.5	1600[b]	100	25	50	18
0.2	4.4	400	50	14	20	10
0.7	3.9	40	10	5	7	4
1.1	3.5	15	6	3	4	3
1.9	2.7	5	3	2	2.4	2
2.6	2.0	3	2	1.6	1.8	1.5
3.2	1.4	2	1.5	1.4	1.4	1.3
4.6	0	1	1	1	1	1

Notes:
[a] normalized to ZAMS age of 4.6 Gyr before present (Table adapted from Güdel, 2007).
[b] Large scatter possible due to unknown initial rotation period of the Sun.

remarkable that the Sun's output increased exclusively in the optical region (by 30% since the zero age main sequence or ZAMS phase) while it has declined by orders of magnitude at all shorter wavelengths as a consequence of the declining dynamo, which in turn is a consequence of rotational braking, i.e. increasing rotation period P (from about 2 to 25 days). The decays are well fitted by power laws that become *steeper toward shorter wavelengths* (see Dorren *et al.*, 1994; Güdel *et al.*, 1997; Guinan *et al.*, 2003; Ribas *et al.*, 2005; Telleschi *et al.*, 2005), where a power-law decay law for the rotation rate with age t, $\Omega \propto t^{-0.6\pm0.1}$, has been assumed (Ayres, 1997). Table 11.1 provides numerical estimates for the solar spectral irradiance at various ages.

UV line radiation from chromosphere:	L (chrom)	$\propto P^{-1.25\pm0.15} \propto t^{-0.75\pm0.1}$
UV line radiation from transition region:	L (TR,UV)	$\propto P^{-1.6\pm0.15} \propto t^{-1.0\pm0.1}$
FUV from transition region (920–1180 Å):	L (TR,FUV)	$\propto P^{-1.4} \propto t^{-0.85}$
Soft-X/EUV radiation from corona (20–360 Å):	L (EUV)	$\propto P^{-2.0} \propto t^{1.2}$
X-rays from corona (1–20 Å):	L_X	$\propto P^{-3.2} \propto t^{-1.9}$

X-ray radiation has thus suffered by far the strongest reduction (by 3 orders of magnitude) in the course of the Sun's history, initially achieving as much as 0.1% of the bolometric output, or 2–4 × 10^{30} erg s^{-1}. These trends are illustrated in Figure 11.3.

Solar and stellar flares

Flares are the most evident manifestation of magnetic-energy release in stellar atmospheres. While they may increase radiation levels – in particular at very short (EUV, X-ray)

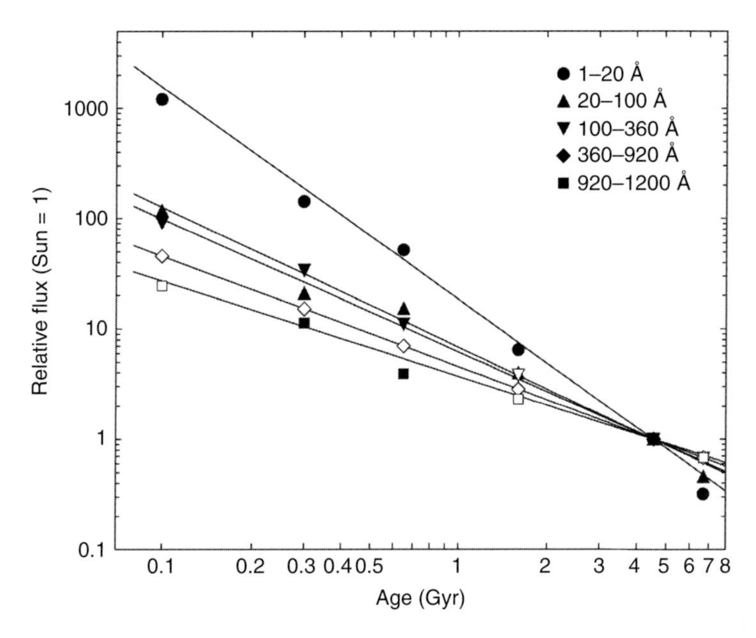

Figure 11.3. Power-law decay laws for solar irradiance in various wavelength bands, normalized to the present-day irradiance (at 4.6 Gyr). Open symbols represent inferred fluxes without direct observations (from Ribas *et al.*, 2005; reproduced by permission of the AAS).

wavelengths and in the radio regime – by up to a few orders of magnitude for up to a few hours, the statistical ensemble of all flares may also be responsible for much of the detected 'non-flaring' electromagnetic radiation from the outer stellar atmospheres. Known as the 'micro-flare hypothesis' in solar physics (e.g. Hudson, 1991), this model requires a sufficiently steep increase of the flare-occurrence rate toward *small* flares; it has found increasing support also from magnetically very active young stars (e.g. Audard *et al.*, 2000; Güdel *et al.*, 2003), thus suggesting that large numbers of high-energy particles are injected continuously into stellar environments.

The solar wind and high-energy particles

Ionized hot winds like the solar wind have not yet been directly detected in other stars; their expected radio or charge-exchange emission is beyond the reach of present-day observing facilities. However, Lyα absorption in so-called astrospheres, large-scale structures created by the interaction between the stellar wind and the interstellar medium, has provided indirect quantitative evidence for stellar winds (Linsky and Wood, 1996; Wood *et al.*, 2002, 2005). The absorption is due to 10^2–10^4 K interstellar H I piling up in the 'heliospheric' shock region. The amount of absorption in these 'hydrogen walls' should scale with the wind ram pressure. Assuming a wind velocity similar to the Sun's, the mass-loss rate can

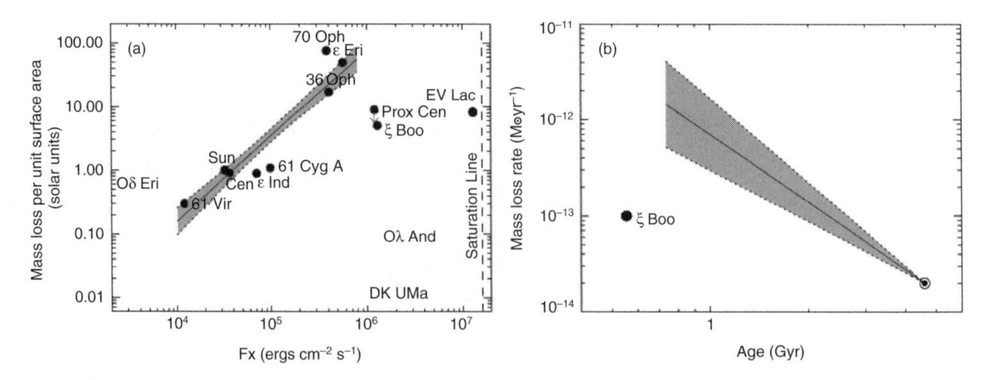

Figure 11.4. Mass-loss rates per unit surface area as a function of stellar X-ray surface flux and age, as inferred from 'astrospheric' Lyα absorption. Note the breakdown of the evolutionary trends for the most active (highest X-ray flux F_X) and youngest objects (from Wood *et al.*, 2005; reproduced by permission of the AAS).

be estimated. The wind mass-loss rate, dM_w/dt, turns out to be another magnetic activity indicator; its evolution is described by:

$$dM_w/dt \propto L_X^{1.34\pm0.18} \quad (11.1)$$

$$dM_w/dt \propto t^{-2.33\pm0.55} \quad (11.2)$$

(Wood *et al.*, 2005). These power laws are reminiscent of those applicable to coronal X-ray emission. This is perhaps not entirely surprising given that the solar wind is a 'coronal wind', heated to similar temperatures as the corona and accelerated and guided along open magnetic-field lines. However, while the above laws hold up to maximum mass-loss rates of 100 times the present solar level, a strong reduction in the wind mass-loss rate is observed for more active stars, with mass-loss rates only 10 times higher than for the present-day Sun (Wood *et al.*, 2005; Figure 11.4). It appears that wind mass loss is inhibited in these stars, perhaps due to the appearance of predominantly polar spot regions and mostly closed global magnetic-field structures.

High-energy non-thermal particles are also immersed in the solar wind. They are either ejected directly by solar coronal flares, or accelerated at interplanetary shock fronts. Measurements in other stellar environments are beyond present-day possibilities. Evidence for the presence of high-energy electrons in stellar coronae (with energies in the MeV range) is available from observations of non-thermal gyrosynchrotron emission at radio wavelengths. Like X-rays, radio emission is strongly enhanced in young active solar-like stars, pointing to flare-like magnetic processes that accelerate particles nearly continuously. Phenomenologically, X-ray and radio luminosities are correlated for the most active stars (Güdel and Benz, 1993), pointing to a physical relationship between coronal heating and particle acceleration and therefore to magnetic activity in general.

High-energy radiation and planetary atmospheres

Atmospheric erosion by thermal and non-thermal escape

High-energy radiation and particles are responsible for several processes in the upper planetary atmospheres that result in atmospheric loss. Irradiation of upper atmospheric layers by high-energy (EUV, X-ray) photons leads to atmospheric heating in the thermosphere (heights above surface of 90–500 km on Earth, 90–210 km on Mars) and eventually to thermal escape from the exosphere (layer above the thermosphere). In the exosphere, the mean free path of the particles is large; particles with thermal energies exceeding the gravitational-binding energy escape into space (Jeans escape). This is predominantly true for hydrogen atoms. If the upper atmosphere is heated to sufficiently high temperatures so that the mean particle energy exceeds the gravitational-binding energy ($kT > mv_{esc}^2/2$, where v_{esc} is the escape velocity), then the exosphere becomes unstable and escapes ('blows off') into space in an uncontrolled manner (for reviews, see Chamberlain and Hunten, 1987; Hunten *et al.*, 1989; Chassefière and Leblanc, 2004; Kulikov *et al.*, 2007). In between these two regimes, there is a stable hydrodynamic escape regime in which the atmosphere escapes in a controlled manner, but in which the pressure force and adiabatic cooling must be considered throughout a wide region of the upper atmosphere. Expanding atmospheres of this kind have been studied by Watson *et al.* (1981) for early Earth and by Kasting and Pollack (1983) for early Venus. A more recent example, in which high EUV fluxes were applied to a modern Earth-type atmosphere, has been modelled by Tian *et al.* (2008). Upper-atmosphere temperature profiles for this case are shown in Figure 11.5. The escape in this case is subsonic because the atmosphere becomes collisionless well below the point at which the flow would become supersonic. This latter regime is probably where the atmospheres of the young terrestrial planets in our Solar System spent most of their time. Alternative non-Maxwellian approaches to studying such atmospheres are also available, for example, the 13-moment approximation used by Cui *et al.* (2008) to study escape of H_2 from Titan. Combining these types of model with detailed parameterizations of upper-atmosphere chemistry and physics is a fruitful area for future research.

Further processes leading to atmospheric escape involve irradiation by high-energy (non-thermal) particles, the solar wind, and planetary magnetic fields; here, the escape process is related to microscopic non-thermal mechanisms, in particular, dissociative recombination or photochemical escape (in which the forming neutrals are energetic); ion pick-up (in which ions produced by photon or particle irradiation are dragged along by magnetic fields); ionospheric outflow (in which particles are accelerated by electric fields in the solar wind); and ion sputtering (in which newly produced atmospheric ions re-impact and eject neutral particles). For reviews, see Lundin and Barabash (2004) and Lundin *et al.* (2007). We now discuss applications to Solar-System objects.

Venus is of particular interest for comparison with the Earth. It shows a very low atmospheric water content (approximately 10^{-5} times the water content of the terrestrial oceans) but a suspiciously high deuterium-to-hydrogen ratio (120–250 times the terrestrial ratio; DeBergh *et al.*, 1991). Both observations suggest that water was initially present in large

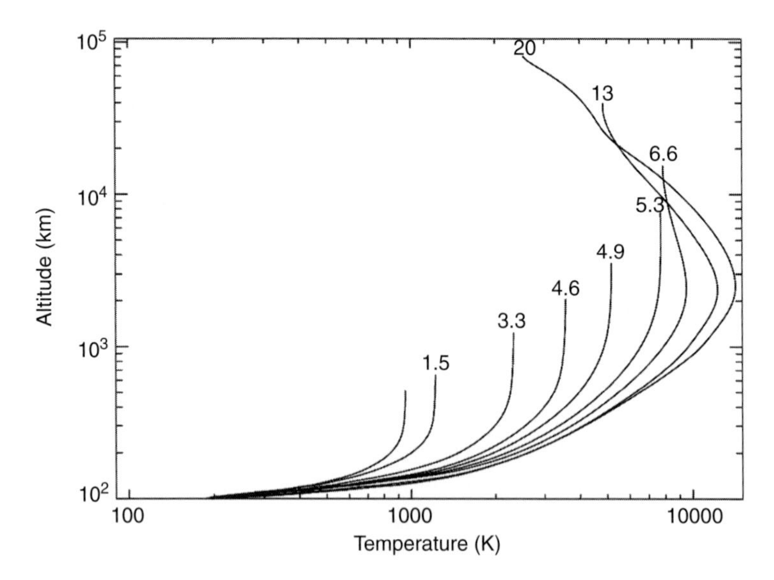

Figure 11.5. Diagram illustrating the effect of enhanced solar EUV/FUV heating on the temperature structure of Earth's (modern) atmosphere. The numbers indicate the increase in solar flux relative to today's solar mean value. At fluxes exceeding 5–6 times present, the atmosphere begins to expand hydrodynamically and cools adiabatically in its upper regions. (From Tian *et al.*, 2008.)

quantities but was removed early in the history of Venus. In relevant models, water evaporated from the initial oceans due to a more efficient greenhouse (Ingersoll, 1969; Kasting, 1988) and then dissociated in the high atmosphere due to strong solar EUV irradiation (Kasting and Pollack, 1983). The resulting hydrogen atoms escaped (Watson *et al.* 1981; Kasting and Pollack, 1983). Hydrodynamic escape conditions for hydrogen applied as long as 250 Myr after the Sun's arrival on the main sequence (Kulikov *et al.*, 2007). Oxygen could also have been removed from the atmosphere, either by being dragged along by the escaping hydrogen (Zahnle and Kasting, 1986) or through ion-pick-up by the solar wind (Lammer *et al.*, 2006; Chassefière, 1997), although the required very strong solar wind with 3–4 orders of magnitude more mass loss than at present has been questioned based on indirect observations (Wood *et al.*, 2005; see 'The solar wind and high energy particle' above).

Mars shows geologic evidence for a warm and wet climate in its early history, with liquid water abundantly present on the surface (Carr, 1986). A strong greenhouse was required to keep temperatures above the freezing point (Chassefière and Leblanc, 2004), especially considering the lower solar flux in early times (see 'The young Sun: activity and radiation'). The precise cause of the greenhouse warming on early Mars is still not fully understood (Kasting, 1991; Forget and Pierrehumbert, 1997), and some authors have suggested that early Mars was warmed only transiently by impacts (Segura *et al.*, 2002). Two factors may have accelerated atmospheric loss on Mars: its much lower gravitational potential and the lack of a strong magnetosphere (at least after the initial few 100 Myr;

Acuña *et al.*, 1998), which allowed the solar wind to interact with the upper atmosphere directly. Hydrodynamic-escape conditions should have applied for hydrogen (Kulikov *et al.*, 2007) and possibly for heavier gases (C, N and O) as well (Tian *et al.*, 2009). Oxygen may also have been lost by the non-thermal processes of sputtering and pick-up (e.g. Lammer *et al.*, 2003) and by reaction with surface material. Water may have been lost by photodissociation and escape and also by formation of ice within the planet's crust. Some water may still be present as a deep subsurface reservoir.

For Earth, one would expect a situation somewhere in between those of Venus and Mars. However, a large water reservoir has been kept intact, possibly thanks to the strong terrestrial magnetosphere that prevented the solar wind from directly interacting with the atmosphere (Kulikov *et al.*, 2007; Lundin *et al.*, 2007). However, the fundamental process that led to the stability of Earth's oceans was the formation of an efficient tropopause cold trap in Earth's atmosphere (Ingersoll, 1969; Hunten *et al.*, 1989). This limited the amount of water vapour in Earth's upper atmosphere, thereby restricting the rate at which hydrogen could be lost to space. That said, there is evidence for early hydrodynamic escape from the terrestrial atmosphere, as suggested by the depletion of light noble gases (e.g. Zahnle and Kasting, 1986). This escape may have occurred during the process of planetary accretion, as large impacts created transient dense steam atmospheres, temporarily eliminating the tropopause cold trap. A ten-fold increase in the EUV flux is sufficient to induce high escape rates for H and several other atmospheric constituents (Kulikov *et al.*, 2007). Some aspects of hydrogen escape from early Earth may have resembled the escape of H_2 from Titan's CH_4/H_2-rich atmosphere (Cui *et al.*, 2008, and references therein).

Propagation of ionizing radiation through planetary atmospheres

Ionizing stellar radiation may be directly relevant for the formation and evolution of life forms on terrestrial planets. Although most of the present-day high-energy solar radiation is absorbed in the high atmosphere of the Earth, the much stronger irradiation regime in the young Solar System may have increased the radiation dose in lower planetary atmospheres. For realistic 'habitable' atmospheres, direct transmission of high-energy radiation (EUV, X-rays and gamma-rays, essentially all radiation shortward of 200 nm) is negligible (Smith *et al.* 2004; Cnossen *et al.*, 2007), but in the 200–320 nm range, the surface irradiation level may have been several orders of magnitude higher than for present-day solar conditions (Kasting, 1987; Cnossen *et al.*, 2007). Figure 11.6 shows the expected surface flux of UV in this range as a function of atmospheric O_2 concentration. This radiation is further enhanced by strong aurora-like UV showers that are produced by molecular excitation by secondary electrons in the high atmosphere (Smith *et al.*, 2004). The transmitted energy in the re-radiated UV photons may amount to a minor percentage of the incident X-ray/gamma-ray energy depending on the atmospheric composition, density and the amount of O_2/O_3 shielding (Smith *et al.*, 2004). The transmitted UV radiation should have been relevant for the formation and evolution of life forms on Earth in the first Gyr (at least).

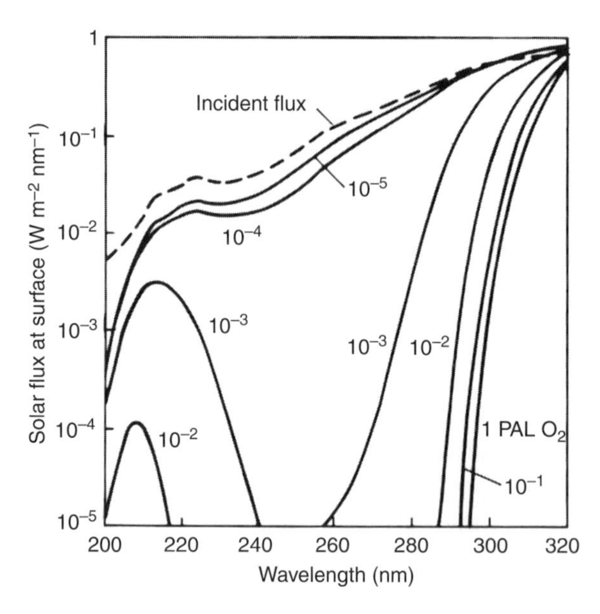

Figure 11.6. Solar flux at the surface of the Earth for different assumed concentrations of atmospheric O_2. The O_2 concentration is in PAL (times the 'present atmospheric level'). The dashed curve gives the UV flux at the top of the atmosphere (for present-day solar output). Present-day concentrations of N_2 and CO_2 were assumed in the calculation. (From Kasting, 1987.)

Atmospheric chemistry induced by high-energy radiation and particles

The increased flux of short-wavelength UV and X-ray radiation from the young Sun should also have affected atmospheric chemistry on the early terrestrial planets. This subject area has not been explored in detail in models, largely because of the lack of data with which to compare during the first few hundred million years of Solar-System history. For the Earth, two particular problems deserve mention here. The first concerns the rate of production of chemical compounds that may have been involved in the origin of life. As discussed elsewhere in this volume, some (but not all) theories of the origin of life make use of compounds that were produced within Earth's atmosphere and oceans. For example, Miller–Urey synthesis involves production of amino acids and other biologically interesting compounds from spark-discharge experiments, simulating lightning in a highly reduced primitive atmosphere (Miller, 1953). That particular model is currently out of favour because the early atmosphere is considered to have been only weakly reduced (Walker, 1977; Kasting, 1993). (But see Hashimoto *et al.*, 2007 and Schaefer and Fegley, 2007 for attempts to revert to Miller–Urey-type models.) By this, we mean that it consisted largely of N_2 and CO_2 (and a little H_2), as opposed to NH_3 and CH_4. But it may still have been possible to produce plausible prebiotic compounds in such an atmosphere, as discussed by Kasting *et al.* (1998). Nucleotides, for example, consist of various bases (e.g. guanine, $C_5H_5N_5O$) attached to a sugar (ribose, $C_5H_{10}O_5$), and linked together with phosphate (PO_4^{3-}). Ribose can, in principle, be made from formaldehyde, H_2CO, through the

Formose reaction (although the steps required to get specifically ribose, as opposed to a mix of various sugars, are not understood). Formaldehyde can be readily synthesized in a weakly reduced atmosphere through the reaction sequence:

$$CO_2 + h\nu \rightarrow CO + O$$
$$H_2O + h\nu \rightarrow H + OH$$
$$H + CO + M \rightarrow HCO + M$$
$$HCO + HCO \rightarrow H_2CO + CO$$

Here, 'M' is a third molecule needed to carry off the excess energy of the reaction. The photolysis reactions that initiate this sequence occur at relatively long wavelengths ($\lambda \sim 200$ nm), and so plenty of photons are available even from the modern UV-weak Sun, but much higher levels should have applied to the young Earth (see 'The long-term evolution of solar irradiance from the optical to x-rays').

Obtaining the starting material for the nucleic acid bases (and for amino acids) is more challenging. Guanine, for example, can be seen to be composed of five molecules of HCN, hydrogen cyanide. Other nucleic acid bases and amino acids also contain C≡N bonds. Forming C≡N bonds in a weakly reduced atmosphere is difficult because the carbon comes from CO_2 and the nitrogen comes from N_2. Shock heating of CO_2 and N_2 (i.e. Miller–Urey-type synthesis) produces NO, not HCN (Chameides and Walker, 1981). The most plausible mechanism for forming HCN in such an atmosphere is that suggested by Zahnle (1986). The reaction sequence is as follows:

$$N_2 + h\nu \rightarrow N_2^+ + e$$
$$N_2^+ + e \rightarrow N + N$$
$$CH_4 + h\nu \rightarrow CH_3 + H$$
$$CH_3 + N \rightarrow HCN + H_2$$

The first two reactions occur high in the atmosphere where ionizing radiation is present. Radiation shortward of ~ 90 nm is required to ionize N_2. This, of course, is precisely the type of short-wavelength radiation that should have been emitted in large quantities by the active young Sun. So, high solar activity may have helped to enable the formation of HCN, and thence the precursors to amino acids and nucleic acid bases. Some methane is required for this pathway to succeed. The methane in this model would have been supplied by water–rock interactions at Earth's surface, perhaps within hydrothermal vents. Even today, methane is thought to be produced by serpentinization reactions within the vent systems (Kelley *et al.*, 2001, 2005). In Zahnle's model, N atoms flowing down from the ionosphere encounter CH_3 radicals produced in the stratosphere, and the resulting HCN diffuses downward and is removed by rainout in the troposphere.

A second example of why the enhanced flux of short-wavelength radiation may have been important concerns the post-biotic Earth. If methanogenic bacteria (methanogens) evolved early, as seems likely, then methane may have been an important atmospheric constituent on the Archean Earth 2.4–3.8 Gyr ago (Kharecha *et al.*, 2005; Haqq-Misra *et al.*, 2008). If the CH_4/CO_2 ratio was ≥ 0.1, which is possible under many circumstances, then photolysis of CH_4 can lead to the formation of Titan-like organic haze (Pavlov *et al.*, 2001; Domagal-Goldman *et al.*, 2008). The wavelength cut-off for methane photolysis is at 145 nm; hence, the solar EUV flux is quite critical in determining the efficiency of the haze-formation process. The early Earth should therefore have been strongly influenced by the activity of the young Sun. The same thought could apply to Earth-like planets orbiting other young solar-type stars.

Conclusions

Habitability of planets and the potential evolution of biological activity are strongly related to the radiative output and mass loss of the central star. We have learned both from studies in the Solar System (geological studies and investigations of meteorites and planetary atmospheres) and statistical studies of stellar samples that the young Solar System was subject to conditions fundamentally different from those we see at present. On the one hand, the bolometric luminosity of the young (near-zero age main-sequence) Sun was 30% below present-day levels, requiring efficient greenhouses to explain the evidently mild climate on young Earth and Mars and to make the formation of life on Earth possible. On the other hand, the short-wavelength (UV, EUV, X-ray and gamma-ray) fluxes were much enhanced in the younger Sun owing to its much higher level of magnetic activity as a consequence of its faster rotation. EUV and X-ray irradiation at levels hundreds to about a thousand times higher than at present led to various processes of atmospheric loss, in particular loss of water from Venus and Mars, although magnetospheric protection and the development of a cold trap kept a strong water reservoir intact on Earth. Radiation also had a fundamental impact on the formation of basic chemical compounds required for the formation of life, such as ribose and starting materials for nucleic acid bases and amino acids. Again, enhanced ultraviolet and ionizing EUV and X-ray radiation were amply present to support the requisite reactions. The same radiation may, on the other hand, have been enhanced also at the terrestrial-surface level, requiring early life forms to have evolved in regions shielded from this radiation, such as in the deep oceans.

References

Acuña, M. H., Connerney, J. E. P., Wasilewski, P., Lin, R. P., Anderson, K. A., Carlson, C. W., McFadden, J., Curtis, D. W., Mitchell, D., Reme, H., Mazelle, C., Sauvaud, J. A., D'Uston, C., Cros, A., Medale, J. L., Bauer, S. J., Cloutier, P., Mayhew, M., Winterhalter, D. and Ness, N. F. (1998). Magnetic field and plasma observations at Mars: initial results of the Mars Global Surveyor mission. *Science*, **279**, 1676–80.

Audard, M., Güdel, M., Kashyap, V. L. and Drake, J. J. (2000). Extreme-ultraviolet flare activity in Tate-type stars. *The Astrophysical Journal*, **541**, 396–409.

Ayres, T. R. (1997). Evolution of the solar ionizing flux. *Journal of Geophysical Research – Planets*, **102**, 1641–52.

Caffee, M. W., Hohenberg, C. M., Swindle, T. D. and Goswami, J. N. (1987). Evidence in meteorites for an active early Sun. *The Astrophysical Journal*, **313**, L31–5.

Carr, M. H. (1986). Mars: a water-rich planet? *Icarus*, **68**, 187–216.
Chamberlain, J. W. and Hunten, D. M. (1987). *Theory of Planetary Atmospheres.* Orlando: Academic Press.
Chameides, W. L. and Walker, J. C. G. (1981). Rates of fixation by lightning of carbon and nitrogen in possible primitive terrestrial atmospheres. *Origins of Life*, **11**, 291–302.
Chassefière, E. (1997). Loss of water on the young Venus: the effect of a strong primitive solar wind. *Icarus*, **126**, 229–32.
Chassefière, E. and Leblanc, F. (2004). Mars atmospheric escape and evolution; interaction with the solar wind. *Planetary and Space Science*, **52**, 1039–58.
Cnossen, I., Sanz-Forcada, J., Favata, F., Witasse, O., Zegers, T. and Arnold, N. F. (2007). Habitat of early life: solar X-ray and UV radiation at Earth's surface 4–3.5 billion years ago. *Journal of Geophysical Research – Planets*, **112**, E02008.
Cui, J., Yelle, R. V. and Volk, K. (2008). Distribution and escape of molecular hydrogen in Titan's thermosphere and exosphere. *Journal of Geophysical Research – Planets*, **113**, 16, E10004.
DeBergh, C., Bezard, B., Owen, T., Crisp, D., Maillard, J. P. and Lutz, B. L. (1991). Deuterium on Venus – observations from Earth. *Science*, **251**, 547–9.
Domagal-Goldman, S. D., Kasting, J. F., Johnston, D. T. and Farquhar, J. (2008). Organic haze, glaciations and multiple sulphur isotopes in the mid-Archean Era. *Earth and Planetary Science Letters*, **269**, 29–40.
Dorren, J. D. and Guinan, E. F. (1994). HD 129333: The Sun in its infancy. *The Astrophysical Journal*, **428**, 805–18.
Dorren, J. D., Guinan, E. F. and DeWarf, L. (1994). The decline of solar magnetic activity with age. In *Eighth Cambridge Workshop on Cool Stars, Stellar Systems, and the Sun*, ed. J.-P. Caillault. San Francisco: ASP, pp. 399–401.
Forget, F. and Pierrehumbert, R. T. (1997). Warming early Mars with carbon dioxide clouds that scatter infrared radiation. *Science*, **278**, 1273–6.
Giampapa, M. S., Rosner, R., Kashyap, V., Fleming, T. A., Schmitt, J. H. M. M. and Bookbinder, J. A. (1996). The coronae of low-mass dwarf stars. *The Astrophysical Journal*, **463**, 707–25.
Gilliland, R. L. (1989). Solar evolution. *Global and Planetary Change*, **1**, 35–55.
Gough, D. O. (1981). Solar interior structure and luminosity variations. *Solar Physics*, **74**, 21–34.
Güdel, M. (2007). X-ray and radio emission from stellar coronae. *Memorie della Società Astronomica Italiana*, **78**, 285–292.
Güdel, M. and Benz, A. O. (1993). X-ray/microwave relation of different types of active stars. *The Astrophysical Journal*, **405**, L63–6.
Güdel, M., Guinan, E. F. and Skinner, S. L. (1997). The X-ray Sun in time: a study of the long-term evolution of coronae of solar-type stars. *The Astrophysical Journal*, **483**, 947–60.
Güdel, M., Audard, M., Kashyap, V. L., Drake, J. J. and Guinan, E. F. (2003). Are coronae of magnetically active stars heated by flares? II. Extreme ultraviolet and X-ray flare statistics and the differential emission measure distribution. *The Astrophysical Journal*, **582**, 423–42.
Guinan, E. F. and Ribas, I. (2002). Our changing Sun: the role of solar nuclear evolution and magnetic activity on Earth's atmosphere and climate. In *The evolving Sun and its Influence on Planetary Environments*, eds. B. Montesinos, A. Gimenez and E. F. Guinan. San Francsico: ASP, pp. 85–106.
Guinan, E. F., Ribas, I. and Harper, G. M. (2003). Far-ultraviolet emissions of the Sun

in time: probing solar magnetic activity and effects on evolution of paleoplanetary atmospheres. *The Astrophysical Journal*, **594**, 561–72.

Haqq-Misra, J. D., Domagal-Goldman, S. D., Kasting, P. J. and Kasting, J. F. (2008). A revised, hazy methane greenhouse for the early Earth. *Astrobiology*, **8**, 1127–37.

Hashimoto, G'L., Abe, Y. and Sugita, S. (2007). The chemical composition of the early terrestrial atmosphere: formation of a reducing atmosphere from CI-like material. *Journal of Geophysical Research – Planets*, **112**, E05010.

Hempelmann, A., Schmitt, J. H. M. M. and Stepien, K. (1996). Coronal X-ray emission of cool stars in relation to chromospheric activity and magnetic cycles. *Astronomy and Astrophysics*, **305**, 284–95.

Hudson, H. S. (1991). Solar flares, microflares, nanoflares, and coronal heating. *Solar Physics*, **133**, 357–69.

Hunten, D. M., Donahue, T. M., Walker, J. C. G. and Kasting, J. F. (1989). Escape of atmospheres and loss of water. In *Origin and Evolution of Planetary and Satellite Atmospheres*, eds. S. K. Atreya, B. J. Pollack and M. S. Matthews. Tucson: University of Arizona, pp. 386–422.

Ingersoll, A. P. (1969). The runaway greenhouse: a history of water on Venus. *Journal of the Atmospheric Sciences*, **26**, 1191–8.

Kasting, J. F. (1987). Theoretical constraints on oxygen and carbon dioxide concentrations in the Precambrian atmosphere *Precambrian Research*, **34**, 205–229.

Kasting, J. F. (1988). Runaway and moist greenhouse atmospheres and the evolution of Earth and Venus. *Icarus*, **74**, 472–94.

Kasting, J. F. (1991). CO_2 condensation and the climate of early Mars. *Icarus*, **94**, 1–13.

Kasting J.F. (1993). Earth's early atmosphere. *Science*, **259**, 920–6.

Kasting, J. F. and Pollack, J. B. (1983). Loss of water from Venus. I – Hydrodynamic escape of hydrogen. *Icarus*, **53**, 479–508.

Kasting, J. F., Brown, L. L. and Brack, A. (1998). Setting the stage: the early atmosphere as a source of biogenic compounds. In *The molecular origins of life: assembling the pieces of the puzzle*, ed. A. Brack. New York: Cambridge University Press, 5–56.

Kelley, D. S., Karson, J. A., Blackman, D. K., Früh-Green, G. L., Butterfield, D. A., Lilley, M. D., Olson, E. J., Schrenk, M. O., Roe, K. K., Lebon, G. T., Rivizzigno, P. and the AT3–60 Shipboard Party. (2001). An off-axis hydrothermal vent field near the mid-Atlantic ridge at 30° N. *Nature*, **412**, 145–9.

Kelley, D. S., Karson, J. A., Früh-Green, G. L., Yoerger, D. R., Shank, T. M., Butterfield, D. A., Hayes, J. M., Schrenk, M. O., Olson, E. J., Proskurowski, G., Jakuba, M., Bradley, A., Larson, B., Ludwig, K., Glickson, D., Buckman, K., Bradley, A. S., Brazelton, W. J., Roe, K., Elend, M. J., Delacour, A., Bernasconi, S. M., Lilley, M. D., Baross, J. A., Summons, R. E. and Sylva, S. P. (2005). A serpentinite-hosted ecosystem: the Lost City hydrothermal vent field. *Science*, **307**, 1428–34.

Kharecha, P., Kasting, J. F. and Siefert, J. L. (2005). A coupled atmosphere–ecosystem model of the early Archean Earth. *Geobiology*, **3**, 53–76.

Kulikov, Y. N., Lammer, H., Lichtenegger, H. I. M., Penz. T., Breuer, D., Spohn, T., Lundin, R. and Biernat, H. K. (2007). A comparative study of the influence of the active young Sun on the early atmospheres of Earth, Venus, and Mars. *Space Science Reviews*, **129**, 207–43.

Lammer, H., Lichtenegger, H. I. M., Kolb, C., Ribas, I., Guinan, E. F. and Bauer, S. J. (2003). Loss of water from Mars: implications for the oxidation of the soil. *Icarus*, **165**, 9–25.

Lammer, H., Lichtenegger, H. I. M., Biernat, H. K., Erkaev, N. V., Arshukova, I. L., Kolb, C., Gunell, H., Lukyanov, A., Holmstrom, M., Barabash, S., Zhang, T. L., and Baumjohann, W. (2006). Loss of hydrogen and oxygen from the upper atmosphere of Venus. *Planetary and Space Science*, **54**, 1445–6.

Lin, R. P., Dennis, B. R., Hurford, G. J., Smith, D. M., Zehnder, A., Harvey, P. R., Curtis, D. W., Pankow, D., Turin, P., Bester, M., Csillaghy, A., Lewis, M., Madden, N., Van Beek, H. F., Appleby, M., Raudorf, T., Mctiernan, J., Ramaty, R., Schmahl, E., Schwartz, R., Krucker, S., Abiad, R., Quinn, T., Berg, P., Hashii, M., Sterling, R., Jackson, R., Pratt, R., Campbell, R. D., Malone, D., Landis, D., Barrington-Leigh, C. P., Slassi-Sennou, S., Cork, C., Clark, D., Amato, D., Orwig, L., Boyle, R., Banks, I. S., Shirey, K., Tolbert, A. K., Zarro, D., Snow, F., Thomsen, K., Henneck, R., Mchedlishvili, A., Ming, P., Fivian, M., Jordan, J., Wanner, R., Crubb, J., Preble, J., Matranga, M., Benz, A., Hudson, H., Canfield, R. C., Holman, G. D., Crannell, C., Kosugi, T., Emslie, A. G., Vilmer, N., Brown, J. C., Johns-Krull, C., Aschwanden, M., Metcalf, T. and Conway, A. (2002).The Reuven Ramaty High-Energy Solar Spectroscopic Imager (RHESSI). *Solar Physics*, **210**, 3–32.

Linsky, J. L. and Wood, B. E. (1996). The Alpha Centauri line of sight: D/H ratio, physical properties of local interstellar gas, and measurement of heated hydrogen (the 'hydrogen wall') near the heliopause. *The Astrophysical Journal*, **463**, 254–70.

Lundin, R. and Barabash, S. (2004). Evolution of the Martian atmosphere and hydrosphere: solar wind erosion studied by ASPERA-3 on Mars Express. *Planetary and Space Science*, **52**, 1059–71.

Lundin, R., Lammer, H. and Ribas, I. (2007). Planetary magnetic fields and solar forcing: implications for atmospheric evolution. *Space Science Reviews*, **129**, 245–78.

Messina, S. and Guinan, E.F. (2002). Magnetic activity of six young solar analogues i. Starspot cycles from long-term photometry. *Astronomy and Astrophysics*, **393**, 225–37.

Miller, S. L. (1953). A production of amino acids under possible primitive Earth conditions. *Science*, **117**, 528–9.

Pavlov, A. A., Kasting, J. F. and Brown, L. L. (2001). UV-shielding of NH_3 and O_2 by organic hazes in the Archean atmosphere. *Journal of Geophysical Research*, **106**, 23267–23288.

Preibisch, T., Kim, Y.-C., Favata, F., Feigelson, E. D., Flaccomio, E., Getman, K., Micela, G., Sciortino, S., Stassun, K., Stelzer, B. and Zinnecker, H. (2005). The origin of T-Tauri X-ray emission: new insights from the Chandra Orion Ultradeep Project. *Astrophysical Journal Supplement Series*, **160**, 401–22.

Radick, R. R., Lockwood, G. W. and Baliunas, S. L. (1990). Stellar activity and brightness variations – a glimpse at the Sun's history. *Science*, **247**, 39–44.

Ribas, I., Guinan, E. F., Güdel, M. and Audard, M. (2005). Evolution of the solar activity over time and effects on planetary atmospheres. I. High-energy irradiances (1–1700 Å). *The Astrophysical Journal*, **622**, 680–94.

Sackmann, I. J. and Boothroyd, A. I. (2003). Our Sun. V. A bright young Sun consistent with helioseismology and warm temperatures on ancient Earth and Mars. *Astrophysical Journal*, **583**, 1024–39.

Schaefer, L. and Fegley, J. B. (2007). Outgassing of ordinary chondritic material and some of its implications for the chemistry of asteroids, planets, and satellites. *Icarus*, **186**, 462–83.

Schüssler, M. and Solanki, S. K. (1992). Why rapid rotators have polar spots. *Astronomy and Astrophysics*, **264**, L13–16.

Segura, T. L., Toon, O. B., Colaprete, A. and Zahnle, K. (2002). Environmental effects of large impacts on Mars. *Science*, **298**, 1977–80.

Smith, D. S., Scalo, J. and Wheeler, J. C. (2004). Transport of ionizing radiation in terrestrial-like exoplanet atmospheres. *Icarus*, **71**, 229–53.

Stauffer, J. R., Caillault, J.-P., Gagné, M., Prosser, C. F. and Hartmann, L. W. (1994). A deep imaging survey of the Pleiades with ROSAT. *Astrophysical Journal Supplement Series*, **91**, 625–57.

Strassmeier, K. G. and Rice, J. B. (1998). Doppler imaging of stellar surface structure. VI. HD 129333 = EK Draconis: a stellar analog of the active young Sun. *Astronomy and Astrophysics*, **330**, 685–95.

Telleschi, A., Güdel, M., Briggs, K., Audard, M., Ness, J.-U. and Skinner, S. L. (2005). Coronal evolution of the Sun in time: high-resolution X-ray spectroscopy of solar analogs with different ages. *The Astrophysical Journal*, **622**, 653–79.

Tian, F., Kasting, J. F. and Solomon, S. C. (2009). Thermal escape of carbon from the early Martian atmosphere. *Geophysical Research Letters*, **36**, 10.1029/2008GL036513.

Tian, F., Kasting, J. F., Liu, H.-L. and Roble, R. G. (2008). Hydrodynamic planetary thermosphere model: 1. Response of the Earth's thermosphere to extreme solar EUV conditions and the significance of adiabatic cooling. *Geophysical Research Letters – Planets*, **113**, E05008.

Walker, J. C. G. (1977). *Evolution of the Atmosphere*. New York: Macmillan.

Watson, A. J., Donahue, T. M. and Walker, J. C. G. (1981). The dynamics of a rapidly escaping atmosphere – applications to the evolution of Earth and Venus. *Icarus*, **48**, 150–66.

Wood, B. E., Müller, H.-R., Zank, G. P. and Linsky, J. L. (2002). Measured mass-loss rates of solar-like stars as a function of age and activity. *The Astrophysical Journal*, **574**, 412–25.

Wood, B. E., Müller, H.-R., Zank, G. P., Linsky, J. L. and Redfield, S. (2005). New mass-loss measurements from astrospheric Lyα absorption. *The Astrophysical Journal*, **628**, L143–6.

Zahnle, K. J. (1986). Photochemistry of methane and the formation of hydrocyanic acid (HCN) in the Earth's early atmosphere. *Journal of Geophysical Research*, **91**, 2819–34.

Zahnle, K. J. and Kasting, J. F. (1986). Mass fractionation during transonic escape and implications for loss of water from Mars and Venus. *Icarus*, **68**, 462–80.

12
Climates of the Earth

Gilles Ramstein

Introduction

Since the beginning of this century, astronomers have discovered hundreds of exoplanets, almost all of them being giant planets. However, we expect the possibility of detecting relatively smaller exoplanets similar in size to our Earth soon. Among the thousands of exoplanets that will be discovered by the end of this century, some may host life. Obviously the possibility of finding life on another planet is not only a function of the number of discovered planets, but also of the stability of life on those planets: if life is only a glimpse, the search for life will be much harder! We have under our feet a marvellous example demonstrating that one kind of life may be hosted on a planet (Earth) for billions of years. This relative stability of life on Earth seems to be strongly correlated with the stable environmental conditions that have prevailed on the surface of our planet for several billion years (1 Ga = 10^9 years). This is the reason why this chapter is devoted to deciphering and understanding the 'stable' climate conditions on Earth since 3.8 Ga. Observation of our neighbouring planets in the Solar System teaches us that the conditions for the development of life (habitability) and sustainability at the surface of a planet are not widespread – at least in the Solar System. Nowadays, Mars is a very cold desert experiencing dust storms, whereas Venus is a burning hell whose surface is totally hidden by a thick greenhouse-gas atmosphere. In comparison, the conditions that prevail at the surface of our planet are obviously much more favourable to life. But has it been the case throughout Earth's entire history?

Deep-past climate

There is, of course, a common forcing factor, a provider of energy: the Sun for 4.6 billion years. The Sun – this is certainly not the most poetic definition – is 'just' a nuclear reactor based on a fusion process, which burns its hydrogen and this will be followed by its helium. Therefore, 4.6 Ga ago, the young Sun was also a 'weak' Sun (Figure 12.1). The implications for Earth (and indeed also for Venus and Mars) are huge. First, because solar radiation is the main source of energy for the Earth System (> 99%), and second, because we know that even the weak variations of solar radiation at the top of the atmosphere, due to orbital parameters

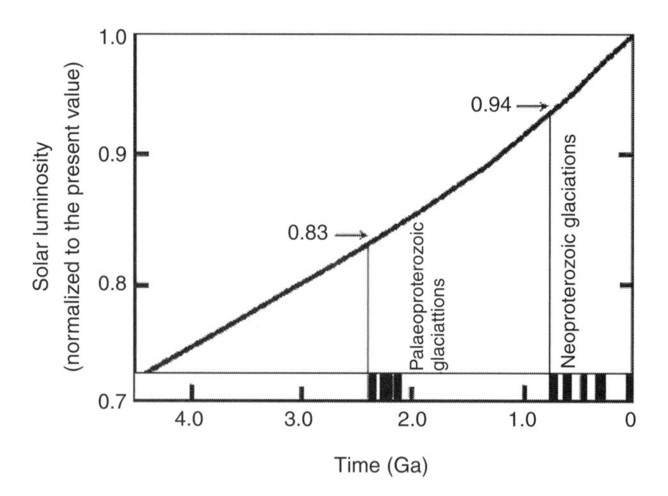

Figure 12.1. Evolution of Sun luminosity with time. (From Gough, 1981.)

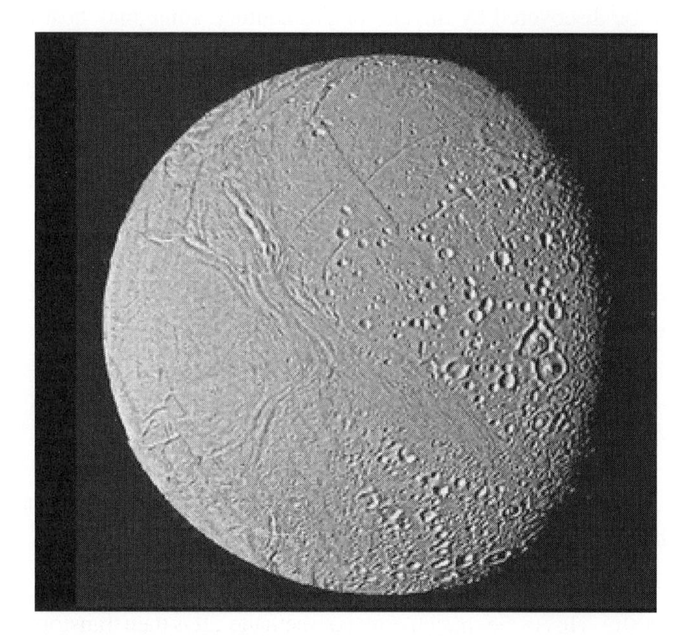

Figure 12.2. Enceladus, the frozen satellite of Saturn.

of the Earth, induced during the Quaternary period's large climatic variations (see 'The last 500 Ma' section). Therefore, the Earth should have experienced a drastic cooling due to this weak Sun. The Earth should look like these frozen satellites (Figure 12.2). But geologic and isotopic measurements demonstrate that, in fact, during the Archaean Era (4.54–2.5 Ga) and Proterozoic Era (2.5–0.54 Ga) (cf. a geological timescale, Gradstein and Ogg, 2004), the Earth was warm with the temperature of the oceans ranging from 70 to 50°C (Robert and

Chaussidon, 2006). Indeed, no measurement is possible before 3.8 Ga ago. Moreover, do these temperatures really represent the Earth's surface temperature? This remains an open question. Nevertheless, the existence of warm conditions raises a deep contradiction in the so-called faint young Sun paradox, that greenhouse gases can help to solve.

Which greenhouse gas? After an initial controversial period (Sagan and Mullen, 1972), most scientists proposed that carbon dioxide appeared to be the most likely one. The volcanic source could have been enhanced during early Earth, whereas the sink of carbon dioxide, especially through silicate weathering, was much reduced because there was no or very little continental surface. Therefore, a high level of carbon dioxide in the atmosphere was possible during early Earth. But too much atmospheric carbon dioxide would have produced siderite ($FeCO_3$) on the continental surface (Rye *et al.*, 1995). None of the oldest rocks show any siderite, which leads several authors (Kasting, 2004) to suggest that the carbon dioxide, by itself, could not 'do the job' and that another greenhouse gas should be invoked. Methane has been suggested as an appropriate candidate, but what about the possible sink and source of methane?

Something we often forget concerning Earth's atmosphere is that for 4.6 billion years the atmosphere was anoxic during half of that time. The oxygen increase in the atmosphere only occurred 2.2–2.4 billion years ago (Catling and Claire, 2005; Bertrand, 2005). This feature is largely accepted by the whole scientific community, even if there are still some controversies (Ohmoto *et al.*, 2006). Several lines of evidence (Knoll, 2004) point to the fact that *Methanogene archae* appeared very early in the Tree of Life. Kasting developed a scenario where the source of methane – which, in contrast to Titan, can only be biologic on Earth – was provided by these archae, which were very abundant in the ocean.

What about the methane sink? In the geologic timescale, methane behaves as a match. It can produce strong warming because it is a powerful greenhouse gas, but its duration in the oxic atmosphere is very short: about 10 years. Methane is oxidized and transformed into carbon dioxide, which is a less powerful greenhouse gas but lasts much longer in the atmosphere: tens of kilo years (1 Ka = 1000 years). This is illustrated by the scenario developed to explain the brutal warming of the Palaeocene/Eocene Era boundary 55 million years ago (1 Ma = 1 000 000 years) (Dickens and Francis, 2004). Many authors point to the large methane emissions in the atmosphere from an enormous quantity of methane hydrate stored within oceanic sediments (Dickens, 2003). The release of methane from the dissociation of the hydrate of methane (highly depleted in $\delta^{13}C$ around – 60‰) explains both the negative carbon excursion observed at the Palaeocene–Eocene Era boundary and the abrupt warming (Zachos *et al.*, 2001). This is the 'match effect of methane'. It is then transformed into carbon dioxide, which is less efficient than methane as a greenhouse gas. Of course, in an anoxic atmosphere, the sink of methane is much less efficient. Therefore, both sink and source provide the possibility of a strong accumulation of methane in an anoxic atmosphere.

Regulation of Earth's surface temperature and major accidents

As described below, in a context where the Sun was fainter by a considerable degree (−30% at 4.6 Ga ago and −6% at 0.6 Ga ago– see Figure 12.1), the carbon dioxide and, perhaps

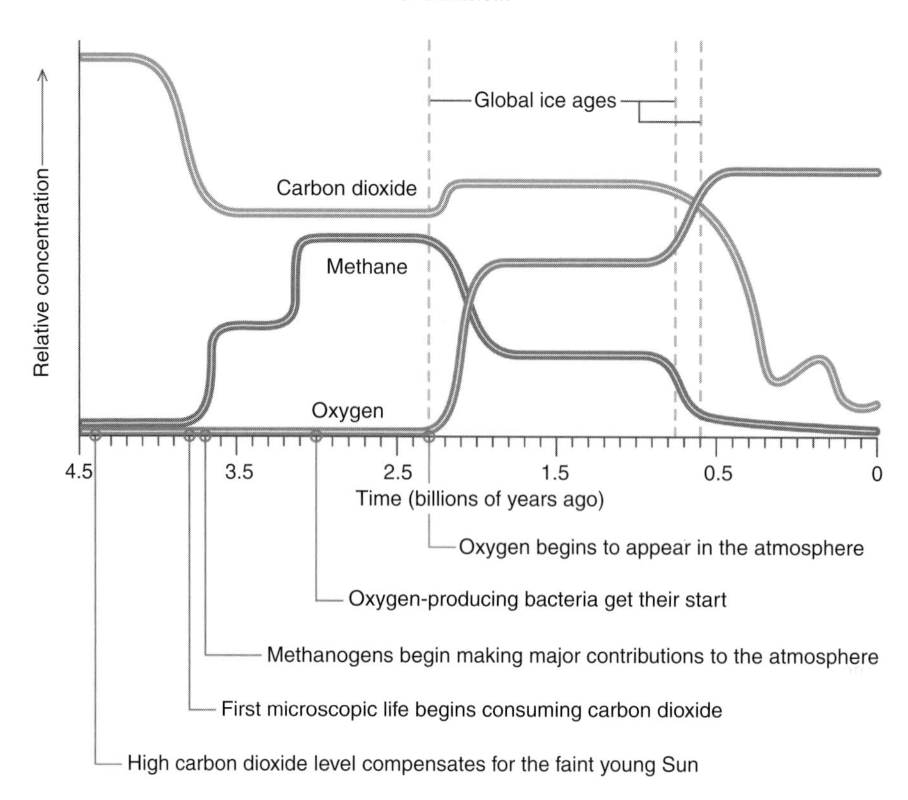

Figure 12.3. Temporal evolution of CO_2, CH_4 and O_2; adapted From Kasting (2004) and Anbar and Knoll (2002).

the methane, helped to largely overcompensate for the cooling due to the weakness of the young Sun to explain warm climates on Earth.

Surprisingly, there is, during the Archaean and Proterozoic Periods, very little evidence of glaciation. Recently, a Pongolian glaciation (~ 2.7 Ga ago) has been pointed out (Crowell, 1999; Young *et al.*, 1998). The next glaciation occurred, accounting for error bars, synchronously with the oxygen rise in the atmosphere (−15% solar reduction – 2.4 to 2.2 Ga ago, known as the Huronian glaciation) (Figure 12.3). The third glaciation occurred more than 1.5 Ga later, during the Neoproterozoic Era. One can always argue that 'the absence of evidence (of other glaciations) is not the evidence of absence'. This idea has indeed to be accounted for when dealing with a deep past for which we only have scarce data. Nevertheless, it appears that glaciations are only accidents on Earth, which is, most of the time, regulated to a warm temperature. But how does this regulation work and why did these accidents occur?

This regulation is far from a new idea. Lovelock carried out pioneering work pointing out how many processes, including biology, make the Earth able to 'protect' itself against perturbations (Lovelock, 1979). Nevertheless, similarly to Wegener with the continental-drift theory, as long as appropriate and quantified mechanisms cannot explain this hypothesis, many scientists will remain sceptical. However, long-term interactions between climate, tectonics and the carbon cycle may explain the regulation of the Earth System (Walker *et al.*, 1981).

It is indeed difficult, maybe impossible, to infer the climate before the late bombarding event at 3.8 Ga ago (Morbidelli *et al.*, 2001), but from 3.8 Ga ago to the Precambrian Period (540 Ma ago), this regulation was very powerful in maintaining a warm climate despite a weaker Sun. We shall now describe the interactions between carbon dioxide and the climate and tectonics, which may produce long-term thermoregulation. This mechanism is closely related to the carbon cycle on the geologic timescale. Concerning the source of carbon dioxide, vulcanism, the most recent publications show (Cogné and Humler, 2004) that for at least the last 100 Ma there is no evidence that the source quantity of carbon dioxide produced by vulcanism has changed. It is indeed a very short time period compared to 3 Ga, but in the absence of recorded variation on this timescale, we shall be conservative and consider the source as constant and equal to the present-day value. In contrast, silicate weathering, which is the most powerful sink, is dependent on tectonics. For instance, if most of the continental mass is located in the equator/tropics, the continents experience huge precipitation. The raindrops, which include atmospheric carbon dioxide, become acidic and dissolve silicate, which is transported by runoff to the ocean, and then sinks with sediments to the bottom of the ocean. This process leads to an atmospheric carbon dioxide decrease. (if most of the continental mass is located at a high latitude, it experiences a dry climate producing weak alteration and erosion. Therefore, it is not a context in favour of 'washing out' – through precipitation and erosion processes – the atmospheric carbon dioxide, which therefore increases). More than 80 years ago, when Wegener explained his plate-drift theory (Wegener, 1915), he also understood the long-term climate impact of his new theory. In collaboration with Köppen, he also published a book that describes how climate change is due to latitudinal drift of continents (Köppen and Wegener, 1924). However, one piece was missing in their understanding. Not only climate, but also carbon dioxide, which acts on climate, was modified by tectonics. This is the extraordinary feedback loop produced by the triptych climate/carbon dioxide/tectonics that is able to explain a warm Earth for most of the time. When temperature increases, weathering also increases and carbon dioxide decreases, which leads to a cooling, and vice versa. Therefore, carbon dioxide is a powerful thermoregulator (Walker *et al.*, 1981). We have described the feedback and regulation, but why have some very rare accidents occurred? The main idea is not to forget that the Earth's climate is a relatively fragile system which depends on its two greenhouse gases (carbon dioxide and methane) to avoid major glaciation due to low energy coming from a 'faint young Sun'. If a destabilization of one of these gases occurs, a drastic cooling may happen and transform the planet into 'snowball Earth'.

A very good point in Kasting's scenario (Kasting, 2004) is the occurrence of the methane drop, which was accompanied by an atmospheric oxygen rise and this ended in the so-called Huronian glaciation (2.2–2.4 Ga ago; Figure 12.3). We have seen in the deep past that carbon dioxide by itself could not have overcome the decrease in short-wave energy due to a faint Sun. In Figure 12.3, J. Kasting describes the evolution, through geological time, of greenhouse gases and oxygen. Even if there are large uncertainties on both axes for the amplitude of these gases in the atmosphere and for the timing, a robust feature is that a large drop in levels of greenhouse gas has occurred with major glaciations. As we have seen previously, the Earth depends on methane and carbon dioxide to maintain a warm temperature and, in fact, overcome the cooling effect of a young Sun; this context makes possible large accidents which drive the system to major glaciations if levels of one of these gases suddenly decrease. What does 'suddenly' mean? What are the causes and the timing of these accidents?

The first one, which occurred 2.2–2.4 Ga ago, synchronously with the rise of oxygen in the atmosphere (see Figure 12.3) could be related to biology. The prevailing methanogenic archae in the Earth's surface ocean may have been around since 3.8–3.5 Ga ago, leading to the production of methane in Earth's atmosphere, whereas oxygen producers (cyanobacteria) were totally marginal. On the other hand, although *Methanogene archae* cannot survive in an oxic environment, they survive in anoxic refugia. Therefore, it is easy to understand, and this is also a strong point of Kasting's theory, that when oxygen levels rise and oxygen producers shift from marginal to prevailing in numbers, the archae disappear, which then leads to a sudden drop in methane levels in the atmosphere. Because on Earth a methane source could only be produced through a biological process, the first large glaciation (Huronian) was driven by a biological crisis, and therefore was 'instantaneous' in terms of the geological timescale.

Neoproterozoic Era global glaciation

Certainly the best documented 'snowball Earth' episodes are the Neoproterozoic Era ones. They consist of three main glaciations: the two older ones, the Sturtian (715 Ma ago) and Marinoan (635 Ma ago) glaciations, are supposed to have been global (Hoffman *et al.*, 1998), whereas the third and most recent one, the Rapitan glaciation (550 Ma ago), appears to have been essentially regional and linked to the Appalachian uplift (Donnadieu *et al.*, 2004a). Here too, such a global glaciation results from a 'rapid' decrease of atmospheric CO_2 content. But let us first summarize the different lines of evidence that support the snowball hypothesis. They consist of four major and 'mysterious' observations: (1) palaeomagnetism evidence for glaciations at low latitude; (2) a strong decrease in $\delta^{13}C$; (3) the reappearance, after 1.5 Ga of absence, of banded-iron formation (BIF); and (4) huge cap-carbonate formations overlaying glacial diamictites (tillites), which are typical glacial deposits.

These observations were made by different research teams and untill the end of the last century, no reliable global explanation was available which could help us to understand what was happening during the Neoproterozoic Era. In fact, concerning the first point, which is based on a few reliable palaeomagnetic data, an attractive explanation (to infer why ice caps were located at the equator and not the poles) was proposed by G. Williams (Williams, 1975) and later on, theoretically demonstrated by D. Williams (Williams *et al.*, 1998). Indeed, if the Earth did have a much larger obliquity (> 60°), it is easy to show (Figure 12.4) that the lowest annual temperature occurred over the equator. Unfortunately, this explanation, very simple and therefore very appealing, was wrong for two main reasons.

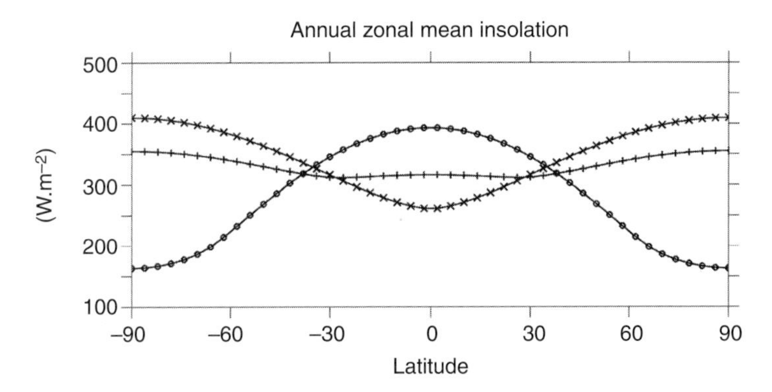

Figure 12.4. Variation of annual mean insolation at the top of the atmosphere for high obliquity and low obliquity (present day).

The first issue was tackled by Levrard and Laskar (2003), who demonstrated that no realistic mechanism existed to produce a decrease in Earth's obliquity from a high value (> 60°) to its present-day low angle (~ 23°).

The second argument came from Donnadieu *et al.* (2002), who showed that such a high obliquity leads to a huge seasonal cycle, which was inconsistent with a large mid-latitude ice sheet during Marinoan glaciations.

Further pioneering work was provided in the 1960s by the American geologist Harland (Harland, 1964). He was the first to point out that glaciations in the tropics (that he observed) would be easily explained in the context of global glaciation. Unfortunately, modellers explained to Harland that, if such an event did occur, it would have led to an increased albedo and the Earth, instead of being a 'blue planet' with a global albedo of 0.3, would become a 'white planet' reflecting much more solar radiation with a global albedo of around 0.6–0.8. A direct consequence is that to escape from such a glaciation, the only opportunity is to increase the Sun's luminosity by a factor of 1.5, which is not possible. Therefore, this hysteresis has the consequence that, if such a snowball did exist, it should still be here!

It was 30 years later that Kirschvink (1992) provided quite a simple explanation: during a snowball Earth episode, vulcanism continues and the carbon dioxide emitted into the atmosphere, in contrast with the normal situation where it interacts with the biosphere and the ocean, remains in the atmosphere. Because continents were covered by ice sheets and oceans by sea ice, carbon dioxide was stored in the atmospheric reservoir. Over millions of years, the carbon dioxide remained in the atmosphere and increased, until it was able, through a super-greenhouse effect, to melt the snowball Earth.

This theory also provides very good explanations for all of the 'mysterious' observations listed above.

The fact that the ocean was covered by sea ice and land by ice sheets completely inhibits the carbon cycle. Therefore, carbon was no longer fractionated and the $\delta^{13}C$ value remained very similar to the original one emitted by the volcanoes. Ventilation of the ocean covered by sea ice completely declined, and therefore oxygenation of the ocean drastically decreased.

These features may explain the reappearance of banded iron formations (BIF) that prevailed in Archaean and early Proterozoic oceans. Finally, the carbon dioxide was stored for a very long time (millions of years) in the atmosphere, until it reached a threshold value when the radiative greenhouse effect could eventually produce a huge deglaciation. Huge precipitation and erosion took place in a very warm climate that brought a large amount of carbonate into the oceans. This feature explains the paradox that tillite was overlaid by cap carbonate. The snowball theory therefore became very popular due to the talent of Hoffman (Hoffman and Schrag, 2002). Most scientists did agree to admit that it explains most of the 'mysterious' features that occurred during the Neoproterozoic Era. Nevertheless, many questions remain unsolved, the first one being: why did the carbon dioxide level largely decrease to produce such a snowball? Why were there such drastic perturbations and why so few during the first billions of years of Earth history? The answers to these questions were resolved by Donnadieu *et al.* (Donnadieu *et al.*, 2004a), thanks to a fruitful collaboration between geochemists and climate modellers (Goddéris *et al.*, 2007).

If, indeed, the snowball theory seems to solve many intriguing questions, no explanation was provided to aid in the understanding as to why such an event occurred (Ramstein *et al.*, 2004a). The cause of the large decrease in carbon dioxide at the origin of the large cooling that led to snowball Earth remains mysterious. Less than ten years ago, the influence of palaeogeography on CO_2 was first invoked. It has been known for a long time that continents were oscillating between building up a large supercontinent and breaking up into several smaller continents that drift away from each other. The last supercontinent was Pangaea and it broke up from the Triassic to the Cretaceous Period. The previous one, which we are interested in here, was Rodinia. It was sitting on the equatorial band and spread to 40° (Figure 12.5A). It was aggregated before 1 Ga ago and began to break up only around 800–750 Ma ago (Figure 12.5B).

There are two very peculiar things in this break-up: The initial supercontinent is located at low latitudes (Figure 12.5A) and most of the fragments remain in the tropical belt (Figure 12.5B). Why are these features so important? If you remember the mechanism we described in previous sections to explain how CO_2–climate–tectonics act together to regulate temperature at the surface of the Earth, you may understand that the configuration, in which all the fragments remain in the tropics, is the most favourable to large atmospheric CO_2 decreases. Indeed, all these continental masses, because it was located in the tropics, experienced a huge precipitation that led to massive erosion and alteration. We showed that this process provides a large sink of atmospheric CO_2 through silicate weathering.

Donnadieu *et al.* (2004b) demonstrated, by using a carbon/climate model, that such a break-up would lead to a tremendous CO_2 decrease – which is consistent with the drastic cooling that finally shifted the Earth towards a snowball Earth (Figures 12.5 and 12.6). To escape such a situation, vulcanism and CO_2 stored during several million years in the atmosphere is certainly a robust mechanism (Kirschvink, 1992; Hoffman *et al.*, 1998). Nevertheless, it has been shown recently (Le Hir *et al.*, 2008a; Le Hir *et al.*, 2008b) that, first, the concept of no interaction of atmospheric CO_2 with oceans is difficult to maintain, and second, that acidification of surface oceans due to enormous quantities of CO_2 suddenly in contact with oceans depleted of CO_2 is inconsistent with biology (Le Hir *et al.*, 2008b).

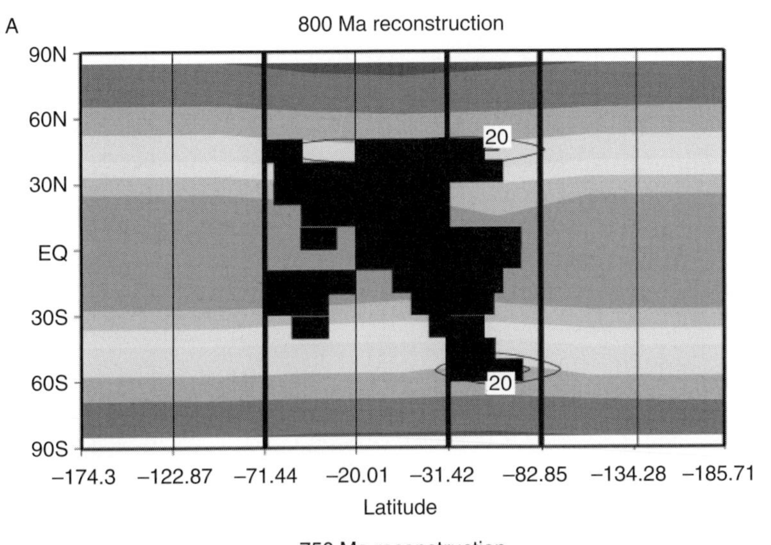

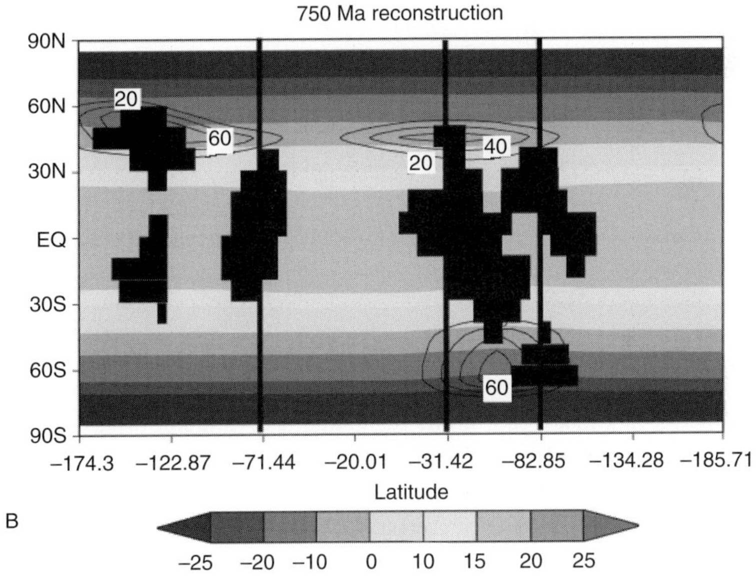

Figure 12.5. A: Distribution of temperature and computation of atmospheric CO_2 by GEOCLIM for 800 Myr. B: Distribution of temperature and computation of atmospheric CO_2 by GEOCLIM for 750 Myr.

For these reasons, the behaviour of the carbon cycle during and after a snowball Earth has been deeply modified these last few years (Le Hir *et al.*, 2008c; Pierrehumbert, 2004).

The amount of atmospheric carbon dioxide, after these glaciations, is able to vary with tectonics during these last 500 Ma. Nevertheless, because of the slow increase in Sun luminosity, which reaches a value of only 5 % weaker than present-day during the Cambrian

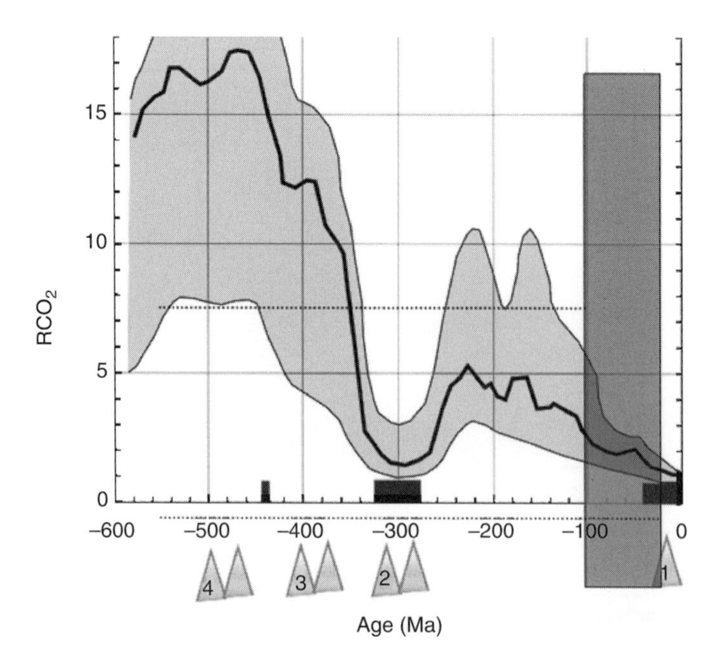

Figure 12.6. Most recent evolution of atmospheric CO_2 derived from GEOCARB Berner's model. Triangles at the foot of the figure correspond to evidence of large glaciations. The thick black line represents the 'best guess' and the shaded area either side of this represents 95% CI.

period, the variation of CO_2 content of the atmosphere can no longer produce any big accidents such as snowball Earths.

The last 500 Ma

Indeed, as will be illustrated in this section, the relationship between tectonics, climate and carbon dioxide plays a major role in the regulation and variation of atmospheric carbon dioxide: most of the time and again, the regulation tends to maintain a rather warm climate. Ice-sheet build-ups are therefore rare events (Ramstein, 2004b).

The last important glaciations occurred at the Permo–Carboniferous boundary due to both a low carbon dioxide content in the atmosphere (Figures 12.6 and 12.7) and a palaeogeography where a large fraction of the continental mass was located at high latitudes in the southern hemisphere (Figure 12.8). These two factors have been used to perform simulations of the Permo–Carboniferous glaciation. The results, shown in Figure 12.8, show that annual temperatures are negative at high latitudes of Pangaea continents. Moreover, we plot on this figure the albedo value corresponding to the summer season for the Southern hemisphere, which shows that perennial snow cover extends through South Gondwana (South Africa, South America, India, Australia and Antarctica). This result is consistent with moraine deposits on these continents, which led Wegener to imagine all these areas were at this time at the same location. Later on, when Pangaea drifts towards the northern hemisphere, the continental mass slowly shifts from high to low latitudes and alteration and

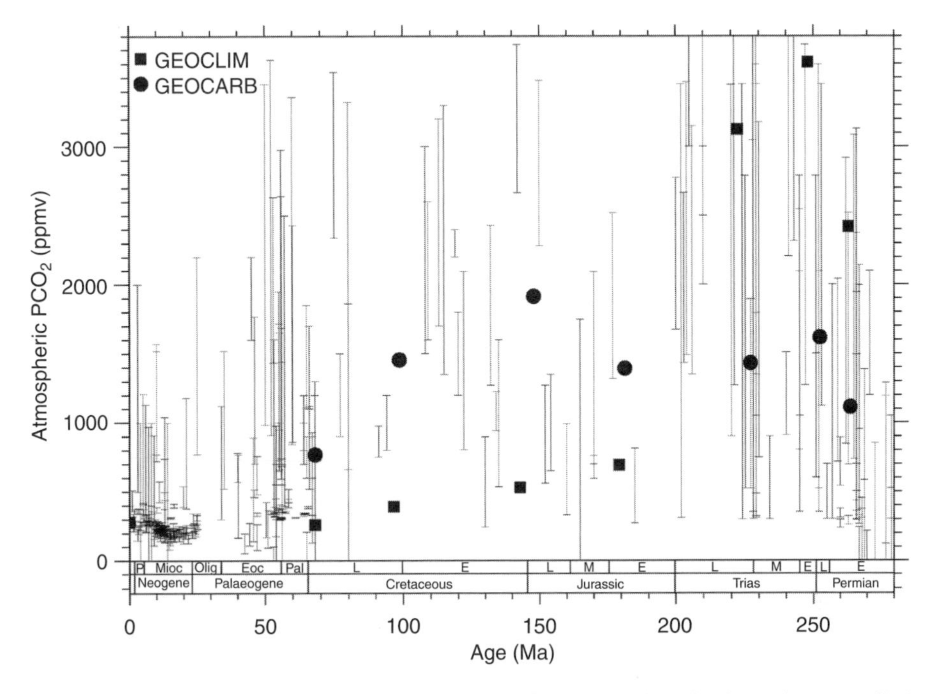

Figure 12.7. Cenozoic and Mesozoic reconstruction of atmospheric CO_2 from data compilation (Roger, 2006) and from models by GOECARB and GEOCLIM.

erosion increase, implying a severe drop in atmospheric carbon dioxide, which is simulated using the GEOCLIM (climate–carbon cycle) model (Donnadieu *et al.*, 2006). Interestingly, the appearance of calcareous plankton corresponds to that period of large carbon dioxide decrease (Figure 12.9) (Rhaetian: 200 Ma ago).

Another interesting period is the Mid-Cretaceous Cenomanian period corresponding to a high level of carbon dioxide (Figures 12.6 and 12.7) and high sea level, leading to a warm and uniform climate with rather weak seasonal cycles and a large decrease in equator-to-pole temperature gradients (Figure 12.10) (Fluteau *et al.*, 2007). In contrast to good agreement between models and available data for previous periods (Permo–Carboniferous and Rhaetian Periods), the models cannot reproduce the rather flat temperature gradient deduced from $\delta^{18}O$ readings of calcite (Barron and Washington, 1982). This disagreement has been shown to be linked with data, since when $\delta^{18}O$ is measured from the apatite of fish teeth (Pucéat *et al.*, 2007), this gradient in temperature is in much better agreement with the model results. A major shift from warm to cold climate occurred 34 Ma ago with the onset of Antarctic glaciations. Different model simulations have shown that the most important feature that explains this shift is the decrease in atmospheric carbon dioxide (DeConto and Pollard, 2003), see Figure 12.11 (Pagani *et al.*, 2005). Moreover, the opening of the Drake Passage also played a role in this transition. Greenland's glaciation only occurred 3 Ma ago, and the most recent results which quantify the role of the different forcing factors

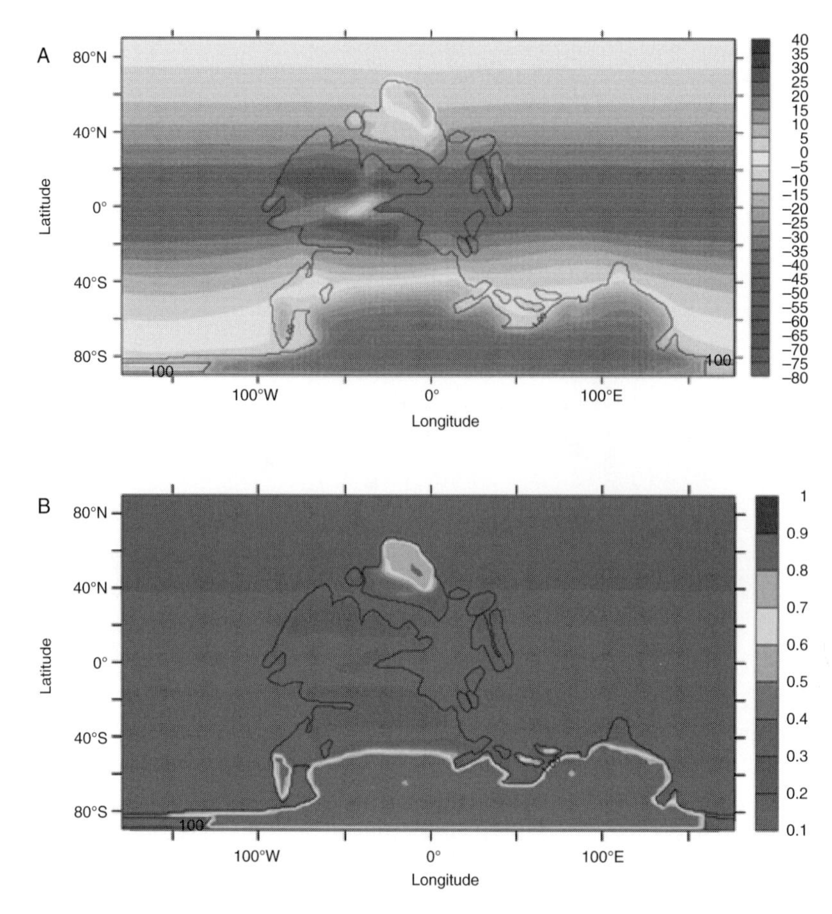

Figure 12.8. Permo–Carboniferous distribution of (A) temperature and (B) summer albedo in the Southern hemisphere when no ice cap is accounted for in South Gondwana.

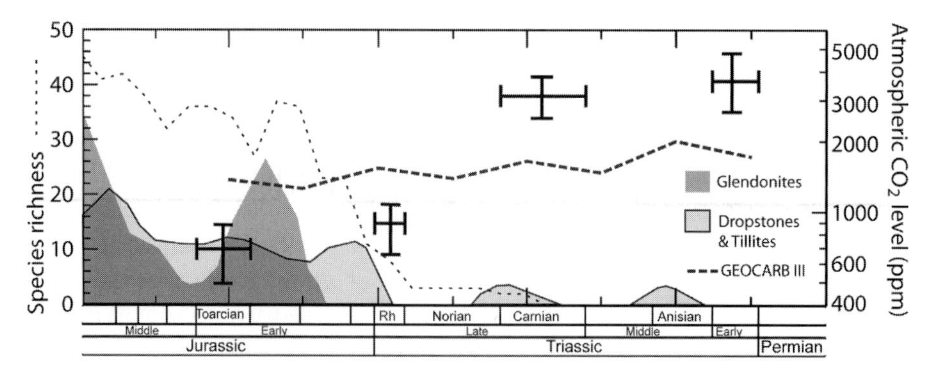

Figure 12.9. Evolution of biodiversity and atmospheric CO_2 during Triassic and Jurassic Periods. The cross corresponds to CO_2 level in the atmosphere, whereas the curve corresponds to calcareous plankton evolution during the same period. (After Donnadieu *et al*,. 2006.)

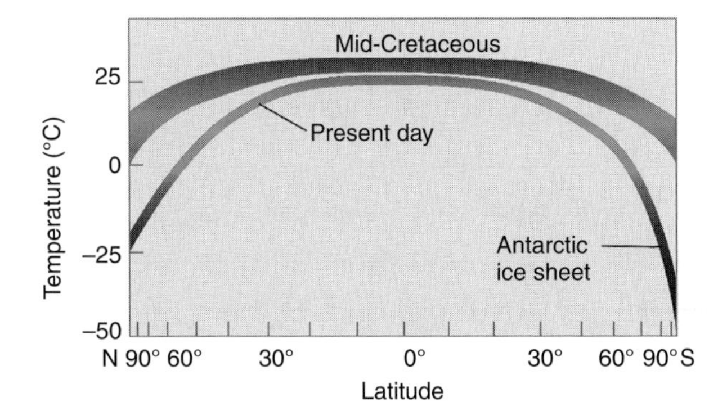

Figure 12.10. Equator-to-pole temperature gradient derived from data for Mid-Cretaceous and compared to present day.

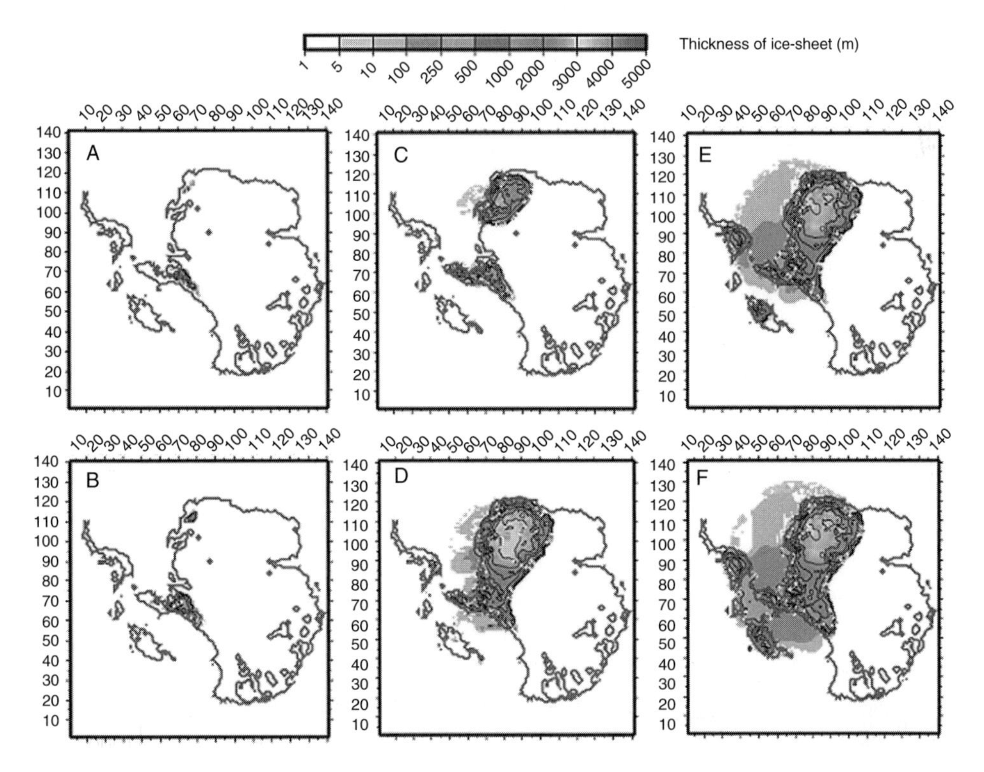

Figure 12.11. Antarctic ice sheet (AIS) topography for different atmospheric CO_2 concentrations and Drake Passage (DP) configurations (fully coupled CLIMBER–GRISLI experiments). In the left panels, the AIS obtained for PCO_2 is represented, set to 4 PAL in the case of shut DP (top, A) and open DP (bottom, B). In the middle panels, the figure shows the AIS topography for PCO_2 set to 3 PAL and, respectively, shut DP (C) and open DP (D), while in the right panels the AIS obtained for PCO_2 set to 2 PAL with shut (E) and open (F) configurations of the DP. The ice-sheet thickness is expressed in m. It is possible to note the presence of ice shelves for the simulations (C) to (F). The scales represent longitude (along the top) and latitude (down the left hand side).

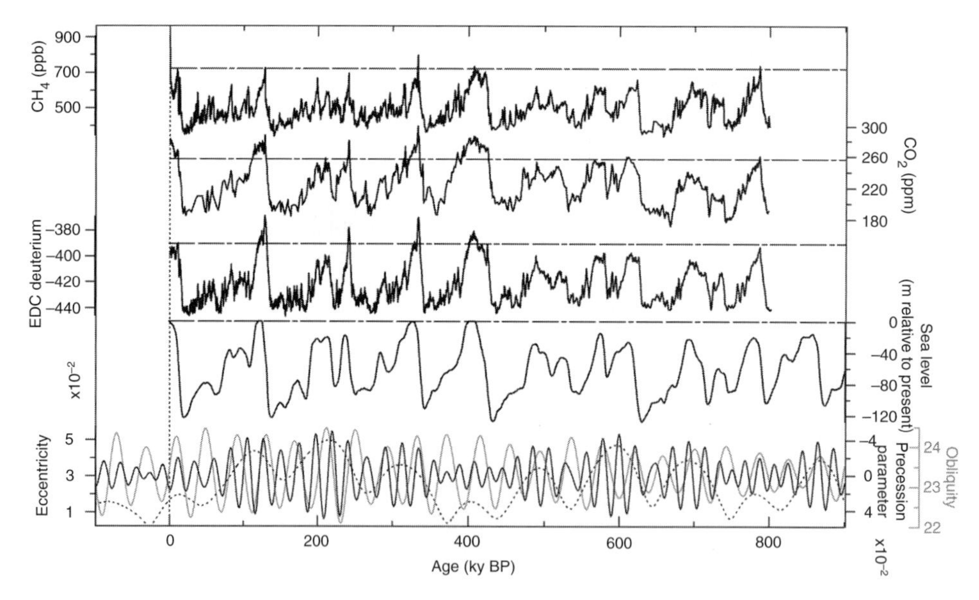

Figure 12.12. From top to bottom: Antarctic records of atmospheric methane concentration (ppbv) (Loulergue, 2008), atmospheric carbon dioxide concentration (ppmv) (Petit, 1999; Siegenthaler, 2005; Lüthi, 2008), and deuterium, a proxy of Antarctic temperature (Jouzel, 2007; Masson-Delmotte *et al.*, 2010). They are compared to the estimate of past changes in sea level (Bintanja *et al.*, 2005) derived from a stack of marine records (Lisiecki and Raymo, 2005) and to the past variations in the Earth's orbital parameters, eccentricity, precession parameter and obliquity. Data are displayed as a function of time in thousands of years before 1950 (ky Before Present). From Pol K. *et al.* 2010.

show that carbon dioxide decrease, Rockies uplift, Panama Passage closure and changes in tropical dynamics were all forcing factors, but that carbon dioxide was the major player (Lunt *et al.*, 2008).

The last million years' climate

Since the Greenland glaciation around 3 Ma BP, the climate in the northern hemisphere has changed drastically, with a big ice-sheet build up for a period of 40 Ka, and since then around 800 000 years for a period of 100 Ka (Paillard, 2008).

An extraordinary feature of this period is that we can measure temperature, methane and carbon dioxide levels in Antarctica's ice sheets – glaciologists hope even to be able to go back as far as 1.2 million years ago. In contrast, for the deep past, measurements of PCO_2 (partial pressure of carbon dioxide) are indirect through different indicators such as boron isotopes or stomata, and therefore uncertainties are much larger (Figures 12.6 and 12.7). As shown in Figure 12.12, the different cycles are indeed not similar, and there are still many unsolved questions that require answers in order for us to understand the climate of the Quaternary Period.

The last glacial–interglacial cycle from 130 Ka BP to Present is certainly the best documented in the different areas. From the modelling point of view, using a model of

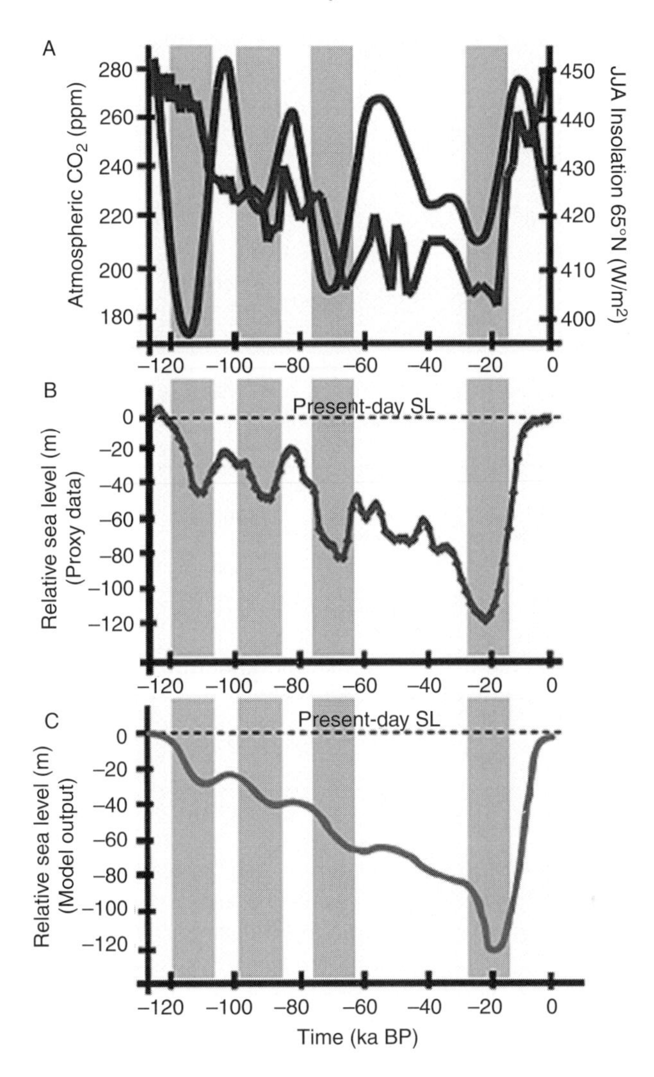

Figure 12.13. On panel (A) both forcing factors used for the simulation from 126 Ka BP to the present day are represented at insolation at 65° N along with CO_2 evolution from the Antarctica ice cores. On panels (B) and (C) are shown the sea-level evolution from data (B) and from the model simulation of Bonelli *et al.*, 2009.

intermediate complexity that allows us to run long simulations (Petoukhov *et al.*, 2000) and ice sheet models (Ritz *et al.*, 1997) has made it possible to simulate the onset of glaciations, the paroxysm at the last glacial maximum (21Ka BP) and the decay of the ice sheets and more stable climate of the Holocene Period (Bonelli *et al.*, 2009).

Figure 12.13 shows the associated sea-level variations simulated versus those reconstructed from data (Waelbroeck *et al.*, 2002). Indeed, being able to simulate a global cycle correctly leads to being able to investigate the long-term behaviour of present-day ice

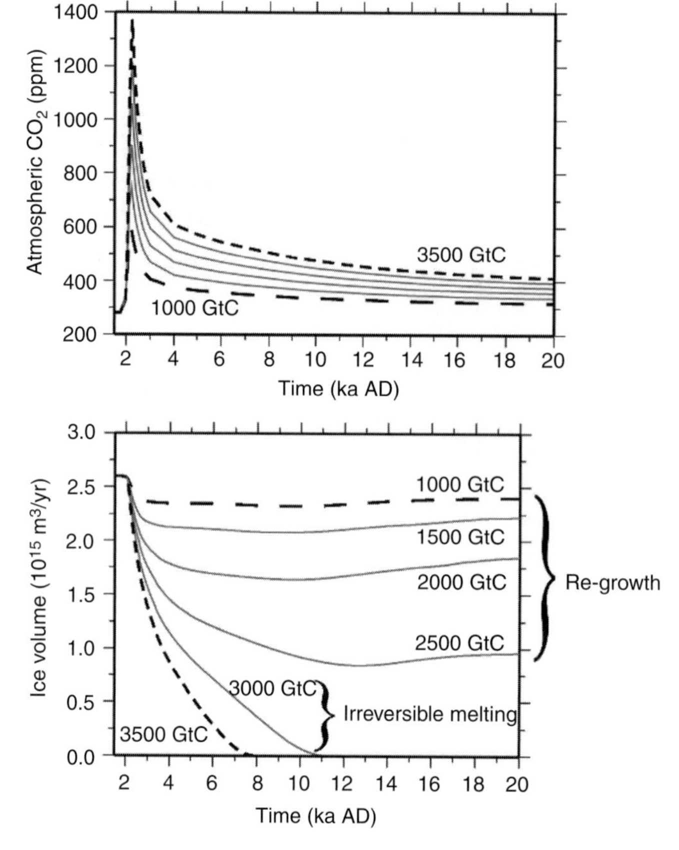

Figure 12.14. Atmospheric CO_2 concentration scenarios (top) and ice sheet response. Above 3000 Gt C (Gigaton of carbon), the Greenland melting is irreversible (bottom). (From Charbit *et al.*, 2008.)

sheets. Using the same models and different scenarios, we point out that for accumulated carbon-dioxide emission larger than 3000 gigatons of carbon in the atmosphere, the irreversible melting of Greenland's ice cap will ensue (Charbit *et al.*, 2008) (Figure 12.14). Moreover, the value is certainly overestimated, because many processes are parameterized in our model because the spatial resolution (40 km) is too large to include explicit physics of ice streams, for instance. Moreover, different types of sediments under the ice-sheet are important for sliding and not yet accounted for in our models. Comparison of our simulation and satellite data shows an underestimation of the ice streams' motion.

Conclusions

In the future, it would not be very surprising if human activity leads to the ice-sheet melting and the disappearance of the ice from at least Greenland and west Antarctica. First because industrial development has induced huge carbon-dioxide and methane perturbations, and second, because, as we show in this contribution, ice-sheets are not very frequent and only exist for low CO_2 levels in Earth's history. Industrial development has deeply perturbed the CO_2 cycle, transferring into the atmosphere huge amounts of carbon that were previously stored on the continents or beneath the ocean. This perturbation has been proved by the CO_2 measured since 1950 in the atmosphere. This increase from 280 ppm (pre-industrial value) to 380 ppm in 2009 is already largely out of the last million years' levels (190–280 ppm), as measured in Antarctica's ice-sheets. In a few decades, the atmospheric CO_2 concentration will reach 420 ppm, a level that in the past was reached tens of Ma ago, a period when no Greenland ice-sheets existed on Earth's surface.

The question is perhaps rather different as the transition does not happen on a geological timescale, but 'instantaneously' with all reservoirs (atmosphere, ocean, vegetation, ice-sheet) not in equilibrium. Therefore, this scenario is more similar to a biodiversity crisis, because of its intensity and abruptness, than to the 'smooth' climate evolution we have described in this chapter. On the other hand, because the perturbation is associated with human activity it can be brought under control and hope lies in the fact that the emission scenarios of the twenty-first century are not yet fully written.

References

Anbar, A.D. and Knoll, A.H. (2002). Proterozoic ocean chemistry and evolution: A bioinorganic bridge? *Science*, **297** (5584): 1137–1142

Barron, E. J. and Washington, W. M. (1982). Cretaceous climate: a comparison of atmospheric simulations with the geologic record. *Palaeogeography, Palaeoclimatology, Palaeoecology*, **40**, 103–33.

Bertrand, P. (2005). Vers une régulation planétaire. In *Des Atomes aux Planètes Habitables*, eds. M. Gargaud, P. Claeys and H. Martin. Bordeaux: Presses Universitaires de Bordeaux, pp. 135–54.

Bintanja, R., van de Wal, R., and Oerlemans, J. (2005). Modelled atmospheric temperatures and global sea levels over the past million years. *Nature*, **437**, 125–128.

Bonelli, S., Charbit, S., Kageyama, M., Woillez, M.-N., Ramstein, G., Dumas, C. and Quiquet, A. (2009). Investigating the evolution of major northern hemisphere ice sheets during the last glacial–interglacial cycle. *Climate of the Past*, **5**, 329–45.

Catling, D. C. and Claire, M. W. (2005). How Earth's atmosphere evolved to an oxic state: a status report. *Earth and Planetary Science Letters*, **237**, 1–20.

Charbit S., Paillard D. and Ramstein G. (2008). Amount of CO_2 emissions irreversibly leading to the total melting of Greenland. *Geophysical Research Letters*, **35**, L12503, 10.1029/2008GL033472.

Cogné, J. P. and Humler, E. (2004). Temporal variation of oceanic spreading and crustal production rates during the last 180 Myr. *Earth and Planetary Science Letters*, **227**, 427–39.

Crowell, J. C. (1999). Pre-Mesozoic ice ages: their bearing on understanding the climate system. Geological Society of America, Boulder. *Geological Society of America Memoir*, **192**.

DeConto, R. and Pollard, D. (2003). Rapid Cenozoic glaciation of Antarctica induced by declining atmospheric CO_2. *Nature*, **421**, 245–9.

Dickens, G. R. (2003). Rethinking the global carbon cycle with a large, dynamic and microbially mediated gas hydrate capacitor. *Earth and Planetary Science Letters*, **213**(3–4), 169–83.

Dickens, G. R. and Francis, J. M. (2004). Comment: a case for a comet impact trigger for the Paleocene/Eocene thermal maximum and carbon isotope excursion. *Earth and Planetary Science Letters*, **217**, 197–200.

Donnadieu, Y., Goddéris, Y., Ramstein, G. and Fluteau, F. (2004a). Global tectonic setting and climate of the Late Neoproterozoic: a climate-geochemical coupled study. In *Multidisciplinary Studies Exploring Extreme Proterozoic Environment Conditions*, eds. G. S. Jenkins, M. McMenamin, C. P. McKay and L. E. Sohl. Geophysical Monograph Series, 79–89.

Donnadieu, Y., Goddéris, Y., Ramstein, G., Nédelec, A. and Meert, J. G. (2004b). Snowball Earth triggered by continental break-up through changes in runoff. *Nature*, **428**, 303–6.

Donnadieu, Y., Ramstein, G., Fluteau, F., Besse, J. and Meert, J. (2002). Is high obliquity a plausible cause for Neoproterozoic glaciations? *Geophysical Research Letters*, **29**(23), 2127.

Donnadieu, Y., Godderis, Y., Pierrehumbert, R., Dromart, G., Fluteau, F. and Jacob, R. (2006). A GEOCLIM simulation of climatic and biogeochemical consequences of Pangea breakup. *Geochemistry Geophysics Geosystems*, **7**, Q11019, 21.

Fluteau, F., Ramstein, G., Besse, J., Guiraud, R. and Masse J. P. (2007). Impacts of palaeogeography and sea level changes on Mid-Cretaceous climate. *Palaeogeography, Palaeoclimatology, Palaeoecology*, **247**(3–4), 357–81.

Goddéris, Y., Donnadieu, Y., Dessert, C., Dupré, B., Fluteau, F., François, L. M., Meert, J., Nédélec, A. and Ramstein, G. (2007). Coupled modeling of global carbon cycle and climate in the Neoproterozoic: links between Rodinia breakup and major glaciations. *Comptes Rendus Geosciences de Paris*, **339**(3–4), 212–22.

Gradstein, F. M. and Ogg, J. G. (2004). Geologic time scale 2004: why, how and where next! *Lethaia*, **37**(2), 175–82.

Harland, W. B. (1964). Evidence of Late Precambrian glaciation and its significance. In *Problems in Palaeoclimatology*, ed. A. E. M. Nairn. London, UK: Interscience, John Wiley and Sons, pp. 119–49.

Hoffman, P. F. and Schrag, D. P. (2002). The snowball Earth hypothesis: testing the limits of global change. *Terra Nova*, **14**, 129–55.

Hoffman, P. F., Kaufman, A. J., Halverson, G. P. and Schrag, D. P. (1998). A Neoproterozoic snowball Earth. *Science*, **281**(5381), 1342–6.

Jouzel, J., Masson-Delmotte, V., Cattani, O., Dreyfus, G., Falourd, S., Hoffmann, G., Minster, B., Nouet, J., Barnola, J.M., Chappellaz, J., Fischer, H., Gallet, J.C., Johnsen, S., Leuenberger, M., Loulergue, L., Luethi, D., Oerter, H., Parrenin, F., Raisbeck, G., Raynaud, D., Schilt, A., Schwander, J., Selmo, E., Souchez, R., Spahni, R., Stauffer, B., Steffensen, J.P., Stenni, B.S., T.F., Tison, J.L., Werner, M., and Wolff, E (2007), Orbital and millennial Antarctic climate variability over the past 800,000 years. *Science*, **317**, 793–796.

Kasting, J. F. (2004). When methane made climate. *Scientific American*, **291**, 78–85.

Kirschvink, J. L. (1992). Late Proterozoic low-latitude glaciation: the snowball Earth. In *The Proterozoic Biosphere*, ed. J. W. Schopf and C. Klein. Cambridge: Cambridge University Press, pp. 51–2.

Knoll, A. H. (2004). *Life on a Young Planet: The First Three Billion Years of Evolution on Earth*. Princeton: Princeton University Press.

Köppen, W. and Wegener, A. (1924). Die klimate der geologischen vorzeit. *Astronomische Nachrichten*, **262**, 393–410.

Le Hir, G., Goddéris, Y., Donnadieu, Y. and Ramstein, G. (2008a). A geochemical modelling study of the evolution of the chemical composition of seawater linked to a 'snowball' glaciation. *Biogeosciences*, **5**, 253–67.

Le Hir, G., Goddéris, Y., Donnadieu, Y. and Ramstein, G. (2008b). A scenario for the evolution of the atmopsheric PCO_2 during a snowball Earth. *Geology*, **36**(1), 47–50.

Le Hir, G., Donnadieu, Y., Goddéris, Y., Pierrehumbert, R. T., Halverson, G. P., Macouin, M., Nédélec, A. and Ramstein, G. (2008c). The snowball Earth aftermath: Exploring the limits of continental weathering processes. *Earth and Planetary Science Letters*, **277**(3–4), 453–63.

Levrard, B. and Laskar, J. (2003). Climate friction and the Earth's obliquity. *Geophysical Journal International*, **154**, 970–90.

Lisiecki, L.E. and Raymo, M.E. (2005). A Pliocene-Pleistocene stack of 57 globally distributed benthic δ18O records. *Paleoceanography*, **20**, PA2007.

Loulergue, L., Schilt, A., Spahni, R., Masson-Delmotte, V., Blunier, T., Lemieux, B., Barnola, J.M., Raynaud, D., Stocker, T. and Chappelaz, J. (2008). Orbital and millenial-scale features of atmospheric CH4 over the last 800 000 years. *Nature*, **453**, 383–386.

Lovelock, J. (1979). *Gaia: A New Look at Life on Earth*. Oxford: Oxford University Press.

Lunt, D. J., Foster, G. L., Haywood, A. M. and Stone, E. J. (2008). Late Pliocene Greenland glaciation controlled by a decline in atmospheric CO_2 levels. *Nature*, **454**, 1102–5.

Masson-Delmotte, V., Stenni, B., Pol, K., Braconnot, P., Cattani, O., Falourd, S., Kageyama, M., Jouzel, J., Landais, A., Minster, B., Barnola, J.M., Chappellaz, J., Krinner, G., Johnsen, S., Rothlisberger, R., Hansen, J., Mikolajewicz, U. and Otto-Bliesner, B. (2010). EPICA Dome C record of glacial and interglacial intensities. *Quaternary Science Reviews*, **29**, 113–128.

Morbidelli, A., Petit, J.-M. and Gladman, B. (2001). A plausible cause of the late heavy bombardment. *Meteoritics and Planetary Science*, **36**, 371–80.

Ohmoto, H., Watanabe, Y., Ikemi, H., Poulson, S. R. and Taylor, B. E. (2006). Absence of mass-independent fractionation of sulphur isotopes: an oxic Archean atmosphere? *Nature*, **442**, 908–11.

Pagani, M., Zachos, J. C., Freeman, K. H., Tipple, B. and Bohaty, S. (2005). Marked decline in atmospheric carbon dioxide concentrations during the Paleogene. *Science*, **309**, 600–3.

Paillard, D. (2008). From atmosphere, to climate, to Earth System science. *Interdisciplinary Science Reviews*, **33**(1), 25–35.

Petit, J. R., Jouzel, J., Raynaud, D., Barkov, N. I., Barnola, J.-M., Basile, I., Bender, M., Chappellaz, J., Davisk, M., Delaygue, G., *et al.* (1999). Climate and atmospheric history of the past 420,000 years from the Vostok ice core, Antarctica. *Nature*, **339**, 429–436.

Petoukhov, V., Ganopolski, A., Brovkin, V., Claussen, M., Eliseev, A., Kubatzki, C. and Rahmstorf, S. (2000). CLIMBER-2: a climate system model of intermediate complexity. Part I: model description and performance for present climate. *Climate Dynamics*, **16**, 1–17.

Pierrehumbert, R. T. (2004). High levels of atmospheric carbon dioxide necessary for the termination of global glaciation. *Nature*, **429**, 646–9.

Pol, K., Masson-Delmotte,V., Johnsen, S., Bigler, M., Cattani, O., Durand, G., Falourd, S., Jouzel, J., Minster, B., Parrenin, F., Ritz, C., Steen-Larsen, H.C. and Stenni B. (2010). New MIS 19 EPICA Dome C high resolution deuterium data: Hints for a problematic preservation of climate variability at sub-millenial scale in the "oldest ice". *Earth and Planetary Science Letters*, **298**, 95–103.

Pucéat, E., Lécuyer, C., Donnadieu, Y., Naveau, P., Cappetta, H., Ramstein, G., Huber, B. T. and Kriwet, J. (2007). Fish tooth $\delta_{18}O$ revising Late Cretaceous meridional upper ocean water temperature gradients. *Geology*, **35**(2), 107–10.

Ramstein, G., Donnadieu, Y. and Goddéris, Y. (2004a). Proterozoic glaciations. *Comptes Rendus Geoscience*, **336**, 639–46.

Ramstein, G., Khodri, M., Donnadieu, Y. and Goddéris Y. (2004b). Impact of the hydrological cycle on past climate changes: three illustrations at different time scales. *Comptes Rendus Geoscience*, **337**, 125–37.

Ritz, C., Fabre, A. and Letreguilly, A. (1997). Sensitivity of a Greenland ice sheet model to ice flow and ablation parameters: consequences for the evolution through the last climatic cycle. *Climate Dynamics*, **13**, 11–24.

Royer, DL. (2006). CO_2-forced climate thresholds during the Phanerozoic. *Geochimica et Cosmochimica* Acta, **70**, 5665–5675

Robert, F. and Chaussidon, M. (2006). A palaeotemperature curve for the Precambrian oceans based on silicon isotopes in cherts. *Nature*, **443**, 969–72.

Rye, R., Kuo, P. and Holland, H. D. (1995). Atmospheric carbon dioxide concentrations before 2.2 billion years ago. *Nature*, **378**, 603–5.

Sagan, C. and Mullen, G. (1972). Earth and Mars: evolution of atmospheres and temperatures. *Science*, **177**, 52–6.

Siengenthaler, U., Stocker, T.F., Monnin, E., Lüthi, D., Schwander, J., Stauffer, B., Raynaud, D., Barnola, J.-M., Fischer, H., Masson-Delmotte, V. and Jouzel, J. (2005). Stable carbon cycle-climate relationship during the last Pleistocene. *Science*, **310**, 1313–1317.

Waelbroeck, C., Labeyrie, L., Michel, E., Duplessy, J. C., McManus, J. F., Lambeck, K., Balbon, E. and Labracherie, M. (2002). Sea-level and deep water temperature changes derived from benthic foraminifera isotopic records. *Quaternary Science Reviews*, **21**, 295–305.

Walker, J. C. G., Hays, P. B. and Kasting, J. F. (1981). A negative feedback mechanism for the long-term stabilization of Earth's surface temperature. *Journal of Geophysical Research*, **86**, 9776–82.

Wegener, A. (1915). *Die Entstehung der Kontinente und Oceane*. Braunschweig: Friedrich Vieweg & Sohn Akt. Ges.

Williams D., Kasting, J. F. and Frakes, L. A. (1998). Low-latitude glaciation and rapid changes in the Earth's obliquity explained by obliquity-oblateness feedback. *Nature*, **396**, 453–5.

Williams, G. E. (1975). Late Precambrian glacial climate and the Earth's obliquity. *Geological Magazine*, **112**, 441–65.

Young, G. M., von Brunn, V., Gold, D. J. C. and Minter, W. E. L. (1998). Earth's oldest reported glaciation; physical and chemical evidence from the Archean Mozaan Group (2.9 Ga) of South Africa. *Journal of Geology*, **106**, 523–38.

Zachos, J., Pagani, M., Sloan, L., Thomas, E. and Billups, K. (2001). Trends, rhythms, and aberrations in global climate 65 Ma to Present. *Science*, **292**, 686–93.

Part III

The role of water in the emergence of life

13

Liquid water: a necessary condition for all forms of life?

Kristin Bartik, Gilles Bruylants, Emanuela Locci and Jacques Reisse

Introduction

In a paper entitled 'The prospect of alien life in exotic forms on other worlds' published in 2006, the authors write:

> The nature of life on Earth provides a singular example of carbon-based, water-borne, photosynthesis-driven biology. Within our understanding of chemistry and the physical laws governing the universe, however, lies the possibility that alien life could be based on different chemistries, solvents, and energy sources from the one example provided by Terran biology. (Schulze-Makuch and Irwin, 2006)

Similar comments can be found in several papers (Bains, 2004).

We are not planning to address the possibility of an alien life, but wish to focus on the issue of the solvent in order to try to demonstrate that water is an essential component of all living systems. Living systems are complex both at the molecular and supra-molecular levels (Schulze-Makuch and Irwin, 2006). Water plays, at both these levels, a role which is crucial for the structure, the stability and the biological function of all molecules that are essential for life, a role that cannot be played by any other solvent or any other molecule.

A solvent is never an inert medium and always interacts with the solute molecules. These interactions affect not only the solute but also the solvent. Water is a unique solvent because solute-induced modifications are very important in this medium. Even if certain properties of pure liquid water are still not fully elucidated and remain the subject of intense research in physical chemistry (Brezin, 2004), it is possible to explain why liquid water behaves differently to any other liquid.

In this chapter we will first introduce the important general concept of solvophobicity, called hydrophobicity when the solvent is water, before focusing on the unique role that liquid water plays on the association of solute molecules. We will also explain why the presence of liquid water seems a prerequisite to reach the organization level of matter necessary to achieve a living and functioning cell.

Solvophobicity – hydrophobicity

Despite the large number of books (Franks, 1973; Ben-Naim, 1980; Tanford, 1980) and papers (Pratt and Chandler, 1977; Silverstein *et al.*, 2000; Southall *et al.*, 2002; Gallagher and Sharp, 2003; Chandler, 2005; Lynden-Bell and Head-Gordon, 2006; Graziano, 2008; Qvist and Halle, 2008) devoted to hydrophobicity, this concept is still frequently misunderstood. In this chapter we intend to present the topic by first insisting on the very general phenomenon called solvophobicity, which is a property common to all solvents (Moura Ramos *et al.*, 1977) before focusing on hydrophobicity.

Solvophobicity can be described in a semi-quantitative manner by considering the transfer of a solute molecule, **A**, from the gas phase into a solvent. This process can be described as the sum of two virtual steps (Pierotti, 1965; Blokzijl and Engberts, 1993): (1) the creation of a cavity in the solvent which can accommodate **A**, and (2) the placement of **A** into this cavity. If we consider the whole process, starting from a mole of **A** in a perfect gas phase and finishing with a mole of **A** dissolved, at infinite dilution, in the solvent, the free energy of transfer, ΔG_{tr}, can be expressed as the sum of two terms:

$$\Delta G_{tr} = \Delta G_{cav} + \Delta G_{\text{int}} \tag{13.1}$$

As for all free energy changes, each of these terms is the sum of an enthalpic (ΔH) and an entropic (ΔS) contribution:

$$\Delta G = \Delta H - T\Delta S \tag{13.2}$$

To create the cavity it is necessary to act against the cohesion of the solvent. This requires a cost in terms of free energy and ΔG_{cav} will consequently be positive. As the cavity has been created to accommodate **A**, the attractive interactions between **A** and the surrounding solvent molecules will necessarily lead to a favourable enthalpy change and ΔH_{int} will be negative. Obviously, the mobility of **A** will be reduced when it is transferred from the gas phase to the liquid phase, which leads to a decrease in its entropy, and ΔS_{int} will be negative. Consequently, the sign of ΔG_{int} will depend on the relative contributions of the enthalpic and entropic terms.

The cavity-formation step, in which the eventual reorganization of the solvent molecules at the cavity interface is included, is always endergonic (absorbing energy in the form of work) and ΔG_{cav} is a measure of the solvophobicity of **A** when immersed in the solvent. It is indeed easy to understand why, for the same solvent, the value of ΔG_{cav} increases with the size of the cavity and why for a defined cavity size, ΔG_{cav} increases with the cohesion of the solvent. The cohesion of a liquid can be quantified by using different descriptors such as the surface tension (γ) or the cohesion energy density (CED). If we exclude mercury, water has the highest surface tension among 200 pure liquids (71.99 mN/m at 25°C) (*Handbook of Chemistry and Physics*, 2001–2002). Hydrocarbons are characterized by a low surface tension, which is furthermore more or less independent of the molecular weight of the hydrocarbon (21.14 mN/m for octane and 27.87 mN/m for octadecane). We can infer from these values that the cohesion of liquid methane is certainly much lower than the cohesion of water and we can safely conclude that the free energy cost of creating a cavity of a definite size in water is higher than in any other solvent. The high cohesion of liquid water is,

Table 13.1. Thermodynamic parameters associated with xenon dissolution (25°C, 1 atm) (Clever, 1979).

	ΔH_{tr} (kJ mol^{-1})	$T\Delta S_{tr}$ (kJ mol^{-1})	ΔG_{tr} (kJ mol^{-1})	Solubility ° (M^{-1})
H_2O	−19.4	− 42.8	23.4	4.3×10^{-3}
C_8H_{18}	−10.4	− 19.3	8.9	150×10^{-3}

of course, related to the small size of the water molecule and to the fact that liquid water is a highly organized solvent, characterized by an extended network of hydrogen bonds.

The solvophobicity of a particular solute immersed in water, called hydrophobicity, is qualitatively different from solvophobicity in a solvent like a hydrocarbon or a ketone. There is also a quantitative difference as ΔG_{cav} is generally markedly larger for water than for common organic liquids (Graziano, 2008). In a non-structured liquid, such as octane, the creation of a cavity does not require the reorganization of solvent molecules. The ΔS_{cav} term will be negligible and the ΔH_{cav} term, corresponding to the loss of favourable non-covalent interactions between the solvent molecules, will always be predominant. In water, it is generally the entropy contribution to ΔG_{cav} which dominates. Liquid water is a highly structured medium which can be described as a fluctuating lattice of water molecules linked to each other by H-bonds. As discussed by Chandler (2005), if the cavity is small (nanometric), it can be created without breaking many H-bonds, through a reorganization of the water molecules in the immediate vicinity of the cavity. Water molecules adopt orientations that allow hydrogen-bonding patterns to accommodate the solute: ΔH_{cav} will be negligible but not ΔS_{cav}. If the cavity is large, it becomes impossible for adjacent water molecules at the interface to maintain a complete hydrogen-bonding network. As water molecules 'do not want to sacrifice' any of their hydrogen bonds, they lose some of their entropy in order to minimize this loss: on average less than one hydrogen bond per water molecule is sacrificed compared with that in the bulk liquid (Chandler, 2005; Graziano and Lee, 2005). Consequently, both the enthalpic and entropic contributions to ΔG_{cav} will be unfavourable. In summary, whereas ΔS_{cav} is negligible in most solvents, it is always important in water.

The comparison of the solubility of xenon in octane and water allows us to exemplify the difference between solvophobicity in the two solvents and to illustrate the specificity of hydrophobicity for a small hydrophobic solute. Table 13.1 gives the ΔH_{tr}, $T\Delta S_{tr}$ and ΔG_{tr} values associated with the transfer of a mole of xenon from the gas phase under a pressure of one atmosphere (which will be considered as a perfect gas state) and at a temperature of 25°C, into a large amount of solvent, leading to an infinitely diluted solution.

Xenon is undoubtedly a hydrophobic solute which is only able to interact with a solvent through London and Debye interactions. Its solubility in octane is much higher than in water, which is expected for a hydrophobic solute. The entropy term is negative in both solvents, partially due to the fact that xenon atoms have more translational freedom in the gas phase than in solution. The loss of entropy is, however, much higher in water than in octane. It is difficult to imagine that the translation of xenon is more hindered in water than in octane. Most of the difference between the entropic contributions (−19.3 kJ. $mole^{-1}$ in octane and −42.8 in water) results from the entropy loss of water when xenon is dissolved, i.e. to $T\Delta S_{cav}$. The transfer enthalpies for the two solvents are the sum of a cavity contribution, ΔH_{cav}, and a solute–solvent interaction contribution, ΔH_{int}. It is impossible to measure independently these two contributions but on the basis of transfer enthalpy alone, xenon 'prefers' water when compared to octane. This is a spectacular illustration of the importance of the entropy contribution in dissolution processes.

The role of hydrophobicity in intermolecular associations

In the gas phase, when a molecule **A** interacts with a molecule **B** to form a complex **AB**, a loss of entropy is always associated with the process. For an equilibrium characterized by an association constant, K ($\Delta G° = -RT\ln K$ with R the ideal gas constant):

$$A + B \rightleftharpoons AB \tag{13.3}$$

the $\Delta S°$ term is negative: the entropy of the complex ($S°_{AB}$) is smaller than the sum of the entropies of the two isolated partners ($S°_A + S°_B$). Indeed, if **A** and **B** are non-linear molecules containing N_A and N_B atoms, respectively, the total number of vibrational degrees of freedom is $(3N_A - 6)$ for **A**, $(3N_B - 6)$ for **B** and $(3N_A + 3N_B - 6)$ for the complex **AB**. When the **AB** complex is formed, there is a loss of three translational degrees of freedom and three rotational degrees of freedom but a gain of six vibrational degrees of freedom. In terms of entropy change it is easy to show, using classical statistical thermodynamics, that the loss of translational and rotational entropy is not compensated for by the gain of vibrational entropy (Knox, 1978). We can therefore conclude that when a complex is formed in the gas phase, it is necessarily an enthalpy-driven process.

The situation is different in solution. First of all, the translation and rotation of solute molecules are hindered in solution due to their interaction with solvent molecules and the corresponding contributions to the entropy of the molecules will consequently be much smaller than in the gas phase. Their vibrational modes are, however, essentially the same in the gas phase and in solution. If we only consider the interacting partners and the complex (**A, B** and **AB**) the loss of entropy associated with the complex formation in solution is necessarily lower than in the gas phase.

An additional factor that must be taken into account when association occurs in solution is the fact that, due to the formation of a contact zone between **A** and **B**, the solute–solvent interface is generally reduced. In water, as the ΔG_{cav} term depends on the area of the solute–solvent interface due to the reorganization of the solvent molecules around the solute, the ΔG_{cav} term for the complex **AB** will be smaller than the sum of the corresponding terms for **A** and B:

$$\Delta G_{cav}(AB) < \Delta G_{cav}(A) + \Delta G_{cav}(B) \tag{13.4}$$

ΔG_{cav} is always positive, which implies that, all other things being equal, hydrophobicity favours association by 'pushing **A** and **B** together' ($\Delta\Delta G_{cav} < 0$). As the increase of water entropy is generally greater than the entropy decrease of the solutes, the association of hydrophobic molecules in water is promoted by entropy. This is called the 'hydrophobic effect'.

It must be noted that this presentation of the role played by entropy in the formation of a complex in water is oversimplified. Indeed, as the complex is generally larger than either **A** or **B**, it is possible that when **A** and **B** are small molecules the term ΔH_{cav} is negligible for the isolated entities but not for the larger **AB** complex. A relevant question, of course, concerns the concept of 'small molecule'. The large number of entropy-driven associations described in the literature and the fact that they also concern associations which involve proteins (Luke *et al.*, 2005), suggests that is it not necessarily the overall size of the molecule that is important but the region directly involved in the association process. Furthermore, as far as proteins are concerned, it is important to point out that their solvent-exposed surface is highly heterogeneous and presents hydrophilic and hydrophobic regions.

Entropy-driven associations play a role in many processes of major importance in the living world. This behaviour can only be explained because water is the solvent: nothing similar could occur in liquid methane or liquid nitrogen and this is one of the reasons why liquid water is indeed a fundamental component of all living cells.

At the end of this section devoted to the role of hydrophobicity in association processes, it is important to point out why the expression 'hydrophobic interaction' is misleading. Indeed, it is frequently confused with van der Waals interactions. It is true that hydrophobic molecules generally interact which each other or with water molecules mainly through van der Waals interactions, but hydrophobicity is essentially due to induced changes in the solvent surrounding the hydrophobic solute.

Why water is essential for life

Obviously, today the presence of life on Earth depends on the availability of liquid water. In living systems water is pervasive and ubiquitous and cannot be considered as a simple diluter. It performs many functions: it transports, structures, stabilizes, lubricates, reacts and partitions. The finely tuned involvement of water in life processes is confirmed by the fact that heavy water is toxic to these processes.

Membranes and water

A living cell is a system that is physically distinct from its environment: a semi-permeable frontier acts as a barrier between the 'inside' and the 'outside'. In living cells as we know them today, this frontier is a layer made of amphiphilic molecules. Phosphoesters of fatty acids (N-acyl phospholipids) are probably the best-known examples of membrane components.

Several authors (Ourisson and Nakatani, 1996; Pozzi *et al.*, 1996; Berti *et al.*, 1998; Bloechliger *et al.*, 1998; Luisi *et al.*, 1999; Takajo *et al.*, 2001) consider that the spontaneous formation of membranes is one of the primordial steps in prebiotic evolution, not only because they separate an 'in' and an 'out' but also because membranes represent an apolar medium in which condensation reactions can take place. These reactions – which involve the elimination of a water molecule – would be impossible in water.

In water, the spontaneous formation of vesicles or liposomes is a well-known and reproducible process. The auto-association of lipid molecules is accompanied by an entropy decrease that is overcompensated for by the entropy increase of the water molecules released from the surface of the hydrophobic tail of the amphiphilic molecules. The spontaneous evolution of the overall system, from isolated phospholipidic molecules dissolved in water to vesicles suspended in water, is governed by the entropy increase of water upon desolvatation of hydrophobic surfaces. This suggests that, without liquid water, life based on cells separated from the external world by organic membranes could never have appeared on Earth.

Ions and water

It is essentially impossible to imagine a living system which does not require the presence of ionic species and ionic gradients between the intracell and the extracell media. Obviously, the role of ions is highly dependent on the presence of water. Water is a unique solvent in ionic chemistry, not only because of the high value of its dielectric constant, but also because water molecules interact efficiently with many anions via H-bonds and with many cations via coordination between the lone pairs on the oxygen atom and the empty orbitals of the cations. In particular, water has a singular affinity for protons and it is known that hydroxonium ions play a prominent role in cell machinery (Rawn, 1989). It is difficult to imagine any catalytic or any energy-exchange processes which do not involve ionic species and, consequently, water.

Proteins and water

Proteins also take part in important biological functions: they catalyse and regulate reactions and transport substrates. It is widely appreciated that water molecules play an invaluable role in governing the structure, stability, dynamics and function of these biomolecules (Poornima and Dean, 1995a; Franks, 2002; Pal *et al.*, 2002; Takano *et al.*, 2003; Halle, 2004). Indeed, they lack activity in the absence of water (Pal and Zewail, 2004). Even if, in some cases, the possibility of life without proteins seems conceivable (Orgel, 2004), proteins were probably selected very early on by the first living cells as efficient catalysts.

Proteins are polymers of α-amino acids which, under reasonable pH, temperature and ionic strength conditions, generally adopt a preferred three-dimensional (3D) fold. This conformation, and especially the conformation of the active site in the case of enzymes, is essential to ensure that the protein can play its biological role. A protein reaches its native conformation through a very complex process called folding. Starting from the so-called denatured state, which has no preferred conformation, folding towards the native conformation is spontaneous and must therefore be associated with a free-energy decrease.

This free-energy decrease has been estimated to be around 40 to 60 kJ mole^{-1} (Finney, 1979). This value is high enough to explain why proteins adopt their native conformation while still maintaining enough flexibility to fulfil their diverse roles.

All studies devoted to the protein-folding problem emphasize the role of water, even if it would be an oversimplification to consider that the increase in water entropy associated with the process is the unique driving force of the folding. There are many enthalpic and entropic contributions which are different in the denatured and in the native protein and it is a subtle balance between these terms which leads to the lower free energy of the native conformation. Even if electrostatic interactions and van der Waals (London) interactions play important roles in determining the stability of proteins, the hydrophobic effect, which leads to the burial of the hydrophobic amino-acid side-chains in the core of the protein, is generally considered to be the main driving force for the folding of globular proteins. In the late 1950s Kauzmann introduced the concept of 'hydrophobicity' to account for the complexities of protein folding (Kauzmann, 1959).

If water is an essential solvent for protein folding, molecular water is also essential for protein stability and activity. Internal water molecules occupy cavities within the protein and are present in most globular proteins (Williams *et al.*, 1994; Halle, 2004). These trapped water molecules are, strictly speaking, part of their structure. Some of them are required in order for the protein to fulfil its biological role while others contribute to the structure and stability of proteins by bridging, via hydrogen bonding, different functional groups. Water molecules have been found to be conserved and occupy the same position in homologous proteins (Sreenivasan and Axelsen, 1992; Poornima and Dean, 1995b). The residence times of internal water molecules have been estimated by nuclear magnetic resonance studies to range between 10^{-2} and 10^{-8} s (Otting, 1997). Hydrogen-bonding networks, involving several water molecules which link distinct parts of a protein or of an association complex, have also been observed experimentally (Sharrow *et al.*, 2005). Proof of the importance of water for proteins is that mutations can affect the number of structural water molecules within the protein core and disrupt essential interactions mediated by water H-bonding, resulting in the destabilization of the 3D structure of the protein (Covalt *et al.*, 2001).

Nucleic acids and water

Even if it is clear that the 'RNA world' must have been preceded by other stages (Orgel, 2004) it is widely accepted that ribonucleic acid (RNA) acted as a precursor of both protein and deoxyribonucleic acid (DNA), in the sense that it can serve both as a catalyst (like protein enzymes) and as a carrier of genetic information (like DNA).

The basic repeating unit in nucleic acids is the nucleotide, which is composed of a cyclic sugar that is phosphorylated in one position and substituted by one of five different heterocyclic aromatic bases in another. The nucleotides are linked together via phosphate groups (phosphodiester bonds). The specific functions of DNA and RNA molecules are modulated by their 3D structure, which depends not only on nucleotide sequence and base composition but also greatly on the environmental conditions.

DNA is double stranded and is usually thought of in terms of a right-handed double helix with complementary bases pairing via H-bonds. The base pairs, located in the centre of the double helix, are coplanar and perpendicular to the helix axis. This 3D structure is capable of considerable variations and can undergo structural transitions between several different conformations. Whilst some of these changes are quite subtle, others are very dramatic and can, for example, involve a complete reversal of the handedness of the double helix. RNAs are not double stranded but adopt regular folds and usually more than half of the nucleotides participate in base pairing. Consecutive stacking of the base pairs gives helices connected by single-stranded stretches.

Along with hydrogen bonding between base pairs and London dispersion forces between the stacked bases, water contributes to the stabilization of RNA and of DNA structures (Berman, 1997; Wahl and Sundaralingam, 1997; Hermann and Patel, 1999; McConnell and Beveridge, 2000). The role of water in the stabilization of the 3D structure of nucleic acids is even more important than in proteins because of the presence of negatively charged phosphate groups. Phosphate–phosphate electrostatic repulsion is diminished in water.

As in proteins, water is an integral part of nucleic-acid structures and the degree of hydration of nucleic acids plays a key role in the helical conformation that is adopted. A shell of hydration is located around DNA double helices (Saenger, 1984). These water molecules are specifically bound to the phosphate groups and to the bases. X-ray studies show that there are six hydration sites per phosphate and that the positions and occupancies of these sites are dependent on the conformation and type of nucleotide (Berman, 1997; Wahl and Sundaralingam, 1997; Schneider *et al.*, 1998). Hydration is more persistent around the bases than around the phosphate groups as the electron distribution of the latter is more diffuse. The paired bases are capable of hydrogen bonding to water within the grooves. A spine of hydration is, for example, observed by X-ray crystallography in the minor groove of B-DNA, the conformation most frequently encountered *in vivo.* RNA is more hydrated than DNA due to its extra oxygen atoms (the ribose O2′) and unpaired base sites. Stable hydration patterns are observed, as in DNA, around the double-stranded regions (Auffinger and Westhof, 1997; Berman, 1997; Wahl and Sundaralingam, 1997; Schneider *et al.*, 1998; Hermann and Patel, 1999; McConnell and Beveridge, 2000). Trapped long-residency water molecules have also been observed in the catalytic core of ribozymes (Rhodes *et al.*, 2006).

Role of water in molecular recognition involving biomolecules

Entropy-driven binding processes, as discussed in the earlier section on hydrophobicity, constitute a phenomenon widely encountered in molecular biology and play a role in many processes of major importance in the living world. Biological macromolecules present hydrophobic and hydrophilic surfaces. The formation of an association complex will be most favourable when the hydrophobic and hydrophilic surfaces of the binding partners are complementary. A high loss of enthalpy due to the dehydration of polar groups which is not compensated for by a favourable enthalpy of interaction with another polar group can of course compensate for the benefits of the dehydration of hydrophobic surfaces.

Most binding processes are highly specific and molecular water plays a key role in the formation of the association complexes (Ladbury, 1996; Tame *et al.*, 1996; Ben-Naim, 2002). The fact that water molecules are abundantly observed experimentally at the interface between (bio)molecules suggests that water is indispensable for biomolecular recognition and self-assembly. Molecular-recognition studies focused for a long time on the geometrical complementarity between the partners, eventually after conformational changes, assuming that the driving forces for binding originate from direct interaction between the partners (attractive London interactions and presence of matching functional groups leading to H-bonding, salt bridges etc.) as it is now generally accepted that water molecules are part of the association complex and must be explicitly considered when undertaking molecular-recognition studies. Water is a highly versatile component at the interface of biomolecular complexes. It can act both as a hydrogen-bond donor and acceptor, imposes few steric constraints on bond formation and can take part in multiple hydrogen bonds. Water can thus confer a high level of adaptability to a surface, allowing promiscuous binding. It can also provide specificity and increased affinity to an interaction. The water molecules involved in protein–ligand interactions can be seen as an extension of the protein structure, allowing optimization of the fit at the interface (Bhat *et al.*, 1994; Hamelberg and McCammon, 2004; Cohen *et al.*, 2005) or the accommodation of varied ligands (Tame *et al.*, 1996). For these water molecules, the energetic gain from water-mediated contacts is greater than the entropic cost resulting from their immobilization.

Binding processes are ubiquitous in living systems as we know them today; they range from the binding of small molecules to the binding of proteins to DNA. It appears impossible to imagine any form of life that could be sustainable without molecular interactions in which water molecules are involved.

Conclusions

In this chapter we have tried to demonstrate that water is essential for all living systems. Water is a unique molecule: it can act both as a hydrogen-bond donor and acceptor, imposes few steric constraints on bond formation and can take part in multiple hydrogen bonds. Water molecules can interact efficiently with anions via H-bonds and with cations via coordination between the lone pairs on the oxygen atom and empty orbitals of the cations. The unusual properties of liquid water, such as its high cohesion-energy density, high dielectric constant and surface tension, are a consequence of the small size of the molecule and of the extensive hydrogen-bonding network between the molecules. Even if some water-like liquids might show comparable properties to water at other thermodynamic state points, these would be far from the ambient conditions under which most biological activity has been optimized (Lynden-Bell and Head-Gordon, 2006).

The living world should be thought of as an equal partnership between proteins, nucleic acids, lipids, ions … and water. The description of water as the solvent and the other (macro)molecules as the solutes does not mean that the solutes are 'more important' than the solvent. In a hydroxonium ion, it would be absurd to consider that the proton is more important than the water molecule; it is the same for all solvated ions.

All living species, from bacteria to roses, from archea to man, follow the laws of physics and chemistry of open systems exchanging matter end energy with their surroundings. One must reflect on why structural complexity, such as membrane formation, was able to appear spontaneously through a 'simple' free-energy decrease. Clearly, liquid water was a prominent actor in this near-equilibrium evolution: the entropy decrease associated with the complexity increase being overcompensated for by the entropy increase of water. Even if the formation of protometabolic molecules is possible in other solvents, it is highly improbable that these media could play similar roles to that of water.

It is not unwarranted to put forward the idea that water has allowed the progressive transition from non-living to living matter. No other molecule, abundant on the primitive Earth, abundant in the Solar System, could have played this role, and certainly not methane or nitrogen. It is even highly probable that any other form of life in the Universe, if it exists, is based on liquid water.

Acknowledgements

G. B. thanks the Belgian Fonds de la Recherche Scientifique (FNRS) for his post-doctoral fellowship.

References

Auffinger, P. and Westhof, E. (1997). RNA hydration: three nanoseconds of multiple molecular dynamics simulations of the solvated tRNAAsp anticodon hairpin. *Journal of Molecular Biology*, **269**, 326–41.

Bains, W. (2004). Many chemistries could be used to build living systems. *Astrobiology*, **4**, 137–67.

Ben-Naim, A. (2002). Molecular recognition – viewed through the eyes of the solvent. *Biophysical Chemistry*, **101–102**, 309–19.

Ben-Naim, B. (1980). *Hydrophobic Interactions*. New York: Plenum Press.

Berman, H. M. (1997). Crystal studies of B-DNA: the answers and the questions. *Biopolymers*, **44**, 23–44.

Berti, D., Baglioni, P., Bonaccio, S., Barsacchi-Bo, G. and Luisi, P. L. (1998). Base complementarily and nucleoside recognition in phosphatidylnucleoside vesicles. *Journal of Physical Chemistry B*, **102**, 303–8.

Bhat, T. N., Bentley, G. A., Boulot, G., Greene, M. I., Tello, D., Dall'Acqua, W., Souchon, H., Schwarz, F. P., Mariuzza, R. A. and Poljak, R. J. (1994). Bound water molecules and conformational stabilization help mediate an antigen–antibody association. *Proceedings of the National Academy of Sciences of the United States of America*, **91**, 1089–93.

Bloechliger, E., Blocher, M., Walde, P. and Luisi, P. L. (1998). Matrix effect in the size distribution of fatty acid vesicles. *Journal of Physical Chemistry B*, **102**, 10383–90.

Blokzijl, W. and Engberts, J. B. F. N. (1993). Hydrophobic effects: opinions and facts. *Angewandte Chemie, International Edition in English*, **32**, 1545–79.

Brezin, E. (2004). *Demain la Physique*. Paris: Odile Jacob.
Chandler, D. (2005). Interfaces and the driving force of hydrophobic assembly. *Nature*, **437**, 640–7.
Clever, H. L. (1979). *IUPAC Solubility Data Series, vol. 2: Krypton, Xenon and Radon – Gas Solubilities*. Oxford, New York: Pergamon Press.
Cohen, G. H., Silverton, E. W., Padlan, E. A., Dyda, F., Wibbenmeyer, J. A., Willson, R. C. and Davies, D. R. (2005). Water molecules in the antibody–antigen interface of the structure of the Fab HyHEL-5 – lysozyme complex at 1.7 Å resolution: comparison with results from isothermal titration calorimetry. *Acta Crystallographica, Section D: Biological Crystallography*, **D61**, 628–33.
Covalt, J. C. J., Roy, M. and Jennings, P. A. (2001). Core and surface mutations affect folding kinetics, stability and cooperativity in IL-1b: does alteration in buried water play a role? *Journal of Molecular Biology*, **307**, 657–69.
Finney, J. L. (1979). The organization and function of water in protein crystals. In *Water: A Comprehensive Treatise*, vol. 6, ed. F. Francks. New York: Plenum Press Corp., pp. 47–122.
Franks, F. (1973). The solvent properties of water. In *Water: A Comprehensive Treatise*, vol. 2, ed. F. Franks. New York: Plenum Press, pp. 38–43.
Franks, F. (2002). Protein stability: the value of 'old literature'. *Biophysical Chemistry*, **96**, 117–27.
Gallagher, K. R. and Sharp, K. A. (2003). A new angle on heat capacity changes in hydrophobic solvation. *Journal of the American Chemical Society*, **125**, 9853–60.
Graziano, G. (2008). Hydrophobicity in modified water models. *Chemical Physics Letters*, **452**, 259–63.
Graziano, G. and Lee, B. (2005). On the intactness of hydrogen bonds around nonpolar solutes dissolved in water. *Journal of Physical Chemistry B*, **109**, 8103–7.
Halle, B. (2004). Protein hydration dynamics in solution: a critical survey. *Philosophical Transactions of the Royal Society of London, Series B*, **359**, 1207–24.
Hamelberg, D. and McCammon, J. A. (2004). Standard free energy of releasing a localized water molecule from the binding pockets of proteins: double-decoupling method. *Journal of the American Chemical Society*, **126**, 7683–9.
Handbook of Chemistry and Physics. (2001–2002). Boca Raton, New York, London, Tokyo: CRC Press.
Hermann, T. and Patel, D. J. (1999). Stitching together RNA tertiary architectures. *Journal of Molecular Biology*, **294**, 829–49.
Kauzmann, W. (1959). Some factors in the interpretation of protein denaturation. *Advances in Protein Chemistry*, **14**, 1–63.
Knox, J. H. (1978). *Molecular Thermodynamics: An Introduction to Statistical Mechanics for Chemists*. New York: Wiley.
Ladbury, J. E. (1996). Just add water! The effect of water on the specificity of protein–ligand binding sites and its potential application to drug design. *Chemistry and Biology*, **3**, 973–80.
Luisi, P. L., Walde, P. and Oberholzer, T. (1999). Lipid vesicles as possible intermediates in the origin of life. *Current Opinion in Colloid and Interface Science*, **4**, 33–9.
Luke, K., Apiyo, D. and Wittung-Stafshede, P. (2005). Dissecting homo-heptamer thermodynamics by isothermal titration calorimetry: entropy-driven assembly of co-chaperonin protein 10. *Biophysical Journal*, **89**, 3332–6.
Lynden-Bell, R. M. and Head-Gordon, T. (2006). Solvation in modified water models: towards understanding hydrophobic effects. *Molecular Physics*, **104**, 3593–605.

McConnell, K. J. and Beveridge, D. L. (2000). DNA structure: what's in charge? *Journal of Molecular Biology*, **304**, 803–20.

Moura Ramos, J. J., Lemmers, M., Ottinger, R., Stien, M. L. and Reisse, J. (1977). Calorimetric studies in solution. Part III. Experimental determination of the activated complex-solvent interaction enthalpy: cis-trans-isomerization of azobenzene. *Journal of Chemical Research, Synopses*, **2**, 56–7.

Orgel, L. E. (2004). Prebiotic chemistry and the origin of the RNA world. *Critical Reviews in Biochemistry and Molecular Biology*, **39**, 99–123.

Otting, G. (1997). NMR studies of water bound to biological molecules. *Progress in Nuclear Magnetic Resonance Spectroscopy*, **31**, 259–85.

Ourisson, G. and Nakatani, Y. (1996). Can the molecular origin of life be studied seriously? *Comptes Rendus de l'Academie des Sciences, Serie* IIB*: Mecanique, Physique, Chimie, Astronomie*, **322**, 323–34.

Pal, S. K. and Zewail, A. H. (2004). Dynamics of water in biological recognition. *Chemical Reviews*, **104**, 2099–123.

Pal, S. K., Peon, J. and Zewail, A. H. (2002). Biological water at the protein surface: dynamical solvation probed directly with femtosecond resolution. *Proceedings of the National Academy of Sciences of the United States of America*, **99**, 1763–8.

Pierotti, R. A. (1965). Aqueous solutions of nonpolar gases. *The Journal of Physical Chemistry*, **69**, 281–8.

Poornima, C. S. and Dean, P. M. (1995a). Hydration in drug design. 1. Multiple hydrogen-bonding features of water molecules in mediating protein–ligand interactions. *Journal of Computer-aided Molecular Design*, **9**, 500–12.

Poornima, C. S. and Dean, P. M. (1995b). Hydration in drug design. 3. Conserved water molecules at the ligand-binding sites of homologous proteins. *Journal of Computer-aided Molecular Design*, **9**, 521–31.

Pozzi, G., Birault, V., Werner, B., Dannenmuller, O., Nakatani, Y., Ourisson, G. and Terakawa, S. (1996). Single-chain polyprenyl phosphates form 'primitive' membranes. *Angewandte Chemie, International Edition in English*, **35**, 177–80.

Pratt, L. R. and Chandler, D. (1977). Theory of hydrophobic effect. *The Journal of Physical Chemistry*, **67**, 3683–704.

Qvist, J. and Halle, B. (2008). Thermal signature of hydrophobic hydration dynamics. *Journal of the American Chemical Society*, **130**, 10345–53.

Rawn, J. D. (1989). *Biochemistry*. Burlington, NC: Neil Paterson Publishers.

Rhodes, M. M., Réblová, K., Šponer, J. and Walter, N. G. (2006). Trapped water molecules are essential to structural dynamics and function of a ribozyme. *Proceedings of the National Academy of Sciences of the United States of America*, **103**, 13380–5.

Saenger, W. (1984). *Principles of Nucleic Acid Structure*. Berlin: Springer-Verlag.

Schneider, B., Patel, K. and Berman, H. M. (1998). Hydration of the phosphate group in double-helical DNA. *Biophysical Journal*, **75**, 2422–34.

Schulze-Makuch, D. and Irwin, L. N. (2006). The prospect of alien life in exotic forms on other worlds. *Naturwissenschaften*, **93**, 155–72.

Sharrow, S. D., Edmonds, K. A., Goodman, M. A., Novotny, M. V. and Stone, M. J. (2005). Thermodynamic consequences of disrupting a water-mediated hydrogen-bond network in a protein–pheromone complex. *Protein Science*, **14**, 249–56.

Silverstein, K. A. T., Haymet, A. D. J. and Dill, A. K. (2000). The strength of hydrogen bonds in liquid water and around nonpolar solutes. *Journal of the American Chemical Society*, **122**, 8037–41.

Southall, N. T., Dill, K., Dill, A. and Haymet, A. D. J. (2002). A view of the hydrophobic effect. *Journal of Physical Chemistry B*, **106**, 521–33.

Sreenivasan, U. and Axelsen, P. H. (1992). Buried water in homologous serine proteases. *Biochemistry*, **31**, 12785–91.

Takajo, S., Nagano, H., Dannenmuller, O., Ghosh, S., Marie Albrecht, A., Nakatani, Y. and Ourisson, G. (2001). Membrane properties of sodium 2- and 6-(poly)prenyl-substituted polyprenyl phosphates. *New Journal of Chemistry*, **25**, 917–29.

Takano, K., Yamagata, Y. and Yutani, K. (2003). Buried water molecules contribute to the conformational stability of a protein. *Protein Engineering*, **16**, 5–9.

Tame, J. R. H., Sleigh, S. H., Wilkinson, A. J. and Ladbury, J. E. (1996). The role of water in sequence-independent ligand binding by an oligopeptide transporter protein. *Nature Structural Biology*, **3**, 998–1001.

Tanford, C. (1980). *The Hydrophobic Effect*. New York: Wiley.

Wahl, M. C. and Sundaralingam, M. (1997). Crystal structures of A-DNA duplexes. *Biopolymers*, **44**, 45–63.

Williams, M. A., Goodfellow, J. M. and Thornton, J. M. (1994). Buried waters and internal cavities in monomeric proteins. *Protein Science*, **3**, 1224–35.

14

The role of water in the formation and evolution of planets

Thérèse Encrenaz

Introduction

Water is known to be ubiquitous in the Universe, from the dark spots of the Sun up to the most distant galaxies. It is also a major component of Solar-System objects, especially in the outer parts beyond heliocentric distances of a few astronomical units (one astronomical unit = AU = average Earth–Sun distance = 149.6×10^6 km). It should be mentioned, however, that the Earth is the only place in the Solar System where water can be present in its three states: solid, liquid and vapour. So far, outside the Earth, water has always been found in the form of vapour or ice, although there are some indications that liquid water might be – or have been – present elsewhere in the Solar System. Liquid water was probably present in the past on the surface of Mars and also possibly Venus; it might currently exist in the interiors of some satellites of giant planets.

The presence of large amounts of water, vapour or ice in the Universe is a natural outcome of the large cosmic abundances of the hydrogen and oxygen elements which form the water molecule. In addition, the large abundance of water in the outer Solar System is also a natural consequence of the formation scenario of the Solar System, which led to the accretion of two classes of planets, the terrestrial and the giant ones, separated by the 'snow line', which basically corresponds to the heliocentric distance of water condensation in the primordial solar nebula.

In this chapter, we will briefly describe the formation scenario of the Solar System and the role played by water in the discrimination between the telluric and the giant planets. Then we will review the different forms of water known in the outer Solar System and finally we will discuss the role of water in the evolution processes of the terrestrial planets.

The formation scenario of the Solar System

It is generally accepted that the Solar System was formed after the collapse of a spinning interstellar cloud, which fell into a disk as an effect of its own gravity. At the centre, the matter accreted into a proto-Sun while, within the turbulent disk, particles, all subject to differential rotation speeds, accreted through mutual collisions. Planets later grew due to their own gravity field, attracting the surrounding matter. The first concepts of this scenario were proposed as early as the eighteenth century by Immanuel Kant and Pierre-Simon de Laplace. However, it was definitely accepted only a few decades ago, on the basis of observational and theoretical support.

First, the development of infrared and millimetre astronomy made possible the study of star formation. The observation of young stellar objects (Herbig–Haro objects, T-Tauri and Fu-Orionis objects) shows very active early phases of the life of the stars; in many cases, protoplanetary disks were observed around these young objects. It became clear that the formation of stars through the collapse of an interstellar cloud and the formation of a disk was a widespread process in the Universe. This idea has received additional support from the discovery of extrasolar planets around nearby solar-type stars (e.g. Reipurth *et al.*, 2007).

Second, the development of numerical simulations has allowed theoreticians to model the various stages of this process, from the collapse of the cloud through to the formation of the protoplanetary disk and the interaction between the disk and the protoplanets (e.g. Ollivier *et al.*, 2009).

The age of the Solar System is known from cosmochemistry, which provides us with an absolute dating of Solar-System objects. The method is based on the abundance measurement of radioactive elements having very long half-times in terrestrial rocks, lunar samples and meteorites. Lunar and meteoritic measurements converge on an age of 4.56 billion years and show that the Sun and the Solar System were all formed quasi-simultaneously (e.g. Taylor, 2001).

The formation of the giant planets: the snow line

Within the planetary disk, what was the matter available for planet formation? As the disk accreted from interstellar matter, it can be assumed that the elements were present with their cosmic abundances: hydrogen was most abundant, then helium, He (about 10%), then (by order of abundance per volume) oxygen, carbon, nitrogen and then the heavier elements, all below the per cent level. These elements (apart from hydrogen and helium) were all formed within the stars by nucleosynthesis (e.g. Lewis, 1997). As the amount of energy necessary to synthesize an element increases with its atomic weight, consequently, the abundance of heavy elements is anticorrelated with their atomic weight. The most abundant 'heavy' elements (C, N and O) combined with hydrogen to form simple molecules such as H_2O, CH_4 and NH_3, and they also combined together to form CO, CO_2 etc.

How did the planets form? An important constraint is that planets accrete from solid particles. Both hydrogen and helium, the most abundant elements, are in gaseous form. The other ones can be either gaseous or solid, depending upon the temperature of their environment. At the time of its formation, just after the collapse phase, the protosolar disk was warmer than the present-day Solar System; however, as is seen today, its temperature decreased from the centre towards the outer edge. Two cases can be considered (e.g. De Pater and Lissauer, 2001; Encrenaz *et al.*, 2004; Ollivier *et al.*, 2009). The first case considers the vicinity of the Sun, where at heliocentric distances smaller than 4 AUs, the temperature must have been a few hundred K. At such temperatures, the only solid materials were the rocks and the metals, the simple molecules like H_2O remaining gaseous.

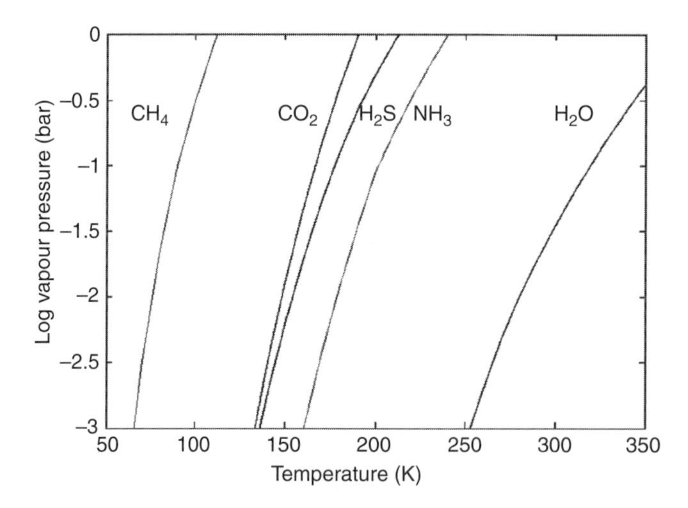

Figure 14.1. Saturation vapour pressures of some condensable gases. From right to left: H_2O, NH_3, H_2S, CO_2, CH_4. It can be seen that water condenses at far higher temperatures than the other species. Saturation laws are taken from Atreya (1986).

The solid matter available for building planetary nuclei was then limited and could only lead to the formation of a few rocky planets – the terrestrial planets. Their high volumic mass (between 3900 and 5400 kg/m^3) indicates that they are mainly constituted of rocks. The second case considers that at greater heliocentric distances, beyond about 4 AUs, the temperature must have been below 160–200 K, i.e. low enough for maintaining all simple molecules (H_2O, CH_4, NH_3, CO_2 etc.) in the form of ice. This solid material was thus available to form large cores with masses as high as about 10–15 terrestrial masses. Theoretical models (Mizuno, 1980; Pollack *et al.*, 1996) show that this is the critical mass that allows the protoplanet to accrete the surrounding matter that falls on it as an effect of its gravity field. As the protoplanetary disk is mostly made of light gases (hydrogen and helium), this process leads to the formation of big objects of low densities: the giant planets. Indeed, the giant planets are characterized by large diameters (from 4 to 11 times the terrestrial ones) and low volumic masses (from 600 to 1700 kg/m^3).

Why is water especially important in this differentiation process? For two reasons: first, water is particularly abundant among the simple molecules, as oxygen is the most abundant element after hydrogen and helium. Second, water is the first molecule to condense as the temperature decreases (Figure 14.1). The level of water condensation, called the 'snow line', marks the separation between the two classes of planet – the terrestrial and the giant ones (Encrenaz, 2008).

Water in the outer Solar System

Water is very abundant in the outer Solar System; indeed, comets are made of about 80% of water by mass. In the case of the giant planets, water vapour is present in the interiors and tropospheres, but also in the stratospheres (although in much lower amounts); their satellites all contain a significant, most often major, amount of water ice.

Water in comets

Comets are small objects (most of them are less than 10 km in size), made of ice and rocks. They travel on very elliptical orbits which occasionally bring them back into the inner Solar System. As they approach the Sun, the ice sublimates, leading to the ejection of gas (the parent molecules) and dust (mostly silicates and carbonaceous material) which form the coma. Because they encounter no thermal or collisional evolution along their trajectories in the outer Solar System, comets are believed to be privileged witnesses of the early processes of Solar-System formation (Crovisier and Encrenaz, 2000; Festou *et al.*, 2004). Some comets are very famous, like Halley (which comes back every 76 years and was approached by 5 spacecraft in 1986, at the time of its last appearance) or Hale Bopp, which appeared in 1997 and was unusually large.

The study of water in comets is very informative. By studying the water rotational transitions at high spectral resolution, it is possible to study the kinematics and the physical properties of the inner coma. By studying the HDO/H_2O ratio, it is possible to infer the D/H ratio, which is a valuable diagnostic of early formation processes. Indeed, deuterium (D) was formed during the Big Bang by primordial nucleosynthesis, but has, since then, been constantly destroyed in stars, where it is transformed into ^4He. The value of D/H in the protosolar disk is thus very low (about 2×10^{-5}). However, ices are enriched in D (resulting in higher D/H ratios), due to fractionation effects occurring during ion–molecule reactions at low temperature; this effect, well known in laboratory experiments, is currently observed in the interstellar medium. Actually, the value of D/H in an outer Solar-System body is an indicator of its formation temperature. The D/H ratio in comets, measured from HDO/H_2O, is about 3×10^{-4}; this high value is consistent with the fact that the objects were formed at low temperature in the outer Solar System, close to the orbits of Uranus and Neptune or beyond. In the case of the terrestrial planets, the D/H ratio in water is diagnostic of other atmospheric processes: the D/H enrichment in the atmospheres of Mars and Venus, as compared to the Earth's oceans, is interpreted as the signature of a differential escape (HDO being heavier than H_2O) and thus provides evidence for more massive, water-rich primitive atmospheres (see 'Comparative evolution of terrestrial planets: the role of water').

The observation of water in comets leads to the determination of another parameter, the ortho–para ratio, which is also a diagnostic of the formation temperature of these objects. Water can be found in two different states, ortho and para, depending upon the spin orientation of the two hydrogen atoms in the water molecule. This ratio depends upon the temperature at the time of its formation. Ortho and para transitions are found in the water spectrum at slightly different frequencies, so the abundances of the two species can be determined and the ortho–para ratio (OPR) can be measured. This has been done on a few comets (Figure 14.2) and in all cases the inferred temperature is very low, below 30 K.

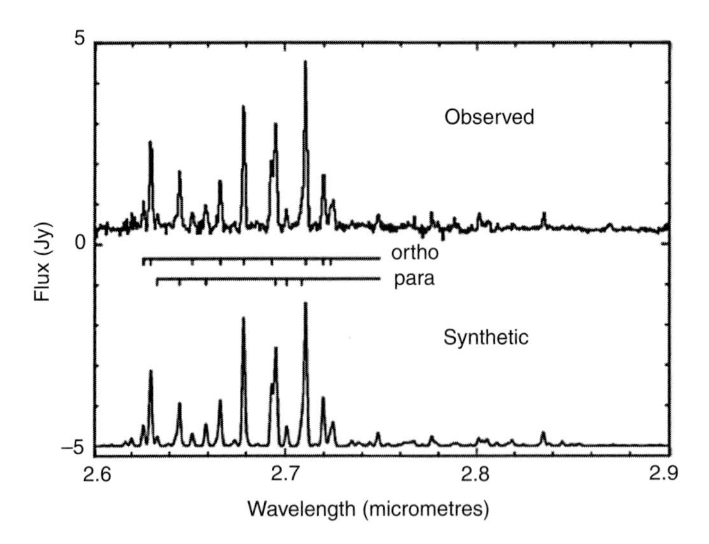

Figure 14.2. Observation of the H_2O near-IR spectrum of Hale-Bopp with the ISO satellite. Top: observations; bottom: synthetic spectrum. Both ortho- and para-transitions are detected, which makes possible a determination of the ortho–para ratio, from which the formation temperature is deduced. The figure is taken from Crovisier *et al.* (1997).

This result confirms the high D/H ratio and suggests that comets were formed at great heliocentric distances, at the orbit of Uranus or beyond.

Water in the giant planets

Before discussing the giant planets we need to introduce a sub-classification between them. On the one hand, Jupiter and Saturn have very huge masses of 318 and 95 terrestrial masses, respectively, while Uranus and Neptune only have 14 and 17 terrestrial masses. Considering that all contain an icy core of about 12 terrestrial masses, it means that Jupiter and Saturn are mostly composed of protosolar gas, while Uranus and Neptune have most of their mass in their icy core. For this reason, Jupiter and Saturn are called the gaseous giants, while Uranus and Neptune are the icy giants.

The atmospheric compositions and structures of the giant planets have been studied since the mid-1970s by space exploration, first with the two Pioneer spacecraft, then by the Voyager mission, which flew by the four giant planets between 1979 and 1989. Later on, Jupiter and its system were extensively explored by the Galileo mission and, in particular, the Galileo probe, which entered the Jovian atmosphere in December 1995. The Cassini mission approached the system of Saturn in 2004 and is still in operation; in January 2005, the Cassini–Huygens probe successfully landed on Titan's surface and sent the first images of Saturn's biggest satellite (see Chapter 29). In addition to space exploration, the atmospheres of the giant planets have been extensively studied using ground-based spectroscopy, especially in the infrared range. Their chemical composition is dominated by hydrogen,

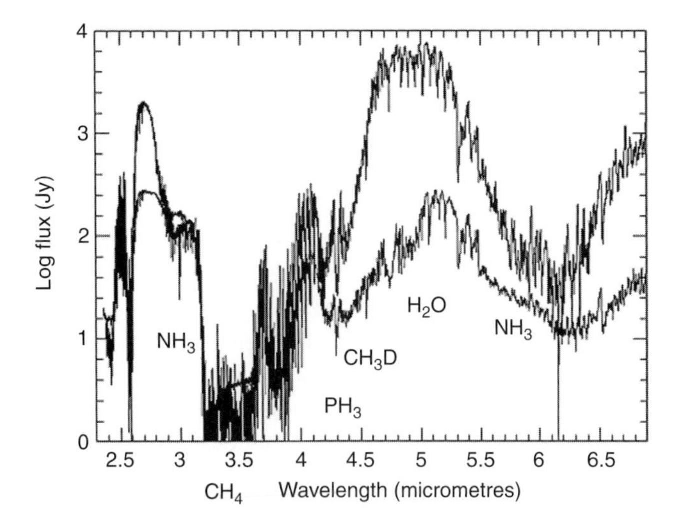

Figure 14.3. The spectrum of Jupiter (upper curve) and Saturn (lower curve) between 2.3 and 6.9 μm, as observed by the short-wavelength spectrometer (SWS) of the Infrared Space Observatory (ISO). Note that, in the 14″ × 20″ aperture of SWS, the spectrum of Saturn at 3 μm is higher than the Jupiter one at this wavelength. The 5-μm region is an atmospheric window where radiation comes from the deep troposphere. Absorption lines of water vapour are detected in the spectrum of Jupiter. The figure is taken from Encrenaz (2003).

then helium, and traces of heavier molecules in reduced form at the per cent level or less (CH_4, NH_3, H_2O, PH_3 etc.; Figure 14.3). The thermal structures are characterized by a convective troposphere, an intermediate region (the tropopause) where the temperature (from 110 K on Jupiter to 50 K for Uranus and Neptune) is at its minimum and then a stratosphere where the temperature increases again with altitude; photodissociation products of methane are found in this region (see Encrenaz *et al.*, 2004, for a review).

There is observational evidence which supports the nucleation model of the giant planets. It is provided by the relative abundances of heavy elements versus hydrogen, as compared to the protosolar ('cosmic') value. This enrichment results from the fraction of heavy elements included in the initial core. In the absence of an icy core, if the giant planets had been directly formed from the collapse of a sub-cloud of protosolar gas, these elemental abundances would reflect the protosolar values. The larger the mass fraction of the core, the larger the enrichment is. Assuming that all elements were equally trapped in the core at the time of the planet's formation, and assuming that they were later rehomogenized within the planet after the collapse phase of the surrounding subnebula, it is possible to calculate the expected enrichment, just knowing the mass fraction of the initial core for each planet. Assuming an initial core of 12 terrestrial masses, the inferred enrichment is 4 for Jupiter, 9 for Saturn and 30–50 for Uranus and Neptune. In the case of Jupiter, the Galileo probe, which entered Jupiter's atmosphere in December 1995, measured the enrichment for several elements, including C, S, N, Ar, Kr and Xe (Niemann *et al.*, 1998; Atreya *et al.*, 1999; Mahaffy *et al.*, 1999). For the other giant planets, only the carbon enrichment could be measured from remote-sensing observations of the atmospheric methane. In all cases, there was remarkable agreement between the predictions and the observations. This diagnostic provides definitive support for the nucleation-model formation of the giant planets (Owen *et al.*, 1999; Owen and Encrenaz, 2006).

In the case of Jupiter, the Galileo probe also measured the oxygen abundance. However, the water content found in the deep troposphere of Jupiter was significantly lower than expected from thermochemical models. On the basis of Jupiter's thermal profile a cloud structure was predicted, including an NH_3 cloud at about 0.5 bar, an NH_4SH cloud at about 2 bars and an H_2O cloud at about 5 bars. The observed cloud structure was much weaker, which implies that the entry region of the Galileo probe was especially dry. This was attributed to a very active convective circulation involving uprising plumes rich in water vapour and ammonia and dry regions of subsidence – the 'hot spots'. These hot spots were known from remote-sensing spectroscopy as being clear regions where the infrared flux comes from deep tropospheric levels. It just happened that the Galileo probe entered one of these hot spots. As a result, the oxygen abundance measured by the Galileo probe (as well as the infrared measurements performed in the hot spots) are not representative of the global oxygen abundance in Jupiter's interior (Atreya *et al.*, 1997). In order to derive this parameter one needs to probe deeper levels down to tens of bars. This is one of the objectives of the NASA space mission Juno, to be launched in 2011, which will be orbiting Jupiter with a microwave sounder.

Water vapour has also been detected in the interior of Saturn (De Graauw *et al.*, 1997) with abundance lower than expected from thermochemical models; this suggests that a dynamical convective process might be at work in Saturn's troposphere, as in the case of Jupiter. Water vapour must also be present inside Uranus and Neptune, but in the lower troposphere, below the water-ice clouds, which are too deep to be observable.

The presence of water in the interiors of the giant planets is a direct consequence of their formation scenario. Its presence in their stratosphere, detected by the Infrared Space Observatory (ISO) satellite in 1997 (Feuchtgruber *et al.*, 1997) was more surprising. Indeed, there is a minimum temperature at the tropopause of all giant planets at a pressure level of 0.1 bar. This minimum temperature is 110 K for Jupiter, 90 K for Saturn and about 52 K for Uranus and Neptune. In all cases, the temperature is too low for water to be in gaseous form, so the water coming from the interior cannot pass this cold trap. As a result, the stratospheric water found in the giant planets must be of external origin. Two kinds of source are possible: a local source (coming from the rings and the satellites) and an interplanetary source including micrometeorites or even comets. In particular, the collision of comet Shoemaker-Levy 9 with Jupiter in July 2004 may have fed Jupiter's stratosphere with water and oxidized material (Noll *et al.*, 1996). It is likely that both sources, local and interplanetary, contribute to the oxygen flux detected on the four giant planets (and also on Titan). More observations will be needed to better understand this phenomenon and its origin; it will be one of the objectives of the key programme 'Water in the Solar System' of the European Herschel satellite, launched in May 2009.

Water ice in the rings and outer satellites

All giant planets exhibit a ring system and a large number of regular satellites. This is a simple consequence of the collapse phase of their formation scenario, which led to the formation of a planetary disk in their equatorial plane. Except for Jupiter, which has a very tenuous internal ring, all ring systems are mostly made of water ice. This is also the case of the outer satellites (Bagenal *et al.*, 2007), with the exception of Io, Jupiter's closest satellite. Subject to strong tidal forces induced by Jupiter's gravity field, this satellite has lost its water ice. Its surface, permanently remodelled by active vulcanism, is mostly made of silicate and sulphur compounds.

Europa, the next Galilean satellite, closest to Jupiter after Io, is of special interest for astrobiology. It is covered with water ice and probably contains an ocean of liquid water below its crust (Tobie *et al.*, 2003). Several facts support this idea: the images of the surface, sent by the Voyager and Galileo spacecraft, suggest the presence of plates covering a viscous or liquid medium, presumably made of water; the Galileo magnetometer has detected the presence of an induced magnetic field which could be generated within salty liquid water (Khurana *et al.*, 1998); finally, models of internal structure predict the presence of a liquid layer within the satellite below the icy crust and above the silicate mantle. Other outer satellites, such as Ganymede and Callisto, are also expected to behave so; however, Europa is close enough to Jupiter to benefit from an additional internal energy force, due (like Io but to a lesser extent) to tidal forces. As a result, the liquid layer could be in direct contact with the silicate surface, which would have important possible consequences for the development of a complex chemistry within liquid water. In addition, the icy crust is expected to be thinner than in the other cases. How thin could it be? According to the models, the crust thickness could reach several tens of kilometres, which means that drilling it would not be an easy task. In any case, Europa appears as a favoured target for searching possible sites for the emergence of life in the outer Solar System. As a first step, the radar instrument planned on the future Europa and Jupiter System Mission (EJSM), currently under study at NASA and ESA, should be able to confirm (or deny) the existence of a subsurface ocean on Europa.

There are two other outer satellites which deserve special attention. Titan of course, Saturn's biggest satellite, is unique with its thick nitrogen atmosphere, which draws an interesting similarity with the Earth (Hunten *et al.*, 1984; Coustenis and Taylor, 2008). As revealed by Voyager and later by the Cassini–Huygens mission, Titan's atmosphere and ionosphere host many complex molecules or ions of hydrocarbons and nitriles. Such molecules are known to be the 'building bricks' for prebiotic molecules like amino acids. For these reasons, Titan is considered as a possible laboratory for prebiotic chemistry, with possible analogies with the primitive Earth. There is a major difference, however: today, the temperature (93 K at Titan's surface) is so low that water cannot be liquid at the surface nor just below it. The boulders identified by the images sent by the Huygens probe when it landed on Titan on 14 January 2005 are most probably made of water ice. Models (Tobie *et al.*, 2005) suggest the presence of liquid water tens of kilometres beneath the surface.

On primordial Titan, the surface temperature may have been above 300 K, allowing the presence of a large abundance of water (possibly liquid) at the surface.

Finally, the Cassini orbiter has made another unexpected discovery: the small satellite Enceladus was found to show active cryovulcanism at its south pole. Plumes, identified as water vapour, were observed in the visible range, while the infrared spectrometer detected a temperature excess in the southern region. The outgassing of Enceladus is believed to feed the E-ring of Saturn, which is at the same distance from Saturn.

Comparative evolution of terrestrial planets: the role of water

Among the four terrestrial planets, three of them have a stable neutral atmosphere. Mercury, the planet closest to the Sun, is too small to retain its atmosphere through its gravity field. The presence of very tenuous traces of water has been reported from several ground-based and space observations in perpetually shadowed craters on the surface of Mercury (Slade *et al.*, 1992).

Since the 1960s, Mars and Venus have been the favoured targets of space exploration, with an important number of failures but also remarkable successes. In the case of Mars, the observations of Mariner 9 in 1972, then the two Viking orbiters and landers from 1975–77, are still used as a reference today. Two decades later, the space exploration of Mars entered a new age with a new generation of orbiters (Mars Global Surveyor, Mars Odyssey, Mars Express, Mars Reconnaissance Orbiter), as well as landers and rovers (Viking, Mars Pathfinder, Spirit, Opportunity, Phoenix). All these missions have provided us with information about the surface topography and mineralogy and the atmospheric composition and structure (Kieffer *et al.*, 1992; Forget *et al.*, 2008).

The case of Venus is different, as its surface is hidden by a thick and opaque cloud layer, mostly composed of sulphuric acid. The atmosphere was probed by several orbiters (the Venera and Pioneer Venus spacecraft) and the surface was first revealed by the Venera 9 and 10 landers in 1975. The best cartography of the surface of Venus was achieved by the radar of the Magellan spacecraft in 1992. In addition, the lower atmosphere of Venus on the night side has been successfully probed by near-infrared ground-based spectroscopy, as the temperature is high enough for the thermal emission to be detectable at these wavelengths (see Bougher *et al.*, 1997 for a review).

The atmospheres of Venus, Earth and Mars show remarkable differences in their physical conditions: the surface temperatures and pressures range respectively from 730 K and 90 bars in the case of Venus to about 230 K and 0.06 bar (with strong variations) in the case of Mars. In between, the Earth, with an average surface temperature of 288 K and a surface pressure of 1 bar, has the unique opportunity to keep water in its three states: vapour, liquid and solid. The primitive atmospheres of the three planets probably had a similar composition, with a large fraction of carbon dioxide (CO_2), a small percentage of N_2 and a large amount of water.

Water vapour on Mars and Venus today is less than one per cent per volume; however, in the past, it is believed to have been another major constituent of both planets, especially Venus. This information comes from the measurement of the HDO/H_2O ratio

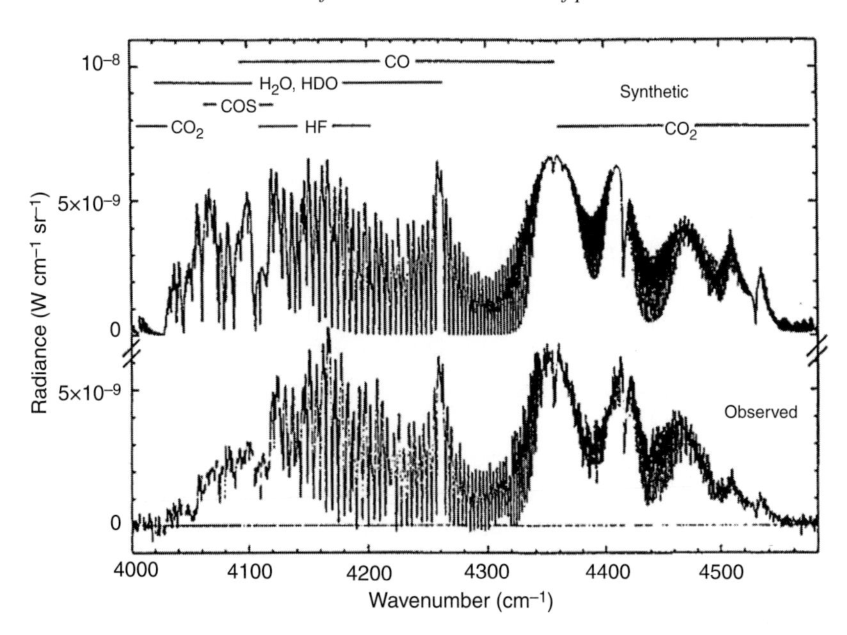

Figure 14.4. The ground-based high-resolution spectrum of Venus' dark side in the 2.35-μm atmospheric window. Upper curve: synthetic spectrum; lower curve: observations. The radiation comes from the deep troposphere where several minor species can be detected, including H_2O and HDO. The figure is taken from Bézard *et al.*, (1990).

of Venus and Mars with respect to the Earth. In both atmospheres, H_2O and HDO are measured in the near-infrared range. On Venus, the observation is made on the dark side of the planet, in the near-infrared range, between the very broad CO_2 absorption bands; measurements made from the ground (Figure 14.4) and from space (Galileo in 1990 and recently Venus Express) have indicated a deuterium enrichment which is about 120 times the terrestrial value (de Bergh *et al.*, 1991). There is, however, some uncertainty in the exact enrichment factor, which must be more precisely measured. On Mars, because of its very tenuous atmosphere, the lines are very narrow. Here also, ground-based infrared spectroscopic measurements have indicated a strong enrichment, although not as spectacular as on Venus: a factor about 5 over the terrestrial value (Owen *et al.*, 1988). In both cases, the deuterium enrichment is interpreted in terms of water escape from the primitive atmospheres of the planets; heavy water (HDO) escaped at a slower rate, which led to the deuterium enrichment observed today.

In the past history of Venus, the fainter radiation of the young Sun probably allowed water to be liquid at the surface. So water was probably in the liquid phase on all three planets, but their surface and atmospheric temperatures, together with their geologic and climate histories, led them on different paths. Why did water disappear from the present atmospheres of Venus and Mars? In the case of Venus it is likely that water has been photodissociated in the upper atmosphere and the hydrogen atoms were light enough to escape. However the fate of the heavier oxygen atoms' enrichment is still unsolved. Was oxygen bound with another molecule like ClO (which would remain to be detected) or did it escape through a non-thermal process? The question is still open. On Mars the situation was different: indeed, it has been recently established that there is a reservoir of water ice below the polar caps, and also possibly a permafrost reservoir below the surface at high latitudes. We do not yet have a reliable estimation of the volume of water contained in this reservoir. Since the Mariner 9 and Viking missions, images of the Martian surface have shown evidence of valley networks and outflow channels. More recently, the presence of hydrated sulphates (in particular gypsum, kieresite and jarosite), found by Mars Express and Opportunity, seems to indicate that water has flowed on the surface in the past, even if we do not know exactly when and how long it lasted. This might indicate that the primitive atmosphere of Mars was warmer, denser and wetter, but this also still remains an open question (see Chapter 15).

In the case of the Earth, atmospheric water condensed very early to form the oceans; indeed, the zircons from Jack Hills in Western Australia demonstrate that liquid water and oceans have been present on Earth since 4.3 Ga ago and very probably since 4.4 Ga ago (Peck *et al.*, 2001; Wilde *et al.*, 2001). In addition, the ocean allowed the trapping of atmospheric carbon dioxide in the form of calcium carbonate. Atmospheric oxygen appeared later as a result of the development of life.

What is the origin of the terrestrial water? Here again, the measurement of D/H provides a reliable answer. As mentioned above, the D/H ratio of different primitive objects of the Solar System is a good diagnostic of the temperature of the place where they formed – a high D/H ratio indicating a cold formation region. In the protosolar disk, the D/H ratio was about 2×10^{-5}. The D/H ratio can be measured in the giant planets – from the HD/H_2 ratio or from the CH_3D/CH_4 ratio (Figure 14.5). Not surprisingly, the protosolar value of D/H is

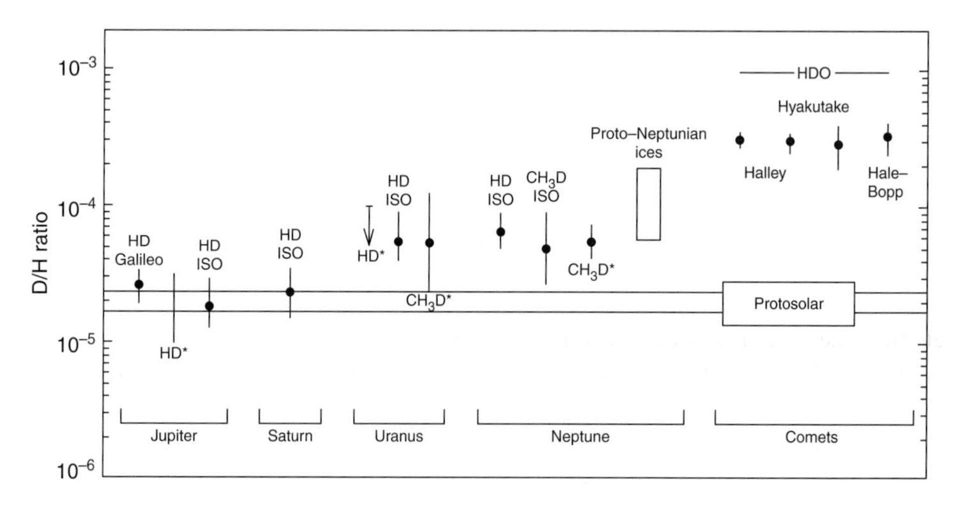

Figure 14.5. D/H ratios in the Solar System. Asterisks indicate ground-based measurements. The Jupiter Galileo measurement is from Mahaffy *et al.* (1999). The figure is adapted from Bockelée-Morvan *et al.* (1998).

also found in Jupiter and Saturn, but the D/H value in Uranus and Neptune (which have a larger mass fraction in their icy cores) is higher (about 6×10^{-5}). The D/H ratio measured in the terrestrial oceans is higher (1.5×10^{-4}). The D/H in comets has been found to be even higher (3×10^{-4}); twice the terrestrial value. Compared with the protosolar value, the D/H in the Earth's oceans is high, which is consistent with its formation scenario: they accreted from solid particles, and their atmospheres were formed at a later stage, partly from outgassing but mostly from meteoritic and cometary impacts. The difference between the terrestrial and the cometary D/H demonstrates that terrestrial water cannot be entirely of cometary origin and rather argues in favour of a source being D-type asteroids coming from the outer part of the main asteroidal belt (Morbidelli *et al.*, 2000).

Why did the three terrestrial planets exhibit such a different evolution? Water is believed to have played a major role in this scenario.

On Venus water was in the form of vapour, which, similar to carbon dioxide, is a very active greenhouse-effect gas. As a result, the surface temperature of Venus increased from its equilibrium temperature (about 350 K) to the high value observed today. The Earth escaped this fate because the water was liquid and formed oceans where carbon dioxide was trapped in solid carbonates.

On Mars, two main factors account for the differences with both Venus and Earth. The greater heliocentric distance of Mars leads to a lower surface temperature. The small mass of the planet (about one-tenth of the terrestrial mass) leads to a weaker gravity field and a smaller amount of internal energy. Mars is believed to have had an intrinsic magnetic field during the first billion years of its history; this is testified by the presence of a remnant crustal magnetic field in the oldest southern terrains, as discovered by MGS (Acuña *et al.*, 1999; Connerney *et al.*, 2001). But due to the lack of sufficient internal energy, the dynamo stopped. The disappearance of Mars's magnetic field may have coincided with the escape of the atmosphere, no longer protected from erosion by the magnetospheric cavity. Another possibility is that Mars encountered major meteoritic impacts, which may have blown off a significant fraction of its atmosphere. Without significant atmosphere, the greenhouse effect was almost nil on Mars, resulting in a low surface temperature such that water remained solid.

Conclusions

The study of water in the Solar System shows that water has played a major role in the formation of Solar-System planets, by separating, at the level of the snow line, the giant planets from the terrestrial ones. In addition, the role played by water, depending on its state (solid, liquid, vapour) was also crucial in the diverging evolution of the terrestrial planets.

There are still many open questions regarding water in the Solar System. In the case of water, what is the D/H ratio in other comets? The objects measured so far all come from the Oort cloud, a distant reservoir where they were expelled after their formation in the giant planet's vicinity. But there is another reservoir of comets, the Kuiper belt; these comets have small inclinations, are weaker and more difficult to observe, but measuring D/H in these objects would help to better constrain their formation processes.

In the case of the giant planets, a parameter to be measured is the O/H ratio in the interior of Jupiter, as well as in the other giant planets, which is difficult as it requires probing at tens or even hundreds of bars. In the case of Jupiter, the Juno mission, to be launched in 2011, could bring an answer by measuring the radio signal from Jupiter's deep interior. A better determination of the D/H ratio in the four giant planets is also needed; this will hopefully be done by the Herschel mission, launched in May 2009.

In the case of Venus, a localized measurement of D/H at the surface would be invaluable for determining the possible presence of active vulcanism. Indeed, if water is currently being outgassed by volcanoes, its D/H ratio should not be enriched with respect to the terrestrial value – it could even be lower if the material inside Venus was formed at high temperature. In contrast, in the absence of outgassing, the D/H ratio should be uniformly enriched by a factor of 120. This relatively simple measurement will, we hope, be performed by the Venus Express spacecraft and its infrared spectrometer VIRTIS.

Finally, on Mars, the search for water in the form of ice, permafrost or a vestige of liquid water, is an ongoing challenge, and future space missions (Mars Science Laboratory, ExoMars and Mars Sample Return) will be devoted to this objective. In addition, mapping the D/H ratio as a function of latitude and season will bring new constraints about possible fractionation effects and will help in determining more accurately the D/H in water ice, the main water reservoir on Mars.

The special role of water in the formation and evolution of the Solar System can probably be understood from the unique physical and chemical properties of the molecule: its high cosmic abundance, its high dipole moment, its condensation at relatively high temperature and the wide temperature range of its liquid state.

It is thus reasonable to consider that water, most likely, has the same key role in extrasolar planetary systems. We already have some evidence that comets exist in the disk of Beta Pictoris, and the existence of a Kuiper belt has been proposed for some evolved stars, from a water excess measured at millimetre wavelengths (Justtanont *et al.*, 2005). The Herschel satellite will be a unique tool for investigating such objects and detecting water emissions from different types of object, including planetary disks, young stellar objects and evolved stars.

Acknowledgements

I am grateful to S. K. Atreya and H. Martin for their helpful comments regarding this paper.

References

Acuña, M. H., Connerney, J. E. P., Ness, N. F., Lin, R. P., Mitchell, D., Carlson, C. W., Mcfadden, J., Anderson, K. A., Reme, H., Mazelle, C., Vignes, D., Wasilewski, P. and Cloutier, P. (1999). Global distribution of crustal magnetization discovered by the Mars Global Surveyor MAG/ER experiment. *Science*, **284**, 790–3.

Atreya, S. K. (1986). *Atmospheres and Ionospheres of the Outer Planets and Their Satellites*. Heidelberg: Springer-Verlag, p. 224.

Atreya, S. K., Wong, M. H., Owen, T. C., Niemann, H. B. and Mahaffy, P. R. (1997). Chemistry and clouds of Jupiter's atmosphere: A Galileo perspective. The Three Galileos: the Man, the Spacecraft, the Telescope. *Proceedings of the Padova Conference*, Italy, 7–10 January 1997. Dordrecht: Kluwer Academic Publishers, Astrophysics and Space Science Library (ASSL), series vol. 220, ISBN 0792348613. Electronic access http//www.wkap.nl/book.htm/0–7923–4861–3, 249–260.

Atreya, S. K., Wong, M. H., Owen, T. C., Mahaffy, P. R., Niemann, H. B., de Pater, I., Drossart, P. and Encrenaz, T. (1999). A comparison of the atmospheres of Jupiter and Saturn: deep atmospheric composition, cloud structure, vertical mixing, and origin. *Planetary and Space Science*, **47**, 1243–62.

Bagenal, F., Dowling, T. E. and McKinnon, W. B. (2007). *Jupiter: The Planet, Satellites and Magnetosphere*. Cambridge: Cambridge University Press.

Bézard, B., de Bergh, C., Crisp, D. and Maillard, J. P. (1990). The deep atmosphere of Venus revealed by high-resolution nightside spectra. *Nature*, **345**, 508–11.

Bockelée-Morvan, D., Gautier, D., Lis, D. C., Young, K., Keene, J., Phillips, T., Owen, T., Crovisier, J., Goldsmith, P. F., Bergin E. A., Despois, D. and Wootten A. (1998). Deuterated water in comet C/1996 B2 (Hyakutake) and its implications for the origin of comets. *Icarus*, **133**, 147–62.

Bougher, S. W., Hunten, D. M. and Phillips, R. J. (1997). *Venus* II. Tucson, AZ: University of Arizona Press.

Connerney, J. E. P., Acuña, M. H., Wasilewski, P. J., Kletetschka, G., Ness, N. F., Rème, H., Lin, R. P. and Mitchell, D. L. (2001). The global magnetic field of Mars and implications for crustal evolution. *Geophysical Research Letters*, **28**, 4015–18.

Coustenis, A. and Taylor, F. W. (2008). *Titan, the Earth-like Moon* (revised edition). Singapore: World Scientific Publishing Company.

Crovisier, J. and Encrenaz, T. (2000). *Comet Science*. Cambridge: Cambridge University Press.

Crovisier, J., Leech, K., Bockelée-Morvan, D., Brooke, T. Y., Hanner, M. S., Altieri, B., Keller, H. U. and Lellouch, E. (1997). The spectrum of Comet Hale-Bopp (C/1995 01) observed with the Infrared Space Observatory at 2.9 AU from the Sun. *Science*, **275**, 1904–7.

De Bergh, C., Bézard, B., Owen, T., Crisp, D., Maillard, J. P. and Lutz, B. L. (1991). Deuterium on Venus – observations from Earth. *Science*, **251**, 547–9.

De Graauw, T., Feuchtgruber, H., Bezard, B., Drossart, P., Encrenaz, T., Beintema, D. A., Grif, M., Heras, A., Kessler, M., Leech, K., Lellouch, E., Morris, P., Roelfsema, P. R., Roos-Serote, M., Salama, A., Vandenbussche, B., Valentijn, E. A., Davis, G. R. and Naylor, D. A. (1997). First results of ISO-SWS observations of Saturn: detection of CO_2, CH_3C_2H, C_4H_2 and tropospheric H_2O. *Astronomy and Astrophysics*, **321**, L13–16.

De Pater, I. and Lissauer, J. (2001). *Planetary Science*. Cambridge: Cambridge University Press.

Encrenaz, T. (2003). ISO observations of the giant planets and Titan: what have we learnt? *Planetary and Space Science*, **51**, 89–103.

Encrenaz, T. (2008). Water in the Solar System. *Annual Review of Astronomy and Astrophysics*, **46**, 57–87.

Encrenaz, T., Bibring, J. P., Blanc, M., Barucci, A. M., Roques, F. and Zarka, P. (2004). *The Solar System*. Heidelberg: Springer.

Festou, M. C., Keller, U. and Weaver, H. A. (2004). *Comets* II. Tucson, AZ: University of Arizona Press.

Feuchtgruber, H., Lellouch, E., de Graauw, T., Bézard, B., Encrenaz, T. and Griffin, M. (1997). External supply of oxygen to the atmospheres of the giant planets. *Nature*, **389**, 159–62.

Forget, F., Costard, F. and Lognonné, P. (2008). *The Planet Mars: story of another world.* Chichester: Praxis Publishing.

Hunten, D.M., Tomasko, M.G., Flasar, F.M, Samuelson, R.E., Strobel, D.F., and Stevenson, D.J. (1984) Titan, in *Saturn*, eds T. Gehrels and M.S. Matthews. Tucson, AZ: University of Arizona Press, pp. 671–759.

Justtanont, K., Bergman, P., Larsson, B., Olofsson, H., Schöier, F. L., Frisk, U., Hasegawa, T., Hjalmarson, Å., Kwok, S., Olberg, M., Sandqvist, A., Volk, K. and Elitzur, M. (2005). W Hya through the eye of Odin. Satellite observations of circumstellar submillimetre H_2O line emission. *Astronomy and Astrophysics*, **439**, 627–33.

Khurana, K.K., Kivelson, M.G., Russel, C.T. (1998). Induced magnetic field as evidence for subsurface oceans in Europa and Callisto. *Nature*, **397**, 777–780.

Kieffer, H. H., Jakosky, B. M., Snyder, C. W. and Matthews, M. S. (1992). *Mars*. Tucson, AZ: University of Arizona Press.

Lewis, J.S. (1997). *Physics and Chemistry of the Solar System*, revised edn. San Diego, CA: Academic Press.

Mahaffy, P. R., Donahue, T. M., Atreya, S. K. and Owen, T. C. (1999). Galileo probe measurements of D/H and 3He/4He in Jupiter's atmosphere. *Space Science Reviews*, **84**, 251–63.

Mizuno, H. (1980). Formation of the giant planets. *Progress of Theoretical Physics*, **64**, 544–57.

Morbidelli, A., Chambers, J., Lunine, J. I., Petit, J. M., Robert, F., Valsecchi, G. B. and Cyr, K. E. (2000). Source regions and time scales for the delivery of water to Earth. *Meteoritics and Planetary Science*, **35**, 1309–20.

Niemann, H. B., Atreya, S. K., Carignan, G. R., Donahue, T. M., Haberman, J. A., Harpold, D. N., Hartle, R. E., Hunten, D. M., Kasprzak, W. T., Mahaffy, P. R., Owen, T. C. and Way, S. H. (1998). The composition of the Jovian atmosphere determined by the Galileo probe mass spectrometer. *Journal of Geophysical Research*, **103**, E10, 22831–46.

Noll, K. S., Weaver, H. A. and Feldman, P. D. (1996). *The Collision of Comet Shoemaker-Levy 9 with Jupiter*. Cambridge: Cambridge University Press.

Ollivier, M., Encrenaz, T., Roques, F., Selsis, F. and Casoli, F. (2009). *Planetary Systems.* Heidelberg: Springer-Verlag.

Owen, T. and Encrenaz, T. (2006). Compositional constraints on giant planet formation. *Planetary and Space Science*, **54**, 1188–96.

Owen, T., Maillard, J. P., de Bergh, C. and Lutz, B. L. (1988). Deuterium on Mars – the abundance of HDO and the value of D/H. *Science*, **240**, 1767–70.

Owen, T., Mahaffy, P., Niemann, H. B., Atreya, S. K., Donahue, T. M., Bar-Nun, A. and de Pater, I. (1999). A low-temperature origin for the planetesimals that formed Jupiter. *Nature*, **402**, 269–70.

Peck, W. H., Valley, J. W., Wilde, S. A. and Graham, C. M. (2001).Oxygen isotope ratios and rare earth elements in 3.3 to 4.4 Ga zircons: ion microprobe evidence for high ^{18}O continental crust and oceans in the Early Archean. *Geochimica et Cosmochimica Acta*, **65**, 4215–29.

Pollack, J. B., Hubickyj, O., Bodenheima, P., Lissauer, J. J., Podolak, M. and Greenzweig, Y. (1996). Formation of the giant planets by concurrent accretion of solids and gas. *Icarus*, **124**, 62–85.

Reipurth, B., Jewitt, D. and Keith, K. (2007). *Protostars and Planets* V. Tucson, AZ: University of Arizona Press.
Slade, M. A., Butler, B. J. and Muhleman, D. O. (1992). Mercury radar imaging – evidence for polar ice. *Science*, **258**, 635–40.
Taylor, S. R. (2001). *Solar System Evolution: A New Perspective*. Cambridge: Cambridge University Press.
Tobie, G., Choblet, G. and Sotin, C. (2003). Tidally heated convection: constraints on Europa's icy shell thickness. *Journal of Geophysical Research*, **108**, (E11), 5124.
Tobie, G., Mocquet, A. and Sotin, C. (2005). Tidal dissipation within large satellites: applications to Europa and Titan. *Icarus*, **177**, 534–49.
Wilde, S. A., Valley, J. W., Peck, W. H. and Graham, C. M. (2001). Evidence from detrital zircons for the existence of continental crust and oceans on the Earth 4.4 Gyr ago. *Nature*, **409**, 175–8.

15

Water on Mars

Jean-Pierre Bibring

Introduction

Three decades after the Viking missions, which failed to detect any biorelics, not even a slight trace of organic activity, the question as to Mars having harboured habitable conditions, if not life, has been dramatically reopened. A key ingredient, liquid water, might have covered large fractions of early Mars over sustained periods, as indicated by the ongoing space missions. This chapter presents our understanding of the evolution over time of the Martian water reservoirs.

It took centuries for Mars to evolve (in human minds) from a 'planet of death' to a 'world of life': its colour no longer referred to blood (thus its being named after the God of war) but to rust; rust: thus water; water: thus life. These later syllogisms have persisted until very recently, translating the transcendental quest of life far beyond the scientific sphere. And yet: is Mars actually covered by ferric material? If so, is liquid water responsible for the oxidation? More importantly still, would that be sufficient for life to have emerged on Mars? Without direct means to address (and possibly answer) such questions, Mars has always been viewed as the closest and most favourable planet to have harboured extraterrestrial life. A variety of similarities between Mars and the Earth could support the 'plurality of worlds' that was conceived as the operational dogma. Mars exhibits two polar caps, the extension of which follows the four seasons, and which trace an obliquity of the rotational axis (25.2°) very close to that of the Earth (23.3°); the durations of the day (24 h 37 min for Mars) are about the same, both very different from (and much shorter than) those of the other planets. It was thus very tempting to interpret seasonal variations of surface colours, derived from telescopic observations, as resulting from the evolution of Martian vegetation. With the space age and the capability to perform *in-situ* measurements, probes were sent to detect extraterrestrial life on Mars. The Apollo programme had not ended when it was decided that the Viking missions were to examine this fundamental objective; astrobiology was entering the scientific era.

The result of these unprecedented missions departed from expectations (Snyder, 1979 and references therein). Mars appeared as an utmost arid and dry desert, at least in the two sites visited by the Viking Landers. No metabolism was detected; the chemical reactivity of highly peroxidic soil alone is sufficient to account for the results obtained. At the same time, the two Viking Orbiters completed a global coverage of the surface units initiated by

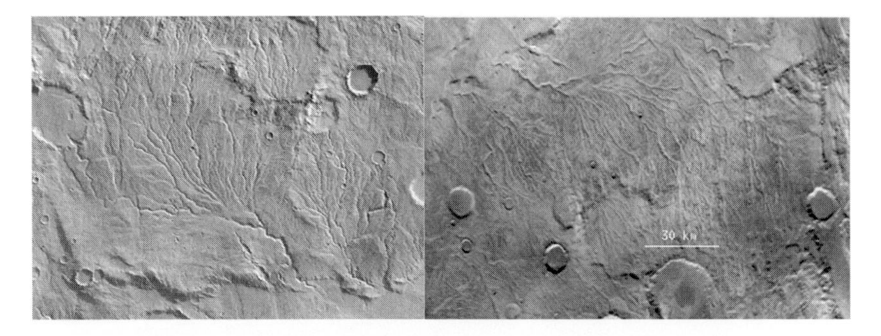

Figure 15.1. Mariner 9 (left) and MOC/MGS (right) images of a valley network.

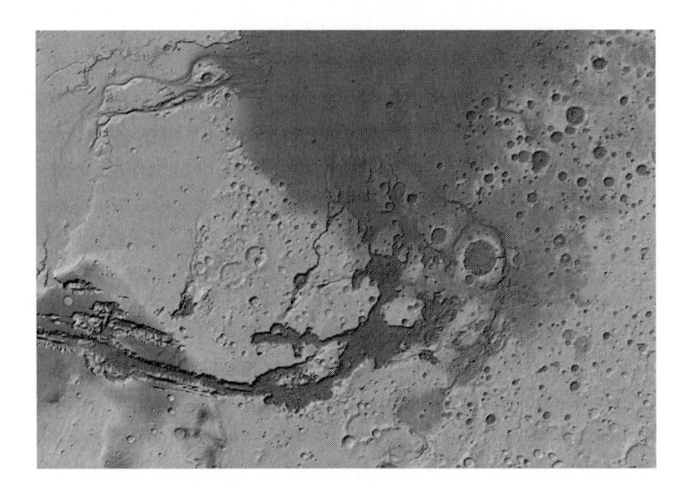

Figure 15.2. MOLA/MGS altimetry map exhibiting outflow channels entering Chryse Planitia.

Mariner 9. Among the variety of surface structures identified, two types associated with fluvial activity were of particular importance: valley networks within the highly cratered terrains (Figure 15.1) and outflow channels ending in the northern plains (Figure 15.2). Although there was no direct evidence (through actual compositional measurement) that they were formed by water, there was no doubt that liquid water once flooded the surface due to their remarkable similarity to terrestrial water-driven features.

The current Martian water reservoirs

Water vapour has been identified and monitored in its space and time variations along several Martian years, through orbital measurements by a number of instruments primarily on board the MGS/NASA (Smith, 2004), Mars Express/ESA (Fouchet *et al.*, 2007) and MRO/NASA missions. Water is a very minor atmospheric constituent: if totally condensed, it would constitute a layer of less than 100 μm in thickness. It exhibits a global planetary

Figure 15.3. In winter (right), water frost deposits at this Viking Landing site (left: summer).

asymmetry, the southern hemisphere being drier than the northern one. Given the low total pressure on Mars, close to or below (depending on the season and location) that of the water triple point (6 mbar), water is not stable in the liquid form, but rather directly condenses (from vapour) or sublimates (from ice) if the temperature crosses 273 K (Figure 15.3). As a consequence, the major current water reservoirs are in the solid state, both as frosts and ices and as hydrated minerals.

The northern polar cap has long been recognized as a massive water-rich glacier, more than 4 km deep at its centre, and hundreds of km wide. Although CO_2 frost condenses and covers most of it during winter, it leaves a CO_2-free H_2O perennial ice cap after spring sublimation is completed. This contrasts with the southern polar cap for which a CO_2 layer remains and actually constitutes the bright surface that shows up in optical images. The spectral imager OMEGA/Mars Express, operating in the near-infrared, has demonstrated that, in fact, this CO_2 ice is only a very thin veneer, some metres deep, covering a wide and thick H_2O-rich glacier, similar to the northern one; this water ice is intimately mixed with dust, which prevents it from shining; it cannot be discriminated from the surrounding water-free soil in the optical images (Bibring *et al.*, 2004).

A quantitative evaluation of the volume of polar water ices has recently been performed using deep radar sounding on board Mars Express/ESA (MARSIS instrument) and MRO/NASA (SHARAD instrument). The principle of these measurements is to send an electromagnetic wave penetrating large depths (kms for MARSIS, tens to hundreds of metres for SHARAD). Whenever these waves cross an interface between media of distinct dielectric constants they are, in part, reflected back to the instrument, where they can be detected. From the time elapsed between emission and reception, the depth of the interface is derived. The interface between the icy cap and the underlying base unit has been clearly detected. The overall icy polar reservoirs amount to 1.4×10^6 km^3 in the north, and 1.6×10^6 km^3 in the south. This would correspond to a layer some 25 metres deep around the entire planet (Plaut *et al.*, 2007).

Actually, the radar sounding was implemented primarily to detect potential deep aquifers: if liquid water was present within the first few kilometres below the surface, it would be detected. So far – the global coverage had not been completed at the time of writing – none have been detected. This 'negative' result does not necessarily deny that liquid water exists, either as deeper (> 5 km) aquifers, or with a water/rock ratio varying with depth without sharp discontinuities. However, this lack of detection may well translate the absence of significant amounts of subsurface liquid water. In fact, the radar probes have failed to detect not only liquid water but also deep water ice, except close to the poles: no thick permafrost has been identified and located so far. The scenario in which the abundant ancient water would now primarily be stored within the subsurface over hundreds of metre- to kilometre-thick layers, is not supported by the measurements.

However, the radar sounding indicates the shallow presence of water ice, down some tens of metres below the surface. This is in agreement with the neutron and gamma-ray measurements on board the Odyssey/NASA mission: their energy spectrum indicates reaction with protons in the upper metre or so, in large abundance, even at low latitudes. This subsurface ice would result from the percolation of the frost over millions to billions of years. This would account for the Phoenix/NASA results, and for the observation made by HIRISE/MRO that shallow craters exhibit bright ejecta when they are formed, which fade out over weeks: the exhumed ice sublimates very rapidly.

Finally, compositional measurements from orbit, by OMEGA/Mars Express and CRISM/MRO, indicate that an important fraction of the water is stored in minerals. The near-infrared is indeed very sensitive to the presence of either or both OH and H_2O, whatever (and discriminating between) their physical state: vapour, frost, ice or within minerals: absorbed, adsorbed, interstitial, structural. The spectral imagers, operating in the near-infrared, have been able to identify, characterize and map most surface constituents, and specifically the hydrated minerals (Figure 15.4). The results are surprising: hydrated minerals are indeed present, but neither those that were previously envisaged nor where they were previously envisaged.

OMEGA has clearly demonstrated that the widespread reddish and bright dust (dark areas in the upper part of Figure 15.4) is actually dominated by ferric oxides: Mars has been rusted. However, liquid water is not responsible for this oxidation. The bright dust is dominated by anhydrous nanophase hematite α-Fe_2O_3, while hydrated ferric oxides, such as goethite with the formula FeO(OH), are extremely rare. The oxidation probably results from a large-scale alteration process in which liquid water did not play a significant role.

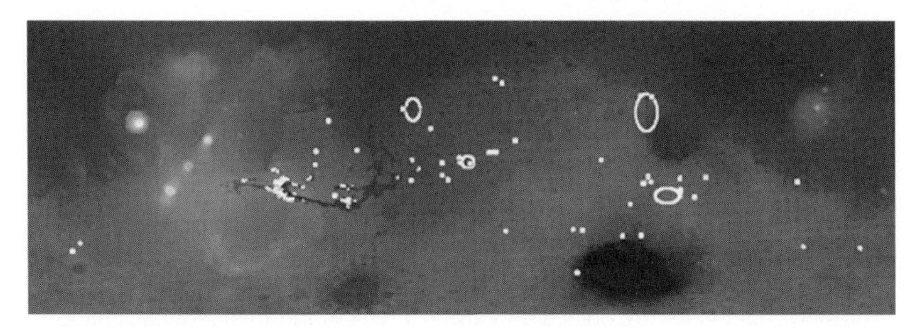

Figure 15.4. Map of hydrated minerals on the surface of Mars.

Figure 15.5. Tracks left by the Spirit rover.

Rather, the alteration results from interaction with the atmosphere, in which trace molecules such as H_2O_2 (a few tens of ppb at most) modify the very superficial grains at a sub-millimetre scale, and on very long (billions of years) timescales. This accounts for the dark tracks left by the rovers on the surface (Figure 15.5): the wheels exhume pristine – and darker – unaltered material.

A second 'negative' mineralogical result is the lack of detection of large carbonate fields. If liquid water was present in the past, thanks to a denser CO_2-rich atmosphere, one could have expected that, similarly to what happened on the Earth, most of the CO_2 would have dissolved, been ionized and precipitated into carbonates, accounting for the present low atmospheric pressure – and resulting lack of sufficient amounts of greenhouse gases to maintain liquid water as stable. The detection of surface carbonates would have both confirmed this hypothesis and indicated where past oceans were present, with possible trapped bio-fossils. An important counter-argument was that if, indeed, the atmospheric CO_2 disappeared through such a process, then the Martian atmosphere should be N_2-rich (as on the Earth), since N_2 cannot dissolve and precipitate. On the contrary, the N_2/CO_2 atmospheric ratio (~ 3%) is similar for Mars and Venus (and for the Earth, if one takes the present carbonates as the reservoir of the primordial CO_2). This is a strong indication that on Mars, both N_2 and CO_2 have been heavily depleted (given the present tenuous atmosphere) with exactly the same efficiency for the two molecules, which requires a single process with no involvement of oceans. As will be emphasized below, all current observations are consistent with a large and early atmospheric escape of all non-condensed species (CO_2 and N_2 primarily, as well as argon and water vapour). Actually, OMEGA/Mars Express has not detected measurable amounts of carbonates above its limit of sensitivity of ~ 1% of carbonates within 99% of other minerals. CRISM/MRO has only detected carbonates in highly restricted spots within Nili Fossae, with no global quantitative significance in terms of CO_2 and H_2O trapped.

The global coverage made by OMEGA has enabled the detection of two classes of hydrated minerals (Bibring *et al.*, 2005), confirmed by CRISM: sulphates and phyllosilicates. Sulphates have been detected both from orbit and on the ground (*in situ*) by the Opportunity Mars Exploration Rover within outcrops close to its landing site, in Terra Meridiani. From orbit, sulphates are detected in a variety of sites, although they are primarily located within terrains in the vicinity of the Tharsis Mons: within Valles Marineris, eastward across Terra Meridiani and up to Aram Chaos towards the north. In chasms within Valles Marineris, sulphates are detected in units of tens of kilometres wide, and kilometres high. Sulphates of diverse compositions have been detected, both *in situ* by Opportunity (jarosite) and from orbit by OMEGA and CRISM: kieserite and other monohydrated sulphates, as well as gypsum and other polyhydrated ones. They all indicate a very acidic environment at the time they formed.

By contrast, phyllosilicates (among which are clays) have been detected all across the ancient heavily cratered crust in small (kilometre-sized) spots. In two locations only, large fields (tens of kilometres wide) of phyllosilicates have been found: in Nili Fossae and in Mawrth Vallis. In this latter area, the depth of the phyllosilicate-rich layer amounts to some 200 m, and their total abundance is higher than 50%. Most phyllosilicates are Mg/Fe-rich smectites, with occurrences of more Al-rich ones, indicating a higher degree of leaching.

Summing up the diverse reservoirs of water already identified, one finds a volume that would correspond to a planetary layer barely in excess of a few tens of metres deep. Even scaled to actual planetary sizes, this is two orders of magnitude less that the terrestrial water reservoir, adding the oceans and magma contents. Again, one cannot totally exclude the possibility that most water is stored at depths or within sinks that are out of reach of current observational tools. However, the lack of detection of hydrated minerals in recent crater ejecta, exhuming deep material, tends to rule out this possibility. This would thus open the questions: were the Martian and terrestrial water reservoirs similar in the past? If so, where has the Martian water gone?

The past Martian water reservoirs

Polar caps have constituted sustained reservoirs of water over Mars's history. Their layered pattern records a traceable evolution of the atmospheric properties. Although these cold traps preserve very crucial information, as the terrestrial Antarctic cap does, these icy reservoirs are not the best remnants of the ancient Mars environment. The chaotic evolution of the Martian obliquity recently simulated by J. Laskar's team in Paris (Laskar *et al.*, 2002), exhibits large variations over timescales of millions of years, by more than ten degrees. As a consequence, the polar ices have frequently been sublimated from their current location, and eventually have redeposited in a variety of sites down to equatorial latitudes, such as the flanks of the giant volcanoes within Tharsis, as demonstrated by F. Forget and his team (Levrard *et al.*, 2004). With the further evolution of either the obliquity or the internal activity triggering transient processes, the ice became unstable in these locations, leaving diagnostic glacial features still visible (Figure 15.6). However, the interpretation of these structures does not require icy reservoirs larger than those currently observed, but rather their geographical redistribution.

The most accurate deciphering of the ancient history of Mars can be derived from the detection of well-preserved hydrated minerals and the characterization of their geological settings: the composition of the minerals gives access to the 'enabling environments' at the time and location of their formation. In particular, the phyllosilicates and sulphates identified, which require very distinct processes and conditions, still record the past aqueous history of Mars; they trace back the evolution of the Martian environment.

The dominant phyllosilicates observed require neutral to alkaline conditions to grow, and abundant perennial surface water to alter their parent magmatic rocks: the implied water to rock ratio of their formation can amount to 1000 and more, indicating a sustained circulation over long durations. Their occurrence in spots spread over the entire cratered

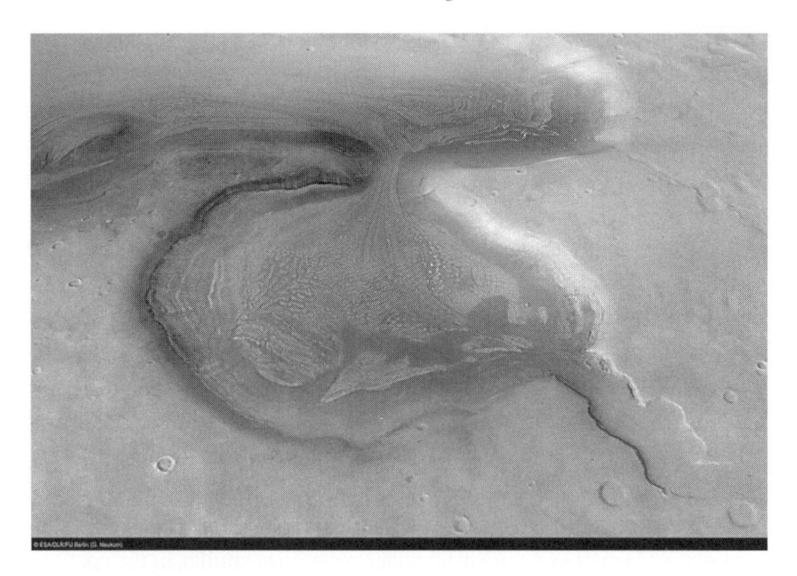

Figure 15.6. Glacial Martian feature.

crust has two implications. First, it tells us that these conditions – for liquid water to remain stable – occurred during the very early Mars ages, some tens to hundreds of millions of years only after the crust solidified. Second, the fact that the crust essentially exhibits its pristine mafic composition instead of being fully covered by phyllosilicates indicates that the era during which water could altered the crust terminated before the heavy bombardment ceased. If the late cratering episode (LHB, late heavy bombardment) (Gomes *et al.*, 2005) had taken place while surface water was still present and stable, the impacts would have mixed up materials and continuously altered them, leading to a widespread surface coverage of hydrated minerals, which is not at all what is observed. The LHB has covered and buried most hydrated material with deeper magmatic minerals, which were and remained unaltered, and which constitute most of the present surface crust. The phyllosilicates are only detected in small sites exhumed by either impact or erosion. They offer access to a very specific and ancient Martian era during which liquid water remained stable: we named it, consequently, the 'Phyllosian Era'. It ended very early, before the LHB, when the environment faced a global climatic change forcing liquid water to leave the surface: some evaporated and escaped and the rest percolated as permafrost.

The sulphates are detected in terrains formed after the bombardment had stopped, exhibiting much lower crater densities. Most of them are located in units whose formation is directly coupled to the building of Tharsis. The building of this huge dome, thousands of km wide and several km high, constitutes the oldest and most intense volcanic activity on Mars. It happened just after the LHB, as dated by its crater density. The lava outflows were accompanied by massive outgassing, in particular of S-rich compounds which spread over the entire planet: this accounts for the high S concentration measured in surface soil at all landing sites (Vikings, PathFinder, MERs). By its mass, Tharsis triggered large structural stresses in the surrounding crust, initiating networks of faults and grabens (e.g. Valles Marineris), and lifting vast areas (Terra Meridiani). The rising of the geothermal front which accompanied the building of Tharsis lifted part of the permafrost (accumulated from the water percolated at the end of the Phyllosian Period) towards the surface: massive supplies of liquid water reached the surface, and cemented the S-rich compounds into vast deposits of sulphates. However, although their formation requires large amounts of water, sulphates do not necessarily trace an era during which liquid water was stable with sustained lakes or oceans. If water vaporizes, it builds mounts of precipitated salts. This is what may well have happened: sulphate deposits grew as water evaporated, since this occurred after Mars's global climatic change. This very specific era of sulphate formation is named the Theiikian Era, from 'theiikos', sulphates in Greek (Bibring *et al.*, 2006).

Mars's global climatic change

The occurrence of phyllosilicates in the most ancient terrains and of sulphates in younger but still old ones, indicates that the Martian environment underwent a global change, from neutral to alkaline conditions with stable liquid water, towards acidic ones when the atmosphere had faded to a level preventing the sustenance of liquid water: one of its major outcomes is the atmospheric loss of most non-condensed species.

This transition happened before the end of the heavy bombardment, and is therefore probably not coupled to this exogenous process, but rather to an endogenous one. One attractive possibility – which remains to be confirmed – is that this climatic change was triggered by the decrease in the Martian dynamo, which very likely occurred in very early Martian times. The MAGER/MGS instrument (Acuña *et al.*, 2008) has shown that Mars has no global magnetic field today, but had one during its early phases, recorded in the remnant magnetization of the ancient cratered crust (Figure 15.7).

A key feature is the absence of remnant magnetized minerals in Tharsis, on the giant volcanoes, in the northern plains and within the basins, indicating that their formation all occurred after the dynamo had dropped: if a magnetic field was still present while their surface materials crystallized, they would also have acquired and would exhibit a remnant magnetization. It is to be emphasized that most large basins, and therefore the onset of craters currently observed, which constitute the LHB, were formed after the global magnetic field had vanished.

The shutdown of the dynamo could have strongly affected the atmospheric properties of Mars. Without an efficient magnetic shield against the lethal effects of the still young Sun (EUV emitter, intense solar wind), most of the planetary atmospheres were highly ionized and sputtered away, decreasing the concentration of greenhouse gases below those required to maintain water above its triple point.

The drop in the dynamo might have originated from a too low global amount of radioactive species, given the planet's size, to maintain mantle convection when most of the accretion energy had faded out. The difference in size between Mars and the Earth, translated into terms of global radioactive content, would have triggered the divergent pathways

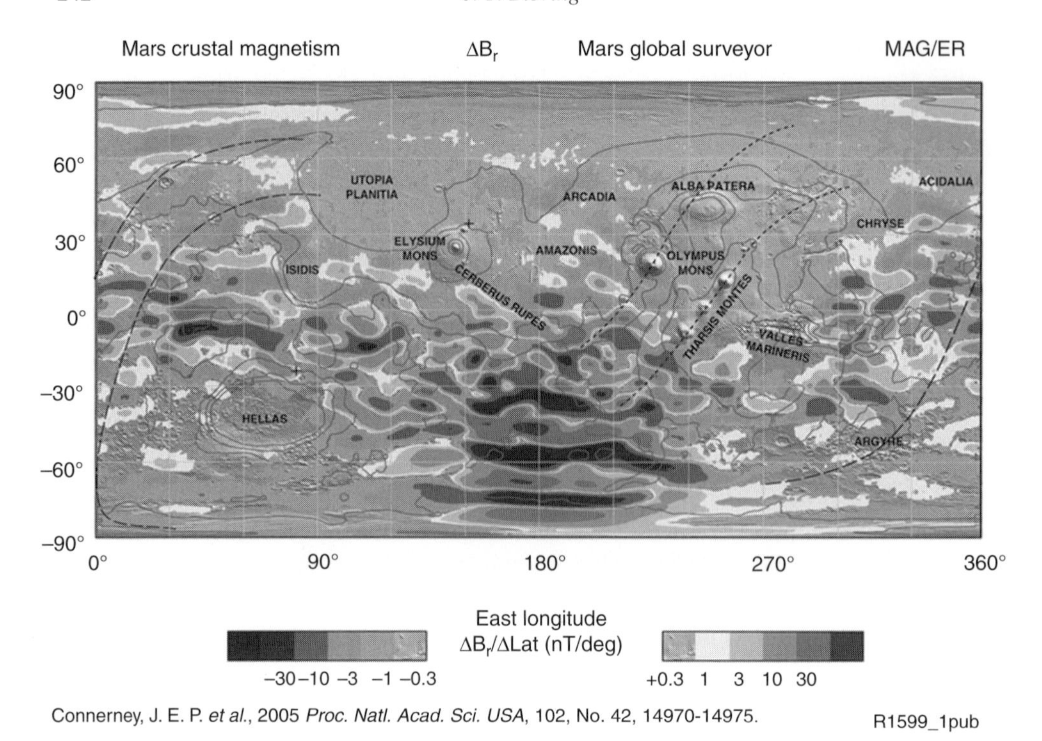

Figure 15.7. Remnant magnetization on the surface of Mars, exhibiting the contrast between the ancient (magnetized) crust and non-magnetized structures such as Tharsis, the giant volcanoes, the northern plains and the large basins.

of these two worlds. The Earth could sustain an efficient internal activity. On Mars, the decrease in mantle convection reduced its capability to efficiently extract energy from the core: consequently, the turbulent core convection disappeared, inhibiting the magnetic dynamo. The further evolution of the mantle dynamics, with downwelling cold plumes initiating instabilities at the core–mantle interface, led to upwelling plumes building Tharsis and filling the northern plains: this thus happened long (tens of millions of years) after the dynamo had vanished and most of the atmosphere had escaped. A single process thus accounts for structural and compositional evidence.

If Mars ever harboured a habitable environment, hosting surface liquid water over geological timescales, it was before Mars's global climatic change, during the Phyllosian Era. Supposedly habitability led to the emergence of living species, relics are to be searched for in still well-preserved Phyllosian terrains. Were these hypothetical species capable of surviving the global climatic change? The Theiikian Era was probably not favourable to the emergence of life; however, as it constituted an era of supply of large amounts of liquid surface water, although not stable, it could have efficiently contributed to the survival of dormant species, trapped in safe niches at the end of the Phyllosian Era.

Figure 15.8. Artist's view of a future lander exploring Mawrth Vallis phyllosilicate-rich terrains.

Mars and Earth: similarities and diverging pathways

Recent discoveries of (highly refractory) terrestrial zircons dated to 4.1 billion years ago or more (prior to the LHB) with varying $^{18}O/^{16}O$ abundances, strongly point towards the existence of an ancient ocean while the terrestrial crust had not yet fully cooled. However, the disappearance of the rocks in which they were included prevents an in-depth study of this era of potential Earth habitability, during which life might have emerged in these primitive oceans: the Hadean–Archaean transition might well have occurred before the LHB. It has been shown very recently that in fact the LHB might not have destroyed more than a few per cent of pre-existing terrestrial oceans. If life had emerged before the bombardment had ceased, it could have lasted and evolved, protected in these oceans.

The Martian phyllosilicates and the terrestrial zircons formed at about the same time, when Mars and the Earth possibly hosted ancient long-standing bodies of water. Phyllosilicates and zircons thus played similar roles, while the former preserved their entire context – including their hosted rocks: they provide a unique means for studying this era of potential emergence of life, on Earth, and possibly on Mars.

Up to now, neither Lander nor Rover have ever explored a Phyllosian site: the landing-site selection, based on imaging primarily, favoured terrains where water did indeed flood, but probably through transient episodes, out of thermodynamic equilibrium. The discoveries of hydrated minerals modify fundamentally the astrobiological exploration of Mars: they identify and locate sites of optimal relevance to the search for habitability, preserving the record of an era during which water might have been stable over long durations. The upcoming MSL/NASA and ExoMars/ESA *in-situ* missions, which will, for the first time, target and explore Phyllosian sites, have, therefore, a dramatic potential in their search for traces of extant/extinct life (Figure 15.8).

References

Acuña, M. H., Kletetschka, M. H. G. and Connerney, J. E. P. (2008). Mars crustal magnetization: a window into the past. In *The Martian Surface: Composition, Mineralogy, and Physical Properties*, ed. J. F. Bell. Cambridge: Cambridge University Press.

Bibring, J. P., Langevin, Y. F., Poulet, F., Gendrin, A., Gondet, B., Berthé, M., Soufflot, A., Drossart, P., Combes, M., Bellucci, G., Moroz, V. and Mangold, N. (2004). Perennial water ice identified in the south polar cap of Mars. *Nature*, **428**, 627–30.

Bibring, J. P., Langevin, Y. Gendrin, A., Gondet, F., Poulet, F., Berthé, M., Soufflot, A., Arvidson, R. E., Mangold, N., Mustard, J. F. and Drossart, P. (2005). Mars surface diversity as revealed by the OMEGA/Mars Express observations. *Science*, **307**, 1575–81.

Bibring, J. P., Langevin, Y., Mustard, J. F., Poulet, F., Arvidson, R. E., Gendrin, A., Gondet, F., Mangold, N., Pinet, P. and Forget, F. (2006). Global mineralogical and aqueous Mars history derived from OMEGA/Mars Express data. *Science*, **312**, 400–4.

Fouchet, T., Lellouch, E., Ignatiev, N. I., Forget, F., Titov, D., Tschimmel, M., Montmessin, F., Formisano, V., Giuranna, M., Maturilli, A. and Encrenaz, T. (2007). Martian water vapour; Mars Express PFS/LW observations. *Icarus*, **190**(1), 32–49.

Gomes, R., Levison, H. F., Tsiganis, K. and Morbidelli, A. (2005). Origin of the cataclysmic late heavy bombardment period of the terrestrial planets. *Nature*, **435**, 466–9.

Laskar, J., Levrard, B. and Mustard, J. M. (2002). Orbital forcing of the Martian polar layered deposits. *Nature*, **410**, 375–7.

Levrard, B., Forget, F., Montmessin, F. and Laskar, J. (2004). Recent ice-rich deposits at high latitudes on Mars by sublimation of unstable ice during low obliquity. *Nature*, **431**, 1072–5.

Plaut, J. J., Picardi, G., Safaeinili, A., Ivanov, A. B., Milkovich, S. M., Cicchetti, A., Kofman, W., Mouginot, J., Farrell, W. M., Phillips, R. J., Clifford, S. M., Frigeri, A., Orosei, R., Federico, C., Williams, I. P., Gurnett, D. A., Nielsen, E., Hagfors, T., Heggy, E., Stofan, E. R., Plettemeier, D., Watters, T. R., Leuschen, C. J. and Edenhofer, P. (2007). Subsurface radar sounding of the south polar layered deposits of Mars. *Science*, **6**, 316, 5821, 92–5.

Smith, M. D. (2004). Interannual variability in TES atmospheric observations of Mars during 1999–2003. *Icarus*, **167**, 148–65.

Snyder, C. W. (1979). The Planet Mars as seen at the end of the Viking missions. *Journal of Geophysical Research*, **84**, B14, 8487–519.

Part IV

From non-living systems to life

16

Energetic constraints on prebiotic pathways: application to the emergence of translation

Robert Pascal and Laurent Boiteau

The origin of life, as with any other process of structure formation, should have been accompanied by a loss of entropy. Since the second law of thermodynamics states that the entropy of an isolated system tends to increase, any self-organizing system must exchange free energy (closed system) and/or matter (open system) with its environment in order that the overall entropy increases (Kondepudi and Prigogine, 1998). This simple observation emphasizes the importance of energy transfers in the origin and development of early life. As far as biochemical systems are concerned, energy exchanges mostly involve chemical energy that is brought about by 'high-energy' carriers so that energy flows through metabolic pathways from free energy-rich compounds towards low-energy molecules, the difference being released in the environment as heat. When the occurrence of a thermodynamically unfavourable reaction makes it necessary, fresh energy is provided to the system through coupled reactions involving a free-energy carrier such as ATP. The principle that energy is brought about by 'high-energy' carriers applies to most metabolic pathways, though some of them do not simply follow this rule. An example is the process of energy collection leading to ATP synthesis, in which 'chemical' energy is generated from a 'physico–chemical' source: a gradient of concentration between two compartments separated by the plasma–cell membrane (Mitchell, 1961). However, these exceptions require very specific devices such as a semi-permeable barrier and the presence of a highly complex membrane protein (ATPase) functioning as a machine made of several subunits cooperating to build one ATP molecule from several discrete events (i.e. the passage of protons through the ATPase).

The requirement that free energy must be provided to the self-organizing system would apply to proto-metabolic chemical networks as well. Then, the principle that energy flows from energy-rich carriers[1] to inactivated molecules is an essential feature of self-organizing proto-metabolic processes and must have been of importance in governing the origin of the metabolism (Weber, 2002). Although 'high-energy' compounds may be formed as unstable and transient intermediates in a given pathway, the preceding principle is mandatory when

[1] A chemical-energy source or a carrier can be defined as a reactant (or more generally as a system of reactants) in far-from equilibrium concentration with respect to products and protected from spontaneous deactivation (isolated from products) by a kinetic barrier. This definition has some connection with the ideas advanced by Eschenmoser (1994; 2007) that kinetic barriers have an essential role in driving a chemical system towards self-organization or with the description of life as a kinetic state of matter (Pross, 2005).

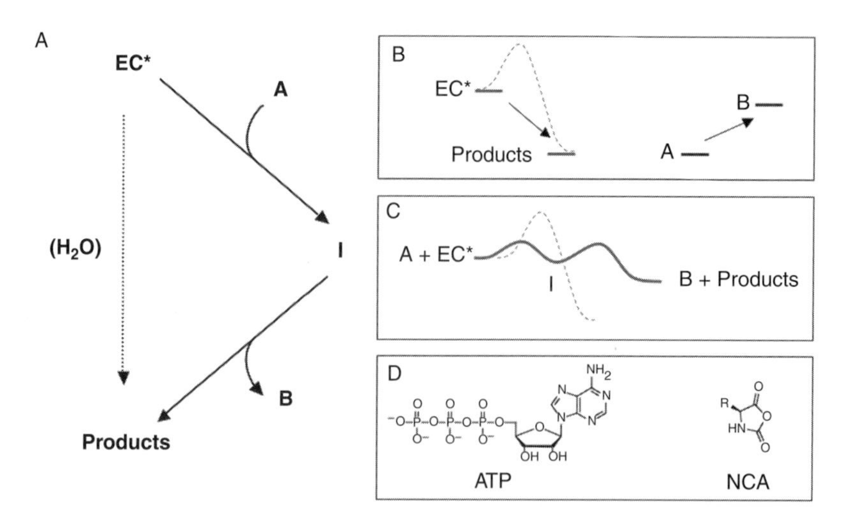

Figure 16.1. Exchange of free energy through a coupled reaction of a free-energy carrier (**EC***). An endergonic reaction (thermodynamically unfavourable with $\Delta G^\circ > 0$) can be coupled to an exergonic reaction ($\Delta G^\circ < 0$). A first requirement for the coupled reaction to take place (**A**) is chemical; a chemical group (often water) has to be exchanged between half-reactions without being released, which necessitates the formation of an intermediate (I). The nature of the second requirement is thermodynamic (box **B**): standard free energies of reactant and intermediates must be compatible. The last requirement is kinetic (box **C**): the kinetic barriers (continuous line) of every step of the coupled reaction must be lower than that of the spontaneous deactivation (dashed line) of the free-energy carrier. Adenosine triphosphate (ATP) is the most common energy carrier in biology (box **D**), whereas amino-acid N-carboxyanhydrides (NCAs) are proposed here to have played this role in early biochemistry.

considering stable products or long-lived intermediates present in significant concentrations. An important improvement of our understanding of these processes lies in the identification of coupled reactions (Figure 16.1) through which energy could be provided to the system or could be circulating among metabolic subsystems. Furthermore, coupled reactions are essential to the development of metabolic cycles (Shapiro, 2006; Plasson, 2008; Blackmond, 2009). On the basis of its universality in biology, it is commonly believed that ATP (or possibly other phosphate anhydrides such as polyphosphates) was initially selected as an energy carrier in early living organisms, but we have no historical evidence in favour of such an early role. The above-mentioned principle requires the free-energy content of the carrier to match the requirements of metabolic pathways. In this chapter, we analyse the origin and development of translation, and after noticing that in the absence of enzymes, α-amino-acid activation would require more free energy than the amount delivered by ATP, we propose that α-amino-acid *N*-carboxy-anhydrides (NCAs) initially constituted the 'high-energy' species of the process.

The translation apparatus and the origin of the genetic code

Studies of the early stages of evolution though molecular-phylogeny approaches suggest that the last universal common-ancestor (LUCA) genome was mainly composed of genes

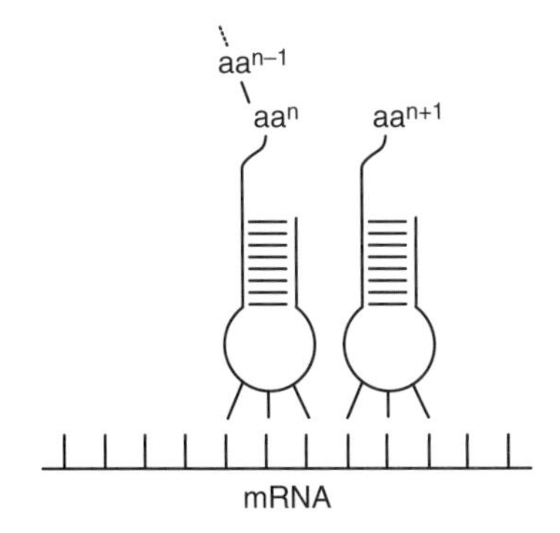

Figure 16.2. Scheme of a primitive translation system originating from small hairpin RNA adaptors.

encoding for components of the translation apparatus (ribosomal proteins and RNAs, transfer RNAs, aminoacyl tRNA synthetases etc.). This predominance has been explained in terms of a (hypothetical) period of evolution called the RNA-protein world, during which RNA played a prominent role in biological processes (Delaye *et al.*, 2005). It also means that translation evolved early from a network of chemical reactions proceeding through spontaneous pathways or through catalytic mechanisms involving simple inorganic or organic molecules and very probably RNA catalysis by ribozymes. The evolution leading to an RNA–protein world involved processes governed by chemical rules. From this perspective, it is worth noting that the ribosome has probably conserved its early function, which corresponds to its mere definition as an entropy trap (Sievers *et al.*, 2004). The principle of a close proximity of tRNAs induced by codon–anticodon recognition on the mRNA strand, which promotes peptide-bond formation, may simply have been improved, without functional change, by the development of the A and P sites on the ribosome. These sites are the places where the amino-acid – tRNA and peptide – tRNA complexes are bound on the ribosome in order to induce the proximity of reacting groups and then facilitate peptide-bond formation. This observation is in agreement with the idea that the function of the ribosome may, in principle, be played by the messenger RNA and simple RNA-made molecular adaptors; which is the basis of the explanations of the emergence of translation. The early development of the translation apparatus very probably required covalent interactions of α-amino acids and/or peptides with RNA adaptors (Figure 16.2). Thus, RNA mini-helices or RNA hairpins covalently bound to amino acids or peptides may have been essential (Di Giulio, 1997, 1998; Tamura and Schimmel, 2003). Then more complex structures may have been formed by hairpin duplication (Widmann *et al.*, 2005).

However, understanding how the ribosome works is not sufficient to provide a full explanation of the origin and development of translation and of the selective pathways allowing activated amino acids to be charged on their cognate tRNA, which is the actual chemical basis of the genetic code rather than the ribosomal process of peptide-bond formation. Initially, the genetic code may have involved a reduced set of prebiotically available amino acids, which has then been expanded by the allocation of new codons to biochemically related amino acids according to the coevolution theory of the genetic code

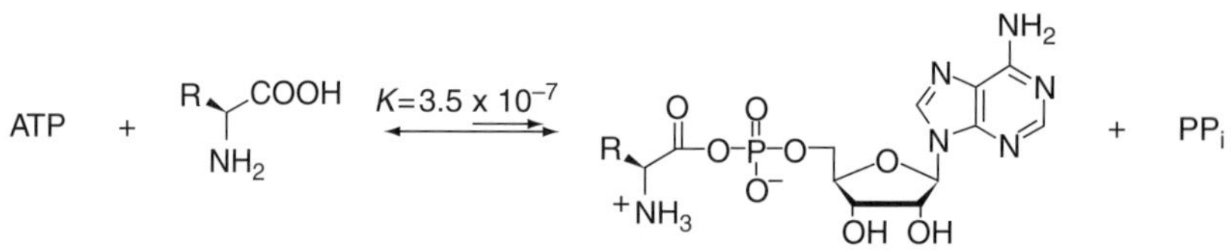

Figure 16.3. The formation of amino-acid adenylates from ATP and free amino acids is thermodynamically unfavourable without enzymatic stabilization of aa-AMP by aaRS. AA = amino acid.

(Wong, 1975, 2005). But there are two different questions related to the organization of the genetic code and the translation apparatus, namely the correspondence of amino acid with codons insured by the set of aminoacyl-tRNA synthetases (aaRSs) ligating activated amino acids to their cognate tRNA and the amino-acid activation reaction that precedes the formation of aminoacyl tRNA. Most investigations have been focused on the selectivity of the loading reaction and on the proofreading mechanisms (Cramer and Freist, 1987). This chapter is devoted solely to the analysis of the chemical processes that were responsible for amino-acid activation and were the chemical basis on which the translation machinery developed.

How did α-amino-acid activation develop?

The thermodynamics of amino-acid activation

The reaction of amino acids with ATP giving the amino-acyl-adenylate (aaAMP) intermediate is highly energetically unfavourable (Figure 16.3). A value of 3.5×10^{-7} has been determined at pH 7.78 for the equilibrium constant in the case of tyrosine activation (Wells *et al.*, 1986) leading to $\Delta G^{\circ\prime}$ = *ca.*–70 kJ mol^{-1} for the value of the standard free energy of hydrolysis of Tyr–AMP at pH 7. Therefore, aa-AMPs can be considered to correspond on a free-energy scale to a level that is *ca.*38 kJ mol^{-1} above that of ATP. Then, almost no aa-AMP could be formed in the absence of an enzyme (aaRS) stabilizing the intermediate, which raises the question: how could this intermediate have been selected in biological pathways if it was not spontaneously present to a significant extent? It could be argued that the reaction of Figure 16.3 could be shifted to the right by pyrophosphate (PP_i) hydrolysis, as observed in living cells. However, it seems highly unlikely that a specific pyrophosphatase (enzyme or more probably ribozyme) could have been present at early stages of evolution with a high activity against PP_i while devoid of deleterious ATPase activity. The most straightforward answer seems to be that aa-AMP synthesis initially took place via an alternative pathway that did not involve activation by ATP (Pascal *et al.*, 2005).

A plausible scenario for the evolution of α-amino-acid activation

This proposal is based on the idea that the formation of the aa-AMPs needed for the development of translation started from a spontaneous chemical process (thermodynamically

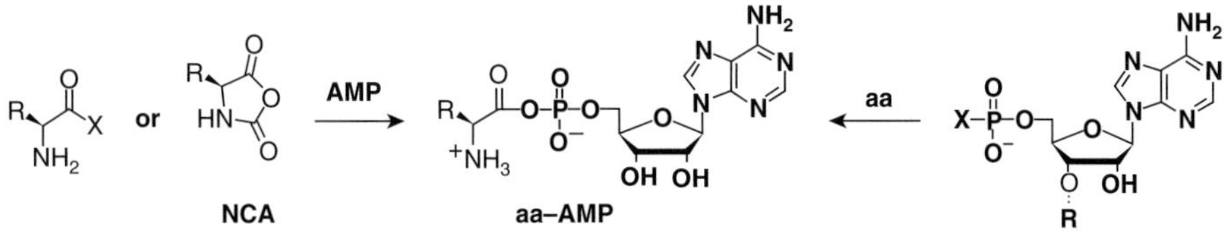

Figure 16.4. Obtaining amino-acyl adenylates from activated nucleotides or activated amino acids (including α-amino-acid *N*-carboxyanhydrides, NCAs).

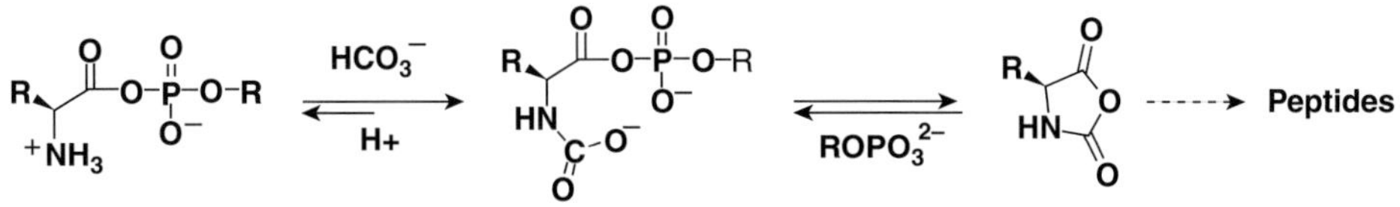

Figure 16.5. The conversion of α-amino-acid phosphoanhydrides into NCAs in aqueous solutions containing CO_2 or hydrogen carbonate.

favoured or close to equilibrium), which was sensitive to catalysis and may then be kinetically improved by enzymes or ribozymes. Proposing an alternative to the reaction of ATP as a precursor of aminoacyl adenylate needs to identify a prebiotic chemical process capable of promoting the synthesis of this intermediate in a better way. Since aa-AMPs are phosphoric–carboxylic mixed anhydrides, in principle the activated forms of both amino acids and nucleotides can be precursors of these intermediates, but we have to look for aminoacyl or phosphoryl transfer agents at least as potent as anhydrides (Figure 16.4).

α-amino-acid N-carboxyanhydrides (NCAs)

A convenient solution could be proposed after the observation that NCAs promote the formation of mixed anhydrides from inorganic phosphate (Biron and Pascal, 2004; Leman *et al.*, 2006) and phosphate monoesters (Biron *et al.*, 2005; Leman *et al.*, 2006) in dilute aqueous solution at pH values close to neutrality. This potential role is also supported by the increasing chemical evidence that NCAs may have been formed in a prebiotic world (Commeyras *et al.*, 2004; Leman *et al.*, 2004; Pascal *et al.*, 2005; Danger *et al.*, 2006). NCAs may thus be considered as good candidates for the formation of aa-AMPs or similar mixed anhydrides before life emerged and at the first stages of evolution. But this role is also supported by another argument: since CO_2 was probably much more abundant in the early Earth's atmosphere than in the contemporary one, hydrogen carbonate was probably present in significant concentrations that are likely to induce NCA formation from aa-AMP according to the following reaction (R = adenosyl), whatever the origin of adenylates could have been (Figure 16.5).

Experimental support in favour of an easy conversion of aa-AMPs into NCAs has been provided by the fact that the length and the yield of peptides formed by amino-acyl-phosphate mixed-anhydride polymerization in aqueous solution is strongly increased by the presence of HCO_3^- (Dueymes 2009).

The fact that NCAs can undergo spontaneous reactions with nucleophiles different from water in aqueous solution is an indication of their ability to behave as free-energy carriers in accordance with the definition that this role must be played by species protected from spontaneous degradation by kinetic barriers. These reactions also constitute evidence of the ability of NCAs to undergo coupled reactions, which is the second feature of an energy carrier.

Adenylates as biochemical fossils testifying to an early role for NCAs?

Any suggestion of an early role for NCAs would get some support by finding indications of their past contribution, which may have been conserved in contemporary biochemical pathways. However, the universality of the genetic code means that it was set before the common ancestor lived so that it is difficult to find pieces of information in relation to earlier stages of evolution in which the current pathway of amino-acid activation was not established. As already mentioned, it is not easy to find a driving force explaining why the high-energy intermediate aa-AMP has been selected by evolution for the aminoacylation of tRNA, since it is much less stable ($\Delta G^{\circ\prime} = -70$ kJ mol^{-1}) than that of ATP ($\Delta G^{\circ\prime} = -32.2$ kJ mol^{-1}; Jencks, 1976). This value means that its equilibrium concentration must have been very low starting from ATP and a free amino acid. In this view the free-energy content of aa-AMPs may be considered as a kind of bioenergetic or metabolic fossil of an early stage during which the formation of aa-AMPs or that of closely related mixed-anhydride intermediates was allowed as free species in solution and not only as a sequestered intermediate in the active site of an aaRS enzyme. Since the value of the free energy of hydrolysis at pH 7, $\Delta G^{\circ\prime} = -70$ kJ mol^{-1} for Tyr–AMP (Wells *et al.*, 1986) matches that of NCA, which is probably close to $\Delta G^{\circ\prime} \approx -60$ kJ mol^{-1} as estimated previously (Pascal *et al.*, 2005), an early role for NCAs as precursors of aa-AMPs is compatible with these results. However, the cytosolic abundance of NCAs in the corresponding organisms is likely to have had deleterious consequences on their metabolism so it is relevant to search for remnants of NCAs in biochemistry attesting not only to the presence of these intermediates, but also to the mechanisms by which cells avoided their deleterious side-effects. Indeed, NCAs are capable of acylating proteins in an uncontrolled way both at the amino terminus and at side-chain amino groups, leading to altered enzymes and proteins likely to have lost the activity of the native form (Figure 16.6). Since there is no indication of a current biochemical role for NCAs, we can look at remnants of these mechanisms of protection against any purported early contribution.

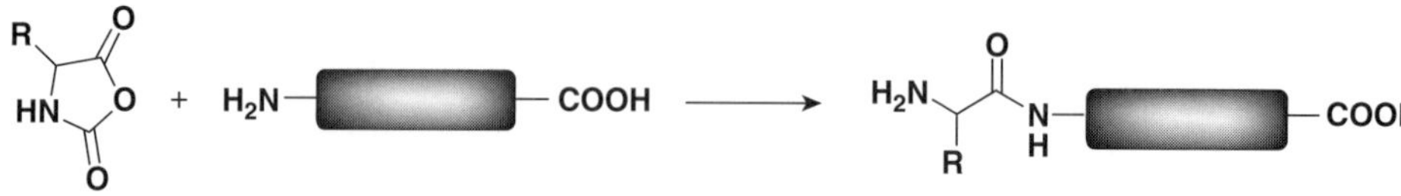

R O HN O O + H₂N— —COOH → H₂N O N H R —COOH

Figure 16.6. Elongation of protein N-terminus by NCA reaction.

aa-AMP sequestration by aaRS

HCO_3^- is usually present in significant concentrations (20–30 mM) in most cells as a product of metabolism, which would be sufficient to induce conversion of aa-AMPs into NCAs. This remark can be associated with the fact that the affinity of aaRSs for adenylates prevents them from being released in solution, which suggests an extra role for the sequestration of aa-AMP in aaRSs, namely that it could prevent NCA formation. We have already mentioned the propensity of aa-AMP and other amino-acyl phosphates to be readily converted into NCA in dilute aqueous solution, which is illustrated by an easy polymerization into oligomers from these intermediates in the presence of hydrogen carbonate (Dueymes, 2009). An unexpected consequence of the absence of free aa-AMPs in the cytoplasm is then the prevention of non-controlled aminoacylation of proteins by NCAs.

Initiation of translation by formyl-methionine transfer RNA

In this discussion it is appropriate to mention that the initiation of protein synthesis in bacteria involves *N*-formyl-methionine (fMet), which prevents any acylation of the amino terminus. Although initiation by fMet-tRNA is not universal (Di Giulio, 2001), the protection of the amino terminus by acylation with the HCO group could be considered as a defence against post-translational non-controlled aminoacylation, which may have resulted from the presence of significant NCA concentrations in the cytoplasm.

The late introduction of lysine in the genetic code

The side-chain amino group of lysine plays an essential role in the active site of a number of enzymes (as, for instance, the catalysis by aldolase involved in glycolysis) by establishing covalent bonds with substrates in essential biochemical pathways. However, it is generally considered that lysine was not among the set of amino acids that were initially selected in the early genetic code and that it was added at a subsequent stage (Wong, 1975, 2005). This late introduction has been attributed to the non-availability of lysine from prebiotic pathways of synthesis – it may also be considered as a consequence of the sensitivity of the lysine side-chain to aminoacylation by NCAs at a stage of evolution when NCAs played a metabolic role. On this view, we can suggest that the possibilities of catalysis allowed by a side-chain amino group became available to cells only when an alternative pathway of synthesis of aminoacyl adenylate was introduced that lowered the formation of NCAs to harmless concentrations because of the sequestration of adenylates in the active sites of aaRSs.

The transition from NCA- to ATP-driven translation

The proposal of a stage in the development of the translation apparatus in which NCAs were used as activated amino-acid forms requires that a scenario explaining how this system could have evolved towards the modern process of amino-acid activation is provided. In this proposal, the need to avoid the deleterious effects of NCAs is considered to have

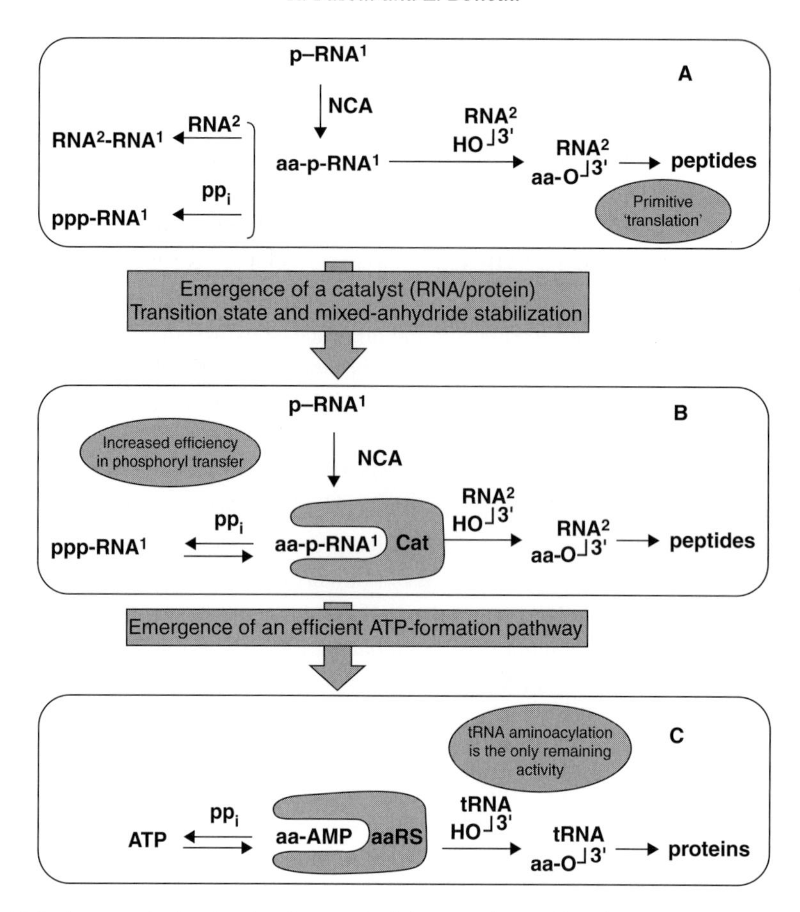

Figure 16.7. A scenario for the development of the α-amino-acid activation pathway. (**A**) NCA-promoted formation of mixed anhydrides with mono- or polynucleotides (p-RNA^1) allows both acyl transfer and phosphoryl transfer. (**B**) The stabilization of the mixed anhydride intermediate (aa-p-RNA^1) by a catalyst (Cat) increases the efficiency of the NCA-promoted primitive translation pathway as well as that of nucleotide activation (polymerization, ligation) and allows the conversion into mono- or poly-nucleotide 5′-triphosphates. (**C**) When the stabilization is sufficient and an independent ATP-synthesis pathway is available, the existing devices can be opportunistically utilized in the reverse direction for an ATP-mediated amino-acid activation pathway, the catalyst evolving into an aminoacyl tRNA synthetase.

driven the change. If we consider that early life mainly evolved by selecting spontaneous chemical pathways that were facilitated by the catalytic action of biomolecules, then any improvement in the efficiency of amino-acid activation would have led to the stabilization of aminoacyl adenylates (Figure 16.7), which is likely to have led to sequestration of this intermediate. But a second factor is likely to have contributed to this change. The reaction giving aa-AMPs and PPi from ATP and free amino acids is strongly unfavourable, which means that it is in principle possible to get ATP by performing this reaction in the reverse direction. This reverse reaction does not work spontaneously, but it could have proceeded owing to biochemical catalysis (by enzymes or ribozymes) leading us towards the hypothesis that the synthesis of ATP was the actual driving force leading to the emergence of biocatalysts for adenylation (Pascal *et al.*, 2005). The abilities of ATP and other nucleotide triphosphates (NTPs) as building blocks for RNA synthesis may explain the utility of this reaction pathway for early living organisms. This system became redundant for ATP synthesis when more efficient pathways for the synthesis of NTPs were available (for instance, after the emergence of chemiosmosis and membrane ATPases). However, the availability of adenylating enzymes capable of sequestrating and stabilizing aa-AMPs gave to evolution the opportunity of using these intermediates for other purposes including protein synthesis. Although highly speculative, this scenario presents the unique advantage that biocatalysts were selected by evolution only to improve the rates of thermodynamically favourable reactions and not to catalyze biochemically useless uphill processes. It is compatible at every stage with the idea that early living organisms evolved to get a selective advantage resulting firstly from improved RNA synthesis and then from the developing pathway of amino-acid activation.

Coevolution or RNA world?

This scenario is consistent with the fact that it is difficult to build a clear separation between amino-acid and nucleotide chemistries at early stages of evolution, which is a strong indication favouring coevolution scenarios (Borsenberger *et al.*, 2004) rather than an unmixed RNA world (Gesteland *et al.*, 1999) in which amino acids and peptides were absent. The dual reactivity (i.e. acyl transfer and phosphoryl transfer) of aa-AMPs and other mixed anhydrides formed from α-amino acids and phosphates or nucleotides (Biron and Pascal, 2004; Biron *et al.*, 2005) constitutes an illustration of this perspective. This property is incorporated in the above-proposed scenario deriving from energetic constraints on the development of translation, which suggests that the current biochemical role of mixed anhydrides with amino acids may have derived from their ability to promote phosphoryl transfer as well as acyl transfer. Then, aa-AMPs or similar nucleotide derivatives may have been efficient in nucleotide ligation or polymerization through triphosphate intermediates. It is even possible to conceive that RNA polymerization started from nucleotides activated as mixed anhydrides with amino acids, which may be supported by the fact that certain polymerases have been shown to accept another family of amino-acid-derived substrates, i.e. phosphoramidates (Adelfinskaya and Herdewijn, 2007; Adelfinskaya *et al.*, 2007).

This NCA-based scenario loses its main original feature within the scope of the 'RNA only' world hypothesis, presuming that translation emerged at a stage at which RNA chemistry was fully developed. In fact, the NCA-induced reaction pathway would have been of poor utility if another pathway of nucleotide activation were already fully established, depriving the system of any evolutionary driving force. Within this approach, a role of NCA might be conceived as an inducer of mixed-anhydride formation. The initially useless mixed-anhydride intermediates would then have been opportunistically utilized in subsequent steps. In the eventuality that NCAs were not present, the selection of mixed anhydrides as intermediates could only be explained by considering ribozymes capable of transiently activating amino acids as phosphate anhydrides.[2] A ribozyme capable of achieving this task (Kumar and Yarus, 2001) has been selected, but the low efficiency of this catalyst and the need to find a source of energy for this reaction (Wolf and Koonin, 2007) show that the development of the amino-acid activation pathway in the RNA-world hypothesis must not be considered as fully understood. This is consistent with the observation that it is not simple to find an evolutionary driving force affording a selective advantage for the evolution of a primitive translation apparatus in an RNA world (Noller, 2004).

Conclusion

Whether the stage in which RNA played an important role corresponded to a true RNA-world or a coevolution process, the discussion of the thermodynamic constraints on amino-acid activation presented here suggests that the emergence of this biochemical pathway could not avoid a contribution of NCAs either as intermediates or at least reagents responsible for a random aminoacylation of translated proteins. Further investigation of the role of NCAs may be devoted to specifying their ability to aminoacyle RNA strands or to promote the ligation of RNA oligomers on templates. It is worth mentioning that considering the prebiotic energy constraints on the emergence of translation leads to a hypothesis on the emergence of the modern amino-acid activation pathway that is consistent with a continuity principle (Wolf and Koonin, 2007) stating that every step of an evolution scenario must have led to an increase in fitness. The resulting scenario also allows us to go beyond the limits of hypotheses encompassing only one kind of polymer and confers an essential role on RNA consistent with the evolution towards an RNA–protein world.

References

Adelfinskaya, O. and Herdewijn, P. (2007). Amino acid phosphoramidate nucleotides as alternative substrates for HIV-1 reverse transcriptase. *Angewandte Chemie International Edition*, **46**, 4356–8.

Adelfinskaya, O., Terrazas, M., Froeyen, M., Marlière, P., Nauwelaerts, K. and Herdewijn, P. (2007). Polymerase-catalyzed synthesis of DNA from phosphoramidate conjugates of deoxynucleotides and amino acids. *Nucleic Acids Research*, **35**, 5060–72.

Biron, J.-P. and Pascal, R. (2004). Amino acid N-carboxyanhydrides: activated peptide monomers behaving as phosphate-activating agents in aqueous solution. *Journal of the American Chemical Society*, **126**, 9198–9.

Biron, J.-P., Parkes, A. L., Pascal, R. and Sutherland, J. D. (2005). Expeditious, potentially primordial, aminoacylation of nucleotides. *Angewandte Chemie International Edition*, **44**, 6731–4.

[2] The claim of a prebiotic formation of aminoacyl adenylates by reaction of ATP with amino acid on clays (Paecht-Horowitz and Katchalsky, 1973), sometimes mentioned in the literature, should not be taken into account, since subsequent attempts to reproduce the experiment failed (Warden *et al.*, 1974), which is consistent with the strongly dissimilar $\Delta G^{\circ\prime}$ values for ATP and aa-AMPs (Wells *et al.*, 1986).

Blackmond, D. G. (2009). An examination of the role of autocatalytic cycles in the chemistry of proposed primordial reactions. *Angewandte Chemie International Edition*, **48**, 386–90.

Borsenberger, V., Crowe, M. A., Lehbauer, J., Raftery, J., Helliwell, M., Bhutia, K., Cox, T. and Sutherland, J. D. (2004). Exploratory studies to investigate a linked prebiotic origin of RNA and coded peptides. *Chemistry and Biodiversity*, **1**, 203–46.

Commeyras, A., Taillades, J., Collet, H., Boiteau, L., Vandenabeele-Trambouze, O., Pascal, R., Rousset, A., Garrel, L., Rossi, J.-C., Biron, J.-P., Lagrille, O., Plasson, R., Souaid, E., Danger, G., Selsis, F., Dobrijevic M. and Martin H. (2004). Dynamic co-evolution of peptides and chemical energetics, a gateway to the emergence of homochirality and the catalytic activity of peptides. *Origins of Life and Evolution of Biospheres*, **34**, 35–55.

Cramer, F. and Freist, W. (1987). Molecular recognition by energy dissipation, a new enzymatic principle: the example of isoleucine–valine. *Account of Chemical Research*, **20**, 79–84.

Danger, G., Boiteau, L., Cottet, H. and Pascal R. (2006). The peptide formation mediated by cyanate revisited. N-Carboxyanhydrides as accessible intermediates in the decomposition of N-carbamoylaminoacids. *Journal of the American Chemical Society*, **128**, 7412–13.

Delaye, L., Becerra, A. and Lazcano, A. (2005). The last common ancestor: what's in a name? *Origins of Life and Evolution of Biospheres*, **35**, 537–54.

Di Giulio, M. (1997). On the RNA world: evidence in favor of an early ribonucleopeptide world. *Journal of Molecular Evolution*, **45**, 571–8.

Di Giulio, M. (1998). Reflections on the origin of the genetic code: a hypothesis. *Journal of Theoretical Biology*, **191**, 191–6.

Di Giulio, M. (2001). The non-universality of the genetic code: the universal ancestor was a progenote. *Journal of Theoretical Biology*, **209**, 345–9.

Dueymes, C. (2009) Contribution à l'étude du rôle des aminoacylphosphates et des acides aminés N-phosphoridês dans la formation prébiotique des peptides. PhD thesis, Université Montpellier.

Eschenmoser, A. (1994). Chemistry of potentially prebiological natural products. *Origins of Life and Evolution of the Biosphere*, **24**, 389–423.

Eschenmoser, A. (2007). Question 1: Commentary referring to the statement 'the origin of life can be traced back to the origin of kinetic control' and the question 'do you agree with this statement; and how would you envisage the prebiotic evolutionary bridge between thermodynamic and kinetic control?' Stated in Section 1.1. *Origins of Life and Evolution of Biospheres*, **37**, 309–14.

Gesteland R. F., Cech T. R. and Atkins J. F. (1999). *The RNA World*, 2nd edn. New York: Cold Spring Harbor Laboratory Press.

Jencks, W. P. (1976). Free energies of hydrolysis and decarboxylation. In *Handbook of Biochemistry and Molecular Biology*, 3rd edn., vol. I, ed. G. D. Fasman. Cleveland: CRC Press, pp. 296–304.

Kondepudi, D. and Prigogine, I. (1998). *Modern Thermodynamics – From Heat Engines to Dissipative Structures*. Chichester, UK: Wiley and Sons Ltd.

Kumar, R. K. and Yarus, M. (2001). RNA-catalyzed amino acid activation. *Biochemistry*, **40**, 6998–7004.

Leman, L., Orgel, L. and Ghadiri, M. R. (2004). Carbonyl sulfide-mediated prebiotic formation of peptides. *Science*, **306**, 283–6.

Leman, L. J., Orgel, L. E. and Ghadiri, M. R. (2006). Amino acid dependent formation of phosphate anhydrides in water mediated by carbonyl sulfide. *Journal of the American Chemical Society*, **128**, 20–1.

Mitchell, P. (1961). Coupling of phosphorylation to electron and hydrogen transfer by a chemi-osmotic type of mechanism. *Nature*, **191**, 144–8.

Noller, H. F. (2004). The driving force for molecular evolution of translation. *RNA*, **10**, 1833–7.

Paecht-Horowitz, M. and Katchalsky, A. (1973). Synthesis of aminoacyl-adenylates under prebiotic conditions. *Journal of Molecular Evolution*, **2**, 91–8.

Pascal, R., Boiteau, L. and Commeyras, A. (2005). From the prebiotic synthesis of α-amino acids towards a primitive translation apparatus for the synthesis of peptides. *Topics in Current Chemistry*, **259**, 69–122.

Plasson, R. (2008). Comment on 're-examination of reversibility in reaction models for the spontaneous emergence of homochirality'. *The Journal of Physical Chemistry B*, **112**, 9550–2.

Pross, A. (2005). On the emergence of biological complexity: life as a kinetic state of matter. *Origins of Life and Evolution of Biospheres*, **35**, 151–66.

Shapiro, R. (2006). Small molecule interactions were central to the origin of life. *The Quarterly Review of Biology*, **81**, 106–25.

Sievers, A., Beringer, M., Rodnina, M. V. and Wolfenden, R. (2004). The ribosome as an entropy trap. *Proceedings of the National Academy of Sciences of the USA*, **101**, 7897–901.

Tamura, K. and Schimmel, P. (2003). Peptide synthesis with a template-like RNA guide and aminoacyl phosphate adaptors. *Proceedings of the National Academy of Sciences of the USA*, **100**, 8666–9.

Warden, J. T., McCullough, J. J., Lemmon, R. M. and Calvin, M. (1974). A re-examination of the zeolite-promoted, clay-mediated peptide synthesis. *Journal of Molecular Evolution*, **4**, 189–94.

Weber, A. L. (2002). Chemical constraints governing the origin of metabolism: the thermodynamic landscape of carbon group transformations under mild aqueous conditions. *Origins of Life and Evolution of the Biosphere*, **32**, 333–57.

Wells, T. N. C., Ho, C. K. and Fersht, A. R. (1986). Free energy of hydrolysis of tyrosyl adenylate and its binding to wild-type and engineered mutant tyrosyl–tRNA synthetases. *Biochemistry*, **25**, 6603–8.

Widmann, J., Di Giulio, M., Yarus, M. and Knight, R. (2005). tRNA creation by hairpin duplication. *Journal of Molecular Evolution*, **61**, 524–30.

Wolf, Y. I. and Koonin, E. V. (2007). On the origin of the translation system and the genetic code in the RNA World by means of natural selection, exaptation, and subfunctionalization. *Biology Direct*, **2**, 14.

Wong, J. T.-F. (1975). A co-evolution theory of the genetic code. *Proceedings of the National Academy of Sciences of the USA*, **72**, 1909–12.

Wong, J. T.-F. (2005). Coevolution theory of the genetic code at age thirty. *BioEssays*, **27**, 416–25.

17

Comparative genomics and early cell evolution

Antonio Lazcano

Introduction

The awareness that genes and genomes are extraordinarily rich historical documents from which a wealth of evolutionary information can be retrieved has widened the range of phylogenetic studies to previously unsuspected heights. The development of efficient sequencing techniques, which now allows the rapid sequencing of complete cellular genomes, combined with the simultaneous and independent blossoming of computer science, has led not only to an explosive growth of databases and new sophisticated tools for their exploitation, but also to the recognition that, in spite of many lateral gene-transfer (LGT) events, different macromolecules are uniquely suited as molecular chronometers in the construction of nearly universal phylogenies.

Cladistic analysis of rRNA sequences is acknowledged as a prime force in systematics and from its very inception had a major impact on our understanding of early cellular evolution. The comparison of small-subunit ribosomal-RNA (16/18S rRNA) sequences led to the construction of a trifurcated unrooted tree in which all known organisms can be grouped in one of three major monophyletic cell lineages, i.e. the domains Bacteria (Eubacteria), Archaea (Archaeabacteria) and Eukarya (Eukaryotes) (Woese *et al.*, 1990), which are all derived from an ancestral form, known as the last common ancestor (LCA).

From a cladistic viewpoint, the LCA is merely an inferred inventory of features shared among extant organisms, all of which are located at the tips of the branches of molecular phylogenies. Some time ago it was surmised that the sketchy picture developed with the limited databases would be confirmed by completely sequenced cell genomes from the three primary domains. This has not been the case: the availability of an increasingly large number of completely sequenced cellular genomes has sparked new debates, rekindling the discussion on the nature of the ancestral entity and its predecessors. As reviewed by Becerra *et al.* (2007), this is shown in the diversity of names that have been coined to describe it, which include, among others, progenote (Woese and Fox, 1977), cenancestor (Fitch and Upper, 1987), LUCA, last universal common ancestor (Kyrpides *et al.*, 1999) or, later on, last universal cellular ancestor (Philippe and Forterre, 1999), universal ancestor (Doolittle, 2000), LCC, last common community (Line, 2002), and MRCA, most recent common ancestor (Zhaxybayeva and Gogarten, 2004). Close analysis shows that these terms are not truly synonymous, and in a way they reflect the controversies on the nature of the universal ancestor and the evolutionary processes that shaped it.

Analysis of homologous traits found among the three major lineages suggests that the LCA was not a direct immediate descendant of the RNA World, a protocell or any other pre-life progenitor system. In fact, given the huge gap existing in current descriptions of the evolutionary transition between the prebiotic synthesis of biochemical compounds and the LCA, it is difficult to see how the universal trees can be extended beyond a threshold that corresponds to a period of cellular evolution in which protein biosynthesis was already in operation (Becerra *et al.*, 2007).

However, from an evolutionary point of view it is reasonable to assume that at some point in time the ancestors of all forms of life must have been less complex than even the simpler extant cells. Although it is naive to attempt to describe the origin of life and the nature of the first living systems from the available rooted phylogenetic trees, molecular cladistics may provide clues to some very early stages of biological evolution. Indeed, the variations of traits common to extant species can be explained as the outcome of divergent processes from an ancestral life form that existed prior to their separation into the three major biological domains. The purpose of this chapter is to discuss such early stages of cellular evolution, i.e. to discuss some of the possible steps that took place between the origin of life and the last common ancestor of all living beings.

Are there fossil records of the cenancestor?

Although no evolutionary intermediate stages or ancient simplified versions of the basic biological processes have been discovered in contemporary organisms, the differences in the structure and mechanisms of gene expression and replication among the three lineages have provided insights into the stepwise evolution of the replication and translational apparatus, including some late steps in the development of the genetic code. We are now in a position in which it is possible to distinguish the origin of life problem from a whole series of other issues, often confused, that belong to the domain of the evolution of microbial life. Accordingly, the most basic questions pertaining to the origin of life relate to simpler replicating entities predating by a long series of evolutionary events the oldest lineages represented by basal branches in universal molecular phylogenies.

Recognition of the significant differences that exist between the transcriptional and translational machineries of the Bacteria, Archaea and Eukarya, which were assumed to be the result of independent evolutionary refinements, led to the conclusion that the primary branches were the descendants of a progenote, a hypothetical biological entity in which phenotype and genotype still had an imprecise rudimentary linkage relationship (Woese and Fox, 1977). However, the conclusion that the LCA was a progenote was disputed when the analysis of homologous traits found among some of its descendants suggested that it was not a direct immediate descendant of the RNA World, a protocell or any other pre-life progenitor system. Under the implicit assumption that LGT had not been a major driving force in the distribution of homologous traits in the three domains, it was concluded that the LCA was already like extant Bacteria (Lazcano *et al.*, 1992).

Variations of traits common to the domains Bacteria, Archaea and Eukarya (Woese *et al.*, 1990) can be easily explained as the outcome of divergent processes from ancestral life forms older than the LCA. No palaeontological remains will bear testimony of such an entity, as the search for a fossil of the cenancestor is bound to prove fruitless. Indeed, there is little or no geological evidence for the environmental conditions on the early Earth at the time of the origin and early evolution of life. It is not possible to assign a precise chronology to the origin and earliest evolution of cells, and identification of the oldest palaeontological traces of life remains a contentious issue. The early archaean geological record is scarce and controversial, and most of the sediments preserved from such times have been metamorphosed to a considerable extent. Although the biological origin of the microstructures present in the 3.5×10^9 year-old Apex cherts of the Australian Warrawoona formation (Schopf, 1993) has been disputed, at this time the weight of evidence favours the idea that life existed 3.5 billion years ago (Altermann and Kazmierczak, 2003).

Isotopic-fractionation data and other biomarkers support the possibility of a metabolically diverse archaean microbial biosphere, which may have included members of the archaeal kingdom. The proposed timing of the onset of microbial methanogenesis based on the low ^{13}C values in methane inclusions found in hydrothermally precipitated quartz in the 3.5-billion-year-old Dresser Formation in Australia (Ueno *et al.*, 2006) has been challenged (Lollar and McCollom, 2006). However, sulphur isotope investigations of the same site indicate biological sulphate-reducing activity (Shen *et al.*, 2001), and analyses of 3.4×10^9 year-old South African cherts suggest that they were inhabited by anaerobic photosynthetic prokaryotes in a marine environment (Tice and Lowe, 2004). These results support the idea that the early archaean Earth was teeming with prokaryotes, which included anoxygenic phototrophs, sulphate reducers and, perhaps, methanogenic Archaea (Canfield, 2006).

Molecular phylogenies and the last common ancestor

In principle, determination of the evolutionary polarity of character states in universal phylogenies should lead to the recognition of the oldest phenotype. Accordingly, the most parsimonious characterization of the LCA can be achieved by proceeding backwards and summarizing the features of the oldest recognizable node of the universal cladogram, i.e. rooting of the universal tree would provide direct information on the nature of the LCA. However, the plesiomorphic traits found in the space defined by rRNA sequences allow the construction of topologies that specify branching relationships but not the position of the ancestral phenotype. This issue was solved independently by Iwabe *et al.* (1989) and Gogarten *et al.* (1989), who analysed paralogous genes encoding (1) the two elongation factors (EF-G and EF-Tu) that assist in protein biosynthesis; and (2) the α- and β-hydrophilic subunits of F-type ATP synthetases. Using different tree-constructing algorithms, both teams independently placed the root of the universal trees between the Eubacteria on the one side, and the Archaea and the eukaryotic nucleocytoplasm on the other. The rooting of universal phylogenies placed the LCA in the bacterial branch of the universal tree (Gogarten *et al.*, 1989; Iwabe *et al.*, 1989).

The rooting of universal cladistic trees determines the directionality of evolutionary change and allows for the recognition of ancestral from derived characters. Determination of the root normally imparts polarity to most or all characters. It is, however, important to distinguish between ancient and primitive organisms. Organisms belonging to lineages located near the root of universal rRNA-based trees are cladistically ancient, but they are not endowed with primitive molecular-genetic apparatus, nor do they appear to be more rudimentary in their metabolic abilities than their aerobic counterparts (Islas *et al.*, 2003). Primitive living systems would initially refer to pre-RNA Worlds, in which life may have been based on polymers using backbones other than ribose–phosphate and possibly bases different from adenine, uracil, guanine and cytosine (Levy and Miller, 1998), followed by a stage in which life was based on RNA as both the genetic material and as catalysts (Joyce, 2002).

Reticulate phylogenies resulting from LGT complicate the reconstruction of cenancestral traits. Driven in part by the impact of lateral gene acquisition, as revealed by the discrepancies of different gene phylogenies with the canonical rRNA tree, and in part by the surprising complexity of the universal ancestor as suggested by direct-backtrack characterizations, Woese (1998) proposed that the LCA was not a single organismic entity, but rather a highly diverse population of metabolically complementary cellular progenotes that occupied as a whole the root of the tree and that were endowed with multiple, small linear chromosome-like genomes engaged in massive multidirectional horizontal-transfer events.

According to Woese's (1998) model, the essential features of translation and the development of metabolic pathways took place before the earliest branching event. As sequence exchange decreased, the resulting genetic isolation led, eventually, to Bacteria, Archaea and Eukarya (Woese, 1998). However, even if the genetic entities that formed such a communal ancestor may have been extremely diverse, their common features, such as the genetic code and the gene-expression machinery, are an indication of their ultimate monophyletic origin (Delaye *et al.*, 2004). It is not necessary to assume that the processes that led to the three domains took place inmediately after the appearance of the code and the establishment of translation. Inventories of LCA genes include sequences that originated in different pre-cenancestral epochs (Delaye and Lazcano, 2000; Anantharaman *et al.*, 2002). The origin of the mutant sequences ancestral to those found in all extant species, and the divergence of the Bacteria, Archaea and Eukarya were not synchronous events, i.e. the separation of the primary domains took place later, perhaps even much later, than the appearance of the genetic components of their last common ancestor (Delaye *et al.*, 2005).

Was the last common ancestor a hyperthermophile?

The examination of the prokaryotic branches of unrooted rRNA trees had already suggested that the ancestors of both Bacteria and Archaea were extreme thermophiles growing optimally at temperatures of 90°C or above (Achenbach-Richter *et al.*, 1987). Rooted universal phylogenies appeared to confirm this possibility, since heat-loving prokaryotes occupied short branches in the basal portion of molecular cladograms (Stetter, 1994). However, the recognition that the deepest branches in rooted universal phylogenies are occupied by hyperthermophiles does not provide, by itself, conclusive proof of a heat-loving LCA, or much less of a hot origin of life. Analysis of the correlation of the optimal growth temperature of prokaryotes and the G + C nucleotide content of 40 rRNA sequences through a complex Markov model has led Galtier *et al.* (1999) to conclude that the universal ancestor was a mesophile. Further refinements using a model that included both rRNA and highly conserved protein sequences suggest more complex evolutionary history, going from a mesophilic LCA into parallel convergent adaptations to high-temperature regimes in the ancestors of both Bacteria and the Archaea. These changes, which may be due to climate changes in the Precambrian Earth, were then lost independently in both lineages (Bousseau *et al.*, 2008).

The chemical nature of the cenancestral genome

Since all extant cells are endowed with DNA genomes, the most parsimonious conclusion is that this genetic polymer was already present in the cenancestral population. Although it is possible to recognize the evolutionary relatedness of various orthologous proteins involved with DNA replication and repair (ATP-dependent clamp-loader proteins, topoisomerases, gyrases and 5′-3′ exonucleases) across the entire phylogenetic spectrum (Olsen and Woese, 1997; Edgell and Doolittle, 1997; Leipe *et al.*, 1999; Penny and Poole, 1999), comparative proteome analysis has shown that (eu)bacterial replicative polymerases and primases lack homologues in the two other primary kingdoms.

The peculiar distribution of the DNA replication machinery has led to suggestions not only of an LCA endowed with an RNA genome, but also of the polyphyletic origins of DNA and many of the enzymes associated with its replication (Leipe *et al.*, 1999; Koonin and Martin, 2005) in which viruses may have played a central role (Forterre, 2006; Koonin, 2009). It has been argued by Koonin and Martin (2005) that the LCA was an acellular entity endowed with high numbers of RNA viral-like molecules that had originated abiotically within the cavities of a hydrothermal mound. This idea, which has little, if any, empirical support, does not take into account the problems involved with the abiotic synthesis and accumulation of ribonucleotides and polyribonucleotides, nor does it explain the emergence of functional RNA molecules.

It has also been suggested that the ultimate origins of cellular DNA genomes lie in viral systems, which gave rise to polyphyletic deoxyribonucleotide biosynthesis (Forterre, 2006). According to a rather complex hypothetical scheme, gene transfers mediated by viral takeovers took place three times, giving origin to the DNA genomes of the three primary kingdoms. The invasion of the ancestor of the bacterial domain by a DNA virus eventually led to a replacement of its cellular RNA genes by DNA sequences, while the archaeal and eukaryal DNA-replication enzymes resulted from an invasion by closely related DNA viruses (Forterre, 2006). This proposal is based not only on the assumption that metabolic pathways can arise in viruses, but also on the possibility of a polyphyletic origin of deoxyribonucleotide biosynthesis. This is unlikely: in sharp contrast with other energetically favourable biochemical reactions (such as phosphodiester-backbone hydrolysis or the transfer of amino groups), the direct removal of the oxygen from the 2′-C ribonucleotide pentose ring to form the corresponding deoxy-equivalents is a thermodynamically much less-favoured reaction, considerably reducing the likelihood of multiple independent origins of biological ribonucleotide reduction.

There are indeed manifold indications that RNA genomes existed during early stages of cellular evolution (Lazcano *et al.*, 1988) but it is likely that double-stranded DNA genomes had become firmly established prior to the divergence of the three primary domains. It is true that the demonstration of the monophyletic origin of ribonucleotide reductases (RNRs) is greatly complicated by their highly divergent primary sequences and the different mechanisms by which they generate the substrate 3′-radical species required for the removal of the 2′-OH group. However, sequence analysis and biochemical characterization of archaeal RNRs have shown their similarities with their bacterial and eukaryal counterparts, confirming their ultimate monophyletic origin (Stubbe *et al.*, 2001).

Strong selection pressures acting over multigene RNA-based mechanisms, such as translation, are responsible for their universal distribution and high conservation, and DNA-based genetic systems were selected for in cells to stabilize earlier RNA replicating systems. This replacement was the outcome of the metabolic evolution of deoxyribonucleotide biosynthesis via the reductive elimination of the 2′-hydroxyl group in ribonucleotides, thymine anabolism and the replacement of uracil, and the selection of editing mechanisms. Indeed, the sequence similarities shared by many ancient large proteins found in all three domains suggest that considerable fidelity existed in the operative genetic system of the LCA. Despite claims to the contrary (Poole and Logan, 2005), such fidelity is unlikely to be found in RNA-based genetic systems (Reanney, 1987; Lazcano *et al.*, 1992), which do not replicate using the multiunit cellular DNA-dependent RNA polymerase, but are based on RNA replicases lacking editing mechanisms.

Gene duplications and the evolution of metabolism

Clues to the genetic organization and biochemical complexity of primitive entities from which the LCA evolved may also be derived from the analysis of paralogous gene families. The number of sequences that underwent such duplications prior to the divergence of the three lineages includes genes encoding for a variety of enzymes that participate in widely different processes such as translation, DNA replication, biosynthetic pathways and energy-producing processes.

Some authors have argued that LCA was an acellular entity, arising directly from abiotic processes (Koga *et al.*, 1998; Koonin and Martin, 2005). However, the high conservation and distribution of many membrane proteins, including, for instance, ATPase hydrophilic subunits (Gogarten *et al.*, 1989), the signal-recognition particles (SRPs) (Gribaldo and Cammarano, 1998) and ABC transporters (Delaye *et al.*, 2005) imply a cellular cenancestor, whose membrane may have been formed by heterochiral lipids composed of a mixture of glycerol-1-phosphate and glycerol-3-phosphate (Wächstershäuser, 2003; Peretó *et al.*, 2004).

The conservation of membrane-bound proton-pump ATPase subunits suggests that the cenancestor was able to produce a chemically driven proton gradient across its cell membrane using a variety of oxidized inorganic molecules as molecular acceptors (Castresana and Moreira, 1999), while the high conservation of manifold ABC transporters involved in the import of metabolic substrates is consistent with the possibility of a heterotrophic LCA that depended on external sources of organic compounds (Becerra *et al.*, 2007).

As discussed elsewhere (Becerra *et al.*, 2007), a survey of the available information shows that sequences that have resulted from early pre-ancestral paralogous expansion may be classified into three major groups:

(1) sequences formed by two tandemly arranged homologous modules that underwent fusion events, such as the (i) protein disulphide oxidoreductase (Ren *et al.*, 1998); (ii) large subunit of carbamoyl phosphate synthetase (Alcántara *et al.*, 2000); and (iii) HisA, an isomerase involved in histidine biosynthesis (Alifano *et al.*, 1996);
(2) gene families that have undergone a major expansion of sequences, such as ABC transporters and other enzymes involved in membrane-transport phenomena (Clayton *et al.*, 1997); and
(3) families formed by a relatively small number of paralogous sequences. These include, among others, the pair of homologous genes encoding the EF-Tu and EF-G elongation factors (Iwabe *et al.*, 1989) as well as the duplicated sequences encoding the F-type ATPase hydrophilic α- and β-subunits (Gogarten *et al.*, 1989).

The identification of sequences formed by tandemly fused homologous modules provides direct evidence of the existence, during early Precambrian times, of smaller functional genes. Moreover, the families of paralogous duplicates also imply that the LCA was preceded by simpler cells with a smaller genome in which only one copy of each of these genes existed, i.e. by cells in which, for instance, protein synthesis involved only one elongation factor, and with ATPases with limited regulatory abilities. Paralogous families of metabolic genes also support the proposal that anabolic pathways were assembled by the recruitment of primitive enzymes that could react with a wide range of chemically related substrates, i.e. the so-called patchwork assembly of biosynthetic routes (Jensen, 1976; Velasco *et al.*, 2002). Such relatively slow unspecific enzymes may have represented a mechanism by which primitive cells with small genomes could have overcome their small coding abilities. How early cells could overcome the bottlenecks imposed by such limitations is an open problem that could be addressed, for instance, by using *in-vitro* systems of anabolic pathways, such as histidine biosynthesis, in which the different homologous enzymes that catalyze several different steps are replaced by one single representative of such a paralogous set.

Genomic evidence for an RNA/protein world

As demonstrated by other analyses, proteins that interact with RNA in one way or another are among the most highly conserved sequences (Delaye and Lazcano, 2000; Anantharaman *et al.*, 2002). A significant percentage of such highly conserved sequences correspond to proteins that interact directly with RNA (such as ribosomal proteins, DEAD-type helicases, aminoacyl tRNA synthetases and elongation factors, among others), or take part in RNA and nucleotide biosynthesis, including the DNA-dependent RNA polymerase β- and β'-subunits, dimethyladenosine transferase, adenyl-succinate lyases, dihydroorotate oxidase and ribose-phosphate pyrophosphokinase, among many others. Few metabolic genes are part of the conserved ORF product set. These include many sugar metabolism-related sequences, such as the enolase-encoding genes noted above, as well as homologues of sequences involved in nucleotide biosynthesis, such as thioredoxin and phosphoribosylpyrophosphate synthase (Delaye *et al.*, 2005; Becerra *et al.*, 2007).

Although RNA hydrolysis is an exergonic process, degradosome-mediated mRNA turnover plays a key role as a regulatory mechanism for gene expression in both prokaryotes and Eukaryotes (Blum *et al.*, 1997). A possible explanation for the vey high conservation of DEAD-type RNA helicases may lie in their roles in protein biosynthesis and in mRNA degradation. This possibility is supported by the phylogenetic relatedness of the RhlB and DEAD sequences (Schmid and Linder, 1992) and by the surprising conservation of the *eno*-like sequences. If this interpretation is correct, then it could be argued that degradosome-mediated mRNA turnover is an ancient control mechanism at the RNA level that was established prior to the divergence of the three primary kingdoms. Together with other lines of evidence, including the observation that the most highly conserved gene clusters in several (eu)bacterial genomes are regulated at the RNA level (Siefert *et al.*, 1997), these results are consistent with the hypothesis that during the early stages of cell evolution RNA molecules played a more conspicuous role in cellular processes (Becerra *et al.*, 2007).

Conclusions

Analysis of the increasingly large database of completely sequenced cellular genomes from the three major domains in order to define the set of the most conserved protein- encoding sequences to characterize the gene complement of the last common ancestor of extant life results in a set dominated by different putative ATPases and by molecules involved in gene expression and RNA metabolism (Delaye *et al.*, 2005). DEAD-type RNA helicase and enolase genes, which are known to be part of the RNA degradosome, are as conserved as many transcription and translation genes. This suggests the early evolution of a control mechanism for gene expression at the RNA level, providing additional support for the hypothesis that during early cellular evolution RNA molecules played a more prominent role. Conserved sequences related to biosynthetic pathways include those encoding putative phosphoribosyl-pyrophosphate synthase and thioredoxin, which participate in nucleotide metabolism. Although the information contained in the available databases corresponds only to a minor portion of biological diversity, these sequences are likely to be part of an essential and highly conserved pool of protein domains common to all organisms (Becerra *et al.*, 2007).

The high levels of genetic redundancy detected in all sequenced genomes imply that duplication has played a major role in the accretion of the complex genomes found in extant cells. They also show that prior to the early duplication events revealed by the large protein families, simpler living systems existed which lacked the large sets of enzymes and the sophisticated regulatory abilities of contemporary organisms. Once it appeared, the LCA would have been in the company of its siblings, a population of entities similar to it that existed throughout the same period. They may have not survived, but some of their genes did if they became integrated via lateral transfer into the LCA genome. The cenancestor is thus one of the last evolutionary outcomes of a series of ancestral events including LGT, gene losses and paralogous duplications that took place before the separation of Bacteria, Archaea and Eukarya (cf. Becerra *et al.*, 2007).

References

Achenbach-Richter, L., Gupta, R., Stetter., K. O., and Woese, C. R. (1987). Were the original eubacteria thermophiles? *Systematic and Applied Microbiology*, **9**, 34–9.

Alcántara, C., Cervera, J., and Rubio, V. (2000). Carbamate kinase can replace in vivo carbamoyl phosphate synthetase. Implications for the evolution of carbamoyl phosphate biosynthesis. *FEBS Letters*, **484**, 261–4.

Alifano, P., Fani, R., Liò, P., Lazcano, A. and Bazzicalupo, M. (1996). Histidine biosynthetic pathway and genes, structure, regulation, and evolution. *Microbiology Reviews*, **60**, 44–69.

Altermann, W. and Kazmierczak, J. (2003). Archean microfossils, a reappraisal of early life on Earth. *Research in Microbiology*, **154**, 611–17.

Anantharaman V., Koonin, E. V. and Aravind, L. (2002). Comparative genomics and evolution of proteins involved in RNA metabolism. *Nucleic Acid Research*, **30**, 1427–64.

Becerra, A., Delaye, L., Islas, A. and Lazcano A. (2007). Very early stages of biological evolution related to the nature of the last common ancestor of the three major cell domains. *Annual Review of Ecology and Evolutionary Systematics*, **38**, 361–79.

Blum E., Py B., Carpousis, A. J., Higgins, C. F. (1997). Polyphosphate kinase is a component of the *Escherichia coli* RNA degradosome. *Molecular Microbiology*, **26**, 387–398.

Bousseau, B., Blanquart, S., Necsulea, A., Lartillot, N. and Gouy, M. (2008). Parallel adaptations to high temperatures in the Archaean Eon. *Nature*, **456**, 942–5.

Canfield, D. E. (2006). Biochemistry, gas with an ancient history. *Nature*, **440**, 426–7.

Castresana, J. and Moreira, D. (1999). Respiratory chains in the last common ancestor of living organisms. *Journal of Molecular Evolution*, **49**, 453–60.

Clayton, R. A., White, O., Ketchum, K. A. and Venter, C. J. (1997). The genome from the third domain of life. *Nature*, **387**, 459–62.

Delaye, L. and Lazcano, A. (2000). RNA-binding peptides as molecular fossils. In *Origins from the Big-Bang to Biology*, eds. J. Chela-Flores, G. Lemerchand and J. Oró. Proceedings of the First Ibero-American School of Astrobiology. Dordrecht: Klüwer Academic Publishers, pp. 285–8.

Delaye, L., Becerra, A. and Lazcano, A. (2004). The nature of the last common ancestor. In *The Genetic Code and the Origin of Life*, ed. L. R. de Pouplana. Georgetown: Landes Bioscience, pp. 34–47.

Delaye, L., Becerra, A. and Lazcano, A. (2005). The last common ancestor, what's in a name? *Origins of Life and Evolution of the Biosphere*, **35**, 537–54.

Doolittle, W. F. (2000). The nature of the universal ancestor and the evolution of the proteome. *Current Opinion in Structural Biology*. **10**, 355–358.

Edgell, R. D. and Doolittle, W. F. (1997). Archaea and the origins of DNA replication proteins. *Cell*, **89**, 995–8.

Fitch, W.M. and Upper, K. (1987). The phylogeny of tRNA sequences provides evidence of ambiguity reduction in the origin of the genetic code. Cold Spring Harbor Symposium. *Quantitative Biology*, **52**, 759–67.

Forterre, P. (2006). Three RNA cells for ribosomal lineages and three DNA viruses to replicate their genomes: A hypothesis for the origin of cellular domain. *Proceedings of the National Academy of Sciences of the USA*, **103**, 3669–74.

Galtier, N., Tourasse, N. and Gouy, M. (1999). A nonhyperthermophilic common ancestor to extant life forms. *Science*, **283**, 220–1.

Gogarten, J. P., Kibak, H., Dittrich, P., Taiz, L. and Bowman, E. J. (1989). Evolution of the vacuolar H^+-ATPase; implications for the origin of eukaryotes. *Proceedings of the National Academy of Sciences of the USA*, **86**, 6661–5.

Gribaldo S., Cammarano P. (1998). The root of the universal tree of life inferred from anciently duplicated genes encoding components of the protein-targeting machinery. *J Mol Evol*, **47**, 508–16.

Islas, S., Velasco, A. M., Becerra, A., Delaye, L. and Lazcano, A. (2003). Hyperthermophily and the origin and earliest evolution of life. *International Microbiology*, **6**, 87–94.

Iwabe, N., Kuma, K., Hasegawa, M., Osawa, S. and Miyata, T. (1989). Evolutionary relationship of archaebacteria, eubacteria, and eukaryotes inferred from phylogenetic trees of duplicated genes. *Proceedings of the National Academy of Sciences of the USA*, **86**, 9355–9.

Jensen, R. A. (1976). Enzyme recruitment in evolution of new function. *Annual Review of Microbiology*, **30**, 409–25.

Joyce, G. F. (2002). The antiquity of RNA-based evolution. *Nature*, **418**, 214–21.

Koga, Y., Kyuragi, T., Nishihara, M. and Sone, N. (1998). Did archaeal and bacterial cells arise independently from noncellular precursors? A hypothesis stating that the advent of membrane phospholipids with enantiomeric glycerophosphate backbones caused the separation of the two lines of descent. *Journal of Molecular Evolution*, **46**, 54–63.

Koonin, E. V. (2009). On the origin of cells and viruses, primordial virus world scenario. *Annals of the New York Academy of Sciences*, **1178**, 47–64.

Koonin E.V. and Martin, W. (2005). On the origin of genomes and cells within inorganic compartments. *Trends in Genetics*, **21**, 647–54.

Kyrpides, N., Overbeek, R. and Ouzonis, C. (1999). Universal protein families and the functional content of the last universal common ancestor. *Journal of Molecular Evolution*, **49**, 413–23.

Lazcano, A., Guerrero, R., Margulis, L., Oró, J., (1988). The evolutionary transition from RNA to DNA in early cells. *Journal of Molecular Evolution*, **27**, 283–290.

Lazcano, A., Fox, G.E. and Oró, J. (1992). Life before DNA, the origin and early evolution of early Archean cells. In *The Evolution of Metabolic Function*, ed. R. P. Mortlock. Boca Raton, FL: CRC Press, pp. 237–95.

Leipe, D. D., Aravind, L. and Koonin, E. V. (1999). Did DNA replication evolve twice independently? *Nucleic Acid Research*, **27**, 3389–401.

Levy, M. and Miller, S. L. (1998). The stability of the RNA bases: implications for the origin of life. *Proceedings of the National Academy of Sciences of the USA*, **95**, 7933–8.

Line, M. A. (2002). The enigma of the origin of life and its timing. *Microbiology*, **148**, 21–7.

Lollar, B. S. and McCollom, T. M. (2006). Biosignatures and abiotic constraints on early life. *Nature*, **444**, E18.

Olsen, G. and Woese, C. R. (1997). Archaeal genomics: an overview. *Cell*, **89**, 991–4.

Penny, D. and Poole, A. (1999). The nature of the last common ancestor. *Current Opinion in Genetics and Development*, **9**, 672–7.

Peretó, J., López-García, P. and Moreira, D. (2004). Ancestral lipid biosynthesis and early membrane evolution. *Trends in Biochemical Sciences*, **29**, 469–77.

Philippe, H. and Forterre, P. (1999). The rooting of the universal tree of life is not reliable. *Journal of Molecular Evolution*, **49**, 509–23.

Poole, A. and Logan, D. T. (2005). Modern mRNA proofreading and repair, clues that the last universal common ancestor possessed an RNA genome? *Molecular Biology and Evolution*, **22**, 1444–55.

Reanney, D. C. (1987). Genetic error and genome design. Cold Spring Harbor Symposium. *Quantitative Biology*, **52**, 751–7.

Ren, B., Tibbelin, G., de Pascale, D., Rossi, M., Bartolucci, S. and Ladenstein, R. (1998). A protein disulfide oxidoreductase from the archaeon *Pyrococcus furiosus* contains two thioredoxin fold units. *Nature Structural and Molecular Biology*, **7**, 602–11.

Schmid, S. R. and Linder, P. (1992). D-E-A-D protein family of putative RNA helicases. *Molecular Microbiology*, **6**, 283–92.

Schopf, J. W. (1993). Microfossils of the early Archaean Apex chert: new evidence for the antiquity of life. *Science*, **260**, 640–6.

Shen, Y., Buick, R. and Canfield, D. E. (2001). Isotopic evidence for microbial sulphate reduction in the early Archaean Era. *Nature*, **410**, 77–81.

Siefert, J. L., Martin, K. A., Abdi, F., Wagner, W. R. and Fox, G. E. (1997). Conserved gene clusters in bacterial genomes provide further support for the primacy of RNA. *Journal of Molecular Evolution*, **45**, 467–72.

Stetter, K. O. (1994). The lesson of archaeabacteria. In *Early Life on Earth*, Nobel Symposium No. 84, ed. S. Bengtson. New York: Columbia University Press, pp. 114–22.

Stubbe, J., Ge, J. and Yee, C. S. (2001). The evolution of ribonucleotide reduction revisited. *Trends in Biochemical Sciences*, **26**, 93–9.

Tice, M.M. and Lowe, D.R. 2004. Photosynthetic microbial mats in the 3,416-Myr-old ocean. *Nature*, **431**, 549–52

Ueno, Y., Yamada, K., Yoshida, N., Maruyama, S. and Isozaki, Y. (2006). Evidence from fluid inclusions for microbial methanogenesis in the early Archaean Era. *Nature*, **440**, 516–19.

Velasco, A. M., Leguina, J. I. and Lazcano, A. (2002). Molecular evolution of the lysine biosynthetic pathways. *Journal of Molecular Evolution*, **55**, 445–9.

Wächstershäuser, G. (2003). From pre-cells to Eukarya – a tale of two lipids. *Molecular Microbiology*, **47**, 13–22.

Woese, C. R. (1998). The universal ancestor. *Proceedings of the National Academy of Sciences of the USA*, **95**, 6854–9.

Woese, C. R. and Fox, G. E. (1977). The concept of cellular evolution. *Journal of Molecular Evolution*, **10**, 1–6.

Woese, C. R., Kandler, O. and Wheelis, M. L. (1990). Towards a natural system of organisms: proposal for the domains Archaea, bacteria, and eucarya. *Proceedings of the National Academy of Sciences of the USA*, **87**, 4576–9.

Zhaxybayeva, O. and Gogarten, P. J. (2004). Cladogenesis, coalescence and the evolution of the three domains of life. *Trends in Genetics*, **20**, 182–7.

18

Origin and evolution of metabolisms

Juli Peretó

Introduction: some conceptual remarks on metabolism

Metabolism is the set of enzymatic reactions that allow living beings to use external energy sources to drive the building of their biochemical components from external chemical sources and also to carry out energy-consuming functions, such as osmotic and mechanical work. The role played by gene-encoded enzymatic catalysts is one of the essential properties of life. Since their stability is finite and there is a need for constant replacement, enzymes are themselves products of metabolism (Cornish-Bowden *et al.*, 2004). Thus, the proteome (i.e. the totality of proteins and their concentrations that exists in a particular cell state) is a product of the metabolome (i.e. the totality of metabolites and their concentrations that exists in a particular metabolic state). This situation gives rise to the 'metabolic circularity' or 'recursivity'; a concept needed for a complete understanding of metabolism (Cornish-Bowden *et al.*, 2007).

Extant metabolic networks are certainly complex, with hundreds or thousands of concatenated enzymatic reactions. Since there is a limited number of coenzymes (i.e. the special reactants that help enzymes to perform their catalytic functions), recurrently used by different enzymes, and some central metabolites are true crossroads between different lines of chemical transformation, complex networks emerge. From a topological perspective, metabolic networks show a power-law distribution of connectivity (Fell and Wagner, 2000). In other words, most metabolites are poorly connected whereas a few of them (coenzymes and metabolic crossroads) support many connections. One of the challenges of the evolutionary study of metabolism is to unravel how these complex networks emerged, from an expectedly simpler chemical network, based on non-enzymatic reactions.

Traditionally, biochemists have identified certain transformations, from sources to sinks, with physiological significance, named 'metabolic pathways'. These pathways can be either 'catabolic' (i.e. using energy- and electron-rich metabolites to produce useful forms of metabolic energy) or 'anabolic' (i.e. energy and electron-consuming pathways aimed at building the cellular components). Obviously, behind the necessary simplification of traditional biochemical studies, many more chemical transformations exist in the cell as a result of the complexity of connections between pathways. Current systems-biology approaches,

Table 18.1. The metabolic modes. Metabolisms are classified taking into account the external sources of energy (visible light and either inorganic or organic oxidative reactions), reducing power (electrons from inorganic or organic compounds) and carbon.

Mode	Energy source	Reducing power	Carbon source	Examples
chemo-litho-autotrophic	oxidation of inorganic compounds	inorganic compounds	CO_2	hydrogen-oxidizing bacteria, methanogens, denitrifying bacteria
photo-litho-autotrophic	visible light	inorganic compounds	CO_2	photosynthetic bacteria, plants
photo-organo-heterotrophic	visible light	organic compounds	organic compounds	nonsulphur purple bacteria, heliobacteria
chemo-organo-heterotrophic	oxidation of organic compounds	organic compounds	organic compounds	animals, fungi, many prokaryotic and eukaryotic microorganisms

which try to understand cell functioning from a holistic perspective, using the data from 'omics' technologies (genomics, transcriptomics, proteomics, metabolomics etc., i.e. the totality of genes, RNAs, proteins, metabolites etc. associated with one biological system) are unveiling, in part, this complexity (Wagner, 2007).

On a global scale, life is sustained by a flux of energy – either visible light or from chemical sources – throughout the metabolic networks. The different combinations of energy, electron and carbon sources offer a diversity of metabolic modes (Table 18.1).

Darwin (1859) cited, as one of the difficulties of his natural-selection theory, the existence of organs of extreme perfection and complication. As a case study he used the eye. But the metabolic machinery is by far more complicated. The eye evolved no fewer than forty times – and probably more than sixty times – independently in various animal lineages (Lane, 2009), whereas oxygenic photosynthesis, for example, was invented only once, by the ancestors of extant cyanobacteria (Barber, 2008). As natural selection fully explains the origin of the eye, we are confident that it will be possible to learn the natural causes for the origin and evolution of metabolism.

The chemical roots of metabolism

Although the problem of the origin of metabolism is intimately intertwined with the problem of the origin of life itself, most authors have instead focused their attention on the origin of the genetic information. Following the Oparin tradition, it is generally assumed that life originated with a heterotrophic metabolic mode, i.e. from a 'primitive soup' rich in organic compounds from which the first organisms gained energy and matter (Oparin,

1924; Haldane, 1929; Lazcano *et al.*, 1992; Peretó *et al.*, 1997). During the last few decades scientists have accumulated many data favouring the contributions of volcanic, atmospheric and cosmic chemistries to the inventories of abiotic compounds and processes in the early Earth. Out of this abiotic chemistry, the emergence of suprachemical (or infrabiological) subsystems exhibiting basic lifelike performances, such as self-reproductive vesicles, self-replicative polymers and self-maintained chemical networks, was a necessary step during the chemical evolution or prebiotic phase. Eventually, the harmonic articulation of those three prebiotic subsystems in the same functional framework could be considered as the beginning of biological evolution. Thus, any evolutionary scenario, to be complete, must give an explicit account of the mechanisms of energy and matter fluxes through these primitive systems.

One of the first questions to be addressed is whether there is a correlation or continuity between prebiotic chemistry and biochemistry. Several authors consider the existence of chemomimetic processes, i.e. the occurrence of organo-chemical mechanisms in abiotic processes that anticipate the kind of chemistry to be used by the first biochemical transformations (Morowitz, 1992; Zubay, 1993; Meléndez-Hevia *et al.*, 1996; Costanzo *et al.*, 2007; Eschenmoser, 2007) leading to a true chemical continuity. Christian de Duve's protometabolic model is a useful theoretical framework for exploring this continuity between the chemical world of a prebiotic phase, basically determined by the participation of non-genetically instructed catalysts (e.g. minerals, short oligomers), and the chemistry of a protometabolism using replicative polymers (e.g. RNA) as catalysts (de Duve, 2005). Since the very moment that replicative polymers appear on the scene, natural selection would automatically emerge and historical contingency would add to the purely chemical processes.

In de Duve's proposal there is an explicit role for thioesters as the first kind of molecule involved in energy transduction that would also serve as a bridge between sulphur and phosphate chemistries. In fact, the low solubility and reactivity of the main mineral source of phosphate (apatite) have obscured the prebiotic origin of the energy-transduction mechanisms based on phosphate-dependent chemical couplings for a long time, whose universality in extant cells is a strong indication of its ancestral character. Nevertheless, interesting candidates for an early bioenergetics based on phosphorus are pyrophosphate and polyphosphate generated in volcanic settings (Baltscheffsky and Baltscheffsky, 1994; Gedulin and Arrhenius, 1994) and meteroritic phosphonates (Pasek, 2008).

However, others prefer a radically different point of departure for early metabolisms. Instead of a heterotrophic scenario (where a chemically-rich environment is the cradle for simple protometabolic networks) an autotrophic origin (i.e. based on the energy and electrons released from redox reactions involving mineral compounds and the synthesis of organic matter from CO_2) is postulated. Thus, the complete organochemical diversity originates from geochemical simplicity (e.g. CO_2 and electron-rich compounds such as H_2S). Morowitz (1992) suggested that a primitive form of the reductive tricarboxylic-acid cycle would be the initial seed of a prebiotic chemistry that prefigured most of modern metabolism. Another model, with partial experimental support, postulates that the anaerobic synthesis of pyrite from H_2S and FeS would serve as a source of electrons and energy for chemical syntheses from CO_2 through the same autotrophic route (Wächtershäuser, 2006). Russell and Martin (2004) postulated a geochemical version of the metabolic synthesis of acetate from CO or CO_2 present in certain anaerobic microorganisms. In these cases, submarine hydrothermal settings would offer the thermal and chemical gradients required by the models.

The coevolution of metabolism and the environment

Living beings are chemical open systems that use chemicals from the environment and produce and release other compounds. Coevolution of microbial metabolisms and the elemental cycles implies the coupling and consistency of any evolutionary scheme with the geochemical record and planetary chemistry. In this sense, chemical and isotopic studies of ancient rocks are fundamental in providing the best estimates for the minimal timing of any metabolic establishment. Nevertheless, current studies show an apparent temporal uncoupling between the origin of metabolic innovations and their signals in the rock record (Nealson and Rye, 2005).

Box 18.1. The bioenergetic processes

Every life form conserves energy using basic and universal mechanisms. (A) Whether the external or primary energy source is visible light or chemical (inorganic or organic) reactions, electrons are channelled to soluble carriers or are associated to membranes. The soluble reduced carriers (e.g. NAD[P]H) are used for biosynthetic purposes, whereas carriers in membranes usually generate electrochemical potential gradients (of either protons or sodium ions), a type of useful energy invested in ATP synthesis, membrane transport, motility etc. Electrochemical potential gradients and ATP are enzymatically interchangeable energy pools. (B) Electron-transport chains are arranged around several common enzymatic complexes that recurrently use the same redox cofactors such as quinones (Q), cytochromes (e.g. *b*, *c*) or iron–sulphur proteins (FeS). In photoelectronic chains, the photochemical reaction centres based on bacteriochlorophyll (BChl) or chlorophyll (Chl) use the electromagnetic energy to drive the uphill electronic transport (gray lines). Peripheral electron-transport proteins are cytochrome *c* (cyt *c*) and plastocyanin (PC). Photosystem II (PSII) has an associated water-oxidative oxygen-releasing enzymatic complex. The archetypal oxygenic photosynthesis present in cyanobacteria (and plastids) could be the result of evolutionary tinkering on previously emerged anoxygenic photosyntheses (as in the red- and green-bacteria versions shown here). On the other hand, respiratory chains accept electrons through oxidative complexes such as NADH dehydrogenase (NDH) and donate electrons to external acceptors such as molecular oxygen (cytochrome oxidase, cyt aa_3), nitrate (nitrate reductase, NAR), nitrite (NO_2^- Rase), nitrous oxide (N_2O Rase), or nitric oxide (NO Rase). The free-quinone pool is indicated as Q inside a dashed-line circle.

Box 18.1. Continued

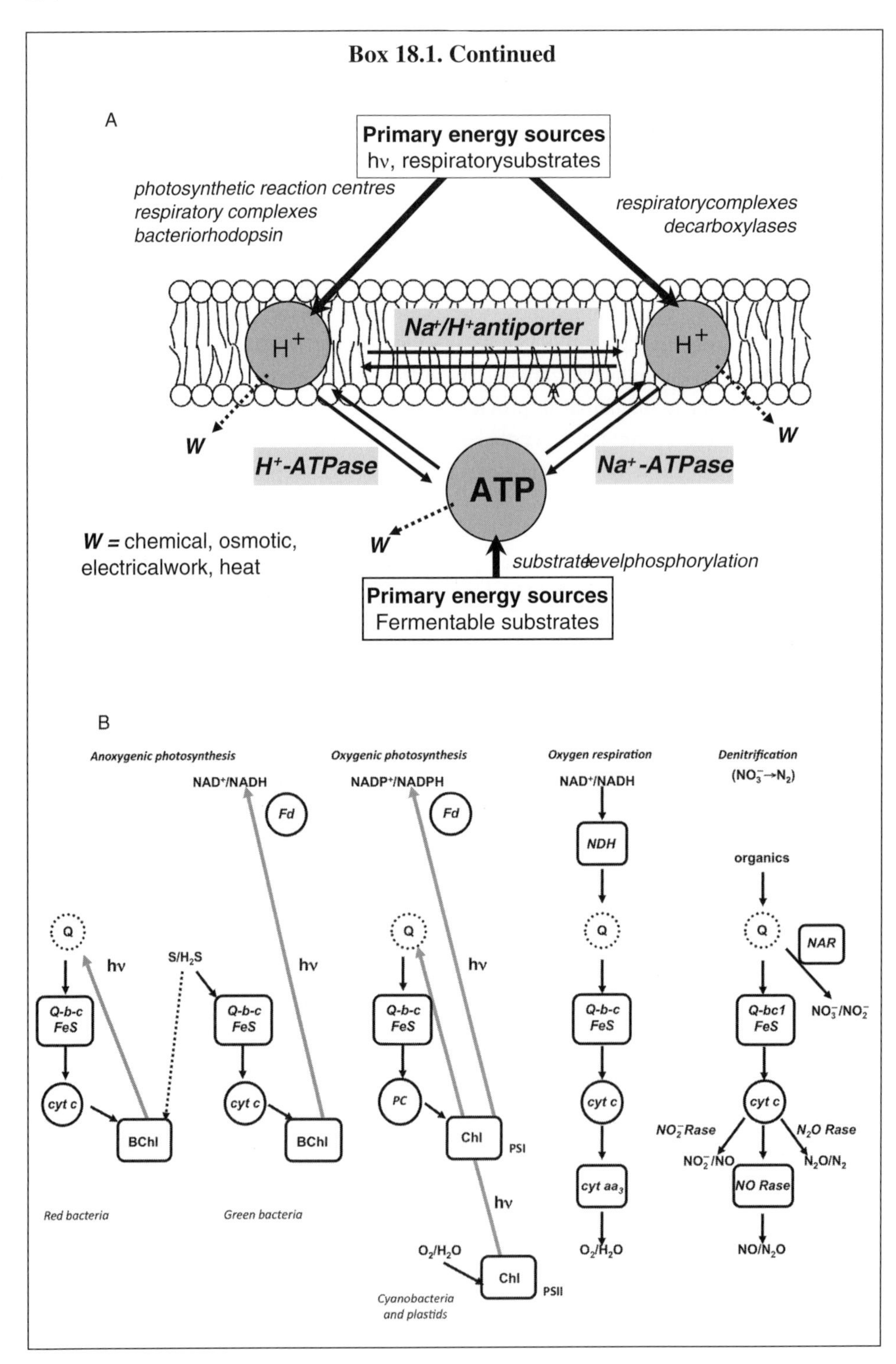

1. C fixation (isotope signatures)
2. Methanogenesis (isotope signatures)
3. Fe oxidation (BIF)
4. Anoxygenic photosynthesis
5. Denitrification
6. Sulphur reduction (isotope signatures)
7. N fixation
8. Oxygenic photosynthesis
9. Oxygen respiration

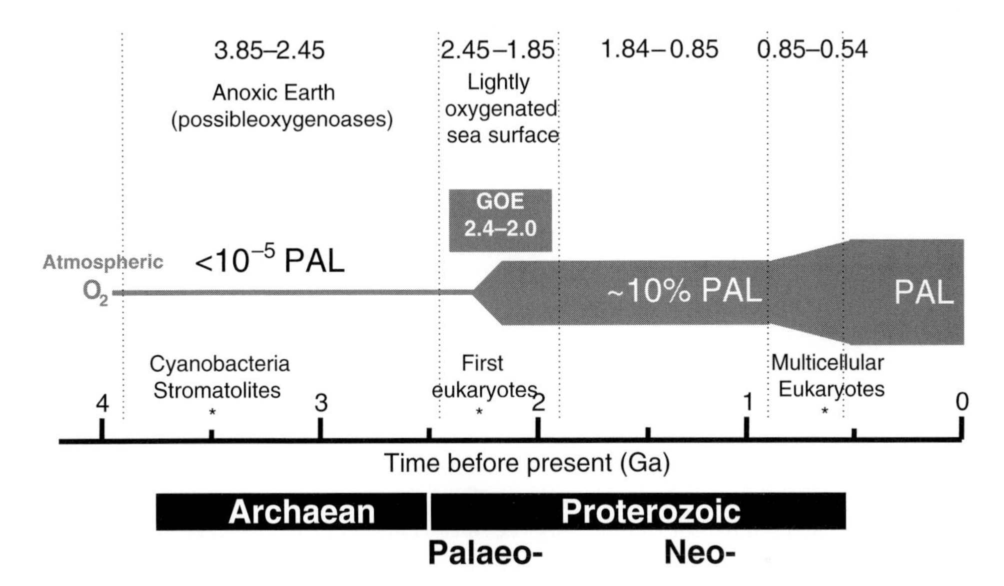

Figure 18.1. A hypothesis on the evolution of metabolisms and planetary history of oxygen. The accumulation of atmospheric oxygen along the geological time (in Ga = 10^9 years) in terms of the fraction of the present atmospheric level (PAL) is a consensus from several sources (Holland, 2006; Falkowski and Isozaki, 2008; Kump, 2008) based on geochemical data. Some palaeontological landmarks (*) and the great oxidation event (GOE) are indicated. The order of appearance of metabolisms is based on different, sometimes controversial, data from geochemistry, biochemistry and molecular phylogeny. BIF: banded-iron formations.

One of the most debated issues is the appearance and atmospheric accumulation of molecular oxygen and its metabolic and geological effects. About 2.4 Ga ago the atmospheric oxygen level rose in what is called the great oxidation event (GOE), to a level reaching about 10% of the present atmospheric level (PAL, Figure 18.1). Most authors agree that the invention of oxygenic photosynthesis (see Box 18.1) by the ancestors of cyanobacteria was the main source of oxygen, so the GOE would set a minimum age for this metabolism. Some authors have postulated an earliest (more than 3.0 Ga ago) origin of cyanobacteria, and thus the lag period for the oxygen increase would be explained by diverse sink reactions, such as from oxygen consumption by methane of biological origin or the oxidation of oceanic iron (with consequent deposition of the banded-iron formations, BIFs). Eventually, the marginalization of methanogenesis (e.g. by the scarcity of an essential metal such as Ni for the enzymes involved in the synthesis of methane; Saito, 2009) would contribute to the accumulation of oxygen. This extremely reactive molecule would also offer the opportunity for the evolution of new metabolic processes, such as oxygen respiration (see Box 18.1) or the synthesis of new metabolites, such as steroids (see below). Many authors also suggest that oxygen accumulation in the atmosphere was a requisite for a further increase in life's complexity, i.e. the origin of biomineralization and multicellularity.

The mechanism of emergence of new metabolic processes would imply recycling and modifying of old molecular devices, e.g. oxygenic from anoxygenic photosyntheses or aerobic from anaerobic respirations. In both cases, the evolutionary schemes proposed remain controversial. At any rate, the crucial step in the origin of oxygenic photosynthesis was the addition of a catalyst able to oxidize water (associated to photosystem II, see Box 18.1) to an older light-dependent anoxygenic electron-transport chain (Allen and Martin, 2007). On the other hand, the enzyme necessary to reduce oxygen to water (cytochrome oxidase) would emerge from the tinkering of older reductases, most likely NO reductase (Ducluzeau *et al.*, 2009), responsible for an early denitrification process (Box 18.1).

The evolution of metabolic pathways

The origin and evolution of metabolic pathways allowed primitive cells to become more chemically independent from the prebiotic sources of essential molecules. It is reasonable to assume that during the early stages of cell evolution, the primitive metabolism was based on a limited number of rudimentary (i.e. unspecific) enzymes. Several mechanisms observed in current cells (Box 18.2) may account for a rapid expansion of metabolic abilities. This 'mode' of evolution would justify a rapid 'tempo' in metabolic innovation, as observed in the spread of antibiotic resistances or the appearance of new metabolic activities within very short periods of time (i.e. from days to months) in experimental evolution with bacterial populations (Blount *et al.*, 2008).

Models of metabolic-pathway evolution

Although we do not know how and when the central metabolic pathways originated, there are several models trying to explain their evolution (for reviews, see Peretó *et al.*, 1997; Fani and Fondi, 2009). The classical hypotheses include (1) the retrograde model (Horowitz, 1945, 1965), (2) the forward model (Granick, 1957, 1965) and (3) the patchwork model (Ycas, 1974; Jensen, 1976). In earlier stages, some combination of enzymatic and non-enzymatic reaction steps could coexist (Lazcano and Miller, 1999).

(1) The first attempt to explain the evolution of metabolic pathways was developed by Horowritz following Oparin's heterotrophic model and the one-gene–one-protein correspondence suggested by

Box 18.2. How do new genes appear?

Let us suppose that we begin with a certain number of (not yet known) 'starter-type' genes (Lazcano and Miller, 1994). Several mechanisms have been discovered in extant cells that could explain a rapid expansion of the enzymatic repertoire.

1. **Duplication and divergence**. DNA duplication may involve from gene fragments (internal duplications) to whole genomes. Two duplicated genes may structurally diverge (paralogs) or may fuse forming an elongated gene (Fani *et al.*, 2007). Regarding their functional fates, gene divergence after duplication may originate new metabolic functions (James and Tawfik, 2003).
2. **Exon shuffling**. Most eukaryotic genes are fragmented, i.e. are organized as mosaics of coding sequences (exons) interrupted by non-coding segments (introns). Since many exons encode functional domains in proteins, new proteins may originate by the shuffling of exons (Long *et al.*, 2003).
3. **Overprinting and *de novo* from non-coding sequences**. Some mutations could give rise to new sequences for transcription and translation initiation and termination, either inside a pre-existing coding sequence (Delaye *et al.*, 2008) or within non-coding regions (Cai *et al.*, 2008). Wide arrows indicate coding regions; boxes, exons; lines, introns; and dashed lines, intergenic regions.

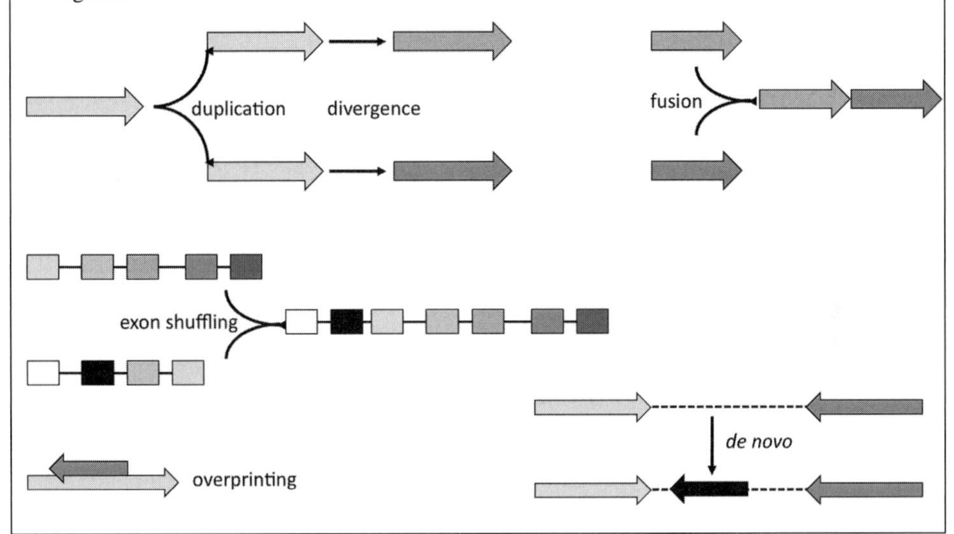

Beadle and Tatum (1941). Horowitz proposed that biosynthetic enzymes were acquired via gene duplication and in the reverse order as found in extant pathways, under the selective pressure of the exhaustion of prebiotic materials. Several criticisms of this model have been addressed elsewhere (Lazcano *et al.*, 1992; Peretó *et al.*, 1997) and, as a matter of fact, it can be applied only to very few cases.

(2) The Granick model assumes that simpler metabolites are older than more complex compounds and, consequently, the enzymes involved in earlier steps in a pathway are older than the later ones. Albeit originally proposed for explaining the evolution of haeme and chlorophyll biosyntheses, this model fits better to the biosynthesis of several membrane components, including steroids (Peretó *et al.*, 1997). In general, the metabolic expansion due to the accumulation of oxygen in the atmosphere could be the result of the extension of an ancestral anaerobic metabolic core (Raymond and Segré, 2006).
(3) According to the patchwork model, metabolic pathways may have been assembled by the recruitment of primitive enzymes that could react with a wide range of chemically related substrates. New enzymes with narrow specificities would result from gene duplication and divergence events. Several pieces of evidence give support to this model, including the analysis of whole genome sequences showing a high proportion of paralogous duplications corresponding to different enzymes evolved from common ancestral sequences and acting today in different pathways, i.e. with diverse substrate specificities, reaction mechanisms or both (Schmidt *et al.*, 2003). Accordingly, the distribution of protein domains across genomes supports the patchwork scenario (Caetano-Anollés *et al.*, 2009). Furthermore, the emergence of new activities during the experimental evolution of bacterial populations also implies the recruitment of old enzymes to serve new functions (Mortlock, 1992).

Enzyme recruitment is a pervasive mechanism invoked to explain the evolution of metabolic pathways. For instance, the urea cycle of terrestrial animals probably evolved by the addition of arginase to the biosynthesis of arginine (Takiguchi *et al.*, 1989). Autotrophy could emerge by adding just one or two new activities to the previous heterotrophic pathways (Peretó *et al.*, 1999). The Krebs cycle might be the result of the assembly of enzymes involved in amino-acid metabolism (Meléndez-Hevia *et al.*, 1996). Patchwork assembly can also explain the evolution of biosynthetic pathways of amino acids, such as histidine (Fani *et al.*, 1995), lysine (Velasco *et al.*, 2002), tryptophan (Xie *et al.*, 2003), and lysine, arginine and leucine (Fondi *et al.*, 2007). The recruitment of stereospecific dehydrogenases from families of less efficient primitive enzymes has been invoked to explain the origin of homochiral lipid membranes (Peretó *et al.*, 2004). Furthermore, the patchwork model offers more explanatory power when studying pathways with chemically complex intermediates, such as in the biosynthesis of cofactors (Holliday *et al.*, 2007). Finally, directed-evolution experiments support this model, since the ability to use new carbon sources (Mortlock, 1992) or to catabolize toxic compounds (Copley, 2000) usually emerges via combinatorial pre-existing pathways, deregulation of transcriptional circuits and selection of enzyme variants with altered catalytic parameters.

Besides the mode of construction of pathways, another question of interest is the order of appearance of the different metabolic modules during evolution. To address this problem one must admit, in agreement with Morowitz (1992), that extant metabolic networks have traces of their history imprinted in their own structure. According to this author, a metabolic ancient core organized around a prebiotic reductive tricarboxylic-acid cycle followed by the addition of ‘enzymatic shells’, successively increases the complexity from central pathways (such as in the Krebs cycle, glycolysis and fatty-acid metabolism) to amino-acid biosynthesis, sulphur metabolism and purine, pyrimidine and cofactor biosyntheses. A strikingly similar structure emerges from studies of protein domain distribution, although in this case the nucleotide metabolism appears as an older module, suggesting a possible chemical link to an early RNA World (Caetano-Anollés *et al.*, 2009). Using a chemical and biochemical retrospective reasoning applied to the history of the appearance of metabolic pathways, Meléndez-Hevia and colleagues (2008) have proposed that early metabolism was directly connected to prebiotic chemistry and was built by radial growth from a central nucleus (glycolysis and an open or horseshoe Krebs cycle) and also by adding other peripheral subnetworks to it.

Enzymatic properties and evolvability

The molecular tinkering associated with protein-function evolution has long been recognized (Jacob, 1977), one classic example being the use of some metabolic enzymes as lens proteins in animals. The conventional view of an extremely specific and proficient enzyme performing a well-defined and unique function must be substituted by the appreciation of several properties – such as 'ambiguity' and 'plasticity' – of crucial importance for the comprehension of enzyme evolvability, defined as 'the ability of proteins to rapidly adopt (i.e. within a few sequence changes) new functions within existing folds or even adopt entirely new folds' (Tokuriki and Tawfik, 2009).

Catalytic ambiguity refers to the ability of one enzyme to promote a secondary activity in the same active centre responsible for the primary one. This is compatible with a view of proteins whereby one primary sequence can adopt a flexible structure with certain conformational diversity (James and Tawfik, 2003). Catalytic ambiguity also implies a certain plasticity or malleability of active sites, i.e. the alteration of activity throughout a small number of amino-acid substitutions. Jensen (1976) emphasized the role of substrate ambiguity in enzyme evolution, since it opens up the possibility of recruiting new activities after duplication and divergence of the gene encoding for the enzyme (Figure 18.2). The evolution of new functions is possible by point mutations with large effects on the secondary activity without compromising the native structure and the primary activity, before the gene duplication, divergence and selection of the new protein (Aharoni *et al.*, 2005; Tokuriki and Tawfik, 2009).

Multifaceted proteins (or 'moonlighting') are a manifestation of the evolution of more than one function by the same protein. Thus, a protein 'moonlights' when it has more than one functional role, e.g. an enzyme with additional structural or regulatory functions. Since amino-acid residues at the catalytic site usually represent a small fraction of the total enzymatic structure, there are ample opportunities to evolve those additional functions.

Adaptive changes in the catalytic performance of enzymes have been reported regarding environmental conditions such as oxygen availability, osmotic relationships and

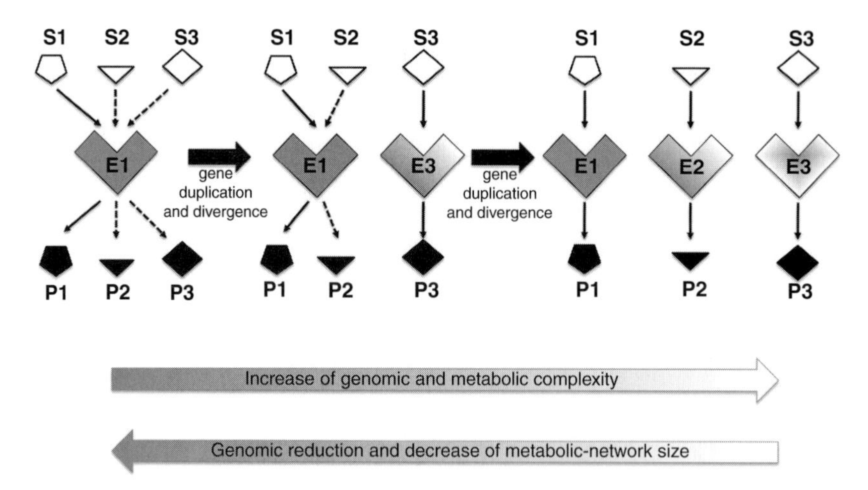

Figure 18.2. Increase of metabolic complexity through gene duplication and divergence. The enzyme E1 shows substrate ambiguity, i.e. it acts on three different substrates S1 to S3 – open pentagon, triangle and rhomboid, the transformation of S1 into P1 being the main or primary activity and the activities on S2 and S3 (dashed arrows) the secondary ones. A first duplication event and divergence specializes E3 in the use of S3 whereas E1 maintains the activity on S1 and, with less extension, on S2. A second duplication event generates three specialized enzymes fully active on each of the three substrates. On the other hand, the reverse phenomenon might occur during genome and metabolic minimization in endosymbionts (Zientz *et al.*, 2004) and pathogens (Yus *et al.*, 2009).

temperature regimes (Hochachka and Somero, 2002). Several authors have also addressed the question of optimality of the enzymatic function, either at the level of the catalytic constants of individual enzymes (Cornish-Bowden, 1976; Heinrich *et al.*, 2002), the architecture of the metabolic pathways (Meléndez-Hevia and Torres, 1988; Heinrich *et al.*, 1999; Cornish-Bowden, 2004) or the fitness of a final metabolic product such as cholesterol (Bloch, 1994) and glycogen (Meléndez *et al.*, 1997) as well-adapted structures able to perform their physiological function.

Examples of mosaic evolution in eukaryotic metabolism

Protein sequence analysis, molecular cladistics and comparative genomics have enhanced our understanding of metabolic evolution. Their applicability cannot be extended far beyond the universal cenancestor, although important insights on early metabolism can be achieved through comparative biochemistry over the broadest representation of genomes. Nevertheless, such backtrack studies are hindered by polyphyletic secondary losses, lateral gene transfers, replacements and redundancies. These difficulties are also patent in the study of less ancient stages of metabolic evolution, e.g. during eukaryotic evolution since about 2 Ga before present. Molecular cladistics and comparative genomics show a mosaic evolution of metabolic pathways during the merging of genomes from different origins, giving rise to the diversity of eukaryotic compartments. Furthermore, ongoing endosymbioses (see Chapter 21) are also the scenario for evolutionary opportunism and molecular tinkering in metabolic innovation. The biochemical and genetic causes of the metabolic mosaicism observed in eukaryotic cells are still poorly understood (Ringemann *et al.*, 2006).

Mosaic pathways in eukaryotic compartments

The bacterial origin of mitochondria and plastids is well established, although some controversies remain (de Duve, 2007). The symbiogenetic acquisition of genomes implies that, initially, many pathways could be redundant between the endosymbiont and the host. The comparative-genomics and proteomics studies by Gabaldón and Huynen (2003, 2004) have reconstructed the ancestral metabolism of the protomitochondrion: a quite complex bacterium whose original metabolism has been almost completely lost or substituted by new functions.

Molecular cladistic analyses of individual enzymes also show that the enzymes involved in a pathway have different phylogenetic origins. For instance, the Calvin cycle, an idiosyncratic pathway from the cyanobacterial ancestor of modern plastids, is catalysed in plants by a mixture of enzymes with different origins, including mitochondrial ones (Martin and Schnarrenberger, 1997). Also, the redundancy of isoenzymes in the cell compartments has been solved differently in the different eukaryotic lineages, as shown by the case of the universally distributed anabolic and catabolic thiolases (Peretó *et al.*, 2005).

Opportunistic metabolic evolution in endosymbionts

Ongoing endosymbioses are excellent case studies for the evolution of metabolic pathways. The reduction of genomes and metabolic networks observed during the endosymbiotic event (Moya *et al.*, 2008) preferentially affects the redundant pathways with the host, e.g. the biosynthesis of fatty acids and membrane lipids. Meanwhile, the biosynthetic pathways for essential metabolites (e.g. amino acids, cofactors) are kept by the endosymbiont and result in a mutual metabolic interdependence with the host. However, sometimes the reduction in the repertoire of enzymatic activities favours a return to a situation similar to an ancestral state, i.e. fewer enzymes have to do the work and reversion to substrate ambiguity could be a solution (Figure 18.2). In fact, some generic enzymes, like transaminases and kinases, have been reduced in number and most probably the substrate specificity has been relaxed (Zientz *et al.*, 2004).

A remarkable aspect of symbiogenetic innovation in metabolism is syntrophy or metabolic complementation. The biochemical integration of hosts and symbionts can be extremely tight since some metabolic pathways are shared and the participating enzymes are encoded by different genomes. In other words, fragments of a metabolic pathway are spatially separated into different compartments and organisms. One striking case is haeme biosynthesis by *Bradyrhizobium japonicum* from a metabolic precursor (δ-aminolevulinic acid) supplied by the host plastid (soybean) (Figure 18.3A).

A

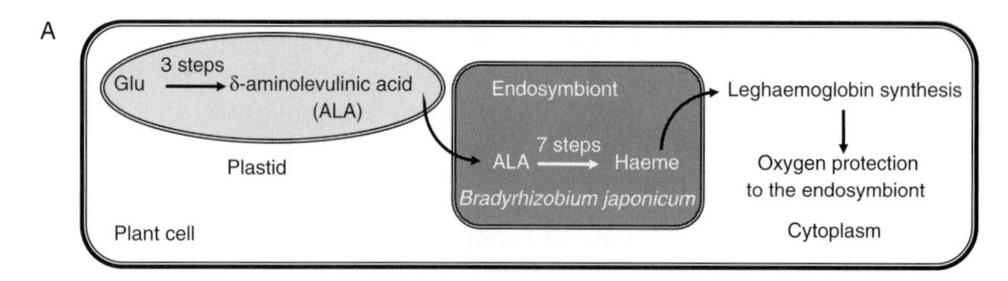

B

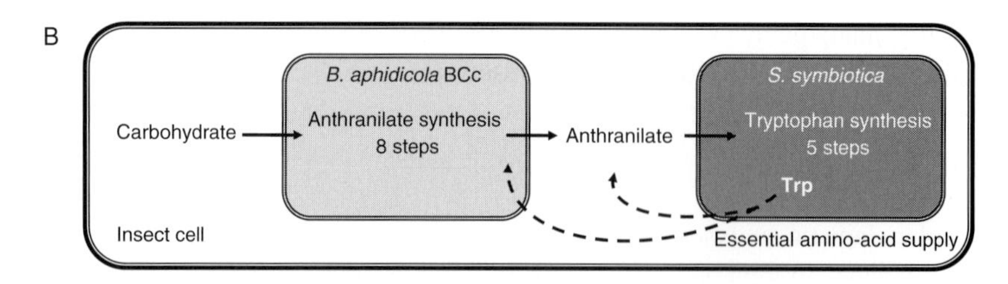

Figure 18.3. Metabolic complementation in endosymbionts. Some metabolic pathways are fragmented between hosts and endosymbionts. (A) The synthesis of haeme in soybean is shared between plastid and the endosymbiont *Bradyrhizobium japonicum*. The plant leghaemoglobin synthesis uses the bacterial haeme. Leghaemoglobin controls oxygen availability to the bacterium. (B) The synthesis of Trp in the cedar aphid is distributed between the two bacterial endosymbionts: *Buchnera aphidicola* supplies the anthranilate necessary for Trp biosynthesis by *Serratia symbiotica*. Trp is an essential amino acid for both the host and the endosymbiont *B. aphidicola*, whereas *S. symbiotica* could be considered an auxotrophic mutant for anthranilate.

Eventually the plant cell uses the haeme group for leghaemoglobin synthesis (Sangwan and O'Brian, 1991).

Another excellent example is provided by the biosynthesis of the essential amino acid Trp in cedar aphids. The common ancestor of animals lacked the complete pathway for Trp biosynthesis. Some 200 million years ago the association of the aphid ancestor with γ-proteobacteria overcame the need for an external supply of Trp, allowing the adaptation of the insect host to new ecological niches and diets devoid of Trp. The association with a second symbiont solved the redundancies in such a way that today there is a need for an interspecific anthranilate supply between the two endosymbionts (Figure 18.3B). *Buchnera aphidicola* contains the biosynthetic pathway up to anthranilate, whereas *Serratia symbiotica* uses this anthranilate to synthesize the Trp needed by all three partners (Gosalbes *et al.*, 2008). Thus, aphids synthesize Trp by using bacterial pathways fully integrated in a vertical inheritance.

Concluding remarks

We still do not know when and how life originated. However, useful hints can be inferred from extant metabolic pathways, as well as from their correlation with environmental changes through planetary history. Although we still lack a narrative for the origin and evolution of metabolic pathways – a true natural history of biochemistry – we are gaining insights from comparative genomics and molecular cladistic analyses of individual enzymes. The fruitfulness of this approach is indebted to the vision of evolution as a tinkerer rather than an engineer. Using the words of Jacob (1977), inspired by Darwin (1862): 'In contrast to the engineer, evolution does not produce innovations from scratch. It works on what already exists ... like a tinkerer who, during millions of years, has slowly modified his products ... using all opportunities to transform and create.'

References

Aharoni, A., Gaidukov, L., Khersonsky, O., McQ Gould, S., Roodveldt, C. and Tawfik, D. S. (2005). The 'evolvability' of promiscuous protein functions. *Nature Genetics*, **37**, 73–6.

Allen, J. F. and Martin, W. (2007). Out of thin air. *Nature*, **445**, 610–12.

Baltscheffsky, H. and Baltscheffsky, M. (1994). Molecular origin and evolution of early biological energy conversion. In *Early life on Earth*, ed. S. Bengston. New York: Columbia University Press, Nobel symposium no. 84, pp. 81–90.

Barber, J. (2008). The photosynthetic generation of oxygen. *Philosophical Transactions of the Royal Society B*, **363**, 2665–74.

Beadle, G.W. and Tatum, E.L. (1941). Genetic control of biochemical reactions in *Neurospora. Proceedings of the National Academy of Sciences of the USA*, **27**, 499–506.

Bloch, K. (1994). *Blondes in Venetian Paintings: The Nine-banded Armadillo, and Other Essays in Biochemistry*. New Haven: Yale University Press.

Blount, Z. D., Borland, C. Z. and Lenski, R. E. (2008). Historical contingency and the evolution of a key innovation in an experimental population of *Escherichia coli. Proceedings of the National Academy of Sciences of the USA*, **105**, 7899–906.

Caetano-Anollés, G., Yafremava, L. S., Gee, H., Caetano-Anollés, D., Kim, H. S. and Mittenthal, J. E. (2009). The origin and evolution of modern metabolism. *The International Journal of Biochemistry and Cell Biology*, **41**, 285–97.

Cai, J., Zhao, R., Jiang, H. and Wang, W. (2008). *De novo* origination of a new protein-coding gene in *Saccharomyces cerevisiae. Genetics*, **179**, 487–96.

Copley, S. D. (2000). Evolution of a metabolic pathway for degradation of a toxic xenobiotic: the patchwork approach. *Trends in Biochemical Sciences*, **25**, 261–5.

Cornish-Bowden, A. (1976). The effect of natural selection on enzymic catalysis. *Journal of Molecular Biology*, **101**, 1–9.

Cornish-Bowden, A. (2004). *The Pursuit of Perfection: Aspects of Biochemical Evolution*. Oxford: Oxford University Press.

Cornish-Bowden, A., Cárdenas, M. L., Letelier, J. C. and Soto-Andrade, J. (2007). Beyond reductionism: metabolic circularity as a guiding vision for a real biology of systems. *Proteomics*, **7**, 839–45.

Cornish-Bowden, A., Cárdenas, M. L., Letelier, J. C., Soto-Andrade, J. and Abarzúa, F. G. (2004). Understanding the parts in terms of the whole. *Biology of the Cell*, **96**, 713–17.

Costanzo, G., Saladino, R., Crestini, C., Ciciriello, F. and Di Mauro, E. (2007). Formamide as the main building block in the origin of nucleic acids. *BMC Evolutionary Biology*, **7**, S1.

Darwin, C. R. (1859). *On the Origin of Species by Means of Natural Selection, or the Preservation of Favoured Races in the Struggle for Life*. London: John Murray, ch. 6.

Darwin, C. R. (1862). *On the Various Contrivances by which British and Foreign Orchids Are Fertilized by Insects, and on the Good Effects of Intercrossing*. London: John Murray, ch. 9.
de Duve, C. (2005). *Singularities: Landmarks on the Pathways of Life*. Cambridge: Cambridge University Press.
de Duve, C. (2007). The origin of eukaryotes: a reappraisal. *Nature Reviews Genetics*, **8**, 395–403.
Delaye, L., Deluna, A., Lazcano, A. and Becerra, A. (2008). The origin of a novel gene through overprinting in *Escherichia coli*. *BMC Evolutionary Biology*, **8**, 31.
Ducluzeau, A.-L., van Lis, R., Duval, S., Schoepp-Cothenet, B., Russell, M. J. and Nitschke, W. (2009). Was nitric oxide the first deep electron sink? *Trends in Biochemical Sciences*, **34**, 9–15.
Eschenmoser, A. (2007). The search for the chemistry of life's origin. *Tetrahedron*, **63**, 12821–44.
Falkowski, P. G. and Isozaki, Y. (2008). The story of O_2. *Science*, **322**, 540–2.
Fani, R., Liò, P. and Lazcano, A. (1995). Molecular evolution of the histidine biosynthetic pathway. *Journal of Molecular Evolution*, **41**, 760–74.
Fani, R., Brilli, M., Fondi, M. and Liò, P. (2007). The role of gene fusions in the evolution of metabolic pathways: the histidine biosynthesis case. *BMC Evolutionary Biology*, **7**, S4.
Fani, R. and Fondi, M. (2009). Origin and evolution of metabolic pathways. *Physics of Life Reviews*, **6**, 23–52.
Fell, D. and Wagner, A. (2000). The small world of metabolism. *Nature Biotechnology*, **18**, 1121–2.
Fondi, M., Brilli, M., Emiliani, G., Pafffetti, D. and Fani, R. (2007). The primordial metabolism: an ancestral interconnection between leucine, arginine and lysine biosynthesis. *BMC Evolutionary Biology*, **16**, S3.
Gabaldón, T. and Huynen, M. A. (2003). Reconstruction of the proto-mitochondrial metabolism. *Science*, **301**, 609.
Gabaldón, T. and Huynen, M. A. (2004). Shaping the mitochondrial proteome. *Biochimica et Biophysica Acta*, **1659**, 212–20.
Gedulin, B. and Arrhenius, G. (1994). Sources and geochemical evolution of RNA precursor molecules: the role of phosphate. In *Early life on Earth*, ed. S. Bengston. New York: Columbia University Press, Nobel symposium no. 84, pp. 91–106.
Gosalbes, M. J., Lamelas, A., Moya, A. and Latorre, A. (2008). The striking case of tryptophan provision in the cedar aphid *Cinara cedri*. *Journal of Bacteriology*, **190**, 6026–9.
Granick, S. (1957). Speculations on the origins and evolution of photosynthesis. *Annals of the New York Academy of Sciences*, **69**, 292–308.
Granick, S. (1965). The evolution of heme and chlorophyll. In *Evolving Genes and Proteins*, eds. V. Bryson and H. J. Vogel. New York: Academic Press, pp. 67–88.
Haldane, J. B. S. (1929). The origin of life. *The Rationalist Annual*, 3–10.
Heinrich, R., Meléndez-Hevia, E., Montero, F., Nuño, J. C., Stephani, A. and Waddell, T. G. (1999). The structural design of glycolysis: an evolutionary approach. *Biochemical Society Transactions*, **27**, 294–8.
Heinrich, R., Meléndez-Hevia, E. and Cabezas, H. (2002). Optimization of kinetic parameters of enzymes. *Biochemistry and Molecular Biology Education*, **30**, 184–8.

Hochachka, P. W. and Somero, G. N. (2002). *Biochemical Adaptation: Mechanism and Process in Physiological Evolution*. Oxford: Oxford University Press.

Holland, H. D. (2006). The oxygenation of the atmosphere and oceans. *Philosophical Transactions of the Royal Society B*, **361**, 903–15.

Holliday, G. L., Thornton, J. M., Marquet, A., Smith, A. G., Rébeillé, F., Mendel, R., Schubert, H. L., Lawrence, A. D. and Warren, M. J. (2007). Evolution of enzymes and pathways for the biosynthesis of cofactors. *Natural Product Reports*, **24**, 972–87.

Horowitz, N. H. (1945). On the evolution of metabolic pathways. *Proceedings of the National Academy of Sciences of the USA*, **31**, 153–7.

Horowitz, N. H. (1965). The evolution of biochemical syntheses: retrospect and propect. In *Evolving Genes and Proteins*. New York: Academic Press, pp. 15–23.

Jacob, F. (1977). Evolution and tinkering. *Science*, **196**, 1161–6.

James, L. C. and Tawfik, D. S. (2003). Conformational diversity and protein evolution: a 60-year-old hypothesis revisited. *Trends in Biochemical Sciences*, **28**, 361–8.

Jensen, R. A. (1976). Enzyme recruitment in evolution of new function. *Annual Review of Microbiology*, **30**, 409–25.

Kump, L. R. (2008). The rise of atmospheric oxygen. *Nature*, **451**, 277–8.

Lane, N. (2009). *Life Ascending: The Ten Great Inventions of Evolution*. London: Profile Books.

Lazcano, A., Fox, G. and Oró, J. (1992). Life before DNA: the origin and evolution of early Archaean cells. In *The Evolution of Metabolic Function*, ed. R. P. Mortlock. Boca Raton: CRC Press, pp. 237–95.

Lazcano, A. and Miller, S. L. (1994). How long did it take for life to begin and evolve to cyanobacteria? *Journal of Molecular Evolution*, **39**, 546–54.

Lazcano, A. and Miller, S. L. (1999). On the origin of metabolic pathways. *Journal of Molecular Evolution*, **49**, 424–31.

Long, M., Deutsch, M., Wang, W., Betrán, E., Brunet, F. G. and Zhang, J. (2003). Origin of new genes: evidence from experimental and computational analyses. *Genetica*, **118**, 171–82.

Martin, W. and Schnarrenberger, C. (1997). The evolution of the Calvin cycle from prokaryotic to eukaryotic chromosomes: a case study of functional redundancy in ancient pathways through endosymbiosis. *Current Genetics*, **l32**, 1–18.

Meléndez, R., Meléndez-Hevia, E. and Cascante, M. (1997). How did glycogen structure evolve to satisfy the requirement for rapid mobilization of glucose? A problem of physical constraints in structure building. *Journal of Molecular Evolution*, **45**, 446–55.

Meléndez-Hevia, E. and Torres, N. V. (1988). Economy of design in metabolic pathways: further remarks on the game of the pentose phosphate cycle. *Journal of Theoretical Biology*, **132**, 97–111.

Meléndez-Hevia, E., Montero-Gómez, N. and Montero, F. (2008). From prebiotic chemistry to cellular metabolism. The chemical evolution of metabolism before Darwinian natural selection. *Journal of Theoretical Biology*, **252**, 505–19.

Meléndez-Hevia, E., Waddell, T. G. and Cascante, M. (1996). The puzzle of the Krebs citric acid cycle: assembling the pieces of chemically feasible reactions, and opportunism in the design of metabolic pathways during evolution. *Journal of Molecular Evolution*, **43**, 293–303.

Morowitz, H. J. (1992). *Beginnings of Cellular Life: Metabolism Recapitulates Biogenesis*. New Haven: Yale University Press.

Mortlock, R. P. (ed.). (1992). *The Evolution of Metabolic Function*. Boca Raton: CRC Press.

Moya, A., Peretó, J., Gil, R. and Latorre, A. (2008). Learning how to live together: genomic insights into prokaryote–animal symbioses. *Nature Reviews Genetics*, **9**, 218–29.

Nealson, K. H. and Rye, R. (2005). Evolution of metabolism. In *Biogeochemistry*, ed. W. H. Schlesinger. Amsterdam: Elsevier, vol. 8, pp. 41–61.

Oparin, A. I. (1924). *The Origin of Life*. Appendix to Bernal, J. D. (1967). *The Origin of Life*. London: Weindenfeld and Nicholson.

Pasek, M. A. (2008). Rethinking early Earth phosphorus geochemistry. *Proceedings of the National Academy of Sciences of the USA*, **105**, 853–8.

Peretó, J., López-García, P. and Moreira, D. (2004). Ancestral lipid biosynthesis and early membrane evolution. *Trends in Biochemical Sciences*, **29**, 469–77.

Peretó, J., López-García, P. and Moreira, D. (2005). Phylogenetic analysis of eukaryotic thiolases suggests multiple proteobacterial origins. *Journal of Molecular Evolution*, **61**, 65–74.

Peretó, J., Fani, R., Leguina, J. I. and Lazcano, A. (1997). Enzyme evolution and the development of metabolic pathways. In *New Beer in an Old Bottle: Eduard Buchner and the Growth of Biochemical Knowledge*, ed. A. Cornish-Bowden. València: Publicacions de la Universitat de València, pp. 173–98.

Peretó, J., Velasco, A. M., Becerra, A. and Lazcano, A. (1999). Comparative biochemistry of CO_2 fixation and the evolution of autotrophy. *International Microbiology*, **2**, 3–10.

Raymond, J. and Segré, D. (2006). The effect of oxygen on biochemical networks and the evolution of complex life. *Science*, **311**, 1764–7.

Ringemann, C., Ebenhöh, O., Heinrich, R. and Ginsburg, H. (2006). Can biochemical properties serve as selective pressure for gene selection during inter-species and endosymbiotic lateral gene transfer? *IEE Proceedings-Systems Biology*, **153**, 212–22.

Russell, M. J. and Martin, W. (2004). The rocky roots of the acetyl-CoA pathway. *Trends in Biochemical Sciences*, **29**, 358–63.

Saito, M. A. (2009). Less nickel for more oxygen. *Nature*, **458**, 714–15.

Sangwan, I. and O'Brian, M. R. (1991). Evidence for an inter-organismic heme biosynthetic pathway in symbiotic soybean root nodules. *Science*, **251**, 1220–2.

Schmidt, S., Sunyaev, S., Bork, P. and Dandekar, T. (2003). Metabolites: a helping hand for pathway evolution? *Trends in Biochemical Sciences*, **28**, 336–41.

Takiguchi, M., Matsubasa, T., Amaya, Y. and Mori, M. (1989). Evolutionary aspects of urea cycle enzyme genes. *Bioessays*, **10**, 163–6.

Tokuriki, N. and Tawfik, D. S. (2009). Protein dynamism and evolvability. *Science*, **324**, 203–7.

Velasco, A. M., Leguina, J. I. and Lazcano, A. (2002). Molecular evolution of the lysine biosynthetic pathways. *Journal of Molecular Evolution*, **55**, 445–9.

Wächtershäuser, G. (2006). From volcanic origins of chemoautotrophic life to bacteria, Archaea and eukarya. *Philosophical Transactions of the Royal Society B*, **361**, 1787–808.

Wagner, A. (2007). *Robustness and Evolvability in Living Systems*. Princeton: Princeton University Press.

Xie, G., Keyhani, N. O., Bonner, C. A. and Jensen, R. A. (2003). Ancient origin of the tryptophan operon and the dynamics of evolutionary change. *Microbiology and Molecular Biology Reviews*, **67**, 303–42.

Ycas, M. (1974). On earlier states of the biochemical system. *Journal of Theoretical Biology*, **44**, 145–60.

Yus, E., Maier, T., Michalodimitrakis, K., van Noort, V., Yamada, T., Chen, W. H., Wodke, J. A., Güell, M., Martínez, S., Bourgeois, R., Kühner, S., Raineri, E., Letunic, I., Kalinina, O. V., Rode, M., Herrmann, R., Gutiérrez-Gallego, R., Russell, R. B., Gavin, A. C., Bork, P. and Serrano, L. (2009). Impact of genome reduction on bacterial metabolism and its regulation. *Science*, **326**, 1263–8.

Zientz, E., Dandekar, T. and Gross, R. (2004). Metabolic interdependence of obligate intracellular bacteria and their insect hosts. *Microbiology and Molecular Biology Reviews*, **68**, 745–70.

Zubay, G. (1993). To what extent do biochemical pathways mimic prebiotic pathways? *Chemtracts Biochemistry and Molecular Biology*, **4**, 317–23.

Part V

Mechanisms for life evolution

19

Molecular phylogeny: inferring the patterns of evolution

Emmanuel Douzery

Preamble: the vertical inheritance of genetic material

Living systems are ephemeral vehicles of their immortal germlines. In the selfish-gene (Dawkins, 1976) or disposable-soma (Kirkwood, 1977) perspectives, genes do have a greater interest in regularly 'buying a new vehicle and throwing away the old one'. Rather than being forever stuck in the same organism genes can reassort themselves with random samples of the genetic-material pool through sexual reproduction. As human beings we are thus familiar with the widespread biparental reproduction procedure. Two individuals – the parents – spawn reproductive cells – the gametes – which merge to produce offspring – the descendants. In diploid populations, each descendant receives a paternal copy and a maternal copy of the genome (Figure 19.1). Because of the recombination occurring in the germline each parental copy is itself a reassortment of the grandmaternal and grandpaternal genomes. From generation to generation gene copies are replicated. If they are transmitted to at least two descendants (Figure 19.1: stars), this will result in branching nodes in what we call 'gene trees' (Maddison, 1997).

This transmission of the genetic material is called 'vertical': DNA of an organism is inherited from its forebears. Vertical gene transfer implies that there is a tree structure describing the history of descent of the genetic material. By contrast, there is horizontal gene transfer when genetic material is passed on from donor organisms to receptors belonging to different species (see Chapter 20).

However, even the high-fidelity DNA replication is prone to errors and the gene copies might change through time, owing to mutations. After the initial occurrence of a mutation in the germline of one single individual, the mutant allele will often be lost. However, it may be transmitted to offspring, which in turn may leave one or more descendants in the next generation. In this way, and after many generations, the mutant allele may progressively replace the wild-type allele in the population. We say that it has been fixed and the overall process is called a 'substitution' (of the wild-type by the mutant allele). During the course of evolution DNA sequences will accumulate such substitutions, and thus, be gradually modified with respect to their progenitors (Figure 19.1). As the different substitutions may be deleterious, nearly-neutral, neutral or advantageous, the rate at which these modifications will occur may vary among the different positions of the same gene and/or among different lineages of organisms (Bromham and Penny, 2003). Molecular-evolution

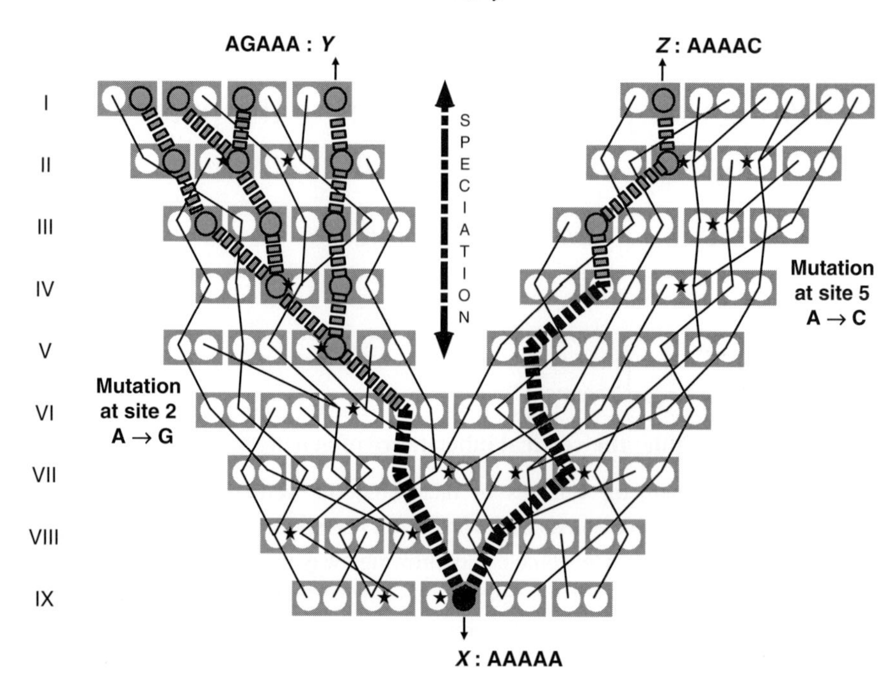

Figure 19.1. Vertical transmission of genetic material among individuals of populations will ultimately generate a branching pattern of DNA sequences through time. Grey rectangles indicate diploid individuals, with white circles denoting maternal and paternal genomes. Thin black lines connecting circles denote successful transmission to offspring of gene copies replicated in the gametes. A black star indicates when an individual is transmitting to the next generation at least two copies from the same gene. An arbitrary timescale graduated from generation I (present-day) to IX (origin) is given on the left. The history of a 5-nt long hypothetical sequence (5′-AAAAA-3′) in individual X is illustrated, with an A → G mutation at time VI, and an A → C mutation at time IV, respectively, on the second and fifth sites. The resulting sequences can be sampled by biologists in individuals Y and Z. Starting from time V, a speciation event occurred that separated two species from which individuals Y and Z were respectively sampled. We have, therefore, gene trees that split into a species tree. For simplicity, recombination is not illustrated here, but it would concur to produce new combinations of maternal and paternal sequences in gametes. This figure is a free modification from Hasegawa *et al.*, 1989.

studies aim, therefore, at describing the phenomenon of sequence divergence by progressive accumulation of substitutions.

Tree-thinking and tree-building approaches to understand life's evolution

A case study: the shift from mesophilic to thermophilic ecological preferences among microorganisms

Let us consider six related microbial species in the primordial world, three of which are mesophilic (Sp. 1, Sp. 3 and Sp. 5 have an intermediate optimal-growth temperature), and

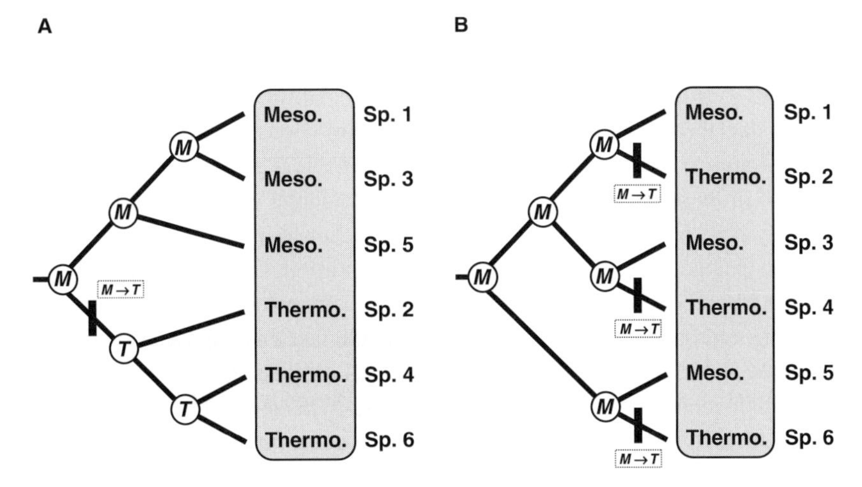

Figure 19.2. Tree-thinking in evolution: knowing the phylogeny of six species helps to understand how their mesophilic and thermophilic ecological preferences evolved. **Panel A**. A first example of phylogenetic relationships among six microbial species (Sp. 1 to Sp. 6). Their optimal-growth temperature is either mesophilic (Meso) or thermophilic (Thermo). Circles indicate that the ancestral preference at nodes of the tree is assumed to be either mesophilic (*M*) or thermophilic (*T*). The black bar indicates that a change from mesophilic to thermophilic preference ($M \rightarrow T$) is likely to have occurred along the corresponding branch of the phylogenetic tree. This interpretation is the most parsimonious one. **Panel B**. An example of alternative phylogenetic relationships among the same six species. The change in the topology of the phylogenetic tree involves at least three shifts from mesophilic to thermophilic preferences are likely to have occurred during microbial evolution. Obviously, other scenarios of changes in ecological preferences are possible, but all will be either equally or less parsimonious than the one illustrated here. In other words, alternative scenarios would involve at least three ecological shifts to explain the pattern observed.

three others are thermophilic (Sp. 2, Sp. 4 and Sp. 6 prefer to live at high temperatures). How can the ecological preferences of these six species be explained? Without any phylogenetic framework it is virtually impossible to understand the evolution of ecological preferences of life forms on Earth, in particular their variability. Now, let us assume a mesophilic last universal common ancestor (LUCA) (Galtier *et al.*, 1999; Boussau *et al.*, 2008). Under the hypothesis of closer affinities of species 1, 3 and 5, relative to species 2, 4 and 6, the most parsimonious scenario would be to assume a single shift from mesophilic to thermophilic preferences along the lineage leading to the hypothetical ancestor of species 2, 4 and 6 (Figure 19.2A). In this case, in spite of several microorganisms being thermophilic, all of them can be traced back to a single shift of optimal-growth temperature, which would thus have been a rare event during the evolution of life on Earth.

By contrast, if we assume a closer relatedness of species 1 and 2, and species 3 and 4, with respect to species 5 and 6, we would infer three independent shifts from mesophilic to thermophilic life-history traits, each one along the branches leading respectively to species 2, 4 and 6 (Figure 19.2B). In this second situation, shifts of optimal-growth temperature among microorganisms would have been more frequent than expected according to the previous scenario. From this theoretical example, alternatives to the evolutionary scenarios here illustrated (Figure 19.2B) are obviously possible. However, all of them would either involve more than three ecological shifts, i.e. less parsimonious explanations of the optimal-growth temperatures, or assume that the ancestral ecological preference was thermophilic.

Of note, all of these scenarios are linked in a fundamental way to the phylogenetic tree on which they have been mapped. In other words, a change in the phylogeny under focus will involve a change in the evolutionary scenarios inferred. Consequently, understanding the patterns and processes of life's evolution requires knowledge of the phylogeny. Such a tree-thinking approach is, however, hampered by the fact that the organismal phylogeny is unknown. Moreover, plenty of distinct phylogenies are possible – more than 34 million bifurcating phylogenetic trees connect only 10 species. This means that we should first find traces of organismal evolution and then use explicit tools to infer from them the tree(s) that best fit our observations.

Characters as witnesses of evolution

Tree-building relies on traces, or witnesses, of the organismal evolution. Most intuitive characters are morphological, anatomical and developmental (e.g. bilaterians are triploblastic animals: they develop from three distinct germ layers, in contrast to the diploblastic radial animals such as cnidarians). However, these characters are often available in limited number, especially when comparing organisms with strikingly contrasted body plans: in a Chinese spring roll, it is easy to distinguish among organisms (the pork marinade), plants (shredded carrots) and fungi (black mushrooms) yet it is more difficult to infer which one is more closely related to the other from an evolutionary point of view. By contrast, molecular characters – DNA, RNA and protein sequences – can always be identified across a wide taxonomic range provided that the corresponding genes are ubiquitous (i.e. present in all species) and that these sequences are not too divergent, so that they can be compared. In this case, from comparison of ribosomal RNA- and protein-coding genes we have learned that fungi are more closely related to animals than to plants (Wainright *et al.*, 1993; Stechmann and Cavalier-Smith, 2002). The solution of the spring-roll phylogenetic conundrum is therefore that animals and fungi are sister taxa, belonging to Opisthokonta, i.e. eukaryotes with cells using a single posterior flagellum to propel themselves (Cavalier-Smith, 1987).

A prerequisite for the analysis of molecular characters is the assessment of their homology. Different copies of a given molecular marker – i.e. DNA, RNA or protein sequences – sampled in different species should be compared to determine positional homology of each nucleotide or amino acid. Moreover, if target sequences are of different length, insertions and deletions should be allowed for. This is done through a step called 'sequence alignment' (Löytynoja and Goldman, 2008). Let us observe the alignment of the universal ribosomal protein RPS5 among representatives of the three domains of life (Figure 19.3). This 26-column alignment corresponds to a portion of constant length across Eukarya, Archaea and Bacteria. Some sites (e.g. sites 12 and 25) are identical in the 12 archaeal, bacterial and

RPS5

Domain	Taxon	1	2	3	4	5	6	7	8	9	10	11	12	13	14	15	16	17	18	19	20	21	22	23	24	25	26
EUK.	*Giardia*	P	R	E	D	S	C	R	I	G	K	G	G	Q	V	R	R	Q	S	V	D	V	S	P	L	R	R
	Trichomonas	A	R	E	D	S	T	R	I	G	V	G	G	T	V	R	R	Q	A	C	D	V	S	P	L	R	R
	Plasmodium	P	R	E	D	S	T	R	I	G	S	A	G	V	V	R	R	Q	A	V	D	V	S	P	L	R	R
	Homo	P	R	E	D	S	T	R	I	G	R	A	G	T	V	R	R	Q	A	V	D	V	S	P	L	R	R
ARC.	*Nanoarchaeum*	P	R	E	E	T	T	T	I	E	Y	G	G	A	R	Y	P	K	A	V	D	C	S	P	Q	R	R
	Aeropyrum	P	R	E	D	T	T	K	I	T	Y	G	G	I	T	Y	R	V	S	V	D	V	A	P	Q	R	R
	Methanobacterium	P	R	E	E	T	T	R	I	K	Y	G	G	I	G	Y	Q	V	A	V	D	I	S	P	Q	R	R
	Thermoplasma	P	R	E	E	V	T	R	L	K	Y	G	G	I	A	V	P	K	S	V	D	V	S	P	S	R	R
BAC.	*Aquifex*	P	E	W	E	V	R	P	T	R	V	G	G	A	T	Y	Q	V	P	I	E	V	P	E	R	R	Q
	Deinococcus	P	R	V	E	V	R	S	R	R	V	G	G	S	T	Y	Q	V	P	V	E	V	G	P	R	R	Q
	Bacillus	P	V	L	E	V	K	A	R	R	V	G	G	A	N	Y	Q	V	P	V	E	V	R	P	E	R	R
	Agrobacterium	P	H	V	E	V	R	S	R	R	V	G	G	A	T	Y	Q	V	P	V	D	V	R	P	E	R	R

Figure 19.3. An example of protein alignment: a portion of the conserved ribosomal protein S5 (RPS5) across the three domains of life. Twenty-six positions are given. Some sites are boxed according to their degree of variability. The three domains are distinguished: Eukarya (EUK), Archaea (ARC) and Bacteria (BAC).

eukaryote taxa, suggesting that they have remained completely conserved during their evolutionary history. Some sites are nearly conserved (e.g. sites 1 and 23): they are conserved in all sampled taxa but one. Some sites are variable, displaying different amino acids (for example D/E at sites 4 and 20; A/G at site 11; or R/Q at site 26). Some other sites are much more variable, showing up to six different amino acids (R/K/T/P/S/A at site 7: Figure 19.3). The generalization of these observations to many protein alignments indicates that most variable sites harbour only a few (2–3) different amino acids. This contraction of the protein alphabet at most sites can be explained by the fact that not all amino acids can occur at a given position because purifying natural selection acts to maintain protein functionality. For example, to facilitate their insertion in the lipid bilayer and increase conformational stability, transmembrane alpha helices contain hydrophobic amino-acid residues, e.g. leucine (L), isoleucine (I), valine (V) and phenylalanine (F), which can hardly be replaced by alternative non-hydrophobic amino acids. Similarly, negatively charged amino acids – aspartic acid (D) and glutamic acid (E) – will often interchange (cf. sites 4 and 20).

In summary, to infer phylogenetic relationships, DNA, RNA and protein alignments are helpful, the latter often involving comparisons at larger evolutionary scales because of the lower rate of amino-acid replacement as compared to nucleotides. Such a lower rate of character-state change is thought to reduce the risk of homoplasy, i.e. the phenomenon of sharing by chance the same amino acid – or the same nucleotide at the gene level – at a given sequence position among evolutionary independent organisms. Homoplasy results from convergences (independent gain of the same derived character state in two distinct lineages) and reversions (return to the ancestral character state). It complicates phylogenetic inference, as undetected convergences and reversions may result in erroneously grouping together unrelated species.

In addition to primary nucleotide substitutions and amino-acid replacements in gene and protein sequences, and as an increasing number of complete genomes are becoming available, additional molecular characters have been proposed. Rare genomic changes have especially focused the attention, for instance indel (insertion or deletion) signatures in conserved (DNA, RNA or protein) regions, retroposon integrations, gene-order changes or genetic-code changes. They have provided useful complementary information for tree-building approaches (Rokas and Holland, 2000). The rarity of such genomic events warrants lower levels of homoplasy; on the other hand, it makes them much more difficult to uncover.

A brief overview of tree-building approaches: distance, cladistics and probability

From a molecular perspective, three main approaches are available to infer phylogenetic trees from sequence alignments. Distance-based approaches were among the first to be used to construct trees. Their principle consists in primarily converting a gene or a protein alignment into a matrix of pairwise genetic distances between sequences. This distance matrix defines a natural scoring function on the space of all possible dendrograms. Intuitively, a reasonable tree will be one such that the distance inferred between any two taxa along the tree is as close as possible to the one specified by the pairwise distance matrix. Therefore, the least-squares criterion is often used to identify the topology and the branch lengths which minimize the distortion between inferred and specified distances (Fitch and Margoliash, 1967). The optimal tree can also be the one with the minimum total branch length (Gascuel *et al.*, 2001). Advantages of distance-based approaches are fast-computing times – noticeably when the number of species is large – and reasonable topology accuracy. However, their caveats are the loss of information caused by compressing alignments into pairwise distances and the difficulty in identifying the correct tree, especially when the evolutionary rate varies among species.

As an alternative to distance methods, character-based approaches have been promoted by cladistics. Here, the primary source of information is the character, i.e. each column of the sequence alignment. Sequences will be grouped into clades according to their shared derived character states ('synapomorphies'). The congruence among characters will then be maximized through the identification of the tree that minimizes the total number of evolutionary changes. This is the maximum-parsimony (MP) principle that stems from Occam's razor, according to which simpler explanations of data are more reasonable. Though widely used in phylogenetics, it has been pointed out that MP – and also distance-based methods – may suffer from the long-branch attraction artefact (Felsenstein, 1978). When two unrelated sequences are faster evolving, or share similar nucleotide or amino-acid composition, they will tend to accumulate homoplasies. For sufficiently fast-evolving taxa, those homoplasies will even dominate the true phylogenetic signal. In this context, MP inference will be more prone to interpret convergences as synapomorphies – as it will save evolutionary changes. This leads to the artefactual grouping in the phylogeny of its fastest-evolving representatives, often pulling them closer to the base of the phylogenetic tree, especially when the outgroup is distantly related (Brinkmann and Philippe, 1999). For example, amitochondriate protists like microsporidia – considered as closely related to fungi (Keeling and Doolittle, 1996) – have long been viewed as early emerging eukaryotes in small-subunit (SSU) rRNA phylogenies (Vossbrinck *et al.*, 1987). This artificially basal position is actually explained by a long-branch attraction phenomenon due to the fast evolutionary rate of their SSU rRNA, combined with the far-away location of the archaean outgroup.

As a way to alleviate some of the above-mentioned problems, the probabilistic approach has been introduced in phylogenetics through the use of likelihood calculation (Felsenstein, 1981). As in cladistics, the primary source of information of probability-based tree inference is the site of the sequence alignment. However, a major difference is the definition of a model (M) that aims at describing the sequence evolution, making explicit the underlying assumptions of the approach (Whelan *et al.*, 2001). The model specifies the process of evolutionary change in mathematical terms, for example, the frequency of each nucleotide or amino acid, and their rate of substitution. Given this predefined model of sequence evolution, we will estimate how likely the occurrence of the data (D) is, i.e. the alignment of sequences, if the evolutionary history of these sequences conforms to a given phylogenetic tree (T). The so-called maximum likelihood (ML) approach will therefore select as the best tree the one that maximizes the likelihood function $\mathrm{P}\,(D|T,M)$. Several parameters will thus be attached to the highest-likelihood tree: topology, branch lengths, nucleotide or amino-acid frequencies, and their relative exchangeabilities. As compared with maximum parsimony, the ML approach has the important advantage of being statistically consistent: ML will converge towards the true tree with increasing certainty as the amount of primary information increases, provided either that the assumptions of the model are met or that the model correctly describes the process that produced the data (Bryant *et al.*, 2005). Of note, all these approaches (distance, parsimony, likelihood) resolve a double-optimization problem, i.e. a 'best among bests' procedure in which all parameters for a given tree (e.g. branch lengths) are estimated under an optimality criterion and then the best tree is identified within the forest of best possible trees.

Because of the evolutionary noise brought by homoplasy in sequence alignments, alternative phylogenetic hypotheses are often similarly reasonable explanations of the data, which makes the choice of any one of them essentially arbitrary. This uncertainty is not directly accounted for by optimality approaches such as ML, which returns only the optimal tree while ignoring the suboptimal slightly less-likely ones. To estimate the phylogenetic uncertainty under ML – and also under MP or distances – additional steps are necessary, for example, the resampling of the original data by bootstrap (Felsenstein, 1985) or the comparison of the likelihood of competing trees (see e.g. the pioneering work by Kishino and Hasegawa, 1989).

To directly cope with the uncertainty during tree inference, the statistical Bayesian approach has been introduced in phylogeny. The Bayes theorem on conditional probabilities indicates that:

$$\mathrm{P}\,(T|D,M) = \mathrm{P}\,(D|T,M) \times \mathrm{P}\,(T)/\mathrm{P}\,(D) \qquad (19.1)$$

$\mathrm{P}\,(T|D,M)$ is the posterior probability, i.e. the probability of the tree given the data and the model. It is the product of the likelihood $\mathrm{P}\,(D|T,M)$ by the a-priori probability $\mathrm{P}(T)$ divided by the normalizing factor P (*D*). Bayesian phylogenetics will focus on the hypotheses having high posterior probability P (*T*|*D*,*M*). However, it will not exclusively focus on the tree of highest posterior probability. Instead, it will produce an estimate by averaging over the posterior distribution. In practice, this is done by (1) sampling trees from the posterior distribution through an algorithm called Markov chain Monte Carlo (MCMC), and (2) computing the consensus of all the trees thus sampled. In this context, Bayesian phylogenetics is a density approach, rather an optimality approach, that will simultaneously return a tree estimate and measures of uncertainty of its nodes (Holder and Lewis, 2003). As for the above-mentioned approaches, Bayesian inference of phylogeny is faced with its own problems, such as prior sensitivity and choice, or the correct interpretation to be given to node posterior probabilities, as compared to confidence measures like bootstrap values (Douady *et al.*, 2003).

Improvements to the probability approach to phylogenetic inference

Ways to improve phylogenetic accuracy

One of the major challenges of the comparative analysis of DNA, RNA and protein sequences is the detection of multiple character-state changes at the same position. For example, let us consider the two archaeal RPS5 sequences of *Methanobacterium* and *Thermoplasma* (Figure 19.3). Both sequences display an arginine (R) at sites 7 and 25, which might suggest that their most recent common ancestor already possessed an R at the same sites. However, site 7 is the most variable of the alignment in terms of the number of different amino acids displayed, whereas site 25 is completely conserved for the 12 taxa here studied. The probability that *Methanobacterium* and *Thermoplasma* share the same amino acid through common ancestry is thus higher for site 25 than for site 7 (on the other hand, site 25 is less informative than site 7). If an ancestral K is, for example, assumed at site 7 for Archaea (i.e. the state observed in *Aeropyrum*), multiple convergent replacements may have occurred on both lineages (such as K → R along the lineage leading to *Methanobacterium*, and K → T → R on the lineage to *Thermoplasma*, among other possibilities).

Unequal rates of evolution among sites imply that faster-evolving characters will concentrate a higher proportion of multiple changes. This saturation phenomenon will generate homoplasy, the misinterpretation of which can yield the artefactual grouping of unrelated sequences. Given the importance of detecting multiple character-state changes for improving phylogenetic accuracy, different strategies have been developed to deal with the problem of saturation.

Improving taxon and sequence sampling should provide more phylogenetic signal and allow for a better resolution of the history of character changes through better detection of site-specific multiple changes (Lecointre *et al.*, 1993; Philippe and Douzery, 1994). By adding more taxa, and thus more sequences, long branches will be broken, thus reducing tree-building artefacts (Poe, 2003). For example, had *Trichomonas* not been sampled, then site 1 would appear fully conserved, displaying only a proline (P) for Eukarya, Archaea and Bacteria (Figure 19.3). However, the presence of *Trichomonas* reveals that the history of this site is more complex, with at least one proline–alanine (A) exchange. This strategy has proven efficient for all tree-building methods, especially parsimony. However, increasing taxon sampling is not always possible, for instance, when clades are naturally depauperate, such as in the ginkgo for plants.

Better selecting sites and sequences should also reduce the detrimental effect of mis-interpreting multiple character-state changes. Because faster-evolving characters are more subject to saturation through time, it has been suggested that the focus should be on only the slowest-evolving ones, which should be enriched in phylogenetic signal pertaining to the deepest nodes of the phylogenetic tree. Accordingly, several studies have documented a reduction of the long-branch attraction phenomenon when removing (1) the fastest-evolving sites of multiple sequence alignments (Brinkmann and Philippe, 1999); (2) the fastest-evolving taxa for a given gene (Aguinaldo *et al.*, 1997); and (3) the fastest-evolving gene(s) and protein(s) (Philippe *et al.*, 2005). The major drawback of these approaches is their heavy reliance on the criteria and statistical tools used to conduct the sorting of sites, genes or taxa. Since these criteria may themselves be (implicitly or explicitly) conditioned on a predefined phylogeny, is it often difficult to rule out any self-reinforcing circularity in such methods.

Observing that the phylogenetic estimates may change depending on the sites, genes or taxa sampled actually suggests that character-selection procedures merely compensate for the weaknesses of the underlying model. Character sorting can thus be viewed as a tool to detect problems, like model mis-specifications, confirming the requirement of refined models of sequence evolution so as to better handle the complexity of the molecular evolutionary processes.

Bias versus variance

Formal models of sequence evolution are tools that facilitate phylogenetic estimation. Even if there are not perfect descriptions of the evolutionary processes that produced the observed sequence alignments, they need to be reasonable approximators capturing the most essential trends present in the data. Parameter-poor models are likely to produce biased estimates of phylogenies. For example, if a model assumes that replacements towards each of the 20 amino acids occur at the same frequency, then the complexity of the data will not be properly handled. In the 26-site zoom on the RPS5 protein of 12 eukaryotes (Figure 19.3), all 20 amino acids are represented except phenylalanine (F) and methionine (M). The most abundant are arginine (R) and valine (V), while the rarest are histidine (H), tryptophan (W) and asparagine (N), all three appearing just once. Adopting a 'horizontal' view on this alignment also shows that the *Aquifex* and *Bacillus* proteins are the most divergent in terms of average amino-acid composition. Obviously, the assumption of equal frequency of the 20 amino acids is here clearly violated.

By contrast, parameter-rich models are likely to produce less biased phylogenetic estimates, but at the expense of a higher variance (greater uncertainty). Estimating an increasing number of parameters from the same finite amount of data will result in an increased variance. Consequently, model selection should ensure a trade-off between bias and variance. This trade-off procedure for model selection has been popularized by the software ModelTest, which implements information-theoretic (e.g. Akaike information criterion [AIC]) and frequentist (e.g. likelihood ratio tests for nested models) criteria of choice (Posada and Crandall, 1998). For instance, the AIC will always allow a model to be selected provided that the likelihood gain is higher than the number of estimable extra parameters involved (Akaike, 1974). It should be noted that more complex parameter-rich models will be better implemented in a Bayesian framework given the inherent handling of uncertainty by this statistical approach.

Time-reversible models of character-state changes

Most current phylogenetic models of DNA, codon and protein evolution share at least six assumptions: (1) tree-like divergence of sequences (see 'Preamble: the vertical inheritance of genetic material'); (2) independence and identical distribution (i.i.d.) of alignment columns; (3) stationarity, i.e. the character-state frequencies remain constant along the tree; (4) homogeneity, i.e. the parameters of the substitution process (e.g. amino-acid frequencies) remain constant through time and at all positions of the biomolecule; (5) reversibility, i.e. equivalence between the process and its own time reversal; and (6) character-state changes are Markovian, or memory free, in the sense that the occurrence of the upcoming nucleotide or amino acid at a given site is not influenced by the previous history of the substitution process (Jayaswal *et al.*, 2005; Kelchner and Thomas, 2006).

Commonly used models in phylogenetics account for unbalanced character-state composition by assuming different frequencies for the 4 nucleotides, the 64 codons or the 20 amino acids. They also account for unbalanced character-state changes by incorporating matrices of exchangeability parameters that describe the rates at which each possible pair of nucleotides, codons or amino acids interchange. While it is not a big problem to estimate the 2×3 parameters of a reversible exchangeability matrix of nucleotides, there is more computational burden in estimating the 10×19 or even 32×63 parameters of amino-acid or codon exchangeability matrices. To attenuate this computational load, empirical exchangeability matrices have been built for proteins and triplet-based DNA sequences by estimating the relative exchangeability rates among amino acids and codons through the comparison of numerous nuclear proteins (e.g. the WAG matrix: Whelan and Goldman, 2001), mitochondrial proteins (e.g. the mtZoa matrix: Rota-Stabelli *et al.*, 2009), and protein-coding DNA alignments (e.g. the ECM matrix: Kosiol *et al.*, 2007).

These empirical general time-reversible (GTR) models better fit the data as compared to simpler models with equality of either character-state frequencies or character-state exchangeabilities. However, these GTR models alone do not handle variations in the vertical and horizontal dimensions of the sequence alignments, i.e. variations in the evolutionary processes, respectively, among sites and among lineages.

Site-heterogeneous models

The assumption of identical distribution of sites in sequence alignments is often violated. Sites often harbour contrasted nucleotide or amino-acid compositions, or display varying evolutionary rates, with sites under stronger purifying selective constraints being less variable than sites under relaxed constraints or positive selection (see 'Preamble: the vertical inheritance of genetic material'). To accommodate for among-site rate variation (ASRV) in character-state exchangeabilities, several site-heterogeneous models have been proposed.

First, partitioning the data, i.e. attributing to each different gene or protein its own predefined model of evolution, will lead to an improved fit, especially if the different molecular markers – or their subsets such as their first/second/third codon positions in protein-coding genes, stems/loops in rRNAs or extra-/trans-membrane regions in proteins – evolve under contrasted selective constraints. In particular, each of the partitions can be given distinct rates of evolution (Swofford *et al.*, 1996; Pupko *et al.*, 2002; Bevan *et al.*, 2007). Though very intuitive, character partitioning is hampered by the fact that different sites from the same partition may behave more differently than sites from different partitions. For example, third-codon positions from the same gene are probably as different from each other as third-codon positions from different genes.

Second, sequence-alignment sites may be separated into variable and invariable partitions rather than by gene or protein boundary (Adachi and Hasegawa, 1995). In this case, the evolutionary processes at variable sites will be i.i.d. Distinguishing a fraction of invariable sites acknowledges the fact that molecular markers with extreme ASRV possess fewer sites that are free to vary, and those sites that do vary may be more saturated owing to their higher evolutionary rate. However, variable sites may themselves represent a very heterogeneous population in which contrasted exchangeability rates should be accounted for. This is the reason why a discrete gamma (Γ) distribution has been introduced in site-heterogeneous models to better handle the heterogeneity of exchangeability rates among sites (Yang, 1996). Schematically, sites are assumed to be distributed among four categories of rate, representing nearly-constant, slowly-evolving, more variable and hypervariable sites. Incorporating gamma-distributed rates into phylogenetic inference has been shown to have a major impact on phylogenetic accuracy, noticeably by reducing the frequency of long-branch attractions (Sullivan and Swofford, 1997). Of note, GTR models of sequence evolution (see 'Time-reversible models of character-state changes') can be refined by simultaneously estimating a proportion of invariable sites combined with gamma-distributed rates for the remaining variable sites (Gu *et al.*, 1995). Even under complex GTR + Γ + invariable models, there remains the assumption that all sites of a given partition of characters evolve under the same model, which is not necessarily true.

Third, so-called mixture models have been developed to accommodate the fact that different sites from the same set of alignment columns may behave differently (Lartillot and Philippe, 2004; Pagel and Meade, 2004). In other words, mixture models will affiliate the sites into a number of categories, each category being characterized by its own set of parameters. Since the affiliation of each site is not known, the sitewise likelihood will be calculated by summing over all possible affiliations, each of them weighted by its prior probability. The number of independent components of the mixture may be specified a priori by the user (Pagel and Meade, 2004). One difficulty with this approach is the pre-specification of the number of processes of the mixture model to be used. As an alternative, it is possible to estimate the number of components – as well as their respective sets

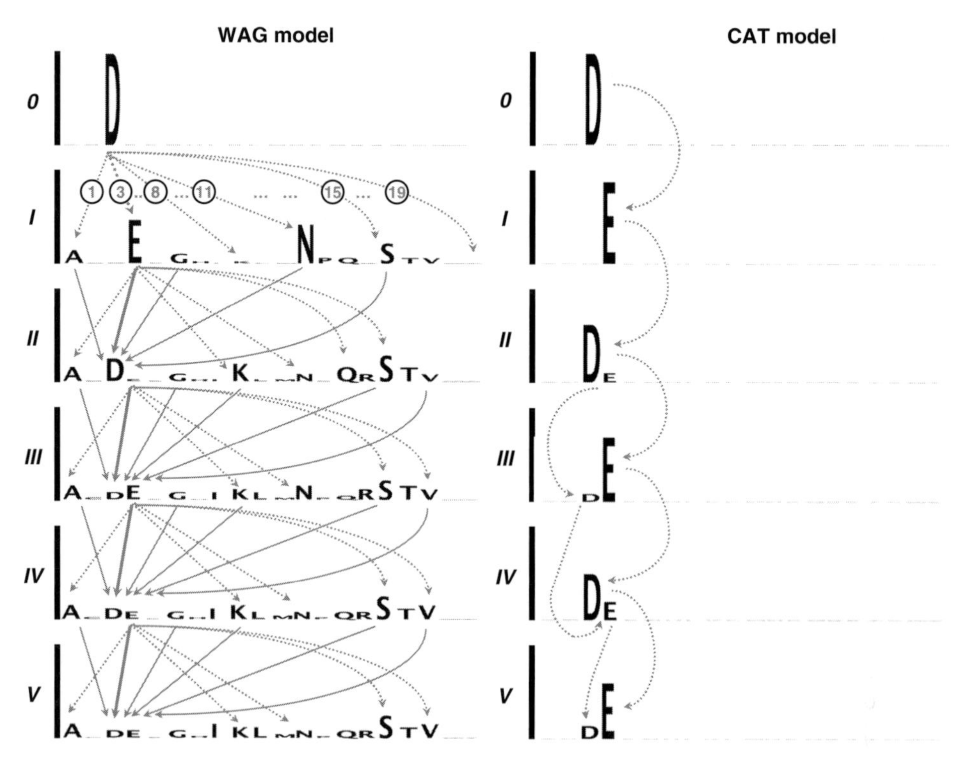

Figure 19.4. Comparison of the site-saturation handling by two models of protein evolution, WAG and CAT. The evolution of a protein site is illustrated for up to five amino-acid replacements (*I* to *V*). Amino acids are given in alphabetical order of their single-letter code. The height of the letters is proportional to the abundance of the corresponding amino acids at the sites considered. Arrows indicate exchangeabilities among amino acids, with focus on the D (aspartic acid)/E (glutamic acid) pair. (Modified from Lartillot and Philippe, 2010.)

of parameters – directly from the data. The CAT model implements this strategy through the use of a Dirichlet prior process, with each model process differing by its relative equilibrium frequencies of character states (Lartillot and Philippe, 2004). The CAT model has been shown to perform well for phylogenetic inference, especially in marked cases of long-branch attraction, by dynamically building a number of amino-acid profiles that individually fit populations of similarly evolving sites (Lartillot *et al.*, 2007; Philippe *et al.*, 2007).

Now, let us explain the relative performances of a site-homogeneous model such as WAG and a site-heterogeneous model such as CAT by considering the case of sites 4 or 20 of the RPS5 protein, both of which display only amino acids D and E (Figure 19.3). The amino-acid frequencies at these sites can be estimated under the WAG model after multiple (here *I* to *V*) replacements, when starting from aspartic acid (D) in each sequence (Lartillot and Philippe, 2010). After the first replacement event, D is replaced by any of the other 19 amino acids (Figure 19.4), proportionally to exchangeability rates. For example, D will be predominantly replaced by glutamic acid (E) as the D $\rightarrow$ E exchange is the fastest one when starting from a D, followed by asparagine (N) as the D → N exchange is the second fastest. Conversely, virtually no cysteine (C) will be generated as the D → C exchanges are among the least likely (occurring at a *c*.200 times lower rate than D → E according to the WAG exchangeability matrix: Whelan and Goldman, 2001). With the second replacement event, E may of course reverse to D (Figure 19.4: thick arrow), but it may also change into any of the 18 remaining amino acids (Figure 19.4: dotted arrows). Even if each of these 18 amino acids may itself be replaced by a D (Figure 19.4: thin arrows), the frequency of D will erode. Reiteration of this erosion after the third, fourth and fifth replacements will lead to a nearly flat distribution over the 20 amino acids, with the most frequent residues being A, I, S or V rather than the initial D and E (Figure 19.4). Interestingly, if the model is now GTR with estimation of amino-acid exchangeabilities instead of using WAG empirical values, very similar equilibrium frequencies will be reached after five replacements (Lartillot and Philippe, 2010). Therefore, it is clear that standard site-homogeneous models overestimate the per-site number of amino acids. By contrast, under the CAT model, the site evolves through near-exclusive exchanges between the D and E residues (Figure 19.4: dotted arrows). The difference in behaviour between site-homogeneous and site-heterogeneous models is striking, especially when the site is saturated due to multiple replacements through time.

The impact on phylogenetic estimates of using either the WAG or CAT models is here illustrated on the tree obtained by Boussau *et al.* (2008) with the concatenation of 56 nearly universal proteins (including RPS5) from the three domains of life. After enforcing the tree topology, branch lengths were estimated from the same amino-acid alignment under the WAG + Γ or CAT + Γ models with the software PhyloBayes (Lartillot *et al.*, 2009). The branch lengths of the phylogram inferred under the CAT + Γ model are on average two times greater than under the WAG + Γ model (Figure 19.5). Overestimation of the number of character states per site under the WAG model increases the size of the possible acceptable changes, which concomitantly decreases the probability of detecting convergences and reversions, whereas the site-heterogeneous CAT model accounts for site-specific contraction of the amino-acid alphabet (see 'Characters as witnesses of evolution'), and allows for better detection of homoplasy at saturated sites. This is reflected by the inference of longer branches.

Time-heterogeneous models

The homogeneity assumption of sequence-evolution models can be violated in a sitewise manner as previously seen, but also branchwise. This is, for instance, reflected in the 'horizontal' dimension of the alignments, with the occurrence of contrasted character-state frequencies among the different sequences. Time-heterogeneous models have therefore been developed whose character-state frequencies and exchangeabilities may vary along each branch of the tree. For example, a non-stationary stochastic DNA model associating a distinct G + C equilibrium frequency to each branch of the tree has been devised in the ML framework, so as to model the fact that the G + C content may diverge with time and among

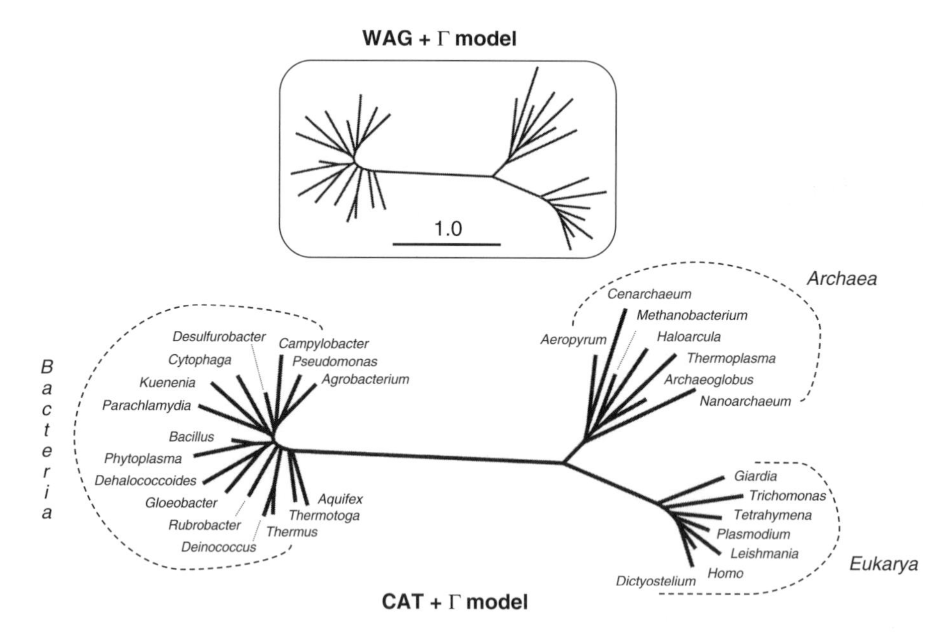

Figure 19.5. Evolutionary divergence of the three domains of life as estimated by two different models of amino-acid replacements. The branch lengths of the phylogram connecting Bacteria, Archaea and Eukarya have been estimated under the WAG (top) and CAT (bottom) models. Both are drawn at the same scale (the bar is proportional to 1 amino-acid replacement per site). Due to the ambiguity about the rooting of the Tree of Life, the phylogram is unrooted. The data (3336 aligned amino acids from 56 proteins) and the tree topology are taken from Boussau *et al.* (2008). Branch lengths have been computed with PhyloBayes (Lartillot *et al.*, 2009). A gamma distribution of the heterogeneity of amino-acid replacement rate among sites has been incorporated for both models.

clades (Galtier and Gouy, 1998). This time-heterogeneous model has been used on SSU and LSU rRNAs sequences, leading to the suggestion that the most recent common ancestor (MRCA) of Bacteria, Archaea and Eukarya did not have the high GC content in RNA stems that would have been expected from a hyperthermophilic ancestor (Galtier *et al.*, 1999; also see 'A case study: the shift from mesophilic to thermophilic ecological preferences among microorganisms'). Failure to account for branch heterogeneity in character-state composition may have the misleading effect of grouping together unrelated sequences that share similar character-state frequencies.

Different time-heterogeneous models have been developed. We have previously seen the importance of accounting for ASRV (cf. 'Site-heterogeneous models'), for example, through a gamma distribution. However, the evolutionary rate of a given site may change over time – a phenomenon called 'heterotachy' – because of potential changes of functional constraints (Fitch, 1971; Lopez *et al.*, 2002). This is illustrated by the RPS5 alignment (Figure 19.3): sites 2 and 3 are conserved in Eukarya plus Archaea while variable in Bacteria, site 10 is conserved in Archaea and in Bacteria while variable in Eukarya, and sites 16 and 17 are conserved in Eukarya and Bacteria while variable in Archaea.

To account for the fact that fast sites may become slowly evolving sites – and vice versa – an ML model that describes the rate of site-specific rate changes has been proposed (Tuffley and Steel, 1998; Galtier, 2001; Huelsenbeck, 2002). This time-heterogeneous model, in which sites can switch from rate categories, allowed the detection of a significant amount of site-specific rate variation in macromolecules.

Previous time-heterogeneous models may be refined by assuming that the changes in evolutionary processes are fully uncoupled from the speciation events: changes may occur at breakpoints anywhere along branches rather than being stuck at nodes of the tree. Because of their complexity, such models have been implemented in the Bayesian framework (see 'Bias versus variance'), allowing them to account for changes in molecular evolutionary rates (Huelsenbeck *et al.*, 2000) or drift of the character-state composition of sequences (Blanquart and Lartillot, 2006). Modelling punctual changes along phylogenetic trees has the advantage of allowing the detection of episodic shifts in molecular processes.

Further model refinements

Because the processes of biomolecule evolution are likely to simultaneously vary among positions and among tree branches, models have been built that account for site and time heterogeneities. The CAT–BP model (Blanquart and Lartillot, 2008) combines two non-homogeneous models: the sitewise CAT model (Lartillot and Philippe, 2004) with the breakpoint (BP) system of timewise character-state changes (Blanquart and Lartillot, 2006). Interestingly, the joint modelling of sitewise and timewise heterogeneities results in a synergistic improvement of phylogenetic inference and ancestral character-state estimation. On the one hand, when phylogenetic trees are reconstructed from a mitochondrial arthropod data set with pervasive evolutionary rate variations, only the CAT–BP recovers a reasonable topology as compared to the less complex GTR, BP or CAT models (Blanquart and Lartillot, 2008). On the other hand, the scenario according to which independent shifts of optimal-growth temperature occurred during microbial evolution (see e.g. the theoretical example of Figure 19.2B) has recently gained some additional support. Indeed, when the ancestral rRNA and protein sequences of the MRCA of extant life forms were reconstructed under site- and time-heterogeneous models, parallel adaptations to high temperatures along the bacterial and archaeal lineages were shown (Boussau *et al.*, 2008). New models of sequence evolution will certainly be devised in the near future, but all of them should hopefully incorporate enough parameters to keep the flexibility required for accounting for complex molecular-evolution processes while keeping in mind the biological questioning under focus.

From gene trees to species trees

In common phylogenetic practice, when reconstructing the phylogeny of a given gene, one hopes that it will perfectly match the phylogeny of the corresponding species (Figure 19.6A). For example, two speciation events generate species X, Y and Z. If the focal-gene copies

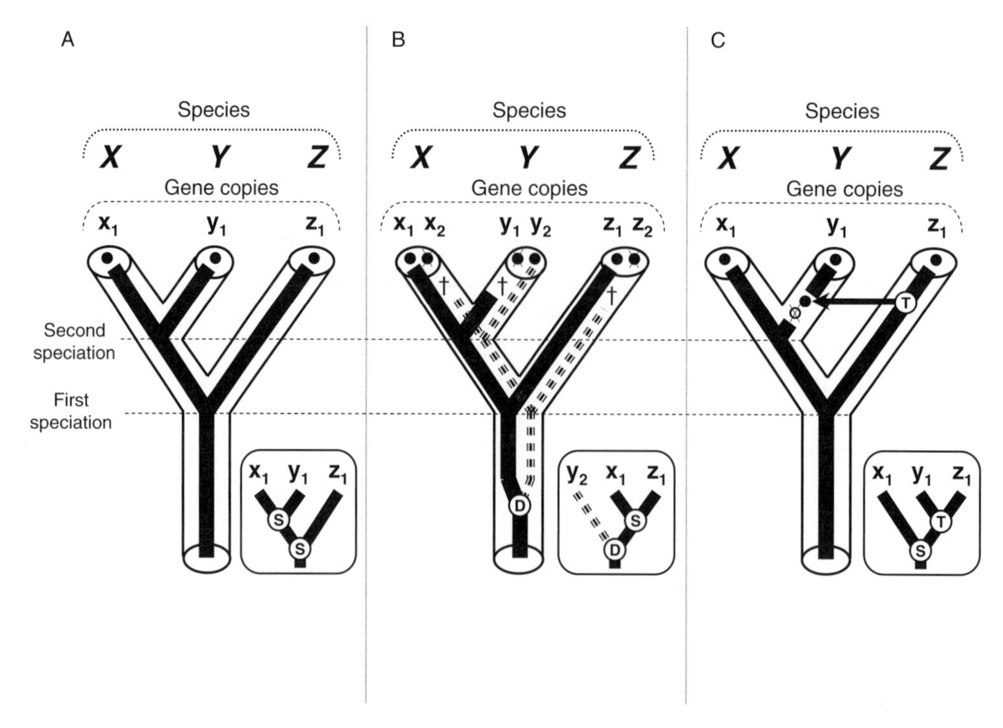

Figure 19.6. Gene trees do not necessarily match the species tree. A tubular species tree is represented with two speciation events: the first isolates species Z and the second separates species X from Y. Three different gene histories occurring in this species tree are illustrated. In the insets, the corresponding gene trees are provided, with circled letters representing gene-lineage splits that may result from speciation (S), duplication (D) or horizontal transfer (T). Three scenarios are depicted. **A**. Gene tree matches species tree. **B**. Hidden paralogy introduces incongruence between gene trees and the species tree. The two genes resulting from the duplication event are represented in black and dashed grey. Some paralogous copies (x_2, y_1 and z_2) have disappeared during the gene evolution. **C**. Horizontal gene transfer introduces incongruence between the gene tree and the species tree. Z is here the donor, and its gene copy (black circle) replaces the one (white circle) of the receptor Y.

split in a near-concomitant way as compared to speciations, the gene tree of x_1, y_1 and z_1 will perfectly match the species tree of X, Y and Z. In addition, the age of the split between x_1 and y_1 will correspond to the speciation event that gave rise to species X and Y, and the same will hold for species Z. However, several phenomena may introduce discrepancies between gene trees and species trees (Maddison, 1997).

Hidden paralogy

Genes can undergo duplications, i.e. molecular events by which two copies of the same gene are generated in the same genome. Homologous copies that arise through a duplication are qualified as paralogous, whereas copies that arise through a speciation are qualified as orthologous. At the molecular level, this means that homology of sequences results from either orthology or paralogy. After a duplication, the two paralogous copies may stand in the same chromosomal vicinity in the case of tandem repeats, but they can also be separated on different chromosomes by translocations. In both cases, substitutions will independently accumulate in the two copies, resulting in the divergence of their primary DNA sequences and of the corresponding proteins. Then, one copy may retain the original function, whereas the second copy will possibly be involved in the acquisition of a new function ('neofunctionalization'). Alternatively, gene duplicates can be partitioned among tissues ('subfunctionalization'). However, loss of gene copies may also occur because of functional redundancy.

Now let us consider a duplication event producing two gene families 1 and 2, each containing three copies deriving from two speciation events (Figure 19.6B). In family 1, the phylogeny of gene copies x_1, y_1 and z_1 therefore conforms to the species tree. The same holds for copies x_2, y_2 and z_2 in family 2. However, independent gene-copy losses can occur in lineages X, Y and Z, leading to a situation in which copies x_2, y_1 and z_2 cannot be sampled by the biologist. Consequently, the gene tree is reconstructed from the paralogous copies x_1, y_2 and z_1, which unfortunately does not reflect the species tree, as copy y_2 is the earliest emerging among present-day sequences. When the paralogy is undetected, the gene tree may provide misleading results regarding the species tree. Of course, this example is a caricature as we can meet an alternative situation in which all copies of family 2 may become extinct, and the gene phylogeny reconstructed from copies of family 1 would accurately represent the species history, provided that efficient tree-building approaches are used (see 'Improvements to the probability approach to phylogenetic inference').

Horizontal gene transfer

Genes can be horizontally transferred, i.e. acquired by one organism from another organism belonging to a different lineage (see Chapter 20). It is opposed to the 'vertical' transfer, i.e. genetic inheritance through reproduction (see 'Preamble: the vertical inheritance of genetic material'). For example, if a horizontal gene transfer (HGT) occurs from Z to Y, the gene copy of the donor (z_1) will replace the copy (y_1) in the genome of the receptor Y (Figure 19.6C). After the HGT, the copy transferred will diverge relative to its progenitor, and the more ancient the HGT, the larger the differences between sequences. Now, if one samples by PCR a gene copy at the locus considered, the species Y will be characterized by a copy (y_1) more similar to the one of Z (z_1) than to the one of X (x_1). Obviously, the gene phylogeny would here suggest that Y and Z are the closest relatives, which contradicts the species tree.

Multi-gene approaches and integrative models

How can we proceed with situations in which gene trees do not necessarily match with the species tree? First, multiple loci can be sampled. A powerful approach is to compare individual gene trees, reconstructed from sequences sampled at independent genetically unlinked loci. If one gene tree is incongruent relative to most of the other gene trees, we can assume that either gene duplication/extinction or HGT did occur during its history. (Other phenomena not described here can contribute to produce incongruences between gene and species trees. For instance, ancestral sequence polymorphism at the focal locus combined with incomplete lineage sorting will disrupt the concomitance of gene splits and species splits: Maddison, 1997.) However, how can we be sure about incongruence? In the ML framework, phylogenetic trees with two different topologies are characterized by different likelihoods. The difference in log likelihoods can be statistically tested in order to evaluate whether or not it is significantly different from its expected variance (Kishino and Hasegawa, 1989; Shimodaira, 2002). If the two trees do not statistically differ, both of them are similarly good explanations of the data under the model considered, indicating that the apparent incongruence just reflects a lack of primary signal in the data. However, if one tree is statistically better than the other, then incongruence is proven. In this case, the molecular marker or the taxa responsible for the discordance in gene trees may be removed, especially for approaches combining multiple independent sources of characters into supermatrices (de Queiroz and Gatesy, 2007) or combining multiple individual topologies into supertrees (Bininda-Emonds, 2004). Yet, this procedure may suffer from the fact that the most frequently observed gene-tree topology does not always correspond to the correct species-tree topology (Degnan and Rosenberg, 2006).

Second, models integrating simultaneously multiple gene histories can be used. The idea of jointly estimating gene and species trees stems from the fact that genetically unlinked loci are actually correlated by a common species history. For example, a Bayesian approach has been proposed in which gene trees are reconciled with the species tree through a process of gene duplications and losses (Arvestad *et al.*, 2003). This allows the number of gene births and deaths to be estimated and the orthologous genes to be identified. Another approach has been developed in which the species tree is estimated through distributions of gene trees by employing standard-substitution models (see ‘Improvements to the probability approach to phylogenetic inference’) combined with the coalescent theory (Liu and Pearl, 2007). With the growing availability of genomic data, and with the increasing use of multiple loci in molecular-phylogenetics studies, new models of joint estimation of gene and species trees will certainly be developed, while requiring protocols to compare their relative performances.

In conclusion, we have seen that the understanding of the evolutionary patterns of biomolecules and species since the origin of life is based on the inference of accurate molecular phylogenies. These investigations routinely rely upon probabilistic approaches that benefit from the development of more realistic, flexible and integrative models of character-state modifications, i.e. exchanges among nucleotides, codons or amino acids, as well as models of multigene inference of species trees.

Acknowledgements

This work greatly benefited from the constructive comments of Nicolas Lartillot. It has been supported by the Agence Nationale de la Recherche Génomique Animale

(ANR-08-GENM-036–01) and Domaines Emergents (ANR-08-EMER-011). This publication is the contribution 2009–126 of the Institut des Sciences de l'Evolution de Montpellier (UMR 5554– CNRS).

References

Adachi, J. and Hasegawa, M. (1995). Improved dating of the human/chimpanzee separation in the mitochondrial DNA tree: heterogeneity among amino acids. *Journal of Molecular Evolution*, **40**, 622–8.

Aguinaldo, A. M. A., Turbeville, J. M., Linford, L. S., Rivera, M. C., Garey, J. R., Raff, R. A. and Lake, J. A. (1997). Evidence for a clade of nematodes, arthropods and other moulting animals. *Nature*, **387**, 489–93.

Akaike, H. (1974). A new look at the statistical model identification. *IEEE Transactions on Automatic Control,* AC–19, 716–23.

Arvestad, L., Berglund, A. C., Lagergren, J. and Sennblad, B. (2003). Bayesian gene/species tree reconciliation and orthology analysis using MCMC. *Bioinformatics*, **19**, i7–i15.

Bevan, R. B., Bryant, D. and Lang, B. F. (2007). Accounting for gene rate heterogeneity in phylogenetic inference. *Systematic Biology*, **56**, 194–205.

Bininda-Emonds, O. R. P. (2004). The evolution of supertrees. *Trends in Ecology and Evolution*, **19**, 315–22.

Blanquart, S. and Lartillot, N. (2006). A Bayesian compound stochastic process for modeling nonstationary and nonhomogeneous sequence evolution. *Molecular Biology and Evolution*, **23**, 2058–71.

Blanquart, S. and Lartillot, N. (2008). A site- and time-heterogeneous model of amino acid replacement. *Molecular Biology and Evolution*, **25**, 842–58.

Boussau, B., Blanquart, S., Necsulea, A., Lartillot, N. and Gouy, M. (2008). Parallel adaptations to high temperatures in the Archean Eon. *Nature*, **456**, 942–5.

Brinkmann, H. and Philippe, H. (1999). Archaea sister group of Bacteria? Indications from tree reconstruction artifacts in ancient phylogenies. *Molecular Biology and Evolution*, **16**, 817–25.

Bromham, L. and Penny, D. (2003). The modern molecular clock. *Nature Reviews Genetics*, **4**, 216–24.

Bryant, D., Galtier, N. and Poursat, M.-A. (2005). Likelihood calculation in molecular phylogenetics. In *Mathematics of Evolution and Phylogeny*, ed. O. Gascuel. Oxford: Oxford University Press, pp. 33–62.

Cavalier-Smith, T. (1987). The origin of Fungi and pseudofungi. In *Evolutionary Biology of the Fungi*, eds. A. D. M. Rayner, C. M. Brasier and D. Moore. Cambridge: Cambridge University Press, pp. 339–53.

Dawkins, R. (1976). *The Selfish Gene*. New York: Oxford University Press.

de Queiroz, A. and Gatesy, J. (2007). The supermatrix approach to systematics. *Trends in Ecology and Evolution*, **22**, 34–41.

Degnan, J. H. and Rosenberg, N. A. (2006). Discordance of species trees with their most likely gene trees. *PLoS Genetics*, **2**, e68.

Douady, C. J., Delsuc, F., Boucher, Y., Doolittle, W. F. and Douzery, E. J. P. (2003). Comparison of Bayesian and maximum likelihood bootstrap measures of phylogenetic reliability. *Molecular Biology and Evolution*, **20**, 248–54.

Felsenstein, J. (1981). Evolutionary trees from DNA sequences: a maximum likelihood approach. *Journal of Molecular Evolution*, **17**, 368–76.

Felsenstein, J. (1985). Confidence limits on phylogenies: an approach using the bootstrap. *Evolution*, **39**, 783–91.
Felsenstein, J. (1978). Cases in which parsimony or compatibility methods will be positively misleading. *Systematic Zoology*, **27**, 401–10.
Fitch, W. M. (1971). Rate of change of concomitantly variable codons. *Journal of Molecular Evolution*, **1**, 84–96.
Fitch, W. M. and Margoliash, E. (1967). Construction of phylogenetic trees. *Science*, **155**, 279–84.
Galtier, N. (2001). Maximum-likelihood phylogenetic analysis under a covarion-like model. *Molecular Biology and Evolution*, **18**, 866–73.
Galtier, N. and Gouy M. (1998). Inferring pattern and process: maximum-likelihood implementation of a nonhomogeneous model of DNA sequence evolution for phylogenetic analysis. *Molecular Biology and Evolution*, **15**, 871–9.
Galtier, N., Tourasse, N. and Gouy, M. (1999). A nonhyperthermophilic common ancestor to extant life forms. *Science*, **283**, 220–1.
Gascuel, O., Bryant, D. and Denis, F. (2001). Strengths and limitations of the minimum evolution principle. *Systematic Biology*, **50**, 621–7.
Gu, X., Fu, Y. X. and Li, W.-H. (1995). Maximum likelihood estimation of the heterogeneity of substitution rate among nucleotide sites. *Molecular Biology and Evolution*, **12**, 546–57.
Hasegawa, M., Kishino, H. and Yano, T. (1989). Estimation of branching dates among primates by molecular clocks of nuclear DNA which slowed down in Hominoidea. *Journal of Human Evolution*, **18**, 461–76.
Holder, M. and Lewis, P. O. (2003). Phylogeny estimation: traditional and Bayesian approaches. *Nature Reviews Genetics*, **4**, 275–84.
Huelsenbeck, J. P. (2002). Testing a covariotide model of DNA substitution. *Molecular Biology and Evolution*, **19**, 698–707.
Huelsenbeck, J. P., Larget, B. and Swofford, D. (2000). A compound Poisson process for relaxing the molecular clock. *Genetics*, **154**, 1879–92.
Jayaswal, V., Jermiin, L. S. and Robinson J. (2005). Estimation of phylogeny using a general Markov model. *Evolutionary Bioinformatics*, **1**, 62–80.
Keeling, P. J. and Doolittle, W. F. (1996). Alpha-tubulin from early-diverging eukaryotic lineages and the evolution of the tubulin family. *Molecular Biology and Evolution*, **13**, 1297–305.
Kelchner, S. A. and Thomas, M. A. (2006). Model use in phylogenetics: nine key questions. *Trends in Ecology and Evolution*, **22**, 87–94.
Kirkwood, T. B. L. (1977). Evolution of ageing. *Nature*, **270**, 301–4.
Kishino, H. and Hasegawa, M. (1989). Evaluation of the maximum likelihood estimate of the evolutionary tree topologies from DNA sequence data, and the branching order in Hominoidea. *Journal of Molecular Evolution*, **29**, 170–9.
Kosiol, C., Holmes, I. and Goldman, N. (2007). An empirical codon model for protein sequence evolution. *Molecular Biology and Evolution*, **24**, 1464–79.
Lartillot, N. and Philippe, H. (2004). A Bayesian mixture model for across-site heterogeneities in the amino-acid replacement process. *Molecular Biology and Evolution*, **21**, 1095–109.
Lartillot, N. and Philippe, H. (2010). Improvement of molecular phylogenetic inference and the phylogeny of Bilateria. In *Animal Evolution, Genomes, Fossils, and Trees*, eds. M. J. Telford and D. T. J. Littlewood. Oxford: Oxford University Press.

Lartillot, N., Brinkmann, H. and Philippe, H. (2007). Suppression of long-branch attraction artefacts in the animal phylogeny using a site-heterogeneous model. *BMC Evolutionary Biology*, **7**, suppl. 1, S4.

Lartillot, N., Lepage, T. and Blanquart, S. (2009). PhyloBayes 3. A Bayesian software package for phylogenetic reconstruction and molecular dating. *Bioinformatics*, **25**, 2286–8.

Lecointre, G., Philippe, H., Lê, H. L. V. and Le Guyader, H. (1993). Species sampling has a major impact on phylogenetic inference. *Molecular Phylogenetics and Evolution*, **2**, 205–24.

Liu, L. and Pearl, D. K. (2007). Species trees from gene trees: reconstructing Bayesian posterior distributions of a species phylogeny using estimated gene tree distributions. *Systematic Biology*, **56**, 504–14.

Lopez, P., Casane, D. and Philippe, H. (2002). Heterotachy, an important process of protein evolution. *Molecular Biology and Evolution*, **19**, 1–7.

Löytynoja, A. and Goldman, N. (2008). Phylogeny-aware gap placement prevents errors in sequence alignment and evolutionary analysis. *Science*, **320**, 1632–5.

Maddison, W. P. (1997). Gene trees in species trees. *Systematic Biology*, **46**, 523–36.

Pagel, M. and Meade, A. (2004). A phylogenetic mixture model for detecting pattern-heterogeneity in gene sequence or character-state data. *Systematic Biology*, **53**, 571–81.

Philippe, H. and Douzery, E. (1994). The pitfalls of molecular phylogeny based on four species as illustrated by the Cetacea/Artiodactyla relationships. *Journal of Mammalian Evolution*, **2**, 133–52.

Philippe, H., Lartillot, N. and Brinkmann, H. (2005). Multigene analyses of bilaterian animals corroborate the monophyly of Ecdysozoa, Lophotrochozoa, and Protostomia. *Molecular Biology and Evolution*, **22**, 1246–53.

Philippe, H., Brinkmann, H., Martinez, P., Riutort, M. and Baguña, J. (2007). Acoel flatworms are not platyhelminthes: evidence from phylogenomics. *PLoS ONE*, **2**, e717.

Poe, S. (2003). Evaluation of the strategy of long-branch subdivision to improve the accuracy of phylogenetic methods. *Systematic Biology*, **52**, 423–8.

Posada, D. and Crandall, K. A. (1998). Modeltest: testing the model of DNA substitution. *Bioinformatics*, **14**, 817–18.

Pupko, T., Huchon, D., Cao, Y., Okada, N. and Hasegawa, M. (2002). Combining multiple data sets in a likelihood analysis: which models are the best? *Molecular Biology and Evolution*, **19**, 2294–307.

Rokas, A. and Holland, P. W. (2000). Rare genomic changes as a tool for phylogenetics. *Trends in Ecology and Evolution*, **15**, 454–9.

Rota-Stabelli, O., Yang, Z. and Telford, M. J. (2009). MtZoa: A general mitochondrial amino acid substitutions model for animal evolutionary studies. *Molecular Phylogenetics and Evolution*, **52**, 268–72.

Shimodaira, H. (2002). An approximately unbiased test for phylogenetic tree selection. *Systematic Biology*, **51**, 492–508.

Stechmann, A. and Cavalier-Smith, T. (2002). Rooting the eukaryote tree by using a derived gene fusion. *Science*, **297**, 89–91.

Sullivan, J. and Swofford, D. L. (1997). Are guinea pigs rodents? The importance of adequate models in molecular phylogenetics. *Journal of Mammalian Evolution*, **4**, 77–86.

Swofford, D. L., Olsen, G. J., Waddell, P. J. and Hillis, D. M. (1996). Phylogenetic inference. In *Molecular Systematics*, eds. D. M. Hillis, C. Moritz and B. K. Mable. Massachusetts: Sinauer, Sunderland, pp. 407–514.

Tuffley, C. and Steel, M. A. (1998). Modelling the covarion hypothesis of nucleotide substitution. *Mathematical Biosciences*, **147**, 63–91.

Vossbrinck, C. R., Maddox, J. V., Friedman, S., Debrunner-Vossbrinck, B. A. and Woese, C. R. (1987). Ribosomal RNA sequence suggests microsporidia are extremely ancient Eukaryotes. *Nature*, **326**, 411–14.

Wainright, P. O., Hinkle, G., Sogin, M. L. and Stickel, S. K. (1993). Monophyletic origins of the Metazoa: an evolutionary link with Fungi. *Science*, **260**, 340–2.

Whelan, S. and Goldman, N. (2001). A general empirical model of protein evolution derived from multiple protein families using a maximum-likelihood approach. *Molecular Biology and Evolution*, **18**, 691–9.

Whelan, S., Lio, P. and Goldman N. (2001). Molecular phylogenetics: state-of-the-art methods for looking into the past. *Trends in Genetics*, **17**, 262–72.

Yang, Z. (1996). Among-site rate variation and its impact on phylogenetic analyses. *Trends in Ecology and Evolution*, **11**, 367–72.

20

Horizontal gene transfer: mechanisms and evolutionary consequences

David Moreira

Introduction

Human common experience tells us that the individuals of a particular species reproduce among themselves to produce a progeny which tends to resemble their parents. This inheritance of characters from one generation to the following one is known as 'vertical inheritance'. Although the nature of the physical support of the genetic information transmitted through generations remained mysterious until 1944 when DNA was shown to constitute such support, Gregor Mendel had already described the laws that control this kind of inheritance in plants in the nineteenth century. The vertical inheritance of favourable modifications is one of the pillars of the Darwinian theory of evolution: natural selection can be effective only if the advantageous characters selected can be transmitted to the progeny. This central role of vertical inheritance in evolution was adopted later on in the twentieth century by the neo-Darwinian evolutionists. For example, they proposed the 'biological species concept', which states that a species is defined by the capacity of its members to reproduce among themselves, namely, a species concept that, instead of being based on morphological characters as in the classical definitions, was based on the capacity of vertical inheritance. In addition, the vertical transmission of the genetic information with changes due to selection defines evolutionary lineages of organisms that are distinct from the rest of the lineages (for example our own *Homo sapiens* lineage). As Darwin stated in his *The Origin of Species*, the evolutionary relationships between those lineages are best represented by a phylogenetic tree. Therefore, in the second half of the twentieth century, the idea that evolution was completely governed by vertical inheritance was the dogma in the biological scientific community.

The discovery of other forms of inheritance was linked to the discovery of the nature of the genetic material itself. In 1944, Avery, MacLeod and McCarty carried out their famous experiment that demonstrated that the genetic information of cells is contained in DNA molecules (Avery *et al.*, 1944). These scientists worked with two strains of the bacterium *Streptococcus pneumoniae*, a virulent one and a non-pathogenic one, and extracted the DNA, the proteins and the lipids from each of them. When the DNA of the pathogenic strain was added to a culture of the non-pathogenic one, some of the cells became pathogenic; a phenomenon not observed when adding the proteins or the lipids of the pathogenic strain.

This proved that the capacity to be a pathogen was encoded in the DNA and that, once the cells had acquired it, it was transmitted to the progeny (i.e. vertical inheritance). The DNA was the mysterious genetic material whose existence was inferred by Mendel one century before. But the historical experiment of Avery and co-workers revealed an additional surprise: organisms of different types (such as their pathogenic and non-pathogenic Bacteria) can exchange genetic information. Nevertheless, this aspect of their research remained secondary behind the spectacular explosion experienced by molecular biology after the discovery of DNA. It was necessary to wait a few decades until a new revolution took place, the advent of the molecular phylogeny era in the 1970s, to realize that the exchange of genetic information between different species, known as 'horizontal gene transfer', could have profound implications. Before discussing those implications, I will give a short overview of the mechanisms that allow the exchange of genetic material between different species.

The mechanisms of horizontal gene transfer

Horizontal gene transfer (HGT) has been studied especially in prokaryotes (Bacteria and Archaea), leading to the discovery of several mechanisms that allow the exchange of DNA between different species (Lawrence, 1999). The simplest one, known as 'transformation', is the acquisition of naked DNA by a recipient cell. As explained above, this is the phenomenon that Avery and co-workers discovered at the same time that they elucidated the chemical identity of the genetic material. Although naked DNA is rapidly degraded in the environment, small amounts can be found, for example, immediately after the death of cells, which lyse and release DNA. A second way to exchange DNA between species relies on the activity of viruses. Many viruses spend part of their infection cycle in an inactive state with their genomes integrated in the chromosomes of their hosts. Owing to certain stimuli, they can activate and excise their genome from the chromosome to begin a lytic phase of viral multiplication that ends up with the death and lysis of the host cell. During the process of excision, the virus can incorporate into its own genetic material a small fragment of the host's chromosome that can contain a few genes, which will be inherited by the next generation of viruses. Once these new viruses invade new hosts and integrate into their genomes, they will also integrate the genes from the previous host carried in their genomes. This process, known as 'transduction', allows the transmission of genes between different hosts infected by the same type of virus. The third major mechanism of HGT is 'conjugation'. It involves the activity of a particular family of genetic elements: the conjugative plasmids. Plasmids are extrachromosomal DNA molecules, most often circular, which contain genes for their own maintenance and, frequently, also genes encoding functions useful for the cell, such as resistance to antibiotics. Conjugative plasmids are a specific type of plasmid that contain genes responsible for the functions involved in conjugation, in particular, the synthesis of proteins that are used to construct a hollow tube between two adjacent cells. This tube allows the transmission of the plasmid between the cells. As in the case of viruses mentioned above, plasmids can also occasionally transfer a portion of the host's genome. Conjugation was initially thought to occur only between prokaryotic cells of the same species, but it was soon demonstrated that it is possible not only between different prokaryotic species but even between Prokaryotes and Eukaryotes (Heinemann and Sprague, 1989).

The organism's view of HGT: a problem or a solution?

Evolution mediated by natural selection leads to the adaptation of organisms to their environment through the optimization of their genomes, both their gene content and sequence. Therefore, it can be imagined that an abrupt change such as the sudden acquisition of exogenous genetic material by HGT has to be most often deleterious. Moreover, as mentioned above, HGT frequently occurs by the activity of genetic elements, viruses or plasmids, which have evolved to perpetuate themselves – in the case of viruses, very often by killing their hosts. Taking these considerations into account, it would appear that HGT is negative for the survival of cells since it can lead to the integration of undesirable genetic material into their genomes. In fact, prokaryotic cells have developed mechanisms to protect themselves against exogenous DNA. For example, restriction endonucleases are enzymes that recognize particular sequences and digest DNA using them as targets. These sequences are absent or modified (e.g. methylated) in the cell genome, so that only a foreign DNA will be digested (Cohan, 2002). Although poorly known, many prokaryotes also possess a defence system against viruses, based on the recognition of particular short sequences present in the viral genomes as well, known as CRISPR (Sorek *et al.*, 2008). Nevertheless, the most important mechanisms limiting the possibilities for foreign DNA to integrate within a new genome is 'mismatch recognition' (Stambuk and Radman, 1998). To integrate within another DNA molecule, a DNA fragment has to recombine with it. Recombination is a complex phenomenon involving several enzymes and it works on the basis of sequence similarity. In normal conditions, only DNA molecules that share a very high sequence similarity can recombine. This is so because the first step in the recombination process between two DNA molecules is the formation of a hybrid double-stranded DNA formed by one DNA strand from one of the molecules and the complementary strand from the second DNA molecule. Subsequently, a complex enzymatic machinery will 'cut and paste' the DNA strands to end up with a product that is a chimeric DNA molecule containing the fragment of exogenous DNA. The crucial step in this whole process is the formation of the hybrid DNA: if the two DNA molecules do not share a perfect sequence identity, mismatches are recognized by a protein called MutS, which can block the rest of the enzymatic machinery and prevent recombination. Only individuals from the same species, or from extremely closely related species or strains, have DNA sequences sufficiently similar to bypass the mismatch-recognition process and recombine.

By avoiding recombination with DNA from distant species, mismatch recognition constitutes the best protection for the integrity of the genome and a powerful barrier against HGT. However, a mechanism that blocks mismatch recognition has been found in many prokaryotic species. It is called the 'SOS response' and it is activated under stress conditions and/or genome damage (Humayun, 1998). The suppression of mismatch recognition opens up the possibility for the cell genome to recombine with exogenous DNA, even if it has a low sequence similarity (Matic *et al.*, 2000). To some extent, this mechanism can be considered as a desperate way for the cell to mutate and/or to acquire genes that can help it to survive under a particularly harsh environmental condition. In contrast with the mechanisms against HGT cited before, the SOS response can be viewed, at least in part, as a process selected to favour HGT under particular conditions.

The consequences of HGT: increase of evolutionary rate

Most often, characters evolve in organisms through the slow accumulation of small changes by natural selection generation after generation. In sharp contrast, HGT allows the sudden appearance of new characters. Sometimes this can allow a species to colonize a completely new niche. For example, the transfer of a plasmid encoding a nitrate reductase in the aerobic bacterium *Thermus thermophilus* has been shown to allow this species to colonize anoxic habitats, gaining energy by respiration using nitrate as its terminal electron acceptor instead of oxygen (Ramirez-Arcos *et al.*, 1998). This dramatic adaptation only takes a few minutes. This is also the case for the resistance to antibiotics encoded by many plasmids. However, the acceleration of evolutionary rate is not only linked to this kind of acquisition of novel functions. At the level of the gene sequences, once a gene has been transferred into the genome of a new species, it has to adapt to the new genetic environment. For example, each species has a particular richness of each of the four nucleotides (A, C, G, T) in its genome and also a particular bias in the use of the codons encoding the different amino acids. When a gene is transferred to a new genome, especially if it comes from a distant species that probably has very different biases in its nucleotide and codon usage, it will progressively adopt the characteristics of the recipient species. This has been called 'gene amelioration' (Lawrence and Ochman, 1997). During this process, the transferred genes will probably change more rapidly than the average of the rest of the genome, namely, they will evolve faster. As a consequence, genes that are frequently transferred (e.g. those involved in antibiotic resistance) tend to show an increased evolutionary rate, although a systematic survey of this phenomenon for all genes acquired by HGT in the genome of a given species is still lacking.

The consequences of HGT: erosion of the species identity

Since the characteristics of a species are encoded in its genome, the integrity of the genome reflects the integrity of the species. If the whole genome of a species is inherited vertically, the evolutionary history of each gene of that genome should be the same as the history of the species itself, namely, all the genes will produce identical phylogenetic trees if they are analysed by molecular-phylogeny techniques. However, if a fraction of genes in the genome has been acquired from other species by HGT, those genes will have an evolutionary history which is different from the rest of the genome. If HGT levels are high, the genome will become a patchwork of genes from different origins, and the species' identity will be lost. In other words, the gene phylogenies will not reflect the evolutionary history of the species (the 'organismal history'). At this point, an important question would be to know whether all genes have the same probability of being transferred between species or if there is a particular class of genes reluctant to undergo HGT, which would be representative of the species' identity and would serve as markers to infer their evolutionary history. Several studies have tried to identify such HGT-free genes. One of the first was based on the 'complexity hypothesis', which states that genes encoding RNAs or proteins that take part in the construction of complex multi-protein structures have to be the least affected by HGT (Jain *et al.*, 1999). The reason for that would be that each one of those RNAs and proteins has to interact with many others in a very precise way. It can be expected that those interactions have been optimized within each species, so that incorporating proteins from other species through HGT events would be deleterious for the recipient species because the acquired proteins would not be able to interact with the others as efficiently as the recipient's own proteins do. The best example for this complexity hypothesis is the ribosome, one of the most complex macromolecular structures found in cells, formed by the interaction of several ribosomal RNA (rRNA) molecules and dozens of different ribosomal proteins. On the other hand, genes encoding proteins involved in a small number of interactions would be more easily transferred. This would be the case for many genes encoding metabolic enzymes, which only interact with a particular metabolite. Supporting this idea, many cases of HGT involving genes encoding metabolic enzymes have been reported, even between very distant species (Jain *et al.*, 1999). In contrast, although not completely impossible, the HGT of genes encoding ribosomal RNAs and proteins appears to be much rarer (Yap *et al.*, 1999; Brochier *et al.*, 2000). Those genes, together with some others involved in essential biological mechanisms such as the synthesis of RNAs (transcription), constitute a set of genes that many authors have called the 'core', which can be used as a data set to infer the evolution of species by phylogenetic reconstruction. However, this core is composed of a relatively small number of genes, so that it is inevitable to ask whether it is actually representative of the evolution of the species, or just a very particular set of genes that do not follow the evolutionary pathways ruling the rest of the genome (Gogarten *et al.*, 2002; Dagan and Martin, 2006).

How much HGT is there in prokaryotes?

It seems logical to think that the degree of erosion of the species' identity is proportional to the frequency of HGT affecting the species. The question of how many of the genes present in a genome have been acquired by HGT is at the centre of a very hot and active debate in the community of evolutionary biologists interested in the evolution of microorganisms. Nevertheless, very few quantitative analyses have been carried out and, moreover, complex methodological problems have made it difficult to get a clear picture, as we will see below.

The accumulation of complete genome sequences has been crucial in the attempts to evaluate the impact of HGT on the evolution of prokaryotic species. In particular, the availability of genome sequences from different strains of the same species allows a precise identification of gene losses and gains in each strain – the latter most likely due to HGT events. One of the first examples of this type of analysis was the comparison of three strains of the bacterium *Escherichia coli*, a common inhabitant of the human intestine and a very well-known laboratory model. This comparison revealed that the three strains share only 2996 genes, which represent only 39.2% of the complete set of genes found in the three genomes (Welch *et al.*, 2002), which is quite a small number taking into account that the average genome size in this species is of ~ 4700 genes. This initial work has been recently confirmed by the analysis of the complete genome sequences of 20 *E. coli* strains, which share ~ 2000 of the ~ 18000 genes that comprise the complete set of genes of this species (Touchon *et al.*, 2009). These findings suggested that a very large proportion of the genome in the strains of this species was acquired by HGT. A closer inspection of the genes shared by all the strains revealed that they were mostly involved in essential cellular processes such as DNA replication and repair, transcription, translation and synthesis of essential cell components (amino acids, lipids, nucleotides etc.). In contrast, the genes differing between the different strains appeared to be related to the adaptation to different environmental conditions and also to virulence. In fact, some of the *E. coli* strains examined were innocuous while others were human pathogens. Subsequent studies have revealed a similar situation for other bacterial species. The term 'pangenome' has been coined to refer to the whole set of genes present in all strains of a species. The pangenome, including all genes acquired by HGT by each strain and also all genes that may have been lost in one or several strains, is always larger than the set of genes shared by all strains, and is very often called the 'core genome'. Usually, it is assumed that the core genome has been inherited vertically by the different strains from its last common ancestor, although this is not necessarily true. In fact, two strains can share a gene even if one of them has acquired it from another species by HGT. For this reason, many authors use the term 'core genome' in a much more restrictive sense, to refer only to the set of genes in a genome that has not been affected by HGT. However, identifying that set of genes is not an easy task.

Several approaches have been used to identify the core genome for diverse samples of bacterial species. A very simple one is based on the fact that different species often have genomes with different nucleotide frequencies, namely, different contents of the four nucleotides adenine (A), cytosine (C), guanine (G) and thymine (T). This is usually expressed as the 'GC content' of a genome. If one species acquires a gene from a different one, it is more or less probable that the transferred gene would have a GC content different from that of the recipient genome. Therefore, it would be possible to detect all genes of a genome that have been acquired by HGT just by looking to the genes that have a GC content different from the average GC value of the complete genome. When this technique was applied to the 4288 genes of the *E. coli* strain MG1655 genome sequence, it was shown that 755 of them had a deviant GC content, suggesting a massive acquisition of genes from other species (Lawrence and Ochman, 1998). However, this approach has important limitations. The GC content of a gene is representative not only of the species to which it belongs, but also of its expression level. Thus, essential

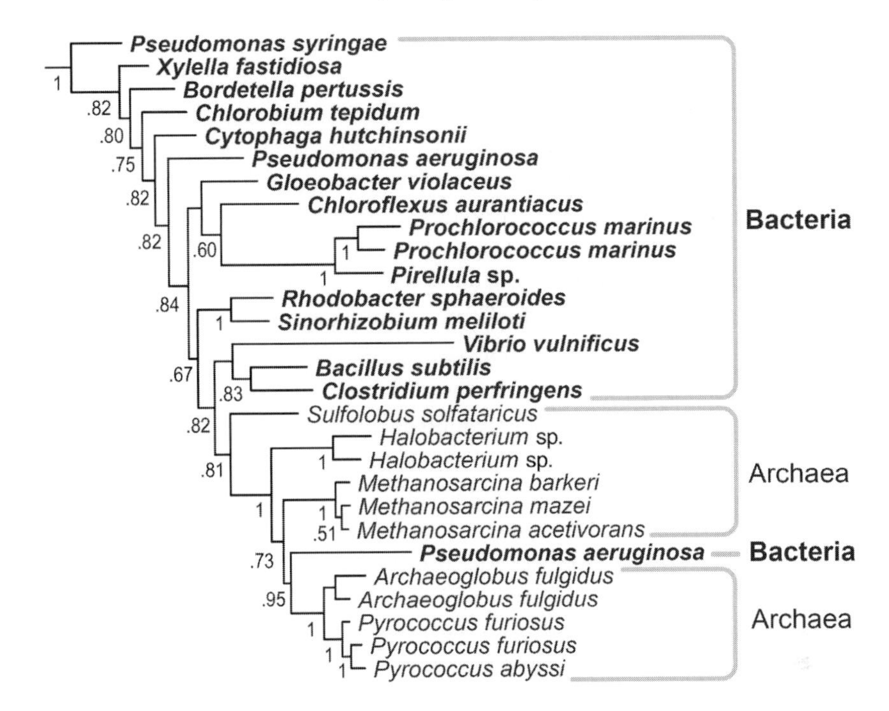

Figure 20.1. Phylogenetic tree of the enzyme UDP-glucose 6-dehydrogenase showing a case of HGT from Archaea to the bacterial species *Pseudomonas aeruginosa*. This species has two copies of the gene encoding this enzyme, one of them branches well-nested within the Archaea, far from the other Bacteria, indicating a relatively recent HGT from an archaeal donor. Bacterial species names are indicated in bold.

genes that are highly expressed tend to have optimal GC contents, whereas accessory weakly-expressed genes may have GC contents far from the optimal value. Therefore, a GC content different from the average is not necessarily synonymous with an exogenous origin by HGT, so that this method may be biased in overestimating the amount of HGT in a genome (Wang, 2001; Daubin *et al.*, 2003a). A widely used alternative method is molecular-phylogenetic analysis. By reconstructing phylogenetic trees of the genes in a genome, it is possible to infer whether they have been acquired by HGT if they branch at abnormal positions in the tree. For example, if a gene of a given bacterial species branches in the middle of a group of Archaea in the corresponding phylogenetic tree, the simplest explanation would be that this bacterium has acquired that gene from an archaeal donor (Figure 20.1). This method is much more reliable than the simple observation of the GC content and, moreover, it provides additional useful information since it can help to identify the gene donors (Archaea in our example). One of the first examples of extensive phylogenetic analysis as a way to quantify HGT concerned the study of the contribution of transferred genes to the genomes of five very distantly related photosynthetic Bacteria. There are 15 possible phylogenetic trees to accommodate five species.

By reconstructing the phylogeny of the genes shared by the five photosynthetic Bacteria, it was found that nearly all the possible 15 topologies were retrieved at more or less similar frequencies (Raymond *et al.*, 2002). This result led to the conclusion that all possible evolutionary histories were retrieved for those genes because they had been exchanged at high rates between the different species. In other words, HGT would be rampant in prokaryotic species, so the evolutionary history of each gene would be different from the others (Raymond *et al.*, 2002).

However, as for the analyses based on GC content, the detection of HGT by means of phylogenetic analysis also has limitations. As gene sequences evolve through time, the information related to the oldest evolutionary events is progressively erased, especially in the case of fast-evolving genes or fast-evolving species. Thus, it is possible to find genes that are only informative for the most recent times. Sequence conservation is essential for the inference of correct multiple-sequence alignments, which are the rough material used for phylogenetic-tree reconstruction. If sequences do not have a minimum of conservation, the inference of the relative order of emergence of the species in the tree will probably be incorrect. To asses the impact of sequence conservation on the ability to detect HGT using phylogenetic analysis, Daubin *et al.* used several bacterial complete genome sequences as their model (Daubin *et al.*, 2003b). When taking into account the quality of the sequence alignment, they realized that only a small fraction of genes with alignments of good quality produced phylogenetic trees supporting HGT events. Often, alignments of inferior quality did not yield the expected phylogenetic relationships, but this could be explained just by the lack of phylogenetic signal and not by HGT. Therefore, in contrast to the analysis of the photosynthetic Bacteria, these authors conclude that HGT is not massive in bacterial species. The huge amount of HGT proposed for the photosynthetic Bacteria could be explained by the fact that the five species studied were very distantly related, so that their relationships can only be inferred from gene sequences that have kept a certain amount of the oldest phylogenetic information. This is a difficult situation since, as mentioned above, that information tends to be progressively lost. Therefore, many of the HGT events proposed for the photosynthetic Bacteria were probably deduced from incorrect phylogenetic trees. In fact, an analysis using the bacterial group of the gammaproteobacteria, applying very strict criteria to select only well-conserved genes, revealed that only 2 genes among the 205 selected supported phylogenies showing HGT events, that is, a very small proportion (Lerat *et al.*, 2000). Therefore, the question about the quantitative importance of HGT in prokaryotes remains open. Many more intensive analyses of all genes in the complete genome sequences available, taking into account the different sources of methodological error, will be necessary to get the answer.

What about Eukaryotes?

The debate about the significance of HGT in the evolution of species has concerned almost exclusively the prokaryotes. In fact, for a long time it has been thought that Eukaryotes were almost completely free of HGT. This was, in part, due to their structural characteristics, in particular the possession of a nuclear membrane separating the genetic material from the rest of the cytoplasm, which would act as an additional barrier hindering the entrance of exogenous DNA into the eukaryotic genome. This led to the curious situation that all phylogenetic trees showing atypical relationships for Eukaryotes were considered as artefactual (which was very important for promoting the study of the different artefacts affecting phylogenetic-tree reconstruction) whereas, in the case of prokaryotes, all atypical trees were considered as evidence for HGT. However, an increasing amount of new data has shown that HGT also exists in Eukaryotes. The first piece of evidence was related to the discovery of a particular type of HGT linked to the presence of organelles of endosymbiotic origin in Eukaryotes. In fact, all Eukaryotes possess – or have possessed – mitochondria, which derive from endosymbiotic alphaproteobacteria, and many of them also possess chloroplasts, which derive from endosymbiotic cyanobacteria. Both mitochondria and chloroplasts have their own genomes, but they are extremely reduced and lack many of the genes required for the functioning of these organelles. Those genes are now encoded in the nuclear genome, though originally they were present in the mitochondrial or chloroplast genomes. In fact, during the evolution of these organelles, most of their genes have been transferred to the nuclear genome, a process of HGT known as 'endosymbiotic gene transfer' (Martin *et al.*, 1998; Nowitzki *et al.*, 1998). The result is that all contemporary eukaryotic genomes are chimeras containing a certain number of genes of bacterial origin.

More recently, it has been suggested that Eukaryotes can also acquire genes from other sources. For example, the 'you are what you eat' hypothesis proposes that phagotrophic Eukaryotes are exposed to a continuous supply of exogenous DNA coming from their food, and that part of that DNA may end up incorporated into the eukaryotic genome (Doolittle, 1998). This could explain the presence of genes of different bacterial origins found in many Eukaryotes (Andersson, 2005). Much rarer are the reports of HGT between different eukaryotic species, but this is probably due to the smaller number of eukaryotic complete genome sequences available compared to prokaryotes. However, some cases are known, affecting, among others, genes that can be involved in the adaptation to parasitism (Richards *et al.*, 2006) or, more surprisingly, the acquisition by an animal of *psb*O, a gene involved in photosynthesis (Rumpho *et al.*, 2008). Thanks to the sequencing of the genomes of a wide variety of Eukaryotes, it will be possible to get a more precise picture of the significance of HGT in Eukaryotes.

Conclusions: HGT, Darwinian evolution and the 'Tree of Life'

As shown above, HGT can lead to the sudden acquisition of new characters that, being coded by genes that are incorporated into the receptor's genome, become inheritable. This has been considered by some authors as a type of Lamarckian 'inheritance of acquired characters', which would be in opposition to the classical Darwinian evolution based on the slow and gradual accumulation of small changes ruled by the process of natural selection. However, this is an artificial antagonism. Although the organisms may experience rapid changes thanks to HGT, the transferred genes themselves evolve in a typical Darwinian way through natural selection. HGT has also been used as a weapon against another of Darwin's most important ideas: the existence of a 'Tree of Life'. It implies that different species originate from their common ancestors following bifurcation patterns and that, if we go sufficiently far back in time, we will see at the end that all species share a unique remote common ancestor (the cenancestor or last universal common ancestor, LUCA). However, if HGT turns out to be rampant, as it has been proposed by many authors to be the case in prokaryotes, the identity of species would be lost very rapidly and a Tree of Life would be impossible to reconstruct. A possible solution might be to use the information contained in the core genome shared by all species to infer a universal tree. Nevertheless, since even the conserved genes of the core genome (e.g. the rRNA genes) can be transferred (Yap *et al.*, 1999), several authors have concluded that it is likely that not a single gene has followed a perfect vertical inheritance from the cenancestor to present-day species (Zhaxybayeva and Gogarten, 2004) so that there is no Tree of Life (Doolittle and Bapteste, 2007). On the other hand, we have seen that other analyses suggest the contrary, namely, that HGT is not so massive (Daubin *et al.*, 2003b; Lerat *et al.*, 2000), so that the phylogenetic signal is kept by the different lineages of living beings throughout their history, and a tree would be an adequate way of representing their evolutionary relationships.

Today, the existence or not of a Tree of Life remains a hotly debated issue (Delsuc *et al.*, 2005; Gogarten and Townsend, 2005). It is certain that HGT has been an important process in shaping the genome content of the different prokaryotic and eukaryotic lineages, but it is also clear that the species of each lineage share a number of characteristics that cannot be explained just by HGT. This is especially clear in the case of Eukaryotes and more difficult to see in prokaryotes, but lineages such as the cyanobacteria or the spirochetes are good examples. Many genes in the genomes of cyanobacteria show that they form an ancient monophyletic group, which is in agreement with the presence of unique characteristics in these organisms, in particular, oxygenic photosynthesis. A similar situation is found in spirochetes, where their monophyly retrieved using many genes is confirmed also by their peculiar morphology, including a very unusual arrangement of their flagella. If HGT was so widespread that species and lineages would lose their identity, we should not expect to observe entire phyla for which both genes and cellular characters clearly support that they have followed a 'vertical' evolution, even if they have also been affected by a certain level of HGT. Likewise, we should not observe that different approaches, such as the reconstruction of phylogenetic trees based on gene content (Snel *et al.*, 1999), the concatenation of proteins involved in protein synthesis (Brochier *et al.*, 2002; Brochier *et al.*, 2005) or the combined analysis of hundreds of genes using supertree methods (Daubin *et al.*, 2001), retrieve the same groups. Perhaps, as proposed by Philippe and Douady (2003), the true evolutionary history of life looks like a mixture of vertical and horizontal transfer of genes, where the core genes depict the major vertical branches (the Tree of Life), immersed in a complex network of horizontal transfers (Figure 20.2).

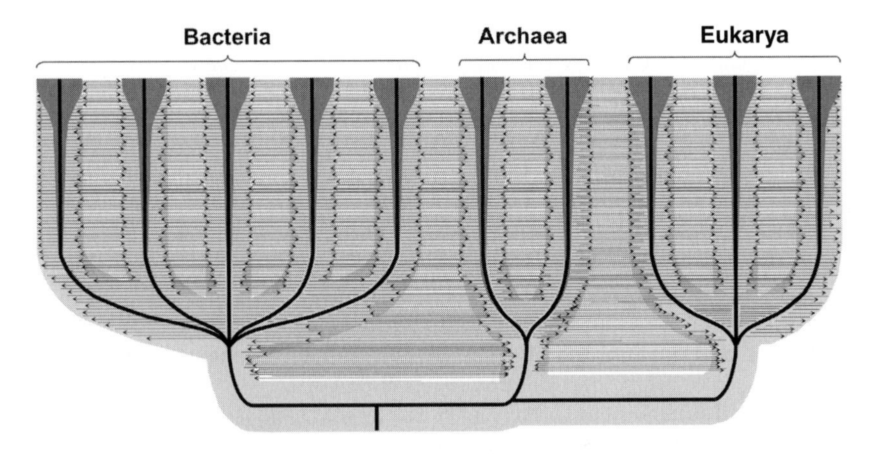

Figure 20.2. Schematic representation of the Tree of Life. Arrows indicate HGT events. The thick black lines correspond to the organismal phylogeny supported by the core genes (those that are never or almost never transferred, such as the ribosomal RNA genes), the dark grey areas correspond to the phylogeny supported by the 'soft core' genes, which may be transferred but very rarely are, and are therefore also useful for reconstructing the evolutionary history of species. Finally, the light grey areas correspond to the highly transferred genes, whose histories are blurred by massive HGT.

References

Andersson, J. O. (2005). Lateral gene transfer in eukaryotes. *Cellular and Molecular Life Sciences*, **9**, 9.

Avery, O. T., Macleod, C. M. and McCarty, M. (1944). Studies on the chemical nature of the substance inducing transformation of pneumococcal types: induction of transformation by a desoxyribonucleic acid fraction isolated from Pneumococcus type III. *Journal of Experimental Medicine*, **79**, 137–58.

Brochier, C., Forterre, P. and Gribaldo, S. (2005). An emerging phylogenetic core of Archaea: phylogenies of transcription and translation machineries converge following addition of new genome sequences. *BMC Evolutionary Biology*, **5**, 36.

Brochier, C., Philippe, H. and Moreira, D. (2000). The evolutionary history of ribosomal protein RpS14: horizontal gene transfer at the heart of the ribosome. *Trends in Genetics*, **16**, 529–33.

Brochier, C., Bapteste, E., Moreira, D. and Philippe, H. (2002). Eubacterial phylogeny based on translational apparatus proteins. *Trends in Genetics*, **18**, 1–5.

Cohan, F. M. (2002). Sexual isolation and speciation in bacteria. *Genetica*, **116**, 359–70.

Dagan, T. and Martin, W. (2006). The tree of one percent. *Genome Biology*, **7**, 118.

Daubin, V., Gouy, M. and Perriere, G. (2001). Bacterial molecular phylogeny using supertree approach. *Genome Information*, **12**, 155–64.

Daubin, V., Lerat, E. and Perriere, G. (2003a). The source of laterally transferred genes in bacterial genomes. *Genome Biology*, **4**, R57.

Daubin, V., Moran, N. A. and Ochman, H. (2003b). Phylogenetics and the cohesion of bacterial genomes. *Science*, **301**, 829–32.

Delsuc, F., Brinkmann, H. and Philippe, H. (2005). Phylogenomics and the reconstruction of the Tree of Life. *Nature Reviews Genetics*, **6**, 361–75.

Doolittle, W. F. (1998). You are what you eat: a gene transfer ratchet could account for bacterial genes in eukaryotic nuclear genomes. *Trends in Genetics*, **14**, 307–11.
Doolittle, W. F. and Bapteste, E. (2007). Pattern pluralism and the Tree of Life hypothesis. *Proceedings of the National Academy of Sciences of the USA*, **104**, 2043–9.
Gogarten, J. P. and Townsend, J. P. (2005). Horizontal gene transfer, genome innovation and evolution. *Nature Reviews Microbiology*, **3**, 679–87.
Gogarten, J. P., Doolittle, W. F. and Lawrence, J. G. (2002). Prokaryotic evolution in light of gene transfer. *Molecular Biology and Evolution*, **19**, 2226–38.
Heinemann, J. A. and Sprague, G. F., Jr. (1989). Bacterial conjugative plasmids mobilize DNA transfer between bacteria and yeast. *Nature*, **340**, 205–9.
Humayun, M. Z. (1998). SOS and Mayday: multiple inducible mutagenic pathways in *Escherichia coli*. *Molecular Microbiology*, **30**, 905–10.
Jain, R., Rivera, M. C. and Lake, J. A. (1999). Horizontal gene transfer among genomes: the complexity hypothesis. *Proceedings of the National Academy of Sciences of the USA*, **96**, 3801–6.
Lawrence, J. G. (1999). Gene transfer, speciation, and the evolution of bacterial genomes. *Current Opinion in Microbiology*, **2**, 519–23.
Lawrence, J. G. and Ochman, H. (1997). Amelioration of bacterial genomes: rates of change and exchange. *Journal of Molecular Evolution*, **44**, 383–97.
Lawrence, J. G. and Ochman, H. (1998). Molecular archaeology of the *Escherichia coli* genome. *Proceedings of the National Academy of Sciences of the USA*, **95**, 9413–17.
Lerat, E., Daubin, V. and Moran, N. A. (2000). From gene trees to organismal phylogeny in prokaryotes: the case of the gamma-proteobacteria. *PLoS Biology*, **1**, 19.
Martin, W., Stoebe, B., Goremykin, V., Hapsmann, S., Hasegawa, M. and Kowallik, K. V. (1998). Gene transfer to the nucleus and the evolution of chloroplasts. *Nature*, **393**, 162–5.
Matic, I., Taddei, F. and Radman, M. (2000). No genetic barriers between *Salmonella enterica* serovar *typhimurium* and *Escherichia coli* in SOS-induced mismatch repair-deficient cells. *Journal of Bacteriology*, **182**, 5922–4.
Nowitzki, U., Flechner, A., Kellermann, J., Hasegawa, M., Schnarrenberger, C. and Martin, W. (1998). Eubacterial origin of nuclear genes for chloroplast and cytosolic glucose-6-phosphate isomerase from spinach: sampling eubacterial gene diversity in eukaryotic chromosomes through symbiosis. *Gene*, **214**, 205–13.
Philippe, H. and Douady, C. J. (2003). Horizontal gene transfer and phylogenetics. *Current Opinion in Microbiology*, **6**, 498–505.
Ramirez-Arcos, S., Fernandez-Herrero, L. A., Marin, I. and Berenguer, J. (1998). Anaerobic growth, a property horizontally transferred by an Hfr-like mechanism among extreme thermophiles. *Journal of Bacteriology*, **180**, 3137–43.
Raymond, J., Zhaxybayeva, O., Gogarten, J. P., Gerdes, S. Y. and Blankenship, R. E. (2002). Whole-genome analysis of photosynthetic prokaryotes. *Science*, **298**, 1616–20.
Richards, T. A., Dacks, J. B., Jenkinson, J. M., Thornton, C. R. and Talbot, N. J. (2006). Evolution of filamentous plant pathogens: gene exchange across eukaryotic kingdoms. *Current Biology*, **16**, 1857–64.
Rumpho, M. E., Worful, J. M., Lee, J., Kannan, K., Tyler, M. S., Bhattacharya, D., Moustafa, A. and Manhart, J. R. (2008). Horizontal gene transfer of the algal nuclear gene *psb*O to the photosynthetic sea slug *Elysia chlorotica*. *Proceedings of the National Academy of Sciences of the USA*, **105**, 17867–71.

Snel, B., Bork, P. and Huynen, M. A. (1999). Genome phylogeny based on gene content. *Nature Genetics*, **21**, 108–10.
Sorek, R., Kunin, V. and Hugenholtz, P. (2008). CRISPR – a widespread system that provides acquired resistance against phages in Bacteria and Archaea. *Nature Reviews Microbiology*, **6**, 181–6.
Stambuk, S. and Radman, M. (1998). Mechanism and control of interspecies recombination in *Escherichia coli*. I. Mismatch repair, methylation, recombination and replication functions. *Genetics*, **150**, 533–42.
Touchon, M., Hoede, C., Tenaillon, O., Barbe, V., Baeriswyl, S., Bidet, P., Bingen, E., Bonacorsi, S., Bouchier, C., Bouvet, O., Calteau, A., Chiapello, H., Clermont, O., Cruveiller, S., Danchin, A., Diard, M., Dossat, C., Karoui, M. E., Frapy, E., Garry, L., Ghigo, J. M., Gilles, A. M., Johnson, J., Le Bouguénec, C., Lescat, M., Mangenot, S., Martinez-Jéhanne, V., Matic, I., Nassif, X., Oztas, S., Petit, M. A., Pichon, C., Rouy, Z., Ruf, C. S., Schneider, D., Tourret, J., Vacherie, B., Vallenet, D., Médigue, C., Rocha, E. P. and Denamur, E. (2009). Organised genome dynamics in the *Escherichia coli* species results in highly diverse adaptive paths. *PLoS Genetics*, **5**, e1000344.
Wang, B. (2001). Limitations of compositional approach to identifying horizontally transferred genes. *Journal of Molecular Evolution*, **53**, 244–50.
Welch, R. A., Burland, V., Plunkett, G., 3rd, Redford, P., Roesch, P., Rasko, D., Buckles, E. L., Liou, S. R., Boutin, A., Hackett, J., Stroud, D., Mayhew, G. F., Rose, D. J., Zhou, S., Schwartz, D. C., Perna, N. T., Mobley, H. L., Donnenberg, M. S. and Blattner, F. R. (2002). Extensive mosaic structure revealed by the complete genome sequence of uropathogenic *Escherichia coli*. *Proceedings of the National Academy of Sciences of the USA*, **99**, 17020–4.
Yap, W. H., Zhang, Z. and Wang, Y. (1999). Distinct types of rRNA operons exist in the genome of the actinomycete *Thermomonospora chromogena* and evidence for horizontal transfer of an entire rRNA operon. *Journal of Bacteriology*, **181**, 5201–9.
Zhaxybayeva, O. and Gogarten, J. P. (2004). Cladogenesis, coalescence and the evolution of the three domains of life. *Trends in Genetics*, **20**, 182–7.

21

The role of symbiosis in eukaryotic evolution

Amparo Latorre, Ana Durbán, Andrés Moya and Juli Peretó

Conceptual and historical framework

Botanists of the late nineteenth century were already familiar with microbial symbioses. In fact, the term 'symbiosis', meaning literally 'living together' was introduced by Anton de Bary and Albert Bernard Frank discussing lichens and mycorrhizae, respectively, at the end of the 1870s. However, until recent times, the idea that microbial associations are central in evolution remained almost marginal. The historian Jan Sapp (1994) proposed several reasons for that situation, including the traditional accent on microorganisms as causative agents of diseases and the prominent concepts of conflict and competition as major evolutionary driving forces. Darwin's metaphoric use of the term 'struggle for existence' included the 'dependence of one being on another', and he actually recognized the existence of species taking advantage of another species, but also emphasized the difficulty for his natural-selection theory to explain the emergence of structures in one species to benefit another (Darwin, 1859, p. 200). Lynn Margulis' contributions since the late 1960s on a symbiotic theory for cell evolution bridged the previous intellectual gap, and proposals made by forgotten biologists that had been dismissed were reopened – especially those of the Russian botany school (Khakhina, 1992; Margulis, 1993). Today there is a wide consensus on the essential role played by symbiosis during the origin and evolution of eukaryotic cells, although very passionate and fundamental debates still persist (de Duve, 2007). Recently, McFall-Ngai has stated that the study of symbiosis is quintessential to systems biology because it integrates all levels of biological analysis, from molecular to ecological, as well as the study of the interplay between organisms in the three domains of life (McFall-Ngai, 2008).

In a broad sense, symbiosis could be defined as a long-term association between two or more organisms of different species at the behavioural, metabolic or genetic level. According to the fitness effects on the members of the association, it is customary to distinguish among 'parasitism', when one species increases its fitness while the fitness of the other species is adversely affected; 'mutualism', when both species increase their fitness; and 'commensalism', when one partner increases its fitness without affecting the other species. However, the fuzzy borders between these terms frequently make their use in real cases very problematic. According to the physical location of the symbiont at the cellular level in the host, we can distinguish between 'ectosymbiosis', when the symbiont lives on the host's body surface, including internal surfaces such as the digestive tube lining and the ducts of glands, and 'endosymbiosis', when the symbiont lives within its host. Among the latter, the most frequent are endosymbioses of prokaryotes or eukaryotic algae within eukaryotic hosts, although there are also prokaryotic endosymbionts in other prokaryotes.

Symbiotic associations have been documented in practically every major branch of the Tree of Life (Moya *et al.*, 2008) and from this observation, together with additional data such us phylogenetic and genomic studies on different associations, we reinforce the view of symbiosis as an important mechanism in the emergence of evolutionary innovations in eukaryotes. Only recently, the genomic era has opened up access to the genetic knowledge of the organisms involved in symbiosis, mainly those non-cultured microbial endosymbionts, allowing a comparison among the different evolutionary innovations carried out by these bacteria on their way from free-living to varied stages of integration with their respective hosts (Moran *et al.*, 2008; Moya *et al.*, 2008). Genomics applied to the study of the symbiotic associations has revealed that the 'symbiogenesis', the process of establishment of new structures, biochemistries or behaviours through symbiosis, is a phenomenon of much more transcendence for evolution than was suspected a few years ago. We are aware, more than ever, of the amplitude of the symbiotic association between bacteria and eukaryotic hosts (McFall-Ngai, 2008). Thus, in addition to its role in the origin and evolution of eukaryotic cells, symbiogenesis is an ongoing phenomenon in the evolution of life. However, some transcendental questions still remain: what evolutionary role does symbiosis play? How has symbiosis contributed to the generation of evolutionary novelties and to the success and diversification of species? It seems premature to state that a correlation between the evolutionary success of some groups of organisms and symbiosis exists, but some examples indicate that this might be the case. At any rate, are mutualistic symbioses adaptive? Are there proofs supporting this statement? Adaptation is a product of natural selection and to answer these questions it is necessary to demonstrate how symbiosis does affect the symbiotic consortium fitness.

Ongoing symbioses

The phenomenon that starts the process of symbiogenesis is a more or less fortuitous event by which two or more organisms live together. The fact that they remain genetically, biochemically or metabolically linked will depend on the evolutionary success of the process, mainly in the mutualistic symbiosis when both members benefit. The differential success of the symbiogenesis process regarding the initial situation where the organisms involved are not in symbiosis, could be a product of natural selection. In general, it is well established and there is a commonality in that symbiotic integration is a process that profoundly changes the gene repertoire of the free-living ancestor and, that depending on the type of symbiotic relationship (i.e. mutualistic or parasitic; facultative or obligate etc.), the age of the association and host necessities (i.e. nutritional, defensive, recycling etc.), the appreciated changes will be more or less dramatic (Wernegreen, 2005).

Insects as a model case

Buchner (1965) stated that microbes and insects not only show an amazing biodiversity in themselves, but they often come together and take evolutionary paths to persistent physical association. Since then, symbiotic associations have been particularly well studied in many insect species and there has been a huge number of genomic, biochemical and physiological studies carried out on their endosymbionts (Baumann, 2005). All these studies have shown that the relationship between insects and endosymbionts has a metabolic foundation. Compared with other much more complex symbiotic associations formed by hundreds or thousands of bacterial species interacting among them and with the host (Guarner and Malagelada, 2003), insect symbioses are very simple and can be taken as models for symbiont biology that could be of great help in understanding how complex symbiotic consortia are formed and have evolved.

Insects constitute 85% of known animal diversity, and it has been estimated that around 15% of them have established a symbiotic relationship with bacteria – probably one of the key factors in their evolutionary success (Douglas, 1998). In general, such insects feed on unbalanced diets, which are supplemented by endosymbionts, called primary endosymbionts. Due to the strict dependence between the host and the symbiont, the bacteria cannot be cultured outside the host and aposymbiotic insects treated with antibiotics do not reproduce. In the mutualistic symbioses a special feature is the host's diet, which is specialized and lacking some nutrients. The role of the bacteria is to supply those nutrients that are lacking in the host's diet. In turn, the host develops specific cells to harbour the bacteria, the bacteriocytes, and must suffer genetic modifications to control the growth of the bacterial population and to ensure their vertical transmission to the offspring. For example, aphids, psyllids, white flies and mealybugs feed on phloem sap that is deficient in nitrogen compounds. All those insects harbour different endosymbionts that provide the absent nutrients, mainly essential amino acids (Baumann, 2005). Tsetse flies feed on blood, deficient in vitamins that are supplied by the primary endosymbiont (Akman *et al.*, 2002). Sharpshooters are xylem-feeding. Since xylem is very poor in nutrients, these are supplied by their co-resident primary symbionts (Wu *et al.*, 2006; McCutcheon and Moran, 2007). A striking case is that of cockroaches and ants, which are omnivores but harbour primary endosymbionts, involved in nitrogen metabolism (Gil *et al.*, 2003, López-Sánchez *et al.*, 2009).

The comparative analysis of the bacterial genes and genomes of endosymbionts and parasites in general, has revealed that adaptation to an intracellular host-associated lifestyle has been accompanied by dramatic changes, both structural and in sequence composition. The most distinctive is the drastic genome shrinkage, where many genes involved in DNA recombination and repair, along with those involved in specific biosynthetic and metabolic pathways, are lost. In contrast, genes involved in informational processes (transcription, translation and replication) are retained. However, different bacteria possess a set of particular genes related to their particular environments, determined by the host's needs, but probably also to the particular process of gene loss, which has occurred throughout adaptation to intracellular life. It is then conceivable that naturally evolved, nearly minimal gene sets may contain substantial differences.

Bacteriocyte-associated endosymbionts of insects are vertically transmitted from mother to offspring (Buchner, 1965). Due to their strictly vertical transmission mode, all obligate intracellular bacteria frequently undergo bottlenecks, resulting in very low effective population sizes, as compared to free-living bacteria. The small effective population size and the inability to recover wild-type phenotypes through recombination (a phenomenon known as Muller's ratchet) led to the accumulation of slightly deleterious mutations in non-essential genes (Moran, 1996). The trend towards genome size reduction in bacteria is associated with large-scale gene loss, reflecting the lack of an effective selection mechanism to maintain genes that are rendered superfluous in the constant and rich environment provided by the host (Andersson and Kurland, 1998; Silva *et al.*, 2001; Wernegreen, 2002).

Genomics and metagenomics provide valuable information on the nature of synthrophic relationships, as they reveal the emergence of the complexity by metabolic complementation between the involved organisms in the symbiotic consortium. In the case of aphids, their deficient phloem-sap diet is complemented by *Buchnera aphidicola*, a γ-proteobacterium that provides the essential amino acids and some vitamins (Shigenobu *et al.*, 2000; Tamas *et al.*, 2002; van Ham *et al.*, 2003). The metabolic bricolage that has been established between aphids and *Buchnera* must be the reason for the evolutionary success of these insects. However, the notion of a symbiotic consortium could be more complex than the one established between the host and one single symbiont. This is the case in the cedar aphid *Cinara cedri*, which harbours, in addition to *B. aphidicola* BCc, a second bacterium called *Serratia symbiotica* that was established before the split of the *Cinara* lineage (Gómez-Valero *et al.*, 2004; Lamelas *et al.*, 2008). Both bacteria live very close to each other in their own bacteriocytes and with a similar density, forming the aphid bacteriome (an organ-like structure that contains the bacteriocytes) (Figure 21.1). The genome of *B. aphidicola* from *C. cedri* probably represents an extreme reduction process with a genome that is about 200 kb smaller than the other three sequenced genomes from *B. aphidicola* strains (Pérez-Brocal *et al.*, 2006). The genome comparisons have shown that this reduction is mainly due to the loss of protein-coding genes and not to a reduction in the size of the intergenic regions or open reading frames. Compared with the already highly compacted genomes of the other strains, *B. aphidicola* BCc has additionally lost the genes responsible for the biosynthesis of nucleotides, cofactors such as riboflavin and most of the transporters, as well as all the genes for peptidoglycan and ATPase subunit biosynthesis. However, despite its extremely reduced genome it still retains the complete machinery for DNA replication, transcription and translation, and a simplified metabolic network for energy production. Thus, with only 362 protein-coding genes, *B. aphidicola* BCc represents a minimal gene set able to support cellular life. It also synthesizes the essential amino acids needed by its aphid host, with the exception of tryptophan. *B. aphidicola* BCc contains the first two genes in the tryptophan synthesis pathway in a plasmid, coding for anthranilate synthase, but lacks the rest of the genes of the pathway. Genome sequencing of *S. symbiotica* has revealed that it contains the rest of the genes needed to complete the synthesis of tryptophan, which is necessary, not only for itself, but also for *Buchnera* and the aphid host.

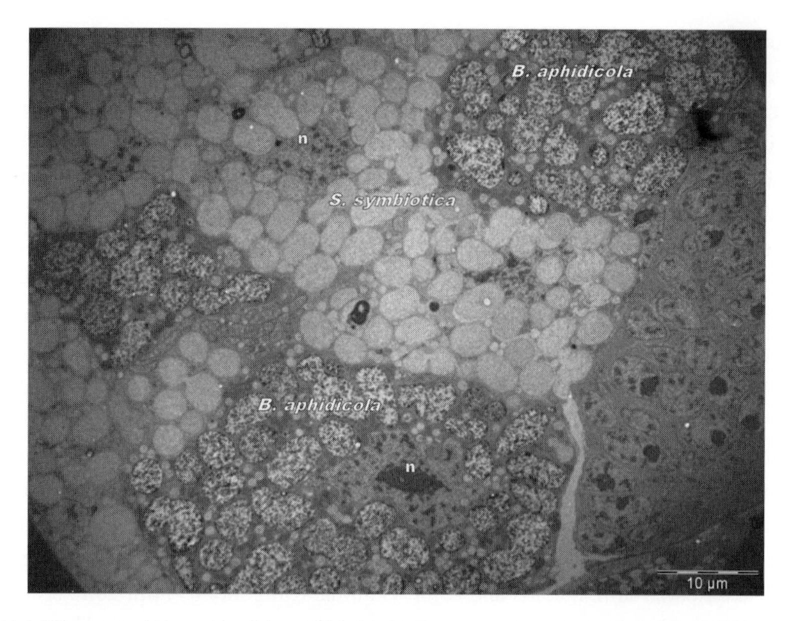

Figure 21.1. Electron micrograph of the aphid *Cinara cedri*. The rounded cells of *Buchnera aphidicola* and *Serratia symbiotica* can be seen in their respective bacteriocytes (specialized symbiont host cells). The nuclei (n) of the bacteriocytes are indicated.

This represents an example of metabolic complementation, where the two symbionts work together to keep the symbiotic consortium alive (Gosalbes *et al.*, 2008).

Another example of metabolic complementation has been found in the sharpshooter *Homalodisca coagulata*, where the two co-resident primary symbionts *Baumannia cicadellinicola* (γ-proteobacterium) and the Bacteroidetes species *Sulcia muelleri* have complementary sets of biosynthetic capabilities needed to provide their host with the nutrients that are lacking in the xylem sap (Wu *et al.*, 2006; McCutcheon and Moran, 2007). Whereas *Baumannia* contains a large number of pathways for the biosynthesis of vitamins, *Sulcia* encodes the enzymes involved in the biosynthesis of most essential amino acids. Apparently, *Sulcia* was ancestrally present in a host lineage that acquired *Baumannia* at the same approximate time as the switch to xylem feeding, consistent with the view that its nutrient-providing capabilities were a requirement for this lifestyle. As in the *Buchnera–Serratia* above-mentioned case, the two symbionts live in close proximity within the host bacteriome and, sometimes, with a *Sulcia* cell surrounded by *Baumannia* cells (Wu *et al.*, 2006).

From endosymbionts to organelles

The complete genome sequencing of endosymbionts with such reduced genomes as the ones shown previously, is cause for reflection on the distinction between 'endosymbiont' and 'organelle'. A common view is that, unlike endosymbionts, organelles have transferred genes to their host and are dependent on a targeting system to re-import their protein products (Keeling and Archibald, 2008). This question has emerged with strength after the genome sequencing of *Carsonella ruddii*, the sap-feeding psyllid *Pachypsylla venusta* symbiont, and the chromatophore of the filose thecamoeba *Paulinella chromatophora*.

The genome of the γ-proteobacterium *C. ruddi* is reduced to a mere 160 kb (so it is within the range of organelles) and only 182 predicted protein-coding genes (Nakabachi *et al.*, 2006). This number is much lower than previous proposals for minimal genomes, and is almost half the number of genes identified in *B. aphidicola* BCc. The small number of genes casts doubt on the character of *C. ruddii* as a living cell, as it lacks many genes for bacterium-specific processes. Moreover, in order to consider that *C. ruddii* is a living organism with a symbiotic relationship with its host, the genes involved in essential living functions, as well as those needed for the maintenance of host fitness, must be preserved. Nakabachi *et al.* proposed that some of the lost genes were transferred from the genome of a *Carsonella* ancestor to the genome of a psyllid ancestor, now being expressed under the control of the host nucleus. This would indicate that *C. ruddii* is on the way to becoming a new type of organelle. However, it would also be possible to consider the implication of the mitochondrial machinery encoded in the insect nucleus. Thus, this strain of *C. ruddii* could be transformed into a new subcellular entity between living cells and organelles (Tamames *et al.*, 2007).

P. chromatophora possesses photosynthetic inclusions, the chromatophores, related to cyanobacteria that might represent a second and independent origin of photosynthetic organelles by symbiogenesis (Keeling and Archibald, 2008). Nowack *et al.* (2008) carried out the genome sequencing of the chromatophore, which is only 1.02 Mb encoding 867 proteins and represents the smallest cyanobacterial genome reported so far. Its close free-living relatives, cyanobacteria of the genus *Synechococcus*, have a genome size of about 3 Mb and 3300 genes, which indicates that similarly to other intracellular bacteria, this *Paulinella* chromatophore has suffered drastic genome shrinkage. It contains a complete set of photosynthetic genes but lacks genes devoted to essential biosynthetic functions (i.e. synthesis of amino acids and cofactors), indicating that the chromatophores are absolutely dependent on their host for growth and survival, but perhaps not for organelle-specific information, in contrast to mitochondria and plastids.

Molecular data from *Paulinella*, *Carsonella* and other endosymbiotic associations at various stages of development can give some clues to an understanding of the symbiogenic events between prokaryotes and primitive eukaryotes that took place around thousands of millions of years ago and gave rise to the nucleocytoplasm, mitochondria and plastids. In the case of *Paulinella*'s chromatophore, the retained genes indicate that the integration of a cyanobacterial symbiont into a eukaryotic host is apparently facilitated by the loss of coding capacity for essential biosynthetic pathways unrelated to photosynthesis – a possible early step in the evolution of a photosynthetic organelle (Nowack *et al.*, 2008).

Symbiosis and gene transfers

Horizontal gene transfer (HGT) refers to the movement of genetic information across normal mating barriers, between more or less distantly related organisms. Extant eukaryotes arose by endosymbiotic gene transfer (EGT) from the endosymbiont's genome to the nucleus, subsequent to the symbiotic events in which the mitochondrial and chloroplast ancestors were engulfed by their hosts.

Primary symbiosis and endosymbiotic gene transfer

All of the examined organellar genomes encode only a small fraction of the organelle's proteins (from 3 to 67 for mitochondria and from 15 to 209 for plastids), with the majority of proteins involved in organelle functions encoded by the nuclear genes, so implying a massive EGT to the host nucleus in the early evolution of mitochondria and plastids (Kurland and Andersson, 2000; Martin *et al.*, 2002).

Following the relocation of organellar genes in the host nucleus, a crucial component of the transformation of the former endosymbionts into cellular organelles was the evolution of targeting sequences in the nuclear-encoded copies and complex-protein import machineries to efficiently target back to the organelle the proteins acting in the organelles but now expressed by the host nucleocytoplasm. This retargeting allowed the endosymbiont-encoded copies to undergo pseudogenization and loss. The process of genome shrinkage was similar to that which is ongoing in obligate intracellular parasites and symbionts. Also, genes of nuclear origin replaced some endosymbiont ones. Comparative proteomic analysis has shown that proteins involved in replication, transcription, cell division and signal transduction were replaced by eukaryotic ones early in the evolution of mitochondria (Gabaldón and Huynen, 2007). For the symbionts, all these processes implied an irreversible loss of autonomy.

Many nuclear genes of organellar origin were able to supply proteins to other cellular compartments and thus became free in terms of being able to evolve new functions (Martin *et al.*, 2002). As a result, enzymes of some biochemical pathways are derived from different sources (Oborník and Green, 2005; Richards *et al.*, 2006).

The similarity in gene content among contemporary plastids and mitochondria suggests that most organellar genes were transferred en masse early in organelle evolution. The subsequent tempo of EGT has been punctuated with bursts of transfer interspersed with long periods of stasis. Many genes also have a patchy distribution across extant organellar genomes, implying recurrent transfers and convergent losses (Keeling and Palmer, 2008). For instance, although mitochondria are considered monophyletic, adaptation to anaerobiosis has resulted in a diversity of metabolisms, including oxygen-independent ATP synthesis, hydrogen-producing mitochondria synthesizing ATP by substrate-level phosphorylation (hydrogenosomes) and extremely reduced organelles without any role in ATP synthesis but probably involved in iron–sulphur cluster assembly (mitosomes). As shown in Figure 21.2, parallel evolution from a putative ancestral facultative anaerobic, metabolically 'pluripotent' mitochondrion has originated eukaryotic lineages with anaerobic mitochondria (e.g. parasitic helminths, ciliates, *Euglena*, *Fusarium*), hydrogenosomes (e.g. trichomonads, ciliates, chytrids) or mitosomes (e.g. diplomonads, *Entamoeba*, Microsporidia) (Tielens *et al.*, 2002; Embley and Martin, 2006).

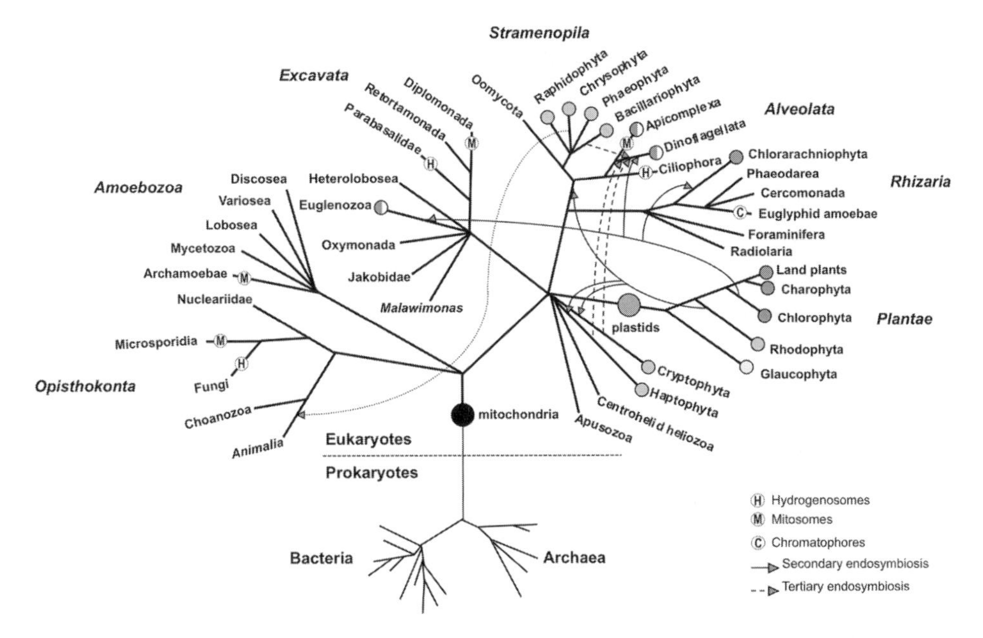

Figure 21.2. The relationships among major groups of eukaryotes are shown. The primary endosymbioses that led to the origin of mitochondria and plastids are indicated, as well as secondary and tertiary endosymbioses involving red and green algal endosymbionts. Lineages in which plastids are known to exist are labelled with full circles. Lineages with a mix of photosynthetic and non-photosynthetic species are partially filled. The acquisition of plastids by the sea slug *Elysia chlorotica* from its algal food source is shown as the arrow that ends in the animal branch. The endosymbiotic event that originated, independently of plastid primary endosymbiosis, the photosynthetic chromatophores of *Paulinella chromatophora* is indicated. Lineages containing species with hydrogenosomes or mitosomes are labelled.

DNA gene transfer from organelles to the nucleus seems to be an ongoing phenomenon, as suggested by the lack of divergence in most nuclear sequences of organellar origin that are postulated to have been acquired recently. Moreover, the transfer of genes from an organelle genome to the nucleus has been observed at high rates in experimental studies (Huang *et al.*, 2003). Almost all present-day nuclear transfers of mitochondrial and plastid DNA give rise to nonfunctional sequences. An organellar-derived gene will become functional if it somehow acquires a eukaryotic-type promoter and regulatory sequences. If the gene product has to go back to the organelle or to another cell compartment it must incorporate the proper targeting sequences. Functionalization of a transferred plastid gene has been observed under laboratory conditions (Stegemann and Bock, 2006).

The continuous transfer of organellar DNA to the nucleus leads to the question as to why organelles retained some genes (Timmis *et al.*, 2004). On one hand, the hydrophobicity hypothesis holds that hydrophobic proteins are poorly imported by organelles and must be encoded in the organellar genomes. In support of this view, the core set of organellar-encoded proteins are almost always hydrophobic proteins, but not all membrane-integral organellar proteins are organelle encoded (e.g. light-harvesting proteins). The redox-control hypothesis holds that individual organelles need to autonomously control the expression of components of their electron-transport chain so that the components can be synthesized when they are needed in order to maintain redox balance. Moreover, the evolution of modified genetic codes and RNA editing in many mitochondria contributes to the mitochondrial genes being locked into place in many lineages (animals prominently).

The DNA flow between genetic compartments in eukaryotic cells seems not to be restricted to organelle-to-nucleus gene transfer, as sequences of chloroplast and nuclear origin have been found in mitochondrial genomes in plants (Unseld *et al.*, 1997; Notsu *et al.*, 2002), providing indirect evidence for plastid-to-mitochondrion and nucleus-to-mitochondrion DNA transfer. There is no evidence for HGT in plant plastids and in animal mitochondria (with one possible exception). However, HGT is very common in plant mitochondria. Among the proposed explanations for this fact is the existence of an active system for the import of double-stranded DNA in plant mitochondria, which makes the uptake of exogenous DNA easier, and allows for a greater propensity for mitochondria to fuse with one another (in the case where HGT was mediated by fusion of the involved organelles) (Keeling and Palmer, 2008). Differences in the streamlining pressures acting on plastid and animal mitochondrial genomes on one hand, and plant mitochondrial genomes on the other hand have been proposed to be responsible for the differences in their genome complexity (Lynch *et al.*, 2006). Equally, these differences could explain the lack of HGT in plastids and animal mitochondria, as they are maintained as compact genomes and probably are less likely to incorporate foreign DNA and more likely to lose it (Keeling and Palmer, 2008).

Symbiosis in the origin and spread of photosynthesis in eukaryotes

The primary endosymbiosis that gave rise to plastids involved the engulfment of a cyanobacterium by a host cell that was a eukaryote with a nucleus, cytoskeleton and mitochondrion. The cyanobacterial genome was greatly reduced by EGT to the host nucleus and gene loss. Primary plastids are characterized by two membranes, both cyanobacterial in nature, and are present in glaucophyte, red and green algae (the latter is the lineage which gave rise to land plants), which together are referred to as the Plantae or Archaeplastida.

Primary plastids subsequently spread to many other eukaryotic lineages by secondary and tertiary endosymbiosis between distantly related groups of eukaryotes, giving rise to the spread of photosynthesis across the eukaryotic tree (Archibald, 2009; Figure 21.2). Secondary endosymbiosis occurs when a primary-plastid-containing alga is engulfed and retained by a heterotrophic eukaryote, and in tertiary endosymbiosis a secondary-plastid-containing alga is taken up by another eukaryote, which may or may not itself possess a plastid. In these cases, in addition to EGT, transfer of genes between the unrelated host and endosymbiont genomes by HGT and loss of genes in the endosymbiont genome occurred as a result of the 'merger' of the two nuclei. These plastids are characterized by the presence of supernumerary membranes of diverse origin. Secondary plastids derived from green algae are found in *Euglena*, which belongs to the Excavata eukaryotic supergroup, and in chlorarachniophytes, members of the Rhizaria supergroup. At least six algal lineages contain secondary plastids derived from red algae: these include cryptophytes (which also have a nucleomorph, the relic of the endosymbiont nucleus), haptophytes, plastid-bearing stramenopiles, apicomplexans, dinoflagellates and *Chromera velia*. Certain dinoflagellates possess a tertiary plastid that has been taken from cryptophytes, haptophytes or stramenopiles.

Thus, cyanobacterial genes that were initially acquired from the endosymbiotic ancestor of plastids have been repeatedly transferred during eukaryotic evolution. First, from plastid to nucleus, and then from the primary host nucleus to the secondary host nucleus in each secondary symbiosis, and again from nucleus to nucleus in each tertiary symbiosis, as the endosymbiont nucleus in most secondary and tertiary symbioses has completely disappeared (Figure 21.2).

EGT from bacterial endosymbionts to the nuclear genome

Recent findings suggest that symbiosis-mediated gene transfers will be found in many more genomes. Until recently, the only documented cases were the transfer of genome fragments of the *Wolbachia* (an α-proteobacterium) endosymbiont to the nucleus of the beetle host *Callosobruchus chinensis* (Kondo *et al*., 2002) and the nematode host *Onchocerca volvulus* (Fenn and Blaxter, 2006). Recent work has provided evidence for gene-transfer events from *Wolbachia* to the nuclear genome of some of its hosts (4 insects and 4 nematode species). In the most extreme case so far, an entire copy of the *Wolbachia* genome was found in the genome of a fruitfly, and it has even been shown that 2% of the inserted genes are transcribed (Hotopp *et al*., 2007; Nikoh *et al*., 2007). Whether these transfers are functional or much less adaptive is unclear. Thus, although the lateral transfer from *Wolbachia* to the eukaryotic hosts can be facilitated by its presence in developing gametes, similar events in other associations cannot be discarded. The sequencing of the genome of the *C. ruddii* host (*P. venusta*) is necessary in order to solve the status of this symbiont.

Eukaryote to eukaryote HGT

HGT has altered our concepts of the genome, the species and the Tree of Life in the bacterial world, and now cases of HGT in eukaryotes are appearing at an increasing rate, accounting for many adaptively important traits.

An example of horizontal gene transfer between eukaryotes is that of the sea slug *Elysia chlorotica* (Rumpho *et al*., 2008), which retains for months in its digestive epithelium photosynthetically active plastids from its algal food source *Vaucheria litorea*, a filamentous secondary alga. Taking into account that most of the proteins required for plastid function are relocated in the algal nucleus, one possible explanation is that these genes have been acquired by the animal from the algal nucleus by horizontal gene transfer and then retargeted to the retained plastids. Sequence and expression analysis have demonstrated that at least the gene *psbO*, encoding a subunit of photosystem II, is provided by the mollusc, as it is absent from the plastid genome, expressed in the mollusc, integrated in its nuclear genome and identical to the algal nuclear homologue. This case of HGT is unusual in two ways: it takes place between two multicellular eukaryotes, an event extremely rarely seen to date, and it has resulted in a photosynthetic animal that needs to acquire 'its' plastids with each generation – another example of the power of HGT in introducing evolutionary novelty.

Ancient symbioses: the symbiogenetic origin of the eukaryotic cell

The origin of eukaryotes remains a matter of controversy (de Duve, 2007). Models of eukaryogenesis must consider the endosymbiotic origin of mitochondria and plastids, the current knowledge of eukaryotic cell biology and that derived from phylogenetic data. Eukaryotic cells differ from prokaryotic ones due to a number of structural features such as a nucleus surrounded by an envelope, and the fact that they contain generally highly structured linear chromosomes, an extensive endomembrane system, a cytoskeleton, mitochondria and, in phototrophic eukaryotes, plastids. Although the nucleus is the defining characteristic of the eukaryotic cell, its origin is uncertain. However, mitochondria and plastids have been proved to have an endosymbiotic origin. Merezhkovsky and Famintsyn were the first supporters of the idea of the endosymbiotic origin of several eukaryotic organelles, in particular chloroplasts (reviewed in Khakhina, 1992), which was revived by Margulis in her theory of the endosymbiotic origin of mitochondria and plastids (Margulis 1970, 1993). Subsequent molecular analysis unequivocally demonstrated that these organelles originally descended from free-living bacteria that became intracellular endosymbionts (Zablen *et al.*, 1975; Schwartz and Dayhoff, 1978). The origin of mitochondria and plastids dates back to at least 1 billion years ago. Subsequent endosymbiosis gave rise to the diverse groups of secondary algae. Sequence comparisons identified α-proteobacteria and cyanobacteria as the bacterial groups that contain the ancestors of mitochondria and plastids, respectively. However, which lineages among those groups gave rise to present-day organelles remains unresolved (Schwartz and Dayhoff, 1978; Gray *et al.*, 1999; Deusch *et al.*, 2008). Both organelles have made a substantial contribution to the complement of genes that are found in the eukaryotic nucleus today, mainly by the early massive relocation of organellar genes to the nucleus, with the subsequent co-evolution of the nuclear and organellar genomes (see 'Conceptual and historical framework').

Phylogenetic analysis based on small subunit ribosomal RNA has split all life forms into three domains (Bacteria, Archaea and Eukarya) (Woese and Fox, 1977). The use of ancestral paralogous genes, like those for the ATPases (Gogarten *et al.*, 1989) or the translation elongation factors (Iwabe *et al.*, 1989) places the root of this universal Tree of Life in the bacterial branch. Such rooting has become widely accepted, although its validity remains controversial. On the other hand, some analyses have shown a genomic chimerism of eukaryotes, that is, informational genes (genes involved in replication, transcription and translation) are most closely related to archaeal genes, whereas operational genes (genes involved in cellular metabolic processes) are most closely related to bacterial genes (Rivera *et al.*, 1998).

Current hypotheses for the origin of eukaryotes fall into two general classes: autogenous models that propose a nucleus-bearing protoeukaryotic cell first, followed by the acquisition of mitochondria, and chimeric models that propose a merging of two typical prokaryotic cells, an archaeon and a bacterium, by physical fusion or by endosymbiosis, followed by the origin of eukaryotic-specific features. In the last class, some models hold that the ancestor of mitochondria is the bacterial partner, whereas in others it is a later incorporation.

Some eukaryogenetic hypotheses have been proposed based on metabolic mutualistic symbiosis (Figure 21.3). These include, among others, the serial endosymbiotic theory (SET) (Margulis, 1970, 1993; Margulis *et al.*, 2000), the symbiosis based on interspecies hydrogen transfer or the hydrogen hypothesis (Martin and Müller, 1998) and the syntrophy hypothesis (Moreira and López-García, 1998; López-García and Moreira, 2006).

The SET proposed by Margulis postulates that eukaryotes evolved from an association between a motile strictly anaerobic *Spirochaeta*-like bacterium and a sulphidogenic thermoacidophilic *Thermoplasma*-like crenarchaeon, which established a sulphur motility syntrophy in warm acidic and sulphurous waters. In this way, the archaeon reduced sulphur and sulphate to sulphide (or O_2 to H_2O in microaerophilic conditions) providing the highly reduced conditions needed for growth by the anaerobic bacterium, which in turn provided fermentation products and electron acceptors to the archaeon, and, importantly, swimming motility to the consortium. The nuclear genome was generated by recombination of bacterial and archaeal DNA, and the eukaryotic undulipodia (motility apparatus based on microtubules) originated from the spirochete. Subsequent endosymbiotic incorporations of the ancestors of α-proteobacteria and cyanobacteria originated mitochondria and plastids, respectively. In favour of her model, the author states that similar consortia can be found nowadays, as in the case of *Thiodendron*, an ectosymbiotic association based on sulphur syntrophy between two bacterial types, *Desulfobacter* and *Spirochaeta*. But this kind of motile symbiosis has never been observed between Archaea and spirochetes, which, in addition, are not known to inhabit extreme acidophilic environments.

In the hydrogen hypothesis, Martin and Müller (1998) proposed that eukaryotes arose through the symbiotic association of a strictly anaerobic and autotrophic methanogenic archaeon (the host) that consumed H_2 and CO_2 to produce methane, with an α-proteobacterium (the symbiont) that was capable of aerobic respiration but generated H_2 and CO_2 as products of anaerobic fermentation. The methanogen acted as a sink for H_2, thus increasing the bacterial fermentative reactions. The host became heterotrophic when it acquired the symbiont's carbohydrate metabolism by endosymbiotic gene transfer, and methanogenesis was lost. In this theoretical framework, Martin and Koonin (2006), following former proposals on the origin of introns (Cech, 1986; Cavalier-Smith, 1991), postulated that the nuclear envelope evolved to allow messenger RNA splicing.

In contrast to the hydrogen hypothesis, the syntrophic hypothesis, presented by Moreira and López-García (1998) and López-García and Moreira (2006), involves two independent symbiotic events. A symbiosis based on interspecies hydrogen transfer was

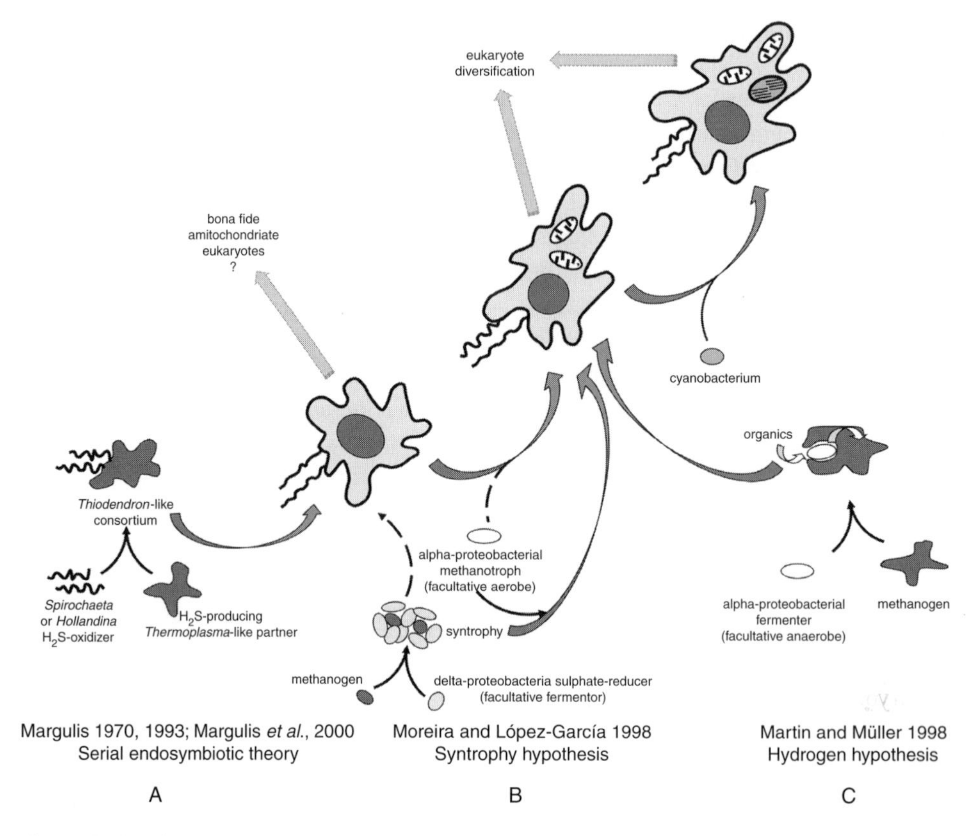

Figure 21.3. Three hypotheses on the origin of eukaryotic cells: (A) the serial endosymbiostic theory (Margulis 1970, 1993; Margulis *et al.*, 2000), (B) the syntrophy hypothesis (Moreira and López-García, 1998; López-García and Moreira, 2006), and (C) the hydrogen hypothesis (Martin and Müller, 1998). According to (A), at least three successive symbiotic events originated the nucleocytoplasm, the mitochondrion and the plastid, respectively. Model (C) postulates that the consortium nucleocytoplasm *and* mitochondrion are the very origin of the eukaryotic cell, whereas model (B) may accept the existence of an amitochondriate stage. As a matter of fact, the discovery of bona fide descendants of this nucleated cell devoted to mitochondria would falsify model (C).

initially established between a methanogenic archaeaon and an ancestral myxobacterium (δ-proteobacterium) under anaerobic conditions. At an early stage, a highly versatile α-proteobacterium, a facultative aerobe able to oxidize methane, among other things, became a member of the consortium and, later, became the mitochondrion. The methanogen endosymbiont became the nucleus, since myxobacteria transferred most of their genes to the archaeal genome. The methanogen provided most of the genetic machinery, while the myxobacteria, characterized by complex life cycles and cell-to-cell communication, would have provided most cytoplasmic features. The nuclear envelope and the endoplasmic reticulum are thought to derive from invaginations of the myxobacterial plasma membranes, initially involved in secretion and connection of the archaeal endosymbiont with the exterior. Two selective forces in the evolution of the nucleus are proposed (López-García and Moreira, 2006): (1) a metabolic compartmentation to avoid deleterious co-existence of anabolic (autotrophic synthesis by the methanogen) and catabolic (fermentation by the myxobacteria) pathways in the cell; and (2) the avoidance of aberrant protein synthesis due to intron spreading in the ancient archaeal genome following mitochondrial acquisition and loss of methanogenesis. In support of this hypothesis, syntrophies between methanogens and δ-proteobacteria mediated by hydrogen transfer are widespread today in anoxic sediments.

Finally, some authors claim that there is no direct evidence that eukaryotes evolved by genome fusion between Archaea and Bacteria, and that these models do not explain many ancient features in eukaryotes, postulating a common ancestor of the three domains of life from which prokaryotes arose by reductive evolution (Kurland *et al.*, 2006).

Concluding remarks

The symbiotic origin of the eukaryotic cell is now widely accepted. Mitochondria and other derived organelles during parallel adaptation to anaerobiosis have a bacterial origin. The same is true for plastids of plants, algae and protists. We cannot disregard the discovery of new instances of endosymbionts currently en route to becoming organelles or fully fledged eukaryotic compartments of bacterial origin.

We are now closer than before, due to the advent of the omics methodologies, not only to further unravelling of the steps towards the origin of the eukaryotic cell, but also to assessing the question as to how much eukaryotic complexity is originated through evolutionary innovations by symbiogenesis.

References

Akman, L., Yamashita, A., Watanabe, H., Oshima, K., Shiba, T., Hattori, M. and Aksoy, S. (2002). Genome sequence of the endocellular obligate symbiont of tsetse flies, *Wigglesworthia glossinidia*. *Nature Genetics*, **32**, 402–7.

Andersson, S. G. and Kurland, C. G. (1998). Reductive evolution of resident genomes. *Trends in Microbiology*, **6**, 263–8.

Archibald, J. M. (2009). The puzzle of plastid evolution. *Current Biology*, **19**, 81–8.

Baumann, P. (2005). Biology of bacteriocyte-associated endosymbionts of plant sap-sucking insects. *Annual Review of Microbiology*, **59**, 155–89.

Buchner, P. (1965). *Endosymbiosis of Animals with Plant Microorganisms*. New York: Interscience.

Cavalier-Smith, T. (1991). Intron phylogeny: a new hypothesis. *Trends in Genetics*, **7**, 145–8.

Cech, T. R. (1986). The generality of self-splicing RNA: relationship to nuclear mRNA splicing. *Cell*, **44**, 207–10.

Darwin, C. (1859). *On the Origin of Species by Means of Natural Selection, or the Preservation of Favoured Races in the Struggle for Life*, 1st edn. London: John Murray.

de Duve, C. (2007). The origin of eukaryotes: a reappraisal. *Nature Reviews Genetics*, **8**, 395–403.

Deusch, O., Landan, G., Roettger, M., Gruenheit, N., Kowallik, K. V., Allen, J. F., Martin, W. and Dagan, T. (2008). Genes of cyanobacterial origin in plant nuclear genomes point to a heterocyst-forming plastid ancestor. *Molecular Biology and Evolution*, **25**, 748–61.

Douglas, A. E. (1998). Nutritional interactions in insect-microbial symbioses: aphids and their symbiotic bacteria *Buchnera*. *Annual Review of Entomology*, **43**, 17–37.

Embley, T. M. and Martin, W. (2006). Eukaryotic evolution, changes and challenges. *Nature*, **440**, 623–30.

Fenn, K. and Blaxter, M. (2006). *Wolbachia* genomes: revealing the biology of parasitism and mutualism. *Trends in Parasitology*, **22**, 60–5.

Gabaldón, T. and Huynen, M. A. (2007). From endosymbiont to host-controlled organelle: the hijacking of mitochondrial protein synthesis and metabolism. *PLoS Computational Biology*, **3**, e219.

Gil, R., Silva, F. J., Zientz, E., Delmotte, F., González-Candelas, F., Latorre, A., Rausell, C., Kamerbeek, J., Gadau, J., Hölldobler, B., van Ham, R. C., Gross, R. and Moya, A. (2003). The genome sequence of *Blochmannia floridanus*: comparative analysis of reduced genomes. *Proceedings of the National Academy of Sciences of the USA*, **100**, 9388–93.

Gogarten, J. P., Kibak, H., Dittrich, P., Taiz, L., Bowman, E. J., Bowman, B. J., Manolson, M. F., Poole, R. J., Date, T. and Oshima, T. (1989). Evolution of the vacuolar H^+-ATPase: implications for the origin of eukaryotes. *Proceedings of the National Academy of Sciences of the USA*, **86**, 6661–5.

Gómez-Valero, L., Soriano-Navarro, M., Pérez-Brocal, V., Heddi, A., Moya, A., García-Verdugo, J. M. and Latorre, A. (2004). Coexistence of *Wolbachia* with *Buchnera aphidicola* and a secondary symbiont in the aphid *Cinara cedri*. *Journal of Bacteriology*, **186**, 6626–33.

Gosalbes, M. J., Lamelas, A., Moya, A. and Latorre, A. (2008). The striking case of tryptophan provision in the cedar aphid *Cinara cedri*. *Journal of Bacteriology*, **190**, 6026–9.

Gray, M. W., Burger, G. and Lang, B. F. (1999). Mitochondrial evolution. *Science*, **283**, 1476–81.

Guarner, F. and Malagelada, J. R. (2003). Gut flora in health and disease. *Lancet*, **361**, 512–19.

Hotopp, J. C., Clark, M. E., Oliveira, D. C., Foster, J. M., Fischer, P., Torres, M. C., Giebel, J. D., Kumar, N., Ishmael, N., Wang, S., Ingram, J., Nene, R. V., Shepard, J., Tomkins, J., Richards, S., Spiro, D. J., Ghedin, E., Slatko, B. E., Tettelin, H. and Werren, J. H. (2007). Widespread lateral gene transfer from intracellular bacteria to multicellular eukaryotes. *Science*, **317**, 1753–6.

Huang, C. Y., Ayliffe, M. A. and Timmis, J. N. (2003). Direct measurement of the transfer rate of chloroplast DNA into the nucleus. *Nature*, **422**, 72–6.

Iwabe, N., Kuma, K., Hasegawa, M., Osawa, S. and Miyata, T. (1989). Evolutionary relationship of archaebacteria, eubacteria, and eukaryotes inferred from phylogenetic trees of duplicated genes. *Proceedings of the National Academy of Sciences of the USA*, **86**, 9355–9.

Keeling, P. J. and Archibald, J. M. (2008). Organelle evolution: what's in a name? *Current Biology*, **18**, 345–7.

Keeling, P. J. and Palmer, J. D. (2008). Horizontal gene transfer in eukaryotic evolution. *Nature Reviews Genetics*, **9**, 605–18.

Khakhina, L. N. (1992). *Concepts of Symbiogenesis. A Historical and Critical Study of the Research of Russian Botanists*, eds. L. Margulis and M. McMenamin. New Haven, London: Yale University Press.

Kondo, N., Nikoh, N., Ijichi, N., Shimada, M. and Fukatsu, T. (2002). Genome fragment of *Wolbachia* endosymbiont transferred to X chromosome of host insect. *Proceedings of the National Academy of Sciences of the USA*, **99**, 14280–5.

Kurland, C. G. and Andersson, S. G. (2000). Origin and evolution of the mitochondrial proteome. *Microbiology and Molecular Biology Reviews*, **64**, 786–820.

Kurland, C. G., Collins, L. J. and Penny, D. (2006). Genomics and the irreducible nature of eukaryote cells. *Science*, **312**, 1011–14.

Lamelas, A., Pérez-Brocal, V., Gómez-Valero, L., Gosalbes, M. J., Moya, A. and Latorre, A. (2008). Evolution of the secondary symbiont '*Candidatus Serratia symbiotica*' in aphid species of the subfamily Lachninae. *Applied and Environmental Microbiology*, **74**, 4236–40.

López-García, P. and Moreira, D. (2006). Selective forces for the origin of the eukaryotic nucleus. *Bioessays*, **28**, 525–33.

López-Sánchez, M. J., Neef, A., Peretó, J., Patiño-Navarett R., Pignatelli, M., Latorre, A. and Moya, A. (2009). Evolutionary convergence and nitrogen metabolism in *Blattabacterium stain Bge*, primary endosymbiont of the cockroach *Blattella germanica*. *PLoS Genetics*, **5**, e1000721.

Lynch, M., Koskella, B. and Schaack, S. (2006). Mutation pressure and the evolution of organelle genomic architecture. *Science*, **311**, 1727–30.

Margulis, L. (1970). *Origin of Eukaryotic Cells*. New Haven: Yale University Press.

Margulis, L. (1993). *Symbiosis in Cell Evolution*, 2nd edn. New York: W. H. Freeman.

Margulis, L., Dolan, M. F. and Guerrero, R. (2000). The chimeric eukaryote: origin of the nucleus from the Karyomastigont in Amitochondriate protists. *Proceedings of the National Academy of Sciences of the USA*, **97**, 6954–9.

Martin, W. and Müller, M. (1998). The hydrogen hypothesis for the first eukaryote. *Nature*, **392**, 37–41.

Martin, W., Rujan, T., Richly, E., Hansen, A., Cornelsen, S., Lins, T., Leister, D., Stoebe, B., Hasegawa, M. and Penny, D. (2002). Evolutionary analysis of *Arabidopsis*, cyanobacterial, and chloroplast genomes reveals plastid phylogeny and thousands of cyanobacterial genes in the nucleus. *Proceedings of the National Academy of Sciences of the USA*, **99**, 12246–51.

Martin, W. and Koonin, E. V. (2006). Introns and the origin of nucleus–cytosol compartmentalization. *Nature*, **440**, 41–5.

McCutcheon, J. P. and Moran, N. A. (2007). Parallel genomic evolution and metabolic interdependence in an ancient symbiosis. *Proceedings of the National Academy of Sciences of the USA*, **104**, 19392–7.

McFall-Ngai, M. (2008). Are biologists in 'future shock'? Symbiosis integrates biology across domains. *Nature Reviews Microbiology*, **6**, 789–92.

Moran, N. A. (1996). Accelerated evolution and Muller's ratchet in endosymbiotic bacteria. *Proceedings of the National Academy of Sciences of the USA*, **93**, 2873–8.

Moran, N. A., McCutcheon, J. P. and Nakabachi, A. (2008). Genomics and evolution of heritable bacterial symbionts. *Annual Reviews of Genetics*, **42**, 165–90.

Moreira, D. and López-García, P. (1998). Symbiosis between methanogenic Archaea and delta-proteobacteria as the origin of eukaryotes: the syntrophic hypothesis. *Journal of Molecular Evolution*, **47**, 517–30.

Moya, A., Peretó, J., Gil, R. and Latorre, A. (2008). Learning how to live together: genomic insights into prokaryote–animal symbioses. *Nature Reviews Genetics*, **9**, 218–29.

Nakabachi, A., Yamashita, A., Toh, H., Ishikawa, H., Dunbar, H. E., Moran, N. A. and Hattori, M. (2006). The 160-kilobase genome of the bacterial endosymbiont *Carsonella. Science*, **314**, 267.

Nikoh, N., Tanaka, K., Shibata, F., Kondo, N., Hizume, M., Shimada, M. and Fukatsu, T. (2007). *Wolbachia* genome integrated in an insect chromosome: evolution and fate of laterally transferred endosymbiont genes. *Genome Research*, **18**, 272–80.

Notsu, Y., Masood, S., Nishikawa, T., Kubo, N., Akiduki, G., Nakazono, M., Hirai, A. and Kadowaki, K. (2002). The complete sequence of the rice (*Oryza sativa* L.) mitochondrial genome: frequent DNA sequence acquisition and loss during the evolution of flowering plants. *Molecular Genetics and Genomics*, **268**, 434–45.

Nowack, E. C., Melkonian, M. and Glöckner, G. (2008). Chromatophore genome sequence of *Paulinella* sheds light on acquisition of photosynthesis by eukaryotes. *Current Biology*, **18**, 410–18.

Oborník, M. and Green, B. R. (2005). Mosaic origin of the heme biosynthesis pathway in photosynthetic eukaryotes. *Molecular Biology and Evolution*, **22**, 2343–53.

Pérez-Brocal, V., Gil, R., Ramos, S., Lamelas, A., Postigo, M., Michelena, J. M., Silva, F. J., Moya, A. and Latorre, A. (2006). A small microbial genome: the end of a long symbiotic relationship? *Science*, **314**, 312–13.

Richards, T. A., Dacks, J. B., Campbell, S. A., Blanchard, J. L., Foster, P. G., McLeod, R. and Roberts, C. W. (2006). Evolutionary origins of the eukaryotic shikimate pathway: gene fusions, horizontal gene transfer, and endosymbiotic replacements. *Eukaryotic Cell*, **5**, 1517–31.

Rivera, M. C., Jain, R., Moore, J. E. and Lake, J. A. (1998). Genomic evidence for two functionally distinct gene classes. *Proceedings of the National Academy of Sciences of the USA*, **95**, 6239–44.

Rumpho, M. E., Worful, J. M., Lee, J., Kannan, K., Tyler, M. S., Bhattacharya, D., Moustafa, A. and Manhart, J. R. (2008). Horizontal gene transfer of the algal nuclear gene *psb*O to the photosynthetic sea slug *Elysia chlorotica. Proceedings of the National Academy of Sciences of the USA*, **105**, 17867–71.

Sapp, J. (1994). *Evolution by Association. A History of Symbiosis*. New York: Oxford University Press.

Schwartz, R. M. and Dayhoff, M. O. (1978). Origins of prokaryotes, eukaryotes, mitochondria, and chloroplasts. *Science*, **199**, 395–403.

Shigenobu, S., Watanabe, H., Hattori, M., Sakaki, Y. and Ishikawa, H. (2000). Genome sequence of the endocellular bacterial symbiont of aphids *Buchnera* sp. APS. *Nature*, **407**, 81–6.

Silva, F. J., Latorre, A. and Moya, A. (2001). Genome size reduction through multiple events of gene disintegration in *Buchnera* APS. *Trends in Genetics*, **17**, 615–18.

Stegemann, S. and Bock, R. (2006). Experimental reconstruction of functional gene transfer from the tobacco plastid genome to the nucleus. *Plant Cell*, **18**, 2869–78.

Tamames, J., Gil, R., Latorre, A., Peretó, J., Silva, F. J. and Moya, A. (2007). The frontier between cell and organelle: genome analysis of *Candidatus Carsonella ruddii. BMC Evolutionary Biology*, **7**, 181.

Tamas, I., Klasson, L., Canbäck, B., Näslund, A. K., Eriksson, A. S., Wernegreen, J. J., Sandström, J. P., Moran, N. A. and Andersson, S. G. (2002). 50 million years of genomic stasis in endosymbiotic bacteria. *Science*, **296**, 2376–9.

Tielens, A. G., Rotte, C., van Hellemond, J. J. and Martin, W. (2002). Mitochondria as we don't know them. *Trends in Biochemical Sciences*, **27**, 564–72.

Timmis, J. N., Ayliffe, M. A., Huang, C. Y. and Martin, W. (2004). Endosymbiotic gene transfer: organelle genomes forge eukaryotic chromosomes. *Nature Reviews Genetics*, **5**, 123–35.

Unseld, M., Marienfeld, J. R., Brandt, P. and Brennicke, A. (1997). The mitochondrial genome of *Arabidopsis thaliana* contains 57 genes in 366 924 nucleotides. *Nature Genetics*, **15**, 57–61.

van Ham, R. C., Kamerbeek, J., Palacios, C., Rausell, C., Abascal, F., Bastolla, U., Fernández, J. M., Jiménez, L., Postigo, M., Silva, F. J., Tamames, J., Viguera, E., Latorre, A., Valencia, A., Morán, F. and Moya, A. (2003). Reductive genome evolution in *Buchnera aphidicola*. *Proceedings of the National Academy of Sciences of the USA*, **100**, 581–6.

Wernegreen, J. J. (2002). Genome evolution in bacterial endosymbionts of insects. *Nature Reviews Genetics*, **3**, 850–61.

Wernegreen, J. J. (2005). For better or worse: genomic consequences of intracellular mutualism and parasitism. *Current Opinion in Genetics and Development*, **15**, 572–83.

Woese, C. R. and Fox, G. E. (1977). Phylogenetic structure of the prokaryotic domain: the primary kingdoms. *Proceedings of the National Academy of Sciences of the USA*, **74**, 5088–90.

Wu, D., Daugherty, S. C., Van Aken, S. E., Pai, G. H., Watkins, K. L., Khouri, H., Tallon, L. J., Zaborsky, J. M., Dunbar, H. E., Tran, P. L., Moran, N. A. and Eisen, J. A. (2006). Metabolic complementarity and genomics of the dual bacterial symbiosis of sharpshooters. *PLoS Biology*, **4**, e188.

Zablen, L. B., Kissil, M. S., Woese, C. R. and Buetow, D. E. (1975). Phylogenetic origin of the chloroplast and prokaryotic nature of its ribosomal RNA. *Proceedings of the National Academy of Sciences of the USA*, **72**, 2418–22.

Part VI

Life in extreme conditions

22

Life in extreme conditions: *Deinococcus radiodurans*, an organism able to survive prolonged desiccation and high doses of ionizing radiation

Magali Toueille and Suzanne Sommer

The high stress resistance of the bacterium *Deinococcus radiodurans*

Deinococcus radiodurans (*D. radiodurans*), initially isolated in canned meat that had been irradiated at 4000 grays in order to achieve sterility (Anderson *et al.*, 1956), is a bacterium belonging to a bacterial genus characterized by an exceptional ability to withstand the lethal effects of DNA-damaging agents, including ionizing radiation, ultraviolet light and desiccation (Battista and Rainey, 2001).

Initially, *D. radiodurans* was named *Micrococcus radiotolerans* because of its morphological similarity to members of the genus *Micrococcus*. Subsequent studies led to its reclassification into a distinct phylum within the domain Bacteria and this bacterium was renamed *Deinococcus radiodurans*, the Greek adjective *deinos* meaning strange or unusual. *Deinococcaceae* were isolated from diverse environments after exposure to high doses of ionizing radiation. Among this family, containing to date more than 20 identified members, *D. radiodurans* is by far the best characterized. *D. radiodurans* cells are non-motile, non-spore-forming and are obligate aerobes that grow optimally at 30°C in rich medium. On agar plates, they are pigmented and appear pink-orange. In liquid media, cells divide alternately into two planes, exhibiting pairs or tetrads (Figure 22.1A).

Ionizing radiation, when applied to any living organism, leads to the formation of highly reactive radicals (e.g. hydroxy radicals) and can cause a variety of DNA damage, such as DNA single- and double-strand breaks and base modifications. A 5000 Gy dose introduces approximately 200 DNA double-strand breaks, 3000 DNA single-strand breaks and more than 1000 damaged bases per *D. radiodurans* genome equivalent (Cox and Battista, 2005). The ability of *D. radiodurans* to survive these injuries without loss of viability is exceptional, as such a dose is 1000 times greater than what would kill a human. It is interesting to note that a 5000 Gy dose of ionizing radiation drastically decreases the viability of *Pyrococcus abyssi*, a hyperthermophilic archaeum with optimal-growth temperatures approaching the boiling point of water (Figure 22.2). However, compared to most living species, *P. abyssi* can be considered as radioresistant, showing 100% viability at a dose of 2000 Gy. This property is not surprising since at high temperature, which corresponds to the natural environment of this Archaea, DNA accumulates many types of damage (e.g. hydrolytic depurination, deamination of cytosine and adenine and single- or double-strand breaks) that have to be very rapidly repaired to avoid complete denaturation of DNA.

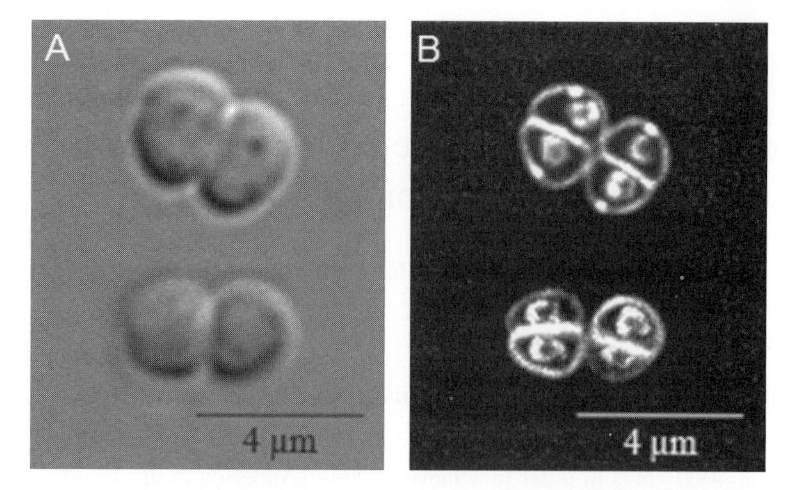

Figure 22.1. (A) Transmission light image of two *D. radiodurans* tetrads. (B) Epifluorescence image of two *D. radiodurans* tetrads. The DAPI-stained nucleoid appears as a ring-like structure in the centre of each cell delineated by the cell membrane stained by the lipophilic dye FM4–64.

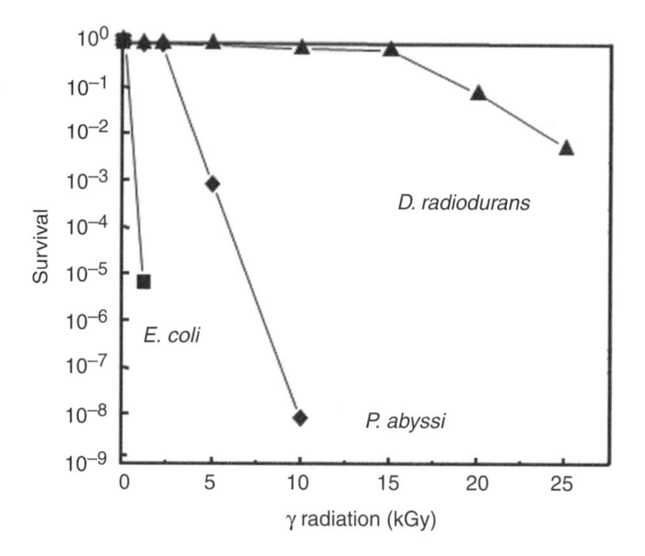

Figure 22.2. Survival curves for *E. coli*, *P. abyssi* and *D. radiodurans* following exposure to gamma radiation.

This further emphasizes the exceptional radioresistance of *D. radiodurans*. *D. radiodurans* was also shown to be extremely resistant to UV radiation, desiccation or repeated cycles of desiccation–rehydration. *D. radiodurans* cell viability is not affected by a dose of 500 J/m^2 UV light that generates up to 5000 covalent bindings of adjacent thymine or cytosine residues in DNA (pyrimidine dimers), and *D. radiodurans* shows 85% viability after 2 years in the presence of less than 5% humidity.

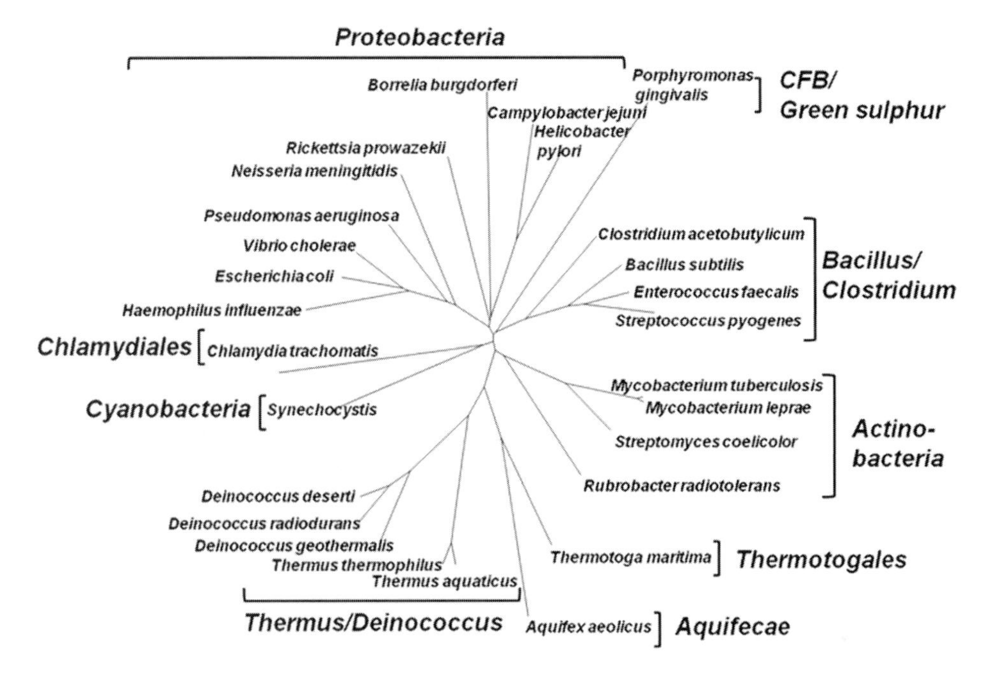

Figure 22.3. A 16S-rRNA gene-sequence-based phylogeny of the main lineages of the domain Bacteria generated by using the neighbour-joining calculation method.

These properties brought some scientists to propose an extraterrestrial origin for this bacterium and its adaptation to constraints for survival on Mars (radiation, cold, vacuum, dormancy) (Diaz and Schulze-Makuch, 2006; Pavlov *et al.*, 2006). Then, the bacterium could have travelled to Earth on pieces of rock that were blasted into space by an impacting asteroid and fallen to Earth as meteorites (Pavlov *et al.*, 2006). The increasing availability of genome sequences and comparative genomics proved that *D. radiodurans* is not extra-terrestrial but is indeed a 'classical' terrestrial bacterium which is located among other terrestrial bacteria in the evolutionary Tree of Life (Cox and Battista, 2005; Omelchenko *et al.*, 2005; Figure 22.3). Indeed, *D. radiodurans* is one of the first bacteria whose genomes were completely sequenced (White *et al.*, 1999). Since this pioneering work, more than 600 bacterial genomes have been sequenced. *D. radiodurans* belongs to the *Deinococcus–Thermus* group, which is deeply branched in bacterial phylogenetic trees, *D. radiodurans* and *Thermus thermophilus*, a radiosensitive thermophile, being considered as sister species belonging to a lineage positioned within a subtree that also includes *Actinobacteria* and *Cyanobacteria* (Omelchenko *et al.*, 2005).

However, the evolution of organisms that are able to grow continuously at 60 Gy/h or survive acute irradiation doses of 15 000 Gy is difficult to explain given the apparent absence of highly radioactive habitats on Earth over geologic time, and thus it seems more likely that the natural selection pressure for the evolution of radiation-resistant bacteria was chronic exposure to non-radioactive forms of DNA damage, in particular those promoted by desiccation

(Mattimore and Battista, 1996). Desiccation leads to DNA hydrolysis and to oxidative damage in a similar way as exposure to ionizing radiation (Potts, 1994). The surface sands of hot arid deserts are exposed to intense ultraviolet (UV) radiation, cycles of extreme temperatures and desiccation. Nevertheless, an extensive diversity of bacterial species has been identified in such extreme and nutrient-poor environments (Rainey *et al.*, 2005). *Deinococcus* species are natural inhabitants of desert environments on Earth; of the 26 species with valid published names, 14 have been isolated from arid environments, i.e. desert soils or rocks (for review see Blasius *et al.*, 2008). Artificial environments, such as radiation and toxic chemical-waste dumps, also provide intense selection pressure for extremophiles. Thus, the high stress-resistance of *D. radiodurans* has probably evolved continuously by gene duplications and through various events of horizontal gene transfer (Makarova *et al.*, 2001; Makarova *et al.*, 2007).

An efficient DNA repair tool box

The radioresistance of *D. radiodurans* cannot be related to prevention of DNA damage, the number of DNA double-strand breaks being formed at the same rate in *Escherichia coli* (*E. coli*) and *D. radiodurans* when cells are irradiated under identical conditions (Gerard *et al.*, 2001). By comparison, only a few DNA double-strand breaks can kill an *E. coli* cell (Krasin and Hutchinson, 1977).

The DNA breakage and repair features of *D. radiodurans* DNA can be monitored using pulsed-field gel electrophoresis, a powerful technique for resolving chromosome-sized DNA fragments. For this purpose, the DNA isolated from *D. radiodurans* cells before and after gamma irradiation is cleaved by a restriction enzyme that recognizes a sequence of eight nucleotides present at only a few sites in the *D. radiodurans* genome, generating 11 resolvable fragments of different size that can be separated by migration in an agarose gel and visualized by the fluorescence under UV-light of ethidium bromide that intercalates between the DNA bases. As shown in Figure 22.4, when cells were exposed to 6800 Gy gamma radiation, the characteristic pattern of restriction is lost, giving rise to hundreds of chromosomal fragments of 50 kb average size. During the first hour post-irradiation incubation, an extensive DNA degradation results in a decrease in the amount of DNA in the cells, and, during the second and third hours post-irradiation incubation, the fragments reassemble to reconstitute an intact genome complement. Further incubation results in an increase in the amount of DNA in the cells, the reconstituted genome being replicated before reinitiation of cell division. These results suggest that *D. radiodurans* possesses a very efficient DNA repair tool box and that a combination of pathways that can deal with DNA double-strand breaks and oxidative damage plays a crucial role in resistance to ionizing radiation, in addition to desiccation.

An efficient protection of proteins against oxidation

In addition to causing DNA damage, elevated doses of ionizing radiation also provoke damage to other cell components such as membranes and proteins, in particular, reactions with water molecules form highly reactive oxygen species. In *D. radiodurans*, the first

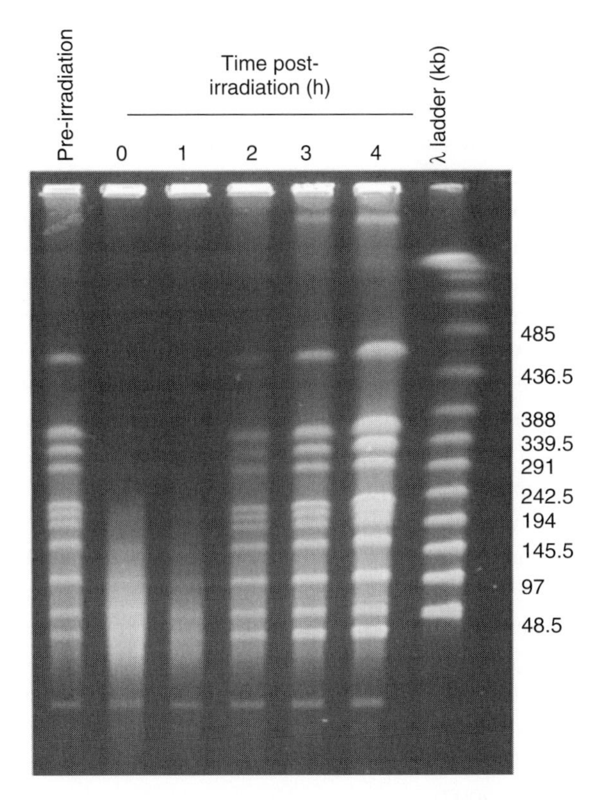

Figure 22.4. Kinetics of genome reconstitution in *D. radiodurans* cells exposed to 6800 Gy gamma radiation (from Blasius *et al*., 2008). *D. radiodurans* bacteria were grown in TGY2X to an A650 of 2, concentrated ten times, exposed to 6800 Gy gamma radiation, diluted in TGY2X to an A650 of 0.2 and then incubated at 30°C with agitation for recovery. To avoid double-strand breaks generated by mechanical manipulation of the DNA and not present *in vivo*, cells were immobilized and lysed in agarose plugs at different post-irradiation times. The plugs were then incubated with proteinase K, an enzyme that destroys proteins, and RNase to eliminate the RNA. Finally, the purified DNA was digested using the restriction enzyme *Not*I before being analysed by pulsed-field gel electrophoresis.

line of protection is the presence of antioxidant enzymes such as superoxide dismutases and catalases which limit oxidation of biomolecules. Inactivation of the corresponding genes results in an increased sensitivity to ionizing radiation as compared to the wild type (Markillie *et al*., 1999). Recently, Daly and colleagues have shown that *D. radiodurans*, as well as other radioresistant bacteria, have high intracellular Mn/Fe concentration ratios (Daly *et al*., 2004; Daly *et al*., 2007). They have reported a relationship between intracellular Mn/Fe concentration ratios and bacterial survival following exposure to ionizing radiation, the most-resistant species containing an intracellular concentration of Mn about 300 times higher, and an intracellular concentration of Fe about 3 times lower than the most-sensitive species (Daly *et al*., 2007). When *D. radiodurans* cells were grown in defined medium without Mn supplementation, cells were depleted in Mn and, at 10 000 Gy, displayed a 1000-fold reduction in survival compared to cells with normal Mn concentrations. Moreover, the amount of protein damage caused by a given dose of gamma radiation for intrinsically resistant and sensitive bacteria is very different. At 4000 Gy, high levels of protein oxidation occured in cells with the lowest intracellular Mn/Fe concentration ratios, whereas no protein oxidation was detected in cells with the highest Mn/Fe ratios (Daly *et al.*, 2007). Thus, protected DNA repair proteins could function with far greater efficiency immediately after irradiation than those unprotected in radiosensitive bacteria, without requiring, as in radiosensitive bacteria, their recycling. It is interesting to note that the desiccation-resistant dry-climate soil bacteria isolated from the shrub-steppe Hanford site in Washington State also accumulate high intracellular Mn and low Fe concentrations compared to desiccation-sensitive bacteria. Moreover, their proteins are protected from oxidation during drying (Fredrickson *et al.*, 2008).

Even though *D. radiodurans* proteins are protected from oxidative damage by elevated Mn/Fe concentration ratios, the removal of detrimental damaged proteins may complement the DNA repair capabilities of *D. radiodurans* to contribute to its extreme radioresistance. *D. radiodurans* encodes an unusually high number of putative proteases (Makarova *et al.*, 2001) and several of these enzymes are induced after irradiation (Tanaka *et al.*, 2004). We have examined the role of these putative proteases in radioresistance and we have shown that two proteases, Lon1 and Lon2, that are mainly involved in abnormal protein degradation are not required for radioresistance, whereas a third one, ClpPX, seems to play an indirect role in the regulation of DNA repair and reinitiation of cell division after gamma irradiation (Servant *et al.*, 2007).

A compact nucleoid that may prevent the dispersion of free DNA ends

In 2003, Abraham Minski proposed that the particular structure of the *D. radiodurans* nucleoid was the clue to its extreme radioresistance (Levin-Zaidman *et al.*, 2003). Bacterial chromosomes usually form condensed structures – the so-called nucleoid. In *D. radiodurans*, doughnut-like DNA structures were detected by transmission electron microscopy (Levin-Zaidman *et al.*, 2003; Englander *et al.*, 2004) and epifluorescence microscopy (Zimmerman and Battista, 2005, Figure 22.1B) and this compact ring-like nucleoid structure remained unaltered after high-dose gamma irradiation, suggesting that this structure may passively contribute to radioresistance by preventing the dispersion of free DNA ends (Levin-Zaidman *et al.*, 2003). A recent examination of nucleoids in members of the radioresistant genera *Deinococcus* and *Rubrobacter* revealed a high degree of genome compaction, whereas the nucleoid appears uniformly distributed in *E. coli* cells (Zimmerman and Battista, 2005). The nucleoid structure of the radioresistant species *Deinococcus radiopugnans, Deinococcus geothermalis* and *Rubrobacter radiotolerans* do not adopt a fixed ring-like shape (Zimmerman and Battista, 2005), suggesting that strong nucleoid compaction, rather than the shape of the nucleoid, may be the common trait among radioresistant organisms. More work is required to characterize the factors involved in the organization of the *D. radiodurans* nucleoid and to investigate the precise role of its compact structure

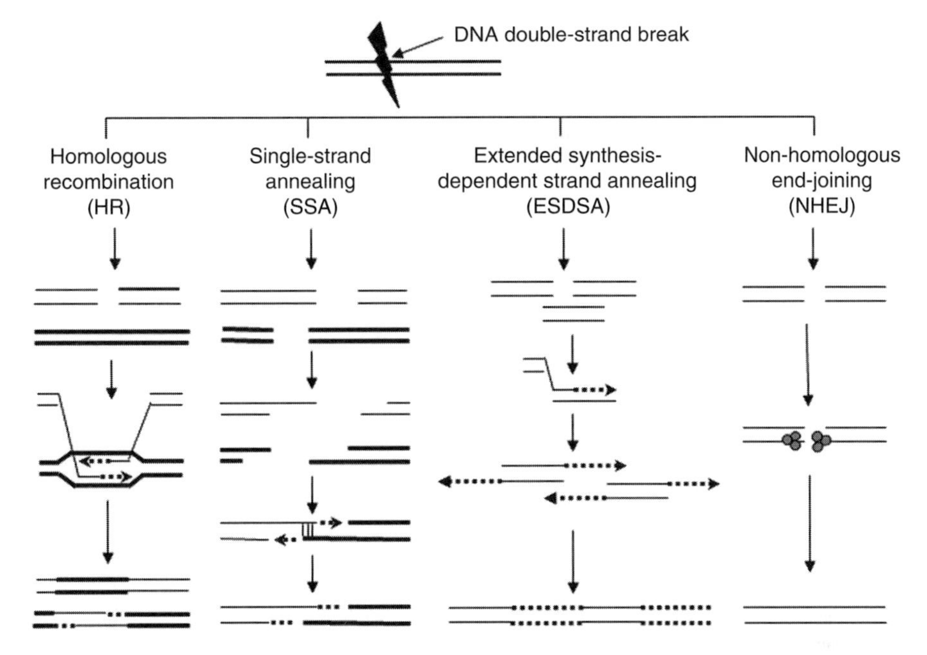

Figure 22.5. Different pathways for DNA double-strand break repair in *D. radiodurans*. Adapted from Blasius *et al.*, 2008. Dotted lines indicate DNA newly synthesized during DNA double-strand break repair.

in *Deinococcus* radioresistance. In particular, it would be interesting to isolate, if viable, mutants with decondensed nucleoids to test their ability to tolerate exposure to high levels of ionizing radiation.

An original pathway of double-strand break repair

Several mechanisms have been proposed to account for efficient repair of DNA double-strand breaks (Figure 22.5). Among these mechanisms, homologous recombination is the main pathway in bacteria (Wyman *et al.*, 2004) and in the yeast *Saccharomyces cerevisiae* (Paques and Haber, 1999). Homologous recombination uses an intact homologous DNA molecule to restore the correct DNA sequence at sites of damage. RecA protein is the recombinase that promotes homologous pairing and strand exchange between the broken and intact homologous DNA molecules.

In addition to homologous recombination, it has been proposed that a single-strand annealing (SSA) reaction occurs at early times in irradiated Deinococcal cells to account for the observation that some of the radiation-induced double-strand breaks can be mended in a recombination-defective *recA* mutant (Daly and Minton, 1996). The process of SSA can occur between different copies of broken chromosomes and be facilitated by the multiplicity of the Deinococcal genome equivalents. It requires maturation of the double-stranded DNA ends to generate single-stranded DNA ends. Then annealing of complementary sequences, followed by gap filling, can ensure the joining of two overlapping fragments. This might constitute a preparatory step for further DNA repair (Daly and Minton, 1996). However, these two DNA double-strand break repair pathways do not correlate with the observation that DNA fragment assembly coincides with a massive DNA synthesis that occurs at a much higher rate in irradiated cells than in unirradiated dividing cells. This massive DNA synthesis begins when the early degradation of DNA stops. It was proposed that an original mechanism, ESDSA (extended synthesis-dependent strand annealing) takes place, involving strand invasion between overlapping fragments belonging to different homologous chromosomes and strand extension proceeding to the end of the invaded fragments (Zahradka *et al.*, 2006).

These extended fragments could then dissociate from the template and anneal with complementary extended single-strand tails. In this model, fragment assembly is RecA-dependent (Slade *et al.*, 2009) and is accompanied by substantial DNA synthesis during the time of repair, the repaired DNA appearing as a patchwork of old and new material, as confirmed experimentally. In contrast to homologous recombination, single-strand annealing or ESDSA, non-homologous end-joining does not require another copy of the DNA strand. In this process, DNA ends are recognized by a protein complex that favours the recruitment of a DNA ligase that joins the two DNA ends (Hefferin and Tomkinson, 2005). It is the major pathway of DNA double-strand break repair in eukaryotes and has recently been identified and characterized in bacteria (Weller *et al.*, 2002; Shuman and Glickman, 2007). It has been proposed that non-homologous end-joining could also play an important role in *D. radiodurans* double-strand break repair (Kobayashi *et al.*, 2004; Lecointe *et al.*, 2004; Narumi *et al.*, 2004). Even though non-homologous end-joining has never been experimentally established in *D. radiodurans* bacteria, it could take place in *D. radiodurans* because of the presence of proteins consistent with this process. In particular PprA, whose expression is highly induced by ionizing radiation or desiccation (Tanaka *et al.*, 2004), has been shown to preferentially bind to DNA double-strand ends and to stimulate the DNA end-joining reaction catalysed by ATP-dependent and NAD-dependent DNA ligases (Narumi *et al.*, 2004). Moreover, cells devoid of PprA are highly radiosensitive (Narumi *et al.*, 2004; Tanaka *et al.*, 2004).

A condensed structure of the *D. radiodurans* nucleoid may provide suitable scaffolds for DNA repair through all of these four DNA double-strand break repair processes.

Original *Deinococcaceae* specific proteins involved in radioresistance

Broad-based bioinformatics and experimental studies have converged on the conclusion that *D. radiodurans* uses a relatively conventional set of DNA repair and protection functions, but with far greater efficiency than radiation-sensitive bacteria (Makarova *et al.*, 2001; Daly *et al.*, 2007; Blasius *et al.*, 2008). The *D. radiodurans* genome was sequenced in 1999 (White *et al.*, 1999). So far, the genomes of two relatives of *D. radiodurans* have been sequenced as well, and their genomes have been compared to the genome of *D. radiodurans*. One was *Thermus thermophilus* (*T. thermophilus*), a radiosensitive thermophile from the same phylum (Henne *et al.*, 2004). The other, *Deinococcus geothermalis* (*D. geothermalis*), is a radioresistant moderate thermophile from the same genus as *D. radiodurans* (Makarova *et al.*, 2007). Comparative genomic analysis suggests that horizontal gene transfer played a major role in the evolution of Bacteria and Archaea (Nelson *et al.*, 1999; Makarova *et al.*, 2001). Both *D. radiodurans* and *D. geothermalis* do not seem to have acquired novel DNA repair systems, but progressively expanded their systems involved in cell cleaning and salvage, in addition to genes involved in transcriptional regulation and signal transduction (Makarova *et al.*, 2000). However, analysis of the transcriptome of *D. radiodurans* revealed an exciting group of genes that are upregulated in response to either desiccation or ionizing radiation (Tanaka *et al.*, 2004). Among these genes, only a limited number of well-defined DNA repair genes were found and the genes that are most highly induced in response to each stress encode *Deinococcaceae*-specific proteins of unknown function. Inactivation of the corresponding genes indicates that they play a role in radioresistance (Tanaka *et al.*, 2004). A novel regulatory protein IrrE, specific to the *Deinococcaceae*, has been shown to be a positive effector of the expression of some of these induced genes. This protein appears to play a crucial role in regulating multiple DNA repair and protection pathways to radiation exposure and to be part of a putative signal-transduction pathway in response to DNA damage in *D. radiodurans* (Earl *et al.*, 2002; Hua *et al.*, 2003). Makarova *et al.* (2007) reported the presence of a potential common radiation-response regulon in *D. radiodurans* and *D. geothermalis* identified by a palindromic motif. This radiation/desiccation response motif is found in the upstream regions of a conserved set of radiation-induced genes.

Conclusions

Bacteria belonging to the *Deinococcaceae* phylum are known for their exceptional resistance to ionizing radiation but other radioresistant bacteria from different phyla or the Archaea have been isolated and their dispersion in the phylogenetic tree suggests that they have acquired radioresistance independently (Cox and Battista, 2005). Under these conditions, it seems rather unlikely that one repair or protection system can be at the origin of radioresistance in all of these organisms, and we can propose that different strategies have evolved in parallel. Considering all the factors and mechanisms that contribute to the radioresistant phenotype of *D. radiodurans* (an ESDSA DNA repair pathway, a high Mn/Fe ratio, a condensed nucleoid, etc.), it seems that radioresistance is achieved by a combination of various proteins and mechanisms. Further work is required to understand the complete survival-kit components that led to the full intricacy of *D. radiodurans* radioresistance.

Acknowledgements

The authors thank A. Bailone, J. Battista, M. Cox and U. Hübscher for their stimulating discussions. P. Servant, C. Bouthier, C. Pasternak, E. Bentchikou, H. H. Nguyen, F. Vannier, G. Coste, F. Lecointe, E. Jolivet and S. Mennecier are acknowledged for their important contribution in the laboratory to a better understanding of the mechanisms involved in *D. radiodurans* radioresistance. We thank E. Prestel for his valuable help in constructing phylogenetic trees, E. Bentchikou and E. Jolivet for their help in providing pictures, and M. Cox, the Editor of *Critical Reviews in Biochemistry and Molecular Biology* for the authorization to use a figure previously published in this journal. M. Toueille is supported by a post-doctoral fellowship from the Agence Nationale de la Recherche (ANR-07-BLAN-0106) and S. Sommer's laboratory is supported by the Centre National de la Recherche Scientifique, the University Paris-Sud XI, the Commissariat à l'Energie Atomique (CEA LRC42V), Electricité de France and the Agence Nationale de la Recherche (ANR-07-BLAN-0106).

References

Anderson, A. W., Nordon, H. C., Cain, R. F., Parrish, G. and Duggan, G. (1956). Studies on a radioresistant *Micrococcus*. I. Isolation, morphology, cultural characteristics, and resistance to gamma radiation. *Food Technology*, **10**, 575–8.

Battista, J. R. and Rainey, F. A. (2001). Family 1. Deinococcaceae. In *Bergey's Manual of Systematic Bacteriology*, eds. D. R. Boone, R. W. Castenholz. New York: Springer, pp. 395–414.

Blasius, M., Sommer, S. and Hubscher, U. (2008). *Deinococcus radiodurans*, what belongs to the survival kit? *Critical Reviews in Biochemistry and Molecular Biology*, **43**, 221–38.

Cox, M. M. and Battista, J. R. (2005). *Deinococcus* radiodurans – the consummate survivor. *Nature Reviews Microbiology*, **3**(11), 882–92.

Daly, M. J., Gaidamakova, E. K., Matrosova, V. Y., Vasilenko, A., Zhai, M., Leapman, R. D., Lai, B., Ravel, B., Li, S. M., Kemner, K. M. and Fredrickson, J. K. (2007). Protein oxidation implicated as the primary determinant of bacterial radioresistance. *PLoS Biology*, **5**(4), e92.

Daly, M. J., Gaidamakova, E. K., Matrosova, V. Y., Vasilenko, A., Zhai, M., Venkateswaran, A., Hess, M., Omelchenko, M. V., Kostandarithes, H. M., Makarova, K. S., Wackett, L. P., Fredrickson, J. K. and Ghosal, D. (2004). Accumulation of Mn(II) in *Deinococcus radiodurans* facilitates gamma-radiation resistance. *Science*, **306**, 1025–8.

Daly, M. J. and Minton, K. W. (1996). An alternative pathway of recombination of chromosomal fragments precedes recA-dependent recombination in the radioresistant bacterium *Deinococcus radiodurans*. *Journal of Bacteriology*, **178**(15), 4461–71.

Diaz, B. and Schulze-Makuch, D. (2006). Microbial survival rates of *Escherichia coli* and *Deinococcus radiodurans* under low temperature, low pressure, and UV-irradiation conditions, and their relevance to possible Martian life. *Astrobiology*, **6**(2), 332–47.

Earl, A. M., Mohundro, M. M., Mian, I. S. and Battista, J. R. (2002). The IrrE protein of *Deinococcus radiodurans* R1 is a novel regulator of recA expression. *Journal of Bacteriology*, **184**(22), 6216–24.

Englander, J., Klein, E., Brumfeld, V., Sharma, A. K., Doherty, A. J. and Minsky, A. (2004). DNA toroids, framework for DNA repair in *Deinococcus radiodurans* and in germinating bacterial spores. *Journal of Bacteriology*, **186**(18), 5973–7.

Fredrickson, J. K., Li, S. M., Gaidamakova, E. K., Matrosova, V. Y., Zhai, M., Sulloway, H. M., Scholten, J. C., Brown, M. G., Balkwill, D. L. and Daly, M. J. (2008). Protein oxidation, key to bacterial desiccation resistance? *ISME Journal*, **2**(4), 393–403.

Gerard, E., Jolivet, E., Prieur, D. and Forterre, P. (2001). DNA protection mechanisms are not involved in the radioresistance of the hyperthermophilic Archaea *Pyrococcus abyssi* and *P. furiosus*. *Molecular Genetics and Genomics*, **266**(1), 72–8.

Hefferin, M. L. and Tomkinson, A. E. (2005). Mechanism of DNA double-strand break repair by non-homologous end joining. *DNA Repair*, **4**(6), 639–48.

Henne, A., Bruggemann, H., Raasch, C., Wiezer, A., Hartsch, T., Liesegang, H., Johann, A., Lienard, T., Gohl, O., Martinez-Arias, R., Jacobi, C., Starkuviene, V., Schlenczeck, S., Dencker, S., Huber, R., Klenk, H. P., Kramer, W., Merkl, R., Gottschalk, G. and Fritz, H. J. (2004). The genome sequence of the extreme thermophile *Thermus thermophilus*. *Nature Biotechnology*, **22**(5), 547–53.

Hua, Y., Narumi, I., Gao, G., Tian, B., Satoh, K., Kitayama, S. and Shen, B. (2003). PprI, a general switch responsible for extreme radioresistance of *Deinococcus radiodurans*. *Biochemical and Biophysical Research Communications*, **306**(2), 354–60.

Kobayashi, Y., Narumi, I., Satoh, K., Funayama, T., Kikuchi, M., Kitayama, S. and Watanabe, H. (2004). Radiation response mechanisms of the extremely radioresistant bacterium *Deinococcus radiodurans*. *Biological Sciences in Space*, **18**(3), 134–5.

Krasin, F. and Hutchinson, F. (1977). Repair of DNA double-strand breaks in *Escherichia coli*, which requires *recA* function and the presence of a duplicate genome. *Journal of Molecular Biology*, **116**(1), 81–98.

Lecointe, F., Shevelev, I. V., Bailone, A., Sommer, S. and Hubscher, U. (2004). Involvement of an X family DNA polymerase in double-stranded break repair in the radioresistant organism *Deinococcus radiodurans*. *Molecular Microbiology*, **53**(6), 1721–30.

Levin-Zaidman, S., Englander, J., Shimoni, E., Sharma, A. K., Minton, K. W. and Minsky, A. (2003). Ringlike structure of the *Deinococcus radiodurans* genome, a key to radioresistance? *Science*, **299**(5604), 254–6.

Makarova, K. S., Aravind, L., Daly, M. J. and Koonin, E. V. (2000). Specific expansion of protein families in the radioresistant bacterium *Deinococcus radiodurans*. *Genetica*, **108**(1), 25–34.

Makarova, K. S., Aravind, L., Wolf, Y. I., Tatusov, R. L., Minton, K. W., Koonin, E. V. and Daly, M. J. (2001). Genome of the extremely radiation-resistant bacterium *Deinococcus radiodurans* viewed from the perspective of comparative genomics. *Microbiology and Molecular Biology Reviews*, **65**(1), 44–79.

Makarova, K. S., Omelchenko, M. V., Gaidamakova, E. K., Matrosova, V. Y., Vasilenko, A., Zhai, M., Lapidus, A., Copeland, A., Kim, E., Land, M., Mavrommatis, K., Pitluck, S., Richardson, P. M., Detter, C., Brettin, T., Saunders, E., Lai, B., Ravel, B., Kernner, K. M., Wolf, Y. I., Sorokin, A., Gerasimova, A.V., Gelfand, M. S., Fredrickson, J. K., Koonin, E. V., Daly, M. J. (2007). *Deinococcus geothermalis*: the pool of extreme radiation resistance genes shrinks. *PLoS ONE*, **2**(9), e955.

Markillie, L. M., Varnum, S. M., Hradecky, P. and Wong, K. K. (1999). Targeted mutagenesis by duplication insertion in the radioresistant bacterium *Deinococcus radiodurans*, radiation sensitivities of catalase (katA) and superoxide dismutase (sodA) mutants. *Journal of Bacteriology*, **181**(2), 666–9.

Mattimore, V. and Battista, J. R. (1996). Radioresistance of *Deinococcus radiodurans*, functions necessary to survive ionizing radiation are also necessary to survive prolonged desiccation. *Journal of Bacteriology*, **178**(3), 633–7.

Narumi, I., Satoh, K., Cui, S., Funayama, T., Kitayama, S. and Watanabe, H. (2004). PprA, a novel protein from *Deinococcus radiodurans* that stimulates DNA ligation. *Molecular Microbiology*, **54**(1), 278–85.

Nelson, K. E., Clayton, R. A., Gill, S. R., Gwinn, M. L., Dodson, R. J., Haft, D. H., Hickey, E. K., Peterson, J. D., Nelson, W. C., Ketchum, K. A., McDonald, L., Utterback, T. R., Malek, J. A., Linher, K. D., Garrett, M. M., Stewart, A. M., Cotton, M. D., Pratt, M. S., Phillips, C. A., Richardson, D., Heidelberg, J., Sutton, G. G., Fleischmann, R. D., Eisen, J. A., White, O., Salzberg, S. L., Smith, H. O., Venter, J. C.,

Fraser, C. M., (1999). Evidence for lateral gene transfer between Archaea and Bacteria from genome sequence of *Thermotoga maritima*. *Nature*, **399**(6734), 323–9.

Omelchenko, M. V., Wolf, Y. I., Gaidamakova, E. K., Matrosova, V. Y., Vasilenko, A., Zhai, M., Daly, M. J., Koonin, E. V. and Makarova, K. S. (2005). Comparative genomics of *Thermus thermophilus* and *Deinococcus radiodurans*: divergent routes of adaptation to thermophily and radiation resistance. *BMC Evolutionary Biology*, **5**, 57.

Paques, F. and Haber, J. E. (1999). Multiple pathways of recombination induced by double-strand breaks in *Saccharomyces cerevisiae*. *Microbiology and Molecular Biology Reviews*, **63**(2), 349–404.

Pavlov, A. K., Kalinin, V. L., Konstantinov, A. N., Shelegedin, V. N. and Pavlov, A. A. (2006). Was Earth ever infected by Martian biota? Clues from radioresistant bacteria. *Astrobiology*, **6**(6), 911–18.

Potts, M. (1994). Desiccation tolerance of prokaryotes. *Microbiology Reviews*, **58**(4), 755–805.

Rainey, F. A., Ray, K., Ferreira, M., Gatz, B. Z., Nobre, M. F., Bagaley, D., Rash, B. A., Park, M. J., Earl, A. M., Shank, N. C., Small, A. M., Henk, M. C., Battista, J. R., Kampfer, P. and da Costa, M. S. (2005). Extensive diversity of ionizing-radiation-resistant bacteria recovered from Sonoran Desert soil and description of nine new species of the genus *Deinococcus* obtained from a single soil sample. *Applied and Environmental Microbiology*, **71**(9), 5225–35.

Servant, P., Jolivet, E., Bentchikou, E., Mennecier, S., Bailone, A. and Sommer, S. (2007). The ClpPX protease is required for radioresistance and regulates cell division after gamma-irradiation in *Deinococcus radiodurans*. *Molecular Microbiology*, **66**(5), 1231–9.

Shuman, S. and Glickman, M. S. (2007). Bacterial DNA repair by non-homologous end joining. *Nature Reviews Microbiology*, **5**(11), 852–61.

Slade, D., Lindner, A. B., Paul, G. and Radman, M. (2009). Recombination and replication in DNA repair of heavily irradiated *Deinococcus Radiodurans*. *Cell*, **136**, 1044–55.

Tanaka, M., Earl, A. M., Howell, H. A., Park, M. J., Eisen, J. A., Peterson, S. N. and Battista, J. R. (2004). Analysis of *Deinococcus radiodurans's* transcriptional response to ionizing radiation and desiccation reveals novel proteins that contribute to extreme radioresistance. *Genetics*, **168**(1), 21–33.

Weller, G. R., Kysela, B., Roy, R., Tonkin, L. M., Scanlan, E., Della, M., Devine, S. K., Day, J. P., Wilkinson, A., d'Adda di Fagagna, F., Devine, K. M., Bowater, R. P., Jeggo, P. A., Jackson, S. P. and Doherty, A. J. (2002). Identification of a DNA nonhomologous end-joining complex in bacteria. *Science*, **297**(5587), 1686–9.

White, O., Eisen, J. A., Heidelberg, J. F., Hickey, E. K., Peterson, J. D., Dodson, R. J., Haft, D. H., Gwinn, M. L., Nelson, W. C., Richardson, D. L., Moffat, K. S., Qin, H., Jiang, L., Pamphile, W., Crosby, M., Shen, M., Vamathevan, J. J., Lam, P., McDonald, L., Utterback, T., Zalewski, C., Makarova, K. S., Aravind, L., Daly, M. J., Minton, K. W., Fleischmann, R. D., Ketchum, K. A., Nelson, K. E., Salzberg, S., Smith, H. O., Venter, J. C., Fraser, C. M. (1999). Genome sequence of the radioresistant bacterium *Deinococcus radiodurans* R1. *Science*, **286**(5444), 1571–7.

Wyman, C., Ristic, D. and Kanaar, R. (2004). Homologous recombination-mediated double-strand break repair. *DNA Repair* (Amst), **3**(8–9), 827–33.

Zahradka, K., Slade, D., Bailone, A., Sommer, S., Averbeck, D., Petranovic, M., Lindner, A. B. and Radman, M. (2006). Reassembly of shattered chromosomes in *Deinococcus radiodurans*. *Nature*, **443**(7111), 569–73.

Zimmerman, J. M. and Battista, J. R. (2005). A ring-like nucleoid is not necessary for radioresistance in the *Deinococcaceae*. *BMC Microbiology*, **5**(1), 17.

23

Molecular effects of UV and ionizing radiations on DNA

Jean Cadet and Thierry Douki

Introduction

Survival of microorganisms in outer space, such as resistant bacterial endospores, is affected by harsh environmental conditions including microgravity, space vacuum leading to desiccation, wide variations in temperature and a strong radiation component of both galactic and solar origins (Nicholson *et al.*, 2000). Solar extraterrestrial UV radiation is mostly deleterious due to its UV component consisting of genotoxic UVC ($200 < \lambda < 280$ nm) and more energetic vacuum–UV photons ($140 < \lambda < 200$ nm) that are able to ionize biomolecules but exhibit very low penetrating features. The galactic cosmic radiation (CGR) is composed predominantly of high-energy protons (85%), electrons, α-particles and high-charge (Z) and energy (E) nuclei (HZE). In addition, solar particle radiation that mostly consists of protons with very small amounts of α-particles and HZE ions is emitted during solar wind and erratic solar flares (Nicholson *et al.*, 2000; Cucinotta *et al.*, 2008). UVC and UVB photons ($280 < \lambda < 320$ nm) are, in the absence of shielding, the main lethal components of space radiation. However, an efficient protection against molecular effects of UV radiations is likely to occur when spores are embedded in micrometeorites according to the scenario that has been proposed for allowing interplanetary or interstellar transfer of microorganisms (Mileikowski *et al.*, 2000; Nicholson *et al.*, 2000). In contrast, under the latter conditions, protection of microorganisms against the damaging effects of CGR, and more precisely, of highly penetrating HZE particles, is at best very limited. The biological effects of UV and ionizing radiations are mostly accounted for by chemical modifications of biomolecules with special attention to nucleic acids, at least to explain the mutagenicity and carcinogenicity of the latter physical agents on humans. Abundant and comprehensive data are now available on the degradation pathways mediated by UV photons (Taylor, 1994; Cadet *et al.*, 2005) and high-energetic ionizing radiation (Cadet *et al.*, 1999, 2008) as inferred from detailed model studies. However, there is still a paucity of information on molecular effects concerning cellular DNA, at least for those effects generated by high-energy photons and heavy ions. Emphasis is placed in this short survey on the description of the main decomposition reactions of cellular DNA that have been identified so far upon exposure to UV light and ionizing radiation, including HZE particles that exhibit high ionization density reflected by a high linear-energy transfer (LET).

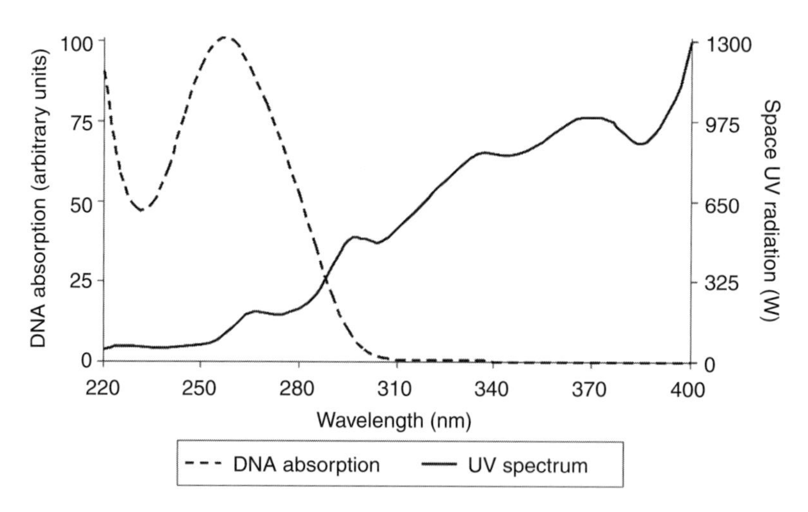

Figure 23.1. Comparison of UV spectrum of cosmic radiation and DNA absorption spectrum.

UV radiation

UV radiation is probably the most efficient genotoxic agent in outer space. Indeed, cosmic rays are spread over the entire UV range where DNA exhibits significant absorption (Figure 23.1). As a result, DNA exposed to space radiation may undergo significant photoreaction triggered by UVC photons. This contrasts with the situation at the Earth's surface where UVC is absent and where only less energetic and less efficiently absorbed UVB photons account for the genotoxicity of sunlight.

Basic photochemistry of DNA

The photochemistry of DNA in the UVC and UVB ranges is dominated by the dimerization of adjacent pyrimidine bases (Cadet *et al.*, 2005). Two main types of photoproducts are induced, namely the cyclobutane pyrimidine dimers (CPDs) and the pyrimidine (6–4) pyrimidone photoproducts (64PPs). A few other photoproducts are produced upon UV-irradiation of DNA but in yields that are at the most one or two orders of magnitude lower than that of pyrimidine dimers. These lesions include adenine dimers, adenine–thymine dimers and cytosine photohydrates. DNA photodamage is also induced by UVA photons, mostly through photosensitization reactions.

Cyclobutane–pyrimidine dimers

CPDs are produced by (2 + 2) cycloaddition between the C5–C6 double bonds of two adjacent pyrimidines (Figure 23.2). Although several diastereomers can be produced, only the *cis,syn* one, with the two bases in a parallel orientation and located on the same side of the cyclobutane ring, is produced in double-stranded DNA. Among other properties, CPDs have been shown to undergo efficient reversion upon exposure to UVC radiation.

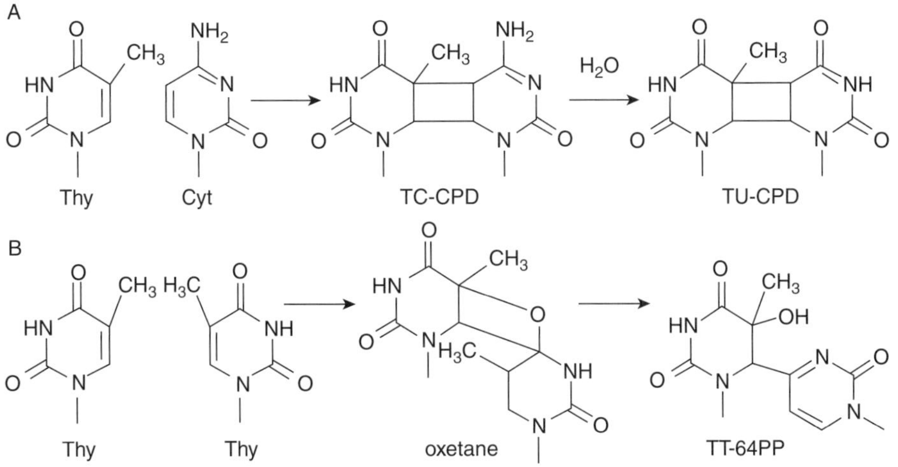

Figure 23.2. (A) Formation and deamination of the thymine–cytosine CPD, and (B) formation of the thymine–thymine 64 PP through an oxetane intermediate. Both types of photoproducts can be produced at TT, TC, CT and CC sites. TC–CPD: thymine–cytosine cyclobutane dimer; TU–CPD: thymine–uracil cyclobutane dimer; TT–64PP: thymine–thymine pyrimidine (6–4) pyrimidone photoproduct.

Consequently, exposure of DNA to UVC leads to a plateau and even a decrease in the frequency of CPDs at high fluxes. Another interesting feature of CPDs is related to the saturated C5–C6 bond. Indeed, cytosine derivatives exhibiting this structure are known to be efficiently hydrolyzed into the corresponding uracil derivative through the so-called deamination reaction (Figure 23.2).

Pyrimidine (6–4) pyrimidone photoproducts

64PPs represent the second major class of UVC- and UVB-induced bipyrimidine lesions in DNA. They arise from a (2 + 2) Paterno–Bücchi cycloaddition between the C5–C6 double bond of the 5′-end pyrimidine and the C4 carbonyl of a thymine (Figure 23.2) or the imine group of a cytosine. The 64PPs arising from a 5-end cytosine may undergo deamination like CPDs. In addition, the pyrimidone ring of the 3′-end base can be converted by UVB photons into a Dewar valence isomer because 64PPs have a maximum absorption around 320 nm. Interestingly, UVA photons are also efficient, especially at low doses in double-stranded DNA and cells where normal bases compete for UVB photon absorption.

UVA-induced DNA damage

DNA absorbs UVA photons very poorly. However, DNA damage may be produced in this wavelength range by photosensitization (Cadet *et al.*, 2005), probably a minor pathway proportion compared to UVC photochemistry in space. In these reactions, light is absorbed by endogenous chromophores that then trigger genotoxic pathways. Oxidation reactions are well documented and may first involve electron abstraction, as discussed below for ionization of DNA. The second pathway involves transfer of excitation energy to molecular oxygen. In both cases guanine is the main target. It should be emphasized, though, that these processes involve molecular oxygen, which is likely to be absent in space or on most planets. UVA has also been reported to induce the formation of cyclobutane thymine dimers in cells (Tyrrell, 1973) and even isolated DNA (Quaite *et al.*, 1992). Although the underlying mechanism remains unclear, this reaction may be of biological relevance, since CPDs were found to be produced in larger yields than oxidative lesions (Douki *et al.*, 2003a; Mouret *et al.*, 2006). Similar data are still lacking for microorganisms.

Formation and fate of bipyrimidine photoproducts in cells

Formation of bipyrimidine photoproducts in cells

UVC radiation has been reported to generate between 2 and 10 CPDs per 10^6 bases per J/m^2 in cultured mammalian cells (Mitchell *et al.*, 1991). The ratio between CPDs and 64PPs ranges between 3 and 5. The yield of damage induced by UVB radiation is approximately one order of magnitude lower than with UVC. Interestingly, the ratio between CPD and 64PP, as well as the overall photoreactivity, varies from one bipyrimidine doublet to the other (Douki and Cadet, 2001). As a general trend, TT and TC are more sensitive than CT and CC. In addition, CPDs are produced in a 10-fold higher yield than 64PPs at TT sites, whereas a ratio closer to 1 is obtained for TC. The quantitative data in microorganisms are more difficult to summarize because they often produce photoprotective compounds and exhibit different membrane composition that shield UV photons to different extents. Yields in the range of 8 bipyrimidine photoproducts/10^6 normal bases per J/m^2 have been reported for *Deinococcus radiodurans* (Pogoda de la Vega *et al.*, 2005) and of 42 for *Bacillus subtilis* (Moeller *et al.*, 2007a). In addition, the composition of DNA in terms of G + C content strongly varies from one species to the other (from 25 to 70%). Variation of this parameter is much more limited in eukaryotic cells (40–45%). The G+C content of the genomes of bacteria was found to have a huge impact on the distribution of photoproducts (Matallana-Surget *et al.*, 2008), with the TT CPDs being a minor lesion at 70% G + C, while it is the main one in mammalian cells that all exhibit a G + C content close to 42%.

Mutagenicity of DNA photoproducts

Because they represent bulky damage within the double helix, UV-induced photoproducts can strongly interfere with basic cellular processes such as replication. Therefore, photoproducts may lead to cell death, for instance, after induction of apoptosis in mammalian cells. If cells survive and divide, mutational events may occur. Indeed, the DNA polymerases may misincorporate the bases in the synthesized strand when encountering a photoproduct (Lawrence *et al.*, 1993; Taylor, 1994). This risk is enhanced for deaminated cytosine photoproducts because the uracil residues code like a thymine. Accordingly, C to T at TC sites and CC to TT double mutations constitute the mutational signature of UV radiation (Brash *et al.*, 1991). It may be added that specialized polymerases, referred to as translesion-synthesis polymerases, are expressed in cells from different domains of life (Bacteria, Archaea, Eukarya) in order to overcome the blocking events associated with the presence of the lesions and to reduce risks of cell death.

DNA repair

A first repair mechanism used by cells to remove bipyrimidine photoproducts from DNA is nucleotide-excision repair (NER). This multi-enzymatic process is present in eukaryotes and prokaryotes, as well as Archaea. It may involve different proteins in these different cell types but follows the same general mechanism (Gillet and Scharer, 2006; Truglio *et al.*, 2006). NER begins with the recognition of the structural modification induced by the photoproduct. Proteins are then recruited that cleave the damaged strand on both sides of the lesion. This step leaves a portion of single-stranded DNA that is further filled by polymerases and then ligated to restore the initial undamaged sequence. In addition to NER, plants, numerous bacteria, yeasts and even a few animals like some fish and opossums, are equipped with an additional repair mechanism absent in humans: the photoreversion of the photoproducts into the initial bases. The involved enzymes are photolyases (Sancar, 2003) specific for either CPDs or 64PPs and are activated by visible or UVA light.

The specific DNA photobiology in bacterial spores

Specific features of DNA in spores

Spores are dormant forms of some bacterial species such as *Bacillus subtilis* (Nicholson *et al.*, 2005) produced under unfavourable conditions. These microorganisms are resistant to a wide series of stresses, including UV radiation (Setlow, 2001). In that respect, they are interesting models for life forms surviving in space vacuum or on planets lacking proper atmospheric UV shielding. Spores are composed of a dehydrated core surrounded by multiple layers of membranes and walls. The core contains large amounts of small acid-soluble proteins (SASP) that bind DNA and modify its structure into an A-like form. The core also exhibits a very high concentration of calcium dipicolinate (Ca-DPA) that further dehydrates DNA. Ca-DPA was found to photosensitize the formation of the spore photoproduct (SP) upon exposure to UVC (Setlow and Setlow, 1993), while protecting DNA against UVB and UVA radiation (Slieman and Nicholson, 2000).

Photochemistry of spore DNA

A consequence of the above-mentioned properties of the spore core is a drastic change in DNA photochemistry. Indeed, the only bipyrimidine photoproduct found in the DNA of UVC-irradiated spores is 5,6-dihydro-5-(α-thyminyl)-thymine (the so called SP), a lesion resulting from the formation of a covalent bridge between two thymines through one of their methyl groups (Figure 23.3). Exposure to UVC radiation leads to the formation of roughly 12 SP/10^6 normal bases per J/m^2 in *Bacillus subtilis* spores (Moeller *et al.*,

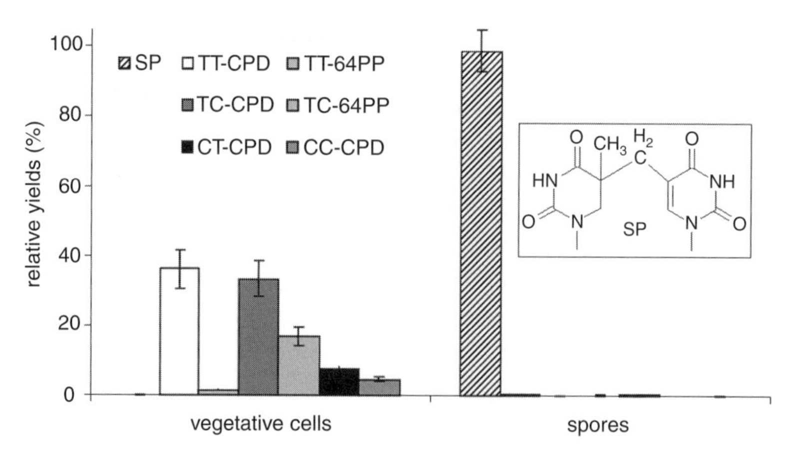

Figure 23.3. Comparison of the distribution of bipyrimidine photoproducts in vegetative cells and in spores of *Bacillus subtilis* upon exposure to UVC radiation. The inset shows the structure of the spore photoproduct. TT: thymine–thymine; TC: thymine–cytosine; CT: cytosine–thymine; CC: cytosine–cytosine; CPD: cyclobutane dimer; 64PP: pyrimidine (6–4) pyrimidone photoproduct; SP: spore photoproduct.

2007b). Dehydration partly explains this photoreactivity, since exposure of dry films of isolated DNA to UV radiation leads to the formation of SP in addition to CPDs and 64PPs. Interestingly, dehydration also modifies the ratio between CPDs and 64PPs (Melly *et al.*, 2002), especially at TC dinucleotides where 64PP becomes the major lesion (Douki and Cadet, 2003b). This favours UV-induced formation of SP upon dehydration is amplified by the addition of SASP and Ca-DPA. Conversely, CPDs and 64PPs are found in significant amounts in spores lacking either SASP or DPA. Interestingly, the overall amount of photoproducts remains approximately constant in these different spores.

The spore photoproduct lyase

Spores are metabolically inactive and no repair of DNA can take place therein. However, a very efficient repair protein loaded into the spore during sporulation, the spore photoproduct lyase (SPL) (Fajardo-Cavazos *et al.*, 1993) reverts the accumulated SP into the starting thymines in the very early stages of germination. This enzyme belongs to the radical-SAM family of proteins and repairs the photoproduct through a radical mechanism. S-adenosylmethionine (SAM) is converted into an adenosine radical (Ado•) in the active site of the enzyme. Ado• then abstracts the H6 hydrogen of the saturated thymine moiety of SP, thereby inducing the reversal of the photoproduct into unmodified thymine (Slieman *et al.*, 2000; Cheek and Broderick, 2002; Chandor-Proust *et al.*, 2008). It thus appears that the photoresistance of spores is explained by their unique DNA photochemistry selecting a specific photoproduct, combined with the presence of an efficient repair enzyme for this lesion that restores undamaged DNA much faster than NER would eliminate CPDs and 64PPs. The importance of SPL is unambiguously shown by the observation that its inactivation drastically reduces spore survival. It may be added that non-homologuous end-joining, a repair mechanism for double-strand breaks, was reported to play a role in the photoresistance of spores (Moeller *et al.*, 2007c).

Ionizing radiation

Space ionizing radiation is characterized by the predominance of high-energy protons and densely ionizing high-LET HZE particles ranging from helium to uranium, whereas terrestrial radiation mostly consists of low-LET photons (X-, β- or γ-rays) (Durante & Cucinotta, 2008). Interaction of ionizing radiation with biological matter is suitably described using a microdosimetry model that highlights the lack of homogeneity and stochastic aspects of energy deposition (Goodhead, 2006). Relevant information is available mostly from Monte Carlo calculations on the effects of track structure following energy deposition. It was thus predicted that clustered damage to DNA (Goodhead, 1994) is expected to occur almost exclusively within single radiation tracks as the result of multiple ionization and excitation events. Ionization of molecules of water bound or in close vicinity to DNA gives rise to highly reactive •OH that is able to efficiently react without diffusion with nucleobases and/or the sugar moiety as the result of the so-called indirect effects (O'Neill and Wardman, 2009). Interaction of high-energy photons and HZE is able to oxidize nucleobases and the 2-deoxribose units through the loss of an electron that gives rise to reactive radical cations. Both processes lead in most cases to the generation of single damage that may consist of modified bases, oxidized abasic sites and strand breaks (Cadet *et al.*, 1999; Dedon, 2008). However, as already discussed, complex lesions initially called 'multiply damaged sites' (Ward, 1988) that may consist of double-strand breaks, several modified bases or single-strand breaks associated with base lesion(s) within one or two helix turns are also expected to be generated with a frequency and complexity that increase with LET.

Isolated lesions

Several radiation-induced base modifications that may also be part of clustered lesions have been recently detected and their formation quantified in cellular DNA using the accurate and sensitive HPLC-electrospray ionization tandem mass spectrometry operating in the accurate multiple-reaction monitoring mode (Frelon *et al.*, 2000).

Hydroxyl radical-mediated oxidation of thymine

Exposure of THP-1 human monocytes to heavy ions provided on Earth by accelerators has been shown to generate six oxidation products of thymidine including 5-(hydroxymethyl)-2′-deoxyuridine (5-HmdUrd), 5-formyl-2′-deoxyuridine (5-FordUrd) and the four *cis* and *trans* diastereomers of 5,6-dihydroxy-5,6-dihydrothymidine (ThdGly) (Pouget *et al.*, 2002; Douki *et al.*, 2006). The yields of the radiation-induced oxidized nucleosides whose formation is linear with the applied dose within the dose range 90–450 Gy are relatively low and

Table 23.1. Yields[a] of degradation products of thymine, guanine and adenine in the DNA of THP-1 malignant human monocytes upon exposure to γ–rays and $^{12}C^{6+}$ particles[b,c].

DNA lesions	$^{12}C^{6+}$ heavy ions	γ–rays
Cis and trans ThdGly	62	97
5-HmdUrd	12	29
5-FordUrd	11	22
8-OxodGuo	10	20
FapyGua	22	39
8-oxodAdo	3	3
FapyAde	1	5

[a] Expressed in lesions per 10^9 nucleobases and per Gy.
[b] Linear energy transfer: 31.5 keV/μm.
[c] From Pouget *et al.*, 2002.

are listed in Table 23.1 as being comprised of between 29 and 97 lesions per 10^9 bases Gy^{-1}. As a striking feature, high-LET $^{12}C^{6+}$particles were found to be less efficient than low-LET γ-radiation at inducing thymidine degradation products. A further decrease in the formation efficiency of ThdGly, 5-HmdUrd and 5-FordUrd is noted upon exposure to $^{36}Ar^{18+}$ that exhibits a higher LET value than that of $^{12}C^{6+}$ heavy ions (Douki *et al.*, 2006).

These observations are probably explained by the decrease in the radiation-induced yield of •OH as LET increases (Goodhead, 1994). As depicted in Figure 23.4, generation of the four diastereomers of ThdGly is explained by initial addition of •OH at C5, and to a lesser extent at C6, of the thymine moiety. In a subsequent step, fast addition of molecular oxygen takes place with reducing 6-yl and oxidizing 5-yl pyrimidine radicals (Cadet *et al.*, 1999). The resulting peroxyl radicals are then converted by reactions with the superoxide radical and subsequent protonation into the corresponding diastereomeric 5- and 6-hydroperoxides that, upon reduction of the peroxidic bond, give rise to ThdGly. 5-(Hydroperoxymethyl)-2′-deoxyuridine that arises from •OH-mediated hydrogen abstraction at the methyl group of thymidine and subsequent O_2 addition to the resulting 5-(2′-deoxyuridilyl) methyl radical is the probable precursor of the stable methyl oxidation products (Cadet *et al.*, 1999, 2008). Loss of a water molecule from the peroxidic function leads to the generation of 5-FordUrd, whereas competitive reduction gives rise to 5-HmdUrd (Figure 23.4).

•OH-mediated degradation pathways of purine bases

8-Oxo-7,8-dihydro-2′-deoxyguanosine (8-oxodGuo) together with 2,6-diamino-4-hydroxy-6-formamidopyrimidine (FapyGua), the related opened imidazole ring compound, were found to be generated in cellular DNA upon exposure to gamma rays and high-LET-$^{12}C^{6+}$ and $^{36}Ar^{18+}$ heavy ions (Pouget *et al.* 2002; Douki *et al.*, 2006). Interestingly, as for

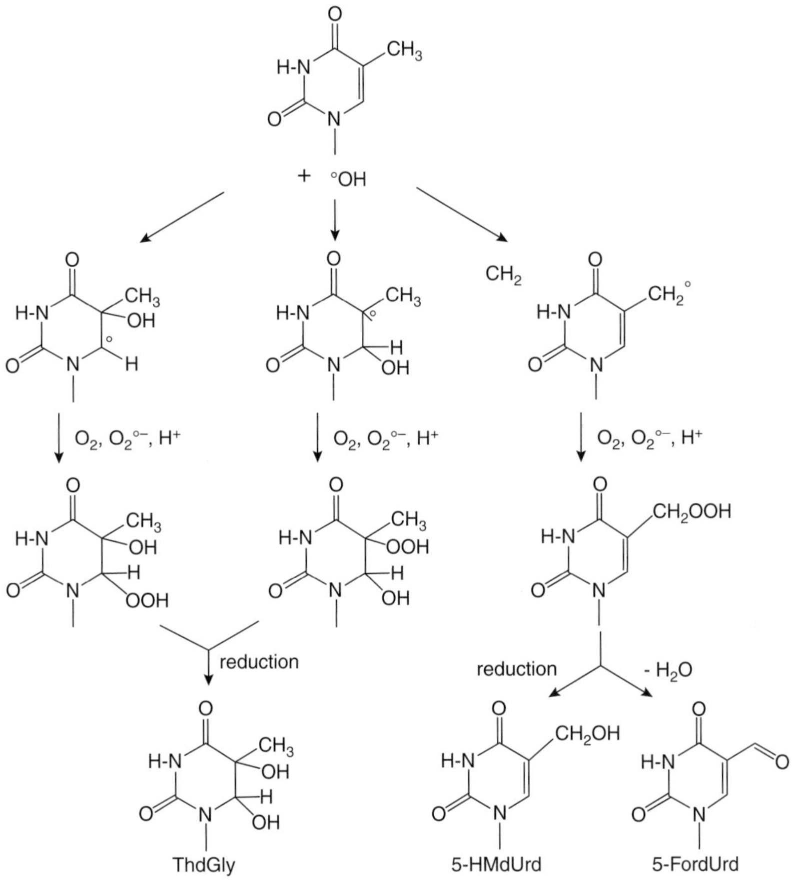

Figure 23.4. Hydroxyl radical-mediated oxidation of the thymine moiety in DNA.

thymidine oxidation products, the radiation-induced formation yields of both 8-oxodGuo and FapyGua were found to decrease with the increase in the LET of the incident photon or particle (Table 23.1). This again is suggestive of the major implication of the •OH radical in the molecular effects of ionizing radiation on the guanine moiety of cellular DNA. In agreement with previous mechanistic studies, addition of •OH to the purine ring at C8 leads to the formation of reducing 8-hydroxy-7,8-dihydro-7-yl radicals (Figure 23.5) which, in the presence of oxidants such as O_2, give rise to 8-oxo-7,8-dihydroguanine. A competitive reaction of the latter radical is one-electron reduction that leads to the formation of FapyGua through the scission of the C8–N9 imidazole bond (Cadet *et al.*, 2008).

Similar radiation-induced degradation pathways of the adenine moiety of cellular DNA have been shown to occur through initial formation of the •OH addition adduct at C8 of the purine ring. This leads to the formation of 8-oxo-7,8-dihydro-2′-deoxyadenosine (8-oxodAdo) and the related imidazole ring open compound, namely 4,6-diamino-5-formamidopyrimidine

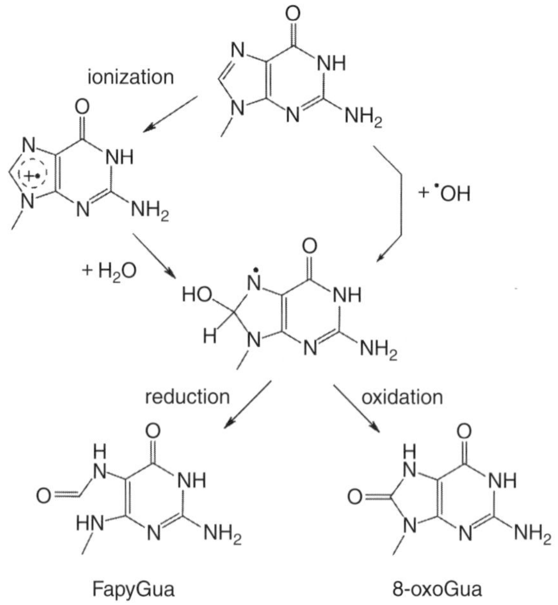

Figure 23.5. Degradation pathways of the guanine moiety in DNA through the indirect (•OH) and direct (ionization) effects of ionizing radiations.

(FapyAde) through oxidation and reduction, respectively, of the 8-hydroxy-7,8-dihydroadenyl radical thus formed. It may be noted that the yields of radiation-induced adenine-degradation products are much lower than those of guanine (Table 23.1).

Ionization reactions of cellular DNA

One-electron oxidation of DNA is involved in the direct effect of ionizing radiation and could therefore be implicated in the biological effects of high-energy UV photons ($\lambda < 200$ nm). However, ionization processes that cannot be easily investigated in DNA, at least in aqueous solution and model studies, have to be designed. Two-quantum photo-ionization provided by 266-nm nanosecond laser pulses has been shown to be an efficient way to mimic the direct ionization effects of at least the purine and pyrimidine bases of free nucleosides and isolated DNA (Douki *et al.*, 2001). Depletion of the initially generated triplet excited-state nucleobases leads upon absorption of a second UV pulse to related radical cations and subsequent chemical reactions. This photochemical approach has been successfully applied to cellular DNA by searching for dedicated final one-electron oxidation products of the nucleobases (Douki *et al.*, 2006). The main oxidized nucleoside was found to be 8-oxodGuo whereas thymidine oxidation products including ThdGly, 5-HmdUrd and 5-FordUrd were generated in much lower yields, the relative ratio between the two groups of degradation products being five. The formation of the latter degradation products may be explained in terms of initial generation of the thymidine radical cation, whose chemical reactions in aerated aqueous solutions have been previously assessed in model studies (Cadet *et al.*, 1999). Thus, the formation of the four diastereomers of ThdGly is rationalized by the transient formation of the 6-hydroxy-5, 6-dihydrothym-5-yl radical upon the specific nucleophilic addition of OH^- at C6 of the thymine radical cation, whereas competitive deprotonation of the latter transient species give rises to the 5-(2′-deoxyuridilyl) methyl radical, the precursor of 5-HmdUrd and 5-FordUrd in aqueous aerated solutions (Figure 23.4). The formation of 8-oxodGuo is explained in terms of hydration of the guanine radical cation (Figure 23.5) that may be produced through direct one-electron oxidation of a guanine residue or, subsequent to hole migration, to a guanine base that acts as a sink from distant adenine and/or pyrimidine radical cations (Cadet *et al.*, 2008). This constitutes the first example so far discovered of the occurrence of charge-transfer reactions within cellular DNA that have been shown to take place within double-stranded oligonucleotides according to several mechanisms including multistep hopping, phonon-assisted polaron-like hopping and coherent super-exchange. The observation of a larger yield of thymine than guanine degradation products in the DNA of cells exposed to ionizing radiation indicates that ionization through direct effects is a relatively minor damaging pathway, even for heavy ions (Pouget *et al.*, 2002; Douki *et al.*, 2006).

Complex DNA damage

Recent progress in the elucidation of the mechanisms of oxidative reactions to DNA have emphasized the probable role of complex lesions that may be produced through several pathways. First, complex DNA damage may be generated by one initial single oxidation event mediated by •OH. For instance, complex radiation-induced DNA lesion formation involves initial hydrogen-atom abstraction at C4′, and subsequent cycloaddition reactions to cytosine were detected in THP1 monocytes (Regulus *et al.*, 2007). It has also been shown that the •OH-induced 5-(uracilyl) methyl radical is able to covalently attach to a vicinal guanine, generating a tandem base lesion in human HeLa-S3 cells (Jiang *et al.*, 2007). However, the main source of radiation-induced complex DNA modifications in cells is the accumulation of damage along the photon or the particle track within one or two helix turns as the result of multi radical and/or excitation events (Hada and Georgakilas, 2008). The clustered lesions thus formed may be double-strand breaks (DSBs) consisting of two closely spaced and opposed single-strand breaks and non-DSB oxidatively generated clustered DNA lesions (OCDLs) containing multiple breaks and/or modified bases.

Double-strand breaks

DNA double-strand breaks constitute a broad class of deleterious nuclear damage that arises from the radiation-induced formation of two nicks on the two opposite DNA strands and that are separated at a maximum by 15 base pairs. Monte Carlo track-structure simulations of the physical and chemical processes have predicted that low-LET radiation such as 4.5-keV electrons, through direct and indirect effects, induce about 20% of complex DSBs

Table 23.2. Yields[a] of DSBs and non-DSB oxidatively generated clustered lesions in human monocytes upon exposure to ^{56}Fe ions and ^{137}Cs γ-rays[b].

Radiation	DSBs	Fpg clusters	Endo III clusters	Endo IV clusters
^{56}Fe ions	10.9	8.5	7.1	5.5
γ-rays	11.9	11.9	10.7	9.5

[a] Expressed as the number of lesions per 10^9 base pairs and per Gy.
[b] From Tsao *et al.*, 2007.

that exhibit at least one additional DNA nick and/or modified base. This proportion has been estimated to increase up to 70% for 2-MeV particles (Nikjoo *et al.*, 2001). Evidence for a higher density of generated DSBs along the ion trajectories with the LET increase of applied HZE was provided by the observation of a larger frequency of fluorescent γ–H2AX clusters for iron ions (LET = 176 keV/nm) with respect to lower LET silicon (LET = 54 keV/nm) heavy particles (Desai *et al.*, 2005). It was also shown that the DSBs generated in human fibroblasts upon exposure to iron ions (150 and 236 keV/nm) are far less subject to repair that the less complex DSBs produced by less energetic silicon (44 keV/nm) and oxygen (14 keV/nm) ions (Asaithamby *et al.*, 2008). This may provide an explanation for the observation of the persistence of chromosome aberrations in mice that were exposed to $^{56}Fe^{26+}$ ions (Tucker *et al.*, 2004).

Non-double-strand break clustered lesions

A repair-based assay has been developed for the detection of non-DSB clustered lesions that consist of one single-strand break together with one or several base lesions and/or apurinic sites on the two complementary DNA strands within one or two DNA helix turns. The DSBs that are generated by enzymic excision of oxidized bases or abasic sites are then detected and quantified using the neutral comet assay of an adaptation of the pulsed-field gel electrophoretic technique (Georgakilas *et al.*, 2004; Hada and Georgakilas, 2008). Three classes of bi-stranded clustered lesions were measured in cellular DNA upon exposure to high-LET heavy ions and gamma rays. These included modified purine bases, oxidized pyrimidine bases and normal and/or oxidized abasic sites that were revealed upon incubation with bacterial formamidopyrimidine glycosylase (Fpg), endonuclease III (endo III) and endonuclease IV (endo IV), respectively. Thus, it was found that the formation of any of the three types of clustered lesions was slightly lower than that of DSBs in the DNA of human monocytes upon exposure to either γ-rays from ^{137}Cs or ^{56}Fe particles (Table 23.2). Similar observations were made for leukaemia Pre-B NALM-6 cells and human breast cancer MCF-7 and related non-malignant cells (Francisco *et al.*, 2008). One may note that the values of the yield of clustered lesions, whose formation efficiency decreases with increasing LET, are similar to those of single oxidized bases in irradiated cells (Table 23.1). This is surprising and perhaps indicative of the occurrence of artefactual oxidation of DNA during its isolation and subsequent processing, as recently suggested (Boucher *et al.*, 2006). It should be noted that up to 15 altered bases have been predicted to be generated within a cluster in cellular DNA (Semenenko and Stewart, 2004), and only one, or at best two of the oxidized bases can be detected using the above assay. It is clear that further work is required to better assess clustered DNA modifications.

Conclusion and perspectives

This short review illustrates the complexity of the molecular mechanisms that are implicated in the degradation of cellular DNA upon exposure to space radiations. Relevant information is available on the main DNA photoreactions that are mediated by UVC and UVB on spores. It remains to be established, however, what the contribution of photons of higher energy – within the range $140 < \lambda < 220$ nm – to the overall degradation effects of vacuum UV light could be. The situation is more complex for ionizing radiation, since identification of clustered lesions remains a challenging issue. It may be added that there is a paucity of data on the radiation chemistry of DNA in spores under vacuum that is likely to be strongly affected by the lack of molecular oxygen by comparison with the DNA degradation pathways established so far in living cells.

References

Asaithamby, A., Uematsu, N., Chatterjee, A., Story, M. D., Burma, S. and Chen, D. J. (2008). Repair of HZE-particle-induced double-strand breaks in normal human fibroblasts. *Radiation Research*, **169**, 437–46.

Boucher, D., Testard, I. and Averbeck, D. (2006). Low levels of clustered oxidative DNA damage induced at low and high LET irradiation in mammalian cells. *Radiation and Environmental and Biophysics*, **45**, 267–76.

Brash, D. E., Rudolph, J. A., Simon, J. A., Lin, A., McKenna, G. J., Baden, H. P., Halperin, A. J. and Ponten, J. (1991). A role for sunlight in skin cancer: UV-induced p53 mutations in squamous cell carcinoma. *Proceedings of the National Academy of Sciences of the USA*, **88**, 10124–8.

Cadet, J., Delatour, T., Douki, T., Gasparutto, D., Pouget, J.-P., Ravanat J.-L. and Sauvaigo, S. (1999). Oxidative damage to DNA: hydroxyl radicals and DNA base damage. *Mutation Research*, **424**, 9–21.

Cadet, J., Sage, E. and Douki, T. (2005). Ultraviolet radiation-mediated damage to cellular DNA. *Mutation Research*, **571**, 3–17.

Cadet, J., Douki, T. and Ravanat, J.-L. (2008). Oxidatively generated damage to the guanine moiety of DNA: mechanistic aspects and formation in cells. *Accounts of Chemical Research*, **41**, 1075–83.

Chandor-Proust, A. Berteau, O., Douki, T., Gasparutto, D., Ollagnier-de-Choudens, S, Atta, M. and Fontecave, M. (2008). DNA repair and free radicals. New insights into the mechanism of spore photoproduct lyase revealed by single mutation. *Journal of Biological Chemistry*, **283**, 36361–8.

Cheek, J. and Broderick, J. B. (2002). Direct H atom abstraction from spore photoproduct C-6 initiates DNA repair in the reaction catalyzed by spore photoproduct lyase: evidence for a reversibly generated adenosyl radical intermediate. *Journal of the American Chemical Society*, **124**, 2860–1.

Cucinotta, F. A., Kim, M.-H. Y., Willingham, V. and George, K. A. (2008). Physical and biological organ dosimetry analysis for International Space Station astronauts. *Radiation Research*, **170**, 127–38.

Dedon, P. C. (2008). The chemical toxicology of 2-deoxyribose oxidation in DNA. *Chemical Research in Toxicology*, **21**, 206–19.

Desai, N., Davis, E., O'Neill, P., Durante, M., Cucinotta, F. A. and Wu, H. (2005). Immunofluorescence detection of clustered γ-H2AX foci induced by HZE-particle radiation. *Radiation Research*, **164**, 518–22.

Douki, T. and Cadet, J. (2001). Individual determination of the yield of the main UV-induced dimeric pyrimidine photoproducts in DNA suggests a high mutagenicity of CC photolesions. *Biochemistry*, **40**, 2495–501.

Douki, T., Angelov, D. and Cadet, J. (2001). UV Laser photolysis of DNA: effect of duplex stability on charge-transfer efficiency. *Journal of the American Chemical Society*, **123**, 11360–6.

Douki, T., Reynaud-Angelin, A., Cadet, J. and Sage, E. (2003a). Bipyrimidine photoproducts rather than oxidative lesions are the main type of DNA damage involved in the genotoxic effect of solar UVA radiation. *Biochemistry*, **42**, 9221–6.

Douki, T. and Cadet, J. (2003b). Formation of the spore photoproduct and other dimeric lesions between adjacent pyrimidines in UVC-irradiated dry DNA. *Photochemical and Photobiological Sciences*, **2**, 433–6.

Douki, T., Ravanat, J.-L., Pouget, J.-P., Testard, I. and Cadet, J. (2006). Minor contribution of direct ionization to DNA base damage induced by heavy ions. *International Journal of Radiation Biology*, **82**, 119–27.

Durante, M. and Cucinotta, F. A. (2008). Heavy ion carcinogenesis and human space exploration. *Nature Reviews*, **8**, 465–72.

Fajardo-Cavazos, P., Salazar, C. and Nicholson, W. L. (1993). Molecular cloning and characterization of the *Bacillus subtilis* spore photoproduct lyase (spl) gene, which is involved in repair of UV radiation-induced DNA damage during spore germination. *Journal of Bacteriology*, **175**, 1735–44.

Francisco, D. C., Peddi, P., Hair, J. M., Flood, B. A., Cecil, A. M., Kalogerinis, P. T., Sigounas, G. and Georgakilis, A. G. (2008). Induction and processing of complex DNA damage in human breast cancer cells MCF-7 and non-malignant MCF-10A cells. *Free Radical Biology and Medicine*, **44**, 558–69.

Frelon, S., Douki, T., Ravanat, J.-L., Pouget, J. P., Tornabene, C. and Cadet, J. (2000). High performance liquid chromatography – tandem mass spectrometry measurement of radiation-induced base damage to isolated and cellular DNA. *Chemical Research in Toxicology*, **13**, 1002–10.

Georgakilas, A. G., Bennett, P. V., Wilson III, D. M. and Sutherland, B. M. (2004). Processing of bistranded abasic DNA clusters in γ-irradiated human hematopoietic cells. *Nucleic Acids Research*, **32**, 5609–20.

Gillet, L. C. and Scharer, O. D. (2006). Molecular mechanisms of mammalian global genome nucleotide excision repair. *Chemical Reviews*, **106**, 253–76.

Goodhead, D. T. (1994). Initial events in the cellular effects of ionizing radiations: clustered damage in DNA. *International Journal of Radiation Biology*, **65**, 7–17.

Goodhead, D. T. (2006). Energy deposition stochastics and track structure: what about the target? *Radiation Protection Dosimetry*, **122**, 3–15.

Hada, M. and Georgakilas, A. G. (2008). Formation of clustered DNA damage after high-LET irradiation: a review. *Journal of Radiation Research*, **49**, 203–10.

Jiang, Y., Hong, H., Cao, H. and Wang, Y. (2007). In vivo formation and in vitro replication of a guanine–thymine intrastrand cross-link lesion. *Biochemistry*, **46**, 12757–63.

Lawrence, C. W., Gibbs, P. E. M., Borden, A., Horsfall, M. J. and Kilbey, B. J. (1993). Mutagenesis induced by single UV photoproducts in *E. coli* and yeast. *Mutation Research*, **299**, 157–63.

Matallana-Surget, S., Meador, J. A., Joux, F. and Douki, T. (2008). Effect of the GC content of DNA on the distribution of UVB-induced bipyrimidine photoproducts. *Photochemistry and Photobiological Sciences*, **7**, 794–801.

Melly, E., Genest, P. C., Gilmore, M. E., Little., S., Popham, D. L., Driks, A. and Setlow, P. (2002). Analysis of the properties of spores of *Bacillus subtilis* prepared at different temperatures. *Journal of Applied Microbiology*, **92**, 1105.

Mileikowsky, C., Cucinotta, F. A., Wilson, J. W., Gladman, B., Horneck, G., Lindegren, I., Melosh, H. J., Rickman, H., Valtonen, M. and Zheng, J.Q. (2000). Natural transfer of viable microbes in space, part 1: from Mars to Earth and Earth to Mars. *Icarus*, **145**, 391–427.

Mitchell, D. L., Jen, J. and Cleaver, J. E. (1991). Relative induction of cyclobutane dimers and cytosine photohydrates in DNA irradiated in vitro and in vivo with ultraviolet-C and ultraviolet-B light. *Photochemistry and Photobiology*, **54**, 741–6.

Moeller, R., Stackebrandt, E., Douki, T., Cadet, J., Rettberg, P., Mollenkopf, H.-J., Reitz, G. and Horneck, G. (2007a). DNA bipyrimidine photoproduct repair and transcriptional response of UV-C irradiated *Bacillus subtilis*. *Archives in Microbiology*, **188**, 421–31.

Moeller, R., Douki, T., Cadet, J., Stackebrandt, E., Nicholson, W. L., Rettberg, P., Reitz, G. and Horneck, G. (2007b). UV-radiation-induced formation of DNA bipyrimidine photoproducts in *Bacillus subtilis* endospores and their repair during germination. *International Microbiology*,**10**, 39–46.

Moeller, R., Stackebrandt, E., Reitz, G., Berger, T., Rettberg, P., Doherty, A. J., Horneck, G. and Nicholson, W. L. (2007c). Role of DNA repair by nonhomologous-end joining in *Bacillus subtilis* spore resistance to extreme dryness, mono- and polychromatic UV, and ionizing radiation. *Journal of Bacteriology*, **189**, 3306–11.

Mouret, S., Baudouin, C., Charveron, M., Favier, A., Cadet, J. and Douki, T. (2006). Cyclobutane pyrimidine dimers are predominant DNA lesions in whole human skin exposed to UVA radiation. *Proceedings of the National Academy of Sciences of the USA*, **103**, 13765–70.

Nicholson, W. L., Munakata, N., Horneck, G., Melosh, H. J. and Setlow, P. (2000). Resistance of *Bacillus* endospores to extreme terrestrial and extraterrestrial environments. *Microbial and Molecular Biological Reviews*, **64**, 548–72.

Nicholson, W. L., Schuerger, A. C. and Setlow, P. (2005). The solar UV environment and bacterial spore UV resistance: considerations for Earth-to-Mars transport by natural processes and human spaceflight. *Mutation Research*, **571**, 249–64.

Nikjoo, H., O'Neill, P., Wilson, W. E. and Goodhead, D. T. (2001). Computational approach for determining the spectrum of DNA damage induced by ionizing radiation. *Radiation Research*, **156**, 577–83.

O'Neill, P. and Wardman, P. (2009). Radiation chemistry comes before radiation biology. *International Journal of Radiation Biology*, **85**, 9–25.

Pogoda de la Vega, U., Rettberg, P., Douki, T., Cadet, J and Horneck, G. (2005). Sensitivity to polychromatic UV-radiation of strains of *Deinococcus radiodurans* differing in their DNA repair capacity. *International Journal of Radiation Biology*, **81**, 601–11.

Pouget, J. P., Frelon, S., Ravanat, J.-L., Testard, I., Odin, F. and Cadet J. (2002). Formation of modified DNA bases in cells exposed either to gamma radiation or to high-LET particles. *Radiation Research*, **157**, 589–95.

Quaite, F. E., Sutherland, B. M. and Sutherland, J. C. (1992). Action spectrum for DNA damage in alfalfa lowers predicted impact of ozone depletion. *Nature*, **358**, 576–8.

Regulus, P., Duroux, B., Bayle, P. A., Favier, A., Cadet, J. and Ravanat, J.-L. (2007). Oxidation of the sugar moiety of DNA by ionizing radiation or bleomycin could induce the formation of a cluster DNA lesion. *Proceedings of the National Academy of Sciences of the USA*, **104**, 14032–7.

Sancar, A. (2003). Structure and function of DNA photolyase and cryptochrome blue-light photoreceptors. *Chemical Reviews*, **103**, 2203–37.

Semenenko, V. A. and Stewart, R. D. (2004). A fast Monte Carlo algorithm to simulate the spectrum of DNA damages formed by ionizing radiation. *Radiation Research*, **161**, 451–7.

Setlow, B. and Setlow, P. (1993). Dipicolinic acid greatly enhances the production of spore photoproduct in bacterial spores upon ultraviolet irradiation. *Applied and Environmental Microbiology*, **59**, 640–3.

Setlow, P. (2001). Resistance of spores of *Bacillus subtilis* to ultraviolet light. *Environmental Molecular Mutagenesis*, **38**, 97–104.

Slieman, T. A. and Nicholson, W. L. (2000). Artificial and solar UV radiation induces strand breaks and cyclobutane pyrimidine dimers in *Bacillus subtilis* spore DNA. *Applied and Environmental Microbiology*, **66**, 199–205.

Slieman, T. A., Rebeil, R. and Nicholson, W. L. (2000). Spore photoproduct (SP) lyase from *Bacillus subtilis* specifically binds to and cleaves SP (5-thyminyl-5,6-dihydrothymine) but not cyclobutane pyrimidine dimers in UV-irradiated DNA. *Journal of Bacteriology*, **182**, 6412–17.

Taylor, J.-S. (1994). Unraveling the molecular pathway from sunlight to skin cancer. *Accounts of Chemical Research*, **27**, 76–82.

Truglio, J. J., Croteau, D. L., Van Houten, B. and Kisker, C. (2006). Prokaryotic nucleotide excision repair: the UvrABC system. *Chemical Reviews*, **106**, 233–52.

Tsao, D., Kalogerinis, P., Tabrizi, I., Dingfelder, M., Stewart, R. D. and Georgakilas, A. G. (2007). Induction and processing of oxidative clustered DNA lesions in ^{56}Fe-ion-irradiated human monocytes. *Radiation Research*, **168**, 87–97.

Tucker, J. D., Marples, B., Ramsey, M. J. and Lutze-Mann, M. H. (2004). Persistence of chromosome aberrations in mice acutely exposed to $^{56}Fe^{26+}$ ions. *Radiation Research*, **161**, 648–55.

Tyrrell, R. M. (1973). Induction of pyrimidine dimers in bacterial DNA by 365 nm radiation. *Photochemistry and Photobiology*, **17**, 69–73.

Ward, J. F. (1988). DNA damage produced by ionizing radiation in mammalian cells: identities, mechanisms of formation and reparability. *Progress in Nucleic Acid Research and Molecular Biology*, **35**, 95–125.

24

Molecular adaptations to life at high salt: lessons from *Haloarcula marismortui*

Giuseppe Zaccai

The origin of life and its existence elsewhere than on planet Earth

Studies of the origins of life are closely interwoven with exobiology (Raulin-Cerceau *et al.*, 1998). It is highly probable that the full range of conditions present on Earth since its formation are present elsewhere. On a virtual trip through the Universe, we would travel not only in space, but also back in time into the Earth's biological history. The search for past, dormant or currently existing extraterrestrial life is one of the most thought-provoking challenges for biology. It is based on the certainty that liquid water and other key chemical and physical environmental conditions for the development of living organisms, as *we* know them, were, or are, present elsewhere in the Universe than on our planet. Any evidence of extraterrestrial life, from Mars sample analysis for example, would be of major interest for all biology. It would contribute to an understanding not only of the definition and origin of life, but also of the evolution and adaptation of molecular mechanisms in living cells, or of how organisms adapt and develop within ecosystems.

Why study life in extreme environments?

Life on Earth is almost everywhere! And because it is *almost* everwhere around us, we can hope to define the extreme limits for its existence by studying it here on Earth. Life has adapted to grow or just survive in what, in our anthropocentric way, we call the most extreme environments: from the deep oceans to mountain peaks way above the treeline, from very hot volcanic springs to glacial Arctic waters, from dry deserts and salt seas to underground rocks and from highly acidic springs to highly alkaline ponds.

López-García and Moreira (2008) reviewed the application of molecular ecology and metagenomics to study the vast microbial biodiversity of the biosphere. An impressive diversity of Archaea, Bacteria and protists has been uncovered by these techniques. And the correlation of function with the phylogenetic diversity observed in natural environments is leading to the discovery of novel metabolisms and to a re-evaluation of the global ecological impact of known ones. All known living organisms have properties in common. They are made up of cells or have passed through a cellular stage. The cell is the fundamental component of all life forms. The common properties of all known life forms, however, extend to the molecular level. Genetic information is encoded in cellular DNA, and passed from one generation to the next by 'replication'. For the information in the DNA to be acted upon, it is 'transcribed' into RNA. The RNA, in turn, is 'translated' into proteins, the molecular machines, which assume the vital specific functions of the cell. Each of replication, transcription and translation is, in fact, a highly catalysed and finely controlled process, involving a large set of enzymes, various repressor and activator proteins and transcription factors. Ribosomes, large protein-RNA molecular machines, are the main seats of translation, which also brings into play aminoacyl-tRNA synthases, tRNA, and a number of initiation, elongation and other factors.

The control and regulation of gene expression represent an essential function of proteins, which they ensure in large part through specific protein–nucleic acid interactions. The genetic information flows from DNA to RNA to protein, but protein is required in all steps, showing that the process is, in fact, already highly evolved. And since it is found in all known organisms, from the prokaryotic Archaea and Bacteria to human beings, it follows we are all already highly evolved and require the full set of macromolecules, ions, small molecules and water in order to be alive. There are no known examples of really primitive life forms and we can only speculate about them.

In the middle of the nineteenth century, Theodor Schwann, Matthias Jakob Schleiden and Rudolf Virchow proposed the three fundamental axioms of 'cell theory'. They remain valid today:

1. All living things are composed of one or more cells.
2. The cell is the most basic unit of life.
3. All cells come from pre-existing cells.

The third point is especially significant. It implies that only if its history is understood can we hope to understand the very existence of a cell, its place in an organism and the reactions that take place within it. Cells are often discussed as wonderfully synchronized extremely complex machines. In a broad sense, cells all perform similar tasks, based on similar basic macromolecular and molecular units. The miracle of adaptation is that, in different cells, at the molecular level these similar reactions take place in a wide variety of physico-chemical conditions – including the *extreme* conditions mentioned above. In the context not only of the search for extraterrestrial life, but also of fundamental biological science, there is a substantial research effort to understand adaptation mechanisms to extreme conditions.

Adaptation of a living organism to an extreme environment occurs at all levels of organization, from the sub-molecular, through the metabolic/physiological level to the ecological level, where it is reflected in biodiversity, which is itself an illustration of the richness of adaptation mechanisms (López-García and Moreira, 2008). This chapter is focused on structural and dynamic molecular features of adaptation to high salt levels in a Dead Sea archaeon, which represents but one aspect of adaptation to extreme conditions. Are the structural and dynamic features involving salt ions and water molecules, which have been identified in *Haloarcula marismortui*, representative of general molecular-adaptation mechanisms or are they specific to this organism? There are indications from genomic and structural analysis that, similarly to the *H. marismortui* enzymes, proteins from other extreme halophile Archaea and from the halophilic bacterium *Salinbacter ruber* also present the strongly acidic surface which participates in water and ion-binding stabilization and solubility in high salt (see the section below on 'Archaea, Bacteria and Eukarya adaptation to high salt'). On the other hand, the thermohalophile *Halothermothrix orenii* has evolved other molecular mechanisms for stability and solubility in a high-salt environment (Sivakumar *et al.*, 2006; Mavromatis *et al.*, 2009).

Water, salt and the Dead Sea that isn't

Life forms have colonized every ecological niche possible. These environments all have in common, however, that they contain liquid water, be it with a low activity coefficient. Organisms have been discovered that live at temperatures above 100°C, where at atmospheric pressure water would vaporize. They live in a deep ocean environment, however, where high pressure maintains the water in its liquid state. Not only is there no evidence of a life form adapted to the total absence of water, but also, organisms that are exposed to very dry conditions have evolved mechanisms to go into suspended animation during drought periods and to recover their usual metabolic state when water becomes available again. The excess production of the disaccharide trehalose, which coats and protects cellular structures, is one such mechanism (Cordone *et al.*, 1999 and references therein). By studying myoglobin dynamics using neutron scattering, Cordone *et al.* have shown that the trehalose coating inhibits large-amplitude denaturing motions, in a way similar to freezing. Salt also can have a protective effect on biological macromolecules. Tehei *et al.* (2002) have shown that an enzyme from an extreme halophile (salt-loving) archaeon, which had been trapped in a salt crystal, was significantly protected against denaturation by heat and desiccation.

Liquid water and various dissolved salts are essential to life. The cell membranes of all organisms contain pumps to exclude Na^+ and to concentrate K^+ in the cytoplasm; they generate significant chemical potential gradients for these ions. It has been recognized for millennia, nevertheless, that saturated salt conditions are not propitious for life. In fact, salt has been used from ancient times to preserve food, because it stopped it 'going bad' – by microorganism infection (but the cause was discovered much more recently). In *Voyage of HMS Beagle*, Charles Darwin wrote: 'Parts of the lake seen from a short distance appeared of a reddish colour, and this perhaps was owing to some infusorial *animalcula*... How surprising it is that any creatures should be able to exist in brine, and that they should be crawling among crystals of sulphate of soda and lime! ... Thus we have a little living world within itself, adapted to these inland lakes of brine'. As observed by Darwin, salt lakes and saline works ponds are often tinted red by the halophile organisms that thrive in their hypersaline waters. The main pigment is beta-carotene, which enters the food chain and is responsible, for example, for the pink colour of flamingos and salmon. The biogeochemistry of hypersaline environments is discussed in a book edited by Aharon Oren (Oren, 1998).

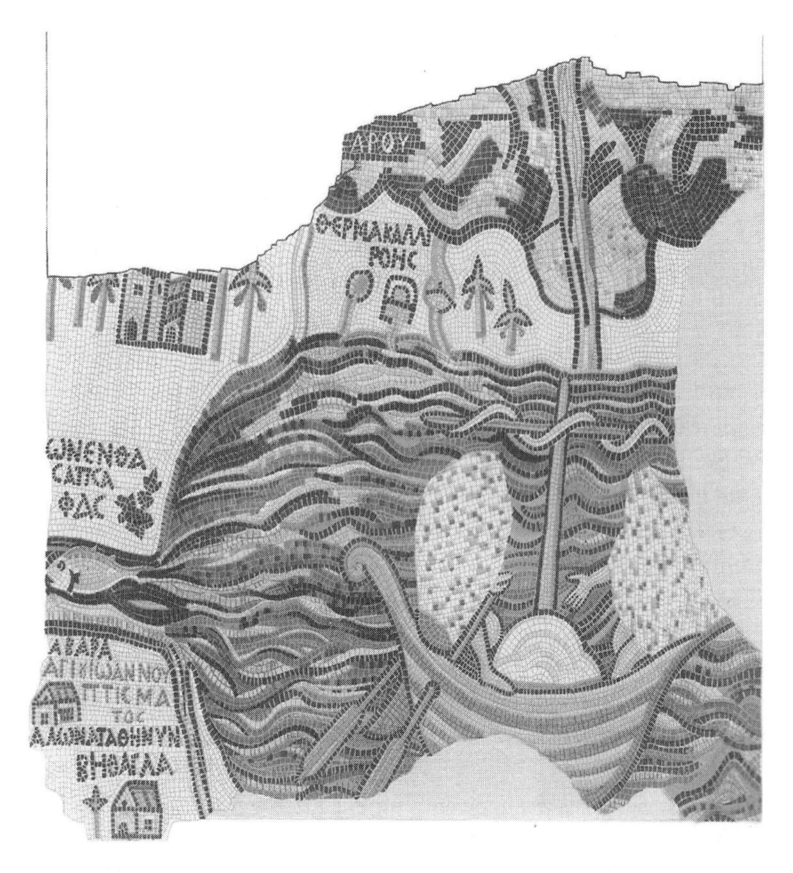

Figure 24.1. The Dead Sea played an important role in the description of halophile life. The Madaba map is a 6th-century floor map from the Church of Madaba, Jordan. A small part of the map shows a boat on the Dead Sea at the point of junction with the River Jordan (see fish turning back). The lumps on the boat represent salt obtained by evaporation from Dead Sea water. From 'The Madaba Mosaic Map' by Michael Avi-Yonah; used with permission (copyright 1954, Israel Exploration Society, Jerusalem).

The first halophile organisms were discovered at the end of the nineteenth century when scientists isolated microorganisms that infected salted cod. The biophysics experiments described further in the chapter, which led to an appreciation of the molecular mechanisms of adaptation to high-salt environments, originated in work done on the biology of the Dead Sea. The Dead Sea (Figure 24.1) is anything but dead. Benjamin Elazari Volcani pioneered the study of its microbial life in the 1930s, and continued its characterization until he died in 1999. In a letter to *Nature*, published in 1936 with the title 'Life in the Dead Sea', Volcani, whose name at the time was Wilkansky, described an indigenous microbial community adapted to the extreme conditions found in the Dead Sea (Wilkansky, 1936). In 1999, Oren and Ventosa published an article dedicated to Volcani's memory describing sixty-three years of studies of the microbiology of the Dead Sea (Oren and Ventosa, 1999). The author of the present chapter became involved in the work on halophile molecular adaptation through close collaborations with scientists studying Dead Sea organisms.

The state of intracellular water in halophile Archaea

The extreme halophile Archaea counterbalance the osmotic pressure due to the external multimolar NaCl concentration by accumulating multimolar KCl in their cytoplasm. There are 55 moles in a kilogram of water. In a six-molal solution of a fully ionized monovalent salt, therefore, there are about 4.6 water molecules per ion. Six water molecules may coordinate each ion, so that each and every water molecule has at least one ion as one of its nearest neighbours, leading to reduced water activity. Considering, furthermore, that the inside of a cell is already very crowded with macromolecules, which occupy close to 30% of its volume, what is the state of water inside halophile Archaea? Is it very different from that of the bulk liquid? And if so, how does this affect the biochemistry of the cell? It must be kept in mind that, in the laboratory, biochemistry is predominantly studied in dilute aqueous solution.

Following previous work on the Dead Sea archaeon, *Haloarcula marismortui* (Hm), which had suggested the existence of non-bulk intracellular water (Ginzburg and Ginzburg, 1975), water dynamics was measured *in vivo*, directly in the organism, by neutron scattering (Tehei *et al.*, 2007). In such experiments, neutrons are bounced off the protons in the hydrogen nuclei of the water molecules (Serdyuk *et al.*, 2007). Like in a billiard-ball collision experiment, the changes in energy and momentum of the neutrons provide information on the rotational and translation motions of the water molecules. The signal from deuterium nuclei is much weaker than that from hydrogen and specific isotope labelling is used to distinguish between the motions of different molecules. Fully deuterated Hm was cultured and then introduced into natural-abundance H_2O in order to observe the motions of the water molecules and not those of deuterons in the cellular structures. It was found that an important fraction of Hm intracellular water had translational motions that were almost two orders of magnitude slower than those of bulk. Was this due to confinement by molecular crowding in the cell, or to the saturated KCl environment? There had been reports that water in all cells was slowed down due to strong confinement by the crowded environment. This is not the case. Neutron-scattering experiments on *E. coli*, similar to the Hm experiments, established that on the atomic scale most of the intracellular water flowed as freely as in bulk (Jasnin *et al.*, 2008). Water motions in saturated KCl and NaCl solutions were also examined by neutron-scattering experiments; some slowing down was found, but only by a small factor compared to bulk (Frölich *et al.*, 2009). Crowding and the salt concentration, therefore, cannot account for the observed effect in Hm, and it appeared that something special was happening to the intracellular water in the extreme halophile. A possible explanation was proposed in terms of the 'solvation-stabilization model' discussed in the following section.

The solvation-stabilization model for halophile proteins

Hm MalDH, the tetrameric malate dehydrogenase from Hm, is the most-studied and best-characterized halophile protein (Eisenberg *et al.*, 1992; Richard *et al.*, 2000; Costenaro *et al.*, 2002; Ebel *et al.*, 2002; Irimia *et al.*, 2003). The protein from Hm is itself a halophile in that it requires very high salt concentrations in its solvent in order to be stable, soluble and active. The Hm MalDH tetramer concomitantly dissociates and unfolds when solvent NaCl or KCl concentration falls below about 2 M (still a *high* salt concentration for most biochemical studies). At first it was thought that under physiological conditions a particularly robust hydration shell, excluding solvent salt ions, surrounded halophile proteins. Careful biophysical experiments have shown, however, that this is not the case. In fact, the opposite situation occurs: the protein recruits solvent ions into its structure that stabilize it. The tetrameric structure is stabilized by chloride ions bridging amino-acid residues between the subunits (Figure 24.2) (Madern and Ebel, 2007; Irimia *et al.*, 2003). The solvation shell is made up of a very high local ionic concentration (equivalent to 5 molal NaCl, for example) and participates effectively in protein stabilization and solubility.

The amino-acid composition of proteins in halophile genomes is particularly rich in the acidic residues, aspartate and glutamate – to the extent that this was considered a signature of halophily in Archaea. The crystal structure of Hm MalDH illustrates how these residues cluster on the protein surface with the carboxyl groups pointing out into the solvent environment. Unfortunately, the ions and water molecules in the hydration shell are not sufficiently ordered to permit their observation in the crystal structure. Carboxyl groups, however, are known to form strong hydrogen bonds with water molecules. They are good candidates for the organization of a relatively strongly bound solvation shell through cooperative interactions with hydrated salt ions. As discussed by Tehei *et al.* (2007), such interactions between carboxyl groups, water molecules and ions have been observed in the crystal structure of the potassium channel, and may well account for the *slow* water found in Hm, which was discussed in the previous section.

Interestingly, Hm MalDH is stabilized *in vitro* by different thermodynamic mechanisms depending on the nature of the solvent salt. Under cosmotropic conditions, in high concentrations of sulphate or phosphate, for example, Hm MalDH stabilization is dominated by the hydrophobic effect with a solvent-excluding hydration shell, similarly to non-halophile proteins. In contrast to stabilization mechanisms in non-halophile proteins, in high concentrations of NaCl or KCl, Hm MalDH thermodynamic stabilization is dominated by enthalpic terms arising from extensive ion binding and hydration interactions. These observations led Bonnete *et al.* (1994) to suggest that stabilization mechanisms in Hm MalDH *adapted* to solvent conditions.

Every protein has evolved to form a stable active particle in its physiological environment. The solvation shell is an integral part of the particle, even though it is composed of solvent components in more or less rapid exchange with their bulk solvent partners. Solvation-shell interactions participate actively in the energetics of the particle, which defines its stability and dynamics (see next section). The solvation shell may be composed

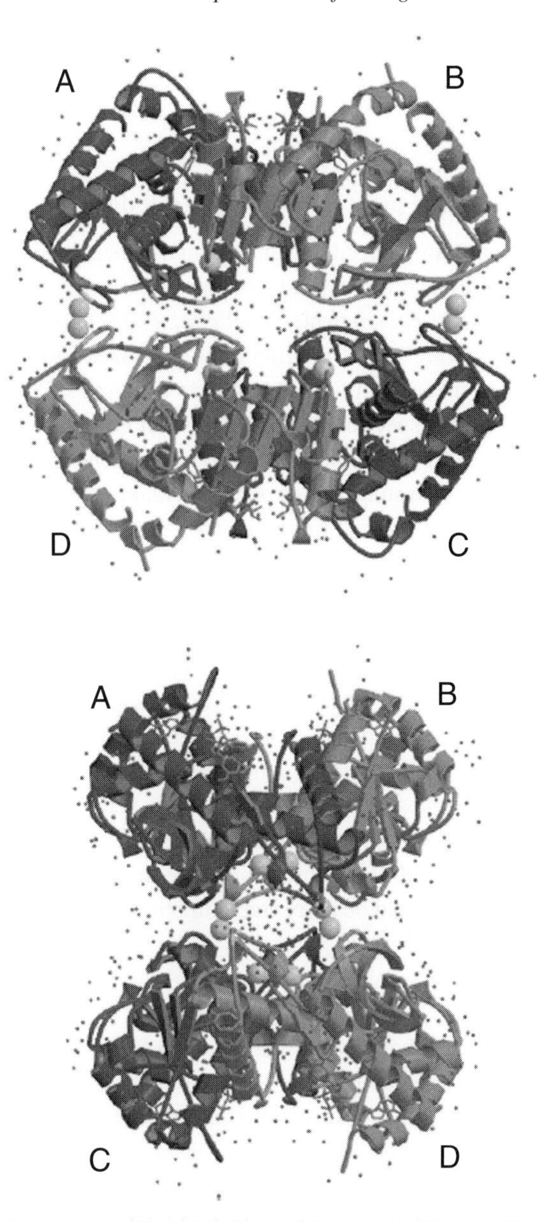

Figure 24.2. Quaternary structure of the (R207S, R292S) mutant of Hm MalDH and ion-binding to the interface. The quaternary structure of (R207S, R292S) MalDH is a tetrameric complex. The structure is shown in two orthogonal views. The monomers are labelled A to D. Secondary structures are shown as thick ribbons. The NADH cofactor is bound to the catalytic site of each monomer. Chloride ions are localized at subunit interfaces. The water molecules form networks on the external surface and at internal interfaces between dimers. The two tight AB and CD dimeric assemblies delimit an elongated cavity at their interface.

of only water in some types of H-bonding network with the protein surface (as in most soluble proteins), or of water and other solvent components, such as the ions observed for the halophile case.

Halophile protein dynamics is salt dependent

There is a strong relationship between protein thermal dynamics and activity. For a protein to be active it should have the appropriate internal motions. These are affected by temperature and also solvent environment. Gabel *et al.* (2002) reviewed protein dynamics measured by neutron scattering and its relation to biological function.

The molecular thermal dynamics of Hm MalDH in different solvents was examined by neutron scattering on the IN13 spectrometer, at the Institut Laue Langevin high-flux reactor in Grenoble (Tehei *et al.*, 2001). IN13 is sensitive to neutron energy and momentum ranges that are perfectly adapted to the study of internal protein dynamics in solution. The results indicated a correlation between the stability of the protein as measured by CD spectroscopy, and the 'resilience' of its structure (expressed as an effective force constant) measured by neutron scattering. The resilience was found to be higher and the mean amplitude of motions (flexibility) lower in concentrated NaCl, in which Hm MalDH is more stable, than in concentrated KCl. The authors suggested that halophile protein resilience was higher in NaCl because of the stronger hydration interaction of Na^+ compared to K^+. The state with the more strongly associated solvation shell would be more stable and more resilient. Recall, however, that the physiological environment of Hm MalDH is high KCl concentration. In this context, the results would suggest that there is a trade-off between stability (resilience), on the one hand, and activity (flexibility, which was higher in KCl than in NaCl), on the other. In other words, Hm MalDH in KCl has the appropriate flexibility for activity, and is sufficiently stable, even though it would be more stable in other salts.

Archaea, Bacteria and Eukarya adaptation to high salt

The discovery of Archaea and their shared features with Bacteria and Eukarya proved that the separation between the three domains of life is not hard and fast. A prokaryote retinal binding protein, for example, was first isolated in the1960s from the extreme halophile Archaea and named bacteriorhodopsin (reviewed by Haupts *et al.*, 1999). Bacteriorhodopsin is still studied very usefully as a tractable model, which almost became a paradigm for many important aspects in biophysics, including light transduction, ion translocation against the membrane potential and alpha-helical membrane protein structure and dynamics (e.g. Wood *et al.*, 2007, 2008). In a search for physiological adaptation mechanisms in halophile Archaea, Franzetti and his colleagues isolated and characterized novel complexes with salt-dependent ribonuclease, chaperone and protease activities (Franzetti *et al.*, 2001, 2002a, 2002b, 1997).

There are also halophile eukaryotes. The genus *Dunaliella*, the unicellular alga that is mainly responsible for primary production in practically all hypersaline environments, was described early in the twentieth century. In a historical survey (Oren, 2005), one hundred years of research on *Dunaliella* was summarized and reviewed. Unlike the extreme halophile Archaea, which accumulate intracellular KCl as the main osmolyte, *Dunaliella* and other Eukarya counterbalance the high osmotic pressure of their hypersaline environment by 'compatible solutes'. Compatible solutes are small organic osmolytes that are compatible with usual biochemical reactions and cellular metabolism even at very high (molar) concentrations, and have a protecive effect on macromolecules (da Costa *et al.*, 1998). The osmolyte in the case of *Dunaliella* is glycerol. Other natural compatible solutes include betaine, proline, ectoine and the disaccharide trehalose, mentioned above in the section on 'Water, salt and the Dead Sea that isn't'. Until a few years ago, it was believed that only the extreme halophile Archaea used salt, mainly KCl, as their osmolyte – while halophile and halotolerant Bacteria and Eukarya used compatible solutes. Then, *Salinibacter ruber* turned up (Anton *et al.*, 2002; Oren *et al.*, 2002).

Salinibacter ruber is an extreme halophile bacterium, which was isolated from saltern crystallizer ponds, and requires at least 150 g/litre of salt for growth. The cells were found to have an extremely high potassium content (the ratio of K^+/protein is in the same range as in extreme halophile Archaea) and only small amounts of potential intracellular compatible solutes were detected (Anton *et al.*, 2002; Oren *et al.*, 2002). These observations suggested that *S. ruber* uses a mode of haloadaptation similar to the extreme halophile Archaea, and accumulates salt and not organic osmolytes such as are used by all other halophilic and halotolerant aerobic Bacteria known at the time. The genome of *S. ruber* was published in 2005 (Mongodin *et al.*, 2005). Its analysis suggested that the resemblance between *S. ruber* and extreme halophile Archaea phenotypes arose through convergence at the physiological *and* the molecular levels (different genes producing similar overall phenotype, and independent mutations yielding similar sequences or structures, respectively). Furthermore, several genes and gene clusters were found to be related by lateral transfer from or to HaloArchaea, including a few for light-sensitive rhodopsins and photoactive yellow protein (Memmi *et al.*, 2008).

Recall from the previous sections that Hm MalDH is the most-studied protein from an extreme halophile. Its study led to the discovery of a new lactate dehydrogenase-like (LDH-like) group of MalDH (Madern and Zaccai, 2004). In recent and current work, members of the LDH-like MalDH enzyme family are used as models by Dominique Madern and his colleagues in a very fruitful comparative study of extremophile adaptation (Madern, private communication, to be published; Coquelle, 2008). In order to identify and compare features of halophile adaptation, the MalDH from *S. ruber* (Sr MalDH) was purified and characterized by analytical ultracentrifugation as a tetramer, indicating that it belongs to the LDH-like MalDH group. Sr MalDH is soluble at high salt concentrations, but in contrast to most other halophilic enzymes, however, which unfold at low salt concentrations, it remains stable in very low salt concentrations. Its activity is reduced by high salt concentration, but remains sufficient for the enzyme to sustain catalysis in 3 M KCl at a significant fraction of the maximum rate. The Sr MalDH crystal structure was resolved to show rich clustering of acidic groups on the surface of the protein (similar to Hm MalDH), but no salt–ion binding between subunits (unlike the case of Hm MalDH). The acidic features and lack of ion-binding in SR MalDH correlate, respectively, with high solubility at high salt and stability of the tetramer at low salt; they account for the activity of the enzyme in a broad range of salt conditions. A comparison of salt (KCl concentration)-dependent features in tetrameric MalDH from *Chlorflexus auranticus* (Ca, a thermophile bacterium), Sr and Hm is summarized as follows (Madern, private communication, to be published; Coquelle, 2008):

(1) Solubility: low surface acidity in Ca correlates with low solubility at high salt; high surface acidity in Sr and Hm correlates with high solubility at high salt.
(2) Stability: high stability at low and high salt in Ca and Sr correlates with no ion-binding between protein subunits; in Hm, the requirement of high salt for stability correlates with ion-binding between subunits.
(3) Activity: peaks at low salt in Ca and effectively disappears in high salt; peaks at low salt in Sr and is maintained at about 50% of the maximum value in high salt; requires high salt in Hm.

Salt, heat, the RNA World and exobiology

Macromolecules are the smallest signatures of life. They encode information and are catalytically active with great specificity. In most cases, the tertiary and quaternary structures of a macromolecule assuring a particular biological function evolved or converged to be very similar in different organisms. Structural and dynamic features associated with stability and activity in different environments have to be strongly adapted, however, and will be quite different for psychrophile, halophile and hyperthermophile proteins, to take extreme examples.

The genome of *S. ruber* (see previous section) is an excellent example of how a variety of survival and thriving strategies to cope with a common extreme environment are easily borrowed and transferred across domain boundaries. These are important points to bear in mind both in origin-of-life studies and in the search for extraterrestrial traces of living organisms.

The RNA World hypothesis in origin-of-life studies was inspired by the discoveries that RNA could have both catalytic and information-coding properties (review by Meli *et al.*, 2001). The halophile endonuclease RNase E, discovered by Franzetti *et al.* (1997) is similar in having the same RNA specificity as its homologue in *E. coli*. Contrary to the *E. coli* enzyme, however, the halophile archaeal enzyme requires a high salt concentration for cleavage specificity and stability. These data indicate that a halophile RNA-processing enzyme can specifically recognize and cleave messenger RNA from *E. coli* in an extremely salty environment (3M KCl). With this discovery, RNase E-like activity has now been identified in all three evolutionary domains: Archaea, Bacteria and Eukarya, strongly suggesting that messenger RNA-decay mechanisms are highly conserved despite quite different environmental conditions. The clustering of thermophile and hyperthermophile organisms close to the putative root of the phylogenetic tree has been interpreted to argue in favour of a 'hot' origin of life. Since RNA is very heat sensitive, would the RNA World scenario be excluded in this case? RNA tertiary structure and stability, however, are sensitively salt type and concentration-dependent (Zaccai and Xian, 1988). Tehei *et al.* (2002) examined the effects of salt on MalDH proteins from different sources, and transfer RNA. Trapping Hm MalDH in *dry* salt crystals protected the enzyme against thermal denaturation. Similar protection was not observed for the homologous mesophile. In the case of transfer RNA, high salt concentration played a protective role against thermal degradation, allowing activity to be recovered. The authors discussed their results in the context of orienting the search for traces of life in planetary exploration towards areas with evaporite structures, suggesting the existence of ancient salt seas that have dried down.

Acknowledgements

The ideas discussed in this chapter have been published in part in *Bioessays* by Maurel and Zaccai (2001) and in the *Proceedings of a Royal Society Meeting on Water and Life* (Zaccai, 2004). They were developed during two decades of work on halophile proteins in Grenoble at the Institut Laue Langevin and in the Laboratoire de Biophysique Moléculaire of the Institut de Biologie Structurale. I would like to thank, in particular, Christine Ebel, Bruno Franzetti, Dominique Madern and Moeava Tehei for many stimulating discussions on extremophile adaptation.

References

Anton, J., Oren, A., Benlloch, S., Rodriguez-Valera, F., Amann, R. and Rossello-Mora, R. (2002). *Salinibacter ruber* gen. nov., sp. nov.: a novel, extremely halophilic member of the Bacteria from saltern crystallizer ponds. *International Journal of Systematic and Evolutionary Microbiology*, **52**, 485–91.

Bonnete, F., Madern, D. and Zaccai, G. (1994). Stability against denaturation mechanisms in halophilic malate dehydrogenase 'adapt' to solvent conditions. *Journal of Molecular Biology*, **244**, 436–47.

Coquelle, N. (2008). *Mécanismes Moléculaires D'adaptation aux Conditions Physico-Chimiques Extrêmes dans la Famille des Lactate–Malate Déshydrogénases: Biologie Structurale et Nanobiologie*. Grenoble: Université Joseph Fourier.

Cordone, L., Ferrand, M., Vitrano, E. and Zaccai, G. (1999). Harmonic behavior of trehalose-coated carbon-monoxy-myoglobin at high temperature. *Biophysics Journal*, **76**, 1043–7.

Costenaro, L., Zaccai, G. and Ebel, C. (2002). Link between protein–solvent and weak protein–protein interactions gives insight into halophilic adaptation. *Biochemistry*, **41**, 13245–52.

da Costa, M. S., Santos, H. and Galinski, E. A. (1998). An overview of the role and diversity of compatible solutes in Bacteria and Archaea. *Advances in Biochemistry, Engineering and Biotechnology*, **61**, 117–53.

Ebel, C., Costenaro, L., Pascu, M., Faou, P., Kernel, B., Proust-De Martin, F. and Zaccai, G. (2002). Solvent interactions of halophilic malate dehydrogenase. *Biochemistry*, **41**, 13234–44.

Eisenberg, H., Mevarech, M. and Zaccai, G. (1992). Biochemical, structural, and molecular genetic aspects of halophilism. *Advances in Protein Chemistry*, **43**, 1–62.

Franzetti, B., Schoehn, G., Ebel, C., Gagnon, J., Ruigrok, R. W. and Zaccai, G. (2001). Characterization of a novel complex from halophilic archaebacteria, which displays chaperone-like activities in vitro. *Journal of Biological Chemistry*, **276**, 29906–14.

Franzetti, B., Schoehn, G., Garcia, D., Ruigrok, R. W. and Zaccai, G. (2002a). Characterization of the proteasome from the extremely halophilic archaeon *Haloarcula marismortui*. *Archaea*, **1**, 53–61.

Franzetti, B., Schoehn, G., Hernandez, J. F., Jaquinod, M., Ruigrok, R. W. and Zaccai, G. (2002b). Tetrahedral aminopeptidase: a novel large protease complex from Archaea. *EMBO Journal*, **21**, 2132–8.

Franzetti, B., Sohlberg, B., Zaccai, G. and Von Gabain, A. (1997). Biochemical and serological evidence for an RNase E-like activity in halophilic Archaea. *Journal of Bacteriology*, **179**, 1180–5.

Frölich, A., Gabel, F., Jasnin, M., Lehnert, U., Oesterhelt, D., Stadler, A. M., Tehei, M., Weik, M., Wood, K. and Zaccai, G. (2009). From shell to cell: neutron scattering studies of biological water dynamics and coupling to activity. *Faraday Discussions*, **141**, 117–30; discussion 175–207.

Gabel, F., Bicout, D., Lehnert, U., Tehei, M., Weik, M. and Zaccai, G. (2002). Protein dynamics studied by neutron scattering. *Quarterly Reviews of Biophysics*, **35**, 327–67.

Ginzburg, M. and Ginzburg, B. Z. (1975). Factors influencing the retention of K in a Halobacterium. *Biomembranes*, **7**, 219–51.

Haupts, U., Tittor, J. and Oesterhelt, D. (1999). Closing in on bacteriorhodopsin: progress in understanding the molecule. *Annual Review of Biophysics and Biomolecule Structure*, **28**, 367–99.

Irimia, A., Ebel, C., Madern, D., Richard, S. B., Cosenza, L. W., Zaccai, G. and Vellieux, F. M. (2003). The Oligomeric states of *Haloarcula marismortui* malate dehydrogenase are modulated by solvent components as shown by crystallographic and biochemical studies. *Journal of Molecular Biology*, **326**, 859–73.

Jasnin, M., Moulin, M., Haertlein, M., Zaccai, G. and Tehei, M. (2008). Down to atomic-scale intracellular water dynamics. *EMBO Reports*, **9**, 543–7.

López-garcía, P. and Moreira, D. (2008). Tracking microbial biodiversity through molecular and genomic ecology. *Research in Microbiology*, **159**, 67–73.

Madern, D. and Ebel, C. (2007). Influence of an anion-binding site in the stabilization of halophilic malate dehydrogenase from *Haloarcula marismortui*. *Biochimie*, **89**, 981–7.

Madern, D. and Zaccai, G. (2004). Molecular adaptation: the malate dehydrogenase from the extreme halophilic bacterium *Salinibacter ruber* behaves like a non-halophilic protein. *Biochimie*, **86**, 295–303.

Maurel, M. C. and Zaccai, G. (2001). Why biologists should support the exploration of Mars. *Bioessays*, **23**, 977–8.

Mavromatis, K., Ivanova, N., Anderson, I., Lykidis, A., Hooper, S. D., Sun, H., Kunin, V., Lapidus, A., Hugenholtz, P., Patel, B. and Kyrpides, N. C. (2009). Genome analysis of the anaerobic thermohalophilic bacterium *Halothermothrix orenii*. *PLoS ONE*, **4**, e4192.

Meli, M., Albert-Fournier, B. and Maurel, M. C. (2001). Recent findings in the modern RNA World. *International Microbiology*, **4**, 5–11.

Memmi, S., Kyndt, J., Meyer, T., Devreese, B., Cusanovich, M. and Van Beeumen, J. (2008). Photoactive yellow protein from the halophilic bacterium *Salinibacter ruber*. *Biochemistry*, **47**, 2014–24.

Mongodin, E. F., Nelson, K. E., Daugherty, S., Deboy, R. T., Wister, J., Khouri, H., Weidman, J., Walsh, D. A., Papke, R. T., Sanchez Perez, G., Sharma, A. K., Nesbo, C. L., Macleod, D., Bapteste, E., Doolittle, W. F., Charlebois, R. L., Legault, B. and Rodriguez-Valera, F. (2005). The genome of *Salinibacter ruber*: convergence and gene exchange among hyperhalophilic Bacteria and Archaea. *Proceedings of the National Academy of Sciences of the USA*, **102**, 18147–52.

Oren, A. (ed.) (1998). *Microbiology and Biogeochemistry of Hypersaline Environments*. Boca Raton: CRC Press.

Oren, A. (2005). A hundred years of *Dunaliella* research: 1905–2005. *Saline Systems*, **1**, 2.

Oren, A., Heldal, M., Norland, S. and Galinski, E. A. (2002). Intracellular ion and organic solute concentrations of the extremely halophilic bacterium *Salinibacter ruber*. *Extremophiles*, **6**, 491–8.

Oren, A. and Ventosa, A. (1999). Benjamin Elazari Volcani (1915–1999): sixty-three years of studies of the microbiology of the Dead Sea. *International Microbiology*, **2**, 195–8.

Raulin-Cerceau, F., Maurel, M. C. and Schneider, J. (1998). From panspermia to bioastronomy, the evolution of the hypothesis of universal life. *Origin of Life and Evolution of the Biosphere*, **28**, 597–612.

Richard, S. B., Madern, D., Garcin, E. and Zaccai, G. (2000). Halophilic adaptation: novel solvent protein interactions observed in the 2.9 and 2.6 A resolution structures of the wild type and a mutant of malate dehydrogenase from *Haloarcula marismortui*. *Biochemistry*, **39**, 992–1000.

Serdyuk, I. N., Zaccai, N. and Zaccai, G. (2007). *Methods in Molecular Biophysics: Structure, Dynamics, Function*. Cambridge, UK: Cambridge University Press.

Sivakumar, N., Li, N., Tang, J. W., Patel, B. K. and Swaminathan, K. (2006). Crystal structure of AmyA lacks acidic surface and provide insights into protein stability at poly-extreme condition. *FEBS Letters*, **580**, 2646–52.

Tehei, M., Franzetti, B., Maurel, M. C., Vergne, J., Hountondji, C. and Zaccai, G. (2002). The search for traces of life: the protective effect of salt on biological macromolecules. *Extremophiles*, **6**, 427–30.

Tehei, M., Franzetti, B., Wood, K., Gabel, F., Fabiani, E., Jasnin, M., Zamponi, M., Oesterhelt, D., Zaccai, G., Ginzburg, M. and Ginzburg, B. Z. (2007). Neutron scattering reveals extremely slow cell water in a Dead Sea organism. *Proceedings of the National Academy of Sciences of the USA*, **104**, 766–71.

Tehei, M., Madern, D., Pfister, C. and Zaccai, G. (2001). Fast dynamics of halophilic malate dehydrogenase and BSA measured by neutron scattering under various solvent conditions influencing protein stability. *Proceedings of the National Academy of Sciences of the USA*, **98**, 14356–61.

Wilkansky, B. (1936). Life in the Dead Sea. *Nature*, **138**, 467.

Wood, K., Grudinin, S., Kessler, B., Weik, M., Johnson, M., Kneller, G. R., Oesterhelt, D. and Zaccai, G. (2008). Dynamical heterogeneity of specific amino acids in bacteriorhodopsin. *Journal of Molecular Biology*, **380**, 581–91.

Wood, K., Plazanet, M., Gabel, F., Kessler, B., Oesterhelt, D., Tobias, D. J., Zaccai, G. and Weik, M. (2007). Coupling of protein and hydration-water dynamics in biological membranes. *Proceedings of the National Academy of Sciences of the USA*, **104**, 18049–54.

Zaccai, G. (2004). The effect of water on protein dynamics. *Philosophical Transactions of the Royal Society London B, Biological Sciences*, **359**, 1269–75; discussion 1275, 1323–8.

Zaccai, G. and Xian, S. Y. (1988). Structure of phenylalanine-accepting transfer ribonucleic acid and of its environment in aqueous solvents with different salts. *Biochemistry*, **27**, 1316–20.

Part VII

Traces of life and biosignatures

25

Early life: nature, distribution and evolution

Frances Westall

Introduction

The first two thirds of the history of life on Earth are dominated by single-celled microorganisms with prokaryotes characterizing the time period up to at least the Palaeoproterozoic Period (from 2.5 to about 1.8 billion years (Ga) ago). The oldest recognizable eukaryotes appear in the Mesoproterozoic Era (and are dated at between 1.6 to 1.8 Ga (Javaux *et al.*, 2001, 2004; see also review in Knoll *et al.*, 2006). This chapter on early life will concentrate on the traces of life contained in the oldest crustal rocks potentially capable of hosting well-preserved biosignatures, i.e. Early to Mid-Archaean, 3.5 to 3.0 Ga-old sediments and volcanic rocks from greenstone belts in both the Pilbara (NW Australia) and the Barberton (East South Africa) Greenstone Belts. The fossil traces of early microorganisms in these rocks resemble prokaryotes in terms of their morphology, metabolic processes and interactions with the environment. Life is directly influenced by its environment and, reciprocally, it can also influence its immediate environment. On the microbial scale, this influence is in proportion to the size of the microbial colonies, biofilms or mats, which can range from tens of microns to several metres or more (sometimes up to kilometres) for well-developed mats. For instance, if one takes into consideration the probable microbial control on the rise of oxygen in the atmosphere (between 2.4 and 2.0 Ga; Bekker *et al.*, 2004; Canfield, 2005), this influence also reaches the planetary scale.

Despite differences in scale, the planetary environment of the early Earth was significant for the small-scale habitats of the early microorganisms, in the sense that global conditions were of primordial importance for the establishment of *habitable* conditions (Southam and Westall, 2007). There were significant differences between the global environment of the early Earth and the Earth of today, especially in terms of atmospheric composition and perhaps the chemistry and temperature of the world's oceans. Extremely low oxygen contents assured an anaerobic atmosphere (Kasting, 1993), whereas ocean temperatures were significantly higher (<55°C according to van der Boorn *et al.*, 2007, although higher temperatures, of up 70–80°C, had been previously hypothesized by Knauth and Lowe, 2003, and Robert and Chaussidon, 2006). The high CO_2 concentration of the atmosphere would have produced a neutral to mildly acidic ocean pH (Grotzinger and Kasting, 1993) and ocean salinities appear to have been higher (de Ronde *et al.*, 1997). In other respects, the early Earth can be considered to have been globally habitable, with a wide variety of habitable environments, ranging from the deep sea to littoral environments, and hydrothermal vents to niches on and in the surfaces of cooled lavas (Westall and Southam, 2006; Westall, 2009). Within the limits of the available rock record, there are indications that all these habitats had been colonized by microbial life by 3.5 to 3.3 Ga ago (Westall and Southam, 2006; Westall, 2008).

The most substantial information regarding early life comes from two areas of well-preserved ancient terrane, the greenstone belts of the Pilbara in NW Australia and Barberton in South Africa, both ranging in age from about 3.55 to about 3.2 Ga old. These terranes have undergone low-grade metamorphism (prehnite–pumpellyite to lowermost greenschist facies) and have not been subjected to high-grade deformation and alteration, although many of the rocks were influenced by (pene)contemporaneous hydrothermal activity (Hofmann and Bohlar, 2007; Pinti *et al.*, 2009). Older crustal materials including sediments (~3.8 Ga) occur in SW Greenland in the Akilia and Isua Greenstone Belts but very high-grade metamorphism has overprinted original protolith signatures to such a degree that the search for traces of early life is difficult and the purported findings highly controversial (Pflug, 1979, 2001; Mojzsis *et al.*, 1996; Rosing, 1999; van Zuilen *et al.*, 2002; Westall and Folk, 2003; Rosing and Frei, 2004). We will restrict this review to the Early to Mid-Archaean Period (3.5 to 3.0 Ga) as represented by well-preserved rocks that contain a diversity of biosignatures and, despite their great age, much information about the nature, distribution and early evolution of life. Since the early Earth was dominated by vulcanism and igneous rocks, the oldest rock record mainly consists of volcanic and plutonic magmatic rocks. Thin layers of volcanically derived sediments were deposited on top of the volcanics between eruptions (Westall, 2005). At least in the Early Archaean Era until about 3.3 Ga, strong hydrothermal activity affected some parts of the upper crust, which, due to silica-saturated seawater, resulted in the silicification of sediments as well as volcanic rocks. Thus, many of the rock formations that contain early traces of life are referred to as 'cherts', whether they are sedimentary or magmatic in origin. It is in these cherts that the oldest traces of life are found.

Archaean biosignatures

Nature of biosignatures

The search for traces of life in some of the oldest rocks on Earth requires investigation of the fossil remains of life in the rocks, i.e. the signatures of life or 'biosignatures'. There are different types of biosignature that provide different kinds of information about the microorganisms that once inhabited the environment represented by a particular rock (Table 25.1; see also reviews in Westall, 2008 and Westall and Cavalazzi, 2009). 'Physical biosignatures' are those related to (1) physical microbial structures, such as the fossilized remains of the organisms themselves, their colonies or biofilms; or (2) the physical structures produced in sediments or on rocks and minerals by microorganisms. The latter are termed microbially induced fabrics and can include laminations in sediments produced by the alternate formation of microbial mats on a sediment surface and burial of the mats by an influx of sediment, or corrosion features produced by microbes on rock and mineral

Table 25.1. Signatures of life.

Major microbial components	Biosignature	Specific measurement, component or structure	
Cell components	Carbon molecules	composition, structure	complexity, non-random distributions, chirality
Cell metabolic activity	Biocatalysis	evidence of catalysis in sluggish reactions: depletion/ accumulation of potential metabolic substrates/ products	Fe, Mn, S, ...; H_2, O_2, CO_2, CH_4, H_2S......
	Biominerals*	direct (magnetite), indirect (aragonite, dolomite)	
	Elements	concentration, co-occurrence, stoichiometric ratios	C, H, N, O, P, S, Ni, Cu, Mn, Co, Mo, Se, V, Fe........
		fractionated isotopic ratios*	C, O, S, N, Fe...
	Microbial influence on:	mineral composition, mineral habit, mineral dissolution	
Physical structures	Cells, colonies, mats, microbialites	cells, colonies, mats, microbialites	clotted fabric, microbial mounds, biolaminites

* Note that minerals that are formed directly or indirectly by microbial processes may also be formed abiogenically.

surfaces. Vertical constructions produced by the interaction of physical and microbial processes, such as domical or conical stromatolites, are also physical biosignatures. 'Organic chemical biosignatures' are the organic molecules that make up the organisms (or their degradation products, known as molecular fossils or biomarkers). 'Metabolic biosignatures' represent traces of the metabolic activities of microorganisms, such as fractionation of the isotopes of certain elements (carbon, sulphur, nitrogen etc.) or precipitation of minerals as a by-product of microbial metabolic processes.

Unfortunately, analysis of biosignatures in very ancient rocks is not straightforward. Many processes act to change the biosignatures since their 'initial' formation, such as alteration of the rocks by metamorphism and physical destruction. Older rocks may also be contaminated by younger organisms, especially those with endolithic habitats, or by the migration of the volatile components of degraded organic matter that can confuse and

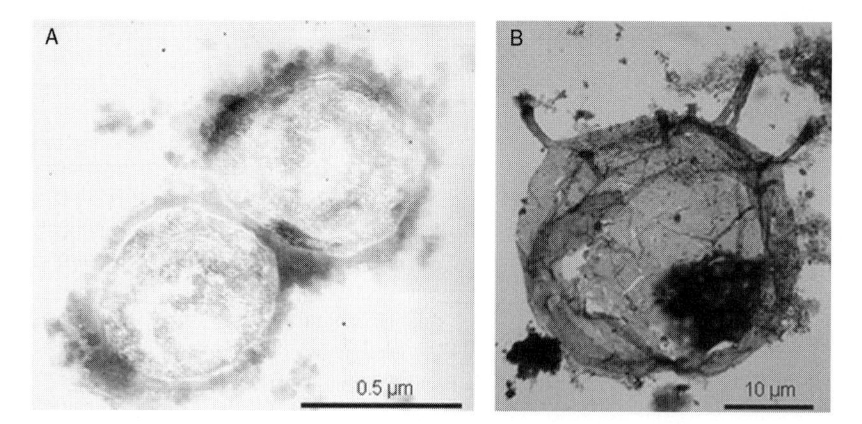

Figure 25.1. Modes of morphological preservation of microorganisms. (A) Silica-replacement of the outer cell envelope and some infilling of the hollow structure after lysis (experimental silicification of microorganisms, Westall *et al.*, 1995). (B) Microorganisms preserved as carbonaceous vesicles in anaerobic fine-grained sediments: the protist *Tappania Plana*, from the ~1.45 Ga-old Roper Formation (Javaux *et al.*, 2001).

disturb the interpretation of some biosignatures. Furthermore, many natural abiotic mineralogical (or organo/mineralogical) precipitations can imitate simple biological structures, thus confusing interpretation of physical biosignatures.

Preservation of biosignatures

The search for biosignatures implies that some trace of past life is conserved in a mineral matrix. In order to become microfossils, living or recently dead microorganisms need to be preserved rapidly before they are completely degraded and used as a nutritional source for other organisms (heterotrophic microorganisms obtain their carbon from organic matter), and this depends upon many factors, including the immediate environment. For instance, in present-day surface hydrothermal systems, organic matter is rapidly oxidized, although the physical structures of the microorganisms inhabiting the environment may be preserved by minerals precipitated on their surfaces. Experiments show that preservation by mineral precipitation is a relatively rapid process, occurring within days to months (Figure 25.1A) (Westall *et al.*, 1995; Westall, 1999; Toporski *et al.*, 2002; Orange *et al.*, 2009). However, some microorganisms are so delicate that they lyse soon after the exponential phase of growth of the microbial colony and there is little time for them to become coated with a mineral and thus physically preserved (Orange *et al.*, 2009). Moreover, fossilization of microorganisms by minerals is also controlled by the composition and the structure of the outer cell envelope (Westall, 1997; Orange *et al.*, 2009). This is because fossilization proceeds through the fixation of minerals in solution to specific functional-group sites of the organic molecules exposed to the solution (see Chapter 29). For instance, Gram-positive bacteria, characterized by a relatively thick outer layer of the compound peptidoglycan that contains many carboxyl groups, can fix mineral ions in solution more easily than Gram-negative bacteria. The thick mineral crusts that form around Gram-positive bacteria contrast with delicate crusts forming around the Gram-negatives because the latter have only a thin layer of peptidoglycan sandwiched between two other layers containing fewer functional groups (Westall, 1997). Microorganisms with a thick and robust mineral crust will have a better chance of being preserved over geological time than those with extremely delicate and fragile mineral crusts. Recent experiments have shown that microorganisms from the Archaea group can also be fossilized, but that the resulting crust is thin and delicate, similar to the mineral crusts around Gram-negative bacteria (Orange *et al.*, 2009).

In the absence of minerals precipitating on the surfaces of microorganisms, the microorganisms can still be preserved as compressed organic-walled vesicles in anaerobic fine-grained sediments (Figure 25.1B) (e.g. Javaux *et al.*, 2001, 2004), as can the degraded organic carbon molecules from such organisms (Brocks *et al.*, 1999; Summons *et al.*, 1999). Gradual breakdown of the organic molecules through time and metamorphism continues to degrade their structure. Up to a certain point the degraded molecules can be related back to specific molecular components of microbial cells. Such biomarkers or molecular fossils have been traced back in time to 2.7 Ga (Brocks *et al.*, 1999; Summons *et al.*, 1999), although it has been contested that the latter are more recent contaminants (Rasmussen *et al.*, 2008). Whereas the labile fraction of the molecules can easily result from contamination, the non-labile residue cannot. Organic molecules in Early to Mid-Archaean rocks are so degraded that they differ little from abiotic polycyclic aromatic hydrocarbons found in meteorites. However, analyses using pyrolysis CG–MS of the non-labile organic fraction in Early Archaean cherts were able to document the biological origin of the molecules based on the non-random distribution of specific structural components (Derenne *et al.*, 2008). Electromagnetic resonance measurements of the same molecules can be used to determine the age of the compounds, and Derenne *et al.* (2008) were thus able to demonstrate that they formed at the same time as the host rock.

Destruction of biosignatures occurs at various levels. Microbial structures preserved by mineral precipitation or microbially precipitated biominerals can be dissolved if there is a change in the physico-chemical environment. An increase in pH will dissolve structures that have been replaced by silica or iron biominerals, whereas a decrease in pH will dissolve those consisting of carbonates. Physical structures produced by microorganisms, such as biolaminated sediments including stromatolites, may be physically destroyed by storms or other natural processes before lithification (i.e. compaction and consolidation) of the sediments. Likewise, physical erosion of the lithified structures, as rocks, is a natural and widespread destructive process. On another level, orogenic processes acting on rocks, as for instance in continental-plate collisions, can lead to heat and pressure stresses that may completely transform the rocks and any biosignatures they may have contained. This is the case for the most ancient sediments on Earth in SW Greenland where ~3.8 Ga-old rocks in the Isua and Akilia areas have been subjected to high-grade metamorphism. Metamorphic effects on carbonaceous materials cause a loss in functional groups, loss of hydrogen, and increase in crystal ordering to graphite, as well as fractionation of carbon isotopes in favour of heavier ones. Morphological structures, such as microfossils, may be completely obliterated by recrystallization during metamorphism. The Early- to Mid-Archaean rocks with which we are concerned in this chapter have undergone low-grade metamorphism in the pumpellyite–prehnite/lowermost greenschist facies (i.e. temperatures of $< 250°C$), and are considered to be well preserved (however, some localized areas in the Pilbara and Barberton Greenstone Belts have undergone higher-grade metamorphism). Thus, both the Pilbara and Barberton Greenstone Belts appear as the best places able to have preserved biosignatures in ancient rocks because (1) they contain rocks formed in habitable environments; (2) at least some of the microorganisms in the communities were susceptible to fossilization; (3) the sediments and rocks in which, and on which the microorganisms lived were lithified before the biosignatures could be destroyed by other processes; (4) the rocks containing these biosignatures did not undergo major metamorphic alteration; and (5) the overlying rocks have been eroded, thus exposing today the Early to Mid-Archaean crust (NB underlying crust can be accessed by drill core, but the resulting microbial-scale context information is lacking in such cases).

Significance of biosignatures

As a result of the processes of preservation and destruction of biosignatures related to individual organisms or to their multispecies communities, those that have survived in rocks will most likely reflect only a portion of the original microbial community inhabiting a particular microenvironment at a particular moment in time. Given the spatial and analytical limits of instrumentation and methods for studying biosignatures, one particular preserved biosignature may actually represent a community rather than a specific species of organism. Microbial mats, for instance, consist of a complex consortium of co-existing microorganisms, the end result of which is a single physical structure with heterogeneous compositional and structural characteristics that cannot be disentangled, at least in the fossilized form. The physical, chemical and metabolic biosignatures in a rock containing one or more microbial mats will therefore reflect the entire community (or communities). The information that can be obtained from isotopic measurements would also represent a mixed signal due to the fact that the carbon necessary for isotopic analysis must be extracted from a large portion of rock, probably comprising many different communities in different microscopic habitats. Even when measurements are made *in situ*, such a large area is measured (150 to 250 μm^2) that, again, the result is not necessarily that of a single species. Interpretations of isotopic signatures alone can therefore be problematical. Furthermore, the composition of microbial communities in one type of habitat can change with time. Thorseth *et al.* (2001) showed that endolithic microorganisms colonizing fractures within the cooled rinds of subsea pillow lavas first consisted of primary chemolithotrophic communities, i.e. organisms obtaining carbon from an inorganic CO_2 source and energy from redox reactions at mineral/rock surfaces, which were later replaced by heterotrophs that extract their carbon from organic sources. Thus, biosignatures in an individual rock will not relate to one particular species of microorganism, but will represent an average over a microbiologically relevant period of time.

Given the complexities of preservation and interpretation of biosignatures, especially in very ancient rocks, it is therefore prudent that studies of ancient microbial life should not be based only on a single biosignature, which is almost always ambivalent. Such analyses should use several independent approaches incorporating many types of biosignature in order to produce complementary evidence that is unambiguous with respect to both the biogenic origin of the biosignatures, as well as to their originality or syngenicity (Westall and Southam, 2006).

The nature of early life and its distribution

This chapter will address the nature and distribution of early life in terms of habitat, including (1) volcanic habitats, (2) shallow water/littoral habitats, (3) planktonic and deep-water habitats, and (4) hydrothermal habitats. Volcanic habitats, i.e. the surfaces of volcanic lavas or particles, can occur in both shallow and deep-sea environments. Shallow water to littoral areas are characterized by exposure to sunlight. Deep-water habitats represent sediments deposited below wave base (which can be 10–100 m), but are basically out of the photic zone. Pelagic habitats are open-water areas that are within the photic zone. The remains of organisms living in pelagic areas may be deposited onto the seafloor, buried and preserved, or destroyed by heterotrophic organisms. Hydrothermal habitats can occur in association with any of the former. (Microorganisms can colonize the surfaces of metamorphic rocks in a sedimentary environment. However, no such occurrences have yet been described from the time period covered in this chapter.)

Volcanic habitats

Putative microbial-corrosion structures have been interpreted in fractures within the glassy rinds of pillow basalts in Early Archaean volcanic sequences from the Barberton Greenstone Belt (Figures 25.2A and 25.2B) (Furnes *et al.*, 2004, 2007; Banerjee *et al.*, 2006, 2007). The structures, sometimes segmented, consist of titanium oxide-lined tubes1–9 μm in width and up to 200 μm in length that extend away from fractures or grain boundaries through which presumably seawater once flowed. The structures were then embedded by precipitated phyllosilicates, now chlorite. Biologically important elements, such as C, N and P, can be traced along the linings of the tubes (Furnes *et al.*, 2004; Banerjee *et al.*, 2007). Similar tubes have been found in the altered surfaces of modern pillow lavas, although most corrosion features in younger lavas are granular rather than tubular (Figure 25.2A) (Furnes *et al.*, 2007, and references therein). One example of putative granular-corrosion signatures was described from the 3.42–3.41 Ga-old EuroBasalt in the Pilbara (Banerjee *et al.*, 2006; Furnes *et al.* 2004, 2007) and Banerjee *et al.* (2006, 2007) concluded that the filamentous traces were indeed biological because the range of morphologies, their distribution, and the association of biologically important elements with the tubes do not have known abiogenic counterparts. Syngenicity is supported by the fact that the tubules are embedded by phyllosilicates and therefore pre-date the early diagenetic phase of alteration.

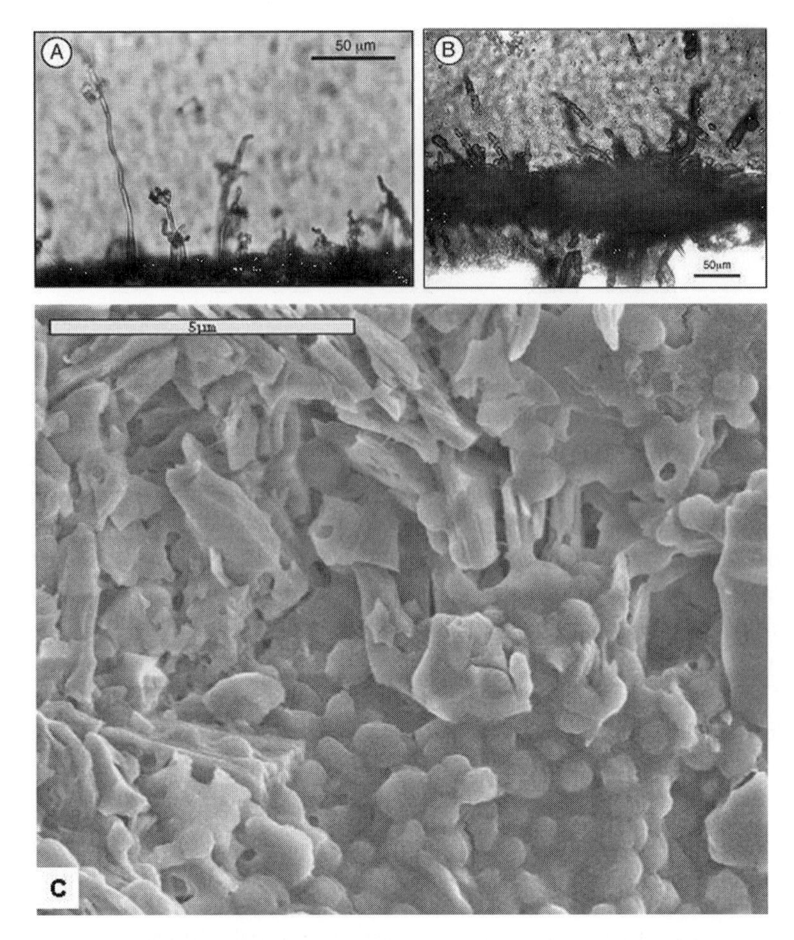

Figure 25.2. Biosignatures associated with volcanic habitats. (A) Tubular microbial corrosion traces in the surface of a modern pillow basalt lava (Furnes *et al.*, 2007). (B) Interpreted tubular microbial corrosion traces in the surface of an Early Archaean pillow basalt lava from the Barberton Greenstone Belt (Furnes *et al.*, 2007). (C) Multispecies colony of silicified coccoidal microorganisms at the surface of a phyllosilicate-altered volcanic particle (upper part of the figure) in Early Archaean (3.466 Ga) mud-flat sediments from the Kitty's Gap Chert, Pilbara (cf. Westall *et al.*, 2006a).

The organisms infiltrating the cracks and vesicles of the volcanic glasses can be classed as endolithic organisms of the chasmolith type, i.e. microorganisms living in fractures or cavities within rocks (Golubic *et al.*, 1981). Both autotrophic and chemoorganotrophic microorganisms use glassy volcanic materials as a substrate. The first communities to colonize the glasses are dominated by autotrophs that use CO_2 as their sole carbon source. As the glasses 'age', the autotrophs are replaced by chemoorganotrophs that use the organic carbon of the previous communities as a source. In addition to carbon, numerous biologically useful elements and compounds would have been available in this microenvironment to the microorganisms: reduced Fe^{2+} and Mn^{2+} in the volcanic glasses would have served as electron donors, whereas oxidized compounds from the circulating seawater, such as SO_2^{2-} and CO_2, as well as Fe and Mn could have acted as electron acceptors. Furnes *et al.* (2007) note that the evidence for biogenicity and syngenicity could be further strengthened by *in situ* demonstration of the depletion of microbially important elements, such as Mg, Fe, Ca and Na in the rocky materials immediately adjacent to the purported corrosion features, as well as by the *in situ* U–Pb dating of the Ti-oxides (Banerjee *et al.*, 2007).

Glassy volcanic lava-flow tops were not the only volcanic habitats of life in the Early Archaean Era. The volcanic sands themselves provided a variety of microenvironments ranging from the surfaces of detrital particles (of volcanic origin) to the pore spaces between the particles, as well as pore spaces within scoriaceous volcanic material, such as pumice. Westall *et al.* (2006a) investigated a chert in the Coppin Gap Greenstone Belt of the Pilbara consisting of volcanic sands deposited in a littoral mud-flat environment (de Vries *et al.*, 2006) from the point of view of individual microenvironments, analysing each layer (a few millimetres thick) in a 4-cm section of the silicified volcanic sediments. In this way, they were able to identify multispecies colonies of coccoidal organisms that had colonized individual volcanic grains in some layers, and pockets of similar colonies in other layers of very fine volcanic dust (Figure 25.2C). *In-situ* documentation of biologically important elements, such as C and N, around the volcanic grains in the same rock sample (Orberger *et al.*, 2006) is consistent with the presence of the microbial colonies observed by Westall *et al.* (2006a). Observation of short tunnels, 5–10 μm in length, filled with a polymer-like substance in the surfaces of some volcanic grains was interpreted by Foucher *et al.* (2009) as corrosion features produced by microbial metabolic activity, the polymer representing extracellular polymeric substances (EPSs). Stepped combustion of the kerogen of the individual layers documented the presence of mature kerogen having carbon isotopic signatures of −25.9 to −25.7‰ that are consistent with fractionation by microorganisms (but, as noted above, this cannot provide information about specific microbial metabolic strategies). Raman spectrometry and HR-TEM analysis of the fringes of the kerogen confirmed its maturity and demonstrated that it consisted of small aromatic molecules, thus showing that the composition and structure of the kerogen were consistent with both the age and metamorphic grade (prehnite–pumpellyite) of the host rocks. The biogenicity of the structures is supported by fine-scale morphological details, including contemporaneous evidence of cell birth and death (cell division and lysis), wrinkled cell surfaces, evidence of flexibility and hollowness and the narrow size range of the individual morphological classes.

The fact that these rocks contain evidence of physical, compositional and metabolic biosignatures is strong support for a biogenic origin. Syngenicity of the structures is documented by the distribution of different types of biosignatures with different types of microenvironments, for instance, the coccoidal colonies around volcanic particle surfaces and a biofilm including locally transported mat fragments associated with what had clearly been a stabilized bedding plane. Westall *et al.* (2006a) originally interpreted the colonies of ~0.4- and ~0.8-μm-diameter coccoids around the volcanic particles as chemolithotrophic microorganisms, on the basis of their intimate relationship with a volcanic substrate. However, recent studies showing that chemolithic colonies in chasmolithic habitats in glassy rinds of pillow lavas can be replaced in (microbial) time by chemoorganotrophs indicate that this may not necessarily be the case (Thorseth *et al.*, 2001).

Shallow water/littoral habitats in the photic zone

Most of the sedimentary deposits from the Early Archaean Era that are preserved were formed in shallow-water environments. The sediment and rock surfaces exposed to sunlight offered an ideal potential habitat for photosynthetic microorganisms that use sunlight as a source of energy. In these generally nutrient-poor (oligotrophic) environments, photosynthetic microorganisms are the primary producers of organic matter (cells, sheaths, EPS) in the uppermost layer of a microbial mat. The organic matter then provides a carbon and nutrient source for heterotrophic microbes below the surface of the mat (Visscher and Stolz, 2005). Because microbial mats occur at the interface between an underlying rocky or sedimentary substrate and either water or the atmosphere, they are characterized by steep physico-chemical gradients. This results in a wide range of microenvironments over a limited vertical scale all hosting a multitude of microbially catalysed biogeochemical processes that take place on different spatial and temporal scales (Reid *et al.*, 2000). Photosynthesizing microbial mats are thus complex in structure and composition.

Structures produced by photosynthesizing mats have been described from both the Pilbara and the Barberton Greenstone Belts (Figure 25.3). Some of them are large enough to be observed with the naked eye, forming flat-lying laminations within a sedimentary deposit (e.g. Figures 25.4A and 25.4B), or vertical constructions ranging from low-amplitude domes to small conical structures up to 10 cm in height. The vertical structures have attracted much attention because of their similarity to present-day columnar stromatolites, formed by *in-situ* lithification of microbial mats and sediment trapping on the sticky mat surfaces. Vertical stromatolite-like structures were first identified in the 1980s in the Pilbara (Lowe, 1980, 1983; Walter *et al.*, 1980; Walter 1983) and in Barberton (Byerly *et al.*, 1986), but it was later hypothesized that the finely laminated conical to domical structures were, in fact, abiogenic artefacts formed by inhomogeneous precipitation of minerals such as, for example, hydrothermal deposits (Lowe, 1994; Grotzinger and Rothman, 1996; Brasier *et al.*, 2002). No microfossils have yet been identified in association with these structures. However, detailed field mapping of conical stromatolite-like structures in the 3.443 Ga-old North Pole Dome of the Pilbara, combined with laboratory microscopic (Figure 25.3) and *in-situ* Raman studies of the organic matter, have documented an intimate relationship between the purported microbial mat structures and changes in microenvironmental parameters that could not have been produced by abiogenic processes (Allwood *et al.*, 2006, 2009). Allwood *et al.* (2009) noted that the conical structure of the stromatolites is, in itself, not evidence for phototactic behaviour, since similar structures can be produced by the influence of various physical processes on microbial mat formation. Their interpretation in terms of photosynthesis is therefore based on the context of the shallow-water environment of deposition.

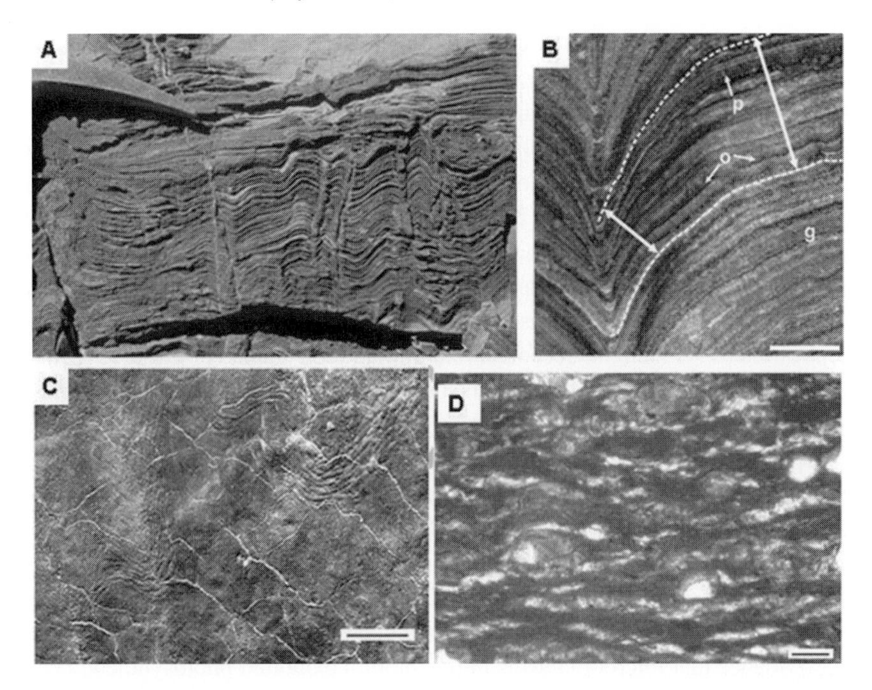

Figure 25.3. Biosignatures formed in shallow-water environments. (A) Field photograph of vertical section through small conical stromatolite-like structures from the 3.42 Ga-old Trendall locality, North Pole Dome, Pilbara (Photo Westall/Marchesini). Hammer head 9 cm. (B) Microscopic view of the fine alternating organic and mineral laminae of these structures (Allwood *et al.*, 2009). (C) 'Elephant-skin' wrinkle structures in the surface of a 3.2 Ga-old microbial mat from the Moodies group, Barberton (Noffke, 2009). (D) Microscopic image of the laminated vertical structure of the lithified microbial mats (Noffke, 2009).

Low-amplitude domical stromatolites on top of basaltic lavas were described from the Fig Tree Formation (3.3–3.2 Ga) in the Barberton Greenstone Belt (Byerly *et al.*, 1986). The fine, wavy non-conformable laminae are suggestive of microbial mats, but Walsh (2004) notes that neither carbon nor microfossils are associated with the structures.

The formation of tabular photosynthetic microbial mats on sediment surfaces in specific environments can leave traces that can be observed both macroscopically and microscopically. These structures have been termed 'microbially induced sedimentary structures' (MISS) by Noffke (2009), who has recognized them in 3.2 Ga-old coastal sedimentary deposits from the Moodies Group in the Barberton Greenstone Belt (Figures 25.3C and 25.3D) (Noffke *et al.*, 2006). Since recognition of these structures is based on morphological criteria and similarities with modern MISS structures, correct interpretation of the palaeoenvironment is fundamental, as the microbial mats only grow in specific microenvironments within tidally influenced littoral habitats. Noffke (2009) noted that sediment grain size is important, since the microbial mats only form on top of fine-grained sediments that, although influenced by wave–current action, are not subjected to strong hydraulic forces that would destroy them. A variety of microtextures related to the interaction of the microbial mats with their immediate sedimentary habitat aid determination of biogenicity.

On a microscopic scale, microbial mats of purported photosynthetic origin have been described from cherts representing shallow-water littoral sediments in a number of formations in the Pilbara and Barberton Greenstone Belts. From the 3.3446 Ga-old 'Kitty's Gap Chert' in the Pilbara, Westall *et al.* (2006a) interpreted a delicate multispecies biofilm that had developed on a stabilized sediment surface in a mud-flat environment as the remains of a possible photosynthetic mat. The biofilm conformably coated the underlying irregular pumice-encrusted surface and consisted of degraded filaments ~0.3 μm in diameter and some tens of microns in length, coccoidal (0.5 μm) and rod-shaped (1 μm in length) microfossils, as well as EPSs. Associated with the biofilm are fragments of reworked robust mats containing filaments similar to those occurring in the biofilm. These fragments must have been transported from the vicinity (the environment is a tidal flat) because they would otherwise have exhibited a more degraded or completely unrecognizable appearance. Carbon isotope ratios ($\delta^{13}C = -27.8‰$) of that particular layer are consistent with microbial processes. Biogenicity is suggested by morphological similarity with modern microorganisms, and syngenicity by the fact that the structure is clearly conformable with an undulating underlying surface and that it contains detrital mat fragments identical to portions of the *in-situ* mat. The evidence that microorganisms in this biofilm exhibited photosynthetic behaviour is, however, circumstantial, based partly on textural details, on the carbon isotope signature (admittedly a 'mixed' signature) and partly on the environmental context.

Evidence of photosynthetic microbial-mat formation is stronger for the remains of a well-preserved 3.334 Ga-old microbial mat that formed on a sediment bedding-plane surface in finely laminated back-barrier/lagoonal volcanic sediments in the 3.334 Ga-old Josefsdal Chert in the Barberton Greenstone Belt (Figure 25.4) (Westall *et al.*, 2001, 2006b, 2008). The excellent preservation of this silicified mat permitted these authors to make a detailed structural investigation showing that it is 5–10 μm thick and has an upper surface consisting of 0.25-μm-thick filaments thickly coated with EPS. The upper surface is interlayered with a suite of ~ 1-μm-sized pseudo-evaporitic minerals (aragonite, Mg–calcite, gypsum and halite). Detrital quartz and volcanic particles are trapped within the mat. The mat shows strong evidence of interaction with the local environment, especially the local hydraulic regime. The filaments on the surface of the mat are all oriented in the same direction, portions of the mat are overturned in the same direction as the filament orientation and there are areas where the mat has been mechanically torn away. This suggests formation under a moderate water flow. The mat was subsequently subjected to subaerial exposure and drying in evaporitic conditions. These are exactly the MISS features described by Noffke (2009). The three-dimensional preservation permitted vertical chemical and structural analysis on a submicron scale (Westall *et al.*, 2008). Sectioning by a focused ion beam (FIB) showed that the interior of the mat consisted of amorphous kerogen having an alveolar texture that was partially replaced by nanocrystalline calcium carbonate. Such alveolar textures are typical of degraded organic matter within microbial mats (Défarge *et al.*, 1999). The structure, environment of formation of the mat plus the

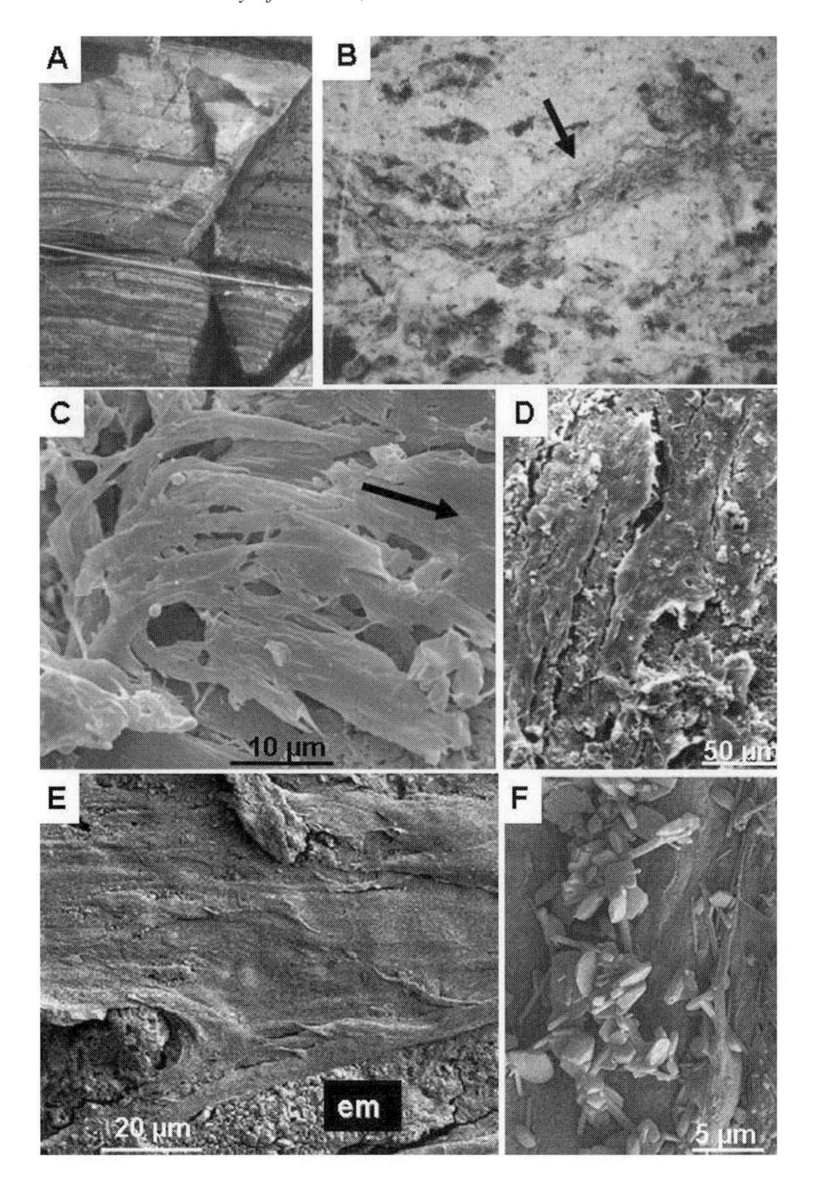

Figure 25.4. Macroscopic and microscopic views of biolaminated sediments from the Josefsdal Chert locality, Barberton (Westall *et al.*, 2006b). (A) Field view of fine carbonaceous laminae in silicified volcanic sediments. (B) Transmitted light microscopic view of one of the black layers in (A) showing that it consists of very fine wavy layers (arrow). (C, D) High-resolution scanning electron microscope micrographs of a specific mat in the Josefsdal Chert (Westall *et al.*, 2006b). (C) Overturned microbial filaments. (D) Desicccated aspect of the microbial mat. (E) Interlayering of the flow oriented mat and pseudo-evaporitic minerals (em). (F) Suite of pseudo-evaporitic minerals including aragonite, Mg–calcite and gypsum.

carbon isotopic signature ($\delta^{13}C = -22.6$ to -27.8‰) are consistent with a photosynthetic origin of the mat (Westall *et al.*, 2006b, 2008). Biogenicity is suggested very strongly by the textural characteristics, interaction with the immediate environment and a consistent $\delta^{13}C$ signature. As noted above, photosynthetic mats are structurally and compositionally complex, with photosynthetic organisms providing the primary biomass for heterotrophic microorganisms that degrade the relics of the photosynthesizers beneath the active surface. The degraded kerogen below the surface of this 3.3 Ga-old mat may be the result of such heterotrophic activity. Again, the evidence of photosynthetic activity in this case is circumstantial, although strongly supported by the complexity of structure and composition of the mat, which are very similar to those of modern photosynthesizing mats.

Other studies have also suggested the existence of photosynthetic microbial-mat communities in the Early Archaean rocks from Barberton, basing their interpretations on observation of mat-like laminations by optical microscopy, consistent carbon isotopic fractionation values and evidence from the sedimentological environment of deposition (Walsh, 1992, 2004; Walsh and Lowe, 1999; Tice and Lowe, 2004, 2006). As with the previous interpretations, although feasible, these interpretations are also circumstantial.

On another cautionary note, van Zuilen *et al.* (2007) noted that small-scale transport of carbon by hydrothermal fluids could produce linear features in these ancient silicified volcanic sediments. The carbon particles deposited in such veins, however, do not exhibit the structural and compositional complexity of the mats described above.

An intriguing study of the sulphur isotopes in microscopic pyrite crystals within bedded shallow-water baryte deposits from the Dresser Formation (3.490 Ga) of the Pilbara has been undertaken by Philippot *et al.* (2007). These authors found $\delta^{34}S = -22.6$‰; this kind of fractionation was previously interpreted by Shen *et al.* (2001) as evidence for sulphur fractionation by sulphate-reducing bacteria (SRBs). However, the range of $\delta^{34}S$ values obtained by Philippot *et al.* (2007) is much larger than those normally obtained from SRBs, leading these authors to hypothesize that a novel form of microbial fraction could have been involved, i.e. microbial disproportionation of elemental sulphur. They suggest that repeated microbial disproportionation of S^0 to H_2S and SO_4^{2-} by a consortium of sulphur-metabolizing microorganisms could have resulted in the wide range of values obtained.

The shallow-water sediments in the Pilbara and Barberton Greenstone Belts have revealed the existence of possible microbial structures of highly uncertain affinity that, for some, have no modern counterparts. In the 3.0 Ga-old Farrel Quartzite of the Gorge Creek Group in the Pilbara, Sugitani *et al.* (2007, 2009) document a range of carbonaceous thread- and film-like structures associated with relatively large, hollow spherical- and spindle-shaped features. The hollow spheres exhibit a wide range in size, from 2.5 μm to 80 μm, whereas the spindle structures range from 20–40 μm in length and 15–35 μm in width. Some of the spheres appear to contain smaller spherical carbonaceous bodies within them. The $\delta^{13}C$ value for the bulk rock is -30‰. The biogenicity of the structures is interpreted from their carbonaceous composition, hollow nature (for the spheres and spindle structures), narrow size distribution, apparent flexibility, association in colonies and morphological variability related to differing degrees of preservation. These structures are far larger than any found in the Early Archaean rock formations (cf. Westall *et al.*, 2001, 2006a, 2006b), although rare spindle-shaped structures and large spheres had been described from the Early Archaean cherts by Walsh (1992) and by Schopf and Walter (1983; and references therein). The affinities of the structures are unknown; spherical cells are common morphologies among microorganisms, but the spindle-shaped features have no known modern counterpart. They are difficult to explain by any abiological processes and do not appear to be younger infiltrations. These enigmatic structures warrant further investigation.

Planktonic and deep-water habitats

For the purposes of this chapter, the term 'deep water' implies sediments deposited below wave base, out of the photic zone. Such sediments are relatively rare in the Early Archaean Era. Generally, the carbonaceous components of these fine-grained sediments have been attributed to the transport of wave-ripped microbial-mat fragments into the deep-water environments from nearby shallower regions. Thus, Tice and Lowe (2004, 2006) identify detrital microbial-mat fragments and clots of carbonaceous matter in deeper water sedimentary facies in the 3.42 Ga-old Buck Reef Chert, in the Barberton Greenstone Belt. The mat fragments were interpreted to be photosynthetic, based on their morphological structure. These shallow-water mat-containing sediments have a $\delta^{13}C = -31.9$ ‰. A similar conclusion regarding the transport and deposition in deeper waters of detrital biological materials originating from photosynthetic microorganisms was reached by de Ronde and Ebbesen (1996). Their study of carbon isotope ratios of fine-grained sediments older than 3.3 Ga in the Barberton Greenstone Belt gave $\delta^{13}C$ values ranging from −9.5 to −28.1 ‰ (Tice and Lowe, 2006, documented an average $\delta^{13}C$ value of −27.2 ‰ for the deeper water facies of the Buck Reef Chert). De Ronde and Ebbesen (1996) also proposed that some of the carbon could have originated from sedimented pelagic photosynthetic microorganisms.

Hydrothermal veins and deposits

Hydrothermal activity appears to have affected the sedimentary horizons and immediately underlying igneous rocks in the Barberton (Hofmann and Bohlar, 2007) and Pilbara (van den Boorn *et al.*, 2007) Greenstone Belts. The hydrothermal activity was, together with the effects of silica-saturated seawater, also the cause of the intense silicification of the Early Archaean sedimentary and igneous rocks. A number of studies of the fractionation of carbon and sulphur isotopes in sedimentary formations from the Barberton Greenstone Belt are indicative of the presence of microorganisms living in the vicinity of hydrothermal vents, such as sulphur-reducing bacteria and methanogens. Light $\delta^{13}C$ signatures up to −40.8 ‰ were measured by Walsh and Lowe (1999), whereas Kakegawa (2001) reported $\delta^{34}S$ values of −4 to +1 ‰.

One of the most famous debates concerning the oldest traces of life is related to some of the earliest descriptions of microfossils in a 3.46 Ga-old chert from the Pilbara: the Apex Chert. Schopf (1993) described a variety of filamentous structures consisting of carbon as

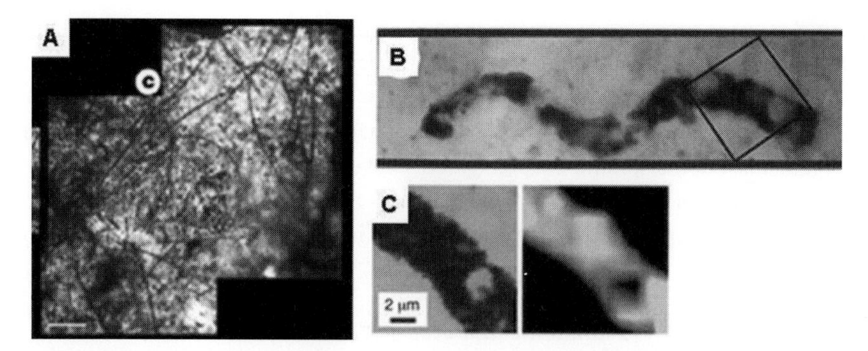

Figure 25.5. Interpreted microfossils from hydrothermal vent environments. (A) Pyritized filaments from a 3.2 Ga-old VMS deposit from the Pilbara (Rasmussen, 2000). (B, C) Carbonaceous filament occurring in a hydrothermal chert vein in the Apex Chert (3.46 Ga), with detail and corresponding map of the Raman G band (Schopf *et al.*, 2002).

the fossil remains of Early Archaean cyanobacteria and bacteria (Figures 25.5B and 25.5C). This interpretation had profound implications for the evolution of life, as well as for the evolution of the Earth's atmosphere. Cyanobacteria are highly evolved organisms, as far as prokaryotes are concerned. They are photosynthesizers that use a complicated but highly efficient method of obtaining energy that releases oxygen as a by-product. Oxygenic photosynthesis is believed to be the main source of oxygen in the atmosphere. Moreover, these organisms are often characterized by differentiated cells. The existence of oxygenic photosynthesis by ~3.5 Ga would imply very rapid evolution of life. It also produced a conundrum: if oxygen was already being produced by 3.5 Ga, why was there no evidence for a globally oxygenated atmosphere until about a billion years later (Rye *et al.*, 1995)?

Examination of the site from which Schopf's samples were obtained showed that the chert was, in fact, a hydrothermal vein, not silicified sedimented materials. Further re-evaluation of the materials with which Schopf worked showed that many of the structures that he described were not simple filaments and, since the original chert was hydrothermal, it was hypothesized that the carbon could have been of abiogenic hydrothermal origin (Brasier *et al.*, 2002). Recently, De Gregorio *et al.* (2009) used *in-situ* Raman spectrometry coupled with a high-resolution TEM investigation as well as a NEXAFS (Near-Edge X-Ray Fluorescence Spectrometry) synchrotron method to determine that the kerogen from the Apex Chert pseudofossils had the same characteristics as those of uncontested microfossils of the 1.87 Ga-old Gunflint Chert, and that the kerogen was indeed of biological origin, as well as being syngenetic with the rock. De Gregorio *et al.* (2009) concluded that, while the context was incompatible with the presence of photosynthetic life, it was possible that chemolithotrophs living close to a hydrothermal vent could have fallen into the vent during periods of inactivity and been silicified during renewed activity. Alternatively, Westall and Southam (2006) suggested that the kerogen could have been entrained by hydrothermal fluids transiting through strata containing kerogenous remains and redeposited further up in the vein. Both hypotheses are consistent with the data. To further complicate matters, Pinti *et al.* (2009) noted the presence of filamentous and filmy structures of biological appearance that appear to have infiltrated the Apex Chert in Caenozoic times. These structures, however, would not exhibit the mature Raman or high-resolution TEM signatures of the carbonaceous structures described by Schopf (1993), Brasier *et al.* (2002) and De Gregorio *et al.* (2009).

As noted above, certain chert veins containing carbonaceous material in the Early Archaean formations from the Pilbara and Barberton have been interpreted as having been infilled with detrital matter from above (Walsh, 2004; De Gregorio *et al.*, 2009) or as containing locally derived organic matter (van Zuilen *et al.*, 2007). Ueno *et al.* (2004) found carbonaceous filaments in 3.46 Ga-old chert veins from the Pilbara, whose $\delta^{13}C$ was –36‰, that they interpreted as indicating the presence of chemoautotrophic microorganisms. A similar interpretation was made for pyritized filaments (Figure 25.5A) in what were interpreted as 3.2 Ga-old VMS (very massive sulphide) deposits, supposedly the equivalent of present-day deep-sea black smokers (Rasmussen, 2000).

Evolution

The fossil record in Early-Mid Archaean sediments is relatively sparse, but recent investigations using multidisciplinary approaches are increasing our knowledge of the nature of early life, its distribution and its evolutionary significance. The early microorganisms were relatively simple anaerobes: coccoids, filaments or rod-shaped structures occurring in colonies, as biofilms or as mats. These organisms do not appear to have developed cellular differentiation. However, larger spherical to spindle-shaped structures from the 3.0 Ga-old Farrel Quartzite, Pilbara, consist of cells contained within a larger envelope-like feature. Rare observations of these features in Early Archaean shallow-water cherts provide a tantalizing possibility of more discoveries to come. Thus, in terms of morphological variety, these earliest traces of life show that it was relatively diverse by 3.5–3.0 Ga ago.

Further identification of the early development of metabolic strategies is constrained by interpretations of isotopic ratios and the immediate microenvironment of a particular type of microorganism. As noted above, interpretations of the isotopic signatures are limited because the analyses are made on bulk rock samples that most likely host a range of microorganisms. However, given these qualifiers, the fossil record appears to consist of (1) chemolithotrophic and/or heterotrophic microorganisms living on or in volcanic/hydrothermal materials and environments (deep- and shallow-water environments); (2) heterotrophs in fine-grained deep-sea (sub-wave base) sediments; and (3) possibly anaerobic photosynthesizers in the photic zone, the latter forming microbial mats on the sunlight-bathed fine sands of the Archaean littoral zones. Microbial communities in modern microbial mats comprise a wide range of organisms using different metabolisms depending on the microenvironments provided by the steep geochemical gradients within the thickness of a single mat. This was probably also the case in the Early-Mid Archaean environment.

Conclusions

The fossil record of the oldest, well-preserved sediments on Earth, dating from 3.5–3.0 Ga ago is both difficult to find and difficult to decipher. Early fossilization by silica of many of the sedimentary sequences ensured preservation of at least some biosignatures, but many may not have survived. This means that the fossil record is patchy and represents probably only some of the original microbial communities that existed on the early Earth. Some of the difficulties in interpretation of the physical and chemical biosignatures lie in the fact that the most widely used biosignatures, fractionated isotopes, relate to bulk rock analyses and, thus, record a mixed signal from different microhabitats and different communities of microorganisms.

Biosignatures occur in all the habitats preserved in the early crustal rocks – volcanic habitats, sunlight-bathed shallow-water environments and deep-water and hydrothermal ecosystems – attesting to the widespread distribution of life forms on the early Earth. Information on the lifestyles of the different life forms is more difficult to obtain for the reasons mentioned above. Correlation of the biosignatures with a specific microenvironment is, for the time being, the most reliable source of information, with the isotopic signatures being indicative rather than absolute. Thus, biosignatures associated with volcanic materials formed in both deep- and shallow-water environments could have been produced by either chemolithotrophic organisms or consecutively, chemolithotrophs and heterotrophs. Microbial mats on the surfaces of shallow-water littoral sediments may have been produced by primary anaerobic photosynthesizers, possibly in a consortium with other kinds of heterotrophs, if the mat communities were constructed in a similar fashion to modern mats. The hydrothermal vent environments may have hosted chemolithotrophs, but other types of microorganism may have been transported from the surface into the vents during periods of inactivity.

Biosignatures in the Early-Mid Archaean rocks demonstrate that life was abundant and possibly even more diverse than currently believed. With new methods of *in-situ* analysis available, it will be possible to verify the biogenicity and syngenicity of physical and/or chemical signatures that are currently problematic. This will greatly aid understanding of the diversity of early life and provide more information regarding its early evolution.

References

Allwood, A. C., Walter, M. R., Kamber, B. S., Marshall, C. P. and Burch, I. W. (2006). Stromatolite reef from the Early Archaean Era of Australia. *Nature*, **441**, 714–18.

Allwood, A. C., Grotzinger, J. P., Knoll, A. H., Burch, I. W., Anderson, M. S., Coleman, M. L. and Kanik, I. (2009). Controls on development and diversity of Early Archean stromatolites. *Proceedings of the National Academy of Sciences of the USA*, **106**, 9548–55.

Banerjee, N. R., Furnes, H., Muehlenbachs, K., Staudigel, H. and de Wit, M. J. (2006). Preservation of microbial biosignatures in 3.5 Ga pillow lavas from the Barberton Greenstone Belt, South Africa. *Earth and Planetary Science Letters*, **241**, 707–22.

Banerjee, N. R., Simonetti, A., Furnes, H., Staudigel, H., Muehlenbachs, K., Heaman, L. and van Kranendonk, M. (2007). Direct dating of Archean microbial ichnofossils. *Geology*, **35**, 487–90.

Bekker, A., Holland, H. D., Wang, P. L., Rumble D., Stein, H. J., Hannah, J. L., Coetzee, L. L. and Beukes, N. J. (2004). Dating the rise of atmospheric oxygen. *Nature*, **427**, 117–20.

Brasier, M. D., Green, O. R., Jephcoat, A. P., Kleppe, A. K., van Kranendonk, M., Lindsay, J. F., Steele, A. and Grassineau, N. (2002). Questioning the evidence for Earth's oldest fossils. *Nature*, **416**, 76–81.

Brocks J. J., Logan, G. A., Buick, R. and Summons, R. E. (1999). Archean molecular fossils and the early rise of eukaryotes. *Science*, **285**, 1033–6.

Byerly G. R., Walsh, M. M. and Lowe, D. L. (1986). Stromatolites from the 3300–3500 Myr Swaziland Supergroup, Barberton Mountain Land, South Africa. *Nature*, **319**, 489–91.

Canfield, D. E. (2005). The early history of atmospheric oxygen. *Annual Reviews of Earth and Planetary Sciences*, **33**, 1–36.

De Gregorio, B. T., Sharp, T. G., Flynn, G. J., Wirick, S. and Hervig, R. L. (2009). Biogenic origin for Earth's oldest putative microfossils. *Geology*, **37**, 631–4.

Défarge, C., Issa, O. M. and Trichet, J. (1999). Field emission cryo-scanning electron microscopy of organic matter and organomineral associations. Application to microbiotic soil crusts. *Comptes Rendus de l'Académie des Sciences – Series iiA – Earth and Planetary Science*, **328**, 591–7.

Derenne, S., Robert, F., Skrzypczak-Bonduelle, A., Gourier, D., Binet, B. and Rouzaud, J. N. (2008). Molecular evidence for life in the 3.5 billion-year old Warrawoona chert. *Earth and Planetary Science Letters*, **272**, 476–80.

de Ronde, C. E. J. and Ebbesen, T. W. (1996). 3.2 billion years of organic compound formation near sea-floor hot springs. *Geology*, **24**, 791–4.

de Ronde, C. E. J., Channer, D. M. D., Faure, K., Bray, C. J. and Spooner, E. T. (1997). Fluid chemistry of Archean seafloor hydrothermal vents: implications for the composition of circa 3.2 Ga seawater. *Geochimica et Cosmochimica Acta*, **61**, 4015–42.

de Vries, S. T., Nijman, W., Wijbrans, J. R. and Nelson, D. R. (2006). Stratigraphic continuity and early deformation of the central part of the Coppin Gap Greenstone Belt, Pilbara, Western Australia. *Precambrian Research*, **147**, 1–27.

Foucher, F., Westall, F., Brandstätter, F., Demets, R., Parnell, J., Cockell, C. S., Edwards, H. C. M., Bény, J. M. and Brack, A. (2009). Testing the survival of microfossils in artificial Martian sedimentary meteorites during entry into the Earth's atmospheric: the STONE 6 experiment. *Icarus*, **207**(2), 616–30.

Furnes, H., Banerjee, N. R., Muehlenbachs, K., Staudigel, H. and de Wit, M. (2004). Early life recorded in Archean pillow lavas. *Science*, **304**, 578–81.

Furnes, H., Banerjee, N. R., Staudigel, H., Muehlenbachs, K., McLoughlin, N., de Wit, M. and van Kranendonk, M. (2007). Comparing petrographic signatures of bioalteration in recent to Mesoarchean pillow lavas: tracing subsurface life in oceanic igneous rocks. *Precambrian Research*, **158**, 156–76.

Golubic, S., Friedmann, I. E. and Schneider, J. (1981). The lithobiontic ecological niche, with special reference to microorganisms. *Journal of Sedimentary Petrology*, **51**, 475–8.

Grotzinger, J. P. and Kasting, J. F. (1993). New constraints on Precambrian ocean composition. *Journal of Geology*, **101**, 235–43.

Grotzinger, J. P. and Rothman, D. H. (1996). An abiotic model for stromatolite morphogenesis. *Nature*, **383**, 423–5.

Hofmann, A. and Bohlar, R. (2007). Carbonaceous cherts in the Barberton Greenstone Belt and their significance for the study of early life in the Archean Record. *Astrobiology*, **7**, 355–388.

Javaux, E. J., Knoll, A. H. and Walter, M. R. (2001). Morphological and ecological complexity in early eukaryotic ecosystems. *Nature*, **412**, 66–9.

Javaux, E. J., Knoll, A. H. and Walter, M. R. (2004). TEM evidence for eukaryotic diversity in mid-Proterozoic oceans. *Geobiology*, **2**, 121–32.

Javaux, E. J., Marshall, C. P. and Bekker, A. (2010). Organic-waller microfossils in 3.2-billion-year-old shallow-marine siliciclastic deposits. *Nature*, **463**, 934–938.

Kakegawa, T. (2001). Isotopic signatures of early life in the Archean oceans: influence from submarine hydrothermal activities. In *Geochemistry and the Origin of Life*, eds. S. Nakashsima, S. Maruyama, A. Brack and B. F. Windley. Tokyo: Universal Academy Press, pp. 237–49.

Kasting, J. F. (1993). Earth's early atmosphere. *Science*, **259**, 920–6.

Knauth, L. P. and Lowe D. R. (2003). High Archean climatic temperature inferred from oxygen isotope geochemistry of cherts in the 3.5 Ga Swaziland Supergroup, South Africa. *Geological Society of America Bulletin*, **115**, 566–80.

Knoll, A. H., Javaux, E. J., Hewitt, D. and Cohen, P. (2006). Eukaryotic organisms in Proterozoic oceans. *Philosophical Transactions of the Royal Society, Biological Sciences*, **361**, 1023–38.

Lowe, D. R. (1980). Stromatolites 3,400-Myr old from the Archean of Western Australia. *Nature*, **284**, 441–3.

Lowe, D. R. (1983). Restricted shallow water sedimentation of early Archean stromatolitic and evaporitic strata of the Strelley Pool Chert, Pilbara Block, Western Australia. *Precambrian Research*, **19**, 239–83.

Lowe, D. R. (1994). Abiological origin of described stromatolites older than 3.2 Ga. *Geology*, **22**, 387–90.

Mojzsis, S. J., Arrhenius, G., McKeegan, K. D., Harrison, T. M., Nutman, A. P. and Friend, C. R. L. (1996). Evidence for life on Earth before 3800 million years ago. *Nature*, **384**, 55–9.

Noffke, N. (2009). The criteria for the biogenicity of microbially induced sedimentary structures (MISS) in Archean and younger, sandy deposits. *Earth Science Reviews*, **96**(3), 173–80.

Noffke, N., Hazen, R. M., Eriksson, K. and Simpson, E. (2006). A new window into early life: microbial mats in a siliclastic Early Archean tidal flat (3.2 Ga Moodies Group, South Africa). *Geology*, **4**, 253–6.

Orange, F., Westall, F., Disnar, J. R., Prieur, D., Bienvenu, N., Le Romancer, M. and Défarge, C. (2009). Experimental silicification of the extremophilic Archaea *Pyroccus abyssi* and *Methanocaldococcus jannaschii*. Applications in the search for evidence of life in early Earth and extraterrestrial rocks. *Geobiology*, **7**, 403–18.

Orberger, B., Rouchon, V., Westall, F., de Vries, S. T., Wagner, C. and Pinti, D. L. (2006). Protoliths and micro-environments of some Archean Cherts (Pilbara, Australia). In *Processes on the Early Earth*, eds. W. U. Reimold and R. Gibson. Geological Society of America Special Paper, **405**,133–52.

Pflug, H. D. (1979). Archean fossil finds resembling yeasts. *Geologie und Palaeontologie*, **13**, 1–8.

Pflug, H. D. (2001). Earliest organic evolution: essay to the memory of Bartholomew Nagy. *Precambrian Research*, **106**, 79–92.

Philippot, P., van Zuilen, M., Lepot, K., Thomazo, C., Farquahr, J. and van Kranendonk, M. (2007). Early Archaean microorganisms preferred elemental sulphur, not sulphate. *Science*, **317**, 1534–7.

Pinti, D. L., Mineau, R. and Clement, V. (2009). Hydrothermal alteration and microfossil artefacts of the 3,456-million-year-old Apex Chert. *Nature Geosciences*, **2**, 640–3.

Rasmussen, B. (2000). Filamentous microfossils in a 3,235-million-year-old volcanogenic massive sulphide deposit. *Nature*, **405**, 676–9.
Rasmussen, B., Fletcher, I. R., Brocks, J. J., and Kilburn, M. R. (2008). Reassessing the first appearance of eukaryotes and cyanobacteria. *Nature*, **455**, 1101–1104.
Reid, R. P., Visscher, P. T., Decho, A. W., Stolz, J. F., Bebout, B. M., Dupraz, C., Macintyre, I. G., Paerl, H. W., Pinckney, J. L., Prufert-Bebout, L., Steppe, T. F. and DesMarais, D. J. (2000). The role of microbes in the accretion, lamination and early lithification of modern marine stromatolites. *Nature*, **406**, 989–92.
Robert, F. and Chaussidon, M. (2006). A palaeotemperature curve for the Precambrian oceans based on silicon isotopes in cherts. *Nature*, **443**, 969–72.
Rosing M. T. (1999). ^{13}C depleted carbon microparticles in > 3700 Ma seafloor sedimentary rocks from West Greenland. *Science*, **283**, 674–6.
Rosing, M. T. and Frei, R. (2004). U-rich Archaean sea-floor sediments from Greenland: indications of > 3700 Ma oxygenic photosynthesis. *Earth and Planetary Science Letters*, **217**, 237–44.
Rye, R., Kuo, P. H. and Holland, H. D. (1995). Atmospheric carbon dioxide concentrations before 2.2 billion years ago. *Nature*, **378**, 603–5.
Schopf, J. W. (1993). Microfossils of the Early Archean Apex Chert: new evidence of the antiquity of life. *Science*, **260**, 640–6.
Schopf, J. W. and Walter, M. R. (1983). Archean microfossils: new evidence of ancient microbes. In *Earth's Earliest Biosphere: Its Origin and Evolution*, ed. J. W. Schopf. Princeton: Princeton University Press, pp. 214–39.
Schopf, J.W., Kudryavtsvev, A.B., Agresti, D.G., Wdowiak, T.J. and Czaja, A.D. (2002). Laser-Raman imagery of Earth's earliest fossils. *Nature*, **416**, 73–76.
Shen, Y., Buick, R. and Canfield, D. E. (2001). Isotopic evidence for microbial sulphate reduction in the Early Archaean Era. *Nature*, **410**, 77–81.
Southam, G. and Westall, F. (2007). Geology, life and habitability. In *Treatise on Geophysics: Planets and Moons*, ed. T. Spohn. Amsterdam: Elsevier, vol. 10, pp. 421–38.
Sugitani, K., Grey, K., Allwood, A., Nagaoka, T., Mimura, K., Minami, M., Marshall, C., van Kranendonk, M. and Walter, M. (2007). Diverse microstructures from Archaean chert from the Mount Goldsworthy–Mount Grant area, Pilbara Craton, Western Australia: microfossils, dubiofossils or pseudofossils? *Precambrian Research*, **158**, 228–62.
Sugitani, K., Grey, K., Nagoaka, T., Mimura, K. and Walter, M. (2009). Taxonomy and biogenicity of Archaean sphereoidal microfossils (ca. 3.0 Ga) from the Mount Goldsworthy–Mount Grant area in the northwestern Pilbara Craton, Western Australia. *Precambrian Research*, **173**, 50–9.
Summons R. E., Jahnke, L. L., Hope, J. M. and Logan, J. H. (1999). 2-Methylhopanoids as biomarkers for cyanobacterial oxygenic photosynthesis. *Nature*, **400**, 554–7.
Thorseth, I., Torsvik, T., Torsvik, K., Daae, F. L., Pedersen, R. B. and Keldysh-98 Scientific Party. (2001). Diversity of life in ocean floor basalts. *Earth and Planetary Science Letters*, **194**, 31–7.
Tice, M. and Lowe D. R. (2004). Photosynthetic microbial mats in the 3,416-Myr-old ocean. *Nature*, **431**, 549–52.
Tice, M. M. and Lowe, D. R. (2006). The origin of carbonaceous matter in pre-3.0 Ga greenstone terrains: a review and new evidence from the 3.42 Buck Reef Chert. *Earth Science Reviews*, **76**, 259–300.
Toporski, J., Steele, A., Westall, F., Thomas-Keprta, K. and McKay, D. S. (2002). The simulated silicification of bacteria: new clues to the modes and timing of bacterial

preservation and implications for the search for extraterrestrial microfossils. *Astrobiology*, **2**, 1–26.

Ueno Y., Yoshioka, H., Maruyama, S. and Isozaki, Y. (2004). Carbon isotopes and petrography of kerogens in *c*.3.5 Ga hydrothermal silica dikes in the North Pole area, Western Australia. *Geochimica et Cosmochimica Acta*, **68**, 573–89.

Van den Boorn, S., Van Bergen, M. J., Nijman, W. and Vroon, P. Z. (2007). Dual role of seawater and hydrothermal fluids in Early Archean chert formation: evidence from silicon isotopes. *Geology*, **35**, 939–42.

van Zuilen, M., Lepland, A. and Arrhenius, G. (2002). Reassessing the evidence for the earliests traces of life. *Nature*, **418**, 627–30.

van Zuilen, M., Chaussidon, M., Rollion-Bard, C. and Marty, B. (2007). Carbonaceous cherts of the Barberton Greenstone Blet, South Africa: isotopic, chemical and structural characteristics of individual microstructures. *Geochimica et Cosmochimica Acta*, **71**, 655–69.

Visscher, P.T. and Stolz, J. (2005). Microbial communities as biogeochemical reactors. *Palaeogeography, Palaeoclimatology, Palaeoecology*, **119**, 87–100.

Walsh, M. M. (1992). Microfossils and possible microfossils from the Early Archean Onverwacht Group, Barberton Mountain Land, South Africa. *Precambrian Research*, **54**, 271–93.

Walsh, M. M. (2004). Evaluation of Early Archean volcanoclastic and volcanic flow rocks as possible sites for carbonaceous fossil microbes. *Astrobiology*, **4**, 429–37.

Walsh, M. M. and Lowe, D. R. (1999). Modes of accumulation of carbonaceous matter in the Early Archaean: a petrographic and geochemical study of carbonaceous cherts from the Swaziland Supergroup. In *Geologic Evolution of the Barberton Greenstone Belt, South Africa*, eds. D. R. Lowe and G. R. Byerly. Geological Society of America Special Paper, **329**, 115–32.

Walter, M. R. (1983). Archean stromatolites: evidence of the Earth's earliest benthos. In *Earth's Earliest Biosphere*, ed. J. W. Schopf. Princeton: Princeton University Press, pp. 187–213.

Walter, M. R., Buick, R. and Dunlop, J. S. R. (1980). Stromatolites 3,400–3,500 Myr old from the North Pole area, Western Australia. *Nature*, **284**, 443–5.

Westall, F. (1997). The influence of cell wall composition on the fossilisation of bacteria and the implications for the search for early life forms. In *Astronomical and Biochemical Origins and the Search for Life in the Universe*, eds. C. B. Cosmovici, S. Bowyer and D. Werthimer. Bologna: Editrici Compositori, pp. 491–504.

Westall, F. (1999). The nature of fossil bacteria: a guide to the search for extraterrestrial life. *Journal of Geophysical Research, Planets*, **104**, 16437–51.

Westall, F. (2005). Life on the early Earth: a sedimentary view. *Science*, **434**, 366–7.

Westall, F. (2008). Morphological biosignatures in terrestrial and extraterrestrial materials. *Space Science Reviews*, **135**, 95–114.

Westall, F. (2009). Life on an anaerobic planet. *Science*, **232**, 471–2.

Westall, F., Boni, L. and Guerzoni, M. E. (1995). The experimental silicification of microbes. *Palaeontology*, **38**, 495–528.

Westall, F., De Wit, M. J., Dann, J., Van Der Gaast, S., De Ronde, C. and Gerneke, D. (2001). Early Archaean fossil bacteria and biofilms in hydrothermally influenced, shallow water sediments, Barberton Greenstone Belt, South Africa. *Precambrian Research*, **106**, 91–112.

Westall, F. and Folk, R. L. (2003). Exogenous carbonaceous microstructures in Early Archaean cherts and BIFs from the Isua greenstone belt: implications for the search for life in ancient rocks. *Precambrian Research*, **126**, 313–30.

Westall, F., de Vries, S. T., Nijman, W., Rouchon, V., Orberger, B., Pearson, V., Watson, J., Verchovsky, A., Wright, I., Rouzaud, J. N., Marchesini, D. and Anne, S. (2006a). The 3.466 Ga Kitty's Gap Chert, an Early Archaean microbial ecosystem. In *Processes on the Early Earth*, eds. W. U. Reimold and R. Gibson. Geological Society of America Special Paper, **405**, 105–31.

Westall, F, de Ronde, C. E. J., Southam, G., Grassineau, N., Colas, M., Cockell, C. and Lammer, H. (2006b). Implications of a 3.472–3.333 Ga-old subaerial microbial mat from the Barberton Greenstone Belt, South Africa for the UV environmental conditions on the early Earth. *Philosophical Transactions of the Royal Society of London Series B*, **361**, 1857–75.

Westall, F. and Southam, G. (2006). Early life on Earth. In *Archean Geodynamics and Environments*, eds. K. Benn, J. C. Marechal and K. C. Condie. American Geophysical Union, Geophysical Monograph, **164**, 283–304.

Westall, F., Lemelle, L., Salomé, M., Simionovici, A., Southam, G., LacLean, L., Wirick, S., Toporksi, J. and Jauss, A. (2008, March). Vertical geochemical profile across a 3.33 Ga microbial mat from Barberton. *39th Lunar and Planetary Sciences Conference*, Houston. Abstract **1636**.

Westall, F. and Cavalazzi, B. (2009). Biosignatures in rocks and minerals. In *Encyclopedia of Geobiology*, eds. V. Thiel and J. Reitner. Berlin: Springer (in press).

26

Early eukaryotes in Precambrian oceans

Emmanuelle Javaux

Origin and diversity of eukaryotes

Origin

Life on Earth is classed into three phylogenetic domains: the Archaea, the Bacteria and the Eucarya. Eukaryotic cells are less diverse than prokaryotes metabolically, but have a complex cellular architecture comprising a nucleus, a cytoskeleton (a proteinic network structuring the cytoplasm to facilitate intracellular traffic, endo- and exo-cytosis and amoeboid locomotion; Cavalier-Smith, 2002), an endomembrane system (a system of internal membranes subdivided into several organelles, and used for synthesis, processing, packaging and transport of macromolecules such as lipids and proteins) and organelles such as mitochondria (or derived organelles) and chloroplasts in photosynthetic eukaryotes. Archaea and Bacteria are called prokaryotes because their cells do not contain a nucleus or organelles, and because transcription and translation are coupled, i.e. they do not occur in different cellular compartments. The prokaryotes have diverse and complex metabolisms, but simpler cellular architecture. They possess proteins playing the role of cytoskeleton, but not the motor proteins involved in intracellular transport as in eukaryotes (Moller-Jensen and Lowe, 2005; Cabeen and Jacobs-Wagner, 2005) and some (the planctomycetales) may possess endomembrane systems (e.g. Fuerst, 2005).

The eukaryotic genome is seen as a mosaic of bacterial genes (involved in energy and carbon metabolism) and archaeal genes (related to DNA replication, transcription and translation), but it also contains a core set of genes and proteins unique to eukaryotes (e.g. Kurland *et al.*, 2006), while the membrane lipids are closer to those of the bacteria (Rivera *et al.*, 1998). The somewhat chimeric nature of eukaryotes has led several authors to propose a range of hypotheses involving a symbiosis between prokaryotes (one archaeon and one bacterium) as the origin of eukaryotic cells. Others suggest the existence of a nucleated proto-eukaryote, or even a viral origin for the nucleus (see review by López-García *et al.*, 2006; Embley and Martin, 2006; Kurland *et al.*, 2006). At present, the origin of the domain Eucarya is an important and still unresolved question, and the status of this domain in the Tree of Life varies from a monophyletic group, to the exclusion of Bacteria and Archaea (Woese *et al.*, 1990; Pace, 2006; Kurland *et al.*, 2006), a group with an archaeal origin (the 'eocyte' hypothesis revisited by Cox *et al.*, 2008), to the root of the Tree itself (Philippe and

Forterre, 1999). Because of the intense lateral gene transfer between and within the three domains of life, a 'Net of Life' has been proposed to better represent the full diversity of life on Earth (Doolittle, 1999).

Based on ultrastructural and molecular observations, the bacterial origin of the eukaryotic organelles is now well established (e.g. Bhattacharya *et al.*, 2007). The mitochondrion ancestor was an α-proteobacterium that was once acquired in the evolution of the eukaryotic cells and which enabled them to live in oxic environments. The chloroplast ancestor was a cyanobacterium, and its acquisition permitted the radiation of several clades of photosynthetic eukaryotes. In the last decades, a consensus has emerged among biologists studying protists (unicellular eukaryotes): all *extant* protists have mitochondria or mitochondrial genes in their genome and derived organelles such as the hydrogenosome or mitosome, suggesting that the ancestor of *extant* eukaryotes was a mitochondriate aerobic protist (review in Embley and Martin, 2006). It is possible, however, that the very first eukaryote (ancestor of all extinct and extant eukaryotes) was an anaerobe (without mitochondria), as suggested by Kurland *et al.* (2006), who proposed that a primitive eukaryote acquired the endosymbiotic ancestor of the mitochondrion by phagocytosis. According to Cavalier-Smith (2009), the first eukaryotes were aerobic phagotrophs, probably benthic heterotrophic amoeboflagellates. The origins of the other constituents of eukaryotic cells, such as the nucleus and associated endomembrane system, the cytoskeleton and the cilium still raise questions (see Jekely, 2006 for a review).

Based on molecular clocks or genetic and ultrastructural studies, the antiquity of the domain Eucarya ranges from the early Palaeoproterozoic (*c.*2.3–1.8 Ga for the origin of the mitochondria; Hedges *et al.*, 2004, 2006), to the Palaeoproterozoic (Yoon *et al.*, 2004) or early Neoproterozoic (*c.*850 Ma; Cavalier-Smith, 2009). The origin of the chloroplast is estimated at 1.5–1.6 Ga (Hedges *et al.*, 2004; Yoon *et al.*, 2004). The diversification of eukaryotes is estimated at the late Mesoproterozoic (1259–950 Ma; Douzery *et al.*, 2004). (See Roger and Hug, 2006, for a discussion on problems with molecular-phylogenetic and molecular-clock estimations of the origin and diversification of eukaryotes.) In principle, nothing precludes the appearance of the ancestor of extant eukaryotes, an aerobic heterotrophic protist, as soon as the ancestor of mitochondria, an alpha-proteobacterium, and cyanobacterium (producing oxygen) had evolved. Cyanobacteria appeared no later than 2.45 Ga (Bekker *et al.*, 2004), but the origin of the particular proteobacterium that gave rise to the mitochondria is unknown, although biomarkers of the purple-sulphur bacteria, another member of the proteobacteria, occur at 1.640 Ga (Brocks *et al.*, 2005).

Diversity

Molecular phylogenies divide the domain Eucarya into several supergroups that diverged rapidly, making it difficult to resolve their relationships and the order of branching events (Baldauf, 2008). Although land plants, animals and fungi are the most accessible and studied eukaryotes, most of the eukaryotic diversity is microscopic (Baldauf, 2008). Characters such as biomineralization (Knoll, 2003a) and multicellularity (Butterfield, 2009a) appeared independently in several clades. Photosynthesis also appeared in several clades, but originated from a single endosymbiotic event involving a cyanobacterium, followed by successive endosymbioses (secondary endosymbioses) of photosynthetic protists (except for one clade of filose *Thecamoeba Paulinella*; Nowak *et al.*, 2008; Baldauf, 2008). A large part of the diversity might still be undiscovered, as reports of picoeukaryotes might suggest (Moreira and López-García, 2002; Liu *et al.*, 2009).

A recent review (Baldauf, 2008) divides the known diversity of the domain into four major groups (Figure 26.1):

(1) the unikonts (primitively uniflagellate cells – although this ancestral state is open to debate). Roger and Simpson (2009) include two supergroups: the Opisthokonts (animals, fungi, choanoflagellates and some other protistan groups) and the Amoebozoa (slime moulds and traditional amoebae);
(2) the Archaeplastida (in which eukaryotic photosynthesis first arose) include red and green algae as well as land plants;
(3) the RAS clade includes three of the most diverse divisions of eukaryotes: the Rhizaria, Alveolates and Stramenopiles (formely Heterokonts), and possibly also a fourth group of algae (the haptophytes [coccolithophorids] and the cryptophytes). The Rhizaria include the Foraminifera,

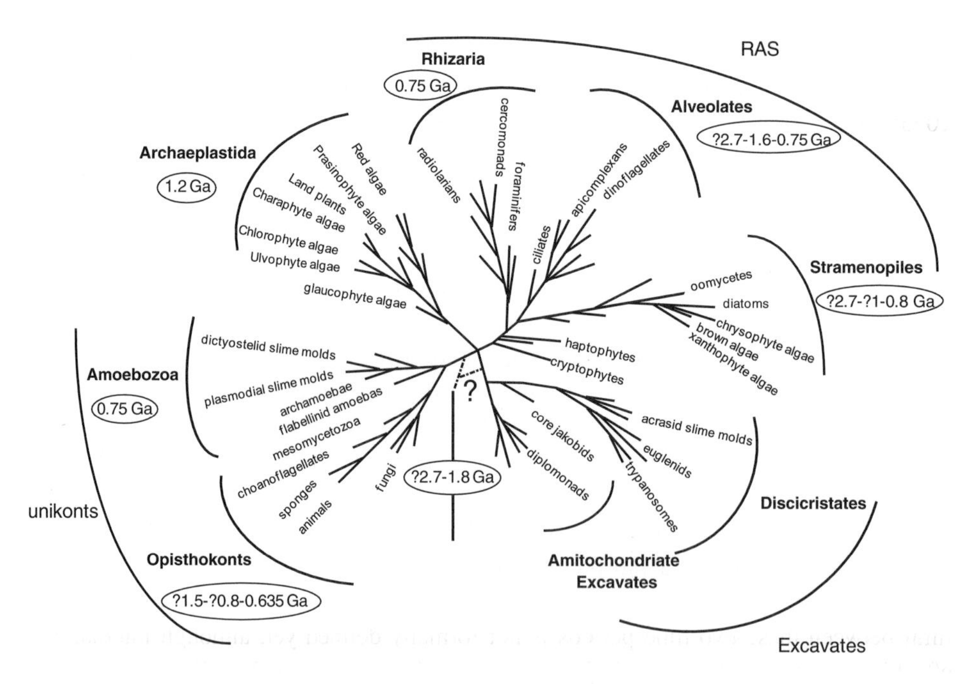

Figure 26.1. Phylogeny of the domain Eucarya. Phylogeny of the domain Eucarya based on molecular phylogenies and ultrastructural data (modified from Baldauf, 2008). Dashed lines are the two currently proposed positions for the eukaryote root (see text for details). The root of the eukaryotic tree would lie between the unikonts and the bikonts, based on gene-fusion and myosin gene data (Stechmann and Cavalier-Smith, 2002; Richards and Cavalier-Smith, 2005), or between the Excavates and the other supergroups. Data for minimum age of the supergroups with a fossil record come from the biomarker and body fossil record (see text).

Radiolarians, Cercozoa – including euglyphids – and other protists; the Alveolates include the ciliates, dinoflagellates and other protists; and the Stramenopiles include the diatoms, other protists and the brown and xanthophyte algae; and

(4) the Excavates include amitochondriate excavates (Metamonads) and mitochondriate excavates (discicristates, such as euglenids and some parasitic protists and Jakobids).

Recent phylogenomic analyses suggest that the supergroup Excavates forms the deepest branch within a megagroup of bikonts (primitively biflagellates), that also includes the supergroup Archaeplastida and the supergoups included in the RAS (Hampl *et al.*, 2009). The root of the eukaryotic tree would lie between the unikonts and the bikonts, based on gene-fusion and myosin gene data (Stechmann and Cavalier-Smith, 2002; Richards and Cavalier-Smith, 2005), although the position of the root and the support of a bikont megagroup are very much discussed (Hedges *et al.*, 2006; Roger and Simpson, 2009) (Figure 26.1).

In the following pages, I will summarize the Precambrian record of some of these supergroups which originated early and have a fossil record. I will then examine the diversification pattern of the domain Eucarya and the appearance of biological innovations in the domain, as well as hypotheses regarding the possible causes of this pattern. Palaeobiological data are essential for testing hypotheses about relationships between clades and order of branching in phylogenetic trees and for understanding the timing of diversification, when fossils can be related to extant clades (e.g. Javaux *et al.*, 2003; Javaux, 2006a; Porter, 2004, 2006). Fossils may also record ancestral forms (and steps in evolution) that might not have any extant relatives. Regardless of taxonomy, fossils may display morphological attributes related to major biological innovations, improving our understanding of the evolution of the domain Eucarya (Javaux *et al.*, 2003, 2006a; Huntley *et al.*, 2006; Knoll *et al.*, 2006b; Xiao and Dong, 2006; Butterfield, 2009a).

The Precambrian fossil record

Major environmental changes in the Precambrian

The Precambrian includes: the Hadean (4.6 to 4 Ga [Ga = billion years]), the Eon starting with the Solar System formation and Earth accretion; the Archean (4 to 2.5 Ga) when life appeared; and the Proterozoic (2.5 to 0.56 Ga), subdivided into the Palaeo-, Meso- and Neoproterozoic Era. The Neoproterozoic Era can be subdivided into three periods: the Early Neoproterozoic (1000–*c*.750 Ma), the Middle Neoproterozoic (*c*.750–635 Ma) (the limit between these two time periods is not formally defined yet, although international effort is ongoing to determine official stratigraphic boundaries), and the Ediacaran, the latest geological period formally defined by the IUGS, from 635 Ma to 542 Ma (Knoll *et al.*, 2004, 2006a). During this large portion of Earth's history, major changes occurred in the environment, as shown by the geological record. These events, such as the step-wise oxygenation of the atmosphere and oceans, meteoritic impacts, supercontinent formation and break-up, and severe glaciations, might have had a profound effect on the early eukaryote evolution (Figure 26.3).

The Archean–Proterozoic transition is marked by the so-called great oxidation event (GOE) around 2.45 Ga, when molecular oxygen accumulated sufficiently in the ocean and atmosphere to be detected widely by a variety of geological and geochemical proxies (Bekker *et al.*, 2004; Sessions *et al.*, 2009). This minimum age for oxygen accumulation of course does not prevent the earlier existence of local oxygenated oases in shallow-water near cyanobacterial mats. Estimates for the possible origin of oxygenic photosynthesis (by cyanobacteria) range between 3.7 Ga based on carbon and Pb isotopic memory of past U/Th ratios (Rosing and Frei, 2004; but see Shen and Buick, 2004); 3.2 Ga based on thick and widespread deposits of kerogenous black shales (Buick, 2008); 2.9 Ga based on S isotopes (Ono *et al.*, 2006); 2.8 Ga based on chromium isotopes (Frei *et al.*, 2009); 2.7 Ga based on biomarkers (Brocks *et al.*, 2003; but see Rasmussen *et al.*, 2008); and at 2.5 Ga based on molybdenum and rhenium data (Anbar *et al.*, 2007). Distinctive microfossils of cyanobacteria occur at 2.1 Ga (Amard and Bertrand-Sarfati, 1997; Hofmann, 1976). Older microfossils could represent cyanobacteria, but their simple morphology does not permit identification. Evidence for photosynthesizing microbial mats occurs at 3.4 Ga (Tice and Lowe, 2004), 3.2 Ga (Noffke *et al.*, 2006) and 2.9 Ga (Noffke *et al.*, 2003), but these may originate from the activity of anoxygenic photosynthesizers.

The Archean ocean was iron-rich and dominantly anoxic. In the mid-Proterozoic (~ 1.9–1.8 to ~ 0.8 Ga), these conditions changed into sulphidic (H_2S-rich) anoxic conditions (euxinia) in the deep ocean (Canfield, 1998). Euxinia also occurred in the lower part of the photic zone, in an oxygen minimum zone (OMZ), while the bottom waters might have been euxinic, or only anoxic or dysoxic (Johnston *et al.*, 2009). The surface waters of the photic zone were moderately oxygenated (1–10% PAL: present atmospheric level of oxygen) since the GOE, although some oxygenation probably occurred earlier, as suggested above. The redox state of the deep ocean between the GOE (2.4 Ga) and 1.8 Ga is not well known, but might also have been euxinic. A transient return to an iron ocean triggered by a decline in oxygen might have occurred possibly around 1.9 Ga (see review in Lyons *et al.*, 2009). The demise of subsurface sulphidic conditions to a transient ferruginous ocean, coeval of the break-up of the supercontinent Rodinia around 750 Ma, and prior to the Sturtian glaciation, is documented by S, C and Fe data (Johnston *et al.*, 2010). A second sharp rise in oxygen levels close to present atmospheric levels (1% PAL) occurred around 750 Ma, but the Ediacaran ocean might have been persistently oxygenated only from 580 Ma, although transient iron-rich and euxinic intervals are suggested (review in Lyons *et al.*, 2009). Transient anoxia occurred prior to the Middle Neoproterozoic glaciations (around 750 Ma) (Nagy *et al.*, 2009) and again at the Proterozoic–Cambrian boundary (Knoll *et al.*, 2006b).

Glass *et al.* (2009) suggested that the history of transition metal availability through time might explain the evolution of primary producers in oceans with changing chemistry. Indeed, these metals, such as Fe, Mo, Ni and Cu are used in prokaryotic and eukaryotic metallo-enzymes to assimilate nitrogen (N). N is essential to all life and is abundant in the atmosphere in the form of the not very soluble N_2. The oceanic concentration of N is therefore low, and poses a major challenge for marine life. Anoxic Fe-rich Archean oceans would be favourable to the evolution of Fe-using enzymes. When oxygen started to accumulate around the GOE, Mo would be mobilized and available for Mo-using enzymes. The following euxinic conditions of the Mid-Proterozoic deep ocean would have limited Mo and Fe availability (forming insoluble precipitates with sulphides) (Anbar and Knoll, 2002). These conditions would disadvantage eukaryotic N_2 fixation because eukaryotic Mo-enzymes need more Mo than the prokaryotic enzymes, thereby favouring prokaryotes such as cyanobacteria. However, Ni and Cu are also important metals used in N assimilation and the evolution of their availability through time is not known yet, although Ni might have been higher in the Archean Eon (Konhauser *et al*., 2009) and Cu, Zn and Cd lowest during the Middle Proterozoic (Anbar and Knoll, 2002; Buick, 2007).

Johnston *et al*. (2009) argued that N would be scarce in the Mid-Proterozoic oxygenated waters above the OMZ, because while ascending from deep waters, it would have been consumed by microbial denitrification, anammox and anoxygenic photosynthesis in the OMZ (Johnston *et al*., 2009). Therefore N-deficiency would have favoured N_2-fixing bacteria over eukaryotic algae. Primary production would be dominated by anoxygenic photosynthesizers (including some cyanobacteria able to perform anoxygenic photosynthesis in anoxic conditions) in the OMZ (although oxygenic photosynthesis still occurred in the surface waters above), maintaining euxinic conditions until about 750 Ma.

Besides a complex (and still unresolved) history of changing ocean and atmosphere chemistry, other important environmental changes included tectonics, meteor impact and major glaciations.

The Early Palaeoproterozoic Period includes evidence for three ice ages (≤ 2.45 to > 2.22 Ga) (the Huronian glaciations) believed to be associated with assembly and rifting of a Late Archean supercontinent (Kenorland) (Bekker *et al*., 2005). Impact layers of spherules also occur.

The Early Neoproterozoic Era (1000–*c*.750 Ma) starts with the amalgamation of the Rodinia supercontinent and increased variability in $\delta^{13}C$ after a prolonged Mesoproterozoic interval of stable values (review in Halverson, 2006). Rodinia break-up ended around 750 Ma.

The Middle Neoproterozoic Period (*c*.750–635 Ma) includes two widespread (Sturtian and Marinoan) glaciations separated by an interglacial. The age and the global extent of the Sturtian glaciation are highly uncertain (see discussion by Halverson, 2006). Datings from different areas range from a maximum age *c*.746 Ma to 670 Ma for the end of the glaciation (Halverson, 2006) or 726–660 Ma (Hoffman and Li, 2009). The bottom age of the interglacial is therefore difficult to constrain, but appears better defined in a few areas (Namibia, Australia and northern Canada). The Marinoan glaciation is well-dated at 635.5±1.2 Ma in Namibia (Hoffman *et al*., 2004) and well-correlated with glacial deposits and overlying cap carbonates elsewhere (Halverson, 2006). Its global extent is well-documented in low latitudes by palaeomagnetic data.

The Ediacaran Period (635–542 Ma) is the latest formally defined geological period (Knoll *et al*., 2004, 2006a). It forms the upper part of the Neoproterozoic Era, and its basis and top are correlated worldwide by carbon isotopic incursion in cap carbonates topping the tills of the Marinoan glaciations (635±0.2 Ma) and by Early Cambrian trace fossils, respectively. Halverson (2006) subdivided the Ediacaran Period into three time slices: the Early Ediacaran, the Gaskiers glaciation and the terminal Proterozoic (Late Ediacaran). Fossil assemblages differ through the Ediacaran but precise datings, correlations and taxonomy are needed to define a reliable biostratigraphy. The timing and extent of the Gaskiers glaciation around 575 Ma are not well constrained. Around 580 Ma, a meteor impact called the Acraman impact is recorded in South Australia (Grey *et al.*, 2004). After the Gaskiers glaciation, the Late Neoproterozoic ocean was highly ^{13}C-depleted just prior to the first appearance of the Ediacara macroscopic biota, possibly due to oxidation of the large organic ^{13}C-depleted reduced carbon reservoir (Halverson, 2006), resulting in stabilization or increase in atmospheric O_2 concentrations. A brief but pronounced C-isotopic event coupled to transient shallow-water anoxia occurred near the Proterozoic–Cambrian boundary (Knoll *et al.*, 2006b).

Methods and limitations of the study of the early eukaryotic fossil record

Palaeontologists have to rely on information other than the genome and internal cellular organization to decipher the record of early eukaryotes. To determine the biological affinities of microfossils at the level of domain or beyond, a set of criteria to differentiate prokaryote from eukaryote microfossils (Javaux *et al.*, 2003, 2004; Knoll *et al.*, 2006b) and a methodology combining microscopy and microchemistry of single organic-walled microfossils (Arouri *et al.*, 1999; Javaux and Marshall, 2006; Marshall *et al.*, 2005) have been developed. Size is not a good criterion to differentiate fossil prokaryotes from eukaryotes, since in both groups, extant pico- and large microorganisms are documented (review in Javaux *et al.*, 2003). Fossils can display morphological and ultrastructural features showing a degree of complexity and/or particular features unknown in prokaryotic organisms, therefore pointing to a eukaryotic affinity. Indeed, the wall structure and ornamentation, the presence of processes that extend from the vesicle wall, the presence of excystment structures (openings through which cysts liberate their content), the wall ultrastructure and the wall chemistry can clarify the eukaryotic affinities of organic-walled microfossils, sometimes even at the level of class. Microchemical analyses such as micro-infra-red (e.g. Marshall *et al.*, 2005) and Raman spectroscopy (e.g. Marshall *et al.*, 2006), scanning transmission X-ray microscopy (STXM; e.g. Bernard *et al.*, 2009) and other techniques applicable to very small sample size such as one microfossil, can be used to characterize the chemistry of organic microfossils and might even reveal biomolecules specific to extant clades. One limitation of this approach is the limited knowledge that we have of extant organisms producing fossilizable structures and their morphological, ultrastructural and chemical properties. This approach requires investigation of preservable biological properties and comparative actualistic studies of taphonomic processes affecting diverse organisms in a range of natural and experimental environmental conditions (Javaux and Marshall, 2006; Javaux and Benzerara, 2009).

Even when a precise identification cannot be achieved, either because the fossil lacks taxonomically informative features permitting to relate it to an extant clade, or because it represents an extinct clade, fossils do provide direct evidence of early organisms, and document steps in biological and biochemical innovations. The original biological properties (morphology, chemistry, division pattern, behaviour) may be well preserved, altered or erased by fossilization processes that vary depending on the mode of preservation (and thus the physico-chemical conditions of the preservational environment) and the organism's composition. Therefore, a good understanding of potential artefacts and biases is essential to decipher the original biology.

The early eukaryotic fossil record includes: carbonaceous compressions (the organisms are preserved as a thin film of carbon); acritarchs (organic-walled vesicles with unknown biological affinities – they can be extracted from shales using strong acids or observed in thin sections through shale, chert or phosphorite); multicellular organic-walled organisms (chert, shale); vase-shaped microfossils; moulds and casts in sandstone, carbonate or shale; mineralized skeletons, walls or scales preserved in carbonates or phosphorite; and phylogenetically informative molecules. The fossil molecules include membrane lipids and pigments, called biomarkers, and recalcitrant cell-wall lipids and sheath or extracellular polysaccharides, called biopolymers (Versteegh and Blokker, 2004). Biomarkers can be preserved in the rocks even in the absence of fossils and provide information about past ecosystems and the evolution of biosynthetic pathways, and in some cases of the occurrence of particular clades. However, several problems, such as contamination by younger molecules, lateral gene transfer (between contemporary unrelated organisms exchanging genes and acquiring the ability to synthesize a particular molecule) and our limited knowledge of the diversity and origin of molecules in extant organisms, limit the interpretation of the biomarker fossil record.

Traces of eukaryotes in the Precambrian oceans

In the following sections, the biomarker and the body fossil records of early eukaryotes are presented in stratigraphic order and the earliest appearance (according to present knowledge) of major extant clades and important biological innovations are underlined. This paper does not intend to give a complete list of all taxa reported as early eukaryotes (for recent compilations, see e.g. Sergeev, 2009), but only the most biologically or stratigraphically informative.

The biomarker record

Archean record (4 to 2.5 Ga)

The Archean record for eukaryotes is questioned and limited to biomarkers: fossil molecules (steranes) derived from sterol lipids found in eukaryotic membranes (Volkman, 2003, 2005; Summons *et al.*, 2006). Steranes from 2.7-Ga bitumens of the Fortescue Group, Australia (Brocks *et al.*, 1999, 2003), may indicate that contemporaneous cells were able to synthesize sterols, requiring a minimum of dioxygen, although the data have been reinterpreted recently as contamination (Rasmussen *et al.*, 2008). Among these biomarkers, dinosterane, derived from dinosterol produced by dinoflagellates and their ancestors, also occurs in the ~2.78–2.45-Ga Mount Bruce Supergroup (Australia) (Brocks *et al.*, 2003). However, dinosterol is also synthesized by diatoms (Rampen *et al.*, 2010). If confirmed, the occurrence of this biomarker could indicate the early evolution of members of the RAS clade (Alveolates or Stramenopiles) or their ancestors. However, this early age may suggest another non-dinoflagellate (and unknown) origin or a contamination (review in Porter, 2006). Other studies report eukaryotic biomarkers in the Late Archean, such as in the 2.67 to 2.46-Ga Transvaal Supergroup (Waldbauer *et al.*, 2009). So if contamination can be ruled out unambiguously by further studies (Fisher, 2008), it is possible that the distribution and preservation of eukaryotic steranes may be dictated by ecology and preservation conditions (Eigenbrode, 2008).

Palaeoproterozoic record (2.5 to 1.6 Ga)

Low abundances of eukaryotic steranes occur in several Palaeoproterozoic successions such as in 2.45-Ga fluid inclusions of the Matinenda Formation, Canada (Dutkiewicz *et al.*, 2006), in bitumen of the 2.1-Ga FA Formation, Franceville Basin, Gabon (George *et al.*, 2009), in the 1.64-Ga Barney Creek Formation, Australia (Summons *et al.*, 1988) and in shales of the ~1.7-Ga estuary environment at the base of the Chuanlinggou Formation of China (Li *et al.*, 2003). Gammacerane, derived from tetrahymenol produced by ciliates, has been found in the 1.7-Ga Tuanshanzi Formation of China (Peng *et al.*, 1998), however it may also be derived from bacteria (Kleeman *et al.*, 1990).

Mesoproterozoic record (1.6–1 Ga)

Steranes, including dinosterane, start to be more abundant in the Mesoproterozoic and occur in the 1.4-Ga McArthur Group (Moldowan *et al.*, 2001), the ~1.4-Ga McMinn Formation (Roper Group, Australia; Summons and Walter, 1990), the ~1.3-Ga Ruyang Group, China (Meng *et al.*, 2005) and the 1.1-Ga Nonesuch Formation, United States (Pratt *et al.*, 1991).

Neoproterozoic record (1 Ga to 542 Ma)

Upper Mesoproterozoic/Lower Neoproterozoic rocks have yielded low abundances of gammacerane and dinosterane in the 830 Ma Bitter Springs Formation, Australia and the 590–570-Ma Pertatataka Formation, also in Australia (Summons and Walter, 1990; Summons *et al.*, 1992). Gammacerane, found in ciliates but also bacteria, occurs in 750-Ma rocks of the Chuar Group, Arizona (Summons *et al.*, 1988).

Love *et al.* (2009) report the occurrence of biomarker 24-isopropylcholestane, a biomarker derived from sterols found only in extant Demosponges, in a complete sedimentary succession throughout the Ediacaran Era down to below the 635-Ma Marinoan glacial rocks from Oman. This report represents the earliest record of the sponge lineage, and possibly of (stem or crown-group) animal life, but not necessarily, as these biomarkers could, in principle, be found in stem-group Ctenophores, the most primitive animals, or even in stem-group protist choanoflagellates, which are closely related but not direct ancestors of true sponges (Brocks and Butterfield, 2009).

The fossil record

Archean record (4 to 2.5 Ga)

There are no unambiguous eukaryotic microfossils reported so far in the Archean rock record. Large organic-walled vesicles with acid-resistant walls have been recently discovered in the Archean (Grey and Sugitani, 2009; Javaux *et al.*, 2010) but their biological affinities are unknown at this point.

Palaeoproterozoic record (2.5 to 1.6 Ga)

The oldest unambiguous eukaryotic microfossils are large organic-walled vesicles with striated walls (*Valeria lophostriata*) (Figures 26.2.A and 26.2B). SEM shows that these striations are 1-μm-spaced ridges on the inner side of the vesicle (Javaux *et al.*, 2004; Figure 26.2B). They occur in the 1.7-Ga Chuanlinggou and in the 1.8 Ga Chuanlinggou Formation (Changcheng Supergroup, China) (Zhang, 1986; Peng *et al.*, 2009; Lamb *et al.*, 2009; Javaux, personal observation), in the 1.65-Ga Mallapunyah Formation (McArthur Supergroup, Australia) (Javaux, 2006a) and in most younger Proterozoic siliciclastic successions.

In Palaeoproterozoic and younger rocks, large, smooth organic-walled vesicles (up to a few 100 μm) are common and may display a medial split suggestive of excystment structures, but the absence of any wall ornamentation prevents their attribution to the eukaryotic domain. They could represent early protists or large prokaryotes such as cyanobacterial envelopes. Peng *et al.* (2009) and Lamb *et al.* (2009) reported putative multilayered wall ultrastructures and medial splits in large Changcheng sphaeromorphs, tentatively relating those to early eukaryotes. As mentioned above, detailed studies of the wall ultrastructure and chemistry may permit their identification in some cases.

Among macroscopic carbonaceous compressions, the coiled filaments *Grypania spiralis* from the 1.87-Ga Negaunee Iron Formation, Michigan (Han and Runnegar, 1992; redated by Schneider *et al.*, 2002) has a diameter up to 30 mm (across the whole coil). Samuelsson and Butterfield (2001) have questioned the eukaryotic nature of these structures, but observations by Knoll (in Knoll *et al.*, 2006a) suggest it was an organism, and not a colony or composite of much smaller prokaryotic filaments. Reassessment of Palaeo- and Mesoproterozoic, macroscopic, coiled filamentous compressions suggests a cyanobacterial affinity for some of them, while others might be dubiofossils and ‘tissue-grade organisms’ (Sharma and Shukla, 2009) or macroalgae (Xiao and Dong, 2006). Bedding-plane structures in sandstones from Montana and Western Australia (Horodyski, 1982; Grey and Williams, 1990; Yochelson and Fedonkin, 2000) called *Horodyskia moniliformis* consist of 1–4-mm spheroidal bodies connected by thin cylindrical strings to form uniseriate, pearl-necklace-like structures up to 10 cm long. These structures have been compared to seaweeds, colonial metazoans, prokaryotic associations or non-biological structures.

Other traces consisting of U-shaped ridges in sandstone of the 2 to 1.8-Ga Stirling Range Formation, Australia, first interpreted as animal traces (Bengtson *et al.*, 2007), have been reassessed as possible traces of giant amoebae similar to observations in recent sediments (Matz *et al.*, 2008; Pawlowski and Gooday, 2008; Bengtson and Rasmussen, 2009).

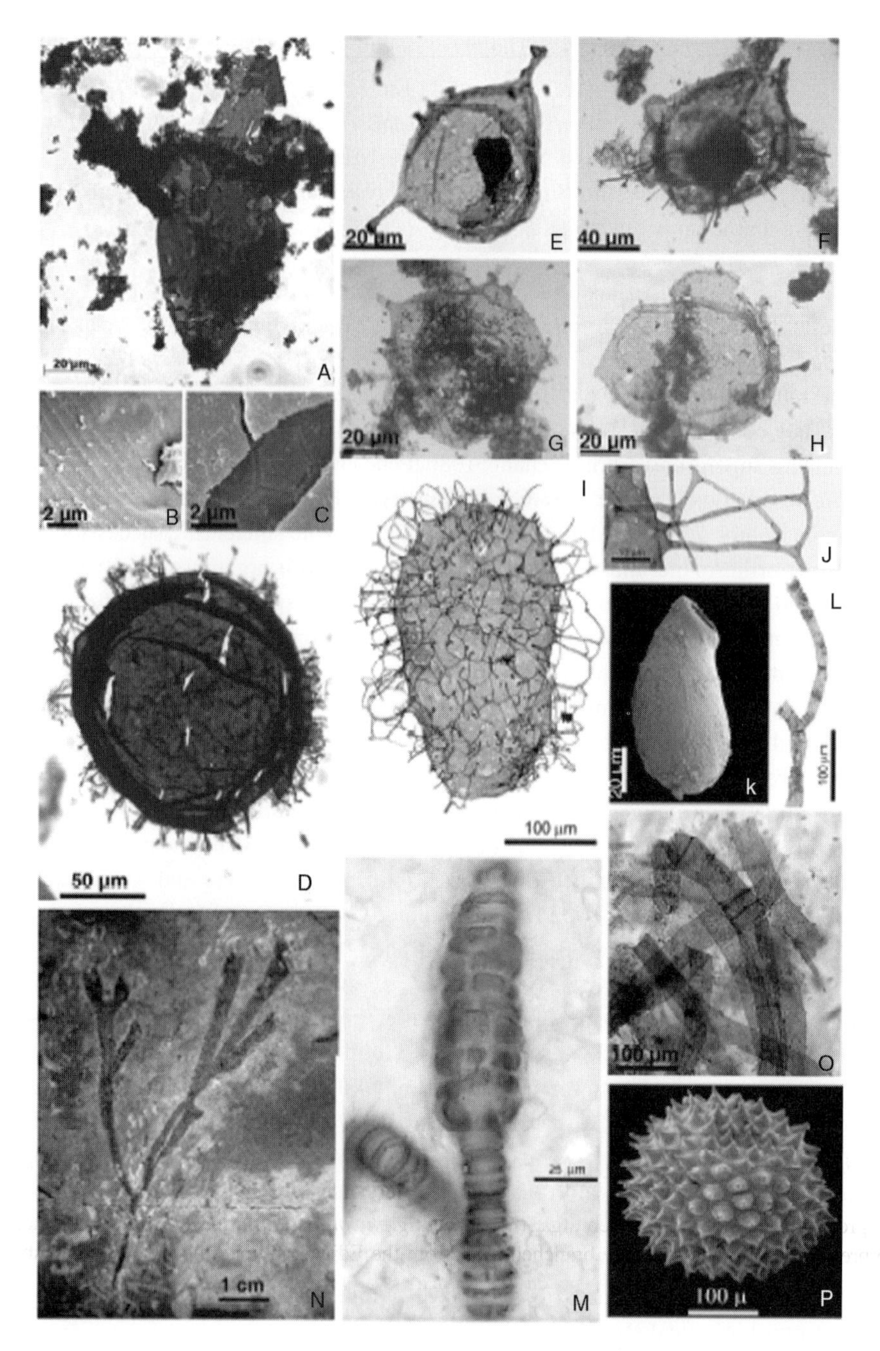

Figure 26.2. Illustration of some representative Precambrian eukaryotic fossils (A, B) *Valeria lophostriata* from the 1.65-Ga Mallapunyah Formation, Australia. (A) Light photograph. (B) Ornamentation of closely spaced parallel ridges on the inner-wall surface in SEM. (C, D) *Shuiyousphaeridium macroreticulatum* from the ~ 1.3-Ga Ruyang Group, northern China. (C) SEM

Mesoproterozoic record (1.6–1 Ga)

Macroscopic compressions of coiled filaments up to 13 mm long and 2 mm wide of *Grypania spiralis* occur in shales of the 1450-Ma Helena Formation, Montana (Walter *et al.*, 1990), as well as in India, where they display millimetre-scale annulation (Kumar, 1995). The segmented *Grypania* have large cells (up to 2 mm wide and 0.2 mm long) that might be interpreted as coenocytic and thus eukaryotic (Knoll *et al.*, 2006b; Butterfield, 2009a). (See Hofman,1994, for a review of macroscopic compressions found in Proterozoic rocks.)

The eukaryotic microfossil record shows higher diversity and more obvious ecological heterogeneity (Javaux *et al.*, 2001). *Tappania plana* is the oldest process-bearing acritarch reported so far (Figures 26.2E–H). It consists of 20- to 160-μm vesicles bearing 0 to 20 (25 to 60 μm) long, closed and sometimes branching processes, distributed irregularly about the vesicle surface. Some vesicles may bear up to 3 bulbous extensions suggestive of budding and some bear neck-like extensions that are sometimes open. One specimen (Figure 26.2G) shows septa at the base of the processes, suggesting that *Tappania* might have been multicellular. Study of more specimens is needed to confirm this observation. *Tappania plana* is interpreted as eukaryotic, based on the combined characters of its complex ornamentation, large size and decay-resistant wall. Its plastic morphology suggests it was a vegetative cell rather than an inert cyst (Javaux *et al.*, 2001). The acritarch *Shuiyousphaeridium macroreticulatum* comprises 50- to 300-μm organic-walled vesicles ornamented with a reticulate surface of ridges, and bearing up to 80 (10–15 μm) long furcating processes (Figures 26.2C and 26.2D). The wall is made of 2μ-wide bevelled hexagonal plates, as seen by SEM and TEM imaging (Javaux *et al.*, 2004; Figure 26.2C). The acritarchs *Tappania plana, Shuiyousphaeridium macroreticulatum*, as well as *Dictyosphaera delicata, Valeria*

Caption for Figure 26.2. Continued

showing inner wall surface of closely packed, bevelled organic hexagonal plates. (D) Light photograph showing specimen with numerous regularly spaced cylindrical processes that flare outward and a reticulated surface. (E–H) Light photographs of *Tappania plana* from the 1.5-Ga Roper Group, Australia. (E) Specimen with heteromorphic processes distributed asymmetrically about the vesicle. (F) Specimen with a branched process and bulbous protrusions interpreted as budding structures. (G) Specimen with septa at the base of processes. (H) Specimen with neck-like extension and processes. (I, J) Light photographs of '*Tappania*' from the ~800-Ma Wynniatt Formation, Arctic Canada; a complex multicellular form interpreted as possible fungi. (I) Specimen with septate, anastomosing processes. (J) Detail showing septae and fusion of processes. (K) *Bonniea dacruchares*, a vase-shaped microfossil from the ~750-Ma Chuar Group, Arizona, interpreted as a lobose testate amoeba. (L) Light photograph of *Proterocladus* sp., from the 750-Ma Svanbergfjellet Formation, Spitsbergen, interpreted as Cladophorale green alga. (M) *Konglingiphyton erecta*, a macroscopic carbonaceous compression of a dichotomously branched alga, from the Ediacaran uppermost Doushantuo shales. (N) Light photograph of *Bangiomorpha pubescens*, from the ~1.2-Ga Hunting Formation, Arctic Canada, preserved tri-dimensionally in chert, showing radial division of cells within uniseriate filaments, interpreted as Bangiophyte red alga. (O) Light photograph of *Palaeovaucheria clavata* from the ~1-Ga Lakhanda Formation, Siberia, interpreted as a possible vaucheriacean alga. (P) SEM photograph of *Meghystrichosphaeridium*, a spiny acritarch preserved tri-dimensionally in phosphorites from the Ediacaran Doushantuo Formation, China. Photographs I, J, L and N courtesy of Nick Butterfield; K courtesy of Susannah Porter; M and P courtesy of Shuhai Xiao.

lophostriata and *Satka favosa* exhibit a complexity of form observed with TEM, SEM and light microscopy that is unknown in prokaryotes. Prokaryotes can have large, ornamented and decay-resistant walls (at least cyanobacteria sheaths), but no prokaryote currently known has all three (large size, ornamentation, preservable acid-resistant walls) at once, in contrast to many eukaryotes. Therefore, these early microfossils display characters of a eukaryotic grade of organization, and are interpreted as eukaryotes with a sophisticated cytoskeleton and a nucleus.

These assemblages of 3 to 8–10 taxa of eukaryotic microfossils co-occurring with prokaryotic filaments and vesicles are preserved in shales of the 1450-Ma Roper Group, Australia (Javaux *et al.*, 2001); the 1.6–1.25-Ga Ruyang Group, China (Yin, 1997; Xiao *et al.*, 1997); the Sarda Formation and the Avadh Formation, India (Prasad and Asher, 2001); and the 1499±43-Ma to 1060±20-Ma Kerpylian Horizon, Siberia (Nagovitsin, 2009). *Valeria* occurs throughout the Proterozoic record and *Satka favosa* is also found in other Mesoproterozoic successions (e.g. Hofmann and Jackson, 1994; Jankauskas, 1989). New Re–Os ages of 1361±21 Ma and 1417±29 Ma of the Upper Roper Group (the Velkerri Formation, Kendall *et al.*, 2009), 200 to 400 metres above the formations containing *Tappania*, which are about 500 m above rocks dated at 1492±3 Ma with U–Pb SHRIMP analyses of zircon, are consistent with earlier Rb–Sr dates (1429±31 Ma) and confirm the early Mesoproterozoic age of these microfossils, as opposed to suggestions by Nagovitsin (2009). The Siberian assemblage that includes *Tappania, Valeria* and *Satka favosa* also contains forms that are more characteristic of early Neoproterozoic age such as *Lophosphaeridum* (acritarchs ornamented with large verrucae or tubercules), *Osculosphaera* (an acritarch with a circular opening), *Obruchevella* (cyanobacterial spiral filaments), *Miroedicha* (vesicles with bulbous protrusions) and putative *Tasmanites* (acritarchs pierced by pores and considered as phycomata of prasinophyte algae) (Nagovitsin, 2009). This assemblage is younger than the Roper Group and Ruyang Group assemblages, extending to the Meso–Neoproterozoic boundary, and its composition and higher diversity are consistent with its intermediary stratigraphic position between the former Groups and the richer 1-Ga Lakhanda assemblage (see below). The Kerpylian assemblage seems to extend the record of *Tappania* to the whole Mesoproterozoic Era.

Taken together, these microfossils record the evolution of biological innovations such as reproduction by budding, vegetative and resting stages, synthesis of resistant polymers, synthesis of various wall ornamentation including processes, striations and polygonal plates, and a moderate diversity and complex ecology (Javaux *et al.*, 2001; Knoll *et al.*, 2006b). Early Mesoproterozoic fossils strongly indicate that eukaryotic organisms of marked cytological and genetic complexity existed at 1.5 Ga.

According to Cavalier-Smith (2002, p. 37), 'cysts with spines or reticulate surface sculpturing would probably have required both an endomembrane system and a cytoskeleton, the most fundamental features of the eukaryotic cell, for their construction' and again, on page 38 of the same paper, 'complex surface sculpturing or spines that hint of the presence of a eukaryotic cytoskeleton in the spiny acritarchs'. The Mesoproterozoic acritarch *Tappania plana* has spines (processes), while the acritarch *Dictyosphaera delicata* has a reticulate surface and *Satka favosa* a wall of interlocking polygonal plates. *Shuiyousphaeridium* has both spines and reticulation (Yin, 1997; Xiao *et al.*, 1997; Javaux *et al.*, 2003; Knoll *et al.*, 2006b). These organic-walled microfossils occur in well-dated 1.5–1.3-Ga successions, as mentioned above. This fact is clearly incompatible with the ~850-Ma origin of eukaryotes suggested by Cavalier-Smith (e.g. 2002, 2009).

The late Mesoproterozoic yields a richer diversity (up to 13 taxa) of eukaryotic organic-walled microfossils. At the end of the Mesoproterozoic Era, multicellular organisms appear, and some can be related to extant clades, such as Bangiophyte red algae and possibly Vaucheriales algae.

The Hunting Formation, Canada, preserves the Bangiophyte red alga *Bangiomorpha pubescens*, the oldest taxonomically resolved eukaryote so far (Figure 26.2N). This fossil includes vertically oriented (15–45-μm wide) uniseriate and (30–67-μm wide) multiseriate unbranched filaments up to 2 mm long, surrounded by a translucent outer wall, and attached to a firm substrate by a lobate multicellular holdfast structure. The ontogeny of *Bangiomorpha* includes single-celled and double-celled stages, and a four-to eight-celled stage where the wedge-shaped cells are arranged along a radial symmetry. This pattern of longitudinal intercalary cell (or pie-like) division is only known in modern Bangia (Bangiales, Rhodophytes). Stigonematales cyanobacteria do produce a comparable thick sheath and multiseriate filaments and differentiated multicellular holdfast, but they are also characterized by branching apical growth and differentiated cells (akinetes and heterocysts), characters not found in *Bangiomorpha* (Butterfield, 2000). No cyanobacteria or green algae show radial intercalary division (one modern prasiolalean chlorophyte does, but the radial cells are not arranged as a fourfold symmetry and they are transient, dividing in three planes into parenchymatous spheres; Butterfield, 2000). Taphonomy also points to a non-cyanobacterial affinity (Knoll *et al.*, 2006b). Therefore, on the basis of a diagnostic fourfold radially symmetrical arrangement of wedge-shaped cells, *Bangiomorpha* is interpreted as a Bangiophyte alga, although it differs in some aspects from modern *Bangia* by having a multicellular holdfast rather than rhizoids (but other filamentous Bangiophytes do have a comparable holdfast) (Butterfield, 2000). *Bangiomorpha* records the evolution of complex multicellularity, cell differentiation, sexual reproduction, eukaryotic photosynthesis and primary endosymbiosis of a chloroplast ancestor by 1.2–1 Ga. Published radiometric dates constrain *Bangiomorpha* only to the interval 1267±2 to 723±3 Ma, but an unpublished Pb–Pb date of 1198±24 Ma and physical stratigraphic relationships strongly suggest a late Mesoproterozoic age (Butterfield, 2000, 2001). This ~1.2-Ga multicellular photosynthetic eukaryote implies the earlier evolution of unicellular algae.

In the > 1005±4-Ma (Rainbird *et al.*, 1998) Lakhanda Formation, Siberia (Herman, 1990), the filamentous fossils *Palaeovaucheria clavata* (Figure 26.2O) are tentatively interpreted as vegetative phases of a xanthophyte alga similar to the extant *Vaucheria*. Populations of *Palaeovaucheria* display morphological traits characteristic of vaucherian xanthophytes such as branching at right angles, two sizes of filaments on the same individual and terminal pores and septae at filament ends (Woods *et al.*, 1998). However, no reproductive phases of *Palaeovaucheria* are known, and this fossil also resembles a freshwater oomycete or a siphonous green alga (Butterfield, 2009a).

The Lakhanda assemblage also includes large (400–1000 μm) and long (up to 5 mm) segmented filaments enclosed in longitudinally striated sheaths in *Eosolenides*, and often including numerous 2–10-μm spheroidal cells (German and Podkovyrov, 2009), networks of cells called *Eosaccharomyces ramosa*, and ornamented acritarchs that are eukaryotic, such as the spiny *Trachyhystrichosphaera aimika*, prokaryotic filamentous sheaths and cellular colonies.

Various grades of eukaryotic multicellularity have evolved since the Mesoproterozoic Era (Butterfield, 2009a). *Eosaccharomyces* colonies show cells in an oriented anastomosing arrangement, suggesting intercellular molecular signalling and localized cell adhesion (Knoll *et al.*, 2006b; Butterfield, 2009a). *Palaeovaucheria* has coenocytic branching filaments while the red algae *Bangiomorpha* shows the most complex developmental history.

Neoproterozoic record (1 Ga to 542 Ma)

Early Neoproterozoic (1000–c.750 Ma) Xiao and Dong (2006) reviewed the diversity of macroalgae in the Proterozoic and described 13 groups of carbonaceous compressions (e.g. Figure 26.2M). In rare cases, some of these algae are also preserved three-dimensionally by permineralization (phosphatization or silicification). Their diversity and morphological complexity have increased since their first occurrence, ranging from mostly spherical, ellipsoidal, tomaculate or cylindrical forms in the Palaeoproterozoic Era (e.g. centimetric spherical *Chuaria*, sausage-shaped *Tawuia*, coiled filamentous *Grypania*); to similar shapes with transverse annulations (*Grypania*) or development of a discoidal holdfast (*Tawuia*-like fossils); to similar morphologies and the presence of stipes in the Early Neoproterozoic; to a higher diversity of thallus morphologies (conical and fan-shaped thalli in addition to previous forms) and thallus differentiation with discoidal and rhizoidal holdfast, stipe, blade (e.g. blade *Longfengshania*) and other features like dichotomous or monopodial branching and apical meristem in the Ediacaran Period (Xiao and Dong, 2006). A significant increase in the surface: volume ratio and canopy height of these macroalgae is also observed, ranging from millimetric to centimetric height in the Palaeoproterozoic up to decimetric size in the Ediacaran Period. Macroalgae are important elements in benthic ecosystems because they offer new habitats where a diverse ecology may develop, and they represent a large biomass (Xiao and Dong, 2006). This ecosystem tiering started already in the Late Mesoproterozoic Era with the evolution of millimetric Bangiophyte algae (Butterfield, 2000), and became more complex through the Proterozoic Period.

The diversity of organic-walled microfossil assemblages increases to about 10 to 25 eukaryotic taxa (Knoll *et al.*, 2006b). Besides a moderate diversity of acritarchs ornamented by pustules (*Kildinosphaera verrucata*), striations (*Valeria*), thick wrinkles (*Cerebrosphaera*), processes (*Trachyhystrichosphaera, Vandalosphaeridium, Cymatiosphaeroides*), equatorial membranes (*Simia*) or enveloping membranes (*Pterospermopsimorpha*), pores (*Tasmanites*) and chagrination (*Trachysphaeridium, Lophosphaeridium*), the record of multicellular microfossils also increases (Butterfield, 2009a), producing a new biological environment with complex ecological interactions. The acritarch *Cerebrosphaera buicki* has been proposed as a potential index fossil for pre-Sturtian rocks older than *c*.777 Ma (Hill *et al*., 2000).

Populations of *Vaucheria*-like fossils assigned to *Jacutianema* (Figure 26.2L) preserving several life cycle stages occur in the 750–800-Ma Svanbergfjellet Formation, Spitsbergen (Butterfield, 2004). The age of the succession is based on microfossils, chemostratigraphy and correlation with the better dated Shaler Group of NW Canada (a minimum of 723±4 Ma, referenced in Butterfield *et al*., 1994). *Jacutianema* has been related to a 'desiccation-induced *Gongrosira*-phase of vaucheriacean algae in which the coenocyte is compartmentalized to form a series of weakly interlinked, sometimes branching, and eventually germinating "akinetes"' (Butterfield, 2009a). 'The key to the identification lies in the large, laterally positioned connections between contiguous akinetes, which, at least among extant plant protists, are unique to *Vaucheria*.' (Butterfield, 2004). *Jacutianema* (and possibly the earlier *Palaeovaucheria*) indicates the appearance of Stramenopiles and of secondary symbiosis (involving a red alga-like endosymbiont).

Co-occurring with *Jacutianema* are populations of *Proterocladus* interpreted as a multicellular siphonocladalean chlorophyte alga based on long (up to 50 μm wide and 1000 μm long) and irregular coenocytic cell lengths, branching, intercellular thick-walled discoidal septae and reproductive structures (Butterfield *et al*., 1994). A complex multicellular form called *Valkyria* is interpreted as eukaryotic, based on the presence of six different cellular types, but its taxonomic affinity is unknown. Another multicellular organism, *Palaeastrum*, is formed by monostromatic sheets of cells arranged as enclosed vesicles, and closely compares to coenobial green algae of the Hydrodictyaceae (Butterfield, 2009a; Butterfield *et al*., 1994).

The 850–750-Ma Wynniatt Formation, Canada, includes fossils assigned to '*Tappania*' by Butterfield (2005a) (Figures 26.2I, J). These exquisitely preserved microfossils exhibit morphological complexity, including features such as serial septae in the hollow, branched processes capable of secondary fusion (interpreted as hyphal fusion) and possibly forming a character combination synapomorphic of the higher fungi. These fossils have been compared to the older 1.5-Ga Roper Group, 1.3-Ga Ruyang Group and the 1.25–1.06-Ga Kerpylian Horizon *Tappania*, but these have neck-like extensions unknown in the Wynniatt '*Tappania*' and do not show filament fusion (Javaux *et al*., 2001; Yin *et al*., 2007; Knoll *et al*., 2006b; Nagovitsin, 2009). One Australian and some Siberian specimens show a septum at the base of processes. At present, the synonymy is not proven. More detailed work on the older material is needed to uncover the range of morphologies displayed by the Neoproterozoic fossils, as well as microchemical analyses (C isotopes, biopolymers) of both populations. Although the early evolution of fungi is plausible since the Opisthokonts are recorded in the Early Neoproterozoic period (see below), the earliest records of fungi include microendolithic borings in Ordovician invertebrate skeletons (Vogel and Brett, 2009) and ascomycetes in Early Devonian cherts (Taylor *et al*., 1999), although the presence of lichens has been suggested in the Doushantuo Formation, China (Yuan *et al*., 2005). The Wynniatt Formation also preserves coenocytic filaments with 'distinctively recurved circumferential flanges' emanating from 'a basal vesicle with one or more flanged openings' assigned to *Cheilofilum* (Butterfield, 2005b, 2009a). Acritarchs with a meshwork of organic cruciate structures reminiscent of stauract sponge spicules are tentatively identified as an early sponge or choanoflagellate (Butterfield, 2009a). *Germinosphaera* is a coenocytic (13–90 μm) vesicle with 1 to 6 open-ended, tubular, hollow, occasionally branched (up to 185 μm long) processes (Butterfield *et al.*, 1994).

Other pre-Ediacaran eukaryotic microfossils include the problematic *Parmia* and *Protoarenicola* and other annulated macroscopic tubes (Vorob'eva *et al.*, 2009).

In the Chuar Group (Grand Canyon, Arizona), the vase-shaped microfossils (VSMs) (e.g. Figure 26.2K) *Melanocyrillium hexodiadema, Trigonocyrillium horodyskii* and *Trachycyrillium pudens* have been related to filose and lobose amoebae ('thecamoebians'), based on a character combination (aperture and test morphology) only found in testate amoebae (Porter and Knoll, 2000). These microfossils provide a firm calibration point for the amoebozoans, and direct evidence for heterotrophic eukaryotes (Porter and Knoll, 2000; Porter *et al.*, 2003). Hemispherical holes in some VSM walls were tentatively interpreted as predation marks. The fossils lie two metres beneath an ash bed dated by U–Pb zircon chronology as 742±6 Ma (Karlstrom *et al.*, 2000). Another VSM, *Melicerion poikilon*, has a pyritized wall perforated by regularly distributed honeycomb-patterned holes interpreted as the former emplacement of siliceous scales, by comparison with modern euglyphids (Porter *et al.*, 2003) and may record, indirectly, the evolution of eukaryotic biomineralization and of Cercozoans (Rhizaria).

Aggregates of micrometre-sized hollow organic spheres called *Sphaerocongregus* (or *Bavlinella*) are interpreted as bacterial remains and occur abundantly in Neoproterozoic glaciogenic strata around the world, and in the Chuar Group in Arizona, before the Sturtian glaciation (Nagy *et al.*, 2009). The distinctive acritarch *Cerebrosphaera* is also present (Nagy *et al.*, 2009).

Middle Neoproterozoic (c.750–635 Ma) Few microfossil assemblages have been reported through and between the glacial deposits, and include low-diversity acritarchs (review in Knoll *et al.*, 2006b) and discoid macrofossils called the Twitya discs (Hofmann *et al.*, 1990). However, this drop in diversity might only reflect sampling bias (Knoll *et al.*, 2006b; Butterfield, 2007) as the occurrence of the relatively diverse 780–620-Ma Tindir assemblage (Butterfield, 2007) may suggest. Early protist biomineralization is recorded by the presence of several types of 5–70-μm siliceous imperforate and perforate scales attributed to chrysophycean and haptophyte algae (Allison and Hilgert, 1986, redated by Kaufman *et al.*, 1992). The Tindir assemblage also includes smooth and spiny acritarchs (*Cymatiosphaera* and *Trachyhystrichosphaera* known in the Early Neoproterozoic), VSMs and filamentous and coccoidal prokaryotes (Allison and Awramik, 1989).

Ediacaran (635 to 542 Ma) During this period, the biosphere diversified substantially in increasingly oxygenated oceans and atmosphere (see above). The appearance of metazoans (possibly already in the Middle Neoproterozoic) added a new dimension to ecosystems, and a new predation pressure (Butterfield, 2007), although proterozoic protists had already evolved heterotrophy during the Early Neoproterozoic (Porter *et al.*, 2003) and probably much earlier. About 60 eukaryotic organic-walled taxa (not counting metazoans) record a high increase in acritarchs' diversity, the presence of animal embryos and eggs, and are contemporary with macroscopic compressions of brown and red algae (Knoll *et al.*, 2006b; Xiao *et al.*, 1998a; Xiao *et al.*, 2002; Xiao and Dong, 2006) (Figure 26.2M).

The Doushantuo Formation in the Yangtze Gorges area is of particular importance due to its exceptional preservational conditions, yielding exquisite 3D-preserved microfossils (Figure 26.2P) as well as macroscopic algal compressions (Figure 26.2M). The formation is bracketed by U–Pb ages between 635±1 and 551±1 Ma (Condon *et al*; 2005), and direct Pb–Pb dating of Upper Doushantuo phosphorite indicates that the fossils are probably 599±4 million years old (Barfod *et al.*, 2002); however, Condon and colleagues (2005) argue that the fossiliferous Upper Doushantuo Formation may be between 580 and 551 million years old. The Doushantuo Formation preserves phosphatized fossils containing anatomical information at the cellular level. Fossils such as *Wengania*, *Thallophycoides* and *Gremiphyca* are interpreted as early, and possibly stem group, branches within florideophyte red algae, an interpretation supported by their simple pseudoparenchymatous thalli, apical growth and lack of cortex–medulla differentiation. Florideophyte algae record the evolution of a tissue-grade organization (coordinated cell differentiation in three dimensions to form a tissue with a specific function). Other Doushantuo fossils, *Thallophyca* and *Paramecia*, are attributed to soft (not calcifying) stem-group corallinaleans, based on their pseudoparenchymatous thallus, apical growth, differentiated medullary and cortical tissues and possible reproductive structures (Xiao *et al.*, 2004). Macroscopic compressions in Upper Doushantuo shales include coenocytic green algae and regularly bifurcating thalli comparable to red and brown algae (Xiao *et al.*, 2002) (e.g. Figure 26.2M).

Among the Doushantuo microfossils, 500-μm phosphatized spheres with an ornamented envelope containing 1, 2, 4, 8 or more closely packed internal bodies with faceted sides and decreasing in size as their number increases, are suggestive of cells dividing by successive binary divisions in cleaving animal embryos (Xiao and Knoll, 2000; Xiao *et al.*, 1998b). Further examination of these Doushantuo embryos using phase-contrast X-ray microtomography suggests putative affinity to poriferan, cnidarian and bilaterian animals (Chen *et al.*, 2009). Bailey *et al.* (2007a) have challenged this metazoan interpretation and proposed an alternative affiliation to giant sulphur bacteria (but see Xiao *et al.*, 2007; Bailey *et al.*, 2007b), but their interpretation cannot explain the presence of an ornamented outer layer around many Doushantuo specimens (Yin *et al.*, 2007).

Ediacaran acritarchs are large and bear regularly distributed processes (e.g. Zang and Walter, 1992; Grey, 2005; Vorob'eva *et al.*, 2009). Some are interpreted as metazoan eggs based on their large size (more than 100 μm) walls ornamented with processes and multilayered wall ultrastructure (Cohen *et al.*, 2009). An algal affinity has also been suggested based on wall ultrastructure (Moczydlowska and Willman, 2009), although their large size compared with older and younger algae has not been explained. One species (*Tanarium conoideum*) had a wall made of a biopolymer similar to algaenan, possibly indicating a green-algal affinity (Marshall *et al.*, 2005). Kodner *et al.* (2009) suggested that algaenan might be more rare than previously thought in marine kerogens because it is limited to a few freshwater extant taxa today. Their argument that algaenan might be diagenetic is plausible and aliphatization is known to occur during diagenesis (e.g. Versteegh *et al.*, 2004). Different acritarch taxa from the same sample and same diagenetic history exhibit different biopolymer composition (not all include algaenan-like biopolymers), suggesting a different original composition, algaenan or, conceivably, another biopolymer transforming preferentially into algaenan during diagenesis. However, this chemical analysis dealt only with one of the spiny Ediacaran acritarchs and it would be interesting to analyse the wall chemistry of the taxa interpreted as animal cysts. Their spines were suggested to be a protective adaptation to the evolution of animal predators (Peterson and Butterfield, 2005), but Cohen *et al.* (2009) proposed a different explanation. The latter authors argued that, in fact, predation pressure appeared earlier with heterotrophic protists and that the Ediacaran spiny acritarchs disappeared before the first body and trace fossil evidence of early animals. However, possible spiculate acritarchs (Butterfield, 2009a), sponge biomarkers (Love *et al.*, 2009) and animal embryos (Xiao and Knoll, 2000) are documented in the Early Neoproterozoic, Middle Neoproterozoic and Ediacaran Periods, respectively, and the spiny microfossils themselves are interpreted by Cohen *et al.* (2009) as animal remains, so it seems that both animals and Ediacaran microfossils coexisted, and in some cases were equivalent. Cohen *et al.* (2009) suggested that these resting stages of early animals would have evolved around 635 Ma because of stressful environmental conditions such as transient anoxia in the water column with variable redox conditions. Spines would have helped prevent sinking into the substrate and to enable changes to be felt in environmental conditions. The spiny microfossils would then disappear around 560 Ma with the increased oxygenation of marine bottom waters. The resting cysts' strategy, developed to avoid unfavourable conditions in the Ediacaran ocean (and earlier in other clades, as documented by excystment structures in older acritarchs), has later reappeared independently in several clades among animals (e.g. crustaceans) and protists (e.g. dinoflagellates).

In South Australia, a low-diversity leiosphere (smooth-walled acritarch) assemblage (ELP) is observed below the 580-Ma Acraman impact layers and is followed by a radiation of large spiny acritarchs (ECAP) above (Grey *et al.*, 2003). Partially similar assemblages are observed in older rocks in the Doushantuo Formation in China, as well as in other formations in Siberia, Svalbard, Norway and India (referenced in Vorob'eva *et al.*, 2009), possibly permitting a detailed biostratigraphy of the Ediacaran period, when problems of taxonomy and accurate datings will be solved. Vorob'eva *et al.* (2009) suggest that the lower boundary of the Ediacaran period might be characterized by a moderately diverse assemblage of mostly leiospheres, and smaller coccoidal and filamentous microfossils with rare large acanthomorphs (spiny acritarchs), followed by a radiation of diverse large acanthomorphs that occurred later in Australia (unless there are taphonomic or sampling biases; see discussion in Halverson, 2006).

After the Gaskiers glaciation around 575 Ma, the macroscopic organisms of the Ediacara biota are preserved as sandstone casts of soft bodies (review in Narbonne, 2005) and interpreted as ancestors of modern metazoans showing radial and bilateral symmetry (Runnegar and Fedonkin, 1992), giant protists (putative foraminifera) (Seilacher *et al.*, 2003) or extinct evolutionary experiments (Seilacher, 1992). Calcareous skeletons (e.g. *Cloudina, Namaccalathus*) appear in < 549-Ma microbial reefs of Namibia (Grotzinger *et al.*, 2000) and record the evolution of animal biomineralization, predation (evidenced by boring in skeletons) and a variety of feeding strategies (review in Narbonne, 2005).

The end of the Ediacaran (550–542 Ma) seems to include impoverished assemblages of mostly unornamented sphaeromorphs including large (100s μm) forms, the acritarchs *Chuaria circularis*, fragments of the ribbon-like macrofossil *Vendotaenia*, cyanobacteria filaments, prokaryotic *Bavlinella faveolata* and rare spiny acritarchs (one *Comasphaeridium*-like species) (Germs *et al.*, 1986). A brief but pronounced C-isotopic event coupled to transient shallow-water anoxia near the Proterozoic–Cambrian boundary may have been one of the causes of the disappearance of the highly diverse Ediacaran spiny acritarchs (Amthor *et al.*, 2003; Knoll *et al.*, 2006b), together with rising predation pressure of macroscopic metazoans (Peterson and Butterfield, 2005). A renewed diversity of small acanthomorphs rose again in the Cambrian (Tommotian, 530–520 Ma) (Butterfield, 2007).

Pattern and causes of diversification

Figure 26.3 summarizes the timing of major environmental changes (supercontinent formation and break-up, widespread glaciations, meteor impact in Australia and change in atmosphere oxygenation and ocean chemistry), of early eukaryotic diversification and biological innovations evidenced by the molecular and body fossil record (modified from Javaux, 2006a, 2007; Lyons and Reinhard, 2009).

The number of fossil taxa represents only the diversity of form (morphological diversity) but not necessarily the genetic diversity. Genetic diversity does not need to translate into phenotypic diversity (similar morphotypes might represent different clades). Conversely, a single clade may include a diversity of phenotypes, often translating into diversity of lower taxonomic levels. Moreover, unicellular and especially multicellular organisms may show different morphologies during their life cycle or produce different parts described as different morpho-taxa, thereby leading to an overestimation of the true diversity (Butterfield, 2004, 2005b). In addition, the fossil record is incomplete (not everything gets preserved, either because the organisms do not produce preservable parts, or they did not die in preservational environmental conditions or the sediments that contained them have been eroded or metamorphosed) and the sampling record is biased (not all sedimentary successions have been sampled, and not all sampled ones have been examined in detail).

Despite the limitations above, it is possible to summarize the fossil record in the form of diversity curves based on the number of taxa of organic-walled microfossils, VSMs and macroscopic compressions (Porter, 2004; Knoll *et al.*, 2006a), or on acritarch data and number of cell types (Butterfield, 2007) or on number of morphological characters (Huntley *et al.*, 2006). Knoll *et al.* (2006a) divided the eukaryotic diversification into five

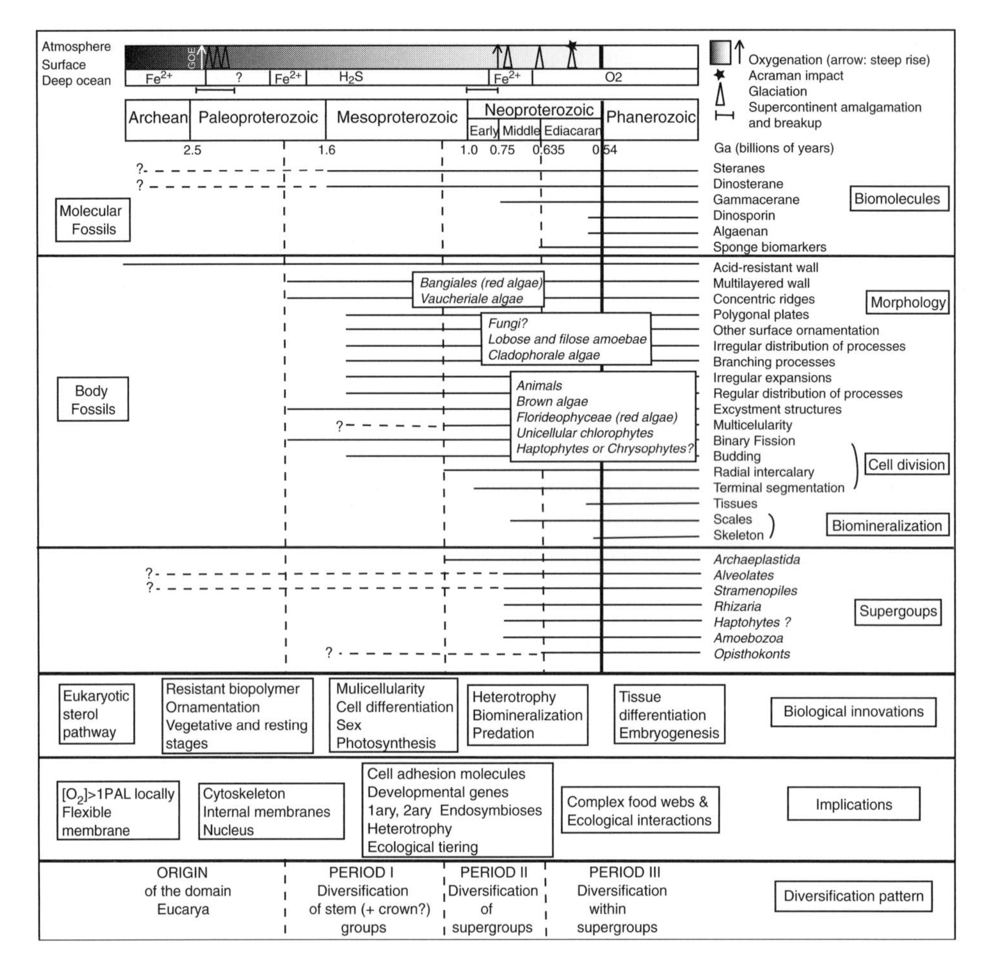

Figure 26.3. Patterns of early eukaryotic diversification, biological innovations and environmental changes. This figure summarizes the timing of major environmental changes (supercontinent formation and break-up, widespread glaciations, a meteor impact in Australia and change in atmosphere and ocean chemistry), molecular fossils, morphological characters displayed by the body fossil record, minimum ages of extant clades and supergroups and the biological innovations and their implications for eukaryotic evolution (modified from Javaux, 2006a, 2007; timing of Palaeoproterozoic glaciations and rifting from Bekker *et al.*, 2005; Neoproterozoic glaciations from Halverson, 2006; atmosphere and ocean chemistry modified from Lyons and Reinhard, 2009). Three periods representing steps in eukaryotic diversification are suggested (see text for details).

time periods before the Cambrian, with a drop during the glaciations before a radiation in the Ediacaran, and a drop again just before a Cambrian radiation, and discussed biological innovations separately from taxonomic diversification. A similar pattern of diversification was described by Porter (2004), who noticed the appearance of most supergoups before the glaciations. Butterfield (2007) argued that there was no extinction in pre-Ediacaran eukaryotes and only modest innovations in assemblages of only long-ranging 'living fossils'; a slow evolution rate also noted by Knoll (1994). Butterfield (2007) proposed that the ecological dynamics of the Ediacaran changed radically from earlier times and became more Phanerozoic-like, in response to the advent of metazoan predation. Huntley *et al.* (2006) observed that half the morphological characters used in their study as a measure of morphological disparity appeared in the early Mesoproterozoic, followed by a plateau possibly related to nutrient stress and an extinction during the glaciations, prior to the Ediacaran diversification.

Indeed, one might wonder why eukaryotes started to diversify so late. After all, they had all the features of the eukaryotic cells since their first occurrence in the fossil record (flexible membrane, cytoskeleton and endomembrane system, nucleus) but the record shows a moderate gradual increase since the Late Palaeoproterozoic, followed by a radiation only in the Ediacaran. The explanations suggested in the literature to tentatively explain the observed pattern focus on changes occurring in the Ediacaran. The suggested motors for diversification that did not exist earlier include: animal evolution and associated predation pressure (Butterfield, 2007; Peterson and Butterfield, 2005), increased oxygenation (Fike *et al.*, 2006), new niches after a meteor impact (Grey *et al.*, 2003; Grey, 2005), new niches after extinction due to glaciations (Knoll, 2003b) and release of nutrients immobilized in a formerly anoxic sulphidic ocean (Anbar and Knoll, 2002).

However, arguments are brought against these suggestions.

Butterfield (2009b) suggests that the answer does not lie in the evolution of permissive environments (especially not oxygenation levels), but rather in the evolution of complex developmental biology. Eukaryotes went from various grades of multicellularity due to the evolution of cell signalling and cell adhesion, to the much more complex control and coordination of organ-grade multicellularity, thanks to unique gene regulatory networks. But, again, why did it take so long?

Rather than glaciations, anoxia due to eutrophication prior to the Middle Neoproterozoic is proposed as the cause of the extinction of pre-Ediacaran organisms (Nagy *et al.*, 2009). Based on microfossil and geochemistry data, Nagy *et al.* (2009) suggest that bacterial blooms of *Sphaerocongregus* evidenced high primary productivity, and resulted in an increased heterotrophic protozoan biomass (the VSMs) and the development of eutrophication of surface waters and water-column anoxia, leading to extinction of pre-Ediacaran diverse acritarch assemblages prior to the glaciations. Other studies (e.g. Moczydlowska, 2008) also argue for the little, or absence of, influence of glaciations on eukaryotic diversity, at least in the photic zone where red, green and xanthophyte algal clades that appeared earlier survived until today.

The evolution of animals was suggested as a major predation pressure causing the diversification of large spiny acritarchs (Butterfield and Peterson, 2005), however, at least some of these are interpreted as metazoan cysts with spines that do not protect them, but may help them to sense when better conditions appear in the environment (Cohen *et al.*, 2009).

The meteor impact cannot offer a global explanation (although a local one might be possible) because it is observed only in Australia, but the Ediacaran radiation seems to occur earlier elsewhere (Halverson, 2006; Xiao *et al.*, 2004).

The euxinic conditions of the Mid-Proterozoic ocean might have had severe consequences on eukaryotic life (Anbar and Knoll, 2002; Johnston *et al.*, 2009; Glass *et al.*, 2009), as explained above. However, the immobilization of transition metals (such as molybdenum) was disadvantageous for algae, but not for heterotrophic eukaryotes (at least those that were not feeding on algae), moreover, it did not restrict diversification in shallow water but in niches further away from the shore (Javaux *et al.*, 2001).

Porter (2004) suggested that it was not the evolutionary innovations of multicellularity, sex or the acquisition of plastids that were key to eukaryotic diversification, because there are many unicellular non-photosynthetic eukaryotic microfossils, but possibly environmental changes such as global tectonics, oxygenation of deep ocean and glaciations. Ecological innovation such as the appearance of microbial predators (the VSMs) around 750 Ma could have fuelled the diversification of primary producers.

Knoll *et al.* (2006a) proposed that several factors linked to genetics, ecology and environmental change, and not a single factor in particular, probably played a role in explaining the pattern of diversification evidenced by the fossil record.

I would like to propose a somewhat different reading of the diversification pattern of early eukaryotes as an hypothesis to be tested: the pattern of diversification of early eukaryotes could be divided into three steps *involving different taxonomic levels*.

Period I (?1.8 to 1.2 Ga) was the time between the origin of the domain (no later than the Late Palaeoproterozoic, possibly in the Archean Eon), through the early Mesoproterozoic, when stem (and possibly but not necessarily crown-group) eukaryotes appeared. These early eukaryotic cells had a flexible lipidic membrane that required > 1 PAL of oxygen concentration locally (for synthesizing eukaryotic sterols; Brocks *et al.*, 2003), a cytoskeleton and associated endomembrane system and a nucleus. These early eukaryotes diversified right next to cyanobacterial mats in local oxygen oases, close to the shores where continental runoff could bring necessary nutrients, but rarer taxa also occurred in the inner and outer shelf and possibly in deeper basin settings (Javaux *et al.*, 2001). Eukaryotic diversity increased from 1 or 2 to about 8 morpho-taxa.

During Period II, from the Late Mesoproterozoic through the Middle Neoproterozoic Era (1.2–0.63 Ga), a major diversification occurred *at the supergroup level*. Members of all extant supergroups: the Opisthokonts, the Amoebozoa, the Archaeplastida, the chromalveolates, and the Rhizaria, except the Excavates, are recorded in the fossil record. Since some of these supergroups include multicellular fossil representatives (the Opisthokonts, the Archaeplastida and the Stramenopiles within the Rhizaria), their origin might actually pre-date Period II. Eukaryotic diversity increased up to 25 morpho-taxa. The glaciogenic Middle Neoproterozoic Era may be included in this period as animal biomarkers, and a moderately diverse assemblage of Chrysophyte and Haptophyte scales, spiny and smooth acritarchs and VSMs have been reported. The occurrence of an extinction and its timing and extension are debated and might have occurred just before the Sturtian (Nagy *et al.*,

2009) or not at all in the pre-Ediacaran (Butterfield, 2007), but anyway, it affected lower taxonomic levels, not the supergroups which still exist today. Major biological innovations such as multicellularity, sex, biomineralization, heterotrophy, predation and eukaryotic photosynthesis appeared, implying ecological tiering, and more complex food webs and ecological interactions.

During Period III, the Ediacaran Period (0.63–0.54 Ga), a second diversification, this time *within the supergroups*, occurred and left more preservable parts and more morphological diversity but within clades. Eukaryotic diversity increased from a few up to about 60 morpho-taxa, different from earlier assemblages, apart from leiospheres. If molecular phylogenies are correct, the Excavates which do not have a fossil record, have a basal position within the eukaryotes, so they should have appeared at least by the time the other supergroups had evolved. Therefore, no new supergroups appeared later than Period II. At the end of the Ediacaran (555–542 Ma), impoverished assemblages of large leiospheres and ornamented acritarchs (3 to 10 taxa), contemporary of mineralized animals, occurred while a transient shallow-water anoxia and animal predation might have caused the extinction of the Ediacara micro- and macro-biota, prior to a new radiation of protists (up to 45 acritarchs) in the Cambrian Period.

The pattern described here is not new, but the focus is not on the number of morphologically different taxa, but on the level at which this diversification might have occurred. What could be the causes for this pattern? Considering Period II as an important diversification period, rather than only the Ediacaran as commonly reported, one may examine changes in environmental conditions, and biological and ecological innovations, linked of course to genetic diversification, which is not directly accessible to palaeobiologists. During this time of supergroup diversification, the amalgamation and break-up of the supercontinent Rodinia (1200–750 Ma) occurred and could be a plausible explanation for the increase in diversity, by comparison with the break-up of Pangaea in the Phanerozoic Era and associated radiations on continents and in epicontinental seas. Indeed, Period II coincides with an increased variability in $\delta^{13}C$ after a prolonged Mesoproterozoic interval of stable values, possibly reflecting changes in organic carbon production and/or burial and increased tectonic activity (review in Halverson, 2006). The chemistry of the ocean and atmosphere was in an intermediary state, with anoxic sulphidic conditions in the bottom waters (Canfield, 1998), and oxygen in surface water, where eukaryotes lived. The rifting of Rodinia has been linked to the transition from the Mid-Proterozoic euxinic ocean to a transient ferruginous ocean around 750 Ma (Johnston *et al.*, 2010). These difficult conditions did not prevent the eukaryotic supergroups evolving early, during Period II. The increased ocean oxygenation and demise of a long period of anoxic sulphidic conditions in deep and subsurface waters, probably led to better ocean and nutrient circulation, as supported by the shift in acritarch ecological distribution from nearshore waters in the Early Mesoproterozoic to the inner shelf in the Neoproterozoic Period (Anbar and Knoll, 2002). However, it is striking to observe little endemism in the Proterozoic Period (if morphological convergence can be excluded), despite the fact that deep oceans might have separated shallower basins. In the Ediacaran, acritarch assemblages show similarities that may allow defining a biostratigraphy, although there are some differences (e.g. Vorob'eva *et al.*, 2009).

More extraordinary were the biological and ecological innovations that evolved during this Period II. The endosymbiosis of the chloroplast ancestor, a cyanobacterium, led to the evolution of eukaryotic photosynthesis and permitted diversification in the photic zone by inventing new pigments and more complex morphologies, new biopolymers to resist bacterial or fungal infection or adverse conditions and flagella or buoyancy control for displacement in the photic zone of the water column. Ecological tiering due to the advent of multicellularity in early algae, possibly fungi, and in other fossils of unknown identity, introduced new niches for early eukaryotes to diversify, as underlined by Butterfield (2000) and Xiao and Dong (2006). These vertical and lateral expansions provided surfaces for attachment, grazing, boring, protection or as sources of oxygen, nutrients or organic detritus. Complex life cycles also developed with vegetative cells and resting cysts, cellular shape and ornamentation control, increased number of cellular types implying cell differentiation within a single organism and thus complex development, and of course sexual reproduction, leading to increased genetic diversity. Autotrophic and heterotrophic protists, including a variety of soft-walled and possibly scaly amoebae, diversified with multicellular green, red and Vaucherialean algae and fungi and other unidentified organisms. The unicellular ancestor of animals or animal themselves could have appeared at that time, as suggested by biomarkers (Love *et al.*, 2009) and putative spiculate microfossils (Butterfield, 2009a), adding a new trophic level and predation pressure. Complex food webs and ecological interactions developed with the evolution of all the eukaryotic supergroups.

Later diversification in the Ediacaran (Period III) led to increased morphological disparity and macroscopic size, possibly resulting from an exponential complexification of ecological interactions, and favoured by environmental conditions (that had started earlier). Episodic glaciations, anoxia and meteor impact freed niches and led to renewed diversification *within* pre-existing supergroups.

At present, it is still difficult to constrain the causes for the observed pattern of early eukaryote diversification, especially because precise timing of environmental changes and biological innovations is lacking. However, it is clear that the diversification of supergroups occurred early, in shallow-water niches above still euxinic deeper waters, during Period II, and coincided with major biological innovations and related ecological complexification. Period II was also a time of major environmental changes, with Rodinia amalgamation (1.2 Ga) and break-up (0.75 Ga) progressively spreading oxygenated niches with declining euxinic conditions (around 0.75–0.6 Ga), and transient anoxia followed by two glaciations at the end of the period. Although the direct possible role of these geological events on eukaryotic evolution is not yet clear, they contrast sharply with the quiet times of Period I. The question remains as to why eukaryotes did not start to diversify earlier since their first fossil occurrence at 1.8 Ga.

Clarifying the biological affinities of organic-walled microfossils using comparative morphology, wall ultrastructure and microchemistry may permit us to identify members of early and later clades or to document biological innovations when identification is not possible. Such a multidisciplinary approach, combined with insights from molecular phylogeny, and improved sampling, datings, geochemical data and taxonomy, is needed to clarify the pattern of diversification and to test hypotheses of its causes.

Chapter summary

The origin of the domain Eucarya is not constrained and might be an Archean event. However, the Archean eukaryotic record is much discussed and limited to biomarkers, possibly indicating the presence of cells able to synthesize eukaryotic sterols in their flexible membranes. A hypothesis is proposed in this paper to offer a new reading of the diversification pattern of early eukaryotes following the origin of the domain, divided into three steps involving *different taxonomic levels*.

Period I (?1.8 to 1.2 Ga) was the time between the origin of the domain (no later than the Late Palaeoproterozoic, possibly in the Archean), through the Early Mesoproterozoic, when stem (and possibly but not necessarily crown-group) eukaryotes appeared. These early eukaryotic cells had a flexible lipidic membrane, a cytoskeleton and associated endomembrane system, and a nucleus. The diversity of organic-walled microfossils is moderate and includes smooth and ornamented sphaeromorphs, and vesicles with asymmetrically distributed processes and one species with symmetrically distributed processes. Macroscopic compressions (interpreted as macroalgae) include few taxa.

During Period II, from the Late Mesoproterozoic through the Middle Neoproterozoic (1.2–0.63 Ga), a major diversification occurred *at the supergroup level*. Members of all extant supergroups: the Opisthokonts, the Amoebozoa, the Archaeplastida, the chromalveolates, and the Rhizaria, except the Excavates, are recorded in the fossil record. The diversity of protists and of macroalgae increases gradually. Major biological innovations such as multicellularity, sex, biomineralization, heterotrophy and eukaryotic photosynthesis appeared, leading to ecological tiering and complex food webs and interactions. During the Middle Neoproterozoic (*c.*750–635 Ma), Chrysophyte or Haptophyte scales, as well as biomarkers of the sponge lineage, document the evolution of a fourth group of algae among the RAS clade and again of Opisthokonts, as well as protist biomineralization, and possibly the first appearance of animals.

During Period III, the Ediacaran (0.63–0.54 Ga), a second diversification, this time *within the supergroups*, occurred and left more preservable parts and more morphological diversity, but within clades. After a drop in diversity (of which the timing and intensity are discussed), the Ediacaran starts with a low-diversity assemblage of leiospheres and rare acanthomorphs (635–555 Ma), then records a burst in diversity of acritarchs with symmetrically distributed processes, microscopic early animals, macroalgal compressions and the macroscopic Ediacara fauna (including cnidarians and bilaterians). Florideophytes and brown algae are documented. The late Ediacaran (555–542 Ma) yields the first mineralized metazoans, and mostly large microscopic smooth vesicles, with rare acanthomorphs. Complex multicellularity (tissue-grade then organ-grade organization), and animal biomineralization and predation evolved during the Ediacaran, leading to more complex ecosystems and diversification within supergroups. A transient shallow-water anoxia and animal predation might have caused the extinction of the Ediacara biota, prior to a Cambrian radiation of much smaller, diverse ornamented and process-bearing acritarchs.

Acknowledgements

I would like to thank the editors for their invitation to write this review chapter. I am also grateful to Nick Butterfield, Susannah Porter and Shuhai Xiao for some of the pictures. Comments by Andy Knoll, Purificación López-García and an anonymous reviewer helped to improve the manuscript and they are gratefully acknowledged.

References

Allison, C. W. and Awramik, S. M. (1989). Organic-walled microfossils from earliest Cambrian or latest Proterozoic Tindir Group rocks, northwest Canada. *Precambrian Research*, **43**, 253–94.

Allison, C.W. and Hilgert, J.W. (1986). Scale microfossils from the Early Cambrian of Northwest Canada. *Journal of Paleontology*, **60**, 973–1015.

Amard, B. and Bertrand-Safarti J. (1997). Microfossils in 2000 Myr old cherty stromatolites of the Franceville Groups, Gabon. *Precambrian Research*, **81**, 197–221.

Amthor, J. E., Grotzinger, J. P., Schröder, S., Bowring, S. A., Ramezani, J., Martin, M. W. and Matter, A. (2003). Extinction of *Cloudina* and *Namacalathus* at the Precambrian–Cambrian boundary in Oman. *Geology*, **31**, 431–4.

Anbar, A. and Knoll, A. H. (2002). Proterozoic ocean chemistry and evolution: a bioinorganic bridge? *Science*, **297**, 1137–42.

Anbar, A. D., Duan Y., Lyons, T. W., Arnold, G. L., Kendall, B., Creaser, R. A., Kaufman, A. J., Gordon, G. W., Garvin, J. and Buick, R. (2007). A whiff of oxygen before the great oxidation event? *Science*, **317**, 1903–6.

Arouri, K., Greenwood, P. F. and Walter, M. R. (1999). A possible chlorophycean affinity of some Neoproterozoic acritarchs. *Organic Geochemistry*, **30**, 1323–37.

Bailey, J. V., Joye, S. B., Kalenetra, K. M., Flood, B. E. and Corsetti, F. A. (2007a). Evidence of giant sulphur bacteria in Neoproterozoic phosphorites. *Nature,* **445**, 198–201.

Bailey, J. V., Joye, S. B., Kalenetra, K. M., Flood, B. E. and Corsetti, F. A. (2007b). Palaeontology: Undressing and redressing Ediacaran embryos (Reply). *Nature*, **446**, E10–E11.

Baldauf, S. (2008). An overview of the phylogeny and diversity of eukaryotes. *Journal of Systematics and Evolution*, **46**, 263–73.

Barfod, G. H., Albarède, F., Knoll, A. H., Xiao S., Telouk, P., Frei, R. and Baker, J. (2002). New Lu–Hf and Pb–Pb age constraints on the earliest animal fossils. *Earth and Planetary Sciences Letters*, **201**, 203–12.

Bekker, A., Holland, H. D., Wang, P. L., Rumble, D., Stein, H. J., Hannah, J. L., Coetzee, L. L. and Beukes, N. J. (2004). Dating the rise of atmospheric oxygen. *Nature*, **427**, 117–20.

Bekker, A., Kaufman, A. J., Karhu, J. A. and Eriksson, K. A. (2005). Evidence for Paleoproterozoic cap carbonates in North America. *Precambrian Research*, **137**, 167–206.

Bengtson, S. and Rasmussen, B. (2009). New and ancient trace makers. *Science*, **323**, 346–7.
Bengtson, S., Rasmussen B. and Krapez B. (2007). The Paleoproterozoic megascopic Stirling biota. *Paleobiology*, **33**, 351–81.
Bernard, S., Benzerara, K., Beyssac, O., Brown, G. E., Jr., Grauvogel Stamm, L. and Duringer, P. (2009). Ultrastructural and chemical study of modern and fossil sporoderms by scanning transmission X-ray microscopy (STXM). *Review of Palaeobotany and Palynology*, **156**, 248–61.
Bhattacharya, D., Archibald, J. M., Weber, A. P. M. and Reyes-Prieto, A. (2007). How do endosymbionts become organelles? Understanding early events in plastid evolution. *Bioessays*, **29**, 1239–46.
Brocks, J. J. and Butterfield, N. (2009). Early animals out in the cold. *Nature*, **457**, 672–3.
Brocks, J. J., Logan, G. A., Buick, R. and Summons, R. E. (1999). Archean molecular fossils and the early rise of eukaryotes. *Science*, **285**, 1033–6.
Brocks, J. J., Buick, R., Summons, R. E. and Logan, G. A. (2003). A reconstruction of Archean biological diversity based on molecular fossils from the 2.78–2.45 billion year old Mount Bruce Supergroup, Hamersley Basin,Western Australia. *Geochimica et Cosmochimica Acta*, **67**, 4321–35.
Brocks, J. J., Love, G. D., Summons, R. E., Knoll, A. H., Logan, G. A. and Bowden, S. A. (2005). Biomarker evidence for green and purple sulphur bacteria in a stratified Palaeoproterozoic sea. *Nature*, **437**, 866–70.
Buick, R. (2007). Did the Proterozoic Canfield Ocean cause a laughing gas greenhouse? *Geobiology*, **5**, 97–100.
Buick, R. (2008). When did oxygenic photosynthesis evolve? *Philosphical Transactions of the Royal Society B*, **363**, 2731–43.
Butterfield, N. J. (2000). *Bangiomorpha pubescens* n. gen., n. sp.: implications for the evolution of sex, multicelluarity and the Mesoproterozoic/Neoproterozoic radiation of eukaryotes. *Paleobiology*, **26**, 386–404.
Butterfield, N. J. (2001). Paleobiology of the late Mesoproterozoic (ca. 1200 Ma) Hunting Formation, Somerset Island, Arctic Canada. *Precambrian Research*, **111**, 235–56.
Butterfield, N. J. (2004). A vaucherian alga from the Middle Neoproterozoic of Spitsbergen: implications for the evolution of Proterozoic eukaryotes and the Cambrian explosion. *Paleobiology*, **30**, 231–52.
Butterfield, N. J. (2005a). Probable Proterozoic fungi. *Paleobiology*, **31**, 165–82.
Butterfield, N. J. (2005b). Reconstructing a complex Early Neoproterozoic eukaryote, Wynniatt Formation, Arctic Canada. *Lethaia*, **38**, 155–69.
Butterfield, N. J. (2007). Macroevolution and macroecology through deep time. *Palaeontology*, **50**, 41–55.
Butterfield, N. J. (2009a). Modes of pre-Ediacaran multicellularity. *Precambrian Research*, **173**, 201–11.
Butterfield, N. J. (2009b). Oxygen, animals and oceanic ventilation: an alternative view. *Geobiology*, **7**, 1–7.
Butterfield, N. J., Knoll, A. H. and Swett, N. (1994). Paleobiology of the Neoproterozoic Svanbergfjellet Formation, Spitsbergen. *Fossils and Strata*, **34**, 1–84.
Butterfield, N. and Peterson, K. (2005). Origin of the Eumetazoa: testing ecological predictions of molecular clocks against the Proterozoic fossil record. *Proceedings of the National Academy of Sciences of the USA*, **102**, 9547–52.
Cabeen, M. T. and Jacobs-Wagner, C. (2005). Bacterial cell shape. *Nature Reviews Microbiology*, **3**, 601–10.
Canfield, D. E. (1998). A new model for Proterozoic ocean chemistry. *Nature*, **396**, 450–3.

Cavalier-Smith, T. (2002). The neomuran origin of archaebacteria: the negibacteria root of the universal tree and bacteria megaclassification. *International Journal of Systematic Microbiology*, **52**, 7–76.

Cavalier-Smith, T. (2009). Megaphylogeny, cell body plans, adaptive zones: causes and timing of eukaryote basal radiations. *Joural of Eukaryotic Microbiology*, **5**, 26–33.

Chen, J-Y., Bottjer, D. J., Davidson, E. H., Li, G., Gao, F., Cameron, R. A., Hadfield, M. G., Xian, D-C., Tafforeau, P., Jia, Q-J., Sugiyama, H. and Tang, R. (2009). Phase contrast synchrotron X-ray microtomography of Ediacaran (Doushantuo) metazoan microfossils: phylogenetic diversity and evolutionary implications. *Precambrian Research*, **173**,191–200.

Cohen, P.A., Knoll, A.H. and Kodner, R.B. (2009). Large spinose microfossils in Ediacaran rocksas resting stages of early animals. *Proceedings of the National Academy of Sciences USA*, **106**, 6519–6524.

Condon, D., Zhu, M., Bowring, S., Jin, Y., Wang, W. and Yang, A. (2005). From the Marinoan glaciation to the oldest bilaterians: U–Pb ages from the Doushantou Formation, China. *Science*, **308**, 95–8.

Cox, C. J., Foster, P. G., Hirt, R. P., Harris, S. R. and Embley, T. M. (2008). The archaebacterial origin of eukaryotes. *Proceedings of the National Academy of Sciences of the USA*, **105**, 20356–61.

Doolittle, W. F. (1999). Phylogenetic classification and the universal tree. *Science*, **284**, 2124–8.

Douzery, E. J. P., Snell, E. A., Bapteste, E., Delsuc, F. and Philippe, H. (2004). The timing of eukaryotic evolution: does a relaxed molecular clock reconcile proteins and fossils? *Proceedings of the National Academy of Sciences of the USA*, **101**, 15386–91.

Dutkiewicz, A., Volk, H., George, S. C., Ridley, J. and Buick, R. (2006). Biomarkers from Huronian oil-bearing fluid inclusions: an uncontaminated record of life before the great oxidation event. *Geology*, **34**, 437–40.

Eigenbrode, J. L. (2008). Fossil lipids for life-detection: a case study from the early Earth record. *Space Sciences Series*, **135**, 161–85.

Embley, T. and Martin, W. (2006). Eukaryotic evolution, changes and challenges. *Nature*, **440**, 623–30.

Fike, D. A., Grotzinger, J. P., Pratt, L. M. and Summons, R. E. (2006). Oxidation of the Ediacaran ocean. *Nature*, **444**, 744–7.

Fisher, W. (2008). Life before the rise of oxygen. *Nature*, **455**, 1051–2.

Frei, R, Gaucher, C., Poulton, S. W. and Canfield, D. E. (2009). Fluctuations in Precambrian atmospheric oxygenation recorded by chromium isotopes. *Nature*, **461**, 250–3.

Fuerst, J. A. (2005). Intracellular compartmentation in planctomycetes. *Annual Review of Microbiology*, **59**, 299–328.

George, S. C., Dutkiewicz, A., Volk, H., Ridley, J., Mossman, D. J. and Buick, R. (2009). Oil-bearing fluid inclusions from the Palaeoproterozoic: a review of biogeochemical results from time-capsules > 2.0 Ga old. *Science in China Series D – Earth Sciences*, **52**, 1–11.

Germs, G. J. B., Knoll, A. H. and Vidal, G. (1986). Latest Proterozoic microfossils from the Nama Group, South West Africa/Namibia. *Precambrian Research*, **32**, 45–62.

German, T. N. and Podkovyrov, V. N. (2009). New insights into the nature of the Late Riphean Eosolenides. *Precambrian Research*, **173**, 154–62.

Glass, J. B., Wolfe-Simon, F. and Anbar, A. D. (2009). Coevolution of metal availability and nitrogen assimilation in caynobacteria and algae. *Geobiology*, **7**, 100–33.

Grey, K. (2005). Ediacaran palynology of Australia. *Memoir of the Association of Australasian Paleontologists*, **31**, 1–432.

Grey, K., Walter, M. R. and Calver, C. R. (2003). Neoproterozoic biotic diversification: Snowball Earth or aftermath of the Acraman impact? *Geology*, **31**, 459–62.

Grey, K. and Sugitani, K. (2009). Palynology of Archean microfossils (c. 3.0 Ga) from the Mount Grant area, Pilpara, Craton, Western Austrlia: Further evidence of Biogenicity. *Precambrian Research*, **73**, 60–69.

Grey, K. and Williams, I. R. (1990). Problematic bedding-plane markings from the Middle Proterozoic Manganese Subgroup, Bangemall Basin, Western Australia. *Precambrian Research*, **46**, 307–27.

Grotzinger, J. P., Watters, W. and Knoll, A. H. (2000). Calcareous metazoans in thrombolitic bioherms of the terminal Proterozoic Nama Group, Namibia. *Paleobiology*, **26**, 334–59.

Halverson, G. P. (2006). A Neoproterozoic chronology. In *Neoproterozoic Geobiology and Paleobiology*, eds. S. Xiao and A. J. Kaufman. Dordrecht: Springer, pp. 231–71.

Hampl, V., Hug, L., Leigh, J. W., Dacks, J. B., Lang, B. F., Simpson, A. G. B. and Roger, A. J. (2009). Phylogenomic analyses support the monophyly of Excavata and resolve relationships among eukaryotic 'supergroups'. *Proceedings of the National Academy of Sciences of the USA*, **106**, 3959–64.

Han, T.-M. and Runnegar, B. (1992). Megascopic eukaryotic algae from the 2.1-billion-year-old Negaunee Iron Formation. *Science*, **257**, 232–5.

Hedges, S. B., Blair, J. E., Venturi, M. L. and Shoe, J. L. (2004). A molecular timescale of eukaryote evolution and the rise of complex multicellular life. *BMC Evolutionary Biology*, **4**, 2.

Hedges, B. S., Battistuzzi, F. U. and Blair J. E. (2006). Molecular timescale of evolution in the Proterozoic. In *Neoproterozoic Geobiology and Paleobiology*, eds. S. Xiao and A. J. Kaufman. Dordrecht: Springer, pp. 199–229.

Herman, N. (1990). *The Organic World One Billion Years Ago*. Leningrad: Nauka.

Hill, A. C., Cotter, K. L. and Grey, K. (2000). Mid-Neoproterozoic biostratigraphy and isotope stratigraphy in Australia. *Precambrian Research*, **100**, 281–98.

Hoffman, P. F. and Li, Z. X. (2009). A palaeogeographic context for Neoproterozoic glaciation. *Palaeogeography, Palaeoclimatology, Palaeoecology*, **277**, 158–72.

Hoffmann, K. H., Condon, D. J., Bowring, S. A. and Crowley, J. L. (2004). A U–Pb zircon date from the Neoproterozoic Ghaub Formation, Namibia: constraints on Marinoan glaciation. *Geology*, **32**, 817–20.

Hofmann, H. J. (1976). Precambrian microflora, Belcher Islands, Canada: significance and systematics. *Journal of Palaeontology*, **50**, 1040–73.

Hofmann, H. J., Narbonne, G. M. and Aitken, J. D. (1990). Ediacaran remains from intertillite beds in northwestern Canada. *Geology*, **29**, 1091–4.

Hofmann, H. J. (1994). Problematic carbonaceous compressions ('metaphytes' and 'worms'). In *Early life on Earth*, ed. S. Bengston. New York: Columbia University Press, pp. 342–58.

Hofmann, H. J. and Jackson, C. D. (1994). Shelf-facies microfossils from the Proterozoic Bylot Supergroup, Baffin Island, Canada. *Paleontological Society Memoir*, **37**, 39.

Horodyski, R. (1982). Problematic bedding-plane markings from the Middle Proterozoic Appekunny Argillite, Belt Supergroup, northwestern Montana. *Journal of Paleontology*, **56**, 882–9.

Huntley, J. W., Xiao, S. and Kowaleski, M. (2006). 1.3 Billion years of acritarch history: an empirical morphospace approach. *Precambrian Research*, **144**, 52–68.

Jankauskas, T. V. (1989). *Mikrofossilii Dokembriya SSSR* (Precambrian microfossils of the USSR). Leningrad: Nauka, pp. 1–190.

Javaux, E. J. (2006a). The early eukaryotic fossil record. In *Origins and Evolution of Eukaryotic Endomembranes and Cytoskeleton*, ed. G. Jekely. Texas, USA: Landes Biosciences, pp. 1–19.
Javaux, E. J. (2006b). Extreme life on Earth–past, present and possibly beyond. *Research in Microbiology*, **157**, 37–48.
Javaux, E. J. (2007). Patterns of diversification in early eukaryotes. In *Recent Advances in Palynology*, eds. P. Steemans and E. Javaux. Carnets de Géologie/Notebooks on Geology, Brest, Memoir 2007/01, pp. 38–42 (CG2007_M01/06).
Javaux, E. J. and Benzerara, K. (2009). Microfossils. *Comptes Rendus Palevol*, **8**, 605–615.
Javaux, E. J. and Marshall, C. P. (2006). A new approach in deciphering early protist paleobiology and evolution: combined microscopy and microchemistry of single Proterozoic acritarchs. *Review of Palaeobotany and Palynology*, **139**, 1–15.
Javaux, E. S., Marshall, C. P. and Bekker, A. (2010). Organic-walled microfossils in 3.2-billion-year-old shallow marine siliciclastic deposits. *Nature*, **463**, 934–938.
Javaux, E. J., Knoll, A. H. and Walter, M. R. (2001). Morphological and ecological complexity in early eukaryotic ecosystems. *Nature*, **412**, 66–9.
Javaux, E. J., Knoll, A. H. and Walter, M. R. (2003). Recognizing and interpreting the fossils of early eukaryotes. *Origins of Life and Evolution of Biospheres*, **33**, 75–94.
Javaux, E. J., Knoll, A. H. and. Walter, M. R. (2004). TEM evidence for eukaryotic diversity in Mid-Proterozoic oceans. *Geobiology*, **2**, 121–32.
Jekely, G. (2006). *Origins and Evolution of Eukaryotic Endomembranes and Cytoskeleton*. Texas, USA: Landes Biosciences, p. 145.
Johnston, D. T., Wolfe-Simon, F., Pearson, A. and Knoll, A. H. (2009). Anoxygenic photosynthesis modulated Proterozoic oxygen and sustained Earth's middle age. *Proceedings of the National Academy of Sciences of the USA*, **106**, 16925–9.
Johnston, D. T., Poulton, S. W., Dehler, C., Porter, S., Husson, J., Canfield, D. E. and Knoll, A. H. (2010). An emerging picture of Neoproterozoic ocean chemistry: insights from the Chuar Group, Arizona. *Earth and Planetary Science of the Letters*, **290**, 64–73.
Karlstrom, K. E., Bowring, S. A., Dehler, C. M., Knoll, A. H., Porter, S. M., Des Marais, D. J., Weil, A. B., Sharp, Z. D., Geissman, J. W., Elrick, M. B., Timmons, J. M., Crossey, L. J. and Davidek, K. L. (2000). Chuar Group of the Grand Canyon: record of breakup of Rodinia, associated change in the global carbon cycle, and ecosystem expansion by 740 Ma. *Geology*, **28**, 619–22.
Kaufman, A. J., Knoll, A. H. and Awramik, S. M. (1992). Biostratigraphic and chemostratigraphic correlation of Neoproterozoic sedimentary successions – Upper Tindir Group, northwestern Canada, as a test case. *Geology*, **20**, 181–5.
Kendall, B., Creaser, R.A., Gordon, G.W., and Anbar, A.D. (2009). Re -Os and Mo isotope systematics of black shales from the Middle Proterozoic Velkerri and Wollogorang Formations, McArthur Basin, northern Australia. *Geochimica et Cosmochimica Acta*, **73**, 2534–2558.
Kleeman, G., Poralla, K., Englert, G., Kjosen, H., Liaaen-Jensen, S., Neunlist, S. and Rohmer, M. (1990). Tetrahymenol from the phototrophic bacterium *Rhodopseudomonas palustris*: first report of a gammacerane triterpene from a prokaryote. *Journal of Genetic Microbiology*, **136**, 2551–3.
Knoll, A. H. (1994). Proterozoic and Early Cambrian protists: evidence for accelerating evolutionary tempo. *Proceedings of the National Academy of Sciences of the USA*, **91**, 6743–50.
Knoll, A. H. (2003a). Biomineralization and evolutionary history. *Review of Mineralogy and Geochemistry*, **54**, 329–56.

Knoll, A. H. (2003b). *Life on a Young Planet*. Princeton, NJ: Princeton University Press.

Knoll, A. H., Walter, M. R. and Christie-Blick, N. (2004). A new period for the geological time scale. *Nature*, **305**, 621–2.

Knoll, A. H., Walter, M. R., Narbonne, G. M. and Christie-Blick, N. (2006a). The Ediacaran Period: a new addition to the geologic time scale. *Lethaia*, **39**, 13–30.

Knoll, A. H., Javaux, E. J., Hewitt, D. and Cohen, P. (2006b). Eukaryotic organisms in Proterozoic Oceans. *Philosophical Transactions of the Royal Society B*, **361**, 1023–38.

Kodner, R. B., Summons, R. E. and Knoll, A. H. (2009). Phylogenetic investigation of the aliphatic, non-hydrolyzable biopolymer algaenan, with a focus on green algae. *Organic Geochemistry*, **40**, 854–62.

Konhauser, K.O., Amskold, L., Lalonde, S.V., Posth, N.R., Kappler, A. and Anbar, A. (2007). Decoupling photochemical Fe(II) oxidation from shallow-water BIF deposition. *Earth and Planetary Science Letters*, **258**, 87–100.

Konhauser, K. O., Pecoits, E., Lalonde, S. V., Papineau, D., Nisbet, E. G., Barley, M. E., Arndt, N., Zahnle, K. and Kamber, B. S. (2009). Oceanic nickel depletion and a methanogen famine before the great oxidation event. *Nature*, **458**, 750–4.

Kumar, S. (1995). Megafossils from the Mesoproterozoic Rohtas Formation (the Vindhyan Supergroup), Katni area, Central India. *Precambrian Research*, **72**, 171–84.

Kurland, C. G., Collins, L. J. and Penny, D. (2006). Genomics and the irreducible nature of eukaryote cells. *Science*, **312**, 1011–14.

Lamb, D. M., Awramik, S. M., Chapman, D. J. and Zhu, S. (2009). Evidence for eukaryotic diversification in the ~ 1800 million-year-old Changzhougou Formation, North China. *Precambrian Research*, **173**, 93–104.

Li, C., Peng, P., Sheng, G. Y., Fu, J. and Yan, Y. (2003). A molecular and isotopic geochemical study of Meso- to Neoproterozoic (1.73–0.85 Ga) sediments from the Jixian section, Yanshan Basin, North China. *Precambrian Research*, **125**, 337–56.

Liu, H., Probert, I., Uitz, J., Claustre, H., Aris-Brosou, S., Frada, M., Not, F. and de Vargas, C. (2009). Extreme diversity in noncalcifying haptophytes explains a major pigment paradox in open oceans. *Proceedings of the National Academy of Sciences of the USA*. (*PNAS* published online before print July 21, 2009, doi:10.1073/pnas.0905841106).

López-García, P., Moreira, D., Douzery, E., Forterre, P., van Zuilen, M., Claeys, P. and Prieur, D. (2006). Ancient fossil record and early evolution (ca. 3.8 to 0.5 Ga). *Earth, Moon and Planets*, **98**, 247–90.

Love, G. D., Grosjean, E., Stalvies, C., Fike, D. A., Grotzinger, J. P., Bradley, A. S., Kelly, A. E., Bhatia, M., Meredith, W., Snape, C. E., Bowring, S. A., Condon, D. J. and Summons, R. E. (2009). Fossil steroids record the appearance of Demospongiae during the Cryogenian period. *Nature*, **457**, 718–22.

Lyons, T. W., Reinhard, C. T. and Scott, C. (2009). Redox redux. *Geobiology*, **7**, 489–94.

Marshall C. P., Carter E. A., Leuko S. and. Javaux E. J. (2006). Vibrational spectroscopy of extant and fossil microbes: relevance for the astrobiological exploration of Mars. *Vibrational Spectroscopy*, **41**, 182–9.

Marshall, C. P., Javaux, E. J., Knoll, A. H. and Walter, M. R. (2005). Combined micro-Fourier transform infrared (FTIR) spectroscopy and micro-Raman spectroscopy of Proterozoic acritarchs: a new approach to palaeobiology. *Precambrian Research*, **138**, 208–24.

Matz, M. V., Frank, T. M., Marshall, N. J., Widder, E. A. and Sönke, J. (2008). Giant deep-sea protist produces bilaterian-like traces. *Current Biology*, **18**, 1849–54.

Meng, F. W., Zhou, C. M., Yin, L. M., Chen, Z. L. and Yuan, X. L. (2005). The oldest known dinoflagellates: morphological and molecular evidence from Mesoproterozoic rocks at Yongji, Shanxi Province. *Chinese Science Bulletin*, **50**, 1230–4.

Moczydlowska, M. (2008). The Ediacaran microbiota and the survival of Snowball Earth conditions. *Precambrian Research*, **167**, 71–92.

Moczydlowska, M. and Willman, S. (2009). Ultrastructure of cell walls in ancient microfossils as a proxy to their biological affinities. *Precambrian Research*, **173**, 27–38.

Moldowan, J. M., Jacobsen, S. R., Dahl, J., Al-Hajji, A., Huizinga, B. J. and Fago, F. J. (2001). Molecular fossils demonstrate Precambrian origins of dinoflagellates. In *The Ecology of the Cambrian Radiation*, eds. A. Y. Zhuravlev and R. Riding. New York: Columbia University Press, pp. 475–93.

Moller-Jensen, J. and Lowe, J. (2005). Increasing complexity of the bacterial cytoskeleton. *Current Opinion in Cell Biology*, **17**, 75–81.

Moreira, D. and López-García, P. (2002). The molecular ecology of microbial eukaryotes unveils a hidden world. *Trends in Microbiology*, **10**, 31–8.

Nagovitsin, K. (2009). *Tappania*-bearing association of the Siberian platform: biodiversity, stratigraphic position and geochronological constraints. *Precambrian Research*, **173**, 137–45.

Nagy, R. M., Porter, S. M., Dehler, C. M. and Shen, Y. (2009). Biotic turnover driven by eutrophication before the Sturtian low-latitude glaciation. *Nature Geoscience*, **2**, 415–18.

Narbonne, G. M. (2005). The Ediacara biota: Neoproterozoic origin of animals and their ecosystems. *Annual Review of Earth and Planetary Sciences*, **33**, 421–42.

Noffke, N., Hazen, R. and Nhleko, N. (2003). Earth's earliest microbial mats in a siliciclastic marine environment (2.9 Ga Mozaan Group, South Africa). *Geology*, **31**, 673–6.

Noffke, N., Eriksson, K. A., Hazen, R. M. and Simpson, E. L. (2006). A new window into Early Archean life: microbial mats in Earth's oldest siliciclastic tidal deposits (3.2 Ga Moodies Group, South Africa). *Geology*, **34**, 253–6.

Nowack, E. C., Melkonian, M. and Glöckner, G. (2008). Chromatophore genome sequence of Paulinella sheds light on acquisition of photosynthesis by eukaryotes. *Current Biology*, **22**, 410–18.

Ono, S., Beukes, N. J., Rumble, D. and Fogel, M. L. (2006). Early evolution of atmospheric oxygen from multiple-sulphur and carbon isotope records of the 2.9 Ga Mozaan Group of the Pongola Supergroup, Southern Africa. *South African Journal of Geology*, **109**, 97–108.

Pace, N. R. (2006). Time for a change. *Nature*, **441**, 289.

Pawlowski, J. and Gooday, A. J. (2008). Precambrian biota: protistan origin of trace fossils? *Current Biology*, **19**, R28–30.

Peng, P., Sheng, G., Fu, J., and Yan, Y. (1998). Biological markers in 1.7 billion year old rock from the Tuanshanzi Formation, Jixian strata section, North China. *Organic Geochemistry*, **29**, 1321–1329.

Peng, Y., Baoa, H. and Yuan, X. (2009). New morphological observations for Paleoproterozoic acritarchs from the Chuanlinggou Formation, North China. *Precambrian Research*, **168**, 223–32.

Peterson, K. J. and Butterfield, N. J. (2005). Origin of the Eumetazoa: testing ecological predictions of molecular clocks against the Proterozoic fossil record. *Proceedings of the National Academy of Sciences of the USA*, **102**, 9547–52.

Philippe, H. and Forterre, P. (1999). The rooting of the universal tree is not reliable. *Journal of Molecular Evolution*, **49**, 509–23.

Porter, S. M. (2004). The fossil record of early eukaryotic diversification. In: *Neoproterozoic-Cambrian biological revolutions. Paleontological Society Papers*, (eds.) J. Lipps and B. Waggoner.10, 35–50.

Porter, S. M. (2006). The proterozoic fossil record of heterotrophic eukaryotes. In *Neoproterozoic Geobiology and Paleobiology*, eds. S. Xiao and A.J. Kaufman. The Dordrecht: Springer, pp. 1–21.

Porter, S. M. and Knoll, A. H. (2000). Testate amoebae in the Neoproterozoic Era: evidence from vase-shaped microfossils in the Chuar Group, Grand Canyon. *Paleobiology*, **26**, 360–85.

Porter, S. M., Meisterfeld, R. and Knoll, A. H. (2003). Vase-shaped microfossils from the Neoproterozoic Chuar Group, Grand Canyon: a classification guided by modern testate amoebae. *Journal of Paleontology*, **77**, 409–29.

Prasad, B. and Asher, R. (2001). Acritarch biostratigraphy and lithostratigraphic classification of Proterozoic and Lower Paleozoic sediments (Pre-Unconformity Sequence) of Ganga Basin, India. *Paleontographica Indica*, **5**, 1–151.

Pratt, L. M., Summons, R. E. and Hieshima, G. B. (1991). Sterane and triterpane biomarkers in the Precambrian Nonesuch Formation, North American Midcontinent Rift. *Geochimica et Cosmochimica Acta*, **55**, 911–16.

Rainbird, R. H., Stern, R. A., Khudoley, A. K., Kropachev, A. P., Heaman, L. M. and Sukhorukov, V. I. (1998). U–Pb geochronology of Riphean sandstone and gabbro from southeast Siberia and its bearing on the Laurentia–Siberia connection. *Earth Planetary Sciences Letters*, **164**, 409–20.

Rampen, S. W., Abbas, B. A., Schouten, S. and Sinninghe-Damste, J. S. (2010). A comprehensive study of sterols in marine diatoms (Bacillariophyta): implications for their use as tracers for diatom productivity. *Limnology and Oceanography*, **55**, 91–105.

Rasmussen, B., Fletcher, I. R., Brocks, J. J. and Kilburn, M. R. (2008). Reassessing the first appearance of eukaryotes and cyanobacteria. *Nature*, **455**, 1101–5.

Richards, T. A. and Cavalier-Smith, T. (2005). Myosin domain evolution and the primary divergence of eukaryotes. *Nature*, **436**, 1113–18.

Rivera, M. C., Jain, R., Moore, J. E. and Lake, J. A. (1998). *Proceedings of the National Academy of Sciences of the USA*, **95**, 6239–44.

Roger, A. J. and Simpson, A. G. B. (2009). Evolution: revisiting the root of the eukaryote tree. *Current Biology*, **19**, 165–7.

Roger, A. J. and Hug, L. A. (2006). The origin and diversification of eukaryotes: problems with molecular phylogenetics and molecular clock estimation. *Philosophical Transactions of the Royal Society B*, **361**, 1039–54.

Rosing, M. T. and Frei, R. (2004). U-rich Archean sea-floor sediments from Greenland – indication of > 3700 Ma oxygenic photosynthesis. *Earth and Planetary Sciences Letters*, **217**, 237–44.

Runnegar B. N. and Fedonkin M. A. (1992). Proterozoic metazoan body fossils. In *The Proterozoic Biosphere; A Multidisciplinary Study*, eds. J. W. Schopf, C. Klein, Cambridge, UK: Cambridge Univ. Press. pp. 999–1007.

Samuelsson, J. and Butterfield, N. J. (2001). Neoproterozoic fossils from the Franklin Mountains, northwestern Canada: stratigraphic and palaeobiological implications. *Precambrian Research*, **107**, 235–51.

Schneider, D. A., Bickford, M. E., Cannon, W. F., Sculz, K. J. and Hamilton, M. A. (2002). Age of volcanic rocks and syndepositional iron formations, Marquette Range Supergroup: implications for the tectonic setting of Paleoproterozoic iron formations of the Lake Superior region. *Canadian Journal of Earth Sciences*, **39**, 999–1012.

Seilacher, A., Grazhdankin, D. and Legouta, A. (2003). Ediacaran biota: the dawn of animal life in the shadow of giant protists. *Paleontological Research*, **7**, 43–54.
Seilacher, A. (1992). Vendobionta and Psammocorallia – lost constructions of Precambrian evolution. *Journal of the Geological Society of London*, **149**, 607–13.
Sergeev, V.N. (2009). The distribution of microfossil assemblages in Proterozoic rocks. *Precambrian Research*, **173**, 212–222
Sessions, A. L., Doughty, D. M., Welander, P. V., Summons, R. E. and Newman, D. K. (2009). The continuing puzzle of the great oxidation event. *Current Biology*, **19**, 567–74.
Sharma, M. and Shukla, Y. (2009). Taxonomy and affinity of Early Mesoproterozoic megascopic helically coiled and related fossils from the Rohtas Formation, the Vindhyan Supergroup, India. *Precambrian Research*, **173**, 105–22.
Shen, Y. and Buick, R. (2004). The antiquity of microbial sulfate reduction. *Earth Science Review*, **64**, 243–72.
Stechmann, A. and Cavalier-Smith, T. (2002). Rooting the eukaryote tree by using a derived gene fusion. *Science*, **297**, 89–91.
Summons, R. E., Powell T. G. and Boreham, C. J. (1988). Petroleum geology and geochemistry of the Middle Proterozoic McArthur Basin, northern Australia: III. Composition of extractable hydrocarbons. *Geochimica et Cosmochimica Acta*, **5**, 1747–63.
Summons, R. E., Thomas, J., Maxwell, J. R. and Boreham, C. J. (1992). Secular and environmental constraints on the occurrence of dinosterane in sediments. *Geochimica et Cosmochimica Acta*, **56**, 2437–44.
Summons, R. E. and Walter, M. R. (1990). Molecular fossils and microfossils of prokaryotes and protists from Proterozoic sediments. *American Journal of Sciences*, **290**, 212–44.
Summons, R. E., Bradley, A. S., Jahnke, L. J. and Waldbauer, J. R. (2006). Steroids, triterpenoids and molecular oxygen. *Philosophical Transactions of the Royal Society B*, **361**, 951–68.
Taylor, T. N., Hass, H. and Kerp, H. (1999). The oldest fossil ascomycetes. *Nature*, **399**, 648.
Tice, M. M. and Lowe, D. R. (2004). Photosynthetic microbial mats in the 3,416-Myr-old ocean. *Nature*, **431**, 549–52.
Versteegh, G. J. M. and Blokker, P. (2004). Resistant macromolecules of extant and fossil microalgae. *Phycological Research*, **52**, 325–39.
Versteegh, G. M. J., Blokker, P., Wood, G. D., Collinson, M. E., Sinninghe Damste, J. H. and de Leeuw, J. W. (2004). An example of oxidative polymerization of unsaturated fatty acids as a preservation pathway for dinoflagellate organic matter. *Organic Geochemistry*, **35**, 1129–39.
Vogel, K. and Brett, C. E. (2009). Record of microendoliths in different facies of the Upper Ordovician in the Cincinnati Arch region USA: the early history of light-related microendolithic zonation. *Palaeogeography, Palaeoclimatology, Palaeoecology*, **281**, 1–24.
Volkman, J. K. (2005). Sterols and other triterpenoids: source specificity and evolution of biosynthetic pathways. *Organic Geochemistry*, **36**, 139–59.
Volkman, J. K. (2003). Sterols in microorganisms. *Applied Microbiology and Biotechnology*, **60**, 495–506.
Vorob'eva, N. G., Sergeev, V. N. and Knoll, A. H. (2009). Neoproterozoic microfossils from the margin of the East European Platform and the search for a biostratigraphic model of lower Ediacaran rocks. *Precambrian Research*, **173**, 163–9.

Waldbauer, J. R., Sherman, L. S., Sumner, D. Y. and Summons, R. E. (2009). Late Archean molecular fossils from the Transvaal Supergroup record the antiquity of microbial diversity and aerobiosis. *Precambrian Research*, **169**, 28–47.

Walter, M. R., Du, R. and Horodyski, R. J. (1990). Coiled carbonaceous megafossils from the Middle Proterozoic of Jixian (Tianjin) and Montana. *American Journal of Sciences*, **290**, 133–48.

Woese, C. R., Kandler, O. and Wheelis, M. L. (1990). Towards a natural system of organisms: proposal for the domains Archaea, Bacteria and Eucarya. *Proceedings of the National Academy of Sciences of the USA*, **87**, 4576–9.

Woods, K. N., Knoll, A. H. and German, T.. (1998). Xanthophyte algae from the Mesoproterozoic/Neoproterozoic transition: confirmation and evolutionary implications. *Geological Society of America*, **30**, 232.

Xiao, S. and Knoll, A. H. (2000). Phosphatized embryos from the Neoproterozoic Doushantuo Formation. *Journal of Paleontology*, **74**, 767–88.

Xiao, S. and Dong, L. (2006). On the morphological and ecological history of Proterozoic macroalgae. In *Neoproterozoic Geobiology and Paleobiology*, eds. S. Xiao and A. J. Kaufman. Dordrecht: Springer, pp. 57–90.

Xiao, S., Knoll, A. H., Kaufman, A. J., Yin, L. and Zhang, Y. (1997). Neoproterozoic fossils in Mesoproterozoic rocks? *Precambrian Research*, **84**, 197–220.

Xiao, S., Knoll, A. H. and Yuan, X. (1998a). *Miaohephyton*, a possible brown alga from the terminal Proterozoic Doushantuo Formation, China. *Journal of Paleontology*, **72**, 1072–86.

Xiao, S., Zhang, Y. and Knoll, A. H. (1998b). Three-dimensional preservation of algae and animal embryos in a Neoproterozoic phosphate. *Nature*, **391**, 553–8.

Xiao, S., Yuan, X., Steiner, M. and Knoll, A. H. (2002). Macroscopic carbonaceous compressions in a terminal Proterozoic shale: a systematic reassessment of the Miaohe biota, South China. *Journal of Paleontology*, **76**, 345–74.

Xiao, S., Knoll, A. H., Yuan, X. and Pueschel, C. M. (2004). Phosphatized multicellular algae in the Neoproterozoic Doushantuo Formation, China, and the early evolution of florideophyte red algae. *American Journal of Botany*, **91**, 214–27.

Xiao, S., Zhou, C. and Yuan, X. (2007).Undressing and redressing Ediacaran embryos. *Nature*, **446**, E10–11, doi: 10.1038/ nature05753.

Yin, L. M. (1997). Acanthomorphic acritarchs from Meso-Neoproterozoic shales of the Ruyang Group, Shanxi, China. *Review of Palaeobotany and Palynology*, **98**, 15–25.

Yin, L. M., Zhu, M. Y., Knoll, A. H., Yuan, X. L., Zhang, J. M. and Hu, J. (2007). Doushantuo embryos preserved inside diapause egg cysts. *Nature*, **446**, 661–3.

Yochelson, E. L. and Fedonkin, M. A. (2000). A new tissue-grade organism 1.5 billion years old from Montana. *Proceedings of the Biological Society of Washington*, **113**, 843–7.

Yoon, H. S., Hackett, J. D., Ciniglia, C., Pinto, G. and Bhattacharya, D. (2004). A molecular time line for the origin of photosynthetic eukaryotes. *Molecular Biology and Evolution*, **21**, 809–18.

Yuan, X. L., Xiao, S. H. and Taylor, T. N. (2005). Lichen-like symbiosis 600 million years ago. *Science*, **300**, 1017–20.

Zang, W. and Walter, M. R. (1992). Late Proterozoic and Cambrian microfossils and biostratigraphy, Amadeus Basin, central Australia. *Memoir of the Association of Australasian Palaeontologists*, **12**, 1–132.

Zhang, Z. (1986). Clastic facies microfossils from the Chuanlinggou Formation (1800 Ma) near Jixian, North China. *Journal of Micropalaeontology*, **5**, 9–16.

27

Biomineralization mechanisms

Karim Benzerara and Jennyfer Miot

Introduction

Biomineralization is the process by which organisms form minerals; this is a widespread phenomenon and more than 60 minerals of biological origin have been identified up to now (e.g. Lowenstam, 1981; Baeuerlein, 2000; Weiner and Dove, 2003). Particular attention has been paid so far to eukaryotic biominerals, including the siliceous frustules of diatoms (e.g. Poulsen *et al.*, 2003; Sumper and Brunner, 2008), the calcitic tests of foraminifers (e.g. Erez, 2003) and the aragonitic skeleton of modern scleractinian corals (e.g. Cuif and Dauphin, 2005; Meibom *et al.*, 2008; Stolarski, 2003). However, prokaryotes can form minerals as well (Figure 27.1; Boquet *et al.*, 1973; Krumbein, 1979). For instance, stromatolites are carbonate deposits that are usually interpreted as the result of bacterial biomineralization. Interestingly too, some bacteria, called 'magnetotactic', can produce intracellular magnetite crystals seemingly aimed at directing their displacements using the local magnetic field (Blakemore, 1982). While eukaryotes obviously synthesize minerals exhibiting very specific structures (although ascertaining quantitatively why it is obvious might be an issue), the biogenicity of prokaryotic biominerals is more difficult to infer. The morphology, the structure (e.g. crystallinity, presence/absence of defects) and the chemistry (including the isotopic composition) of these prokaryote biominerals have, however, frequently been proposed as potential biosignatures (e.g. Konhauser, 1998; Little *et al.*, 2004). Such biosignatures have been used to infer the presence of traces of life not only in ancient terrestrial rocks but also in extraterrestrial rocks such as the Martian meteorite ALH 84001 (McKay *et al.*, 1996). In that case, the similarity of magnetite crystals produced by terrestrial bacteria with magnetite crystals found in the meteorite was used as one of the pieces of evidence for the presence of past traces of life on Mars. Moreover, the finding of zoned carbonate globules harbouring these magnetites in ALH 84001 triggered the study of analogous terrestrial carbonate globules, possibly formed by microorganisms (Kazmierczak and Kempe, 2003). Such interpretations rely on actualism and assume that a mineral in an ancient and/or extraterrestrial rock is a biomineral if it looks similar to a mineral produced by modern terrestrial microbes. In addition to the dubious validity of such reasoning, the major issue is our poor knowledge of the processes involved in the precipitation of minerals by organisms. Therefore, it is difficult to infer whether a given

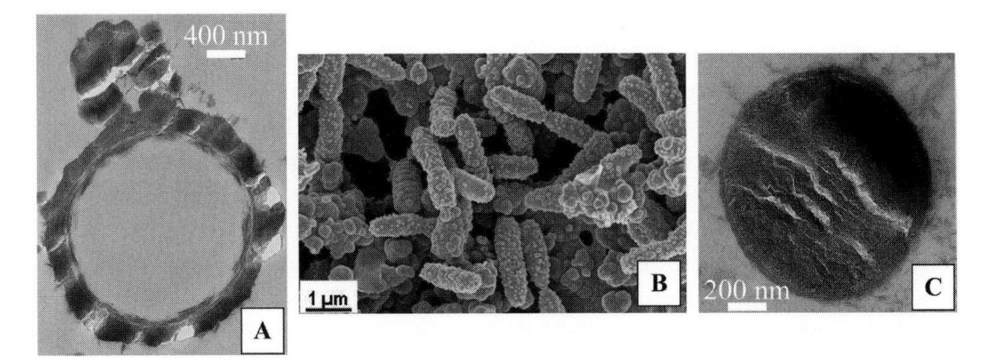

Figure 27.1. Examples of bacterial biomineralization. (A) Transmission electron microscopy (TEM) image of a section prepared by ultramicrotomy across an encrusted bacterial cell from the Carnoules acid-mine drainage (Benzerara *et al.*, 2008). The Fe- and As-containing minerals precipitated around the cell as two electron-dense layers: one more internal, thin and layered, the other external, thick and homogeneous. (B) Scanning electron microscopy (SEM) image of a culture of Fe-oxidizing bacteria (Miot *et al.*, 2009). The Fe minerals precipitated at the surface of the cells forming little globules with a patchy distribution on the cells. (C) TEM image of an ultramicrotomy section of a calcified (here, hydroxyapatite, a calcium phosphate: $[Ca_5(PO_4)_3(OH)]$) bacterial cell (Benzerara *et al.*, 2004). A ~40-nm thick outer rim, corresponding to the encrusted periplasm, can be distinguished from the precipitates filling the cell cytoplasm. These mineralized cells still contain significant amounts of proteins (e.g. Benzerara *et al.*, 2006b).

property of a biomineral is specific of the activity of the organism producing it, and hence corresponds to a biosignature, or results from purely abiotic processes. Understanding biomineralization mechanisms is thus crucial to better assess what trace of biological activity may be left in the geological record. Moreover, biomineralization occurring at the contact of microbial cells can lead to their entombment in minerals and thus fossilize them. The study of biomineralization can also help in inferring the ultrastructural and biochemical details that may be preserved and looked for in fossils.

It is impossible here to review all the different biomineralization systems that have been studied so far (e.g. Baeuerlein, 2000; Mann, 2001; Weiner and Dove, 2003). As our understanding of their biomineralization is generally more advanced, eukaryotic systems are particularly interesting. However, in this chapter, we will focus on a few prokaryotic systems, as they are probably more relevant to the search for ancient and/or extraterrestrial life. There are several steps leading to the formation of a mineral (e.g. De Yoreo and Vekilov, 2003). First, the chemistry of the solution in which it forms requires being favourable to precipitation. Precipitation can be written as a chemical reaction that can be exemplified by the following simplified equation:

$$A + B = \text{Mineral C (Ks)},$$

where A and B are two dissolved species.

The activity product of the reactants is defined as:

$$Q = a(A) * a(B)$$

If the activity product of the reactant exceeds the equilibrium constant (Ks), the solution is said to be supersaturated with mineral C. By modifying the solution chemistry in their vicinity, organisms can significantly impact on that first step (e.g. Posth *et al.*, 2008). Then, nucleation occurs by the formation of very small mineral seeds. However, such very small nuclei are thermodynamically very unstable and overcome dissolution once they are large enough. Organisms can also impact on that step, in particular, by providing large surfaces on which nucleation is favoured by a thermodynamic stabilization of the nuclei. Finally, growth of the mineral phases takes place layer by layer (although sometimes aggregation of subparticles is also inferred). Organisms can again affect that step, favouring crystal growth along preferential directions, which leads to the formation of minerals with particular shapes. This is achieved by modification of the chemistry of the solution, for example, by the production of organic molecules, some of them having an inhibitory effect on the growth of one mineral face over another (e.g. Addadi *et al.*, 1990; Orme *et al.*, 2001). Finally, it should be noted that the chirality of organic molecules can impact on the growth of the resulting biominerals. This has been shown, in particular, with amino acids (e.g. Wolf *et al.*, 2007).

In the following, we will focus on selected prokaryote biomineralization systems that illustrate those different levels of control and discuss how they provide a framework to assess the existence of potential biosignatures in minerals. Three different types of microbially mediated biomineralization processes have been distinguished in the literature (see Dupraz *et al.*, 2009, for an exhaustive review): (1) biologically controlled biomineralization refers to cases in which a specific cellular activity directs the nucleation, growth, morphology and final location of a mineral; (2) biologically induced biomineralization results from the indirect modification of the chemistry of the environment by biological activity; (3) biologically influenced biomineralization is defined as passive mineralization of organic matter, whose properties influence crystal morphology and composition. The term organomineralization encompasses biologically influenced and biologically induced biomineralization (Dupraz *et al.*, 2009).

Triggering of biomineralization by chemical transformations of a solution (biologically induced biomineralization)

The metabolic activity of microorganisms can significantly impact on the chemistry of their environment and possibly lead to the achievement of supersaturation with respect to several mineral phases. This chemical shift has been proposed for almost all biomineralization systems. However, the significance of the impact of microbes on the chemistry of their environment is not always obvious. On one hand, some minerals can unequivocally be attributed to bacterial activity and not to abiotic reactions in environments with a specific chemistry. For instance, in an anoxic environment, the rapid precipitation of Fe(III)-mineral phases can reasonably be attributed to bacterially catalyzed Fe-oxidation, as abiotic processes such as photo-oxidation of Fe by UV are slow (e.g. Konhauser *et al.*, 2007). On the other hand, the precipitation of calcium carbonates in a highly supersaturated environment such as an alkaline lake might not be much influenced by the metabolic activity of cyanobacteria, which under other chemical conditions do trigger this precipitation (e.g. Arp *et al.*, 2001). Systems in which only microbes impact on this chemical shift have been traditionally referred to as 'biologically induced mineralization' systems (e.g. Lowenstam, 1981; Weiner and Dove, 2003). However, sometimes it may be difficult to ascertain that there is no other process involved (such as heterogeneous nucleation on specific organic molecules, for example, and thus biologically *influenced* mineralization). Prokaryotes can catalyse a huge variety of reactions involving many different chemical elements (e.g. Nealson, 1997; Newman and Banfield, 2002), which for some of them can induce the precipitation of mineral phases. Here, we discuss further two particularly interesting biomineralizing systems and focus on the connection existing between a better knowledge of the biomineralization mechanisms and the definition of biosignatures: the formation of dolomite at low temperature by sulphate-reducing bacteria and the formation of stromatolites.

Dolomite, a magnesium–calcium carbonate ($[Ca,Mg]CO_3$) first described by de Dolomieu (1791), is the main component of most ancient carbonate sediments, some of which have been undoubtedly formed under marine conditions. However, dolomitization has long been a geologic enigma (e.g. Machel and Mountjoy, 1986; Burns *et al.*, 2000). Sulphates, which are abundant in seawater, are indeed inhibitors of dolomite formation and laboratory experiments failed for years to synthesize dolomite in seawater at low temperature. Some people have thus proposed that dolomite found in the geological record was mostly formed secondarily through diagenetic processes, i.e. secondary transformations of the sediment by fluids of different chemistry and at higher temperature. However, Vasconcelos *et al.* (1995) showed that primary dolomite could be formed in seawater in the presence of sulphate-reducing bacteria that overcome the kinetic barrier of dolomite nucleation by reducing sulphates into sulphides (see also Wright and Oren, 2005, for a review). Since then, this process has been reproduced experimentally (Warthmann *et al.*, 2000) and the possible involvement of cell surfaces and of extracellular polymers in the nucleation and growth of dolomite crystals has been investigated (e.g. Van Lith *et al.*, 2003; Bontognali *et al.*, 2007). In this system, the aqueous solution is initially oversaturated with dolomite, i.e. precipitation is thermodynamically favoured, but kinetically inhibited, and only microorganisms can overcome this limitation. This example illustrates that biosignatures, if they exist, should be used in a subtle and cautious way requiring a detailed knowledge of the chemistry of the solution in which they formed. Dolomite per se is not a biosignature, as it can form abiotically under many different conditions. However, if it can be independently established that dolomite is primary (and not diagenetic), and formed at low temperature in a sulphate-rich seawater-like environment, then the role of microorganisms might be invoked. In addition, sulphate reduction is coupled to organic carbon mineralization, which provides a source of carbonate ions able to contribute to the formation of dolomite. Hence, several chemical signatures such as C-, O- and S-isotopic compositions have been looked for in such precipitates (e.g. Meister *et al.*, 2007; Wacey *et al.*, 2007). It should be noted that this approach potentially shows the processing of these elements through sulphate reduction instead of directly demonstrating the involvement of a metabolic activity in the nucleation and growth of dolomite.

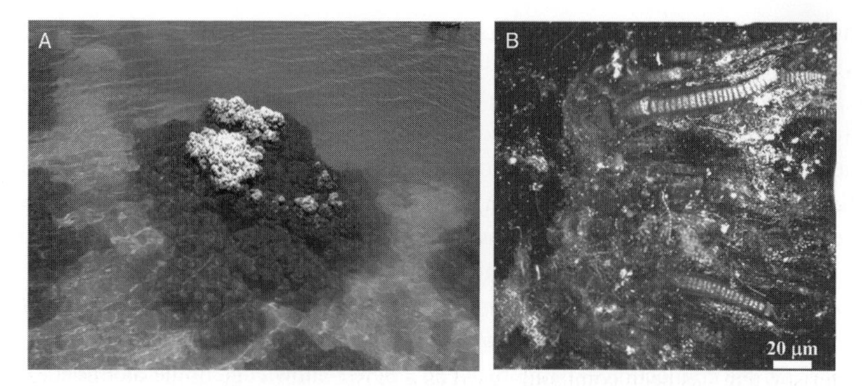

Figure 27.2. Stromatolites at different scales. (A) Lacustrine stromatolites from Alchichica (Mexico). These metre-sized carbonate precipitates formed in an alkaline lake and are inhabited by a highly diverse microbial community (courtesy of P. López-García, D. Moreira, B. Kremer and J. Kazmierczak). (B) Confocal laser scanning microscopy image of a biofilm collected at the surface of Alchichica stromatolites. (Microscope from the GAP team at IPGP, France, courtesy of M. Ibrahimi and E. Gerard.) Extracellular polymeric substances (EPS) with or without calcium carbonates are important components of these biofilms and can be observed in the background. Diverse microorganisms are also observed.

Stromatolites are notorious biomineralization systems, which are often composed of calcium carbonates (although a secondary silicification has been frequently observed), and are usually considered as one of the oldest traces of life in the geological record (Altermann *et al.*, 2006; Allwood *et al.*, 2006). They consist of layered sedimentary structures initiated from a limited surface and forming a variety of morphologies (e.g. Grotzinger and Knoll, 1999; Dupraz *et al.*, 2006). The name stromatolite was first proposed by Kalkowsky (1908), and since the very first studies the involvement of life in their construction has been proposed. Modern stromatolites similar in morphology to Archaean stromatolites are indeed populated by cyanobacteria (Figure 27.2). The traditional biogenetic model very schematically considers that cyanobacteria trigger the precipitation of carbonates by harvesting CO_2 during photosynthesis. When photosynthesis stops, *in-situ* precipitation is overcome by detritic sedimentation; the alternation of both processes results in the formation of laminations. This general framework indirectly supported the interpretation of carbonaceous filaments found in the 3.5-Ga silicified stromatolites from Warrawoona (Pilbara, Western Australia) as microfossils of cyanobacteria (Schopf and Packer, 1987), which would have been a major actor in the formation of these rocks. The processes leading to the formation of modern stromatolites have been, however, significantly reassessed over the last few years. First, other metabolisms that can trigger calcium-carbonate precipitation have been shown to operate in modern stromatolites, in particular, sulphate reduction (e.g. Baumgartner *et al.*, 2006; Dupraz and Visscher, 2005). Other studies have also stressed the inhibitory effect of organic polymers on the precipitation of carbonates and proposed that heterotrophs, including sulphate-reducing, but also oxygen-respiring bacteria, could also favour lithification by degrading organic molecules (e.g. López-García *et al.*, 2005; Braissant *et al.*, 2007, 2009).

Recently, it has even been proposed that anoxygenic photosynthetic bacteria could be involved in the formation of stromatolites during Archaean times when the atmosphere was not yet oxygenated (Bosak *et al.*, 2007). As a result, the relative importance of these different metabolisms in the formation of modern and ancient stromatolites is still not understood (e.g. Aloisi, 2008). Here, we have only discussed the chemical shifts induced by these metabolisms that may trigger calcium-carbonate precipitation. If some chemical signatures of these different metabolisms could be stored in minerals, it would then be possible to better assess their relative contribution. However, this purely chemical view cannot by itself account for the achievement of specific morphologies for stromatolites or for the formation of laminations, although the stromatolitic structure at the macroscale is the morphologic characteristic that has been commonly used as a biosignature. This suggests that regarding our understanding of the biomineralization mechanisms involved in the formation of stromatolites, a biogenic interpretation of these structures remains very empirical.

Nucleation and growth of minerals in association with extracellular organic polymers (biologically influenced mineralization)

Microorganisms also produce a great variety of organic molecules that are exported outside the cells. Some of them can have enzymatic activity, such as the carbonic anhydrases produced by cyanobacteria that catalyse the rapid conversion of carbon dioxide to bicarbonate (e.g. Kupriyanova *et al.*, 2004), thus promoting the precipitation of carbonate minerals; others can facilitate the mobility of toxic or necessary elements between cells and the extracellular medium (e.g. von Canstein *et al.*, 2008). Finally, some form a mostly polysaccharidic gel-like matrix around the cells which compose the framework of what has been called EPS (extracellular polymeric substances; Flemming *et al.*, 2007). EPS can remain attached to the cell surfaces or be excreted. In addition to polysaccharides, they contain a variety of organic molecules including proteins and nucleic acids. They can play multiple roles that have been intensely debated, including the buffering of the chemical conditions prevailing in the vicinity of the cells, the formation of biofilms and the attachment of cells to substrates (e.g. Decho *et al.*, 2005).

Owing to their chemical composition, EPS can obviously affect biomineralization by promoting or modifying the nucleation and growth of minerals. Indeed, these molecules can interact with cations and complex them on negatively charged functional groups. They can also attach on crystal faces and inhibit further growth. All these processes have been abundantly documented in eukaryotic systems and several studies have shown the involvement in biomineralization of specific molecules, such as proteins rich in aspartic and glutamic acids or in acidic or sulphated polysaccharides (e.g. Marie *et al.*, 2008). In contrast, the existence of biochemical molecules specifically involved in extracellular biomineralization is still not clear for prokaryotes, although many studies have suggested that these extracellular polymers affect the nucleation and growth of minerals.

Some studies have focused on the impact of bacterial polysaccharides on the precipitation, nucleation and growth of carbonates. They are indeed pervasive in modern microbial precipitates (e.g. Kawaguchi and Decho, 2000; Kazmierczak and Kempe, 2004; Benzerara *et al.*, 2006a; Braissant *et al.*, 2007). On one hand, they can have an inhibitory effect on nucleation and growth of calcium carbonates by lowering, to some extent, the supersaturation of the solution (e.g. Braissant *et al.*, 2007, 2009) and by adsorbing onto some crystal faces, resulting in the formation of particular crystal shapes (e.g. De Yoreo *et al.*, 2007). On the other hand, it has been proposed that after some structural modifications they may act as a template for the formation of calcium carbonate (Trichet and Défarge, 1995). By promoting the formation of numerous nucleation sites (Grassmann and Lobmann, 2004) and inhibiting crystal growth by poisoning their surfaces (Manoli and Dalas, 2002), polysaccharide matrices may be responsible for the frequently observed clustering of submicrometre-sized calcium carbonate crystals in stromatolites, the so-called micritic texture (e.g. Riding, 2000; Dupraz *et al.*, 2004; Benzerara *et al.*, 2006a). Similar properties have been reported in other biomineralizing systems. For example, EPS have been shown to affect the growth and orientation of iron oxide nano-crystals, precipitating in the presence of iron-oxidizing bacteria (Chan *et al.*, 2004; Figure 27.3). The small size of the crystals

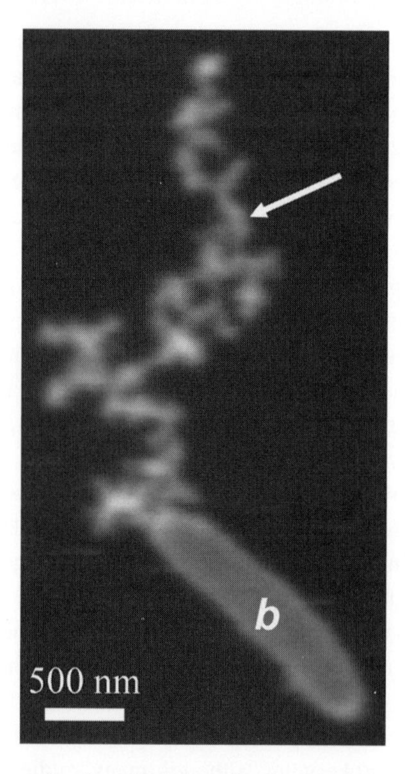

Figure 27.3. Polysaccharide-templated iron biomineralization in cultures of the anaerobic, neutrophilic iron-oxidizing bacterial strain BoFeN (Miot *et al.*, 2009). The bacterial cell is labelled 'b'. Arrow points at the polysaccharide filament. STXM analyses showed that this filament is composed both of Fe minerals and polysaccharides.

is definitely not a biosignature as it can be produced by abiotic processes (e.g. Lin *et al.*, 2002). But their presence in biological systems can be explained by the impact of organics on their nucleation and growth and the resulting mineral–organics associations might be fossilized in very ancient rocks. As an example, we recently reported the presence of aragonite ($CaCO_3$) crystals associated with organic globules exhibiting a baterium-like morphology in 2.7-Ga old stromatolites from Pilbara (Lepot *et al.*, 2008). We noted the similarity of these objects to those that can be observed in modern stromatolites, and thus proposed that they might represent fossils of microorganisms and of aragonitic micrite. Similarly, it has been shown that very specific crystal sizes and shapes are obtained when extracellular precipitation of magnetite crystals occurs in association with polysaccharides in cultures of the iron-reducing bacterium *Geobacter metallireducens* (Vali *et al.*, 2004).

Interestingly, the function of prokaryote extracellular polymers is not always known and the possibility that some of them may have a specific role in biomineralization similarly to what is known in eukaryotes has been proposed for several systems (e.g. Phoenix and Konhauser, 2008; Chan *et al.*, 2004). For instance, Moreau *et al.* (2007) have found proteins associated with zinc sulphide nanoparticles formed extracellularly by sulphate-reducing bacteria. They concluded that these proteins may be specifically produced in order to trigger that biomineralization, and that they consequently play a detoxification role or prevent entombment of the cells. Recently, Barabesi *et al.* (2007) have shown that calcification might be related, in a poorly understood way, to genetic-controlled processes, although this type of biomineralization is usually seen as purely incidental. van Dijk *et al.* (1998), moreover, isolated specific bacterial proteolipids involved in calcification. Finally, outer membrane vesicles (OMVs) produced by Gram-negative bacteria have been inferred as potential important biomineralizing structures (e.g. Aloisi *et al.*, 2006; Benzerara *et al.*, 2008). Much has still to be discovered regarding the mechanisms involved in the nucleation and growth of minerals by extracellular polymers, and whether the latter have an adaptive role in prokaryotes. This is why the traditional view, suggesting that 'biologically induced mineralization' and 'biologically influenced mineralization' are just secondary, not controlled processes, in contrast to 'biologically controlled mineralization', might be misleading, as long as most of the molecular processes involved in these reactions remain unknown. An interesting outcome of biomineralization on extracellular polymers is that some organic molecules can be trapped and protected by a layer of minerals, and might thus be more easily preserved in the geological record. Those organic molecules might be an interesting source of information about the processes of formation of their enclosing minerals.

Biomineralization at the surface and within microbial cell walls (biologically influenced or controlled mineralization)

Although molecular interactions between organic polymers at the surface of bacteria and minerals are the same at the atomic scale as on extracellular polymers, precipitation at the surface of microorganisms has received particular attention (e.g. Beveridge and Murray, 1980). Bacterial surfaces indeed contain a relatively high density of functional groups that are negatively charged over a wide range of pH with very few variations among different species (e.g. Fein *et al.*, 1997; Phoenix *et al.*, 2002; Borrok *et al.*, 2005; Pokrovsky *et al.*, 2008) and that are thought to adsorb metals as a first step prior to mineral nucleation (e.g. Fortin *et al.*, 1998). Moreover, biomineralization at the cell surface results in the formation of microfossils with specific morphologies that can be easily recognized and potentially preserved in the geological record (Ferris *et al.*, 1988).

A first set of studies is particularly interesting to better assess the current challenges in our understanding of biomineralization at cell surfaces. The formation of iron oxides has been studied by oxidizing Fe(II) in solution under oxic conditions and in the presence or absence of non-metabolizing cultured bacteria (e.g. Rancourt *et al.*, 2005; Fakih *et al.*, 2008). It was shown that the structure of the mineral products can change with the addition of bacterial surfaces, and the formation of mineral particles can even be inhibited when high concentrations of bacteria are added (Fakih *et al.*, 2008). This might be partly due to a delay of the oxidation of Fe(II) adsorbed on bacterial surfaces. Moreover, it has been stressed that under those conditions bacterial surfaces are not good nucleation surfaces. While it brings an experimental validation that structural and morphological differences can exist between iron oxides precipitated on the cells and away from the cells, which have been observed in other biomineralizing systems (e.g. Ferris and Magalhaes, 2008), it seems to contradict the common belief that bacterial cells are good nucleation sites.

In some cases, biomineralization proceeds within the cell wall, in particular, within the periplasm of Gram-negative bacteria. This has been shown for several systems, including calcium phosphate-precipitating bacteria (e.g. Benzerara *et al.*, 2004; Goulhen *et al.*, 2006) and iron-oxidizing bacteria (e.g. Miot *et al.*, 2009). A first result of that biomineralization confined to a ~40-nm thick space surrounding the cell is the fossilization of the bacterial morphology (Figure 27.4).

However, in any case (biomineralization at the cell surface or within the cell wall), using the morphology of the resulting microfossils as a biosignature remains ambiguous, as

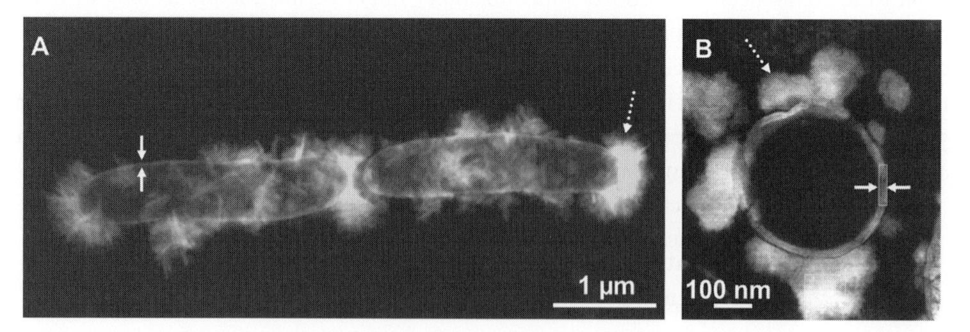

Figure 27.4. Periplasmic biomineralization of iron minerals in anaerobic, neutrophilic iron-oxidizing bacteria, strain BoFeN1. Scanning transmission electron microscopy images of whole cells (**A**) and of a thin section of a single resin-embedded cell (**B**). Solid arrows point at the iron-encrusted periplasm and dashed arrows point to extracellular iron precipitates at the surface of the cells.

bacterium-like morphologies can also be produced through abiotic processes. In contrast, the potential fossilization of organic molecules themselves within these microfossils could provide stronger and more reliable criteria of biogenicity. It has been shown, for instance, that some proteins could be trapped within the fossilized periplasm of iron-oxidizing bacteria (Miot *et al.*, 2009). Finally, looking at the texture of biominerals can provide additional potential biosignatures. For example, in the case of calcium phosphate biomineralization, a particular crystallographic texture has been observed. Calcium phosphate precipitates in the periplasm as nanometre-sized hydroxyapatite crystals ($Ca_5[PO_4]_3[OH]$) along a preferential crystallographic axis parallel to the cell surface. This indicates that some structure within the periplasm has acted as a template for the growth of hydroxyapatite. The significance of this biomineralization pattern, as well as the mechanisms involved in the precipitation confined within the periplasm are still unknown. It should be noted however, that altogether, these very specific features may provide interesting biosignatures if they can be preserved in the geological record.

Intracellular biomineralization (biologically controlled mineralization)

Only a few prokaryote biomineralizing systems are known to produce intracellular precipitates including intracellular calcite ($CaCO_3$) formed by sulphur-oxidizing bacteria (Gray *et al.*, 1997), intracellular iron oxides produced by iron-reducing bacteria (Glasauer *et al.*, 2002) and magnetite crystals formed by magnetotactic bacteria (e.g. Blakemore, 1975). The latter have received huge attention as one of the few established examples of biologically controlled mineralization in prokaryotes and also because magnetite crystals with similar crystallographic features have been found in a Martian meteorite and interpreted as fossils of a Martian ancient life (McKay *et al.*, 1996). In addition, in order to demonstrate the significant role played by magnetotactic bacteria in the magnetization of sediments over the geological record, Vasiliev *et al.* (2008) stressed the need to understand the processes of biomineralization in these microorganisms for a better assessment of their potential signatures in rocks.

In these bacteria, chains of magnetite (Fe_3O_4)) (or in some cases of greigite, $[Fe_3S_4]$) form intracellularly, each one being surrounded by a lipid-bilayer membrane (magnetosome) (Figure 27.5). Chains are fixed within the cell so that the bacterium can passively align in and navigate along geomagnetic fields (Blakemore, 1975). Modern magnetotactic bacteria are found in the oxic–anoxic transition zone of sediments. Sensing the magnetic field thus provides a selective advantage to these bacteria for an efficient search for low-oxygen environments (Komeili, 2007).

Much of what we know of the biochemical processes involved during the formation of magnetite in magnetotactic bacteria is based on the study of cultured strains, which represent only a small fraction of the whole diversity of magnetotactic bacteria (see for example Flies *et al.*, 2005; Isambert *et al.*, 2007). The reader will find a detailed and recent review in Komeili (2007). Although not fully understood yet, the model of magnetite biomineralization in magnetotactic bacteria is by far the most detailed one among

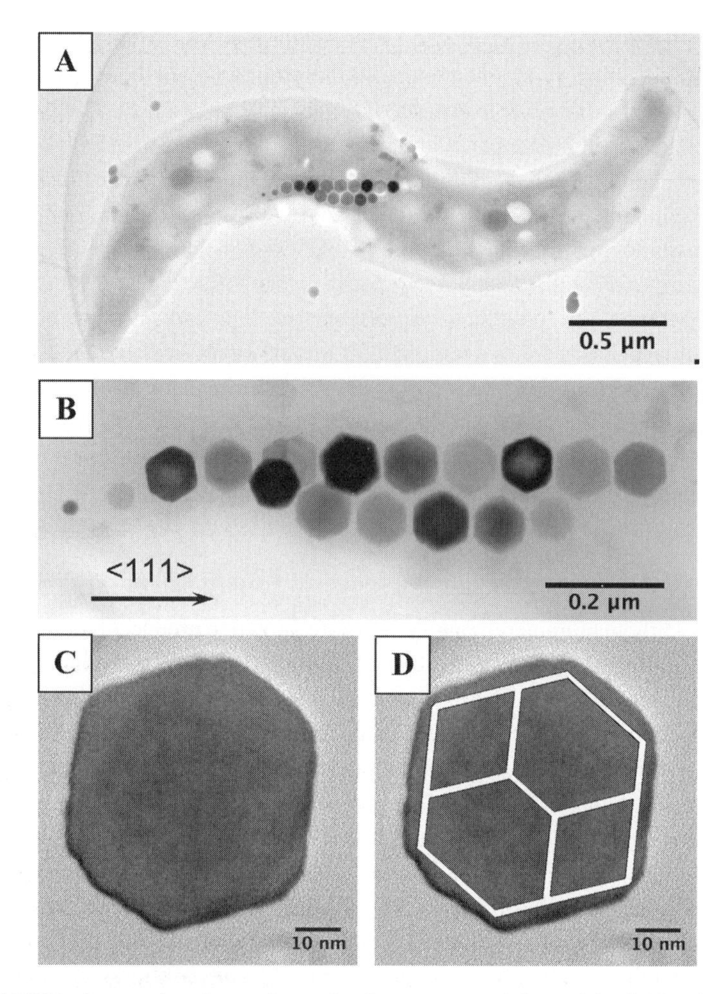

Figure 27.5. TEM micrographs at increasing scale of a magnetotactic bacterial cell (close in morphology to *Magnetospirillum* sp.) collected from the Seine River in France (courtesy of N. Menguy). The magnetite crystals, which have a cubo-octahedral morphology, are crystallographically aligned within the cell along the [111] axis, which is also the easy axis of magnetization. See Isambert *et al.*, 2007 for further information. The whole bacterial cell can be seen on (A). (B) is a close-up of the aligned magnetosomes. (C) is a high-resolution image of a single magnetite crystal and (D) is the cubo-octahedral morphology interpretation superimposed on the same image.

bacterial biomineralizing systems, owing to an interdisciplinary approach including crystallography, biochemistry and genetics. Bacteria import Fe(III) into the periplasm using iron chelators and then reduce it. Dissolved Fe(II) is then transferred into the preformed magnetosomes, recently suggested to result, at least in some cases, from evaginations of the periplasm (Komeili *et al.*, 2006). Fe(II) is then presumably partly re-oxidized leading to the precipitation of magnetite. This step might involve precursor phases such as ferrihydrite (FeOOH, e.g. Frankel and Blakemore, 1984) but there seems to be no consensus on that point (e.g. Staniland *et al.*, 2007; Faivre *et al.*, 2008). The stability of the magnetosome chain is ensured by a bacterial filament cytoskeleton (Komeili *et al.*, 2006; Scheffel *et al.*, 2006). In addition to a control of the chemical conditions (e.g. pH, Eh, activity of trace elements) necessary to grow chemically pure magnetite, there appears to be involvement of molecules, possibly proteins, controlling the growth, hence the final morphology of the magnetite crystals (e.g. Arakaki *et al.*, 2003; Amemiya *et al.*, 2007; Scheffel *et al.*, 2008). Those recent works offer a promising perspective for improving our understanding of the mechanisms leading to the particular morphological features observed in biogenic magnetite. Magnetotactic bacteria thus use very specific, genetically controlled biochemical processes to form magnetite. This supports the general idea that such magnetite has very distinctive features, although the exact relationship between these signatures and the biomineralization processes is still poorly understood (e.g. Devouard *et al.*, 1998; Clemett *et al.*, 2002; Thomas-Keprta *et al.*, 2002): (1) crystals in magnetotactic bacteria are of high chemical purity with concentrations of trace elements such as Ti, Al and Cr lower than in most abiotic experiments (Thomas-Keprta *et al.*, 2000; Bazylinski and Frankel, 2003); (2) they have few crystallographic defects; (3) they display a narrow size range in the single domain (around 35 to 120 nm approximately) and a narrow distribution of shape factor within a single species. Moreover, the size distribution of magnetite produced by some strains is an asymmetric log-normal function skewed towards smaller sizes, while the size distribution in abiotic magnetite generally follows an opposite asymmetry (e.g. Devouard *et al.*, 1998; Faivre *et al.*, 2005); (4) magnetites produced by magnetotactic bacteria can have various shapes depending on the species, some of which have been so far observed only in magnetotactic bacteria; and (5) the magnetites in magnetotactic bacteria are aligned in chains. The possibility of forming chains by abiotic processes has, however, been discussed by Kopp *et al.* (2006). All of those features can be seen as an optimization of the system to sense the magnetic field lines.

However, it should be noticed that these potential mineralogical biosignatures are based on a few strains grown under restricted chemical conditions. Some variations have been documented within a single species cultured under different conditions or between different species (e.g. Taylor and Barry, 2004; Arato *et al.*, 2005; Isambert *et al.*, 2007; Faivre *et al.*, 2008). As a result, magnetites formed by magnetotactic bacteria do not always display the biosignatures presented above. Variations in magnetite morphologies upon varying environmental conditions (e.g. Faivre *et al.*, 2008) and/or species (e.g. Isambert *et al.*, 2007) are still poorly understood, but will be crucial to assess in order to get a more comprehensive view of the distinctive biogenic features that could be preserved and looked for in the fossil record.

Conclusion

Bacteria can affect the formation of minerals by modifying the chemistry of their environment, by producing or degrading biomolecules that may interfere with the nucleation and growth of mineral phases and/or by creating very restricted and chemically controlled volumes favourable to controlled biomineralization within their cell wall or within their cytoplasm. All these processes are believed to affect or even control the structure, chemistry and morphology of the resulting mineral products. However, the specificity of these features compared to those produced through abiotic processes is yet poorly constrained. Possibly, the most direct evidence of biogenicity of such biominerals may be given by the trapping and preservation of organic molecules within or around them.

Biomineralization in general, and in particular in prokaryote systems is still a field in its infancy. The advent of new methodological tools capable of probing the mineral–microbe interface down to the nanoscale is promising for a better understanding of the molecular processes involved in these interactions. The traditional discrimination that has been made between biologically induced mineralization and biologically controlled mineralization may be misleading as long as we have not identified all the biochemical actors playing a role in prokaryote systems. Combinations of mineralogical approaches with biochemistry and genetics would thus also be a crucial way to achieve a broader understanding of those processes at the mechanistic level, as exemplified by the recent advances in the modelling of magnetite biomineralization by magnetotactic bacteria. The potential impact of diverse organic polymers in most of these systems has been clearly established, but the specificity of their involvement in these biomineralization processes is largely unknown. This, in turn, will open new perspectives on the possibility that biomineralization might have been an evolutionary strategy selected over most of geological time and not just since eukaryotes appeared.

References

Addadi, L., Berman A., Moradial-Oldak, J. and Weiner, S. (1990). Tuning of crystal nucleation and growth by protein: molecular interaction at solid-liquid interfaces in biomineralization. *Croatica Chemica Acta*, **63**, 539–44.

Allwood, A. C., Walter, M. R., Kamber, B. S., Marshall, C. P. and Burch, I. W. (2006). Stromatolite reef from the Early Archaean Era of Australia. *Nature*, **441**, 714–18.

Aloisi, G., Gloter, A., Kröger, M., Wallmann, K., Guyot, F. and Zuddas, P. (2006). Nucleation of calcium carbonate on bacterial nanoglobules. *Geology*, **34**, 1017–20.

Aloisi, G. (2008). The calcium carbonate saturation state in cyanobacterial mats throughout Earth's history. *Geochimica et Cosmochimica Acta*, **72**, 6037–60.

Altermann, W., Kazmierczak, J., Oren, A. and Wright, D. T. (2006). Cyanobacterial calcification and its rock-building potential during 3.5 billion years of Earth history. *Geobiology*, **4**, 147–66.

Amemiya, Y., Arakaki, A., Staniland, S.S., Tanaka, T. and Matsunaga, T. (2007). Controlled formation of magnetite crystal by partial oxidation of ferrous hydroxide in the presence of recombinant magnetotactic bacterial protein Mms6. *Biomaterials*, **28**, 5381–9.

Arakaki, A., Webb, J. and Matsunaga, T. (2003). A novel protein tightly bound to bacterial magnetic particles in *Magnetospirillum magneticum* strain AMB-1. *Journal of Biological Chemistry*, **278**, 8745–50.

Arato, B., Szanyi, Z., Flies, C., Schuler, D., Frankel, R. B., Buseck, P. R. and Posfai, M. (2005). Crystal-size and shape distributions of magnetite from uncultured magnetotactic bacteria as a potential biomarker. *American Mineralogist*, **90**, 1233–40.

Arp, G., Reimer, A. and Reitner, J. (2001). Photosynthesis-induced biofilm calcification and calcium concentrations in phanerozoic oceans. *Science*, **292**, 1701–4.

Barabesi, C., Galizzi, A., Mastromei, G., Rossi, M., Tamburini, E. and Perito B. (2007). *Bacillus subtilis* gene cluster involved in calcium carbonate biomineralization. *Journal of Bacteriology*, **189**, 228–35.

Baeuerlein, E. (2000). *Biomineralization: From Biology to Biotechnology and Medical Application*. Weinheim, Germany: Wiley-VCH.

Baumgartner, L. K., Reid, R. P., Dupraz, C., Decho, A. W., Buckley, D. H., Spear, J. R., Przekop, K. M. and Visscher, P. T. (2006). Sulfate reducing bacteria in microbial mats: changing paradigms, new discoveries. *Sedimentary Geology*, **185**, 131–45.

Bazylinski, D. A. and Frankel, R. B. (2003). Biologically controlled mineralization in prokaryotes. *Biomineralization*, **54**(1), 217–47.

Benzerara, K., Menguy, N., Guyot, F., Skouri, F., de Luca, G., Barakat, M. and Heulin, T. (2004). Biologically controlled precipitation of calcium phosphate by *Ramlibacter tataouinensis*. *Earth and Planetary Science Letters*, **228**, 439–49.

Benzerara, K., Menguy, N., López-García, P., Yoon, T. H., Kazmierczak, J., Tyliszczak, T., Guyot, F. and Brown, G. E. (2006a). Nanoscale detection of organic signatures in carbonate microbialites. *Proceedings of the National Academy of Sciences of the United States of America*, **103**, 9440–5.

Benzerara, K., Miller, V. M., Barell, G., Kumar, V., Miot, J., Brown, Jr., G. E. and Lieske, J. C. (2006b). Search for microbial signatures within human and microbial calcifications using soft X-ray spectromicroscopy. *Journal of Investigative Medicine*, **54**, 367–79.

Benzerara, K., Morin, G., Yoon, T. H., Miot, J., Tyliszczak, T., Casiot, C., Bruneel, O., Farges, F. and Brown G. E. (2008). Nanoscale study of As biomineralization in an acid mine drainage system. *Geochimica et Cosmochimica Acta*, **72**, 3949–63.

Beveridge, T. J. and Murray, R. G. E. (1980). Sites of metal-deposition in the cell-wall of *Bacillus subtilis*. *Journal of Bacteriology*, **141**, 876–87.

Blakemore, R. P. (1975). Magnetotactic bacteria. *Science*, **190**, 377–9.

Blakemore, R. P. (1982). Magnetotactic bacteria. *Annual Review of Microbiology*, **36**, 217–38.

Bontognali, T., Vasconcelos, C., Warthmann, R. and McKenzie, J. A. (2007). Dolomite nucleation on extracellular polymeric substances. *Geochimica et Cosmochimica Acta*, **71**, A108.

Boquet, E., Boronat, A. and Ramos-Cormenzana, A. (1973). Production of calcite (calcium carbonate) crystals by soil bacteria is a general phenomenon. *Nature*, **246**, 527–8.

Borrok, D., Turner, B. F. and Fein, A. B. (2005). A universal surface complexation framework for modeling proton binding onto bacterial surfaces in geologic settings. *American Journal of Science*, **305**, 826–53.

Bosak, T., Greene, S. E. and Newman, D. K. (2007). A likely role for anoxygenic photosynthetic microbes in the formation of ancient stromatolites. *Geobiology*, **5**, 119–26.

Braissant, O., Decho, A. W., Dupraz, C., Glunk, C., Przekop, K. M. and Visscher, P. T. (2007). Exopolymeric substances of sulfate-reducing bacteria: interactions

with calcium at alkaline pH and implication for formation of carbonate minerals. *Geobiology*, **5**, 401–11.
Braissant O., Decho A. W., Przekop K. M., Gallagher K. L., Glunk G., Dupraz C. and Visscher P. T. (2009). Characteristics and turnover of exopolymeric substances in a hypersaline microbial mat. *FEMS Microbiology Ecology*, **67**, 293–307.
Burns, S. J., McKenzie, J. A. and Vasconcelos, C. (2000). Dolomite formation and biogeochemical cycles in the Phanerozoic. *Sedimentology*, **47**, 49–61.
Chan, C. S., De Stasio, G., Welch, S. A., Girasole, M., Frazer, B. H., Nesterova, M. V., Fakra, S. and Banfield, J. F. (2004). Microbial polysaccharides template assembly of nanocrystal fibers. *Science*, **303**, 1656–8.
Clemett, S. J., Thomas-Keprta, K. L., Shimmin, J., Morphew, M., McIntosh, J. R., Bazylinski, D. A., Kirschvink, J. L., Wentworth, S. J., McKay, D. S., Vali, H., Gibson, E. K. and Romanek, C. S. (2002). Crystal morphology of MV-1 magnetite. *American Mineralogist*, **87**, 1727–30.
Cuif, J.P. and Dauphin, Y. (2005). The environment recording unit in coral skeletons – a synthesis of structural and chemical evidences for a biochemically driven, stepping-growth process in fibres. *Biogeosciences*, **2**, 61–73.
de Dolomieu, D. (1791). Sur un genre de pierres calcaires très peu effervescentes avec les acides et phosphorescentes par la collision. *Journal de Physique*, **39**, 3–10.
De Yoreo, J. J. and Vekilov, P. G. (2003). Principles of crystal nucleation and growth. *Biomineralization*, **54**(1), 57–93.
De Yoreo, J. J., Wierzbicki, A. and Dove, P. M. (2007). New insights into mechanisms of biomolecular control on growth of inorganic crystals. *CrystEngComm*, **9**, 1144–52.
Decho, A. W., Visscher, P. T. and Reid, R. P. (2005). Production and cycling of natural microbial exopolymers (EPS) within a marine stromatolite. *Palaeogeography, Palaeoclimatology, Palaeoecology*, **219**, 71–86.
Devouard, B., Posfai, M., Hua, X., Bazylinski, D. A., Frankel, R. B. and Buseck, P. R. (1998). Magnetite from magnetotactic bacteria: size distributions and twinning. *American Mineralogist*, **83**, 1387–98.
Dupraz, C., Visscher, P. T., Baumgartner, L. K. and Reid, R. P. (2004). Microbe–mineral interactions: early carbonate precipitation in a hypersaline lake (Eleuthera Island, Bahamas). *Sedimentology*, **51**, 745–65.
Dupraz, C. and Visscher, P. T. (2005). Microbial lithification in marine stromatolites and hypersaline mats. *Trends in Microbiology*, **13**, 429–38.
Dupraz, C., Pattisina, R. and Verrecchia, E. P. (2006). Translation of energy into morphology: simulation of stromatolite morphospace using a stochastic model. *Sedimentary Geology*, **185**, 185–203.
Dupraz, C., Reid, R. P., Braissant, O., Decho, A. W., Norman, S. R. and Visscher, P. T. (2009). Processes of carbonate precipitation in modern microbial mats. *Earth-Science Reviews*, **96**, 141–62.
Erez, J. (2003). The source of ions for biomineralization in foraminifera and their implications for paleoceanographic proxies. *Biomineralization*, **54**, 115–49.
Faivre, D., Menguy, N., Guyot, F., Lopez, O. and Zuddas, P. (2005). Morphology of nanomagnetite crystals: implications for formation conditions. *American Mineralogist*, **90**, 1793–800.
Faivre, D., Menguy, N., Posfai, M. and Schuler, D. (2008). Environmental parameters affect the physical properties of fast-growing magnetosomes. *American Mineralogist*, **93**, 463–9.

Fakih, M., Chatellier, X., Davranche, M. and Dia, A. (2008). *Bacillus subtilis* bacteria hinder the oxidation and hydrolysis of Fe^{2+} ions. *Environmental Science and Technology*, **42**, 3194–200.
Fein, J. B., Daughney, C. J., Yee, N. and Davis, T. A. (1997). A chemical equilibrium model for metal adsorption onto bacterial surfaces. *Geochimica et Cosmochimica Acta*, **61**, 3319–28.
Ferris, F. G., Fyfe, W. S. and Beveridge, T. J. (1988). Metallic ion binding by *Bacillus subtilis* – implications for the fossilization of microorganisms. *Geology*, **16**, 149–52.
Ferris, F. G. and Magalhaes, E. (2008). Interfacial energetics of bacterial silicification. *Geomicrobiology Journal*, **25**, 333–7.
Flemming, H. C., Neu, T. R. and Wozniak, D. J. (2007). The EPS matrix: the 'House of Biofilm cells'. *Journal of Bacteriology*, **189**, 7945–7.
Flies, C. B., Jonkers, H. M., de Beer, D., Bosselmann, K., Bottcher, M. E. and Schuler, D. (2005). Diversity and vertical distribution of magnetotactic bacteria along chemical gradients in freshwater microcosms. *FEMS Microbiology Ecology*, **52**, 185–95.
Fortin, D., Ferris, F. G. and Scott, S. D. (1998). Formation of Fe-silicates and Fe-oxides on bacterial surfaces in samples collected near hydrothermal vents on the Southern Explorer Ridge in the northeast Pacific Ocean. *American Mineralogist*, **83**, 1399–408.
Frankel, R. B. and Blakemore, R. P. (1984). Precipitation of Fe_3O_4 in magnetotactic Bacteria. *Philosophical Transactions of the Royal Society of London Series B-Biological Sciences*, **304**, 567–74.
Glasauer, S., Langley, S. and Beveridge, T. J. (2002). Intracellular iron minerals in a dissimilatory iron-reducing bacterium. *Science*, **295**, 117–19.
Goulhen F., Gloter A., Guyot F. and Bruschi M. (2006). Cr(VI) detoxification by Desulfovibrio vulgaris strain Hildenborough: microbe -metal interactions studies. *Appl. Microbiol. Biotechnol.*, **71**, 892–897.
Grassmann, O. and Lobmann, P. (2004). Biomimetic nucleation and growth of $CaCO_3$ in hydrogels incorporating carboxylate groups. *Biomaterials*, **25**, 277–82.
Gray, N. D., Pickup, R. W., Jones, J. G. and Head, I. M. (1997). Ecophysiological evidence that *Achromatium oxaliferum* is responsible for the oxidation of reduced sulphur species to sulfate in a freshwater sediment. *Applied and Environmental Microbiology*, **63**, 1905–10.
Grotzinger, J. P. and Knoll, A. H. (1999). Stromatolites in Precambrian carbonates: evolutionary mileposts or environmental dipsticks? *Annual Review of Earth and Planetary Sciences*, **27**, 313–58.
Isambert, A., Menguy, N., Larquet, E., Guyot, F. and Valet, J. P. (2007). Transmission electron microscopy study of magnetites in a freshwater population of magnetotactic bacteria. *American Mineralogist*, **92**, 621–30.
Kalkowsky, E. (1908). Oolith und Stromatolith im norddeutschen Buntsandstein. *Zeitschrift der Deutschen geologischen Gesellschaft*, **60**, 68–125.
Kawaguchi, T. and Decho, A. W. (2000). Biochemical characterization of cyanobacterial extracellular polymers (EPS) from modern marine stromatolites (Bahamas). *Preparative Biochemistry and Biotechnology*, **30**, 321–30.
Kazmierczak, J. and Kempe, S. (2003). Modern terrestrial analogues for the carbonate globules in Martian meteorite ALH 84001. *Naturwissenschaften*, **90**,167–72.
Kazmierczak, J. and Kempe, S. (2004). Microbialite formation in seawater of increased alkalinity, Satonda Crater Lake, Indonesia – Discussion. *Journal of Sedimentary Research*, **74**, 314–17.

Komeili, A. (2007). Molecular mechanisms of magnetosome formation. *Annual Review of Biochemistry*, **76**, 351–66.

Komeili, A., Li, Z., Newman, D. K. and Jensen, G. J. (2006). Magnetosomes are cell membrane invaginations organized by the actin-like protein MamK. *Science*, **311**, 242–5.

Konhauser, K. O. (1998). Diversity of bacterial iron mineralization. *Earth-Science Reviews*, **43**, 91–121.

Konhauser, K. O., Amskold, L., Lalonde, S. V., Posth, N. R., Kappler, A. and Anbar, A. (2007). Decoupling photochemical Fe(II) oxidation from shallow-water BIF deposition. *Earth and Planetary Science Letters*, **258**, 87–100.

Kopp, R. E., Weiss, B. P., Maloof, A. C., Vali, H., Nash, C. Z. and Kirschvink, J. L. (2006). Chains, clumps, and strings: magnetofossil taphonomy with ferromagnetic resonance spectroscopy. *Earth and Planetary Science Letters*, **247**, 10–25.

Krumbein W. E. (1979). Calcification by bacteria and algae. In *Biogeochemical Cycling of Mineral-forming Elements*, eds. P. A. Trudinger and D. J. Swaine. Amsterdam: Elsevier, pp. 47–68.

Kupriyanova, E. V., Markelova, A. G., Lebedeva, N. V., Gerasimenko, L. M., Zavarzin, G. A. and Pronina, N. A. (2004). Carbonic anhydrase of the alkaliphilic cyanobacterium *Microcoleus chthonoplastes*. *Microbiology*, **73**, 255–9.

Lepot, K., Benzerara, K., Brown, G. E., Jr. and Philippot, P. (2008). Microbially influenced formation of 2,724-million-year-old stromatolites. *Nature Geoscience*, **1**, 118–21.

Lin, R. Y., Zhang, J. Y. and Zhang, P. X. (2002). Nucleation and growth kinetics in synthesizing nanometer calcite. *Journal of Crystal Growth*, **245**, 309–20.

Little, C. T. S., Glynn, S. E. J. and Mills, R. A. (2004). Four-hundred-and-ninety-million-year record of bacteriogenic iron oxide precipitation at sea-floor hydrothermal vents. *Geomicrobiology Journal*, **21**, 415–29.

López-García, P., Kazmierczak, J., Benzerara, K., Kempe, S., Guyot, F. and Moreira, D. (2005). Bacterial diversity and carbonate precipitation in the giant microbialites from the highly alkaline Lake Van, Turkey. *Extremophiles*, **9**, 263–74.

Lowenstam, H. A. (1981). Minerals formed by organisms. *Science*, **211**, 1126–31.

Machel, H. G. and Mountjoy, E. W. (1986). Chemistry and environments of dolomitization – a reappraisal. *Earth-Science Reviews*, **23**, 175–222.

Mann, S. (2001). *Biomineralization: Principles and Concepts in Bioinorganic Materials Chemistry*. Oxford: Oxford University Press.

Manoli, F. and Dalas, E. (2002). The effect of sodium alginate on the crystal growth of calcium carbonate. *Journal of Materials science-Materials in Medicine*, **13**, 155–8.

Marie, B., Luquet, G., Bedouet, L., Milet, C., Guichard, N., Medakovic, D. and Marin, F. (2008). Nacre calcification in the freshwater mussel *Unio pictorum*: carbonic anhydrase activity and purification of a 95 kDa calcium-binding glycoprotein. *ChemBioChem*, **9**, 2515–23.

McKay, D. S., Gibson, E. K., Thomas-Keprta, K. L., Vali, H., Romanek, C. S., Clemett, S. J., Chillier, X. D. F., Maechling, C. R. and Zare, R. N. (1996). Search for past life on Mars: possible relic biogenic activity in Martian meteorite ALH 84001. *Science*, **273**, 924–30.

Meibom, A., Cuif, J. P., Houlbreque, F., Mostefaoui, S., Dauphin, Y., Meibom, K. L. and Dunbar, R. (2008). Compositional variations at ultra-structure length scales in coral skeleton. *Geochimica et Cosmochimica Acta*, **72**, 1555–69.

Meister, P., McKenzie, J. A., Vasconcelos, C., Bernasconi, S., Frank, M., Gutjahr, M. and Schrag, D. P. (2007). Dolomite formation in the dynamic deep biosphere: results from the Peru Margin. *Sedimentology*, **54**, 1007–31.

Miot, J., Benzerara, K., Morin, G., Kappler, A., Bernard, S., Obst, M., Férard, C., Skouri-Panet, F., Guigner, J. M., Posth, N., Galvez, M., Brown Jr, G. E. and Guyot, F. (2009). Iron biomineralization by anaerobic neutrophilic iron-oxidizing bacteria. *Geochimica et Cosmochimica Acta*, **73**, 696–711.

Moreau, J. W., Weber, P. K., Martin, M. C., Gilbert, B., Hutcheon, I. D. and Banfield, J. F. (2007). Extracellular proteins limit the dispersal of biogenic nanoparticles. *Science*, **316**, 1600–3.

Nealson, K. H. (1997). Sediment bacteria: who's there, what are they doing, and what's new? *Annual Review of Earth and Planetary Sciences*, **25**, 403–34.

Newman, D. K. and Banfield, J. F. (2002). Geomicrobiology: how molecular-scale interactions underpin biogeochemical systems. *Science*, **296**, 1071–7.

Orme, C. A., Noy, A., Wierzbicki, A., McBride, M. T., Grantham, M., Teng, H. H., Dove, P. M. and DeYoreo, J. J. (2001). Formation of chiral morphologies through selective binding of amino acids to calcite surface steps. *Nature*, **411**, 775–9.

Phoenix, V. R. and Konhauser, K. O. (2008). Benefits of bacterial biomineralization. *Geobiology*, **6**, 303–8.

Phoenix, V. R., Martinez, R. E., Konhauser, K. O. and Ferris, F. G. (2002). Characterization and implications of the cell surface reactivity of *Calothrix sp.* strain KC97. *Applied and Environemental Microbiology*, **68**, 4827–34.

Pokrovsky, O. S., Martinez, R. E., Golubev, S. V., Kompantseva, E. I. and Shirokova, L. S. (2008). Adsorption of metals and protons on *Gloeocapsa* sp cyanobacteria: a surface speciation approach. *Applied Geochemistry*, **23**, 2574–88.

Posth, N. R., Hegler, F., Konhauser, K. O. and Kappler, A. (2008). Alternating Si and Fe deposition caused by temperature fluctuations in Precambrian oceans. *Nature Geoscience*, **1**, 703–8.

Poulsen, N., Sumper, M. and Kroger, N. (2003). Biosilica formation in diatoms: characterization of native silaffin-2 and its role in silica morphogenesis. *Proceedings of the National Academy of Sciences of the United States of America*, **100**, 12075–80.

Rancourt, D. G., Thibault, P. J., Mavrocordatos, D. and Lamarche, G. (2005). Hydrous ferric oxide precipitation in the presence of nonmetabolizing bacteria: constraints on the mechanism of a biotic effect. *Geochimica et Cosmochimica Acta*, **69**, 553–77.

Riding, R. (2000). Microbial carbonates: the geological record of calcified bacterial-algal mats and biofilms. *Sedimentology*, **47**, 179–214.

Scheffel, A., Gardes, A., Grunberg, K., Wanner, G. and Schuler, D. (2008). The major magnetosome proteins MamGFDC are not essential for magnetite biomineralization in *Magnetospirillum gryphiswaldense* but regulate the size of magnetosome crystals. *Journal of Bacteriology*, **190**, 377–86.

Scheffel, A., Gruska, M., Faivre, D., Linaroudis, A., Plitzko, J. M. and Schuler, D. (2006). An acidic protein aligns magnetosomes along a filamentous structure in magnetotactic bacteria. *Nature*, **440**, 110–14.

Schopf, J. W. and Packer, B. M. (1987). Early Archean (3.3-billion to 3.5-billion-year-old) microfossils from Warrawoona group, Australia. *Science*, **237**, 70–3.

Staniland, S., Ward, B., Harrison, A., van der Laan, G. and Telling, N. (2007). Rapid magnetosome formation shown by real-time x-ray magnetic circular dichroism. *Proceedings of the National Academy of Sciences of the United States of America*, **104**, 19524–8.

Stolarski, J. (2003). Three-dimensional micro- and nanostructural characteristics of the scleractinian coral skeleton: a biocalcification proxy. *Acta Palaeontologica Polonica*, **48**, 497–530.

Sumper, M. and Brunner, E. (2008). Silica biomineralisation in diatoms: the model organism *Thalassiosira pseudonana*. *ChemBioChem*, **9**, 1187–94.

Taylor, A. P. and Barry, J. C. (2004). Magnetosomal matrix: ultrafine structure may template biomineralization of magnetosomes. *Journal of Microscopy – Oxford*, **213**, 180–97.

Thomas-Keprta, K. L., Bazylinski, D. A., Kirschvink, J. L., Clemett, S. J., McKay, D. S., Wentworth, S. J., Vali, H., Gibson, E. K. and Romanek, C. S. (2000). Elongated prismatic magnetite crystals in ALH 84001 carbonate globules: potential Martian magnetofossils. *Geochimica et Cosmochimica Acta*, **64**, 4049–81.

Thomas-Keprta, K. L., Clemett, S. J., Bazylinski, D. A., Kirschvink, J. L., McKay, D. S., Wentworth, S. J., Vali, H., Gibson, E. K. and Romanek, C. S. (2002). Magnetofossils from ancient Mars: a robust biosignature in the Martian meteorite ALH 84001. *Applied and Environmental Microbiology*, **68**, 3663–72.

Trichet, J. and Défarge, C. (1995). Non-biologically supported organomineralization. *Bulletin de l'Institut Oceanographique de Monaco*, **14**, 203–26.

Vali, H., Weiss, B., Li, Y. L., Sears, S. K., Kim, S. S., Kirschvink, J. L. and Zhang, L. (2004). Formation of tabular single-domain magnetite induced by *Geobacter metallireducens GS-15*. *Proceedings of the National Academy of Sciences of the United States of America*, **101**, 16121–6.

van Dijk, S., Dean, D. D., Liu, Y., Zhao, Y., Chirgwin, J. M., Schwartz, Z. and Boyan, B. D. (1998). Purification, amino acid sequence, and cDNA sequence of a novel calcium-precipitating proteolipid involved in calcification of *Corynebacterium matruchotii*. *Calcified Tissue International*, **62**, 350–8.

Van Lith, Y., Warthmann, R., Vasconcelos, C. and McKenzie, J. A. (2003). Microbial fossilization in carbonate sediments: a result of the bacterial surface involvement in dolomite precipitation. *Sedimentology*, **50**, 237–45.

Vasconcelos, C., McKenzie, J. A., Bernasconi, S., Grujic, D. and Tien, A. J. (1995). Microbial mediation as a possible mechanism for natural dolomite formation at low temperatures. *Nature*, **377**, 220–2.

Vasiliev, I., Frank, C., Meeldijk, J. D., Dekkers, M. J., Langereis, C. G. and Krijgsman, W. (2008). Putative greigite magnetofossils from the Pliocene epoch. *Nature Geoscience*, **1**, 782–6.

von Canstein, H., Ogawa, J., Shimizu, S. and Lloyd, J. R. (2008). Secretion of flavins by *Shewanella* species and their role in extracellular electron transfer. *Applied and Environmental Microbiology*, **74**, 615–23.

Wacey, D., Wright, D. T. and Boyce, A. J. (2007). A stable isotope study of microbial dolomite formation in the Coorong Region, South Australia. *Chemical Geology*, **244**, 155–74.

Warthmann, R., van Lith, Y., Vasconcelos, C., McKenzie, J. A. and Karpoff, A. M. (2000). Bacterially induced dolomite precipitation in anoxic culture experiments. *Geology*, **28**, 1091–4.

Weiner, S. and Dove, P. M. (2003). An overview of biomineralization processes and the problem of the vital effect. *Biomineralization*, **54**(1), 1–29.

Wolf, S. E., Loges, N., Mathiasch, B., Panthöfer, M., Mey, I., Janshoff, A. and Tremel, W. (2007). Phase selection of calcium carbonate through the chirality of adsorbed amino acids. *Angewandte Chemie International Edition*, **46**, 5618–23.

Wright, D. T. and Oren, A. (2005). Nonphotosynthetic bacteria and the formation of carbonates and evaporites through time. *Journal of Geomicrobiology*, **22**, 27–53.

28

Limits of life and the biosphere: lessons from the detection of microorganisms in the deep sea and deep subsurface of the Earth

Ken Takai

Naturally occurring physical and chemical constraints of life and the biosphere

Deep-sea and deep-subsurface environments have been recognized to be among the most extreme biotopes potentially placed very close to an interface between the habitable and the uninhabitable terrains for life on Earth. The concept of habitability appears difficult to define, particularly in terms of an astrobiological perspective. Nevertheless, it is widely accepted that the harshest habitats for life, such as deep-sea and deep-subsurface environments in this 'highly habitable' planet, the Earth, may be approximated to the most plausible environments for extraterrestrial life in some 'hardly habitable' planets and moons of our Solar System. Thus, to understand the limits of life and the biosphere in the deep-sea and deep-subsurface environments of the Earth could be a key for elucidating the potential habitability of extraterrestrial life in the Universe. In this chapter, the possible factors that limit life and the biosphere on the Earth are overviewed and discussed from insights gained from the recent biogeochemical and geomicrobiological explorations in the deep-sea and deep-subsurface biosphere.

In the deep-sea and deep-subsurface environments many physical and chemical parameters limiting the activities of microbial life have been elucidated. The best example is temperature. In the terrestrial and oceanic surface environments, liquid water boils at around 100°C, while with an increasing pressure (hydrostatic), liquid water can be present at up to 373°C for pure water and 407°C for seawater (critical points) (Bischoff and Rosenbauer, 1988). Indeed, the highest temperature record of liquid water (407°C) was found in a deep-sea hydrothermal-vent fluid in the Mid-Atlantic Ridge (Table 28.1). Thus, since the first discovery of superheated hydrothermal vents, called 'black smokers', in 1979 at 21°N on the East Pacific Rise (Spiess and the RISE Group, 1980), deep-sea hydrothermal-vent microbiologists have been interested in empirically determining the upper temperature limit (UTL) for life. The research has caught the interest of scientists from many disciplines, as well as that of the general public, because the deep-sea hydrothermal vents and the subseafloor environments host the only naturally occurring habitats closely situated to the near-critical point of water. In contrast, most of the deep-sea environments are characterized by low temperatures of around 1–4°C, which is not the lowest temperature for life

Table 28.1. Physical and chemical constraints of natural environments on the Earth.

Physical and chemical factors	Representative environment on the Earth	Representative environment in the deep-sea and deep subsurface	Representative environment hosting microbial signatures
Lowest temperature of liquid H_2O	−89°C as ice (Vostok Station in Antarctica)	> 0°C (Abyssal seawater)	In ice of Antarctica or in abyssal seawater
Highest temperature of liquid H_2O	407°C (A deep-sea hydrothermal field in the equatorial Mid-Atlantic Ridge) (News in brief, 2006, *Nature*, **441**, 563)	407°C (A deep-sea hydrothermal field in the equatorial Mid-Atlantic Ridge) (News in brief, 2006, *Nature*, **441**, 563)	365°C (The Kairei Hydrothermal Field in the Central Indian Ridge) (Takai *et al.*, 2008a)
Lowest pressure of liquid H_2O	< 0.6 KPa as ice (Stratosphere)	> 0.1 MPa	0.6 KPa as saturated vapour pressure (Stratosphere of 58 km above) (Imshenetsky *et al.*, 1976)
Highest pressure of liquid H_2O	110 MPa (the Challenger Deep of the Mariana Trench) (Kato *et al.*, 1997)	110 MPa (the Challenger Deep of the Mariana Trench) (Kato *et al.*, 1997)	110 MPa (the Challenger Deep of the Mariana Trench) (Kato *et al.*, 1997)
Lowest pH of liquid H_2O	pH −3.6 (Mine drainage water) (Nordstrom *et al.*, 2000)	pH −3.6 (Nordstrom *et al.*, 2000)	pH 0–1 (Mine drainage water and geothermal water) (Edwards *et al.*, 2000)
Highest pH of liquid H_2O	> pH 12.5 (The subseafloor pore water in the South Chamorro Seamount of the Mariana Forearc) (Salisbury, 2002)	pH 12.5 (The subseafloor pore water in the South Chamorro Seamount of the Mariana Forearc) (Salisbury, 2002)	> pH 12.5 (The subseafloor pore water in the South Chamorro Seamount of the Mariana Forearc) (Salisbury, 2002)
Highest salinity in liquid H_2O	Many saturated soda lakes or crystal salts	Saturated (A deep-sea brine pool in the Red Sea) (Swallow and Crease, 1965)	Many saturated soda lakes or crystal salts (Rainey and Oren, 2006)

Table 28.2. Presently known limits of microbial growth and survival.

Physical and chemical factors	Limit for growth and maintenance	Limit for survival
Lowest temperature in liquid H_2O	< 0°C (Many psychophiles) (Rainey and Oren, 2006)	< 0°C for very long time (Many bacteria and Archaea)
Highest temperature in liquid H_2O	122°C at 20 MPa (*Methanopyrus kandleri* strain 116) (Takai *et al.*, 2008a)	130°C for 180 min at 30 MPa (*Methanopyrus kandleri* strain 116) (Takai *et al.*, 2008a)
Highest presssure in liquid H_2O	130 MPa at 2°C (strain MT41) (Yayanos, 1986)	1.4 GPa at 120°C (*Clostridium botulinum* spore) (Margosch *et al.*, 2006)
Lowest pH of liquid H_2O	pH 0 (*Picrophilus oshimae* and *P. torridus*) (Schleper *et al.*, 1995)	pH 0
Highest pH of liquid H_2O	pH 12.4 (*Alkaliphilus transvaalensis*) (Takai *et al.*, 2001b)	Probably up to pH 14 (*Alkaliphilus transvaalensis* spore)
Highest salinity in liquid H_2O	Saturated (Many extreme halophiles) (Rainey and Oren, 2006)	In crystal for 250 million years (*Bacillus sphaericus*) (Vreeland *et al.*, 2000)

activity, but is close to the freezing temperature of liquid water. Thus, a great variation in temperature is present in the deep-sea and deep-subsurface environments.

Hydrostatic pressure is another important physical parameter in the deep-sea and deep-subsurface environments. The deepest habitat of the Earth explored by researchers is the Challenger Deep, in the Mariana Trench, at a water depth of ~ 10 900 m (Kato *et al.*, 1997), which corresponds to 110 MPa of hydrostatic pressure (Table 28.2). The opposite extreme is the high-altitude atmosphere of the Earth, of which the stratosphere has been microbiologically explored (e.g. Imshenetsky *et al.*, 1976). Probably, this environment is not a place for life activity but of life survival, and astrobiologically bridges between the Earth and the Universe.

Some of the deep-sea and deep-subsurface environments, deep-sea hydrothermal deposits and uranium mines, are exposed to relatively high natural radioactivity as compared with ambient environments, due to the enrichment of radioactive elements such as uranium and thorium (Charmasson *et al.*, 2009; Lin *et al.*, 2005). Although the net dosage of radiation in these natural environments is uncertain, radioactivity may be a physical constraint for life activity and life survival, and in particular, it may be the limit for dormancy in the deep-sea and deep-subsurface environments (McKay, 2001).

Major chemical parameters, extreme conditions of pH and salinity are also found in the deep-sea and deep-subsurface environments. Subaerial volcanic and geothermal fields often host extremely acidic environments in which pH values drop to close to 0 (e.g. Schleper *et al.*, 1995). This extreme acidity is caused by sulphuric acid and hydrochloric acid originally provided from magmatic volatiles associated with volcanic activity. If similar vulcanism takes place in the deep sea, then deep-sea, extremely acidic hydrothermal systems occur. Low pH value (1.6) was reported for hydrothermal fluid from the TOTO Caldera Field in the Mariana Arc (Nakagawa *et al.*, 2006), while several more acidic hydrothermal fluids (< pH 1) have now been observed in the western Pacific submarine volcanoes (Resing *et al.*, 2007). Acid mine drainage environments are often extremly acidified by excess amounts of sulphuric acid produced from the chemical oxydation of sulphide ores (Nordstrom *et al.*, 2000). Indeed, the negative pH value of acid mine water has been reported in a subsurface water environment of Iron Mountain, California (Nordstrom *et al.*, 2000). On the other hand, alkaline environments are also generated in the deep-subsurface water–rock processes. It is well known that the serpentinization reaction of water and olivine, a major mineral of ultramafic rocks abundantly present in the upper mantle, generates highly alkaline water environments (McCollom and Bach, 2009; Takai *et al.*, 2006). The highest pH value ever known in deep-sea and deep-subsurface environments is pH 12.5 in the porewater of the South Chamorro Seamount in the Mariana Forearc (Table 28.1) (Salisbury, 2002). Probably, this pH value is also close to the highest pH naturally occurring, even in the surface environments such as soda lakes. The subaerial soda-lake extreme habitats not only have extreme pH, but also salinity. The evaporation of water finally forms salt crystals and over-saturated brines that produce the highest salinity in known Earth environments. A similar process occurs in deep-sea hydrothermal systems. The high temperature and rapid decompression induce phase-separation of hydrothermal fluids into vapour and brine in the deep subseafloor. Phase-separation-controlled hydrothermal fluids are often observed in global deep-sea hydrothermal systems, and both hypersaline and vapour-rich fluids have been identified. It is expected that the brine phase of fluids generates the local salts-saturated or salts-oversaturated environments associated with subseafloor hydrothermal circulation. Furthermore, deep-sea brine pools and deep-subseafloor salt deposits are often found (Table 28.1) (e.g. Swallow and Crease, 1965; Swart *et al.*, 2000). These environments result from the past dried-up events of seawater induced by palaeoenvironmental changes such as decreasing sea level and tectonics. In contrast, very low salt-concentration environments could also be generated in the deep sea and deep subsurface from the water vapour of hydrothermal phase separation, the input of magmatic volatiles and the dissociation of the gas (CO_2 and CH_4) hydrates that exclude the dissolved salts during hydrate formation. Other than these factors (temperature, pressure, pH and salinity), many other physical and chemical extremes, such as quite low energy sources for life activity and survival, are expected in the deep-sea and deep-subsurface environments (Valentine, 2007). The naturally occurring extreme environmental conditions encompass the physical and chemical constraints for limits of life and the biosphere.

Extremities of physical and chemical constraints on Earth

Constitutive microorganisms of communities occurring in a certain extreme environment might have types of physiological and ecological functions and mechanisms for their habitability under these 'abnormal' environmental conditions. From these extreme environments, microorganisms that can preferentially live and survive under these conditions (called 'extremophiles') have been identified and cultivated. To know the potential limits of microbial habitability on the Earth, it is very helpful to know what kind of extremophiles have been found as 'living evidence' to represent the limits for growth and survival in terms of each of the physical and chemical factors. Summarized in the following sections are the extreme environments in which living microbial signatures have been detected.

Upper temperature limit

Until the discovery of the deep-sea hydrothermal vents, terrestrial hot springs were believed to be the place hosting life having the upper temperature limits for growth and survival, which was established by Tom Brock's pioneering studies in the Yellowstone National Park. From a series of investigations on extreme thermophiles in Yellowstone National Park, the upper temperature limit (UTL) for life was established at up to 91°C (Brock, 1978) (Figure 28.1A). In 1983, a hyperthermophilic archaeon *Pyrodictium occultum* was isolated from a shallow marine hydrothermal vent in Italy by Karl Stetter, and represented the most hyperthermophilic life ever found, growing at up to 110°C (Stetter *et al.*, 1983) (Figure 28.1A). The exploration of hyperthermophiles (microorganisms optimally growing at over 80°C and capable of growth at over 90°C) in deep-sea hydrothermal vents opened a new age of extending the UTL on the Earth. In 1989, a hyperthermophilic methane-producing archaeon *Methanopyrus kandleri*, was isolated from the Guaymas Basin (Huber *et al.*, 1989) that tied with the previous UTL record of *P. occultum* (110°C) isolated from a shallow marine-hydrothermal system (Stetter *et al.*, 1983). An abyssal *Pyrodictium* species was also retrieved from the Guaymas Basin hydrothermal field that had a maximum growth temperature of 110°C (Pley *et al.*, 1991). A few years later, the UTL was elevated with the discovery of *Pyrolobus fumarii* from the TAG Field on the Mid-Atlantic Ridge growing at up to 113°C (Blöchl *et al.*, 1997) (Figure 28.1.A). As the 3 C degree elevation of the UTL took 14 year to discover, many microbiologists believed that a temperature zone of 115–120°C was potentially the upper temperature for microbial growth. It was also considered that this temperature range corresponded to the temperature limit of stability for low-molecular-weight cofactors and metabolites such as AMP, ADP, ATP, NAD/P(H) and acetyl-CoA solely in water *in-vitro* experiments, although several researchers pointed out a variety of potential mechanisms of thermal stabilization by high-molecular-weight cellular components such as proteins, lipids and nucleic acids over this temperature range (e.g. Cowan, 2004).

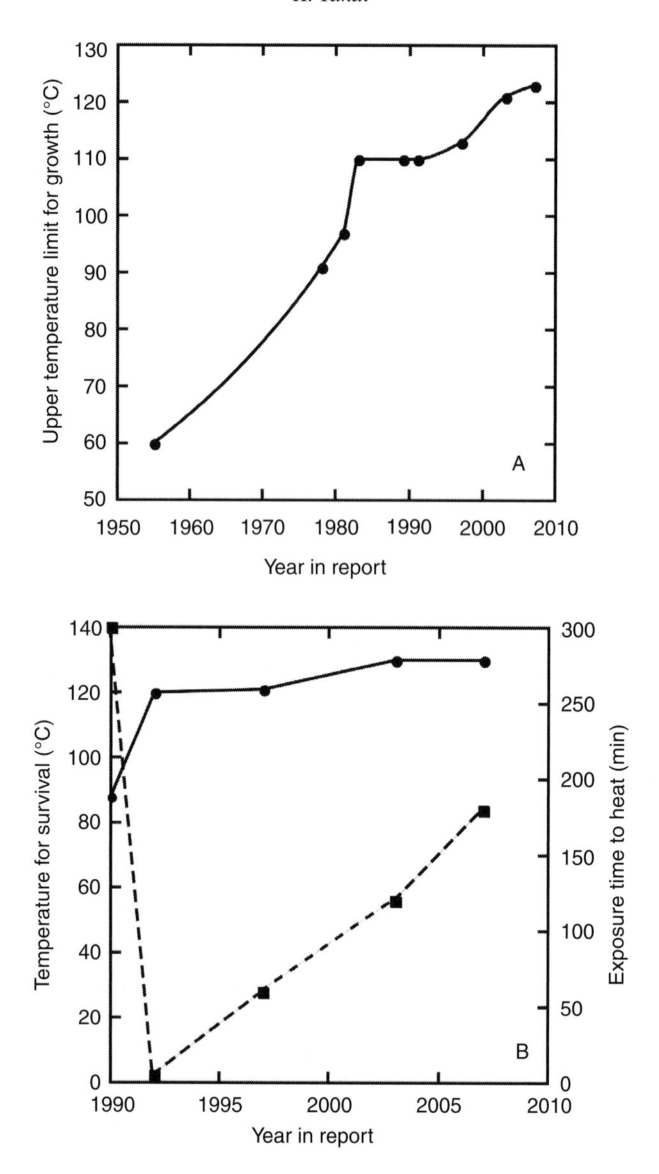

Figure 28.1. History of upper temperature limit for microbial growth (A) and survival potential at high temperatures (B). (A) The record of upper temperature limit for microbial growth and its publication year. (B) The highest temperature (•) and the longest time (■) obtained from the survival-potential experiment, and its publication year.

Then, in 2003, Kashefi and Lovley (2003) reported the isolation of a hyperthermophilic archaeon strain 121 from the Juan de Fuca Ridge, growing at temperatures up to 121°C (Figure 28.1.A). Since this hyperthermophilic archaeon was not available from any of the public culture collections, this record was not verified by other researchers and aroused controversy as to whether the growth at 121°C was reproducible or not (e.g. Cowan,

2004). In 2008, a new strain of *M. kandleri* was isolated from a 365°C black-smoker fluid habitat in the Kairei Hydrothermal Field (water depth = 2450 m) of the Central Indian Ridge (CIR), and this *M. kandleri* strain 116 was found to grow at between 85 and 116°C at near atmospheric pressure (0.3 MPa of gas pressure) (Takai *et al.*, 2008a). This was the second-highest temperature for growth after 121°C in the case of hyperthermophilic archaeon strain 121, and the 365°C of black-smoker fluid of the CIR Kairei Field also represented the highest temperature of habitat known where 'living microorganisms' and 'microbial molecular signatures' (rRNA gene sequences) were identified (Takai *et al.*, 2004, 2008a). However, under the high hydrostatic pressures (20 MPa and 40 MPa), the *M. kandleri* strain 116 was capable of growth at up to 122°C, which was the highest record of UTL for life ever reported (Figure 28.1A) (Takai *et al.*, 2008a). This record of UTL for life provided important insights into understanding the physical and chemical constraints of life and the biosphere. It is very likely that the effects of physical and chemical factors are interactive (e.g. a cooperative temperature and pressure effect for life). One must always keep in mind that almost all the microorganisms collected from the deep-sea and deep-subsurface environments have been cultivated under conventional pressures that are quite different from those in their *in-situ* habitats. In addition, it must be pointed out that most of the living microorganisms in the natural extreme environments cannot be cultured. Thus, 122°C is just a transient UTL, and if new methodological and technological innovations are incorporated in the UTL experiments, the UTL could be further extended.

As shown in the case of the growth temperature of *M. kandleri* strain 116, the deep-sea hydrothermal environments also provide the perfect venue for research on the combined effect of temperature and pressure on the limit of survival for life. In 1992, Jannasch *et al.* (1992) demonstrated that some *Pyrococcus* strains growing at up to 103°C isolated from the Guaymas Basin could be alive after exposure to 120°C for 5 min under a hydrostatic pressure of 20 MPa (Figure 28.1B). This was the first report of cellular survival potential of a deep-sea hyperthermophile at the highest temperature and for the longest time under the combined conditions of temperature and pressure. However, the cellular survival potentials of the *Pyrococcus* strains were at similar levels to the survival potentials of spores produced by mesophilic, Gram-positive bacteria such as *Bacillus* and *Clostridium* members, which have been fundamental controls for autoclave sterilization (121°C, 20 min). The isolation of *P. fumarii* greatly elevated the cellular survival potential at higher temperatures (Figure 28.1B) (Blöchl *et al.*, 1997). It could revive after incubation at 121°C for 60 min (under 0.3 MPa) (Blöchl *et al.*, 1997). Then, the hyperthermophilic archaeon strain 121 extended the cellular survival-potential record to 130°C for 120 min (under 0.3 MPa) (Figure 28.1B) (Kashefi and Lovley, 2003). Finally, the cellular survival potential at 130°C was extended up to 180 min by an experiment using *M. kandleri* strain 116 under 30 MPa (Figure 28.1B) (Takai *et al.*, 2008a). Under conventional gas pressure (0.3 MPa), the survival potential of *M. kandleri* strain 116 at 130°C was the same (120 min) as that of the hyperthermophilic archaeon strain 121, while the elevated hydrostatic pressure extended the survival duration (Takai *et al.*, 2008a). This result represents the present limit of survival for life at the highest temperature and for the longest time.

However, the experimentally determined survival temperatures of the deep-sea hydrothermal-vent microorganisms in the laboratory seem dramatically low as compared to the temperatures of natural habitats in the deep-sea hydrothermal environments where molecular signatures and even viable cells of potentially indigenous living microorganisms are retrieved. It is very likely that the molecular signatures of microbial components detected in the superheated hydrothermal fluids (> 250°C) are derived from the microbial communities hosted in the much the lower temperatures of habitats associated with the subseafloor hydrothermal fluid flows. Indeed, hyperthermophiles have often been cultured from samples of the high-temperature (> 250°C) hydrothermal fluids or materials (e.g. Takai *et al.*, 2004, 2008a, 2008b). The occurrence of living microorganisms in the samples at these extraordinary temperatures was interpreted as follows: (1) these microorganisms might originate from lower temperatures of rich habitats that enabled their active growth somewhere in deep and shallow subseafloor hydrothermal environments as explained above; (2) hydrothermal fluid migration might bring some populations of the microbial components into higher temperatures of hydrothermal fluids; and (3) most of the entrained microorganisms were killed by the heat (molecular signatures were still present) while tiny populations might survive under exposure to extraordinary temperatures during the short time of entrainment by the hydrothermal fluids (Takai *et al.*, 2004, 2008b). *Thermococcus* sp. strain Tc-1–95 is an example of a microorganism isolated from an *in-situ* colonization system deployed at a > 250°C black-smoker fluid in the CIR Kairei Field (Takai *et al.*, 2004). It represented one of the most predominant living microbial populations recovered from the *in-situ* colonization system. However, this strain could not survive at 120°C even for 1.5 s under an *in-situ* hydrostatic pressure in a laboratory experiment (Mitsuzawa *et al.*, 2005). This result strongly suggests that there is still a big gap between observation of the natural environments and laboratory experiments to estimate the temperature limits of survival and even growth for life. However, it is also important to note that this type of laboratory experiment has never reproduced the complete *in-situ* physical and chemical conditions of the deep-sea hydrothermal habitats other than the temperature and pressure. Thus, the previously unexamined factors, such as physical interaction between minerals and microbial cells and the dissolved-gas components of hydrothermal fluids, may drastically increase the temperature limits of survival and growth for microorganisms in the deep-sea hydrothermal environments and even on the surface of the Earth.

Upper pressure limit

A present upper pressure limit for microbial growth is known to be 130 MPa, which was recorded as the highest hydrostatic pressure for growth of deep-sea psychrophilic heterotroph strain MT41 isolated from Challenger Deep, the Mariana Trench (Table 28.2) (Yayanos, 1986). A similar piezophilic response of growth was also found in the *Shewanella* and *Moritella* isolates from a slightly deeper part of the Mariana Trench at 10 898 m (Kato *et al.*, 1998). From the Challenger Deep sediments of the Mariana Trench, a great phylogenetic diversity of rRNA gene sequences and physiological diverse microorganisms have also been identified (Kato *et al.*, 1997; Takami *et al.*, 1997; Takai *et al.*, 1999). Thus, it is already evident that the greatest pressure environment that we can now explore on the Earth is not a pressure margin limiting the habitability for life.

The extreme piezophiles from the Mariana Trench are all aerobic psychrophilic heterotrophs cultivated using the conventional heterotrophic media (Yayanos, 1986; Kato *et al.*, 1998). Thus, if more variable media and cultivation conditions are adopted to cultivate extreme piezophiles from the ultra-deep-sea habitats such as the Mariana Trench, phylogenetically and physiologically more diverse piezophiles, particularly the ones extending the upper pressure limit for growth, could be obtained. Not only the ultra-deep-sea trench systems but also ultra-deep-subsurface environments characterized by high lithostatic pressure in addition to hydrostatic pressure will be good candidates for searching for the microorganisms potentially extending the upper pressure limits for life on this planet in the future.

Pressure (hydrostatic and lithostatic) in deep-sea and deep-subsurface environments also has an impact on survival. The pressure and temperature effects on spore survival have been long investigated in the field of food microbiology. As far as is known, some bacterial spores (*Clostridium* and *Bacillus* spp.) can revive after pressure treatment at > 1 GPa at 25°C and at 1.4 GPa at 120°C for a few minutes (Table 28.2) (e.g. Margosch *et al.*, 2006). Based on the kinetic modelling of temperature and pressure effects on spore viability (Margosch *et al.*, 2006), a possible upper pressure limit for spore survival for a few minutes is expected to be below 2 GPa. This pressure corresponds to > 150 km of water depth in the Earth's ocean or > 50 km of lithospheric depth in the Earth's crust and fully covers all the possible oceanic and crustal environments of the Earth.

On the other hand, the pressure effect on cellular survival has been investigated using a very limited number of non-piezophilic bacterial strains such as *E. coli*, *Listeria monocytogenes* and *Lactobacillus* spp. At 4°C and 700 MPa, most of these bacterial members (if more than 10^9 cells are processed under this pressure) are not completely dead after more than 5 min exposure (Klotz *et al.*, 2007). Since no similar study has been applied to the piezophiles and extreme piezophiles, it is still uncertain how a greater hydrostatic pressure of > 700 MPa would affect the cellular survival of extreme piezophiles dominating in the deep-sea and deep-subsurface environments. However, if a possible upper pressure limit for cellular survival for a few minutes is expected to be similar to that for spore survival, pressure alone is not a factor controlling the spatial limitation of life and the biosphere on this planet.

pH Limit

The pH range of the natural environments on the Earth is from < pH–4 to probably > pH 13, while the range in the deep-sea and deep-subsurface environments is now known to be from pH –3.6 to pH 12.5 (Table 28.1). The acidic pH limit for growth and survival is well established; extremely acidophilic Archaea belonging to the phylum *Thermoplasmata* such as *Picrophilus* and *Ferroplasma* members can grow even at pH 0 (Table 28.2) (Schleper *et al.*, 1995; Edwards *et al.*, 2000), and several eukaryotic species can also grow at near pH 0 and can survive at pH 0 (Doemel and Brock, 1971; Rothschild and Mancielli, 2001). These archaeal and eukaryotic members are all found in the terrestrial surface and subsurface environments. From the TOTO Caldera Hydrothermal Field in the Mariana Volcanic Arc where pH 1.6 hydrothermal fluid was identified, a phylogenetic diversity of bacterial and archaeal rRNA gene sequences were characterized (Nakagawa *et al.*, 2006). Most of the sequences were considered to be the entrained microbial signatures from more moderate pH subseafloor habitats (Nakagawa *et al.*, 2006), while some specific bacterial and archaeal components might be extreme acidophiles. The most acidophilic microorganism isolated from the deep-sea hydrothermal and deep-subseafloor environments is *Acidiprofundum boonei* from the Lau Basin Deep-sea Hydrothermal Field (Reysenbach *et al.*, 2006), and the lowest pH limit for its growth is pH 3.3 (Reysenbach *et al.*, 2006). This thermophilic archaeon is phylogenetically affiliated into the phylum *Thermoplasmata* and is a relative of terrestrial *Picrophilus* and *Ferroplasma* members (Reysenbach *et al.*, 2006; Takai and Horikoshi, 1999; Takai *et al.*, 2001a). Thus, previously identified but uncultivated deep-sea Archaea within the phylum *Thermoplasmata* may be able to grow at near pH 0. Nevertheless, since high temperatures of deep-sea hydrothermal fluids (> 150°C) are expected to reveal an *in-situ* pH of two orders of magnitude higher than that measured at room temperature (due to HCl solubility) (Seyfried *et al.*, 1991), the lowest pH limit for the growth of *Acidiprofundum boonei* (pH 3.3) may be around the lowest limit for growth of many deep-sea hydrothermal-vent microbial components.

The alkaline pH limit for growth and survival is relatively less investigated as compared to the acidic one. The alkaliphilic bacteria are known to be ubiquitous in non-alkaline habitats such as soil, freshwater and ocean environments and are usually able to grow at up to pH 10–11 (Horikoshi, 1999). This pH range is almost equivalent to the alkaline pH limit for life (pH 10–12) that has often been described in the literature (Rothschild and Mancinelli, 2001; Rainey and Oren, 2006). However, to the best of my knowledge, the most alkaliphilic microorganism is known to be *Alkaliphilus transvaalensis* isolated from an ultra-deep South African gold mine (3.2 km deep below land surface), of which the pH limit for its growth is pH 12.4 (Table 28.2) (Takai *et al.*, 2001b). This bacterium also represents the living microorganism retrieved from one of the deepest terrestrial subsurface environments (Takai *et al.*, 2001b). Since *A. transvaaalensis* is a Gram-positive, spore-forming bacterium, the alkaline pH limit for cellular and spore survival could shift to more alkaline conditions, probably up to pH 14. However, this bacterium has never been found in the deep-sea and deep-subseafloor environments.

In the subseafloor serpentine mud environment of the South Chamorro Seamount, the pore water indicates a pH value of > 12.5 (Table 28.1) (Salisbury, 2002). From this hyper-alkaline environment, an approximately 30-m deep-core sample was obtained in Ocean Drilling Program (ODP) Leg#195 (Salisbury, 2002; Takai *et al.*, 2005). Down to 1.5 m below the seafloor (up to a pH value of 11), a living alkaliphilic heterotroph *Marinobacter alkaliphilus* was detected (Takai *et al.*, 2005). However, from deeper parts of the samples (higher than pH 11), none of the living microorganisms were identified, even though some populations of DAPI-stained microbial cells were consistently observed through the 30 m of core (Takai *et al.*, 2005). From the same core samples, considerable amounts of bacterial and archaeal lipids were also obtained (Mottl *et al.*, 2003). Thus, potentially living microbial populations could be present in the hyperalkaline subseafloor environments but this needs to be substantiated with clear evidence. The pH limit for growth of *M. alkaliphilus* (pH 11.4) is probably the present highest pH value ever reported for growth of deep-sea and subseafloor microorganisms.

Another well-known, deep-sea hyperalkaline environment is the Lost City Hydrothermal Field (Kelly *et al.*, 2001, 2005). The alkaline pH of the hydrothermal fluid is generated in a similar process to that of the pore water of the South Chamorro Seamount (McCollom and Bach, 2009). The pH value of the Lost City hydrothermal fluids often exceeds pH 11.5 (Kelly *et al.*, 2001, 2005). In the Lost City hydrothermal fluids, microbial cells and rRNA genes could be identified (Schrenk *et al.*, 2004), while it is still uncertain whether the microscopic and molecular signatures reflect the subseafloor indigenous alkaliphilic microbial communities.

Salinity and other limits

The salinity (NaCl) range of the natural environments on the Earth is substantially covered with the range of microbial growth. Many freshwater microorganisms can grow in distilled water only supplemented with complex organic substrates, while extreme halophiles grow in NaCl-saturated media and can survive even in salt crystals for the geologic timescale (Vreeland *et al.*, 2000). There has been no example of cultivation of extremely halophilic Archaea from the deep-sea environments, although their 16S rRNA gene sequences have been identified in deep-sea hydrothermal-vent chimneys and in deep-sea brine pools (Eder *et al.*, 1999; Takai *et al.*, 2001a). However, since many hypersaline and salt-deposit habitats are present in deep-subseafloor environments, extremely halophilic living Archaea and bacteria might be cultivated if such subseafloor biospheres are explored in the future.

Surviving radioactive irradiation is a significant process in the propagation of life on the Earth and in the Universe. The most radiation-resistant microorganism is known to be *Deinococcus radiodurans*, which can revive after a 150-kGy dose of gamma irradiation (Minton, 1994; Battista, 1997). This species and its relatives are usually isolated from various surface environments and no strain has yet been obtained from deep-sea and deep-subsurface environments (Battista, 1997). However, the 16S rRNA gene signatures of *Deinococcus* spp. were detected in the subseafloor environments of the PACMANUS Hydrothermal Field during the ODP Leg#193 (Kimura *et al.*, 2003). It is still uncertain whether the DNA signatures were truly retrieved from indigenous *Deinococcus* cells or from external contamination, and if they were subsurface *Deinococcus* members, whether they were radioactivity-tolerant or not. However, since deep-sea hydrothermal systems are naturally highly radioactive environments on Earth (Charmasson *et al.*, 2009), they may host many potentially radiation-resistant microorganisms. Indeed, many of the deep-sea hydrothermal-vent Archaea show relatively high radioresistance (Jolivet *et al.*, 2004).

Table 28.3. Various chemolithoautotrophic redox reactions possible in the deep-sea and deep-subseafloor biosphere.

Energy metabolism	Electron donor	Electron acceptor	Redox reaction	Cultivated (C) or Identified (I) in the deep sea or deep subseafloor
Aerobic methanotroph	CH_4	O_2	$CH_4 + 2O_2 = CO_2 + 2H_2O$	I and C
Sulphate-reducing anoxic methanotroph	CH_4	SO_4^{2-}	$CH_4 + SO_4^{2-} = HCO_3^- + HS^- + H_2O$	I but not yet C
Nitrate-reducing anoxic methanotroph	CH_4	NO_3^-	$5CH_4 + 8NO_3^- = 5HCO_3^- + 3OH^- + 6H_2O + 4N_2$	Not I or C
Hydrogenotrophic methanogen	H_2	CO_2	$4H_2 + CO_2 = CH_4 + 2H_2O$	I and C
Hydrogenotrophic acetogen	H_2	CO_2	$4H_2 + 2CO_2 = CH_3COOH + 2H_2O$	I and C
Anoxic carboxydotroph	CO	H_2O	$CO + H_2O = H_2 + CO_2$	I and C
Hydrogenotrophic sulphate-reducer	H_2	SO_4^{2-}	$4H_2 + SO_4^{2-} + 2H^+ = H_2S + 4H_2O$	I and C
Hydrogenotrophic sulphur-reducer	H_2	S^0	$H_2 + S^\circ = H_2S$	I and C
Aerobic sulphide-oxidizer	H_2S	O_2	$H_2S + 2O_2 = SO_4^{2-} + 2H^+$	I and C
Aerobic sulphur-oxidizer	S^0	O_2	$S^\circ + H_2O + 3/2O_2 = SO_4^{2-} + 2H^+$	I and C
Aerobic thiosulphate-oxidizer	$S_2O_3^{2-}$	O_2	$S_2O_3^{2-} + 5/2O_2 = 2SO_4^{2-}$	I and C
Denitrifying thiosulphate-oxidizer	$S_2O_3^{2-}$	NO_3^-	$5S_2O_3^{2-} + 8NO_3^- + H_2O = 10SO_4^{2-} + 4N_2 + 2H^+$	I and C
Ammonifying thiosulphate-oxidizer	$S_2O_3^{2-}$	NO_3^-	$S_2O_3^{2-} + 2H_2O + NO_3^{2-} = 2SO_4^{2-} + NH_4^+$	I and C
Denitrifying sulphur-oxidizer	S°	NO_3^-	$5S^\circ + 2H_2O + 6NO_3^- = 5SO_4^{2-} + 4H^+ + 3N_2$	I and C
Ammonifying sulphur-oxidizer	S°	NO_3^-	$4S^\circ + 7H_2O + 3NO_3^- = 4SO_4^{2-} + 3NH_4^+ + 2H^+$	I and C

Table 28.3. Continued

Energy metabolism	Electron donor	Electron acceptor	Redox reaction	Cultivated (C) or Identified (I) in the deep sea or deep subseafloor
Denitrifying sulphide-oxidizer	H_2S	NO_3^-	$5H_2S + 8NO_3^- = 5SO_4^{2-} + 4N_2 + 4H_2O + 2H^+$	I and C
Hydrogenotrophic oxygen-reducer	H_2	O_2	$2H_2+O_2 = 2H_2O$	I and C
Hydrogenotrophic iron-reducer	H_2	Fe(III)	$H_2 + 2Fe^{3+} = 2Fe^{2+} + 2H^+$	I and C
Aerobic carboxydotroph	CO	O_2	$2CO + O_2 = 2CO_2$	I but not yet C
Aerobic iron-oxidizer	Fe(II)	O_2	$4Fe^{2+} + O_2 + 4H^+ = 4Fe^{3+} + 2H_2O$	I and C
Denitrifying iron-oxidizer	Fe(II)	NO_3^-	$Fe^{2+} + 1/5NO_3^- + 2/5H_2O + 1/5H^+ = 1/10N_2 + Fe^{3+} + OH^-$	Not yet I or C
Aerobic manganese-oxidizer	Mn(II)	O_2	$2Mn^{2+} + O_2 + 4H^+ = 2Mn^{4+} + 2H_2O$	Not yet I or C
Denitrifying manganese-oxidizer	Mn(II)	NO_3^-	$5Mn^{2+} + 2NO_3^- + 12H^+ = N_2 + 5Mn^{4+} + 6H_2O$	Not yet I or C
Hydrogenotrophic manganese-reducer	H_2	MnO_2	$H_2 + MnO_2 + 2H^+ = Mn^{2+} + 2H_2O$	Not yet I or C
Aerobic nitrite-oxidizer	NO_2^-	O_2	$2NO_2^- + O_2 = 2NO_3^-$	Not yet I or C
Aerobic ammonia-oxidizer	NH_3	O_2	$2NH_3 + 3O_2 = 2NO_2^- + 2H_2O + 2H^+$	I and C
Denitrifying hydrogenotroph	H_2	NO_3^-	$2H_2 + 2NO_3^- = N_2 + 4H_2O + 2OH^-$	I and C
Ammonifying hydrogenotroph	H_2	NO_3^-	$4H_2 + NO_3^- = NH_4^+ + H_2O + 2OH^-$	I and C
Anoxic ammonia-oxidizer	NH_3	NO_3^-	$5NH_3 + 3NO_3^- + 3H^+ = 4N_2 + 9H_2O$	I but not yet C
Sulphur-disproportionator	S°	H_2O	$4S + 4H_2O \rightarrow SO_4^{2-} + 3H_2S + 2H^+$	Not yet I or C
Thiosulphate-disproportionator	$S_2O_3^{2-}$	H_2O	$S_2O_3^{2-} + H_2O = SO_4^{2-} + H_2S$	Not yet I or C

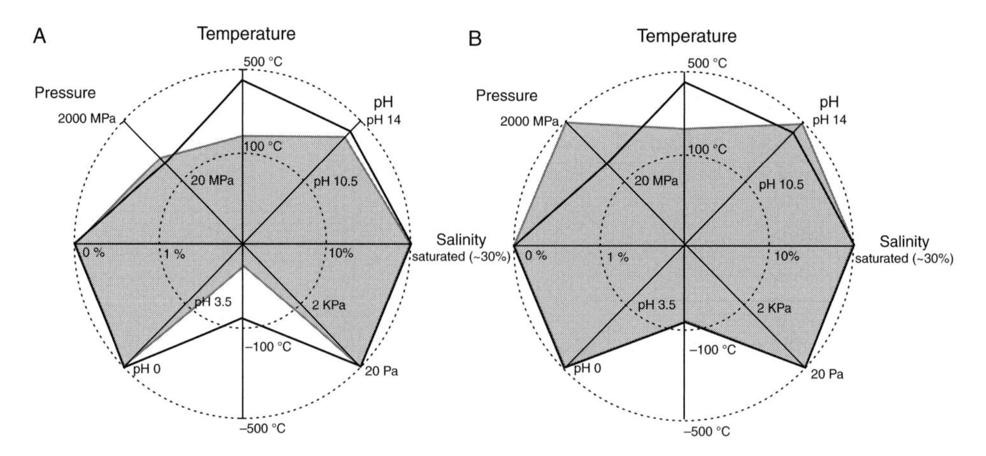

Figure 28.2. Diagrams of physical and chemical limits in natural environments of the Earth and for microbial growth (A) and survival (B). Open black line indicates the limits observed in the natural environments on the Earth. Closed grey line indicates the limits for microbial growth (A) and survival (B).

Among them, *Themococcus gammatolerans* is probably the most radioresistant microorganism from the deep-sea and deep-subsurface environments (Jolivet *et al.*, 2003). It was isolated from enrichment cultures after a 30-KGyr dose of gamma irradiation (Jolivet *et al.*, 2003).

The limitation and balance of energy and essential-element sources are also quite important chemical factors for controlling habitability on the Earth and in the Universe. Based on the recent microbiological exploration in the deep-sea and deep-subsurface environments, it is true that some types of microorganism are always present in any of the explored environments that, energetically, seem extremely barren and harsh. Prokaryotic life has great metabolic potential and can gain the energy for growth, maintenance and survival from numerous thermodynamically feasible redox couples and even from thermodynamically unpractical chemical reactions (Table 28.3). At present, it seems very likely that the balance of energy and essential-element sources has a great impact on the abundance of life and activity in a given habitat, but there has yet been no clear example found of habitats where the energy and elemental fluxes are below the sustainability of a microbial community.

Concluding remarks

In this chapter, the primary physical and chemical factors that constrain the limits of life and the biosphere on the Earth are overviewed, and the importance of exploration in the deep-sea and deep-subsurface biospheres is pointed at to elucidate the practical limits of life and the biosphere on the Earth and in the Universe. Among the primary physical and chemical constraints, temperature seems to be the greatest physical barrier for life to extend habitability (Figure 28.2). Namely, there is still a big gap between the temperature range of naturally present environments, the temperature range of environments hosting detectable cellular and molecular signatures of life and the temperature range of empirically determined microbial growth. This gap may point to the true temperature limit for life and the biosphere, but probably reveals the technical and methodological limits for detection and estimation of life and the biosphere. Not only temperature, but also other physical and chemical constraints should be more widely and intensively explored for clarification of mechanisms associated with restriction of life and activity. The marginal condition between the habitable and the uninhabitable terrains for Earth's life is of great importance in the field of astrobiology. Future deep-sea and deep-subsurface microbiological exploration will provide a key for elucidating the potential habitability of extraterrestrial life in the Universe.

References

Battista, J. R. (1997). Against all odds: the survival strategies of *Deinococcus radiodurans*. *Annual Review of Microbiology*, **51**, 203–24.

Bischoff, J. L. and Rosenbauer, R. J. (1988). Liquid–vapour relations in the critical region of the system NaCl-H_2O from 380 to 415°C: a refined determination of the critical point and two-phase boundary of seawater. *Geochimica et Cosmochimica Acta*, **52**, 2121–6.

Blöchl, E., Rachel, R., Burgraff, S., Hafenbradl, D., Jannasch, H. W. and Stetter, K. O. (1997). *Pyrolobus fumarii*, gen. and sp. nov., represents a novel group of Archaea, extending the upper temperature limit for life to 113°C. *Extremophiles*, **1**, 14–21.

Brock, T. D. (1978). *Thermophilic Microorganisms and Life at High Temperatures*. New York: Springer-Verlag.

Charmasson, S., Sarradin, P. M., Le Faouder, A., Agarande, M., Loyen, J. and Desbruyeres, D. (2009). High levels of natural radioactivity in biota from deep-sea hydrothermal vents: a preliminary communication. *Journal of Environmental Radioactivity*, **100**, 522–6.

Cowan, D. A. (2004). The upper temperature for life – where do we draw the line? *Trends in Microbiology*, **12**, 58–60.

Doemel, W. N. and Brock, T. D. (1971). The physiological ecology of *Cyanidinium caldarium*. *Journal of General Microbiology*, **67**, 17–32.

Eder, W., Ludwig, W. and Huber, R. (1999). Novel 16S rRNA gene sequences retrieved from highly saline brine sediments of Kebrit Deep, Red Sea. *Archives of Microbiology*, **172**, 213–18.

Edwards, K. J., Bond, P. L., Gihring, T. M. and Banfield, J. F. (2000). An archaeal iron-oxidizing extreme acidophile important in acid mine drainage. *Science*, **287**, 1796–9.

Horikoshi, K. (1999). Alkaliphiles: some applications of their products for biotechnology. *Microbiology and Molecular Biology Reviews*, **63**, 735–50.

Huber, R., Kurr, M., Jannasch, H. W. and Stetter, K. O. (1989). A novel group of abyssal methanogenic archaebacteria (*Methanopyrus*) growing at 110°C. *Nature*, **342**, 833–4.

Imshenetsky, A. A., Lysenko, S. V., Kazakov, G. A. and Ramkova, N. V. (1976). On microorganisms of the stratosphere. *Life Sciences and Space Research*, **14**, 359–62.

Jannasch, H. W., Wirsen, C. O., Molyneaux, S. J. and Langworthy, T. A. (1992). Comparative physiological studies on hyperthermophilic Archaea isolated from deep-sea hot vents with emphasis on *Pyrococcus* strain GB-D. *Applied and Environmental Microbiology*, **58**, 3472–81.

Jolivet, E., Corre, E., L'Haridon, S., Forterre, P. and Prieur, D. (2003). *Thermococcus gammatolerans* sp. nov., a hyperthermophilic archaeon from a deep-sea hydrothermal vent that resists ionizing radiation. *International Journal of Systematic and Evolutionary Microbiology*, **53**, 847–51.

Jolivet, E., Corre, E., L'Haridon, S., Forterre, P. and Prieur, D. (2004). *Thermococcus marinus* sp. nov. and *Thermococcus radiotolerans* sp. nov., two hyperthermophilic Archaea from deep-sea hydrothermal vents that resist ionizing radiation. *Extremophiles*, **8**, 219–27.

Kashefi, K. and Lovley, D. (2003). Extending the upper temperature limit for life. *Science*, **301**, 934.

Kato, C., Li, L., Tamaoka, J. and Horikoshi, K. (1997). Molecular analyses of the sediment of the 11 000-m deep Mariana Trench. *Extremophiles*, **1**, 117–23.

Kato, C., Li, L., Nogi, Y., Nakamura, Y., Tamaoka, J. and Horikoshi, K. (1998). Extremely barophilic bacteria isolated from the Mariana Trench, Challenger Deep, at a depth of 11 000 meters. *Applied and Environmental Microbiology*, **64**, 1510–13.

Kelley, D. S., Carson, J. A., Blackman, D. K., Fruh-Green, G. L., Butterfield, D. A., Lilley, M. D., Olson, E. J., Shrenk, M. O., Roe, K. K., Lebon, G. T., Rivizzigno, P. and the AT3–60 Shipboard Party. (2001). An off-axis hydrothermal vent field discovered near the Mid-Atlantic Ridge at 30° N. *Nature*, **412**, 145–9.

Kelley, D. S., Karson, J. A., Früh-Green, G. L., Yoerger, D., Shank, T. M., Butterfield, D. A., Hayes, J. M., Schrenk, M. O., Olson, E., Proskurowski, G., Jakuba, M., Bradley, A., Larson, B., Ludwig, K. A., Glickson, D., Buckman, K., Bradley, A. S., Brazelton, W. J., Roe, K., Elend, M., Delacour, A. G., Bernasconi, S. M., Lilley, M. D., Baross, J. A., Summons, R. E. and Sylva, S. P. (2005). A serpentinite-hosted ecosystem: The Lost City Hydrothermal Field. *Science*, **307**, 1428–34.

Kimura, H., Asada, R., Masta, A. and Naganuma, T. (2003). Distribution of microorganisms in the subsurface of the Manus Basin hydrothermal vent field in Paupa New Guinea. *Applied and Environmental Microbiology*, **69**, 644–8.

Klotz, B., Pyle, D. L. and Mackey, B. M. (2007). New mathematical modeling approach for predicting microbial inactivation by high hydrostatic pressure. *Applied and Environmental Microbiology*, **73**, 2468–78.

Lin, L.-H., Hall, J., Lippmann-Pipke, J., Ward, J. A., Lollar, B. S., DeFlaun, M., Rothmel, R., Moser, D., Gihring, T. M., Mislowack, B. and Onstott, T. C. (2005). Radiolytic H_2 in continental crust: nuclear power for deep subsurface microbial communities. *Geochemistry Geophysics Geosystems*, **6**, Q07003.

Margosch, D., Ehrmann, M. A., Buckow, R., Heinz, V., Vogel, R. F. and Gänzle, M. G. (2006). High-pressure-mediated survival of *Clostridium botulinum* and *Bacillus amyloliquefaciens* endospores at high temperature. *Applied and Environmental Microbiology*, **72**, 3476–81.

McCollom, T. M. and Bach, W. (2009). Thermodynamic constraints on hydrogen generation during serpentinization of ultramafic rocks. *Geochimica et Cosmochimica Acta*, **73**, 856–75.

McKay, C. P. (2001). The deep biosphere: lessons for planetary exploration. In *Subsurface Microbiology and Biogeochemistry*, eds. J. K. Fredrickson and M. Fletcher. New York: Wiley-Liss, Inc., pp. 315–27.

Minton, K. W. (1994). DNA repair in the extremely radioresistant bacterium *Deinococcus radiodurans*. *Molecular Microbiology*, **13**, 9–15.

Mitsuzawa, S., Deguchi, S., Takai, K., Tsujii, K. and Horikoshi, K. (2005). Flow-type apparatus for studying thermotolerance of hyperthermophiles under conditions simulating hydrothermal vent circulation. *Deep-Sea Research* 1, **52**, 1085–92.

Mottl, M. J., Komor, S. C., Fryer, P. and Moyer, C. L. (2003). Deep-slab fluids fuel extremophilic Archaea on a Mariana forearc serpentinite mud volcano: Ocean Drilling Program Leg 195. *Geochemistry Geophysics Geosystem*, **4**, (doi:10.1029/2003GC000588).
Nakagawa, T., Takai, K., Suzuki, Y., Hirayama, H., Konno, U., Tsunogai, U. and Horikoshi, K. (2006). Geomicrobiological exploration and characterization of a novel deep-sea hydrothermal system at the TOTO Caldera in the Mariana Volcanic Arc. *Environmental Microbiology*, **8**, 37–49.
Nordstrom, D. K., Alpers, C. N., Ptacek, C. J. and Blowes, D. W. (2000). Negative pH and extremely acidic mine waters from Iron Mountain, California. *Environmental Science and Technology*, **34**, 254–258.
Pley, U., Schipka, J., Gambacorta, A., Jannasch, H. W., Fricke, H., Rachel, R. and Stetter, K. O. (1991). *Pyrodictium abyssi* sp. nov. represents a novel heterotrophic marine archaeal hyperthermophile growing at 110°C. *Systematic and Applied Microbiology*, **14**, 245–53.
Rainey, F. A. and Oren, A. (2006). Extremophile microorganisms and the method to handle them. In *Methods in Microbiology: Extremophiles*, eds. F. A. Rainey and A. Oren. London, UK: Elsevier, vol. 35.
Resing, J. A., Lebon, G., Baker, E. T., Lupton, J. E., Embley, R. W., Massoth, G. J., Chadwick, W. W. and de Ronde, C. E. J. (2007). Venting of acid-sulfate fluids in a high-sulfidation setting at NW rota-1 submarine volcano on the Mariana Arc. *Economic Geology*, **102**, 1047–61.
Reysenbach, A.-L., Liu, Y., Banta, A. B., Beveridge, T. J., Kirshtein, J. D., Schouten, S., Tivey, M. K., Vom Domm, K. L. and Voytek, M. A. (2006). A ubiquitous thermoacidophilic archaeon from deep-sea hydrothermal vents. *Nature*, **442**, 444–7.
Rothschild, L. J. and Mancinelli, R. L. (2001). Life in extreme environments. *Nature*, **409**, 1092–101.
Salisbury, M. H. (2002). ODP Leg 195 Shipboard Scientific Party Site 1200. *Proceedings of Ocean Drilling Program Initial Reports*, vol. **195**. (www-odp.tamu.edu/publications/195_IR/195TOC.HTM.)
Schleper, C., Pühler, G., Kühlmorgen, B. and Zillig, W. (1995). Life at extremely low pH. *Nature*, **375**, 741–2.
Schrenk, M. O., Kelley, D. S., Bolton, S.A. and Baross, J. A. (2004). Low archaeal diversity linked to subseafloor geochemical processes at the Lost City Hydrothermal Field, Mid-Atlantic Ridge. *Environmental Microbiology*, **6**, 1086–95.
Seyfried, W. E., Jr., Ding, K. and Berndt, M. E. (1991). Phase equilibria constraints on the chemistry of hot spring fluids at mid-ocean ridges. *Geochimica et Cosmochimica Acta*, **55**, 3559–80.
Spiess, F. N. and RISE Group. (1980). East Pacific Rise; hot springs and geophysical experiments. *Science*, **297**, 1421–33.
Stetter, K. O., König, H. and Stackebrandt, E. (1983). *Pyrodictium* gen. nov., a new genus of submarine disc-shaped sulphur reducing archaebacteria growing optimally at 105°C. *Systematic and Applied Microbiology*, **4**, 535–51.
Swallow, J. C. and Crease, J. (1965). Hot salty water at the bottom of the Red Sea. *Nature*, **205**, 165–6.
Swart, P. K., Wortmann, U. G., Mitterer, R. M., Malone, M. J., Smart, P. L., Feary, D., Hine, A. C. and Shipboard Scientific Party. (2000). Hydrogen sulfide-rich hydrates and saline fluids in the continental margin of South Australia. *Geology*, **28**, 1039–42.
Takai, K. and Horikoshi, K. (1999). Genetic diversity of Archaea in deep-sea hydrothermal vent environments. *Genetics*, **152**, 1285–97.

Takai, K., Inoue, A. and Horikoshi, K. (1999). *Thermaerobacter marianensis* gen. nov., sp. nov., an aerobic extremely thermophilic marine bacterium from the 11 000-m deep Mariana Trench. *International Journal of Systematic Bacteriology*, **49**, 619–28.

Takai, K., Komatsu, T., Inagaki, F. and Horikoshi K. (2001a). Distribution and colonization of Archaea in a black smoker chimney structure. *Applied and Environmental Microbiology*, **67**, 3618–29.

Takai, K., Moser, D. P., Onstott, T. C., Spoelstra, N., Pfiffner, S. M., Dohnalkova, A. and Fredrickson, J. K. (2001b). *Alkaliphilus transvaalensis* gen. nov., sp. nov., an extremely alkaliphilic bacterium isolated from a deep South African gold mine. *International Journal of Systematic and Evolutionary Microbiology*, **51**, 1245–56.

Takai, K., Gamo, T., Tsunogai, U., Nakayama, N., Hirayama, H., Nealson, K. H. and Horikoshi, K. (2004). Geochemical and microbiological evidence for a hydrogen-based, hyperthermophilic subsurface lithoautotrophic microbial ecosystem (HyperSLiME) beneath an active deep-sea hydrothermal field. *Extremophiles*, **8**, 269–82.

Takai, K., Moyer, C. L., Miyazaki, M., Nogi, Y., Hirayama, H., Nealson, K. H. and Horikoshi, K. (2005). *Marinobacter alkaliphilus* sp. nov., a novel alkaliphilic bacterium isolated from subseafloor alkaline serpentine mud from Ocean Drilling Program (ODP) Site 1200 at South Chamorro Seamount, Mariana Forearc. *Extremophiles*, **9**, 17–27.

Takai, K., Nakamura, K., Suzuki, K., Inagaki, F., Nealson, K. H. and Kumagai, H. (2006). Ultramafics-Hydrothermalism-Hydrogenesis-HyperSLiME (UltraH^3) linkage: a key insight into early microbial ecosystem in the Archean deep-sea hydrothermal systems. *Paleontological Research*, **10**, 269–82.

Takai, K., Nakamura, K., Toki, T., Tsunogai, T., Miyazaki, M., Miyazaki, J., Hirayama, H., Nakagawa, S., Nunoura, T. and Horikoshi, K. (2008a). Cell proliferation at 122 °C and isotopically heavy CH_4 production by a hyperthermophilic methanogen under high-pressure cultivation. *Proceedings of the National Academy of Sciences of the USA*, **105**, 10,949–54.

Takai, K., Nunoura, T., Ishibashi, J., Lupton, J., Suzuki, R., Hamasaki, H., Ueno, Y., Kawagucci, S., Gamo, T., Suzuki, Y., Hirayama, H. and Horikoshi, K. (2008b). Variability in the microbial communities and hydrothermal fluid chemistry at the newly-discovered Mariner Hydrothermal Field, southern Lau Basin. *Journal of Geophysical Research*, **113**, G02031, doi:10.1029/2007JG000636.

Takami, H., Inoue, A., Fuji, F. and Horikoshi, K. (1997). Microbial flora in the deepest sea mud of the Mariana Trench. *FEMS Microbiology Letters*, **152**, 279–85.

Valentine, D. L. (2007). Adaptations to energy stress dictate the ecology and evolution of the Archea. *Nature Reviews Microbiology*, **5**, 316–23.

Vreeland, R. H., Rosenzweig, W. D. and Powers, D. W. (2000). Isolation of a 250-million-year-old halotolerant bacterium from a primary salt crystal. *Nature*, **407**, 897–900.

Yayanos, A. A. (1986). Evolutional and ecological implications of the properties of deep-sea barophilic bacteria. *Proceedings of the National Academy of Sciences of the USA*, **83**, 9542–6.

Part VIII

Life elsewhere?

29

Titan and the Cassini–Huygens mission

François Raulin and Jonathan Lunine

Titan before Cassini–Huygens

With a diameter of 5150 km, Titan is the largest satellite of Saturn. It was discovered in 1655 by Christiaan Huygens. The period of rotation of Titan around the Sun is that of Saturn, 29.5 years. With its obliquity of 27°, Saturn has seasons, each of 7 years' duration, and Titan's seasonality is the same thanks to the close alignment of its pole with Saturn's. In addition, Titan turns around Saturn – with synchronous rotation – within 16 Earth-days, thus Titan's solid surface rotates slowly; however, its atmosphere presents a super-rotation due to strong zonal winds. Titan's mean distance from the Sun is that of Saturn's – about 9.5 astronomical units (AU). This corresponds to a received solar flux at the top of its atmosphere just slightly more than 1% of the flux at the Earth. Moreover, distant from Saturn by about 20 Saturnian radii, Titan is far enough from the giant planet to avoid interactions with the rings, but still close enough to allow its atmosphere to interact with the electrons of the magnetosphere of Saturn, which thus play a role in its chemical evolution, together with the solar photons.

Titan is the only satellite in the Solar System having a dense atmosphere. The presence of its atmosphere was suggested in 1907 by José Comas-Sola based on his observations of the centre-to-limb darkening of Titan's disk. This was confirmed by Gerard Kuiper in 1944, who detected the IR signature of gaseous methane when observing Titan and deduced a methane partial pressure of around 0.1 bar at Titan's surface. Although the Pioneer 11 and two Voyager spacecraft, in 1979, 1980 and 1981, only flew by Titan, the Voyager 1 mission provided a very important set of data, which established the basic parameters of the atmosphere and set the stage both for future missions, and for important ground-based studies in the decades to follow.

The many pictures taken by the Voyager imaging system show the presence of orange-reddish haze layers in the whole atmosphere, which mask the surface of the satellite at optical wavelengths. The radio-occultation experiment provided the vertical profile of density, pressure and temperaturic of the atmosphere from 200 km altitude down to the surface, indicating a surface pressure and temperature of 1.5 bar and 94 K, respectively. The Voyager UV experiment identified N_2 as the main atmospheric constituent. The IRIS infrared spectroscopy experiment showed that CH_4 is the second most abundant atmospheric molecule, and detected several organic compounds as minor constituents including C_2–C_4 hydrocarbons (ethane, ethylene acetylene, propane, propyne and butadiyne) and C_1–C_4 nitriles (hydrogen cyanide, cyanogen, cyanoacetylene and dicyanoacetylene). It also allowed the detection of carbon dioxide. Thus, the Voyager data provided the first detailed picture of Titan's atmosphere: mainly made of N_2 (> 80%) and CH_4 (initially estimated to a few % up to more than 10%), with around 0.1% H_2 (Coustenis and Taylor, 2008). Argon was not detected and had a generous upper limit from Voyager data, which was re-estimated later on to be 6%. The re-estimated upper limit for methane was then 6%. Such a methane-rich atmosphere suggested the presence of methane reservoirs on Titan's surface, in the crust, or both. Several models were then described on the possibility of an ocean of methane and ethane on Titan's surface (see Lunine, 1993, for a review).

During the two decades which followed the Voyager flyby of Titan, several observations of the satellite were carried out from ground-based or Earth-orbiting observatories, allowing the detection in the IR or microwave spectral regions of new atmospheric species: CO, CH_3CN, and more recently H_2O and benzene. By observing Titan in the near-infrared region, between 1 and 2 micron (see for instance Combes *et al.*, 1997) in spectral windows where methane does not absorb strongly and diffraction by aerosols becomes weaker at longer wavelength, it was also possible to start seeing Titan's surface. These first observations showed a non-homogeneous surface, ruling out the possibility of a global ocean. Radar echo experiments carried out in the 1990s yielded similar conclusions, but suggested the presence of specular reflection in a few low-latitude areas. These specular reflections were interpreted to be caused by liquid (Campbell *et al.*, 2003).

Meanwhile, several photochemical models were published providing more detailed interpretation of Voyager data and insight into the chemical and physical processes involved in the chemical evolution of Titan's atmosphere, including formation of many primarily organic species. In parallel, laboratory experiments were carried out to mimic such processes and to produce the variety of organic compounds expected to be present in Titan's environment. Titan was starting to be a target of interest for exobiology because of the possibilities offered by its potential complex chemistry.

However, many questions still remained unsolved, and it was obvious soon after the first Titan data were obtained from Voyager, that a dedicated mission to the Saturn system was necessary to understand Titan, the many processes involved in its evolution and present state and the potential astrobiological aspects of this planetary body. To study such aspects became one of the main goals of the Cassini–Huygens mission.

The Cassini–Huygens mission

The idea of returning to Saturn and Titan was first initiated as early as 1982 by several scientists (Daniel Gautier, Wing Ip and Tobias Owen) from both sides of the Atlantic. Their efforts resulted in a joint NASA–ESA study from which came out a mission concept where NASA was to take care of a spacecraft to the giant planet, which would become an orbiter around Saturn, and ESA was put in charge of developing an atmospheric probe to explore Titan. After several changes in the mission configuration, the Cassini–Huygens mission was

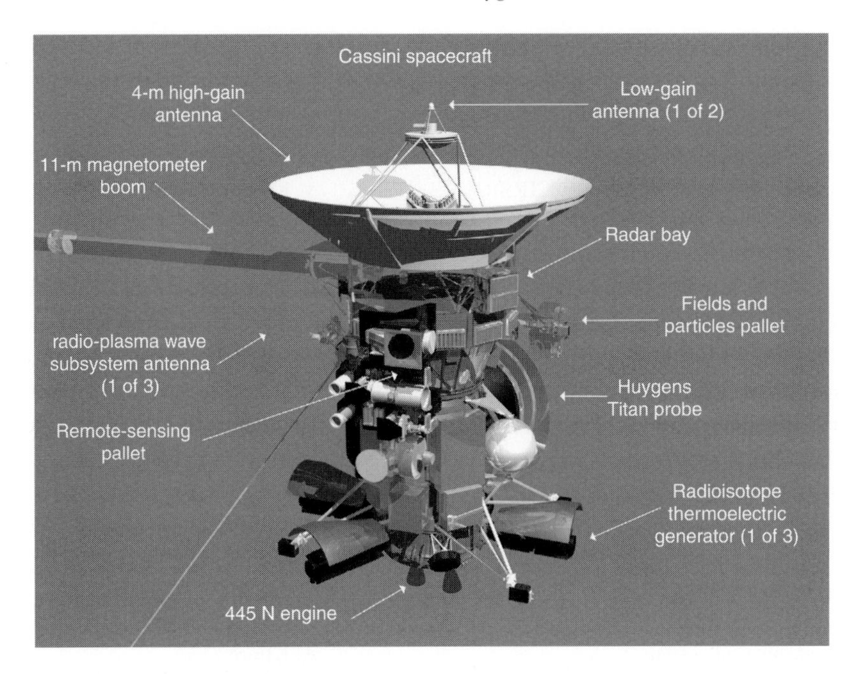

Figure 29.1 The Cassini Spacecraft. Courtesy of NASA/JPL/Caltech.

approved by NASA and ESA in 1989: the Saturn orbiter spacecraft was named 'Cassini' and the Titan probe 'Huygens'. The scientific payload of both components of the mission, and the interdisciplinary scientific investigations (IDS) were selected at the end of 1989 for Huygens and 1990 for Cassini.

Almost 7 metres high, with a 4-metre high-gain antenna and 11-metre-long boom, and a total mass of 5600 kg (including 3100 kg of propellant and 350 kg of Huygens probe) Cassini (Figure 29.1) is the largest spacecraft sent to the outer Solar System (Russell, 2005; JPL/NASA website). Three-axis-stabilized, with a system of gyroscopic wheels, the spacecraft is powered by radioisotope thermoelectric generators. It moves and points to the different targets which are selected for observation, then it turns its high-gain antenna to the Earth to send the data stored in a solid-state recorder system.

Cassini is carrying 12 scientific instruments, representing 335 kg (Table 29.1). Four instruments are devoted to remote-sensing observations in the visible, IR and UV ranges. These include the imaging sciences system (ISS), with a spectral coverage extending from the UV to the near-IR, the visual and mapping spectrometer (VIMS), a near-IR mapper able to observe Titan's surface, the composite infrared spectrometer (CIRS), a high-resolution Fourier transform IR spectrometer for studying Titan's atmospheric composition and temperature, and the ultraviolet imaging spectrometer (UVIS) providing information on the thermal and compositional profiles of the high atmosphere. Two instruments are observing in the radio-wavelength ranges, in particular a radar mapper. Five instruments are performing field-particle and wave measurements, including an ion and neutral mass spectrometer (INMS) able to measure *in situ* neutral molecules and positive and negative ions at very low concentration.

Table 29.1. Cassini spacecraft: scientific payload and interdisciplinary investigations.

Cassini instruments and interdisciplinary programmes	PI, team leader or IDS	
Optical remote-sensing instruments		
Composite infrared spectrometer (CIRS)	V. Kunde/M. Flasar	USA
Imaging science subsystem (ISS)	C. Porco	USA
Ultraviolet imaging spectrograph (UVIS)	L. Esposito	USA
Visual and IR mapping spectrometer (VIMS)	R. Brown	USA
Field particle and wave instruments		
Cassini plasma spectrometer (CAPS)	D. Young	USA
Cosmic dust analysis (CDA)	E. Grün/R. Srama	Germany
Ion and neutral mass spectrometer (INMS)	H. Waite	USA
Magnetometer (MAG)	D. Southwood /M. Dougherty	UK
Magnetospheric imaging instrument (MIMI)	S. Krimigis	USA
Radio and plasma wave spectrometer (RPWS)	D. Gurnett	USA
Microwave remote sensing		
Cassini radar (Radar)	C. Elachi	USA
Radio science subsystem (RSS)	A. Kliore	USA
Interdisciplinary investigations		
Magnetosphere and plasma	M. Blanc	France
Rings and dust	J. N. Cuzzi	USA
Magnetosphere and plasma	T. I. Gombosi	USA
Atmospheres	T. Owen	USA
Origin and evolution	J. Pollack	USA
Satellites and asteroids	L. A. Soderblom	USA
Aeronomy and solar-wind interaction	D. F. Strobel	USA

In addition, Cassini also carried Huygens (Russell, 2003; Lebreton *et al.*, 2009; ESA website), a 350-kg probe, devoted to the exploration of Titan's atmosphere, especially from about 150-km altitude to the surface, and, contingent on surviving landing devoted to the exploration of Titan's surface, with six instruments weighing in total 48 kg for landing. The scientific payload (Table 29.2) is composed of a gas chromatograph and mass spectrometer (GC–MS), with 3 GC columns, a mass range of the MS subsystem from 2 to 141 Daltons and 5 ion sources (one per column, plus two for direct MS analysis of the atmosphere and of ACP samples, respectively). ACP is an aerosols collector and pyrolyzer. It is designed to collect the atmospheric aerosols on a filter in two different zones from the stratosphere to the troposphere. Then the filter is moved into an oven, heated at different temperatures, which includes pyrolysis at 600°C. The resulting gases are analysed by the GC–MS instrument. The descent imager and spectral radiometer (DISR) is an optical remote-sensing instrument to study the main composition and dynamics of Titan's atmosphere, the properties of the aerosols, as well as the physical and chemical nature

Table 29.2. Huygens probe: scientific payload and interdisciplinary investigations.

Huygens instruments and interdisciplinary programmes	PI, team leader or IDS	
Scientific instruments		
Gas chromatograph–mass spectrometer (GC–MS)	H. Niemann	USA
Aerosol collector and pyrolyser (ACP)	G. Israël	France
Huygens atmospheric structure instrument (HASI)	M. Fulchignoni	Italy
Descent imager and spectral radiometer (DISR)	M. Tomasko	USA
Doppler wind experiment (DWE)	M. Bird	Germany
Surface science package (SSP)	J. Zarnecki	UK
Interdisciplinary investigations		
Aeronomy	D. Gautier	France
Atmosphere–surface interactions	J. I. Lunine	USA
Chemistry and exobiology	F. Raulin	France

of the surface. The Huygens atmosphere structure instrument is equipped with different sensors to measure the physical properties of Titan's atmosphere (temperature, pressure and lightning). The surface science package is dedicated to surface measurements, providing information on its physical nature. Finally, the Doppler wind experiment (DWE) is devoted to measurements of the directions and strengths of the zonal winds.

In addition (Tables 29.1 and 29.2), 7 IDSs were selected by NASA and 3 by ESA to carry out scientific investigations based on the use of data from different instruments of the mission in a synergic manner.

Cassini was launched on 15 October 1997 from Cape Canaveral, Florida, and flew a complex trajectory with multiple gravity assists needed to reach Saturn. Indeed, after two flybys of Venus in April 1998 and June 1999, followed by a flyby of the Earth in August 1999, the spacecraft flew by Jupiter in December 2000 and arrived in the Saturn system in 2004. The Saturn orbit insertion occurred on 1 July 2004, making Cassini the first artificial satellite of the Solar System's second giant planet. Its initial orbit allowed a first Titan fly-by in October 2004. On December 24, 2004, Cassini, during its third orbit around Saturn, released the Huygens probe, which entered into Titan's atmosphere on 14 January 2005, while Cassini flew by Titan at a minimum altitude of 60 000 km. The Huygens probe was decelerated with its parachutes, and the scientific instruments analysed the atmosphere, mainly from 150 km down to the surface during 2 hours and 26 minutes of descent, and more than 3 hours at the surface (Figure 29.2). All the atmosphere data – with the exception of DWE, because of the loss of one of the two Huygens transmitting channels – and more than one hour of surface data were recovered until the spacecraft came out of the field of view of the probe. DWE data were partly recovered using the Huygens carrier signal received directly by large radio telescopes on the Earth.

Figure 29.2. Descent and landing of the Huygens probe on Titan on 14 January 2005, as seen by an artist. Credit: NASA/JPL/ESA (PIA06434).

The nominal Cassini mission included 74 orbits around Saturn and 44 encounters with Titan. It was extended to mid-2010, becoming the 'Cassini Equinox Mission'. It was then 'extended-extended', with the 'Cassini Solstice Mission' which will run through 2017, allowing observations of Titan in total for almost half of a Saturn year.

Titan as seen by Cassini–Huygens

Titan, like the other main satellites of Saturn, was formed about 4.6 billion years ago in the sub-nebula of the giant planet, from planetesimals rich in a number of carbon and nitrogen compounds including CH_4–NH_3. The latter two volatile compounds, trapped as clathrate and stoichiometric hydrates (Alibert and Mousis, 2007, and references therein), have been progressively released into the atmosphere, where NH_3 has been converted into N_2. During the first million years of Titan's history, thermal energy produced by gravity processes and decay of radioactive nuclei had been high enough to maintain a large liquid water–ammonia ocean covered by a warm and dense atmosphere at the surface of Titan. The bottom of this ocean was in direct contact with silica bedrock (Tobie *et al.*, 2005). As the thermal energy fluxes decreased, the surface temperature did so also, and surface liquid water progressively froze out of the water–ammonia mixture. A surface water ice or methane clathrate hydrate (Tobie *et al.*, 2006) or mixed-composition (Fortes *et al.*, 2007) crust formed with a thickness increasing with time, covering an internal ocean, the bottom of which became progressively a thick layer of high-pressure water-ice (Lunine and Stevenson, 1987; Grasset and Sotin, 1996; Fortes, 2000). From the current models of the internal structure of Titan, the internal ocean today is 50–150 km below a primarily water-ice crust, is made of liquid water with a small percentage of ammonia and is located between two thick layers of water-ice.

General data

Since 2004, the Cassini–Huygens mission has been providing essential information on the Saturn system, and particularly on Titan. The Cassini orbiter offers global-scale mapping of Titan at moderate spatial resolution (hundreds of metres to kilometres) with potential temporal and spatial variations. The Huygens probe provides very detailed information on one particular location of Titan. The Cassini mission is now in its extended phase, but has the potential to continue for many years. The data already obtained have revealed a new vision of Titan. The first images of Titan taken by the Cassini instruments have shown a very diversified surface (Figure 29.3), with bright and dark zones, and terrains of various morphologies (Porco *et al.*, 2005; Elachi *et al.*, 2005; McCord *et al.*, 2008; Lunine *et al.*, 2008). At higher resolution fluvial, lacustrine and aeolian processes appear to have shaped the surface as they have on our own planet. This is one of the main astrobiological aspects of Titan. Others are related to the presence of a complex organic chemistry in the environments of the satellite, and the question of potential habitability.

The presence of an internal ocean is supported by internal models and the HASI experiment on Huygens. Beghin *et al.* (2009) recently interpreted the extremely low-frequency

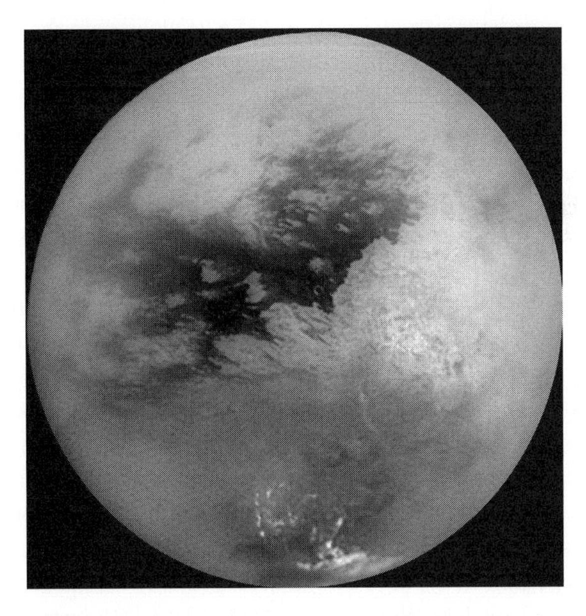

Figure 29.3. Mosaic of Titan's surface from images taken by the narrow-angle camera of the Cassini spacecraft. The large bright white area is named Xanadu. The resolution is about 1.3 km/pixel. Credit: NASA/JPL/Space Science Institute.

electric signal recorded by HASI as a Schumann resonance between the ionosphere and a modestly conducting ocean (since the ice is not conductive) roughly 50 km below the surface.

Cassini and Huygens observations have confirmed the structure of the low atmosphere, determined by Voyager, provided a more precise vertical profile in the stratosphere and troposphere and obtained information in the higher atmospheric zones. The equatorial surface temperature and pressure measured by HASI (Fulchignoni *et al.*, 2005) are, respectively, 93.65±0.25 K and 1.467±1 hPa. Due to its low surface temperature, Titan's atmosphere near the surface is about 4.5 times denser than the Earth's. Moreover, surface temperature measurements by CIRS (Jennings *et al.*, 2009) show temperatures 3 or 4 degrees lower at the poles relative to that at the equator, where Huygens landed).

Similarities with the Earth

Titan has a dense atmosphere which extends well above 1000 km. Like the atmosphere of the Earth, it is mainly composed of N_2. The other main constituents are CH_4 (~ 1.6–2% in the stratosphere as measured by the CIRS on Cassini and 5% near the surface as determined by the GC–MS on Huygens) and H_2 (~ 0.1%). The Cassini–Huygens mission has shown that argon is not present as a major atmospheric constituent, though the presence of small amounts of radiogenic 40-argon suggests outgassing has occurred over Titan's history (Niemann *et al.*, 2005). Titan is much colder than the Earth and has a troposphere with temperatures ranging from ~ 94 K near the surface to 70.4 K at the tropopause, and a stratosphere with temperatures from 70.4 K to 175 K (Fulchignoni *et al.*, 2005). However, Titan's atmosphere presents a complex structure qualitatively similar to that of the Earth. As shown by Voyager and Cassini–Huygens it has a mesosphere and a thermosphere. Indeed, both atmospheres have greenhouse gases and anti-greenhouse species. On Titan, CH_4 (condensable in Titan's conditions) and H_2 (non-condensable) are equivalent to H_2O and CO_2 on Earth. The haze particles and clouds in Titan's atmosphere induce an anti-greenhouse effect similar to that of the aerosols and clouds in the atmosphere of the Earth (McKay *et al.*, 1991).

On Titan, methane plays the role of water on the Earth. There is thus a methane cycle on Titan, which presents several analogies with that of water on Earth (Atreya *et al.*, 2006), but also many differences. Indeed on Earth, water is recycled into the interior at subduction zones (sink), and released at mid-ocean ridges and volcanoes (source), which provides the long-term cycle (one to a few hundred million years). On Titan, methane is photo-dissociated and forms ethane and other organic products in the atmosphere at similar, although shorter timescales (tens of millions of years), which do not recycle methane. The sink of methane on Titan is relatively well understood, but the source of methane is still very model-dependent. Cassini ISS has detected surface features near the South Pole, having the shape of a large lake (Porco *et al.*, 2005). The VIMS instrument has identified the presence of liquid ethane in Ontario Lacus confirming it is liquid (Brown *et al.*, 2008). The Cassini radar has discovered the presence of hundreds of lakes and three large seas

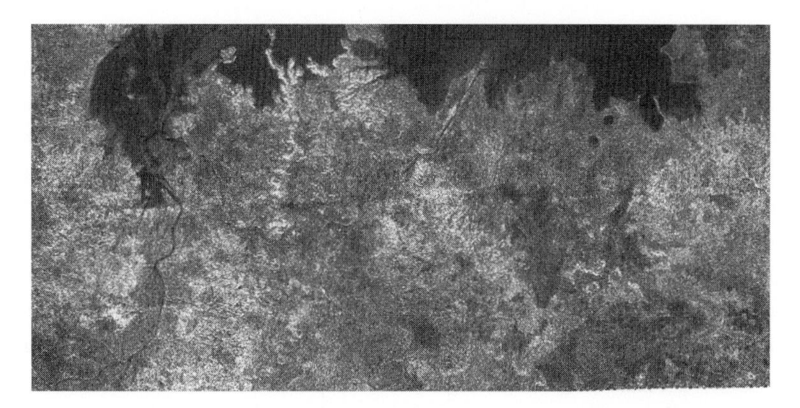

Figure 29.4. Cassini Radar image of Titan's surface in the north polar region taken in October 2006 showing several dark channel lakes of a total length of about 300 km. Credit: NASA/JPL (PIA01942).

Figure 29.5. Dendritic structures on the surface of Titan as seen by the DISR instrument on Huygens at 6.5-km altitude. Credit: ESA/NASA/JPL/University of Arizona (PIA07236).

in the north polar regions (Stofan *et al.*, 2007), also likely to be of liquid methane and ethane (Figure 29.4). The pictures of Titan's surface provided by the DISR instrument on the Huygens probe show dendritic structures (Tomasko *et al.*, 2005) similar to fluvial networks seen on Earth (Figure 29.5). Analogous but larger structures are also observed at coarser resolution on Cassini Orbiter radar images (Lorenz *et al.*, 2008a). These structures are located in relatively young terrains and produced by a recently flowing liquid. Moreover, the Huygens GC–MS (Niemann *et al.*, 2005) has measured the atmospheric CH_4 abundance from 150 km altitude down to the surface and the data show that above the landing site methane reaches the saturation level at an altitude of approximately 8 km, compatible with cloud formation, although they have not been observed at those latitudes. Such clouds could produce rain, which could be the origin of the observed fluvial-like networks. In addition, the GC–MS measurements after the probe landed strongly suggest the presence of condensed methane on the landing site.

Many observations of Titan's surface from the Cassini instruments show features evidently derived from diverse geological processes analogous to the terrestrial ones. There are mountains on Titan of several hundred metres high (Radebaugh *et al.*, 2007), signs of the presence of tectonic activity, fluvial and wind-driven features. There may be cryovolcanoes (Sotin *et al.*, 2005; Lopes *et al.*, 2007); traces of former flows across parts of the surface. However, on Titan fluid water mobilized and made buoyant by ammonia and other materials replaces terrestrial melted silicates. The surface of Titan also shows the presence of sedimentological and meteorological processes, as we see on Earth. There are many large dune areas (Lorenz *et al.*, 2006) where the terrestrial silica sand is probably replaced by the organic material of the aerosols (Lorenz *et al.*, 2008b). There are many features that look like terrestrial dry lakebeds, one of which may have been the flat pebbly area on which Huygens landed (Figure 29.6; Tomasko *et al.*, 2005). But again on Titan, the terrestrial silica pebbles are replaced by water-ice pebbles. There are also lakes and seas on Titan (Stofan *et al.*, 2007) observed in the Polar regions: here the terrestrial liquid water is replaced by liquid hydrocarbons, mainly methane and ethane. And there are clouds in the atmosphere (Griffith *et al.*, 2006; Rannou *et al.*, 2006, and references therein) made here also of methane and/or ethane.

Argon has been detected by INMS (Waite *et al.*, 2005) and GC–MS (Niemann *et al.*, 2005) in Titan's atmosphere. The most abundant isotope is ^{40}Ar, produced by the radioactive decay of ^{40}K, as on Earth; however, its mole fraction in the stratosphere of Titan is much lower: approximately 4×10^{-5}. Primordial argon, ^{36}Ar, has also been detected, at levels about 200 times lower. Other primordial noble gases may also be present, but have not been detected, meaning that their mole fraction is less than 10×10^{-9} (10 ppb). These observations indicate that the atmosphere of Titan is likely to be of secondary origin, formed by the degassing of gases trapped in the icy planetesimals of the Saturnian sub-nebula. Since N_2 is not efficiently trapped but NH_3 could be, this suggests that Titan's primordial atmosphere was made of ammonia, and the low argon abundance is consistent with this hypothesis (Owen, 1982). Then NH_3, through solar UV photolysis and impact-driven chemical processes, was progressively transformed into N_2. GC–MS has measured the stratospheric $^{14}N/^{15}N$ ratio: the value is 183, 1.5 times smaller than that of the primordial nitrogen. This suggests that the atmosphere has been lost several times since its formation, implying the simultaneous transformation of methane into organics and the presence of large deposits of organics on Titan's surface (Niemann *et al.*, 2005).

Figure 29.6. The surface of Titan around Huygens landing site, as seen by DISR instrument. The size of the pebble-like objects in the middle of the image is about 15 cm and they are likely to be made of water-ice. Credit: ESA/NASA/JPL/University of Arizona (PIA07232).

An active organic chemistry

There is an active organic chemistry which is currently going on in the different parts of Titan's environments. In spite of the absence of permanent liquid water bodies on Titan's surface, this organic chemistry shows many analogies with the terrestrial prebiotic chemistry which allowed the emergence of life on Earth. Indeed, several organics considered as key molecules in terrestrial prebiotic chemistry: hydrogen cyanide (HCN), cyanoacetylene (HC_3N) and cyanogen (C_2N_2) have been detected in Titan's atmosphere. They are present in the stratosphere among many compounds including hydrocarbons (with both saturated and unsaturated chains) and N-organics, essentially nitriles, as expected from laboratory simulation experiments. Cassini observations with CIRS have confirmed unambiguously the presence of water and benzene and provided the latitudinal and longitudinal vertical-concentration profiles of many of the stratospheric constituents (Flasar *et al.*, 2005; Coustenis *et al.*, 2007; Lavvas *et al.*, 2008).

A paramount and unexpected discovery on Titan's chemistry was made by INMS, which performed *in-situ* analysis of Titan's ionosphere during the low-altitude Cassini flyby of the satellite. INMS data show (Waite *et al.*, 2005, 2007) the presence of many organics at very high altitude (1100–1300 km). Although the mass range of this instrument is limited to 100 Daltons, extrapolation of its data, and coupling with the data from CAPS, indicates that high-molecular-weight species, up to several 1000 Daltons, may be present in the ionosphere (Waite *et al.*, 2007). This opens new pathways to explain the formation of complex organic compounds in Titan's environment, with an important role of ionospheric chemistry never considered before.

The low stratosphere and troposphere, on the contrary, seem relatively poor in organics, at least in the gas phase, as observed by GC–MS. This can be explained by their condensation on solid particles, very abundant in Titan's atmosphere. Two samplings of these aerosols have been made by the ACP instrument on Huygens: one in the 130–25-km altitude range and the second in the 25–20-km altitude range (Israël *et al.*, 2005), then the samples were heated and the resulting gases analysed by GC–MS. The data obtained indicate that the aerosols are made of refractory organics which release HCN and NH_3 during pyrolysis. These observations strongly support the idea that the aerosols have a molecular composition very close to that of the laboratory tholins and are probably made of a refractory organic nucleus, covered with condensed volatile compounds, with a mean diameter of the order of 1 micron.

The atmospheric aerosols sediment down to Titan's surface where they form a deposit of complex refractory organics and frozen volatiles. DISR data related to Titan's surface indicate the presence of water-ice but not of tholin-like materials (Tomasko *et al.*, 2005). However, the GC–MS data after Huygens landed show the presence of many organics, including C_3 and C_4 hydrocarbons, benzene and N-organics which seem to be released by the soil, due to the heating of the probe (Niemann *et al.*, 2005). This supports the hypothesis that the surface is rich in condensed volatile organic compounds and that most of the organics are condensed in the low atmosphere.

Thus, chemical evolution on Titan starts from the main atmospheric constituents N_2 and CH_4 and the coupling of their chemistries in the high atmosphere. The main products of these atmospheric chemical processes are complex refractory organics, which accumulate on the surface together with condensed volatile organic compounds such as HCN and benzene, but also ethane, which accumulates on the surface or the near subsurface, eventually dissolved in methane–ethane lakes and seas. These lakes are a very interesting depository of these various organics, since their concentration is generally much higher in such an exotic solvent than in the atmosphere (Raulin, 1987, 2008).

Moreover, in spite of the low temperatures, chemical evolution at Titan's surface is not frozen. Titan is still an evolving planetary body and its chemistry is also. Indeed, one can imagine that the aerosols and their complex organics after they accumulate on Titan's surface chemically react with the surface water-ice. Laboratory experiments on the possible evolution of tholins show that these analogues of Titan's aerosols can release a large variety of organics, including compounds of biological interest, such as amino acids, urea and hydroxyl-carboxylic acids, once in contact with liquid water (Khare *et al.*, 1986; Neish *et al.*, 2006; Raulin *et al.*, 2007; Neish *et al.*, 2008). This could even be possible with water-ice, with a longer time duration (Neish *et al.*, 2010; Ramirez *et al.*, 2010). Those processes could occur in areas of Titan's surface with cryovolcanic activity. Thus, in addition to the variety of compounds formed in Titan's atmosphere, one should also take into account those coming from surface evolution. Chemical analysis of Titan's surface, especially by *in-situ* measurements, is a unique opportunity to study some of the many processes which could be involved in prebiotic chemistry, including isotopic (Nguyen *et al.*, 2007) and enantiomeric fractionation.

This complex atmospheric chemistry also has some potential similarities with Earth's prebiotic atmospheric chemistry. New models of hydrogen escape from the primitive Earth atmosphere (Tian *et al.*, 2005) suggest that it has been more reducing than it was estimated previously, and would have included more than 30% H_2 and large amounts of CH_4. Studying Titan's atmosphere and its chemistry today may be a fantastic way to study, in real conditions, many of the prebiotic processes which were occurring on Earth four billion years ago, including in the aerosol phase, which has not yet been much considered.

In spite of the low surface temperature, the possible transient presence of liquid water cannot be excluded. The surface water-ice can be melted by large impactors, such as cometary impacts. Depending on the size of the impactor, water could remain liquid during periods of as long as several thousand years (O'Brien *et al.*, 2005), offering conditions analogous to terrestrial prebiotic syntheses with liquid water. Another location on Titan of great interest for potential prebiotic aqueous chemistry is the possible internal ocean. It could have been a very efficient prebiotic reactor at the beginning of Titan's history, when it was in direct contact with the silicate bedrock. The conditions were similar to those of the prebiotic oceans on Earth, with the possible presence of hydrothermal activity. This could have allowed a complex prebiotic chemistry, involving carbon, hydrogen, nitrogen but also oxygen atoms, to occur and evolve over several million years. Such conditions have permitted the emergence of life on Earth. Thus, even the emergence of life in this early Titan ocean cannot be excluded. If it did appear, was this environment habitable and is it still today?

Habitability of Titan

When considering life as we know it, based on carbon chemistry, liquid water and energy fluxes, the surface of Titan seems much too cold to provide conditions of habitability. However, the possibility of an exotic life, based on carbon but using other solvents than liquid water has already been discussed (Benner *et al.*, 2004; NRC, 2007). In particular, the use of liquid methane–ethane instead of water has been suggested by Schulze-Makuch and Grinspoon (2005) and McKay and Smith (2005). The latter have considered that life on Titan's surface in liquid methane could get the energy needed for its metabolism by the reduction of hydrocarbons, such as acetylene, into methane.

The habitability of the subsurface ocean of Titan has also been considered for a more conventional life, based on liquid water. In particular, Fortes (2000) has reported that there is no real obstacle to the development of living systems in this environment. The possible mean temperature of the ocean could be 260 K and even 300 K near possible cryovolcanic zones, compatible with some life we know on Earth. The high pressure expected in the ocean, at a depth of about 200 km, is of the order of 5 kbar, still in the range of possible values for terrestrial barophiles. The ocean is supposed to be made of liquid water with up to about 15% by weight of ammonia. Its pH would be ~ 11.5, still lower that the maximum value known on Earth, compatible with the development of alkaliphilic bacteria.

Even the energy resources, which may be limited in this ocean, are not incompatible with the development of life. Cassini–Huygens data do not show evidence of a dominant presence of biological activity on Titan's surface. For example, the measurements of the $^{12}C/^{13}C$ ratio in methane by GC–MS in the low atmosphere and near the surface of Titan are consistent with a non-biological origin (Niemann *et al.*, 2005). Nevertheless, this does not exclude the possible presence of a limited biota in the various environments of Titan outlined above, and future missions with specifically designed payloads to search for life are needed to address this exciting possibility.

The future

The Cassini–Huygens mission is in an extended phase and could still give surprising discoveries on Titan, in particular relating to its astrobiologically interesting properties. It has already provided many surprises. Its observations of Enceladus are a perfect illustration of that. This satellite – ten times smaller than Titan – is becoming a new astrobiological planetary target. Indeed, the images of Enceladus obtained by Cassini during its close flyby show large plumes of icy particles, permanently ejected from South-Pole terrains (Figure 29.7). The latter are covered with fractures, named 'tiger stripes', warmer than the surrounding areas, and probably in contact with a warmer internal part. Models developed to explain such phenomena consider the presence of a large pocket of pressurized liquid water inside Enceladus, partially in contact with hot rocks (Matson *et al.*, 2007), a situation of great potential for prebiotic chemistry, and habitability. Chemical analysis of the plumes by INMS shows the presence of ammonia and a variety of hydrocarbons, indicating that some organic chemical processes are going on inside Enceladus (Waite *et al.*, 2009). This supports the hypothesis that the internal water reservoir of the satellite may be a favourable environment for the synthesis of various organic compounds, including the building blocks of life, and even the emergence and persistence of life (McKay *et al.*, 2008).

With the many discoveries related to Titan and Enceladus, the Saturn system is becoming more and more interesting for astrobiology. New concepts of mission to return to Titan and

Figure 29.7. ISS Cassini image of Enceladus showing plumes coming from the south polar region of the satellite. Image credit: NASA/JPL/Space Science Institute (PIA07758).

Enceladus, with scientific instrumentation capable of studying these satellites more precisely, in particular their surfaces, have been considered by NASA (TEM: Titan Explorer Mission) and ESA (TandEM: Titan and Enceladus Mission) (Coustenis *et al.*, 2008). Both space agencies have fused these two concepts and are currently considering a joint mission, TSSM, Titan Saturn System Mission, especially to explore Titan in great detail. TSSM includes an orbiter and two *in-situ* elements, a Montgolfiere for long-term (six months) *in-situ* observation and a lander for *in-situ* exploration of one of Titan's hydrocarbon lakes. If this mission is selected, we could get new data of paramount interest for astrobiology, in particular on Titan's surface and subsurface, in the 2030s time frame. It is not that far away, when considering that the first studies on the Cassini–Huygens mission started in 1982, just after the Voyager data on Titan became available.

Acknowledgements

JIL's contribution was financed within the scope of the programme 'Incentivazione alla mobilita' di studiosi straineri e italiani residenti all'estero.' FR's contribution was financed by a CNES grant and by ESA's travel support.

References

Alibert, Y. and Mousis, O. (2007). Formation of Titan in Saturn's subnebula: constraints from Huygens probe measurements. *Astronomy and Astrophysics*, **465**, 1051–60.

Atreya, S. K., Adams, E. Y., Niemann, H. B., Demick-Montelara, J. E., Owen, T. C., Fulchignoni, M., Ferri, F. and Wilson, E. H. (2006). Titan's methane cycle. *Planetary and Space Science*, **54**, 1177–87.

Beghin, C., Canu, P., Karkoschka, E., Sotin, C., Bertucci, C., Kurth, W. S., Berthelier, J. J., Grard, R., Hamelin, M., Schwingenschuh, K. and Simões, F. (2009). New insights on Titan's plasma-driven Schumann resonance inferred from Huygens and Cassini data. *Planetary and Space Science*, **57**, 1872–88.

Benner, S. A., Ricardo, A. and Carrigan, M. A. (2004). Is there a common chemical model for life in the Universe? *Current Opinion in Chemical Biology*, **8**, 672–89.

Brown, R. H., Soderblom, L. A., Soderblom, J. M., Clark, R. N., Jaumann, R., Barnes, J. W., Sotin, C., Buratti, B., Baines, K. H. and Nicholson, P. D. (2008). The identification of liquid ethane in Titan's Ontario Lacus. *Nature*, **454**, 607–10.

Campbell, D. B., Black, G. J., Carter, L. M. and Ostro, S. J. (2003). Radar evidence for liquid surfaces on Titan. *Science*, **302**, 431–4.

Combes, M., Vapillon, L., Gendron, E., Coustenis, A., Lai, O., Wittemberg, R. and Sirdey, R. (1997). Spatially resolved images of Titan by means of adaptive optics. *Icarus*, **129**, 482–97.

Coustenis, A. and 24 co-authors. (2007). The composition of Titan's stratosphere from Cassini/CIRS mid-infrared spectra. *Icarus*, **189**, 35–62.

Coustenis, A. and 155 co-authors. (2008). TandEM: Titan and Enceladus mission. *Experimental Astronomy*, doi: 10.1007/s10686–008–9103-z.

Coustenis, A. and Taylor, F. (2008). *Titan: exploring an Earthlike World.* Singapore: World Scientific.

Elachi, C. and 34 co-authors. (2005). Cassini radar views the surface of Titan. *Science*, **308**, 970–4.

ESA website on Huygens probe: www.sci.esa.int/huygens/.

Flasar, F. M. and 44 co-authors. (2005). Titan's atmospheric temperatures, winds, and composition. *Science*, **308**, 975–8.

Fortes, A. D. (2000). Exobiological implications of a possible ammonia–water ocean inside Titan. *Icarus*, **146**, 444–52.

Fortes, A. D., Grindrod, P. M., Trickett, S. K. and Vocadlo, L. (2007). Ammonium sulfate on Titan: possible origin and role in cryovolcanism. *Icarus*, **188**,139–53.

Fulchignoni, M. and 42 co-authors. (2005). Titan's physical characteristics measured by the Huygens atmospheric instrument (HASI). *Nature*, **438**, 785–91.

Grasset, O. and Sotin, C. (1996). The cooling rate of a liquid shell in Titan's interior. *Icarus*, **123**, 101–12.

Griffith, C. A. and 13 co-authors. (2006). Evidence for ethane clouds on Titan from Cassini VIMS observations. *Science*, **313**, 1620–2.

Israël G. and 21 co-authors. (2005). Evidence for the presence of complex organic matter in Titan's aerosols by in situ analysis. *Nature*, **438**, 796–9.

Jennings, D. E., Flasar, F. M., Kunde, V. G., Samuelson, R. E., Pearl, J. C., Nixon, C. A., Carlson, R. C., Mamoutkine, A. A., Brasunas, J. C., Guandique, E., Achterberg, R. K., Bjoraker, G. L., Romani, P. N., Segura, M. E., Albright, S. A., Elliott, M. H., Tingley, J. S., Calcutt, S., Coustenis, A. and Courtin, R. (2009). Titan's surface brightness temperatures. *The Astrophysical Journal Letters*, **691**, L103–5.

JPL/NASA/ website on Cassini mission: www.saturn.jpl.nasa.gov/index.cfm

Khare, B. N., Sagan, C., Ogino, H., Nagy, B., Er, C., Schram, K. H. and Arakawa, E. T. (1986). Amino acids derived from Titan tholins. *Icarus*, **68**, 176–84.

Lavvas, P. P., Coustenis, A. and Vardavas, I. M. (2008). Coupling photochemistry with haze formation in Titan's atmosphere, Part ii: results and validation with Cassini/Huygens data. *Planetary and Space Science*, **56**, 67–99.

Lebreton, J. P., Coustenis, A., Lunine, J. I., Raulin, F., Owen, T. and Strobel, D. (2009). Results from the Huygens probe on Titan. *The Astronomy and Astrophysics Review*, **17**,149–79.

Lopes, R. M. C. and 43 co-authors. (2007). Cryovolcanic features on Titan's surface as revealed by the Cassini Titan radar mapper. *Icarus*, **186**, 395–412.

Lorenz, R. D. and 39 co-authors. (2006). The sand seas of Titan: Cassini RADAR observations of longitudinal dunes. *Science*, **312**, 724–7.

Lorenz, R. D. and 14 co-authors. (2008a). Fluvial channels on Titan: initial Cassini RADAR observations. *Planetary and Space Science*, **56**, 1132–44.

Lorenz, R. D., Mitchell, K. L., Kirk, R. L., Hayes, A. G., Aharonson, O., Zebker, H. A., Paillou, P., Radebaugh, J., Lunine, J. I., Janssen, M. A., Wall, S. D., Lopes, R. M., Stiles, B., Ostro, S., Mitri, G. and Stofan, E. R. (2008b). Titan's inventory of organic surface materials. *Geophysical Research Letters*, **35**, doi: 10.1029/2007GL032118.

Lunine, J. I. (1993). Does Titan have an ocean? A review of current understanding of Titan's surface. *Reviews of Geophysics*, **31**, 133–49.

Lunine, J. I. and Stevenson, D. J. (1987). Clathrate and ammonia hydrates at high pressure – application to the origin of methane on Titan. *Icarus*, **70**, 61–77.

Lunine, J. I. and 43 co-authors. (2008). Titan's diverse landscapes as evidenced by Cassini RADAR's third and fourth looks at Titan. *Icarus*, **195**, 415–33.

McCord, T. B., Hayne, P. G., Combe, J. P., Hansen, G. B., Barnes, J. W., Rodriguez, S., Le Mouélic, S., Baines, K. H., Buratti, B. J., Sotin, C., Nicholson, P., Jaumann, R., Nelson, R. and the Cassini VIMS Team. (2008). Titan's surface: search for spectral diversity and composition using the Cassini VIMS investigation. *Icarus*, **194**, 212–42.

McKay, C. P., Pollack, J. B. and Courtin, R. (1991). Titan: greenhouse and anti-greenhouse effects. *Science*, **253**, 1118–21.

McKay, C. P. and Smith, H. D. (2005). Possibilities for methanogenic life in liquid methane on the surface of Titan. *Icarus*, **178**, 274–6.

Matson, D. L., Castillo, J. J. and Johnson, T. V. (2007). Enceladus' plume: compositional evidence for a hot interior. *Icarus*, **187**, 569–73.

McKay, C. P., Porco, C. C., Altheide, T., Davis, W. L. and Kral, T. A. (2008). The possible origin and persistence of life on Enceladus and detection of biomarkers in the plume. *Astrobiology*, **8**, 909–19.

Neish, C. D., Lorenz, R. D., O'Brien, D. P. and the Cassini RADAR Team. (2006). The potential for prebiotic chemistry in the possible cryovolcanic dome Ganesa Macula on Titan. *International Journal of Astrobiology*, **5**, 57–65.

Neish, C. D., Somogyi, Á., Imanaka, H., Lunine, J. I. and Smith, M. A. (2008). Rate measurements of the hyrolysis of complex organic macromolecules in cold aqueous solutions: implications for prebiotic chemistry on the early Earth and Titan. *Astrobiology*, **8**, 273–87.

Neish, C. D., Somogyi, Á. and Smith, M. A. (2010). Titan's primordial soup: formation of amino acids via low-temperature of tholins. *Astrobiology*, **10**, 337–347.

Niemann, H. B. and 17 co-authors. (2005). The abundances of constituents of Titan's atmosphere from the GCMS instrument on the Huygens probe. *Nature*, **438**, 779–84.

Nguyen, M. J., Raulin, F., Coll, P., Derenne, S., Szopa, C., Cernogora, G., Israël, G. and Bernard, J. M. (2007). Carbon isotopic enrichment in Titan's tholins? Implications for Titan's aerosols. *Planetary and Space Science*, **55**, 2010–14.

NRC. National Research Council. (2007). *The Limits of Organic Life in Planetary Systems*. Washington, DC: Space Studies Board.

O'Brien, D. P., Lorenz, R. D. and Lunine, J. I. (2005). Numerical calculations of the longevity of impact oases on Titan. *Icarus*, **173**, 243–53.

Owen, T. (1982). The composition and origin of Titan's atmosphere. *Planetary and Space Science*, **30**, 833–8.

Porco, C. C. and 35 co-authors. (2005). Imaging of Titan from the Cassini spacecraft. *Nature*, **434**, 159–68.
Radebaugh, J., Lorenz, R. D., Kirk, R. L., Lunine, J. I., Stofan, E. R., Lopes, R. M. C., Wall, S. D. and the Cassini Radar Team. (2007). Mountains on Titan observed by Cassini Radar. *Icarus*, **192**, 77–91.
Ramirez, S. I., Coll, P., Buch, A., Brasse, C., Poch, O. and Ranlin, F. (2010). The fate of aerosols on the surface of Titan. *Faraday Discussions*, in press.
Rannou, P., Montmessin, F., Hourdin, F. and Lebonnois, S. (2006). The latitudinal distribution of clouds on Titan. *Science*, **311**, 201–5.
Raulin, F. (1987). Organic chemistry in the oceans of Titan. *Advances in Space Research*, **7**, 571–81.
Raulin, F., Nguyen, M. J. and Coll, P. (2007). Titan: an astrobiological laboratory in the Solar System. In *Proceedings in SPIE 6694, Instruments, Methods, and Missions for Astrobiology x*, eds. R. B. Hoover, G. V. Levin, A. Y. Rozanov and P. C. W. Davies, doi: 10.1117/12.732883.
Raulin, F. (2008). Planetary science: organic lakes on Titan. *Nature*, **454**, 587–9.
Russell, C. T. (2003). *The Cassini-Huygens Mission: Overview, Objectives and Huygens Instrumentarium*. Dordrecht, The Netherlands: Kluwer Academic Publishers.
Russell, C. T. (2005). *The Cassini-Huygens Mission: Orbiter Remote Sensing Investigations*. Dordrecht, The Netherlands: Kluwer Academic Publishers.
Schulze-Makuch, D. and Grinspoon, D. H. (2005). Biologically enhanced energy and carbon cycling on Titan? *Astrobiology*, **5**, 560–7.
Sotin, C. and 25 co-authors. (2005). Release of volatiles from a possible cryovolcano from near-infrared imaging of Titan. *Nature*, **435**, 786–9.
Stofan, E. R. and 37 co-authors. (2007). The lakes of Titan. *Nature*, **445**, 61–4.
Tian, F., Toon, O.B., Pavlov, A.A. and De Sterck, H. (2005). A Hydrogen-rich early Earth atmosphere. *Science*, **308**, 1014–17.
Tobie, G., Grasset, O., Lunine, J. I., Mocquet, A. and Sotin, C. (2005). Titan's internal structure inferred from a coupled thermal-orbital model. *Icarus*, **175**, 496–502.
Tobie, G., Lunine, J. and Sotin, C. (2006). Episodic outgassing as the origin of atmospheric methane on Titan. *Nature*, **440**, 61–4.
Tomasko, M. G. and 38 co-authors. (2005). Results from the descent imager/spectral radiometer (DISR) instrument on the Huygens probe of Titan. *Nature*, **438**, 765–78.
Waite, J. H. and 21 co-authors. (2005). Ion neutral mass spectrometer results from the first flyby of Titan. *Science*, **308**, 982–6.
Waite, J. H. and 15 co-authors. (2009). Liquid water on Enceladus from observations of ammonia and ^{40}Ar in the plume. *Nature*, **460**, 487–90.
Waite, J. H., Young, D.T., Cravens, T. E., Coates, A. J., Crary, F. J., Magee, B. and Westlake, J. (2007). The process of tholin formation in Titan's upper atmosphere. *Science*, **316**, 870–5.

30

The role of terrestrial analogue environments in astrobiology

Richard J. Léveillé

Introduction

Because Earth is the only place where we are certain that life exists, the characteristics of terrestrial life underpin our search for life elsewhere. In essence, the search for extraterrestrial life begins here on Earth. In the mid-twentieth century, early astrobiologists had recognized this reality and began studying life in remote and extreme environments that could be considered as analogues to places on Mars or elsewhere (e.g. Kooistra *et al.*, 1958; Cameron, 1963; Briot *et al.*, 2004). Early work by NASA and the Jet Propulsion Laboratory included studies of arid-soil microbiology in various locations, including the Atacama Desert and the Antarctic Dry Valleys (Cameron *et al.*, 1966; Cameron, 1969; Horowitz *et al.*, 1969; Cameron *et al.*, 1970). Testing of NASA's earliest life-detection instruments also took place at these and other extreme environments (Levin *et al.*, 1962; Levin and Heim, 1965). In parallel, microbiologists were also studying experimentally the survivability and adaptation of microorganisms isolated from desert soils and exposed to Lunar and Martian conditions in the context of forward contamination of the Moon and Mars, as well as towards the possibility of the existence of extraterrestrial life (e.g. Fulton, 1958; Kooistra *et al.*, 1958; Davis and Fulton, 1959; Packer *et al.*, 1963).

In recent years, Earth-based microbiological research, especially in harsh or extreme environments, has greatly expanded our understanding of the nature and limits of life (e.g. Rothschild and Mancinelli, 2001; Steven *et al.*, 2006; Pikuta *et al.*, 2007; Southam *et al.*, 2007). For example, we now know of microorganisms that thrive at, or at least tolerate, temperatures near or below freezing, very high acidity and salinity and very dry conditions. This research has, in turn, influenced the elaboration of astrobiology goals and the planning of space missions, including selecting landing sites, producing measurement requirements, instrument designs and integrated payload concepts (cf. Des Marais *et al.*, 2008). Learning about the range of environments on Earth that are inhabitable by microorganisms and the adaptations that these organisms possess has led to a consideration of a greater diversity of similar environments, past or present, in the Solar System where life could exist (Des Marais *et al.*, 2008). Furthermore, understanding the formation and preservation of microbial biosignatures in these environments directly aids in devising search strategies for life beyond Earth. Terrestrial analogue sites are thus an essential component of planetary exploration and astrobiology, especially given that the focus of missions (particularly for Mars) is currently evolving from global mapping to more detailed analyses at the outcrop, and even microscopic, scales (Farr, 2004).

Various aspects of terrestrial analogues, including analogue missions involving humans (Snook *et al.*, 2007), Canadian (Osinski *et al.*, 2006) and Australian (West *et al.*, 2009) analogue sites, instrumentation testing and validation (Léveillé, 2009) and exploration of Mars (Farr, 2004) have previously been reviewed. The current review focuses specifically on astrobiology. A description of the functions of analogues, and their evaluation, in terms of astrobiology is given here, along with several examples and some recommendations for future studies.

Astrobiology functions of analogues

Terrestrial analogue environments are defined as places on Earth that present one or more geological or environmental conditions similar to those found on an extraterrestrial body, either current or past (Osinski *et al.*, 2006; Léveillé, 2009). For example, the Atacama Desert, Chile, is considered to be one of the driest places on Earth, consequently, it is a useful analogue to the arid surface of Mars and for analogue studies of hydrology, mineralogy and extreme microbiology (Cabrol *et al.*, 2007a; Navarro-González *et al.*, 2003). A number of terrestrial analogue environments with important attributes or features relevant to astrobiology are listed in Table 30.1. In its widest definition, the term 'analogue' includes natural environments, materials (e.g. rocks, meteorites, soils, ice) and laboratory or artificial environments that mimic specific extraterrestrial conditions. For example, various 'Mars chambers' have been devised in order to simulate Martian environmental conditions for mineralogical and microbiological studies (e.g. Cloutis *et al.*, 2008; Jensen *et al.*, 2008).

The term 'analogue missions' refers to simulations of human and/or robotic planetary surface operations that involve an integrated set of activities at analogue field sites, encompass multiple aspects of a target mission and result in system-level interactions. Snook *et al.* (2007) define analogue missions as any analogue activity that integrates two or more high-fidelity (see below for discussion of fidelity) mission elements. Analogue missions can specifically target operations requirements and produce lessons learned in order to reduce costs and risks associated with eventual missions. For example, analogue missions can help to develop and test sampling strategies and measurement requirements, make ground truth measurements, direct training in planning science operations, promote team building and improve communications (Lebeuf *et al.*, 2008). Some examples of recent astrobiology-focused analogue missions include the MARTE drilling project at Rio Tinto, Spain (e.g. Stoker *et al.*, 2008); the Life in the Atacama Desert project (Cabrol *et al.*, 2007a); and the AMASE Svalbard expeditions (Steele *et al.*, 2008). Testing Mars surface operations, both robotic and human, in analogue settings is a high priority of the planetary science community (Farr, 2004; Osinski *et al.*, 2006).

Table 30.1. Important astrobiology-relevant terrestrial analogue environments and their main features.

Location	Main astrobiology features	Selected references
Devon Island, Nunavut, Canada	Haughton impact structure, impact-induced hydrothermal deposits, intracrater palaeolacustrine deposits (palaeo-lakes), polar desert, polygon terrain, gullies, Mars-like alteration minerals	Cockell *et al.* (2006); Glass *et al.* (2008); Lee (2007); Léveillé and Lacelle (2007); Snook *et al.* (2007)
Axel-Heiberg Island, Nunavut, Canada	Permafrost, massive ground ice, perennial cold springs, palaeosprings, gullies, endoliths/hypoliths, pingos/seasonal frost mounds	Andersen *et al.* (2002); Pollard *et al.* (2009)
Ellesmere Island, Nunavut, Canada	Supraglacial sulphur springs (Borup Fjord), endoliths/hypoliths, massive ground ice, patterned ground	Grasby *et al.* (2003); Omelon *et al.* (2007)
Pavilion Lake, British Columbia, Canada	Fresh-water microbialites	Laval *et al.* (2000)
Northern Canadian Mines: Lupin Mine, Nunavut, Canada; Kidd Creek Mine, Ontario	Deep-subsurface microbial ecosystems, deep-subsurface hydrocarbon gases	Pfiffner *et al.* (2008)
Canadian Shield and Abitibi Greenstone Belt, Canada	Archaean basalts and biosignatures, banded-iron formations, stromatolites, Precambrian microfossils	Ohmoto *et al.* (2008)
Svalbard, Norway	Vulcanism, ice, carbonates	Steele *et al.* (2008)
Rio Tinto Basin, Spain	Acidic river and groundwater system, iron- and sulphate-rich deposits, biosignatures, MARTE drilling project	Fernández-Remolar *et al.* (2005); Amils *et al.* (2007); Stoker *et al.* (2008)
Dry Valleys, Antarctica	Extreme aridity, high winds, extreme low temperatures, palaeolacustrine deposits, periglacial features, endoliths/hypoliths	Cameron *et al.* (1970); Dickinson and Rosen (2003)
Atacama Desert, Chile	Extremely arid conditions, dry + oxidizing soils, caves	Navarro-González *et al.* (2003); Cabrol *et al.* (2007a)
Western Deserts, California/Nevada/ Utah/ Arizona, USA	Cold desert, dry + oxidizing soils, iron concretions, endoliths/hypoliths, playas	Dong *et al.* (2007); Douglas *et al.* (2008)
Witwatersrand Basin, South Africa	Deep-subsurface microbial ecosystems, deep-subsurface hydrocarbon gases	Sherwood Lollar *et al.* (2007)
Central and Western Australia	Impact craters, extremely arid desert, playas, clay pans, river networks, oldest known microfossils (Pilbara), microbialites, acid lakes	West *et al.* (2009); Allwood *et al.* (2006); Rasmussen (2000)

Table 30.1. (Continued)

Location	Main astrobiology features	Selected references
Patagonia, Argentina	Proglacial and periglacial features, gullies	Pacifici (2009)
Andean lakes, Chile/ Bolivia	High-altitude and ice-covered lake, volcano-lacustrine sediments	Cabrol *et al.* (2007b)
Iceland	Active and recent basaltic vulcanism, hydrothermal systems, lava tubes and caves, cold-climate basaltic weathering	Preston *et al.* (2008)
Northern Africa	Buried river networks and craters (Egypt), playas and evaporates, episodic discharge (Tunisia)	Barbieri *et al.* (2006)

Modified from Léveillé (2009), Osinski *et al.* (2006) and Farr (2002).

In general, analogue studies can serve a variety of functions, which can be grouped into four main categories (Lee, 2007; Snook, 2008) as described below.

Learn

Planetary scientists regularly perform studies of terrestrial analogues in order to better understand remote or *in situ* robotic observations of other bodies, and to ultimately better understand our Solar System (Chapman, 2007). As planetary missions continually provide new and more detailed information on surface features, composition and structure, analogue studies can, in turn, help to reassess and deepen our current knowledge. With respect to astrobiology, terrestrial analogue sites provide insight into the limits to life on Earth, adaptations of organisms to extreme conditions, and the diversity of habitable environments, such as those colonized by endolithic microorganisms (e.g. Friedmann and Ocampo, 1976; Omelon *et al.*, 2007) or those in rocks affected by impact processes (e.g. Cockell *et al.*, 2006). Analogue studies, of both modern and ancient systems, also provide clues to the origin and early evolution of life on Earth and to the formation and preservation of biosignatures in the geological record (Southam *et al.*, 2007; Ohmoto *et al.*, 2008). In turn, these results directly assist in the formulation of scientific objectives, as well as operational and instrumentation requirements for future planetary missions. In addition, studies in analogue sites, especially those with extreme environmental conditions, can potentially lead to the discovery of novel organisms and metabolisms and the chemical or isotopic signatures of these metabolisms in Mars-like environments (NRC, 2007). For example, the study of marine anaerobic sediments and cold seeps led to the discovery of microbial anaerobic methane oxidation by a syntrophic consortium of microorganisms (Boetius *et al.*, 2000; Raghoebarsing *et al.*, 2006).

Test

Terrestrial analogue environments can be used to test hypotheses about possible extraterrestrial processes, including biosignature formation, degradation and preservation. They can also be used to test exploration strategies, biosignature detection and measurement techniques, sample-handling protocols and mission concepts in a variety of geological and environmental settings. Perhaps most importantly, terrestrial analogues can be used to test and validate analytical instruments, robotic systems and integrated hardware packages in challenging environments and under realistic operating conditions (i.e. representative of actual mission operations) at a relatively low cost and risk (Léveillé, 2009). For example, several Martian drills are currently being developed and many of these have recently been tested in the Atacama Desert, Hawaii, the Canadian Arctic, and at Rio Tinto, Spain (e.g. Glass *et al.*, 2008; Pfiffner *et al.*, 2008; Stoker *et al.*, 2008). Lessons learned from these drill tests can be directly applied to modifying and improving drill designs and related software and hardware systems for an eventual mission to Mars. Similarly, the Mars Exploration Rovers were tested extensively in 'Mars yards', artificial Mars-like terrains. Studies of analogue environments can also help to establish baseline abiotic signatures and provide positive controls for life-detection instruments.

Train

Studies of analogue sites are essential to mission development and execution and to the training of scientists and engineers, mission planners, science teams (backrooms), astronauts, ground operations crew and other stakeholders engaged in space exploration missions in a cost-effective low-risk way (NRC, 2007). Science teams need to learn how to detect and identify planetary features before missions are deployed by studying analogues (Chapman, 2007). Field-based projects also help to train students and researchers in the tools and knowledge needed to participate in future astrobiology research and payload development. In the early years of the Apollo programme, astronauts were trained extensively in field geology and geophysics at various analogue sites (Schaber, 2005). Future astronauts involved in missions to Mars will also benefit greatly from training in analogue environments, including participating in analogue missions (Thirsk *et al.*, 2007).

Engage

Analogue field campaigns, often in remote or exotic locations, are inherently of interest to the media and public, as well as to students of all ages. Such media attention can be positive for stakeholders and may be used as a recruitment tool for students and technical staff. Employing students in analogue research activities also helps to train highly qualified personnel who will become capable of contributing significantly to future astrobiology research and space-exploration programmes.

Analogue field studies also have the potential to bring together integrated teams of scientists, engineers and operations experts in coordinated field campaigns and mission simulations. Such multidisciplinary work can benefit cross-agency relations and international cooperation. Even countries without significant space programmes may be involved in mission preparation and planning through the use of their analogue sites.

Analogue activities are also useful for community building and catalysing research. For example, the Canadian Analogue Research Network (CARN), supported by the Canadian Space Agency, has been instrumental in developing the Canadian astrobiology community by offering funding for field work to be done in various analogue sites (Hipkin *et al.*, 2007). Though not specifically an astrobiology programme, roughly two-thirds of CARN projects to date have had at least a significant astrobiology component to them.

Non-terrestrial analogues

Non-terrestrial analogue environments can also be useful in astrobiology. The Moon may contain early crustal material from Earth, as well as preserve a record of large impacts and long-term exposure to radiation from a time when life on Earth is thought to have originated (Jakosky *et al.*, 2004). Organic material may be preserved in lunar polar regions and these may provide clues to the early evolution of life on Earth and possibly elsewhere in the Solar System. The Moon is also a useful analogue to Mars in terms of technology validation and testing. Finally, the Moon can serve as a negative control for life-detection studies and may provide lessons in forward contamination (planetary protection). Similarly, the Martian moons Phobos and Deimos may also serve as analogues to Mars. These satellites may contain water and carbon-rich rocks, which may inform us on the early environment and surface evolution of Mars. These moons, as well as near-Earth objects (e.g. asteroids), may also serve as testing grounds for future Martian missions, such as sample return missions (Lee *et al.*, 2008).

Fidelity and analogue evaluation

As no site on Earth is identical to another extraterrestrial body in all its aspects, the study of several different analogue environments may be necessary to fully understand an extraterrestrial process or the data returned from a space mission, or to properly test and validate space-mission technologies. The term 'fidelity' is often used to characterize the level of similarity of an analogue with respect to the extraterrestrial environment to which it is compared (Osinski *et al.*, 2006; Snook *et al.*, 2007; Léveillé, 2009). Sites which demonstrate several closely related attributes are said to offer a greater fidelity and are generally more useful as an analogue (though not exclusively). For example, polar deserts like the Antarctic Dry Valleys or Devon Island in the Canadian Arctic are considered to be high-fidelity analogues because they are cold, dry and unvegetated, with a variety of geological and geomorphologic features analogous to those on Mars.

Different methods of analogue fidelity evaluation have been proposed. The 'site-characterization matrix' rates various physical characteristics of an analogue site with respect to various functions (e.g. operations, geology field training; Snook, 2008). Qualitative estimates of different elements of fidelity (e.g. science value and operations, technology development and integration, outreach and education) have also been proposed along with their fidelity parameters (Snook *et al.*, 2007). For example, for the category 'science', fidelity parameters include relevance of the science to science goals determined by advisory committees or the scientific community. However, Snook *et al.* (2007) state that such parameters are subjective and that difficulties exist in comparing one-off field tests with multi-year multi-component testing programmes. Similarly, an 'analogue value index' has been proposed as a relative metric to evaluate and quantify the value of planetary analogues with respect to their fidelity, but also their applications, functions and cost-effectiveness (Lee, 2008). The NASA 'analogues database' (www.external.jsc.nasa.gov/analogs/analogsdb/) features a series of analogue sites, analogue missions and related infrastructure. In some cases, a 'planetary analogue site evaluation card' provides a brief description of the site, as well as figures of merit for various attributes (e.g. topography, biology, terrain and logistics).

The selection of an analogue site will depend on a combination of scientific, technical and logistical considerations. These considerations may be compatible or mutually exclusive depending on both the site and the investigation to be undertaken. Greeley and King (1977) noted that the basaltic Snake River Plain, Idaho, was ideal for analogue studies due to the combination of the relatively good state of preservation, lack of heavy vegetation and presence of a network of trails for access. Similarly, the Mojave Desert–Death Valley area is also considered to be a useful analogue to Mars due to its accessibility and variety of Martian-like landforms and processes there (Farr, 2004). The Canadian Arctic contains diverse analogue environments that are generally more accessible than those found in Antarctica (Osinski *et al.*, 2006). Other attributes of analogues include developed infrastructure (e.g. power, communication, roads and accommodation), easy aircraft overflight access (for remote sensing), cold and/or dry climate and good documentation (including maps, drill logs and photographs; Farr, 2004). However, it could be argued that developed infrastructure at an analogue site could, in fact, reduce the desirability of a site with respect to a surface-mission simulation. In some cases, simply being in a remote location with little infrastructure is an attribute in itself and makes for greater fidelity with respect to analogue mission-type field deployments. Furthermore, a lack of local infrastructure can force participants to utilize novel field-based systems or even space prototypes (e.g. communications, power) and may lead to useful collaborations with respect to support technologies for future space missions.

Future of analogues and astrobiology

An eventual detection of an extraterrestrial biosignature will raise two important questions: is it a true biosignature (i.e. has it been produced by life rather than abiotic processes) and is it real (i.e. has there been contamination along the overall pathway of sample acquisition, handling, processing and analysis)? Both of these questions can benefit greatly from extensive analogue studies, as shown herein. In this light, the US National Research Council recommends that terrestrial analogue studies be included as a fundamental component of Mars astrobiological research and the development of future robotic missions (NRC, 2007). Analogue studies will also be critical for Mars sample-return missions, whereby a complex coordination of novel technologies will be required at very carefully preselected sites (Borg *et al.*, 2008; Lee *et al.*, 2008). Analogue studies will also continue to be essential for selecting appropriate target sites (e.g. Golombek, 1997; Guizzo *et al.*, 2008), sampling and site-characterization strategies and for planning overall mission operations, as well as for developing the multitude of technologies and systems needed for such missions.

Terrestrial analogues have been used in planetary science and astrobiology for well over half a century. However, as new information is continually returned from various orbiter and surface missions, our understanding of extraterrestrial bodies evolves. This leads to the search for new and relevant analogue sites here on Earth. For instance, the discovery of the mineral jarosite ($KFe_3[(OH)_3\ SO_4]_2$) at Meridiani Planum by the MER Opportunity (Klingelhöfer *et al.*, 2004) has led to a search for relevant jarosite-precipitating terrestrial systems (e.g. Lacelle and Léveillé, 2009; Fernández-Remolar *et al.*, 2005). In light of recent discoveries on Mars, sites to further investigate for astrobiological analogue studies include lava tubes and subsurface caves (e.g. Léveillé and Datta, 2009), phyllosilicate-rich deposits in basaltic weathering systems (e.g. Léveillé and Konhauser, 2007), silica-rich hydrothermal deposits (e.g. Allen and Oehler, 2008; Squyres *et al.*, 2008) and cold-climate Mars-like sedimentary mineral localities (e.g. Peterson *et al.*, 2007; Lacelle and Léveillé, 2009).

Currently, there are a wide number and diversity of astrobiology studies taking place in analogue sites around the world (Table 30.1). It is recommended that there be greater coordination at a given site and amongst individual projects, as well as in future collaborative research proposals. Léveillé and Lacelle (2007) recommend coordinated astrobiological investigations and analogue missions at the Haughton-Mars Project Research Station, Devon Island, including integrated studies with a mission-centric (e.g. MSL, ExoMars) targeted approach and interdisciplinary and international collaboration. Farr (2004) recommends the coordinated deployment of remote-sensing instruments, *in situ* field instruments and personnel (including scientist–astronauts) to several sites in order to test instruments and technology, as well as to perform geological process studies. The US National Research Council recently recommended that studies in analogue sites include testing of instrumentation, development of techniques for the detection of biosignatures in conditions approaching the Martian environment and technology-validation studies (NRC, 2007).

Accurately georeferenced data collected from past and future analogue studies should be compiled, packaged and made widely and rapidly available either in an 'analogue atlas' or in existing planetary databases, such as NASA's planetary data system (PDS) or by other means (Farr, 2004). Currently, some useful planetary analogue collections of data are scattered in various locations, such as the geologic remote-sensing field experiment (GRSFE; Arvidson *et al.*, 1991), available at NASA's PDS geoscience node. The Canadian Space Agency is currently compiling and acquiring remote-sensing and ground-based data for analogue field sites as part of the CARN and a planetary WebGIS infrastructure is being developed for terrestrial analogues, as well as for the Moon and Mars (Germain *et al.*,

2008; Williamson *et al.*, 2008). NASA currently hosts a prototype web-based analogue database, but with limited or unequally detailed information.

In analogue studies, it is important to understand what might be similar and what might be different compared to the extraterrestrial site, and to ensure that the differences do not affect the analysis or test for which the analogue is being used (NRC, 2007). Therefore, in order to better assess the usefulness of different analogue sites, international standards for site characterization similar to standards for characterization of soils (e.g. geotechnical), including physical, chemical and mineralogical properties, should be developed. Some qualitative criteria already exist (Snook *et al.*, 2007). This would facilitate comparison of studies and interlaboratory comparison of analogue materials (astromaterials) and their analysis.

In the past, NASA has produced a series of comparative planetary geology field guides (Hawaii vulcanism; Colorado Plateau; Snake River Plain; Aeolian dunes of southern California). These field guides and others like them should be made widely available and updated using new knowledge, both from Earth and Mars, as well as with respect to modern instrumentation. Field guidebooks for analogue workshop field trips in Tunisia and Sicily are available from the International Research School of Planetary Sciences (Ori *et al.*, 2001; Ori *et al.*, 2002). In contrast, no detailed astrobiology field guide exists. Nevertheless, field-based astrobiology workshops (e.g. Stedman and Blumberg, 2005; Ohmoto *et al.*, 2008) greatly benefit the community and help to bring together different stakeholders, including scientists and engineers, as well as communications and operations specialists. Reports and guidebooks from these and future astrobiology workshops should also be published and made widely available.

Coordination with non-space government agencies will also benefit space-related analogue studies. In the USA, the USGS (e.g. Astrogeology programme, Hawaiian Volcano Observatory) and the National Park Service (e.g. Hawaii Volcanoes, Yellowstone and Lava Beds) offer unique expertise and local resources (Farr, 2004). In Canada, the CSA is currently collaborating with various departments of Natural Resources Canada, including the Geological Survey of Canada, the Polar Continental Shelf Programme, and CANMET in several analogue field-based activities. Cross-agency and international coordination is also recommended, as several countries are currently involved in analogue studies or in the development of analogue-related programmes and projects. Recently, an International Analogue Working Group has been started and an International Analogue Network has been proposed (Hipkin *et al.*, 2007).

One way to build expertise with respect to planetary exploration is to focus analogue studies on a few field sites and to acquire diverse remote-sensing and ground-truthing data (Farr, 2004). The CSA's CARN currently exploits three main sites with logistical support and various services for researchers: the Pavilion Lake Research Project, British Columbia; The Haughton-Mars Project Research Station, Devon Island, Nunavut; and the McGill Arctic Research Station, Axel-Heiberg Island, Nunavut (Hipkin *et al.*, 2007; Pollard *et al.*, 2009). For each site, WebGISs are currently being developed and populated with various remote-sensing and ground-based data (Germain *et al.*, 2008; Williamson *et al.*, 2008).

In addition, CSA–CARN-funded research is also taking place at a variety of other sites as well, both in Canada and abroad (Hipkin *et al.*, 2007).

Education and public outreach are important issues and ones where analogue studies can contribute greatly. Workshops for teachers and school students at Hawaii Volcanoes National Park already incorporate aspects of space missions and analogue studies. The CSA hosts an annual educator-training event, where analogue studies could be integrated into the curriculum of the conference. The Canadian Arctic Field Experience for Science Teachers took place on Axel-Heiberg Island in 2008 and enabled two Canadian teachers to become familiar with analogue-related aspects of the High Arctic. This is similar to the previously held MARSFEST professional development workshop (Buxner *et al.*, 2007). SETI's Astrobiology Summer Science Experience for Teachers (ASSET) is a science and curriculum institute for high school science teachers that includes pedagogical materials. Spaceward Bound, developed at NASA AMES, seeks to train the next generation of space explorers by having students and teachers explore and work in remote analogue environments (e.g. Canadian Arctic, Mojave Desert; McKay *et al.*, 2007). Similar opportunities should be continued to be developed.

Conclusions

Terrestrial analogue environments have been a part of astrobiology and planetary science for over fifty years and they will continue to be essential for our search for life in the Solar System and future exploration missions. Terrestrial analogues perform several functions relevant to astrobiology. Specifically, they provide insight into the origin and record of early life on Earth, inform us on search strategies for extraterrestrial habitable environments and biosignatures, and assist in the development, testing and validation of astrobiology instruments and space-exploration hardware and software systems. A more co-ordinated and collaborative approach to studying diverse and well-characterized relevant analogue environments will greatly advance the science of astrobiology in the coming years. Furthermore, testing and validating astrobiology instrumentation in appropriate analogue environments will significantly increase our chances of success in future missions to Mars and beyond.

References

Allen, C. C. and Oehler, D. Z. (2008). A case for ancient springs in Arabia Terra, Mars. *Astrobiology*, **8**, 1093–112.

Allwood, A. C., Walter, M. R., Kamber, B. S., Marshall, C. P. and Burch, I. W. (2006). Stromatolite reef from the Early Archaean Era of Australia. *Nature*, **441**, 714–18.

Andersen, D. T., Pollard, W. H., McKay, C. P. and Heldmann, J. (2002). Cold springs in permafrost on Earth and Mars. *Journal of Geophysical Research – Planets*, **107**, 4–1 to 4–7.

Amils, R., González-Toril, E., Fernández-Remolar, D., Gómez, F., Aguilera, A., Rodríguez, N., Malki, M., García-Moyano, A., Fairén, A. G., de la Fuente, V. and

Luis Sanz, J. (2007). Extreme environments as Mars terrestrial analogs: the Rio Tinto case. *Planetary and Space Science*, **55**, 370–81.

Arvidson, R. E., Dale-Bannister, M. A., Guiness, E. A., Slavney, S. H. and Stein, T. C. (1991). *Archive of Geologic Remote Sensing Field Experiment Release 1.0. NASA Planetary Data System: Geoscience Node*. www.pds-geosciences.wustl.edu/missions/grsfe/index.htm

Barbieri, R., Stivaletta, N., Marinangeli, L. and Ori, G. G. (2006). Microbial signatures in sabkha evaporite deposits of Chott el Gharsa (Tunisia) and their astrobiological implications. *Planetary and Space Science*, **54**, 726–36.

Boetius, A., Ravenschlag, K., Schubert, C. J., Rickert, D., Widdel, F., Gleseke, A., Amann, R. and Pfannkuche, O. (2000). A marine microbial consortium apparently mediating anaerobic oxidation of methane. *Nature*, **407**, 623–7.

Borg, L. E., Des Marais, D. J., Beaty, D. W., Aharonson, O., Benner, S. A., Bogard, D. D., Bridges, J. C., Budney, C. J., Calvin, W. M., Clark, B. C., Eigenbrode, J. L., Grady, M. M., Head, J. W., Hemming, S. R., Hinners, N. W., Hipkin, V., MacPherson, G. J., Marinangeli, L., McLennan, S. M., McSween, H. Y., Moersch, J. E., Nealson, K. H., Pratt, L. M., Righter, K., Ruff, S. W., Shearer, C. K., Steele, A., Sumner, D. Y., Symes, S. J., Vago, J. L. and Westall, F. (2008). Science priorities for Mars sample return. *Astrobiology*, **8**, 489–536.

Briot, D., Schneider, J. and Arnold, L. (2004). G. A. Tikhov, and the beginnings of astrobiology. In *Extrasolar Planets: Today and Tomorrow*, eds. J. P. Beaulieu, A. Lecavelier des Etangs and C. Terquem, Twentieth IAP Colloquium, ASP Conference Series, vol. CS 321, pp. 219–20.

Buxner, S. R., Keller, J. M., Shaner, A. J. and Bitter, C. F. (2007). *MARSFEST (Martian Arctic Regions Science Field Experience for Secondary Teachers) Professional Development Workshop and Teacher Ambassador Program*. 38th Lunar and Planetary Science Conference, Abstract #2050.

Cabrol, N. A., Wettergreen, D., Warren-Rhodes, K., Grin, E. A., Moersch, J., Diaz, G. C., Cockell, C. S., Coppin, P., Demergasso, C., Dohm, J. M., Ernst, L., Fisher, G., Glasgow, J., Hardgrove, C., Hock, A. N., Jonak, D., Marinangeli, L., Minkley, E., Ori, G. G., Piatek, J., Pudenz, E., Smith, T., Stubbs, K., Thomas, G., Thompson, D., Waggoner, A., Wagner, M., Weinstein, S. and Wyatt, M. (2007a). Life in the Atacama: searching for life with rovers (science overview). *Journal of Geophysical Research – Biogeosciences*, **112**, G04S02.

Cabrol, N. A., Grin, E. A. and Hock, A. N. (2007b). Mitigation of environmental extremes as a possible indicator of extended habitat sustainability for lakes on early Mars. *Proceedings of SPIE – The International Society for Optical Engineering*, **6694**, 669410.

Cameron, R. (1963). The role of soil science in space exploration. *Space Science Reviews*, **2**, 297–312.

Cameron, R. (1969). Abundance of microflora in soils of desert regions. *NASA Technical Report*, **32**, 1378.

Cameron, R., Gensel, D. R. and Blank, G. B. (1966). Soil studies – Desert microflora. XII. Abundance of microflora in soil samples from Chile Atacama Desert. In *Space Programs Summary*, Jet Propulsion Laboratory, Pasadena, CA, **37–38**(4), pp. 140–7.

Cameron, R. E., King, J. and David, C.N. (1970). Soil microbial ecology of Wheeler Valley, Antarctica. *Soil Science*, **109**, 110–20.

Chapman, M. G. (2007). *The Geology of Mars: Evidence from Earth-based Analogs*. Cambridge: Cambridge University Press.

Cloutis, E. A., Craig, M. A., Kruzelecky, R. V., Jamroz, W. R., Scott, A., Hawthorne, F. C. and Mertzman, S. A. (2008). Spectral reflectance properties of minerals exposed to simulated Mars surface conditions. *Icarus*, **195**, 140–68.
Cockell, C. S., Lee, P., Broady, P., Lim, D. S. S., Osinski, G. R., Parnell, J., Koeberl, C., Pesonen, L. and Salminen, J. (2006). Effects of asteroid and comet impacts on habitats for lithophytic organisms – a synthesis. *Meteoritics and Planetary Science*, **40**, 1901–14.
Davis, I., and Fulton, J. D. (1959). Microbiologic studies on ecologic considerations of the Martian environment. *Aeromedical Review*, review 2–60, 10p., Brooks Air Force Base, San Antonio, Texas, School of Aviation Medicine, United States Air Force.
Des Marais, D. J., Nuth III, J. A., Allamandola, L. J., Boss, A. P., Farmer, J. D., Hoehler, T. R., Jakosky, B. M., Meadows, V. S., Pohorille, A., Runnegar, B. and Spormann, A. M. (2008). The NASA astrobiology roadmap. *Astrobiology*, **8**, 715–30.
Dickinson, W. W. and Rosen, M. R. (2003). Antarctic permafrost: an analogue for water and diagenetic minerals on Mars. *Geology*, **31**, 199–202.
Dong, H., Rech, J. A., Jiang, H., Sun, H. and Buck, B. J. (2007). Endolithnic cyanobacteria in soil gypsum: occurences in Atacama (Chile), Mojave (United States), and Al-Jafr Basin (Jordan) Deserts. *Journal of Geophysical Research – Biogeosciences*, **112**, G02030.
Douglas, S., Abbey, W., Mielke, R., Conrad, P. and Kanik, I. (2008). Textural and mineralogical biosignatures in an unusual microbialite from Death Valley, California. *Icarus*, **193**, 620–36.
Farr, T. G. (2004). Terrestrial analogs to Mars: the NRC community decadal report. *Planetary and Space Science*, **52**, 3–10.
Farr, T. G. (2002). Terrestrial analogs to Mars. In *Community Contributions to the NRC Solar System Exploration Decadal Survey*, Astronomical Society of the Pacific Conference Series, ed. M. V. Sykes, vol. 272, pp. 35–76.
Fernández-Remolar, D., Morris, R. V., Gruener, J. E., Amils, R. and Knoll, A. H. (2005). The Río Tinto Basin, Spain: mineralogy, sedimentary geobiology, and implications for interpretation of outcrop rocks at Meridiani Planum, Mars. *Earth and Planetary Science Letters*, **240**, 149–67.
Friedmann, E. I. and Ocampo, R. (1976). Endolithic blue-green algae in dry valleys: primary producers in Antarctic desert ecosystem. *Science*, **193**, 1247–9.
Fulton, J. D. (1958). Survival of terrestrial microorganisms under simulated Martian conditions. In *Physics and Medicine of the Atmosphere and Space*, eds., O. O. Benson and H. Strughold. John Wily and Sons Inc., New York, pp. 606–613. 606–13.
Germain, M., Williamson, M-.C., Léveillé, R. and Phaneuf, M. (2008). *Advanced Uses of Geographic Information Systems for Terrestrial Analogues and Planetary Databases*. 6th Canadian Space Exploration Workshop, St-Hubert, QC.
Glass, B., Cannon, H., Branson, M., Hanagud, S. and Paulsen, G. (2008). DAME: planetary-prototype drilling automation. *Astrobiology*, **8**, 653–64.
Golombek, M. (1997). Size-frequency distributions of rocks on Mars and Earth analog sites: implications for future landed missions. *Journal of Geophysical Research – Planets*, **102**, 4117–29.
Grasby, S. E., Allen, C. C., Longazo, T. G., Lisle, J. T., Griffin, D. W. and Beauchamp, B. (2003). Supraglacial sulphur springs and associated biological activity in the Canadian high Arctic – signs of life beneath the ice. *Astrobiology*, **3**, 583–96.

Greeley, R. and King, J. S. (1977). *Volcanism of the Eastern Snake River Plain, Idaho: A Comparative Planetary Geology Guidebook*. Washington DC: National Aeronautics and Space Administration, p. 308.

Guizzo, G. P., Bertoli, A., Torre, A. D., Magistrati, G., Mailland, F., Vukman, I., Philippe, C., Jurado, M. M., Ori, G. G., Macdonald, M., Romberg, O., Debei, S. and Zaccariotto, M. (2008). Mars and Moon exploration passing through the European Precision Landing GNC Test Facility. *Acta Astronautica*, **63**, 74–90.

Hipkin, V. J., Osinski, G. R., Berinstain, A. and Léveillé, R. (2007). *The Canadian Analogue Research Network (CARN): Opportunities for Terrestrial Analogue Studies in Canada and Abroad*. 38th Lunar and Planetary Science Conference, Abstract # 2052.

Horowitz, N. H., Bauman, A. J., Cameron, R. E., Geiger, P. J., Hubbard, J. S., Shulman, G. P., Simmonds, P. G. and Westberg, K. (1969). Sterile soil from Antarctica: organic analysis. *Science*, **164**, 1054–6.

Jakosky, B., Anbar, A., Taylor, J. and Lucey, P. (2004). *Astrobiology Science Goals and Lunar Exploration: NASA Astrobiology Institute White Paper*. Mountain View, CA: NASA Astrobiology Institute, p. 12.

Jensen, L. L., Merrison, J., Hansen, A. A., Mikkelsen, K. A., Kristoffersen, T., Nørnberg, P., Lomstein, B. A. and Finster, K. (2008). A facility for long-term Mars simulation experiments: the Mars environmental simulation chamber (MESCH). *Astrobiology*, **8**, 537–48.

Klingelhöfer, G., Morris, R. V., Bernhardt, B., Schröder, C., Rodionov, D. S., de Souza Jr, P. A., Yen, A., Gellert, R., Evlanov, E. N., Zubkov, B., Foh, J., Bonnes, U., Kankeleit, E., Gütlich, P., Ming, D. W., Renz, F., Wdowiak, T., Squyres, S. W. and Arvidson, R. E. (2004). Jarosite and hematite at Meridiani Planum from Opportunity's Mössbauer spectrometer. *Science*, **306**, 1740–5.

Kooistra, J. A., Mitchell, R. B. and Strughold, H. (1958). The behavior of microorganisms under simulated Martian environmental conditions. *Publications of the Astronomical Society of the Pacific*, **70**, 64–9.

Lacelle, D. and Léveillé, R. (2009). Acid drainage generation and associated Ca-Fe-SO_4 minerals in a periglacial environment, Eagle Plains, northern Yukon, Canada: an analogue for low temperature sulfate formation on Mars. *Planetary and Space Science*, doi: 10.1016/j.pss.2009.06.009.

Laval, B., Cady, S. L., Pollack, J. C., McKay, C. P., Bird, J. S., Grotzinger, J. P., Ford, D. C. and Bohm, H. R. (2000). Modern freshwater microbialite analogues for ancient dendritic reef structures. *Nature*, **407**, 626–9.

Lebeuf, M., Madder, M. M. and Williamson, M. C. (2008). *Analogue missions as an integration mechanism to develop lunar exploration strategies*. Joint Annual Meeting of LEAG-ICEUM-SRR, abstract 4033.

Lee, P., Hildebrand, A. R., Richards, R. and PRIME Mission Team. (2008). *The PRIME (Phobos reconnaissance and international Mars exploration) mission and Mars sample return*. 39th Lunar and Planetary Science Conference, abstract 2268.

Lee, P. (2008). *Planetary analogs: A quantified evaluation*. Geological Association of Canada – Mineralogical Association of Canada Joint Annual Meeting, Québec, QC.

Lee, P. (2007). *Haughton-Mars project 1997–2007: a decade of Mars analog science and exploration research at Haughton Crater, Devon Island, High Arctic*. Second International Workshop – Exploring Mars and its Earth Analogues, Trento, Italy.

Léveillé, R. J. and Datta, S. (2009). Lava tubes and basaltic caves as astrobiological targets on Earth and Mars: a review. *Planetary and Space Science*, doi: 10.1016/j.pss.2009.06.004.

Léveillé, R. J. and Konhauser, K. O. (2007). *Geomicrobiology of clay minerals: implications for life on Early Mars*. 38th Lunar and Planetary Science Conference, abstract 1338.

Léveillé, R. J. and Lacelle, D. (2007). *Astrobiology investigations in and around the Haughton Impact Structure*. Geological Society of America Annual Meeting, Denver, CO, abstract 47–6.

Léveillé, R. J. (2009). Validation of astrobiology technologies and instrument operations in terrestrial analogue environments. *Comptes Rendus Palevol*, **8**, 637–48.

Levin, G. V., Heim, A. H., Clendenning, J. R. and Thompson, M.-F. (1962). 'Gulliver' – a quest for life on Mars. *Science*, **138**, 114–21.

Levin, G. V. and Heim, A. H. (1965). *Gulliver and Diogenes – Exobiological Antitheses*. COSPAR, Life Sciences and Space Research III. Amsterdam: North-Holland Publishing Co., pp. 105–19.

McKay, C. P., Coe, L. K., Battler, M., Bazar, D., Boston, P., Conrad, L., Day, B., Fletcher, L., Green, R., Heldmann, J., Muscatello, T., Rask, J., Smith, H., Sun, H. and Zubrin, R. (2007). *Spaceward Bound: field training for the next generation of space explorers*. LEAG Workshop on Enabling Exploration: The Lunar Outpost and Beyond, abstract 3028.

National Research Council. (2007). *An Astrobiology Strategy for the Exploration of Mars*. Washington DC: The National Academies Press, pp. 118.

Navarro-González, R., Rainey, F. A., Molina, P., Bagaley, D. R., Hollen, B. J., de la Rosa, J., Small, A. M., Quinn, R. C., Grunthaner, F. J., Caceres, L., Gomez-Silva, B. and McKay, C. P. (2003). Mars-like soils in the Atacama Desert, Chile, and the dry limit of microbial life. *Science*, **302**, 1018–21.

Ohmoto, H., Runnegar, B., Kump, L. R., Fogel, M. L., Kamber, B., Anbar, A. D., Knauth, P. L., Lowe, D. R., Sumner, D. Y. and Watanabe, Y. (2008). Biosignatures in ancient rocks: a summary of discussions at a field workshop on biosignatures in ancient rocks. *Astrobiology*, **8**, 883–907.

Omelon, C. R., Pollard, W. H. and Ferris, F. G. (2007). Inorganic species distribution and microbial diversity within high Arctic cryptoendolithic habitats. *Microbial Ecology*, **54**, 740–52.

Ori, G. G., Glamoclija, M. and Rossi, A. P. (2002). *Field Trip Guidebook: Sicily and Mount Etna*. Workshop on Exploring Mars and its Earth Analogues. International Research School of Planetary Sciences, Pescara, Italy, pp. 86.

Ori, G. G., Komatsu, G. and Marinangeli, L. (2001). *Field Trip Guidebook: Chott el Gharsa and Chott el Jerid*. Workshop on Exploring Mars and its Earth Analogues. International Research School of Planetary Sciences, Pescara, Italy, pp. 95.

Osinski, G. R., Léveillé, R., Berinstain, A., Lebeuf, M. and Bamsey, M. (2006). Terrestrial analogues to Mars and the moon: Canada's role. *Geoscience Canada*, **33**, 175–88.

Pacifici, A. (2009). The Argentinean Patagonia and the Martian landscape. *Planetary and Space Science*, **57**, 571–8.

Packer, E., Scher, S. and Sagan, C. (1963). Biological contamination of Mars ii. Cold and aridity as constraints on the survival of terrestrial microorganisms in simulated Martian environments. *Icarus*, **2**, 293–316.

Peterson, R. C., Nelson, W., Madu, B. and Shurvell, H. F. (2007). Meridianiite: a new mineral species observed on Earth and predicted to exist on Mars. *American Mineralogist*, **92**, 1756–9.

Pfiffner, S. M., Onstott, T. C., Ruskeeniemi, T., Talikka, M., Bakermans, C., McGown, D., Chan, E., Johnson, A., Phelps, T. J., Puil, M. L., Difurio, S. A., Pratt, L. M., Stotler, R., Frape, S., Telling, J., Lollar, B. S., Neill, I. and Zerbin, B. (2008). Challenges for coring deep permafrost on Earth and Mars. *Astrobiology*, **8**, 623–8.

Pikuta, E. V., Hoover, R. B. and Tang, J. (2007). Microbial extremophiles at the limits of life. *Critical Reviews in Microbiology*, **33**, 183–209.

Pollard, W., Haltigin, T., Whyte, L., Niederberger, T., Andersen, D., Omelon, C., Nadeau, J. and Lebeuf, M. (2009). Overview of analogue science activities at the McGill Arctic Research Station, Axel Heiberg Island, Canadian High Arctic. *Planetary and Space Science*, **57**, 646–59.

Preston, L. J., Benedix, G. K., Genge, M. J. and Sephton, M. A. (2008). A multidisciplinary study of silica sinter deposits with applications to silica identification and detection of fossil life on Mars. *Icarus*, **198**, 331–50.

Raghoebarsing, A. A., Pol, A., van de Pas-Schoonen, K. T., Smolders, A. J. P., Ettwig, K. F., Rijpstra, W. I. C., Schouten, S., Sinninghe Damsté, J. S., Op den Camp, H. J. M., Jetten, M. S. M. and Strous, M. (2006). A microbial consortium couples anaerobic methane oxidation to denitrification. *Nature*, **440**, 918–21.

Rasmussen, B. (2000). Filamentous microfossils in a 3235-million-year-old volcanogenic massive sulphide deposit. *Nature*, **405**, 676–9.

Rothschild, L. J. and Mancinelli, R. L. (2001). Life in extreme environments. *Nature*, **409**, 1092–101.

Schaber, G. G. (2005). The U.S. Geological Survey, branch of Astrogeology – A Chronology of Activities from Conception through the End of Project Apollo (1960–1973). *U.S.G.S. Open-File Report*, 2005–1190.

Sherwood Lollar, B, Voglesonger, K., Lin, L. H., Lacrampe-Couloume, G., Telling, J., Abrajano, T. A., Onstott, T. C. and Pratt, L. M. (2007). Hydrogeologic controls on episodic H_2 release from Precambrian fractured rocks: energy for deep subsurface life on Earth and Mars. *Astrobiology*, **7**, 971–86.

Snook, K., Glass, B., Briggs, G. and Jasper, J. (2007). Integrated analog mission design for planetary exploration with humans and robots. In *The Geology of Mars: Evidence from Earth-based Analogs*, ed. M. Chapman. Cambridge: Cambridge University Press, pp. 424–55.

Snook, K. (2008). Outpost science and exploration working group (OSEWG). *Presentation made to NASA Planetary Sciences Subcommittee*, March 4, 2008.

Southam, G., Rothschild, L. J. and Westall, F., (2007). The geology and habitability of terrestrial planets: fundamental requirements for life. *Space Science Reviews*, **129**, 7–34.

Squyres, S. W., Arvidson, R. E., Ruff, S., Gellert, R., Morris, R. V., Ming, D. W., Crumpler, L., Farmer, J. D., Des Marais, D. J., Yen, A., McLennan, S. M., Calvin, W., Bell, J. F., III., Clark, B. C., Wang, A., McCoy, T. J., Schmidt, M. E. and de Souza, P. A., Jr. (2008). Detection of silica-rich deposits on Mars. *Science*, **320**, 1063–7.

Stedman, K. and Blumberg, B. S. (2005). The NASA Astrobiology Institute virus focus group workshop and field trip to Mono and Mammoth Lakes, CA, 22–24 June 2004. *Astrobiology*, **5**, 441–3.

Steele, A., Amundsen, H. E. F., Conrad, P. G., Benning, L. and Fogel, M. (2008). *Arctic Mars analogue Svalbard expedition (AMASE) (2007).* 39th Lunar and Planetary Science Conference, abstract 2368.

Steven, B., Léveillé, R. J., Pollard, W. H. and Whyte, L. G. (2006). Microbial ecology and biodiversity of permafrost. *Extremophiles*, **10**, 259–67.

Stoker, C. R., Cannon, H. N., Dunagan, S. E., Lemke, L. G., Glass, B. J., Miller, D., Gomez-Elvira, J., Davis, K., Zavaleta, J., Winterholler, A., Roman, M., Rodriguez-Manfredi, J. A., Bonaccorsi, R., Bell, M. S., Brown, A., Battler, M., Chen, B., Cooper, G., Davidson, M., Fernández-Remolar, D., Gonzales-Pastor, E., Heldmann, J. L., Martínez-Frías, J., Parro, V., Prieto-Ballesteros, O., Sutter, B., Schuerger, A. C., Schutt, J. and Rull, F. (2008). The 2005 MARTE robotic drilling experiment in Río Tinto, Spain: objectives, approach, and results of a simulated mission to search for life in the Martian subsurface. *Astrobiology*, **8**, 921–45.

Thirsk, R., Williams, D. and Anvari, M. (2007). NEEMO 7 undersea mission. *Acta Astronautica*, **60**, 512–17.

West, M. D., Clarke, J. D. A., Thomas, M., Pain, C. F. and Walter, M. R. (2009). The geology of Australian Mars analogue sites. *Planetary and Space Science*, doi: 10.1016/ j.pss.2009.06.012.

Williamson, M.-C., Germain, M., Lavoie, D. and Gulick, V. C. (2008). *Comparative geoscientific and geomatic analysis of hydrothermal zones in volcanic terrain on Earth and Mars*. 39th Lunar and Planetary Science Conference, Houston, TX, abstract 2188.

Index